Water Science and Technology

INSTRUMENTATION, CONTROL AND AUTOMATION OF WATER AND WASTEWATER TREATMENT AND TRANSPORT SYSTEMS 1993

Organising Committee

B. Jank *(Chairman)*	Canada
R. Briggs *(Vice Chairman)*	UK
W. F. Garber *(Secretary)*	USA
J. F. Andrews	USA
A. Di Pinto	Italy
J. Federico	USA
W. Frey	Austria
H. D. Gilman	USA
C. F. Guarino	USA
M. Hiraoka	Japan
A. Klapwijk	The Netherlands
L. H. Lotter	South Africa
A. Lynggaard-Jensen	Denmark
A. Milburn	UK
T. Palmgren	Sweden
Soon Bo Shim	Korea
M. Viitasaari	Finland

Scientific Programme Committee

B. Jank	Canada
R. Briggs	UK
M. T. Garrett	USA
M. Hiraoka	Japan
J. Lohman	Germany
P. Marchandise	France
G. Olsson	Sweden
M. Stenstrom	USA

INSTRUMENTATION, CONTROL AND AUTOMATION OF WATER AND WASTEWATER TREATMENT AND TRANSPORT SYSTEMS 1993

Proceedings of the Sixth IAWQ Workshop, held in Banff and Hamilton, Canada, 17–25 June 1993

Editor

B. JANK

Wastewater Technology Centre, Burlington, Ontario, Canada

with the assistance of

J. Andrews (USA), D. Averill (Canada), M. Barnett (Canada), F. Benoit (Canada), H. Campbell (Canada), D. Chapman (Canada), P. Crawford (Canada), P. Dold (Canada), P. Fowlie (Canada), T. Garrett (USA), B. Hill (USA), M. Hiraoka (Japan), M. Hriblijan (Canada), R. Jones (Canada), G. Lebel (Canada), P. Marchandise (France), B. Monaghan (Canada), D. Nolasco (Canada), S. Nutt (Canada), G. Olsson (Sweden), G. Patry (Canada), W. Schuk, (USA), M. Stenstrom (USA), G. Speirs (Canada), J. Stephenson (Canada), P. Sutton (Canada), I. Takacs (Canada), T. Vassos (Canada), Z. Vitasovic (USA) and G. Zukovs (Canada)

PERGAMON

U.K.	Elsevier Science Ltd, The Boulevard, Langford Lane, Kidlington, Oxford OX5 1GB, U.K.
U.S.A.	Elsevier Science, Inc., 660 White Plains Road, Tarrytown, New York 10591-5153, U.S.A.
JAPAN	Elsevier Science Japan, Tsunashima Building Annex, 3-20-12 Yushima, Bunkyo-ku, Tokyo 113, Japan

First edition 1994

Library of Congress Catalog Card No. 82-645900

ISBN 0 08 042495 3

Water Science and Technology *publishes the proceedings of IAWQ Conferences, Symposia and Workshops. All papers have been selected for publication after refereeing under the auspices of the Editor(s) and the Programme/Organising Committee(s) of each IAWQ event.*

The subjects covered in Water Science and Technology *are defined by IAWQ's mission statement which is "to advance the science and practice of water pollution control and water quality management worldwide". This encompasses five broad areas:*
(1) wastewater treatment processes for domestic, industrial and municipal effluents;
(2) diffuse source pollution;
(3) impacts of pollutants on receiving waters (surface, ground and marine water);
(4) hazardous waste management including source control;
(5) restoration of the aquatic environment.

In order to make this volume available as economically and as rapidly as possible it has been produced when practicable from the author's word processor disks. Every effort has been made to ensure an exact reproduction in the time available.

Printed and bound in Great Britain by BPC Wheatons Ltd, Exeter

CONTENTS

Water Distribution Control Systems

Dynamic Modelling and Simulation

Process Control Strategy Development

Instrumentation Applications and Maintenance

Control Strategies Application

PREFACE

These proceedings are comprised of a selection of manuscripts originally presented as papers or posters at the 6th IAWQ Workshop on Instrumentation, Control and Automation of Water and Wastewater Treatment and Transport Systems, held in Banff, Alberta and Hamilton, Ontario, Canada, June 1993. This Workshop series, sponsored by the IAWQ Specialist Group on Instrumentation, Control and Automation (ICA), has been bringing together the world's leading specialists in water and wastewater ICA since 1973. The 7th in this Workshop series, to be chaired by Ronald Briggs (City University, London, UK), will be held in the United Kingdom in 1997.

Advances in sensor development and computer technology, accompanied by reductions in hardware costs have made it possible to achieve considerable energy savings and upgrade existing facilities by identification and correction of design and operational deficiencies. These technological advances have also provided the tools to optimize the design and operation of new facilities. The 15 technical sessions and 30 poster presentations placed emphasis on the development of control strategies and their applications.

The first day of the Workshop included the presentation of 16 country status reports addressing the Workshop theme "Advancing ICA Technology by Eliminating the Constraints". The representative from each country was asked to address three questions: What are the present constraints to advancing ICA technology in your country? What are the driving forces which are advancing ICA technology in your country? Where do you expect to be in four years time (i.e. what will the state-of-the-art be in your country at the time of the 7th ICA Workshop in 1997)? A panel discussion was held following the presentation to give the other Workshop participants an opportunity to present their views. The results were summarised and presented during the technical programme by Professor Gustaf Olsson (Lund Institute, Sweden). His report summarising the country status reports is included in the proceedings. The report provides an important assessment of the present status and future directions of ICA in the water and wastewater industry.

As Workshop Chairman I would like to thank my Organizing Committee, the industrial sponsors and exhibitors who provided financial support, and Environment Canada's Wastewater Technology Centre whose financial resources and employees made the conference possible. A special thanks is due to Beverly Pasian, Conference Manager, who did the majority of the work. As Editor of these proceedings I would like to thank all of those who so generously donated their valuable time to the original and final review of the manuscripts. I would especially like to thank my Scientific Programme Committee – Professor Ronald Briggs (Specialist Group Vice Chairman), City University, UK; Professor Masakatsu Hiraoka (Specialist Group Past Chairman), Kyoto University, Japan; Professor Gustaf Olsson, Lund Institute, Sweden; Dr Truett Garret, City of Houston, USA; Professor Mike Stenstrom, University of California, Los Angeles, USA; Dr Patrick Marchandise, World Health Organization, France; and Mr Gordon Speirs, Wastewater Technology Centre, Canada.

Finally, I would like to thank all conference participants, especially those who presented papers and posters. It is their generous contribution of time that makes these technology transfer sessions and the subsequent documentation of results possible. The ICA Specialist Group requests that you consider participation in the 7th ICA Workshop in the UK. We look forward to seeing you in 1997.

Bruce Jank
Chairman of the Programme Committee

SUMMARY REPORT

Pergamon

Wat. Sci. Tech. Vol. 28, No. 11-12, pp. 1–7, 1993.

0273–1223/93 $7·00 + 0·00

ADVANCING *ICA* TECHNOLOGY BY ELIMINATING THE CONSTRAINTS

Gustaf Olsson

Department of Industrial Electrical Engineering and Automation (IEA), Lund Institute of technology, P.O. Box 118, S-22100 Lund, Sweden

ABSTRACT

Experts from 16 countries were surveyed to provide an overview of the current status, the constraints and the future prospects for instrumentation, control and automation in water and wastewater treatment and transport. Their responses are summarised.

KEYWORDS

Instrumentation; control; automation; planning; design; legislation; cost.

INTRODUCTION

The title of this report was the theme for the Banff session of the 6th *IAWQ workshop on instrumentation, control and automation of water and wastewater treatment and transport systems.* In preparation for the meeting representatives of 17 countries were invited to provide brief *country status reports.* Each one was asked to address the following questions concerning instrumentation, control and automation (*ICA*):

What are the present constraints to advancing *ICA* technology in your country?

What are the driving forces which are advancing *ICA* technology in your country?

Where do you expect to be in four years time, i.e. what will the state-of-the-art be within your country at the time for the 7th *ICA* workshop in 1997?

Answers were obtained from 16 countries. The answer of each question was documented on one slide and briefly presented at the Banff session. The following countries and individuals have contributed to the content:

Australia	Bob Newell
Austria	Wilhelm Frey
Belgium	Peter Vanrolleghem
Canada	Robert Pickett
Denmark	A. Lynggaard-Jensen
Finland	Matti Viitasaari
France	Patrick Marchandise
Germany	Sigurd Schlegel

Italy	Appio di Pinto
Japan	Masakatsu Hiraoka
Korea	Soon Bo Shim
The Netherlands	Henri Spanjers
South Africa	Laurraine Lotter
Sweden	Torsten Palmgren
UK	Ron Briggs
USA	Truett Garrett

This report has been extracted and structured from the different presentations. A panel discussion was held after the presentations, and some of its content is also reflected in the paper. The panellists were:

Prof. Masakatsu Hiraoka, Japan
Prof. Ron Briggs, UK
Dr. Bruce Jank, Canada
Prof. Gustaf Olsson, Sweden

PRESENT CONSTRAINTS

The constraints look different in the various countries, but still there are a lot of common problems that have to be solved in order to advance *ICA* technology within the wastewater treatment area. Most of these constraints are related to:

legislation,
education – training – understanding,
collaboration,
economy,
measuring devices,
insufficient sewer systems,
plant constraints,
lack of software standardisation.

Legislation

In many countries there is a lack of a tight regulatory standard, which makes it unnecessary to improve the control on plants. The enforcement of the effluent regulations is often quite inadequate. In some European countries the legislation has not been adapted to uniform rules in different parts of a country (such as Belgium) or to a common European standard.

Many regulatory standards are not adapted to the need of the receiving water. Instead they only consider static conditions and do not take dynamic variations into account. It should not be necessary to consider the worst case all the time. Many rules do not recognise geographical differences between the receiving waters. With unclear legislation this results in inconsistent rules for the effluent water quality.

Since legislation asks for off-line data it is not well adapted to the use of *ICA* and on-line measurements. The use of on-line sensors for reporting to the authorities is desired by many users.

We have to remind ourselves that there is hardly any regulation on the influent to the plants, only on the output. Control of the input ought to be considered more.

Education – Training – Understanding

In most countries the public awareness has been small until recently. This has caused delayed investments in better treatment facilities.

Operators are not always adequately educated to deal with on-line instrumentation and control. Many of them do not know where and how to take proper samples and how to condition them. Most environmental engineers (including operating engineers and design engineers) would need more basic education in dynamics in order to appreciate the need and potential for *ICA*.

Communication between users, suppliers, researchers and funding agents is often unsatisfactory and leads not only to misunderstandings but also to poorly designed plants. Manufacturers and suppliers are often not aware of the need for better flexibility and control. It is too late to start thinking about the control system when the design is already fixed.

It has to be realised that the management is part of the overall control of a plant. Many managers are not technically trained, which may lead to misunderstandings and non-consistent operation of a plant.

<u>Economy</u>

Since the wastewater treatment industry is basically a non-profit enterprise there have been quite unclear goals as to what *ICA* could do. Automation has been considered costly and not part of the initial design. Instead, it has been introduced after the design completion. Consequently, a proper weighting between construction and operation costs is often not achieved. An adequate use of *ICA* may in fact be a crucial factor that can make a plant run consistently.

With the recession today many countries postpone the introduction of better operating methods.

<u>Measuring Devices</u>

There is a lack of reliable on-line sensors today. As an example the reliability has to be significantly increased for:

MLSS meters,
ammonia sensors,
phosphorus sensors,
COD monitors.

Redox meters have a considerable drift. There is a lack of fatty acids monitors for the control of anaerobic processes.

In many cases operators are not sufficiently qualified to use and maintain instrumentation systems. Most sensors today do not allow long periods without extensive maintenance.

<u>Plant Constraints</u>

There is still a lot of uncertainty about the efficiency of *ICA* technology. Since there is a lack of good plants demonstrating the success of *ICA*, many designers are still conservative. One looks for tried methods, so plants are constructed with large safety margins. Unfortunately, this also means that too little flexibility and controllability is built into most plants. In fact, today there are disincentives for innovative design!

In most countries the majority of the plants are quite small. In such plants it is even more unrealistic to defend the installation of any advanced *ICA*.

<u>Software</u>

Today there is a lack of software standardisation. Most software systems are proprietary and are often quite unfriendly for the user. It is consequently an educational problem to make the users competent in using the software. However, it is also a need to make software more user-friendly and flexible, so that the specific needs of a plant operation can be met.

DRIVING FORCES

Several significant driving forces were presented. Together they will help to advance the use of *ICA* in wastewater treatment. It was that the development of *ICA* should *not* be driven by technology but by the needs. Some of the driving forces can be categorised as:

effluent quality standards,
economy,
increasing plant complexity,
models for operation.

Moreover, the development in sensor and computer technologies are important driving forces, although probably not the primary forces.

In some countries there is a water shortage which in itself is a driving force for improved wastewater treatment, e.g. for recycling. Moreover, the desire for a municipality to be a leader in environmental consciousness is often a driving force for better wastewater treatment standards. Many process industries are becoming environment-conscious and wish to improve their environmental record.

Effluent Quality Standards

An increasing public awareness is an efficient driving force. The formation of national water quality management strategies will support the development of better wastewater treatment. In many countries there are increasingly severe penalties for pollution offences.

The priorities of plant operation can be defined in the following order (John F Andrews, USA):

(1) keep the plant running;
(2) meet the effluent requirements;
(3) minimise the cost.

The effluent quality and a good performance has to be guaranteed *consistently*, which will have a great impact on instrumentation, control and automation. The legislators often look at the best available technology. The effluent quality standards, however, have to be technically and economically feasible, where the need for the receiving water should be the driving force.

In Europe the EC directives will be enforced, which means that the countries have to bring these demands into practice.

It was mentioned as a constraint that effluent quality has to be measured off-line. The wastewater industry is obliged to show that on-line sensors are good enough to be used for regulatory reports. We would need a co-ordinated international program – on-line versus off-line sensors – with an approved statistically valid experimental design. This could lead to realistic demands and be a good background for legislation.

Economy

This is probably the primary driving force. In most countries the standards are formulated as permits. Then it is a driving force to change the emphasis from compliance to compliance at least cost. In some countries the authorities charge according to the effluent pollution, where different components are given their weighting factors.

Whatever the rules are, there is a driving force to minimise the costs and get more out of existing systems. Often the effluent water quality standard is set according to what is economically feasible.

ICA will not be generally accepted simply because it raises the technological level of the plant operation. It has to be proven that *ICA* provides a more consistently good operation than manual control. In particular one has to look at savings in energy, chemical, and personnel costs. An improved operation by *ICA* may also lead to deferred capital expenditure. This is likely to be the most significant economic factor.

Recycling of process water will become an issue in the future, making wastewater treatment part of an overall production process. This means that wastewater treatment can not be seen as a non-profit activity. Instead this gives further incentives for source control, since raw material and products can be saved in the production industry.

Extracting heat from the effluent water in heat pumps is used today routinely, for example in district heating systems.

Plant Complexity

Plant complexity may be one of the most important driving forces; *ICA* will defer capital expenditures for plant expansion. Since nutrient removal will become more common the plant complexity will still grow. The need for consistent operation further emphasises the importance of *ICA*.

It will become more and more important to understand the impact of load variations.

The increased quantities of sludge make it necessary to develop better control methods for sludge treatment and disposal processes.

Models for Operation

Most models of today, describing the structured behaviour of activated sludge plants, are far too complex to be used for operation. They cannot be verified by on-line plant measurements, and are consequently inadequate for predicting plant behaviour.

For advanced control it is important to obtain models with adequate complexity. They should be updated by on-line measurements and be able to predict the plant behaviour for an adequate time period.

Process simulators are becoming available, but still the operational models have to be further developed in order to make the simulators valuable for operation.

Sensor Technology

There will be an increasing need to measure dynamic changes, not only of the influent wastewater (such as toxic warnings) but of variations in the plant itself. There has to be a proper awareness of the required sampling frequency. This will further emphasise the need for better on-line sensors.

The continuous improvements of sensors (like respirometers) is in itself an incentive to promote the control applications. The evaluation of sensors using approved test protocols is extremely important. Test protocols could be used in procurement specification.

Computer Technology

Software has to be made more modular. There is an increasing awareness among operating engineers that the software has to be flexible. The users do not want to be completely dependent on the vendors in order to make minor or major changes of the software.

More and more users ask for open systems both in terms of hardware and software.

THE STATE OF THE ART IN 1997

Almost surely the **legal demands** will be stricter in most places. One of the consequences will be that **biological nutrient removal** will become more common. At the same time we have to recognise that the rules cannot be the same in all countries.

Already today there is *ICA* technology available that could be applied in many plants. The most important that were identified are:

sensors,
estimation techniques,
control technology,
integration of control, and
open computer systems.

Sensors

The number of available sensor types will be increased, such as for monitoring of:

ammonia,
nitrite – nitrate,
phosphorus,
redox,
respiration rate.

These sensors are already available today, but are expected to be more commonly used. Monitoring of redox potential and pH in anaerobic environment as well as respirometric on-line toxicity detection will be more common. More enzyme based sensors will probably be used.

In process industry in general there is a trend towards **"intelligent" sensors**. There are many reasons to believe that this will be the case also in the wastewater treatment industry. In such a sensor the primary sensing element is complemented with information technology to extract more information from the signals and to interpret the information. Such examples may be image processing and advanced respirometers.

Estimation Techniques

In estimation techniques the sensor signal is "mixed" with model information so as to create a more accurate calculation of a process variable. Typically one may say that "intelligent" sensors use estimation techniques. As an example, automatic image analysis of sludge properties may offer interesting possibilities.

The application of estimation techniques for **diagnosis** purposes will increase. One example may be estimation of floc property changes in the process. The use of **early warning systems** for influent variations or process changes in the plant will probably increase.

Another application of estimation techniques will be on-line **model verification**. An update of dynamical models, based on current data, can be used for diagnostic purposes and as a base for more advanced control.

Control Technology

Having an updated model is crucial in order to apply dynamical process **simulation** as a tool for operation. Then the model could be used to explore different control actions before they are realised in the plant. The operator can then look for different alternatives to *what happens if.*

The state of control technology is very advanced today. Much of it can be applied for wastewater treatment plant control, once the understanding is there and the need is known. However, the controllability of a plant may still not be satisfactory. The possible control actions are simply not sufficient to reject disturbances and to meet the requirements.

New control methods are being explored and will probably be applied more. The use of **neural networks** to identify process behaviour is gaining interest. **Fuzzy logic** as a base for control of wastewater treatment processes has been tried and will probably be more commonly applied in the future.

Integration of control

There are at least three types of integrations that have to be improved:

integration between design, operation and training;
integration between air, water, and sludge quality;
integration in operation of all systems from the sewer to the receiving water.

Such an integration may not be the state of the art already in 1997. However, we can already see the trend towards more integration.

To integrate design, operation and training dynamic models have to be further improved to serve different purposes. Since there are different goals the models for design, operation and training have to differ. Still all of them have to be based on a solid understanding of the processes, and the verification of their accuracy has to be considered. Good models are necessary to get advanced simulators. They are crucial for good design, training, control and operation.

For an advanced and improved *overall* operation the whole system – from the sewer to the receiving water – has to be considered. The sewer operation has to reflect how much load the plant can receive, and bypass decisions have to be made, based on on-line calculations in both the sewer and in the treatment plant. This shows, that it is important to have an adequate organisation in order to operate the whole system.

The operation of a treatment plant has to consider all the interactions between different unit processes. Not only the well known coupling between the aerator and the settler, but all other recycling streams have to be taken into account.

The relation between industrial production processes and their effluent to treatment plants can be further explored. In a systems approach one cannot simply accept the fact that a production process delivers a certain waste stream. The overall design has to consider improvements of the industrial process. Such improvements have come a long way e.g. in the paper and pulp industry.

An integrated approach to control will require advanced information networks. The development in communication technology gives all the technical support that will be needed. The real challenge will be concerned with adequate systems modelling and understanding and proper information exchange and control methods.

Open Computer Systems

There is a clear trend in both software and hardware towards more open systems. The system will become more easily configurable. With modular hardware the user can more easily expand a system. Having software from several vendors a user can also make his own design of software systems so that it better fits his own needs.

CONCLUSIONS

We should set some specific **goals** to be achieved by 1997 with specific countries to report on the development in these areas. The goals should be based on constraints/driving forces.

INVITED PAPER

Pergamon

Wat. Sci. Tech. Vol. 28, No. 11-12, pp. 9–14, 1993.

0273–1223/93 $7·00 + 0·00

FUTURE DIRECTIONS IN INSTRUMENTATION, CONTROL AND AUTOMATION IN THE WATER AND WASTEWATER INDUSTRY

Troy D. Vassos

NovaTec Consultants Inc., Vancouver, British Columbia, Canada

ABSTRACT

The need to optimize treatment plant performance and to meet increasingly stringent effluent criteria are two key factors affecting future development of instrumentation, control and automation (ICA) applications in the water and wastewater industry. Two case studies are presented which highlight the need for dynamic modelling and simulation software to assist operations staff in developing effective instrumentation control strategies, and to provide a training environment for the evaluation of such strategies. One of the limiting factors to date in realizing the potential benefits of ICA has been the inability to adequately interpret the large number of existing instrumentation inputs available at treatment facilities. The number of inputs can exceed the number of control loops by up to three orders of magnitude. The integration of dynamic modelling and expert system software is seen to facilitate the interpretation of real-time data, allowing both quantitative (instrumented) and qualitative (operator input) information to be integrated for process control. Improvements in sensor reliability and performance, and the development of biological monitoring sensors and control algorithms are also discussed.

KEYWORDS

Wastewater treatment instrumentation; process control strategies; dynamic modelling; simulation; expert systems; biological nutrient removal; NADH sensors; oxidation-reduction probes.

INTRODUCTION

While water and wastewater treatment plants are usually designed under steady-state conditions, their performance is sensitive to time-varying conditions and other environmental factors which are usually beyond the control of the plant operators. The performance of wastewater treatment facilities is particularly dependent on the experience of the operations staff, who develop a "feel" for the operation, allowing them to cope reasonably well to changing influent conditions, using conventional treatment technologies. However, operations staff do not have the knowledge and experience base to apply to the increasingly complex wastewater treatment facilities, with the associated inter-dependence of process operations. This is resulting in poor plant performance, and the inability of treatment processes to achieve optimal levels of treatment.

One of the key factors inhibiting the development and incorporation of instrumentation for the control and automation of water and wastewater treatment processes is that sensors are often used primarily for monitoring or status purposes. The number of input/output points to a control system for a large-scale (i.e.

greater than 100 MGD) plant can exceed the number of control loops by up to three orders of magnitude. Little of the monitored data is actually used within a control loop. There is a need for effective control strategies which can take advantage of available real-time information.

The factors which limit the ability to meet performance objectives are directly related to limitations in existing instrumentation, control and automation (ICA) capabilities. The need to optimize plant performance and meet increasingly stringent effluent standards are two key factors which will influence the development of ICA technologies in the future.

ICA TECHNOLOGY NEEDS

The following describes two wastewater treatment process case studies (industrial and municipal) which demonstrate some of the needs in ICA technology development

Pulp and Paper

The pulp and paper industry is faced with increasingly stringent effluent standards, with requirements to reduce toxic organic contaminants such as dioxins and furans to levels approaching the analytical detection limits. While there is no clear understanding regarding the biodegradability of these compounds (e.g. whether the reduction is more dependent on hydraulic retention time or sludge age), it is recognized that conventional aerated lagoon treatment is incapable of effectively achieving the required treatment levels, and that a mechanical process is required. In the absence of specific process guidelines, the mechanical treatment facilities are typically activated sludge processes, which are designed to be as flexible as possible. For example, one such facility (located in British Columbia) incorporates an existing aerated lagoon system, in addition to five new activated sludge bio-reactors which can be operated either in series or parallel, with the added option of step-feed. In the absence of effective dynamic models, the process design has been based on steady state assumptions, with some sensitivity assessment with respect to these design assumptions. The purpose of providing process flexibility is to allow the operations staff to change the process configuration if the plant has difficulty meeting the Discharge Permit quality criteria. The problem lies in determining what process change will improve effluent quality considering: the lack of operations staff experience, and training; the lack of knowledge regarding which process parameters most affect optimal performance; and lack of adequate sensors to reflect the performance levels (i.e. AOX, dioxin and furan levels in the effluent). There is also a need to have the wastewater treatment I&C integrated into the existing pulp and paper process control systems, and to automate where possible to minimize cost.

Biological Nutrient Removal (BNR)

An extensive amount of municipal wastewater treatment plant rehabilitation and new construction has been carried out in recent years as a result of more stringent discharge standards, with particular world-wide emphasis on nutrient removal. Single sludge BNR treatment plants, capable of simultaneously removing both nitrogen and phosphorus, represent the state-of-the-art in wastewater treatment technology. The degree of process complexity can be demonstrated by considering the BNR facility constructed at Penticton, British Columbia. Based on extensive pilot scale research carried out at the University of British Columbia, the Penticton BNR plant is capable of reducing effluent phosphorus and nitrogen concentrations to below 0.1 and 3.0 mg/l, respectively. Although intended to operate primarily in a four stage modified University of Cape Town (UCT) configuration, the plant process configuration, a schematic of which is illustrated in Figure 1, can be readily changed to achieve a wide range of BNR system configurations. This is accomplished by adjusting the discharge gates leading into, and out-of, the recirculation channels, and by varying the mixing and aeration characteristics of each of the reactor cells. Up to 30% of the plant reactor volume (i.e. bioreactor cells 10 and 11) can be taken out of service to accommodate wide seasonal hydraulic variations due to the influx of tourists during the summer months. The relative proportions of each of the anaerobic, anoxic and aerobic reactor zones can also be adjusted. For example, bioreactor cells 6 through 11 can operate with or without an air supply, and the gate positions (open or closed) from the recirculation channels, leading to bioreactors 1 through 5, can be used to adjust the anoxic and anaerobic reactor volumes.

Optimization of such a flexible plant presents a challenge for operations staff. The technology is relatively new and rapidly advancing, requiring operations staff to make process control decisions based on "hands-on" experience, and experimentation. Effective and optimal plant performance is limited by a number of factors including: the lack of operations staff experience, and training; the lack of knowledge regarding which process parameters most affect optimal performance; and the absence of adequate sensors to assess specific environmental conditions in each bioreactor. Despite the large number of inputs to the distributed control system, there are few control loops, and the process requires an extensive amount of operator intervention.

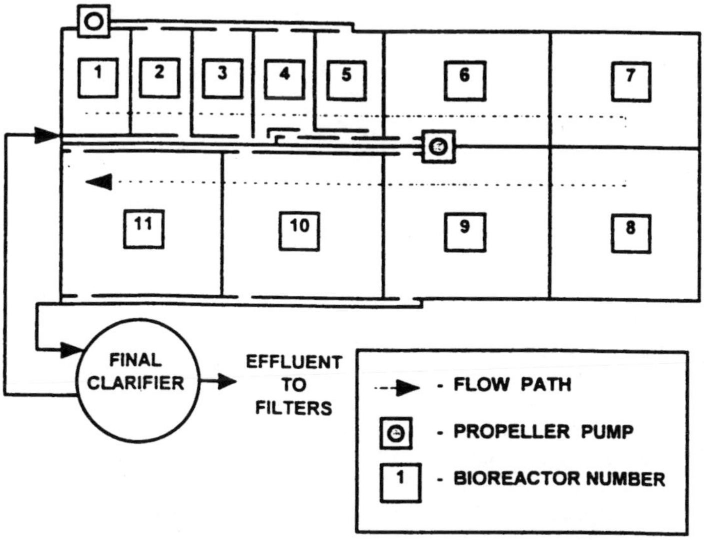

Fig. 1. Process schematic for the penticton, British Columbia, BNR plant.

Both of the above case studies outline a number of potential performance problems which could be addressed through improved ICA technology applications, specifically:

1. Modelling/simulator software, which can incorporate real-time data, to assist in operator training and evaluation of process control alternatives;

2. Expert system (artificial intelligence) software tools to incorporate qualitative information into the control loop, and aid in data interpretation;

3. Instrumentation and sensor developments to provide reliable information for use in control loops.

MODELLING AND SIMULATOR SOFTWARE ADVANCES

Before operations staff can consider alternative configurations, they need to have a much larger experience base upon which to base process decisions, and a method to assess what the outcome of a change will be. Without the ability to verify the outcome of process changes, optimal biological treatment conditions will be difficult to achieve

Up until very recently, the only means by which an operator could determine the effects of process control modifications was experimentation. Recent advances in dynamic models, artificial intelligence, neural network modelling, and computer technology, now make it possible to accurately simulate plant processes for use as an operations control and diagnostic tool.

There are a number of dynamic activated sludge modelling programs available which have great potential for process design, operator training and process control. One such program has been developed over the past eight years by an Ontario based firm (Hydromantis, Inc.), with the financial assistance of Environment Canada. This state-of-the-art dynamic modelling, and simulation, computer program is referred to as the **General Purpose Simulator (GPS)**, and makes use of the latest developments in mathematical modelling and computer simulation. The GPS is designed to assist in the planning, design, operation and control of wastewater treatment plants. The simulator can also be effectively used for operator training, allowing plant personnel to experiment with operational strategies without affecting the normal operation of the plant. Already being applied to a number of large-scale wastewater treatment plants in Canada, the U.S. and U.K., this approach is proving to be very successful, and an important key to improved process automation.

Where there is both limited design and operating experience available for a particular wastewater treatment facility, a simulator could be used to great advantage to prepare contingency plans for emergency conditions, and to train operations staff prior to commissioning the plant. This is particularly important for industrial wastewaters where the feed stock can originate from a number of unit processes with widely varying wastewater characteristics, in addition to fluctuations in hydraulic loading. Application of simulators to wastewater treatment processes has also proven to facilitate the identification of critical process conditions and control loops.

Incorporating a calibrated computer simulation program, operations staff will be able to determine the effect of process changes on effluent quality before they are actually carried out on the plant. Used as a training tool, plant operators will be able to learn about the process and make mistakes without impacting the receiving environment.

For the process modelling and simulation software to be effective, the process model must first be calibrated using reliable on-line data, collected from the plant under existing plant operating conditions. This data is also needed to ensure the model remains calibrated and accurately represents the treatment plant. The calibrated model can then provide information on the time varying performance of the plant, subject to a variety of dynamic events, including:

1. Variations in hydraulic loading to the plant;

2. Changes in influent composition;

3. Operational scenarios of interest, including alternative process configurations, adjusted reactor zones and sizes, and chemical supplementation conditions.

EXPERT SYSTEM ADVANCES

There have been a number of knowledge based expert systems (KBES) for wastewater treatment operation, and control, reported in the literature (Beck *et al.*, 1990; Berkmen *et al.*, 1990; Lai and Berthouex, 1990). Although most of the rule-based systems reported to date are described as being prototype, and few have been used in practice, there is great potential in this approach. The major problems reported concern acquisition of knowledge and subsequent validation of the knowledge base (Barnett and Patry, 1992). These problems are due to the use of inappropriate knowledge representations, where rules alone are insufficient to represent the many different kinds of knowledge required for diagnosis and control of wastewater treatment processes.

One approach to address the deficiencies of past KBES applications is to develop a flexible framework for knowledge, with sufficient expressive power to facilitate the capture of different types of knowledge, as well as the manipulation of knowledge in the inference process. The use of an object-oriented model-based knowledge framework (Kunz *et al.*, 1989) is a recent and promising innovation within the artificial intelligence field. It has the advantages of being flexible, extensible and highly modular in design, and enables a variety of inferencing schemes to be implemented.

The coupling of a comprehensive dynamic treatment plant model with a computer-based expert system, could be an effective process control tool for use in process optimization. Integrated into the instrumentation and control system, such software would have the following characteristics:

1. Adaptability to changing influent and process conditions;

2. Ability to incorporate both qualitative and quantitative data;

3. Auto-calibration and modelling verification capabilities.

In the absence of operating experience, a dynamic model could be used to develop the preliminary knowledge base that will be used in the early phases of the plant operation. This concept of using rule-extraction from simulation has been previously proposed by Patry and Olson (1987) to develop some of the basic operational rules that will make up a significant part of the knowledge base, incorporating qualitative modeling and rule extraction techniques.

Rule extraction techniques can be applied to the dynamic mechanistic model as the basic source of knowledge prior to the plant construction, and serve as one of the major components in the development of the knowledge base. This could be used as a powerful tool for the development and assessment of operational strategies before and after construction of the plant. The resulting process simulator, incorporating both quantitative and qualitative data could be coupled with a real-time operational control system, forming a key component in the computer-based training program.

INSTRUMENTATION IMPROVEMENTS

Improvements in instrumentation equipment are a critical factor in the implementation of effective control strategies for water and wastewater treatment plants. The Water and Wastewater Instrumentation Testing Association (ITA) has played an active role in assessing sensor reliability, and establishing installation, testing, maintenance and calibration protocols for a number of process monitoring components, including: chlorine residual analyzers; suspended solids sensors; dissolved oxygen sensors; and flow measuring devices.

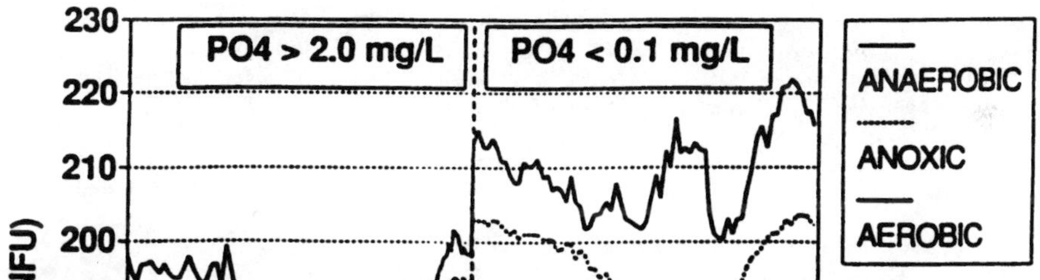

Fig. 2. NADH probe response characteristics under optimal and sub-optimal BNR phosphorus removal conditions.

The development of control strategies for BNR processes which incorporate specific ion probes (i.e., NH_3, NO_X) and bio-activity levels (i.e. respirometers, oxidation reduction potential [ORP], and NADH probes) show great promise for use in process optimization. Figure 2 illustrates the application of NADH probe data in establishing an optimum BNR process "fingerprint". Optimal phosphorus and nitrogen removal depends on maintaining anaerobic, anoxic and aerobic conditions within the process. The relative NADH level in each of the bioreactor zones (i.e. anaerobic, anoxic and aerobic), and the degree of separation in signal levels between the zones, can be used to indicate optimal phosphorus removal conditions. As noted in Figure 2, higher NADH levels and greater bioreactor signal separation is indicative of optimal phosphorus removal efficiencies. Oxidation reduction potential (ORP) sensors can also be used in a similar manner. Consequently, methods of assessing bioreactor conditions (i.e. ORP and NADH) could be used for process control using pattern recognition techniques, such as neural network based modelling.

CONCLUSION

The development of ICA applications for the water and wastewater treatment industry will be directly affected by the development of software which enables operations staff to effectively use the information gathered for process control. In addition, there is a need for the development of effective control strategies which can take advantage of available real-time information. Two key driving factors are the need to optimize the performance of existing treatment facilities, and the need to meet increasingly stringent treatment standards.

Dynamic modelling and simulation software, as an integral component of instrumentation applications, will play a significant role in developing control strategies and in optimizing plant performance. With continuous calibration, such software could be effectively used to train operations staff, and provide a "safe" environment to evaluate alternative process control strategies. Expert system (artificial intelligence) software tools will likely become integrated into dynamic modelling and simulation systems, incorporating qualitative information into the control loop, and assisting in data interpretation. Finally, continued advances are expected in instrumentation and sensor equipment to improve the reliability of inputs, and to provide information pertaining to biological activity within unit processes.

REFERENCES

Barnet, M. W. and Patry, G. G. (1992). Knowledge based (expert) systems for the activated sludge process. In: *Dynamics and Control of the Activated Sludge Process*, Tecnomic Publishing Co., Lancaster, Pennsylvania.

Beck, M. B., Lumbers, J. P., MacKenzie, H. E. C. and Jowitt, P. W. (1990). *A Prototype Expert System for Operational Control of the Activated Sludge Process*. Internal report, Imperial College, Department of Civil Engineering, Imperial College Road, London, SW7 2BU.

Berkman, L., Lennon, M. and Law, K. (1990). An expert system to diagnose performance limiting factors at publicly owned treatment works. In: *Expert Systems for Environmental Applications*, Judith M. Hushon, (ed.), American Chemical Society.

Gall, R. A. B. and Patry, G. G. (1989). Knowledge-based system for the diagnosis of an activated sludge plant. In: *Dynamic Modelling and Expert Systems in Wastewater Engineering*, G. G. Patry and D. Chapman, (eds), Lewis Publishers, Inc., Chelsea, MI 48118.

Kunz, J. C., Stelzner, M. J. and Williams, M. D. (1989). From classic expert systems to models: introduction to a methodology for building model-based systems. In: *Topics in Expert System Design*, G. Guida and C. Tasso, (eds), North-Holland, Amsterdam.

Lai, W. and Berthouex, P. M. (1990). Testing expert system for activated sludge process control. *ASCE, J. of Environmental Engineering*, **116**(5): 890-909.

Patry, G. G. and Olson, G. (1987). Dynamic modelling and expert systems in wastewater engineering. In: *Wastewater Technology Centre Newsletter*, D. T. Chapman, (ed.), Environment Canada.

MANAGEMENT COMMITMENT AND STRATEGIES

Pergamon

Wat. Sci. Tech. Vol. 28, No. 11-12, pp. 15–19, 1993.

0273–1223/93 $7·00 + 0·00

CRITICAL VARIABLES IN SYSTEM SUCCESS: TOP MANAGEMENTS' COMMITMENT IS KEY

Mary H. Winter and Alan W. Manning

EMA, Inc., 1970 Oakcrest Avenue, St. Paul, Minnesota 55113-2624, USA

ABSTRACT

Billions of dollars and vast amounts of human energy go into implementing advanced technology every year. Yet the success rate of automated systems is estimated as low as 25%. Even today's sophisticated technology cannot prevent expensive failure or non-use of automated systems. This suggests that a change in approach is needed to successfully plan and implement automated systems, which raises the following questions.

Are human factors or technical issues more significant to control systems success?

What are the most important non-technical contributors?

Is it possible to assure the success of a control systems project?

To answer these questions a survey was carried out of people who have experience working with computer systems at utilities.

KEYWORDS

Management; automation; control systems; human factors.

SURVEY DESIGN AND DISTRIBUTION

EMA, Inc., a systems control consulting firm, developed a three-part questionnaire and sent 376 copies to water and wastewater agencies in 31 states and Canada.

Although respondents could remain anonymous, they were asked to identify their title, type of organization, plant design capacity and number of employees. Questions related to the agency's control system included: how long had it been in service; did it generally meet expectations; was it justified based on cost or intangible benefits; and had savings or benefits been realized.

Respondents were asked to rank the most important contributors and most serious obstacles to computer control system success. They could either write in their own answers or choose from the provided list (Figure 1). In addition, they were asked to identify the top three from each category. While not identified as such on the form, technical items concerned hardware, software, measurement or technical tasks. Human factors items related to people and organizational considerations pertaining to automation.

PART I

Instructions:

1. Please check (✓) up to 10 items which you believe most significantly contribute to computer control system success.

2. Please circle the 3 items which you believe are the most important.

____ Good planning
____ Realistic expectations
____ Clearly defined needs
____ Measurable system performance goals
____ Plant management backing
____ Staff acceptance
____ Operator acceptance
____ Good consulting engineer training and support
____ Good vendor training and support
____ Assigning a system "champion" or manager
____ Thorough, detailed design and specification
____ Thorough inspection and testing
____ Sufficient staff to operate and maintain system
____ Support staff expertise
____ Reliable computer equipment
____ Working field instruments
____ Easy-to-use operator station hardware
____ Adequate funding
____ Quality in-house support
____ System flexibility
____ System maintainability
____ Other ____
____ Other ____
____ Other ____
____ Other ____
____ Other ____

COMMENTS:

PART II

Instructions:

1. Please check (✓) up to 10 items which you believe are the most serious pitfalls and problems limiting computer control system success.

2. Please circle the 3 items which you believe are the most serious.

____ Inadequate planning
____ Unrealistic expectations
____ Needs not clearly defined
____ No good way to measure system success
____ Inadequate plant management backing
____ Staff resistance
____ Operator resistance
____ Poor consulting engineer training and support
____ Poor vendor training and support
____ Lack of detailed design, vague specification
____ Poor design
____ Inadequate inspection and testing
____ Insufficient staff to operate and maintain system
____ Inadequate staff expertise
____ Unreliable computer equipment
____ Field instruments not working
____ Operator station hardware not easy-to-use
____ Inadequate funding
____ Lack of quality in-house support
____ System inflexible
____ System not maintainable
____ Other ____
____ Other ____
____ Other ____
____ Other ____
____ Other ____

COMMENTS:

Fig. 1. A questionnaire helped water and wastewater managers identify the key factors for computer control system success.

Plant size determined survey recipients in large metropolitan areas across the United States and Canada. Since the survey concentrated on plants with more, in-depth practical control systems experience, very small plants were not surveyed. Respondents identified a variety of plant sizes with 19% in the 11-40 MGD range, 64% in the 41-250 MGD range and 10% over 250 MGD. Ninety percent of these control systems had been operational for at least one year; 39% had been in place for three or more years.

RESPONSES

EMA received 155 responses, a return rate of 41%. Wastewater managers made up 70% of the responses due to the smaller sample group of water utilities.

All respondents by title included:

Division/department manager	23%
Director/administration	20%
Operations manager	17%
Engineer	16%
Maintenance manager	8%
Other	8%

HUMAN FACTORS ARE MOST IMPORTANT

Of the 10 most significant contributors to control systems success listed by respondents, the top three were:

1. Good planning

2. Clearly defined needs

3. Sufficient staff to operate and maintain

Seventy percent of the items were human rather than technical factors (Figure 2). Most human factors included in the top 10 fell into two categories: up-front work (good planning, clearly defined needs, realistic expectations) and staffing and attitude (sufficient staff, operator acceptance, management commitment).

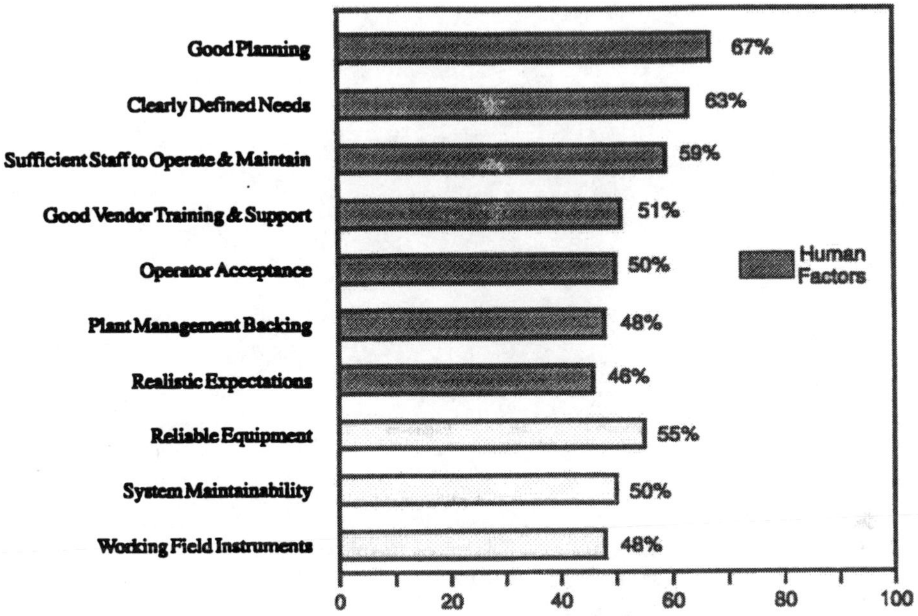

Fig. 2. Seven of top 10 contributors are human factors.

UP-FRONT WORK IS CRITICAL

Good planning and clearly defined needs ranked first and second, respectively, on the managers' top 10 list of critical success variables.

Respondents placed considerably less emphasis on realistic expectations, the third up-front human factor. Surprisingly, it appeared tenth on the list. This, however, could be attributed to unclear communication in the survey. Respondents may not have seen the intended difference between good planning/clearly defined needs and realistic expectations.

Sixty-seven percent of the respondents realized their system's projected savings or benefits. This could be another reason they placed less emphasis on realistic expectations. In other words, they got what they expected from their system.

When respondents first decided to automate, 45% cited intangible benefits, followed by cost savings (30%). Fourteen percent of the respondents marked both items. Intangible benefits, from a managerial viewpoint, are believed to include accurate and timely information, coordination between operations, and positive public relations generated by a state-of-the-art plant.

VIEWS VARY BY POSITION

Managers of both water and wastewater agencies emphasized the importance of human factors in system success, regardless of their position in the organization. Similar emphasis was evident across virtually all organizational positions surveyed. The key was management commitment.

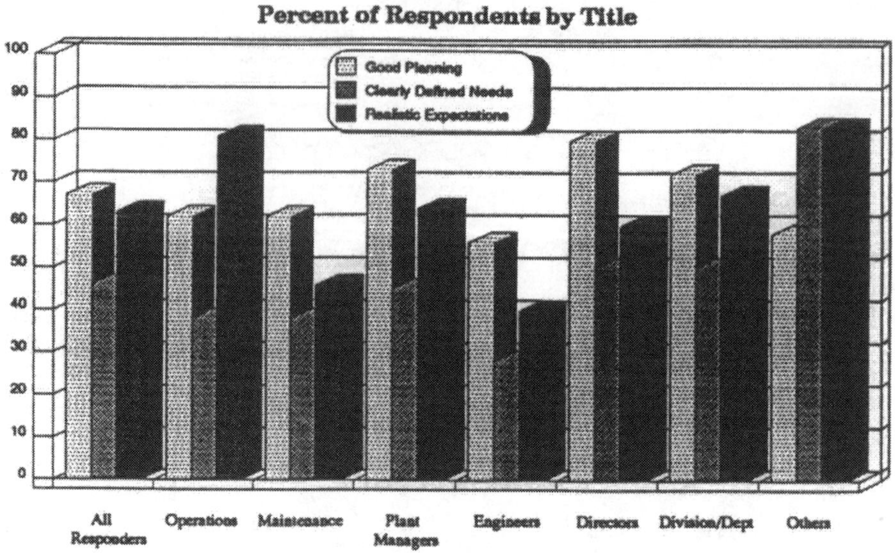

Fig. 3. Up-front work identified as significant.

Figure 3 shows the percentage of respondents in each organizational group and how they ranked the top contributing factors. Operations, maintenance, engineering and administrative managers gave similar answers for their critical success variables.

Human factors comprise half of the top 10 critical variables listed by operations and maintenance managers. Directors and administrators were more emphatic, saying human factors contribute to system success 73% and 60% respectively.

For example, operations managers, supervisors and superintendents value a clear approach to specific objectives. Eighty-two percent of this group ranked clearly defined needs in the top 10. Good planning and detailed designs and specifications are also important to operations people who must keep facilities and projects running efficiently.

Maintenance managers require technical information and staff resources to keep all parts of control systems functioning and to respond quickly to equipment problems. They ranked vendor support and training and adequate staff highest in importance (85% for each variable).

Engineering managers' responses reflected a systems perspective. For them, all parts of a control system need to be functioning reliably, starting with a foundation of accurate process measurement. They ranked working field instruments highest (68%), followed by reliable equipment.

Engineering managers also understood the importance of *humanware* in control systems. They ranked operator acceptance, reliable equipment, and top management commitment as key contributors to system success.

Senior managers value planning. Eighty percent of directors and administrators in water and wastewater agencies say good planning is the most important factor. Staffing ranked second at 67%.

CONCLUSION

Managers know that control systems include hardware, software **and** the critical third element – people. Managers recognize the importance of human factors, however, these factors rarely receive proper attention during system design. More time and skill is required to anticipate, plan and address critical humanware variables. These variables include people and their personalities, expectations and training needs.

Attention to human factors usually occurs during the final implementation phase of a project, if at all. For control systems to be successful, human factors must be addressed even before defining a specific project and then continue throughout all subsequent project phases. Human factors deal with organizational and personal attributes. The key human factor identified in this survey was the commitment of management to automation of any system.

Identifying and addressing human and organizational issues associated with any system is a three-part process – planning, designing and implementing. Addressing human factors begins with a thorough assessment of the status quo to define the current environment into which the new system is being introduced. Next, a definition must be made of the changes likely to occur as a result of the new system. In addition, positive and negative forces which will work for or against the changes are identified. Addressing these issues requires thorough strategy development and step-by-step implementation as the project progresses.

Survey data shows that managers of successful control systems have strong opinions about this topic.

Human factors are more important than technical.

Up-front work is critical to success.

EMA's experience corroborates these findings with more than 200 control systems projects.

REFERENCES

Gray, T. A. (1989). *Computer Systems for Water and Wastewater Application: How Good are They?*, ISA/89 International Conference and Exhibit.

Majchrzak, A. (1988). *The Human Side of Factory Automation*, Jossey-Bass Publishers, San Francisco, CA.

Mainiero, L. A. and DeMichiell, R. L. (1986). Minimizing employee resistance to technological change. *Personnel Magazine*.

Manross, R. C. (1985). Human component extends computer performance half-life. *TAPPI Journal*, **68**(6).

Walton, R. E. and Susman, G. T. (1987). People policies for the new machines. *Harvard Business Review*.

Zuboff, S. (1985). Automate/Informate: the two faces of intelligent technology. *Organizational Dynamics*.

Pergamon

Wat. Sci. Tech. Vol. 28, No. 11-12, pp. 21–27, 1993.

0273–1223/93 $7·00 + 0·00

REALIZING THE BENEFITS OF ENTERPRISE-WIDE COMPUTING

Michael Sweeney* and Alan Manning**

** Greater Cincinnati Metro Sewer District, 1600 Gest Street, Cincinnati, Ohio 45204, USA*
*** EMA, Inc., 1970 Oakcrest Avenue, St. Paul, Minnesota 55113-2624, USA*

ABSTRACT

This paper is a case history on the benefits the city of Indianapolis has been able to realize from incorporating an enterprise-wide model for approaching automation at their two Advanced Wastewater Treatment (AWT) Plants. Emphasized is the importance of strategic planning, performance measures and human factors associated with the suggested implementation of applied technology. Also discussed are the results, thus far, with the implemented portions of the plan. A framework is summarized for other municipalities to use as a guide to applying virtually any automation technology.

KEYWORDS

Wastewater treatment; computer systems; process control; strategic planning; system irrigation; data management; performance measurement; human factors.

INTRODUCTION

The Indianapolis AWT Division faces major challenges. The division operates and maintains two 125 MGD treatment plants which discharge to the White River. The treatment processes involve state-of-the-art unit processes including bioroughing, oxygen activated sludge (nitrification), effluent filtration, ozone disinfection and cryogenic oxygen generation. Sludge treatment includes gravity thickening, belt filter presses and multiple hearth incineration.

The White River has a low $Q_{7,10}$ of 35 MGD. Therefore, during the summer, plant effluent dominates the flow. In addition, there is down-stream aquatic life as well as river water supply intakes on the White River. To meet water quality standards, the Belmont and Southport plants have very stringent effluent limits for dissolved oxygen, ammonia, nitrogen, total suspended solids and BOD. The summer effluent limits for Belmont are 10 mg/l BOD, 10 mg/l total suspended solids, 3.7 mg/l ammonia nitrogen, and 6.0 mg/l dissolved oxygen. Southport AWT Plant limits are similar.

Due to the complexity of the unit processes employed and the stringency of the discharge limits, AWT had no choice but to become more proficient in its use of computers compared to most other wastewater treatment facilities. During the past several years, beginning in 1983 with plant start-up, culminating a $250 million AWT construction project, plant-wide computer controls were implemented, followed by the installation of a maintenance management system, four local area networks, and a laboratory and information management system (LIMS). In 1990 (and revised in 1992) the AWT Division formulated a computer systems integration plan to further coordinate and capitalize on earlier work and existing equipment (Quinn *et al.*, 1992). Since then, these networks have been combined into a wide-area network, the need having been realized to more fully incorporate data and programs into a fully relational database and users have also been trained to use personal computers and develop a number of useful software applications.

This paper is about that integration plan and the results achieved from that planning.

WHY INTEGRATE SYSTEMS?

Why did the AWT feel systems integration was so important? Many reasons for systems integration exist, depending on unique user needs and the current level of applied technology. A basic model is shown in Figure 1. Major systems integration benefits revolve around the strategic value of information. (Brueck and Bennett, 1992). Lets take a look at some of the specific reasons at Indianapolis.

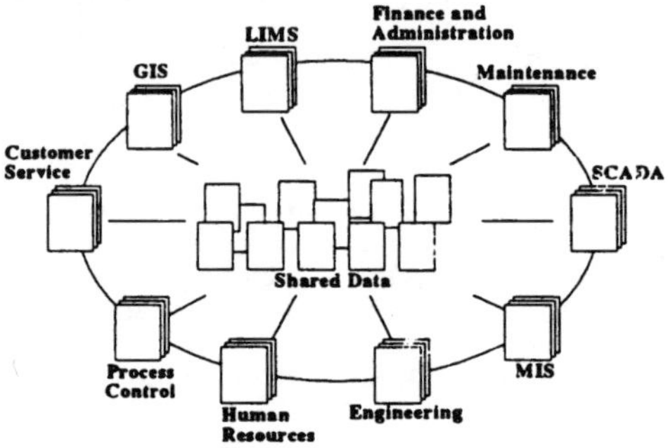

Fig. 1. Systems integration model represents sharing of data among applications and users.

1. Accessibility (appropriate information available to different user groups without human intervention, serves to promote data exchange).

2. Timeliness (responsive information available for analysis or action without delays more time for informed decision making).

3. Quality (captured and verified information close to its source transferred electronically reduces human error).

4. Synergy (multiple-source information combined for responses and decisions based on the widest possible knowledge base serves to synergize the conversion of simple data to information when combined with other data from various sources).

5. Training and Support (maximizes the impact and allows ease of multiple connectivity as well as facilities to easily support).

6. Security (system security is enhanced, data can be read, but is protected, back-ups done more frequently).

7. *Ad Hoc* Query (the database can be queried rapidly, promoting creativity in problem solving).

The base philosophy then is effective decision making having the right information, in the right format, at the right time, easily accessible to enable the manager or operator to make the right decision or do the right thing.

STRATEGIC INTEGRATION PLANNING

Background

Certainly there are success stories for independent systems at other utilities. However, what became apparent to Indianapolis was that the independent systems, no matter how successful, weren't enough. What was needed were integrated successful systems to accomplish the base philosophy. systems needed to allow each user direct access to all functions and data. The objective of the strategic plan was to formulate an ongoing program to keep the computer facilities current with technology, including enhancements for all user interfaces, user inquiries, networking standards and independent from single hardware or software suppliers. The effectiveness of strategic planning is measured by how well it builds on the previous foundation.

While over 90% of the computer systems at Belmont and Southport are networked and provide a variety of functions as shown in Figure 2, the next five years will demand an increase in the degree and effectiveness of the computer network and its member components. The non-transparent, often labor intensive gateways between functions must be identified, removed or improved.

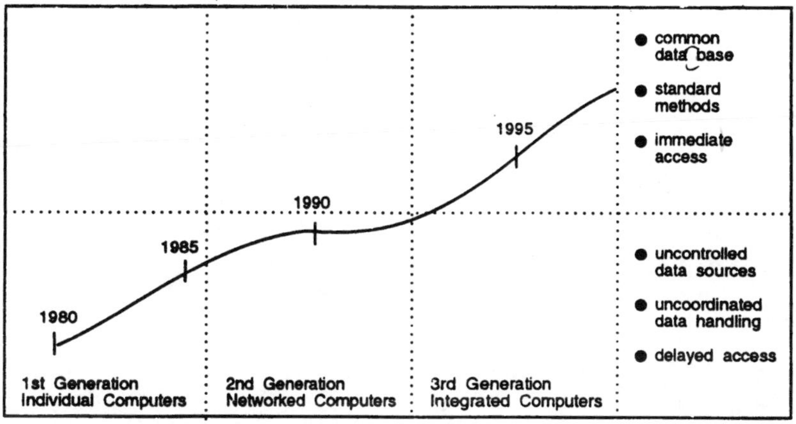

Fig. 2. System development the strategic plan builds on previous work to improve access to common data.

Objectives of Systems Integration

The primary objective of this project was to develop a model and plans for integrating existing AWT computer systems. A review was made of the interconnections required within AWT. The project also reviewed the potential interfaces with other Divisions of DPW and with other City Departments. A review was made of the existing computer hardware and basic systems software. The existing requirements and future requirements were both reviewed. The secondary objective was to provide plans for meeting functional needs that are not met by existing systems.

Summary of Key Points and Recommendations for Indianapolis

The plan yielded the following projects (things to do) to integrate.

Upgrade Local Area Network (LAN) and Include Southport AWT Plant

The Southport Plant LAN was connected to the Belmont LAN. This completed data communications between the two plants and allowed widespread easy access to the City-wide purchasing system and DEC VAX cluster, both remotely located. These capabilities improve organizational effectiveness and efficiency by cutting mail delays, reducing redundant work, and improving the overall information flow between plants and administration.

Relational Database

Initially, AWT needed to share data with the control systems, LIMS, and some key programs such as budget and a plant reporting database where information is required widely with AWT. Selecting the right software tools made a big difference in the cost and time to develop these programs and in the ability of AWT to keep them current, operational and flexible in the long run. Relational database technology is absolutely one of the revolutionary developments in software. The technology is new, but the benefits are so great that it was worthwhile for AWT as well as other plants to go through the learning curve.

Client–Server Configuration

The recommended hardware for supporting the relational database is the PC-based client–server configuration. Traditional minicomputer processing is split into back-end or server portions and front-end or client portions. The server stores the data and on demand sends the necessary data to client PCs. The client PCs then display, manipulate or print reports with the data. The client server provides great power at a lower cost than minicomputers provide. This portion is currently in process with *ad hoc* query and formatted daily reports and graphs depicting process performances being produced.

Inventory Control Budget Monitoring Programs

Several existing applications use freestanding PC programs that should be integrated. An example is the tracking program used by Inventory Control and by the Stock Room Sections. While the existing PC versions are preferable to manual records, they fall far short of the value to be obtained from a networked version of the program. A networked version will provide all authorized users access to the up-to-date information they need and a way of updating information on-line. It will also reduce paperwork, speed requisition entry, tracking, and integration into operations data.

Plant Reporting Database

One of the important applications for operating the wastewater plants is the long-term database for reporting and long-term data access. This, too, is well underway. Considering the value for decision making, engineering, optimization, and month-to-month operations, this is a vital system augmenting the present real-time process control system. The most effective way to handle these needs is the use of a relational database accessed from the wide-area network. A bonus is the saving of about 24-person days per month in data manipulation and report preparation.

Structured Query Language

The standard language for programming with relational databases is the Structured Query Language (SQL) developed originally by IBM. SQL has become a standard for the major relational database from all vendors. By settling on one database language, AWT has been able to train personnel who can work on a wide range of applications. SQL is used directly for accessing a database and embedded in program code in other languages such as C, FORTRAN, and COBOL.

Upgrade Maintenance Management Systems (MMS) to RISC Technology

The existing Hewlett Packard MMS computer is now fully utilized. Additional ports for connecting users and printers are needed that are beyond the capacity of the existing computer. HP has changed their computer technology to Reduced Instruction Set Computing (RISC) architecture. The computer chips operate significantly faster than the previous chip operated. The significance to AWT is the improved hardware performance at a proportionately lower cost. The processors are fast enough to offer better response time and support more users. Existing MMS software can easily be recompiled to operate on these processors. The upgrade will include larger disk drives and faster backup equipment. This technology provides a migration to more powerful computers in the future if they are needed.

Add Enhancements to Maintenance Software

A number of enhancements were proposed by AWT maintenance personnel. After review, some of these enhancements were made to the software. The effect was to make MMS more closely meet AWT needs. Other MMS capabilities not currently used may be implemented.

Use Structured Workshops to Improve Performance

Structured workshops have been and continue to be used to address issues, solve problems, and optimize design tasks. Results are obtained more quickly and are more satisfactory with structured workshops than with traditional techniques. This technique ensures that continuous improvement in quality occurs.

Upgrade Process Control Systems

The sludge processing, monitoring and control system (SPMCS) should be connected to the Belmont LAN. The main control system (MCS) at each plant should be replaced in phases starting in approximately five years. Such a plan is currently in place.

START WITH THE VISION DOING MORE WITH LESS

It is extremely important with any planning effort that right from the start the vision of doing more with less be established in the minds of all of the participants. If you look at any city, historically the O&M costs either per MG or on an annual basis typically go up four to seven percent a year. It is very difficult to control the rising costs because inflation and the costs of materials and medical plans continue to go up. A goal of beating the inflation rate in terms of O&M costs per MG is to be established by all participants very early in the game (see Figure 3 below).

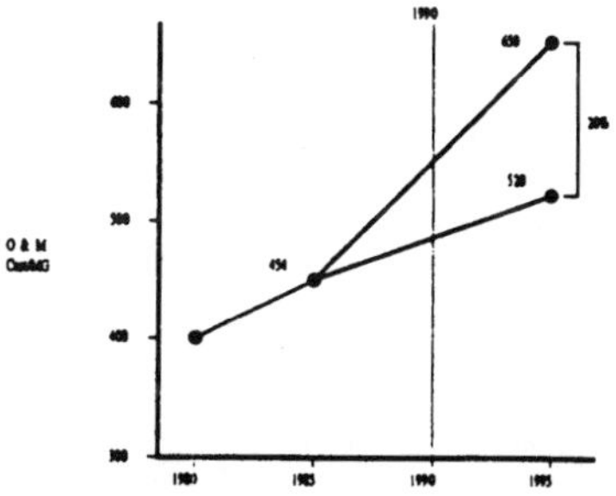

Fig. 3. Vision of doing more with less must be set early.

However, when talking about the vision, that vision must include expenditure of capital. This commitment must be made and the performance of the degree of progress toward the vision be measured against it.

At Indianapolis, the goal (starting in 1985) of ending up 20% less than a linear progression of past increases was stipulated. When you do that, the results you can achieve often far outdo the expectations.

APPLYING TECHNOLOGY ACHIEVES RESULTS

One of the primary mechanisms for reducing costs is to use technology so that manually monitored equipment can be eliminated. The shift in staff hours at Indianapolis from 1985 to 1991 amounted to over

66,000 h saved. This equates to about 32 positions. These 32 positions were eliminated through attrition and that meant that all other positions within the City were impacted. In other words, people were doing different, more productive work. This is one of the outcomes of applied technology. Also, utilizing automation to manage electrical energy and chemical usage contributed $1.1 million per year in avoided costs. When you apply technology to reduce costs, people end up doing different jobs. This 25.6% reduction in O&M costs from $454/MG to $338/MG was the most significant impact of applied technology at Indianapolis (Figure 4).

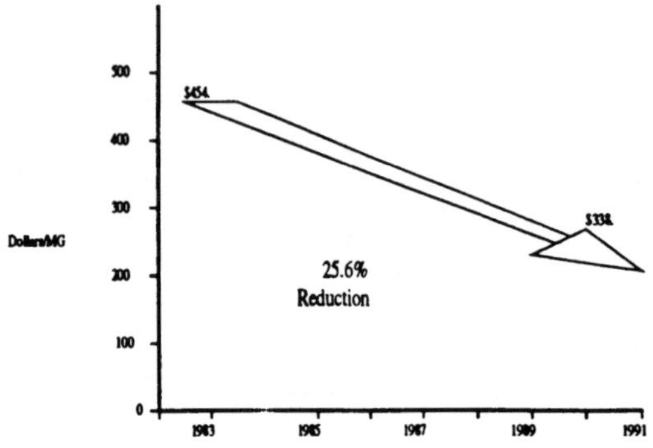

Fig. 4. Optimized work force means less overtime and big savings

APPLIED TECHNOLOGY KEEPS LID ON COSTS

The cost of applied technology systems is decreasing not only because of cheaper hardware, but because of easier-to-apply software. On the other hand, the cost of basic resources, labor and materials such as chlorine and the cost of medical programs, are increasing at about 10% per year. So the reverse nature of these curves offers the opportunity for the benefits which Indianapolis has proven can be derived from applied technology.

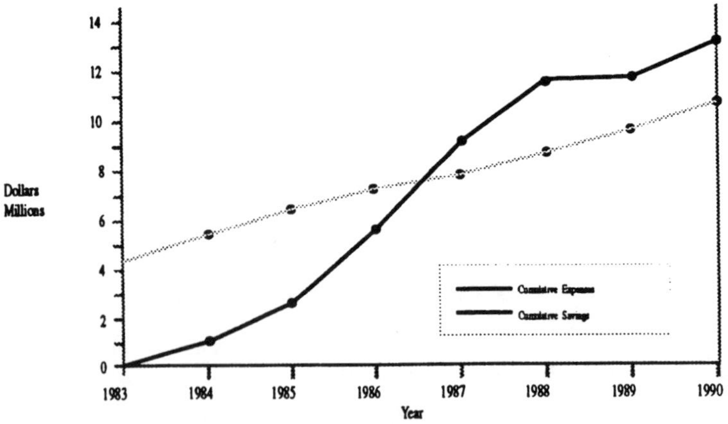

Fig. 5. Cumulative savings vs expenses: applied technology is a strategic opportunity to keep a lid on rising costs.

Figure 5 compares the cumulative costs of applied technology in Indianapolis with cumulated savings since AWT startup in 1983. An overall benefit took until 1987 to be realized but the results were worth the wait. Expenditures toward applied technology allowed expenditures for personnel, power and materials to decline or remain static. Note that the $4.5 million cost for automating the new plants before 1983 was included.

The profit on this long term investment in systems and training continues to the present. The conclusion is that productivity and efficiency have improved with applied technology as the main contributor.

THE BOTTOM LINE RESULTS EXCEEDED GOALS

It is technology that allows, even encourages, synergistic people interaction and allows them to continually improve what they do. This encourages productivity, high morale, shared data, and lower operations and maintenance costs.

At the City of Indianapolis from 1985 to the present, the O&M costs per MG were actually reduced from $454/MG to $338/MG as shown in Figure 6. This far exceeds the goal set at the beginning. In addition to this, it is anticipated that the costs will continue to be reduced through continuous improvement methodologies in the future, through benefits of enterprise-wide computing and the entire planning effort.

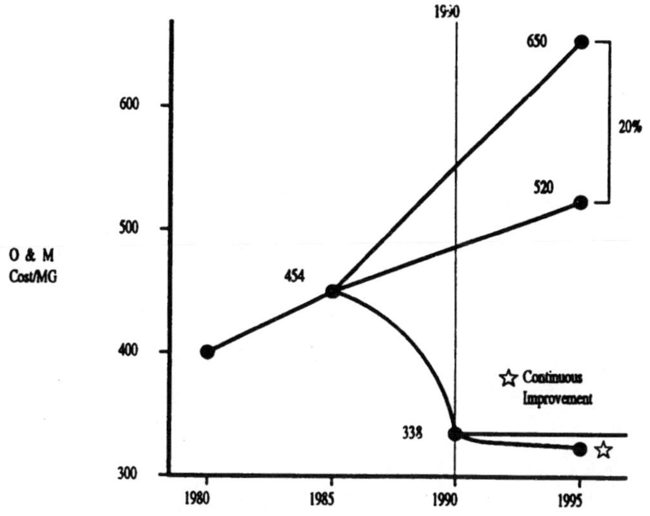

Fig. 6. Vision of doing more with less must be set early.

REFERENCES

Quinn, T. J., Sweeney, M. W. and Manning, A. W. (1992). The human element: getting good results from technology. *Water Environment and Technology*, **4**(7), 47.

Brueck, T. M. and Bennett, M. W. (1992). Integrated systems within your reach. *ISA Transactions*, **31**(1), 17.

Pergamon

Wat. Sci. Tech. Vol. 28, No. 11-12, pp. 29–36, 1993.

0273–1223/93 $7·00 + 0·00

USE OF PART-TIME UNMANNED OPERATION AT A LARGE WASTEWATER TREATMENT PLANT

Doug J. Lumley, Kent-Åke Sahlberg and Jan Haraldsson

GRYAAB – Göteborg Regional Sewage Works, Karl IX:s väg, S-417 22 Göteborg, Sweden

ABSTRACT

The use of part-time unmanned operation at the Göteborg Regional Sewage Works, Göteborg, Sweden (average flow 325 000 m^3/d) is described. Outside normal working hours, a two-man standby crew is responsible for the operation of the plant. The plant's SCADA system and special equipment used by the standby crews (e.g. pagers, cellular telephones, Notebook PC) inform the crews when an alarm has been generated. The standby crew can then react to the alarm, often by logging into the SCADA system and taking appropriate actions. More serious disturbances require visiting the plant. Use of the standby system has only a marginal effect on effluent quality and has not affected effluent compliance. A cost comparison of the standby system with a traditional full-time manned system indicates annual savings of about 1 350 000 SEK.

KEYWORDS

Wastewater treatment plant; SCADA; part-time unmanned operation; standby duty; remote control.

INTRODUCTION

The increasing use of instrumentation, control and automation (ICA) in the water industry has led to a vast improvement in the process information available to plant operators. ICA is used quite extensively at wastewater treatment plants (WWTP) in Sweden (Zagerholm, 1992) and has been a key factor in developing plants that are manned only during normal working hours (7.00 a.m. to 3.30 p.m.) (Bornö, 1990). The rest of the time these plants are operated by standby crews who are on call and responsible for the plant. This paper presents how the standby system is applied at the Rya WWTP and the effects on plant performance, effluent permit compliance and operating costs.

THE RYA WASTEWATER TREATMENT PLANT

The Göteborg Regional Sewage Works (GRYAAB – Göteborgsregionens Ryaverksaktiebolag) is a public utility owned by seven municipalities in the Göteborg Region on Sweden's west coast. GRYAAB owns and operates the Rya WWTP and an extensive tunnel system (total length approx. 90 km) which transports wastewater to the plant (Lumley *et al.*, 1988). The Rya WWTP serves about 550 000 persons and has about 793 000 population equivalents connected including industry. The design flow is 328 000 m^3/d and the maximum flow which can receive full secondary treatment is 518 000 m^3/d. Flows exceeding that level receive only primary treatment. The effluent permit allows a maximum BOD_7 of 15 g O_2/m^3 and total phosphorus of 0.5 g P/m^3 in the treated effluent, calculated as quarterly averages.

The treatment scheme consists of a pumping station with bar screens, primary settling, biological treatment (activated sludge/contact stabilization), chemical treatment (simultaneous precipitation with $FeSO_4$) and secondary settling. Sludge treatment consists of gravity thickening, anaerobic digestion, dewatering with filter belt presses and disposal in a rock cavern located 7 km from the main plant or reuse in land reclamation.

The plant has a staff of 49 employees (expressed in man-years) who are organized into four departments (number of employee man-years in each department indicated): Operations (13); Maintenance (17); Laboratory and Process Control (11) and Administration and Technical Support (4+4). Total operating costs for the facility were 61.2 MSEK in 1992.

INSTRUMENTATION, CONTROL AND AUTOMATION AT THE RYA WWTP

The plant has a distributed supervisory control and data acquisition (SCADA) and data analysis system based on a central process computer and local substation computers. Fourteen local substation computers, rack type modular built stations, are serially linked to the central process computer (HP 1000/A900) via a multiplexer. Each substation is programmable locally or from the central process computer and will operate independently should the central process computer malfunction. The current system has approximately 1 700 analogue I/O and 4 000 digital I/O signals. After the introduction of the SCADA system in 1984, use of the original conventional control room had tapered off until, in 1990, it was completely dismantled and removed. All work with the SCADA system is performed via terminal although it is possible to manoeuvre all machinery by hand. Advanced data analysis and long term data storage are available from a waste data analysis system (Lumley *et al.*, 1992).

Full operator access to the SCADA system is available at 16 colour graphics terminals, five monochrome graphics terminals and 15 PC's running terminal emulation software. The colour and monochrome terminals are placed at easily accessible locations throughout the plant and the PC's are used in the offices.

All alarms generated by events in the processes are classified as A, B or C class, according to the severity of the alarm, by a program run on the central process computer and displayed directly on the operator terminals. There are 20 alarm centres at the plant and to reduce the complexity of the system, these are grouped into six alarm groups. Each terminal can be configured to indicate alarms from the whole plant or from just one of the 20 alarm centres. There are presently about 2800 alarms in the SCADA system.

The alarms are also sent to a local pager system at the main plant which is part of the plant's telephone system. Individual pagers can be configured to receive alarms from any or all of the six alarm groups in the plant depending on where the operator is working. This allows an operator to be notified of a new alarm while working in one part of the plant but also to cover another part if required. The pager indicates activation by a beep, a blinking lamp and a nine character text message (the telephone number to call or the alarm group). Since the local pager system only indicates that an alarm has occurred, the operator must, via an operator terminal, check which alarm has been activated, cancel it and take the appropriate action.

STANDBY DUTY ORGANIZATION

The plant is normally manned five days a week between 7.00 a.m. and 3.30 p.m. (with 1 h flex time). During unmanned hours a two man standby crew is responsible for plant operation. The members of the standby crew are drawn from two groups: Group A consisting of eight senior operators who are responsible for the plant during standby duty; and Group B with ten junior operators who assist the senior operators. One person from each group has standby duty for 1 week on a rotating system. The standby assignments are on top of regular working hours. The senior operators have, on the average, 22 to 25 h of overtime per standby assignment and Group B 15 to 18 h. The average call-in (visit to the plant) takes between 4 and 5 h combined time for both operators.

The head of the operations department is responsible for administering the standby program. The senior operators meet each Monday morning to discuss the incidents which have occurred during the previous week and to plan for the coming week. The current standby crew is thus well informed about any repair or maintenance work which is planned and the group members are kept up to date on process changes, equipment modifications, etc. between their standby assignments. This is useful since several members of Group A are not from the operations department and may not otherwise receive pertinent information. The Group B members are from the operations and maintenance departments.

While on standby assignment, the crew members are required to be able to reach the plant within 1 h of receiving an alarm. A few years ago it was necessary to stay near a telephone so that a manned alarm centre could call up the person on duty. Modern telecommunications equipment has relaxed this requirement considerably and everything the crew needs to maintain contact with the plant fits in a Notebook PC carrying case.

The crew members receive a nominal compensation while on duty but not at the plant and double time compensation of their regular salary when called to the plant during standby hours. The standby crews also have the right to extra compensation after an assignment. This can be in the form of a day of rest after their assignment is over, e.g. Fridays for Group B, or in the form of a pay bonus.

SPECIAL EQUIPMENT USED FOR STANDBY DUTY

To reduce the impact of the standby assignment on the private lives of the crew members, many of the latest developments in communications and information processing have been implemented. Both men on standby have pocket cellular telephones and two pagers. One is a MiniCall Text (MCT) pager which can receive messages of up to 2000 characters sent by the plant process computer. The other is a simple pager used as a reserve system when contact cannot be made via the MCT pager. The senior operator also has a Notebook PC with built-in modem as well as a cellular telephone modem which makes it possible to contact the plant process computer from anywhere. The Notebook PC has the same terminal emulation capacity as the PC's at the plant, thus allowing them to log into the process computer with the same access as the operator terminals at the plant. During standby hours, each crew member also has the use of a company vehicle.

ALARM NOTIFICATION

A flow chart of the alarm response scheme used for the standby duty system at the Rya WWTP is shown in Figure 1. The standby duty alarm system is activated by the Group A operator when he leaves the plant. During standby hours, all A class alarms are sent directly to the senior operator's MCT pager. B class alarms can, on the other hand, be delayed to a given time according to the operator's preference, e.g. all B class alarms generated between 10.00 p.m. and 7.00 a.m. can be sent out at 7.00 a.m. The alarm program signals the MCT system, shown on the right-hand side of Figure 1, to send out an alarm with the appropriate text to the senior operator's MCT pager.

The MCT pager can indicate that it has received an alarm by visual, audio and/or vibrational signals according to user preference. The alarm is displayed with the same descriptive text message that is displayed on the operator terminals at the plant. To acknowledge and cancel the alarm, the operator can dial the MCT modem at the plant or he can log into the process computer with the Notebook PC and cancel the alarm.

If no response is made within a prescribed time limit, then the MCT system uses a second modem to dial two telephone numbers configured by the senior operator (normally his home telephone number and his cellular telephone) and a banded message is played back which simply states that an alarm has been activated at the plant. Now the senior operator has five minutes to respond. If he does not, then the alarm is sent to the junior operator's MCT pager and if no response is received within 10 minutes, the MCT system dials two telephone numbers configured by the junior operator. If the alarm has not been cancelled within five minutes then the sequence starts again.

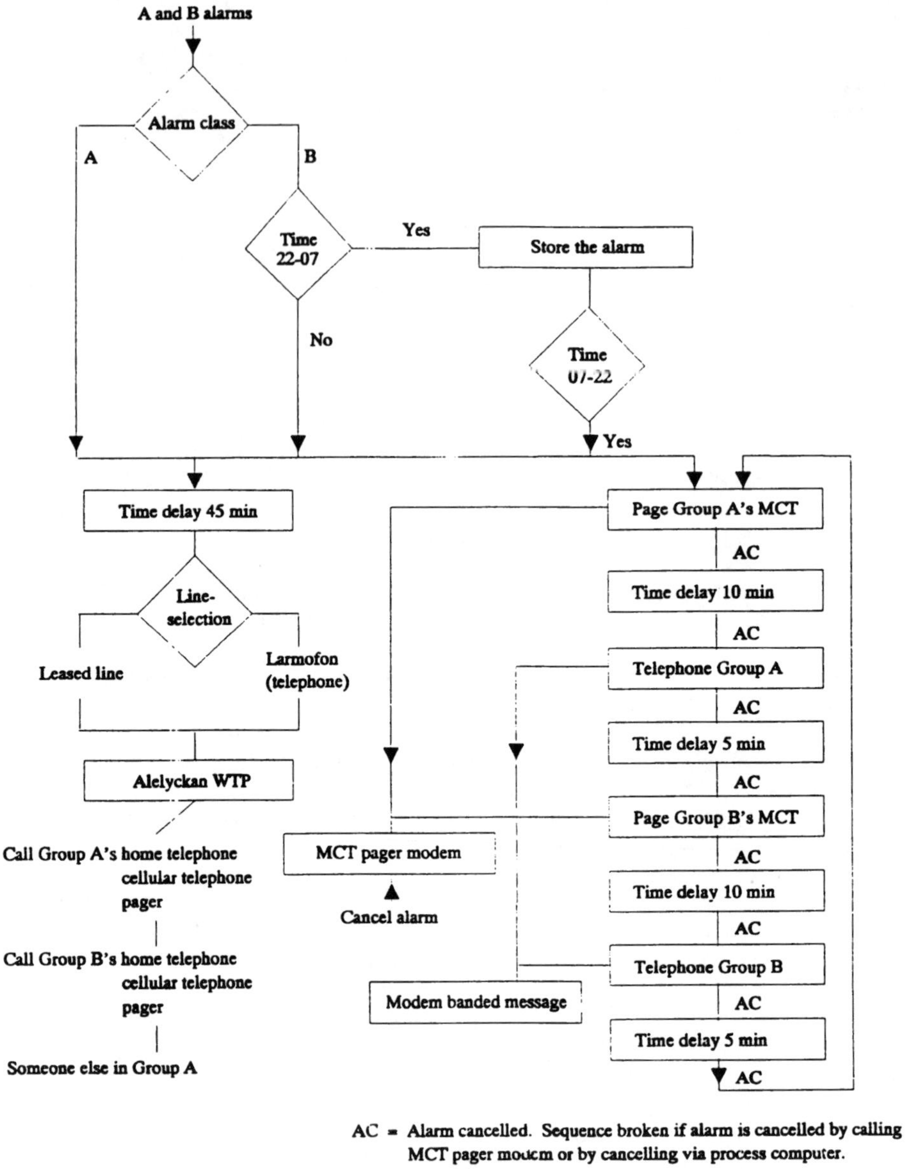

Fig. 1. Flow chart of the standby duty alarm scheme.

Parallel with the MCT system, a second, backup system is activated by the alarm program (shown in the left-hand block in Figure 1). This system waits 45 minutes for the MCT system to attempt to reach the standby crew. Should the alarm not be cancelled after this time, then the alarm message is relayed to the Alelyckan Water Treatment Plant in Göteborg which is manned around the clock and the operator on duty there tries to contact the standby crew directly by calling the senior operator's home telephone, cellular telephone, and then their second pager. If contact is not made then the same is done for the junior operator.

When the operator has received an alarm, he can log into the process computer with the Notebook PC, investigate the cause of the alarm, perhaps see directly any consequences it has on plant performance and often take immediate actions to alleviate the problem. If the problem cannot be resolved remotely, then one or both of the operators must visit the plant.

CONSEQUENCES ON PLANT PERFORMANCE

An analysis of plant performance was made to evaluate the effect that part-time unmanned operation had on effluent quality and plant operation using two types of reports filed by the senior operator: incident reports that register events that may have influenced effluent quality; and operating disturbance reports which are made each time a standby crew responds to an alarm. Since late 1992 a database program has been used to register a combined incident/disturbance report directly in the standby crew's Notebook PC. This has significantly reduced the number of "late" and forgotten reports and makes follow up much easier since SQL type searching can be used. In 1991 there were 23 incident reports registered and 10 of these were found to directly affect effluent quality. There were 395 disturbance reports registered in 1991 and 75 of these were resolved remotely, i.e. the standby crews were able to take measures to correct or circumvent a problem by logging in with their Notebook PC and taking the appropriate action. The other 320 disturbances required visits to the plant although in many cases some actions were first taken remotely before travelling to the plant. For comparison, in 1992 there were 469 disturbance reports, 178 of which were resolved remotely.

An attempt was made to quantify the effect that the 10 incidents had on effluent quality in 1991. The response times to the alarms and the actions taken for each incident were examined and a theoretical comparison of the consequences for standby and fully manned operation was made. It was estimated that about 23 tons of a total of 1056 tons of effluent BOD could be directly attributed to the standby system. For phosphorous, the estimate was 475 kg of a total of 48 tons P. These correspond to 2% and 1%, respectively, of the total BOD and P loads to the recipient. This extra loading did not however affect effluent compliance. Many unknown factors make it difficult to precisely quantify the effects but the estimates should be of the correct order of magnitude.

The most common incidents recorded involved loss of solids from the secondary settlers into the effluent and problems with the blowers to the aeration system. The most common operating disturbances were problems with the primary sludge pumps, the cable drawn scrapers in the secondary settlers and delivery of heated water to the city's district heating system. Another significant disturbance was false alarms that resulted in unnecessary call-ins.

OPERATING AND INVESTMENT COSTS FOR THE STANDBY SYSTEM

Typical annual operating costs for the standby system are given in Table 1. The labour costs are about 7.3% of the total labour costs and about 1.6% of the total operating costs. Table 2 shows the investment costs for the present system. Investments costs are depreciated over 5 years and, after taking into account the rest value of the equipment, the annual investment cost is about 50 000 SEK. Thus, the total annual cost of the standby system is nearly 1 100 000 SEK. The rapidly changing costs for telecommunication and computer equipment, and new developments in these branches make it difficult to give representative investment costs. For comparison purposes 1 USD is approximately 7.3 SEK (April 1993).

TABLE 1. Annual Operating Costs (in SEK) for the Standby System

Labour costs (base standby + overtime)	970 000
Telephone (cellular + pager)	24 000
Vehicles	28 000
Alelyckan WTP (annual service fee for operator assistance)	20 000
TOTAL	1 042 000

TABLE 2. Investment Costs (in SEK) for the Standby System.

2 Company vehicles	167 000
MCT hardware (pagers + modems)	18 000
MCT software	40 000
2 Cellular telephones	28 000
Cellular telephone modem	26 000
Notebook PC (incl. terminal software + built-in modem)	24 000
TOTAL	303 000

Note: One vehicle is used primarily for standby duty (80%) while the other is used for both standby duty (60%) and operations and maintenance (40%). The total investment cost for the vehicles was about 240 000 SEK.

Had the plant been manned full-time, then this would require five shifts of two persons to supervise operation of the plant and eight persons for the day shift of the operations department. The net increase in staff for the operations department would be eight person-years [5 shifts x 2 persons per shift + 8 regular - 10 in operations department at present] at a cost of about 2 450 000 SEK per annum [(180 000 x 1.53 (salary costs) + 24 000 (shift premium)) x 8 persons + 24 000 (shift premium) x 2 persons]. Total personnel costs were 13.2 MSEK in 1991 and thus full time staffing would increase this by about 11% to about 14.6 MSEK after taking into account that there would be no standby. Total operating costs would increase by about 2.4%. A permanent supervisory shift would also require an office and equipment for their use, requiring an investment of approximately 100 000 SEK. The net annual saving with the standby system is therefore about 1.35 MSEK.

STANDBY SYSTEM'S INFLUENCE ON STAFF

The advances made in telecommunications that have been incorporated into the standby system have lessened the impact that a standby assignment has on the crew. The system is quite flexible and by using reserve equipment, one of the others in each respective group can take over for an evening or a weekend if some other commitment should coincide.

The Notebook PC with terminal emulation software is routinely used by the senior operator to log into the process computer to monitor a known problem or just to "keep an eye" on the plant. Often some actions can be taken after receiving an alarm which will correct the problem, lessen the impact of the problem or make some other change in the process. Typical actions include changing pumping rates, switching to another pump, opening or closing valves or gates, shutting down a process or starting up an alternative process.

The standby system is well liked by those involved with it. Shift work is unpopular amongst the staff and the alternative of having standby duty five to seven times a year is more appealing. The extra status and income earned are also positive factors.

FACTORS TO CONSIDER WHEN APPLYING UNMANNED OPERATION

Many factors should be taken into consideration when using the standby system method for the operation of a wastewater treatment facility. Plants in Sweden that are currently fully manned but have considered the standby system as an alternative have found considerable resistance from some staff members who have, of necessity, arranged their lives around shift work (spouse's work, child care, social contacts, etc.) and do not welcome such major changes. Trade unions can also be very reluctant to major changes, which could affect labour relations.

The Rya WWTP has used the standby system since just after startup in 1972. A similar system had also been used at the plant which the Rya WWTP replaced. As the system has evolved, experience has shown that the following should be taken into consideration:

- Thoroughly test new facilities and process modifications before bringing them on-line.

- Make available enough information (alarms, setpoints, historical data, current process state) to the home terminal to allow events to be evaluated.

- Provide enough manoeuvrability from the home terminal to allow the necessary control actions to be taken.

- Input from sensors and instruments should be reliable and well behaved (or at least predictable).

- Eliminate false alarms.

- Automate processes with intelligent controllers that can be manoeuvred remotely.

- Formulate alternative process strategies.

- Have a follow up program. Notify all significant process changes/modifications to the standby crews. Analyze alarm/event lists to detect trends. What is causing the most frequent call-ins and what can be done to reduce them? Could a process change or improved automation resolve a problem?

- It is necessary to have a flexible SCADA system which can be adapted to new developments in control systems, instrumentation and telecommunications.

CURRENT DEVELOPMENTS

A hydrodynamical model of the catchment and tunnel system to the Rya WWTP is being developed (Gustafsson *et al.*, 1993) to provide four to six hour forecasts of flows to the plant pumping station using MouseONLINE. The simulator will be able to compare several control strategies for the appurtenances in the tunnel system and suggest suitable setpoints to the SCADA system.

A new, XWindow based SCADA operator interface has been installed which allows the operator to view several process mimics concurrently and provides easy access to information in the SCADA system. In the near future it should be possible to make this system available remotely to the standby crew.

SUMMARY

The use of part-time unmanned operation has only a minor impact on effluent quality at the Rya WWTP and has not affected effluent compliance. An estimated 1 350 000 SEK in operating costs are saved annually by using the standby system. Modern SCADA systems, telecommunications and computer equipment make the remote control of the plant both feasible and flexible. This has helped to reduced the impact of standby duty on the crews' private lives.

REFERENCES

Bornö, C. (1990). Process Computers in the Swedish Water Industry. In: *Instrumentation, Control and Automation of Water and Wastewater Treatment and Transport Systems* (*Adv. Wat. Pollut. Control* no. 10), R. Briggs (ed.)., Pergamon Press, Oxford, pp. 315-318.

Gustafsson, L.-G., Lumley, D., Lindeborg, C. and Haraldsson, J. (1993). Integrating a Catchment Simulator into Wastewater Treatment Plant Operation. To be presented at the 6th IAWQ Workshop on Instrumentation, Control and Automation of Water and Wastewater Treatment and Transport Systems, Hamilton, Canada. June 21-25, 1993.

Lumley, D., Balmér, P. and Adamsson, J. (1988). Investigations of Secondary Settling at a Large Treatment Plant. *Wat. Sci. Tech.*, **20**(4/5), 133-142.

Lumley, D., Haraldsson, J. and Sahlberg, K.-Å. (1992). Considerations for Part-Time Unmanned Operation of Large Wastewater Treatment Plants. *Proceedings 65th Water Environment Federation Annual Conference and Exposition*, New Orleans. Sept. 20-24, 1992.

Zagerholm, B. (1992). *Inventering av Datorbaserade System för Overvakning Och Styrning inom Kommunal Va-teknik. (Inventory of Computer Based Systems for Monitoring and Control within the Field of Municipal Water and Wastewater Technology)* (In Swedish). VA-FORSK Report 1992-07, Swedish Water and Wastewater Works Association, ISBN 91-88392-00-7.

Pergamon

Wat. Sci. Tech. Vol. 28, No. 11-12, pp. 37–43, 1993.

0273–1223/93 $7·00 + 0·00

FUNCTIONAL APPROACH LEADS TO SUCCESSFUL ICA ACROSS 124 WATER AND 389 WASTEWATER TREATMENT WORKS

M. J. Parker*, R. J. Casey*, L. K. Reynolds*, R. R. de Vries**, T. M. Brueck** and C. N. Williams**

** Thames Water Utilities Limited, London Water Control Centre, Lower Sunbury Road, Hampton, Middlesex, TW12 2ES, UK*
*** EMA United Kingdom Limited, Westerly Point, Market Street, Bracknell, Berkshire, RG12 1QB, UK*

ABSTRACT

Thames Water Utilities Limited operates a complex network of 124 water and 389 wastewater treatment works. Regulatory requirements are driving a massive capital programme of more than £2 Billion over 5 years to improve performance of these facilities. Improvements are planned, engineered, and operated by a variety of internal and external groups and individuals. Shareholders are demanding profitability and the Director General is pushing for improvements to customer service.

To achieve its objectives, Thames has developed an instrumentation, control and automation (ICA) systems master plan. The plan is based on the utility s water and wastewater functional requirements. The functional approach is applied to the large number of diverse facilities by grouping facilities and processes into classifications. The approach is used throughout the plan. Each process and its associated process control strategy are clearly defined to ensure consistency and the cost-effective implementation of instrumentation and control across Thames.

The functional approach was developed for each process control strategy using a series of structured workshops to achieve consensus and company-wide acceptance. Each department of the utility was involved in the workshops (including engineering, operations, maintenance, management systems, research and development, operational science). Top management sponsored the planning effort and committed to making the plan a reality.

KEYWORDS

Water; wastewater; process control; instrumentation; automation; standardisation; structured workshops; master plan; instrumentation, control and automation (ICA).

INTRODUCTION

Most water and wastewater utilities share the common objectives of providing consistent quality potable water to the customer s tap and a consistent acceptable quality of effluent back into the environment. These are difficult enough for utilities with a small number of facilities. It is even more difficult for utilities like

Thames Water which has 389 wastewater treatment works, 124 water treatment works, more than 30,000 km of water mains, and more than 82,000 km of sewers.

To meet its business and operations objectives, Thames prepared an instrumentation control and automation (ICA) master plan (Thames Water Utilities Ltd, 1993). The master plan was developed to ensure the most cost-effective implementation of ICA across Thames over the next decade. It is based on a functional approach which uses business objectives as the driving force and defines each process and its control strategy as the basic building block for operational success. The planning approach uses top-down objective definition and bottom-up strategy definition. To apply these concepts across Thames it was necessary to classify facilities such as treatment works, sewer networks, and water transmission and distribution networks.

Consistent quality across more than 500 locations requires consistent process monitoring and control across all of the locations. With a single bottom line dictated largely by efficiency and performance, it is imperative that Thames have consistent process instrumentation, control, and automation.

FUNCTIONAL APPROACH KEY TO PLANNING

The key to planning was the functional approach. The functional approach determined objectives of business and process functions. By focusing on functions rather than organisational structure, geographical boundaries or other secondary factors, the approach produced automation solutions basic to the company s needs. The intent of the approach was to establish objectives inherent to the business and processes which would be valid for the next ten years.

Objectives were defined starting with the utility's corporate mission statement, translating it into operational targets and objectives, and taking these targets and objectives to the process level on a priority basis. The high level objective of satisfying the customer, for example, would among other things subsequently be expressed in a process such as ozone treatment, where one of the main process objectives was to improve the aesthetic properties of the water. Or, for example, high level objectives would drive the operating objectives of the sewerage system, where minimising basement flooding to a specific limit is clearly a regulatory objective. This was the process of *Top-Down Objectives Definition* as shown in Figure 1.

The process objectives were then translated into functional strategies to form the basic building blocks for any instrumentation, control, and automation solution across Thames. These process strategies form the foundation for cross-function and multi-function ICA strategies. They include, for example, an energy management strategy across liquids treatment or solids treatment functional areas for wastewater (typically a works), or a capacity management strategy across the source and treatment functional areas for water. This was the process of *Bottom-Up Strategy Development.*

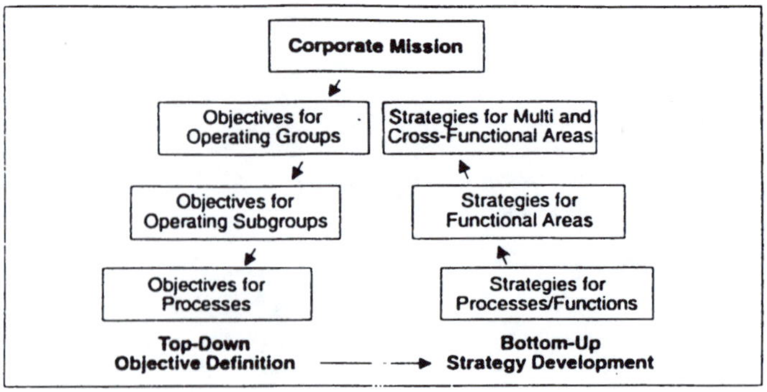

Fig. 1. Functional approach results in consistent success of ICA solutions in meeting business objectives.

The functional approach showed how to get from the current situation (widely varying process applications across Thames) to consistent ICA design, operation and maintenance. A technique that kept the planning effort true to the functional approach was the use of structured workshops. This involved the following steps.

Step 1

Diagnosing the operational needs and problems in meeting objectives from everyone s perspective including operations, maintenance, engineering, and other groups impacted.

Step 2

Bringing all the different perspectives together in structured workshops to create a common understanding of the problem and building collective agreement.

Step 3

Developing a common consensus-based long-term solution across Thames.

Workshops were attended by people from the various business, process, and technical areas of the organisation responsible for applying ICA. The process of gaining their consensus was considered vital to preparing a plan that would be implemented. The workshops were prepared and facilitated by a neutral facilitator to obtain the maximum amount of information in a short amount of time. Besides structuring discussions for analysis of functions, the facilitator made sure that discussions were free of technical and process jargon. The facilitator also helped each person participate in an open dialogue of the issues, and bridge gaps between organisational units.

FUNCTIONAL DEFINITION OF WATER

Definition of objectives and development of strategies were both organised along functional divisions of the business. Functions were first split into water and wastewater. Water was divided into four distinct functional areas as shown in Figure 2: Source, Treatment, Transmission, and Distribution.

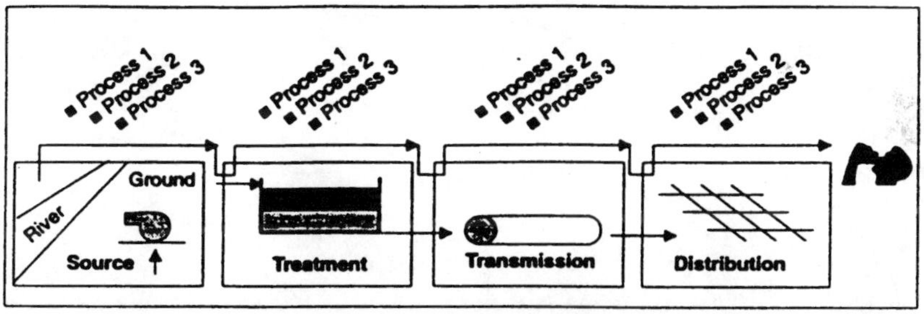

Fig. 2. Water functional definition.

The demarcations between these functional areas were clearly defined both for setting objectives and defining strategies. It was very important, for example, to define whether a borehole (well) pump should actually be considered as part of the Treatment functional area or as part of the Source functional area.

Within each of the functional areas, processes were identified. For example, the Treatment functional area included processes such as rapid gravity filtration, secondary slow sand filtration, and ozone treatment. Certain functions like flow balancing and energy management crossed boundaries of functional areas. To address these functions, multi- and cross-functional areas were defined.

FUNCTIONAL DEFINITION OF WASTEWATER

Wastewater was split into four distinct functional areas as shown in Figure 3: Sewerage, Liquids Treatment, Solids Treatment and Solids Disposal. The demarcations between these functional areas were also clearly defined.

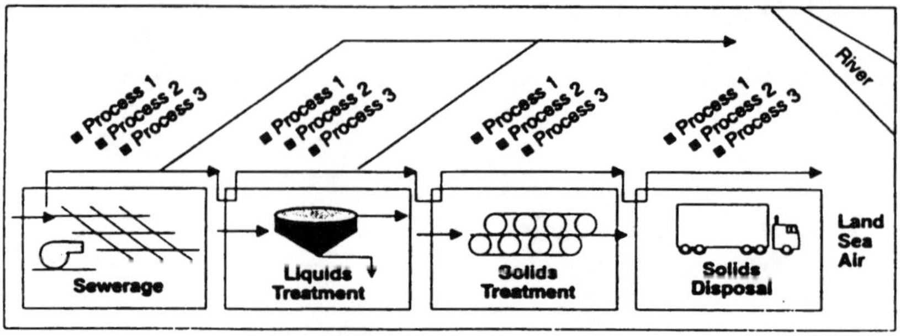

Fig. 3. Wastewater functional definition.

Here it was very important, for example, to define whether the terminal pumping station should actually be considered as part of the Sewerage functional area or as part of the Liquids Treatment functional area. Within each of the functional areas, processes were identified. For example, the Solids Treatment functional area included processes such as centrifuges, belt thickeners and plate presses. Multi- and cross-functional areas were also defined.

FACILITY CLASSIFICATION

Facility classifications were developed for works, sewerage networks, and water transmission and distribution systems. This allowed the functional and process strategies to impact the entire Thames Water asset base without needing location specific strategies for each of the 124 water treatment works, 389 wastewater treatment works, and the vast number of sewage collection and water transmission and distribution systems. This paper only discusses the classification performed for water and wastewater treatment works.

The classification criteria for works were identified as:

- Process complexity
- Production capacity
- Upstream and downstream factors.

Each of these factors was analysed according to their impact on the management and control of the works. Figure 4 shows the factors determining the works classification.

Facility classification was made on process criteria in order to keep planning on a functional basis. Rather than considering factors such as site staffing, maintenance support or similar issues, facilities were classified based on process factors. With this approach, the ICA solutions that were ultimately determined in the plan were considered to be inherent requirements of meeting business and operational objectives.

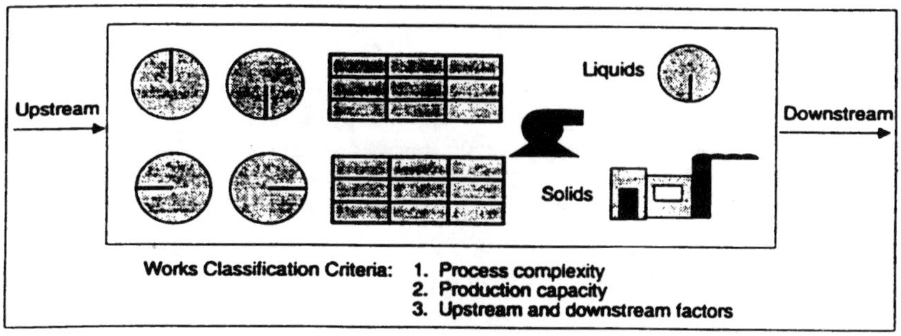

Fig. 4. Water and wastewater treatment works classification criteria.

CLASSIFYING WATER TREATMENT WORKS

Table 1 presents the classification of water treatment works. Process complexity in water treatment works ranges from simple disinfection to process streams including primary rapid gravity filtration, ozone treatment, granular activated carbon (GAC) treatment, and disinfection. Water treatment works range in size from small bore hole sites to large surface water treatment works treating up to 760 Ml/d.

Upstream factors such as variability of source water quality and storage capacity at source were considered in the classification analysis. Downstream factors such as the storage capacity of the transmission and distribution subfunctions were also reviewed for their impact. Analysis of these factors led to classifying the water treatment works into four types.

TABLE 1. Classification of Water Treatment Works

Type	1	2	3	4
Description	Large surface water works	Surface water works	Complex bore holes	Simple bore holes
Production Capacity	> 140 ml/d	≤ 140 Ml/d	All sizes	All sizes
Typical Treatment Processes	Includes advanced water treatment (e.g. Ozone, GAC)	Includes all process streams for conventional water treatment	Includes advanced water treatment (e.g. GAC, nitrate removal	Does not include advanced water treatment processes

CLASSIFYING WASTEWATER TREATMENT WORKS

Table 2 presents the classification of wastewater treatment works. Process complexity in wastewater treatment works ranges from process streams with passive processes such as percolating (trickling) filters, to process streams with active processes such as activated sludge, anaerobic digestion, power generation and incineration. The classification includes both the liquids and solids treatment processes typically found in works. Wastewater treatment works range in size from small package works processing several hundred population equivalents of flow to major treatment works processing over 1,000,000 population equivalents per day.

Upstream factors such as sewer storage capacity were considered in the classification analysis. Downstream factors such as the sensitivity of the effluent receiving waters were also reviewed for their impact. Analysis of these factors led to classifying the wastewater treatment works into four types.

TABLE 2. Classification of Wastewater Treatment Works

Type	1	2	3	4
Description	Large active process works	Medium active process works	Small active process works	Small passive process works
Processing Capacity	> 250,000 p.e.	30-250,000 p.e.	< 30,000 p.e.	All sizes
Typical Treatment Processes	Activated sludge	Activated sludge	Biologically active filter	Percolating (trickling) filter
	Anaerobic digestion	Anaerobic digestion	Activated sludge	Pumping
	Power generation	Tertiary treatment		
	Incineration	Combined heat & power		

STRATEGIES – BUILDING BLOCKS FOR SUCCESSFUL CONTROL

All water and wastewater processes and functions were prioritised according to their impact on achieving regulatory and business targets and objectives. High impact processes were analysed first with less important processes following later. The following items were included for each strategy:

- Process description – an overview of the process characteristics

- Process objectives – a prioritised list of the regulatory, statutory and business and operational objectives specific to the process

- Parameters – a prioritised list of all the parameters necessary to achieve the process objectives

- Process control strategy – a description of the process control and management philosophy as it applies to each of the objectives - a cost benefit analysis is associated with an objectives and strategies which are not absolutely necessary

- Process control diagram – a diagram showing the process and its main components, the location of any instruments and control devices, and the control relationships between components

- Input/output ranges – an indication of the number of inputs and outputs based on the cost-justified level of instrumentation control and automation

The approach used to clearly define and prioritise each process is shown in Figure 5.

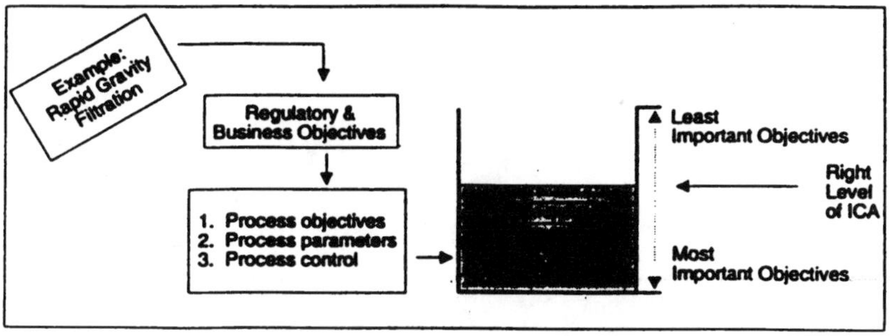

Fig. 5. Defining process objectives, parameters and control.

In keeping with the functional approach and bottom-up strategy development, strategies were first developed for individual processes and functions. Building on these definitions, strategies were developed in increasingly broadening scope for functional, cross-functional, and multi-functional areas. Water strategies included: spanning each functional area (e.g. energy management across the treatment functional area - typically a works), across more than one functional area (e.g. supply/demand balancing across source, treatment and transmission) and across the entire water function (e.g. quality monitoring).

Wastewater strategies included: spanning each functional area (e.g. flow/storm balancing across the sewerage functional area), across more than one functional area (e.g. energy management across liquids treatment and solids treatment) and across the entire wastewater function (e.g. quality monitoring).

This approach resulted in a purely functionally based requirement for any ICA solution – a solution based only on prioritised objectives.

These strategies are used as a consistent base for design and refurbishment of any ICA across Thames.

CONCLUSION – FUNCTIONAL APPROACH RESULTS IN CONSISTENT SUCCESS OF ICA

Thames is well on the way to achieving consistent application of ICA across its complex network of water and wastewater facilities. ICA is consistent in meeting business and operations objectives through cost-effective solutions.

Analysis of needs on a functional basis and facility classifications have set objectives for the control strategies. The process and functionally-based control strategies will, in turn, lead to consistent ICA system solutions across Thames.

The success of the effort to achieve consistency is largely due to the contributions from more than 150 participants in the structured workshops and numerous participants in a "core team" guiding the efforts. The continuation of this success will depend on the continued involvement of Thames employees and the continued support and commitment from management.

REFERENCE

Thames Water Utilities Limited (1993). Instrumentation Control & Automation Master Plan (February), Reading, Berkshire, U.K. Thames confidential internal document.

COLLECTION SYSTEM MONITORING AND CONTROL

Pergamon

Wat. Sci. Tech. Vol. 28, No. 11-12, pp. 45–54, 1993.

0273–1223/93 $7·00 + 0·00

INTEGRATING A CATCHMENT SIMULATOR INTO WASTEWATER TREATMENT PLANT OPERATION

Lars-Göran Gustafsson*, Doug J. Lumley**, Claes Lindeborg*** and Jan Haraldsson**

* *VBB/VIAK Ledningsförnyelse, Gullbergs Strandvägen 3, S-403 14 Göteborg, Sweden*
** *GRYAAB - Göteborg Regional Sewage Works, Karl IX:s väg, S-417 22 Göteborg, Sweden*
*** *Department of Control Engineering, Chalmers University of Technology, S-412 96 Göteborg, Sweden*

ABSTRACT

A catchment model which describes the hydrological and hydrodynamic processes in the catchment to the Rya wastewater treatment plant in Göteborg, Sweden is presented. The model uses hydrological models (MouseNAM) of 20 subcatchments and a hydrodynamic model (MousePIPE) of the tunnel system to make flow and CSO predictions. The catchment model has been used to evaluate several different operating strategies of the plant's pumping station and will be implemented as an operating tool. Efforts are under way to establish hydraulic and mass transport models for the pumping station and primary settling so that a loading model for the activated sludge process can be defined.

KEYWORDS

Catchment simulator; deterministic model; hydrological and hydrodynamic modelling; wastewater treatment plant; SCADA; on-line simulation.

INTRODUCTION

The increasing pressure to reduce pollutant loadings from urban areas on receiving waters has led to more holistic approaches to designing and operating sewage collection systems and wastewater treatment plants (WWTP) (Durchschlag *et al.*, 1991). It is quite common that sewage collection systems are designed and operated without taking into consideration the hydraulic or mass loadings to the WWTP or the total effect of combined sewer overflows (CSO) to the recipient. On the other hand WWTP operators have traditionally been complacent and accepted without questioning whatever comes down the pipe, whenever it comes.

This paper describes the development of a model of the catchment to the Rya WWTP named RAOM (Ryaverkets Avrinnings Områdes Model – Rya WWTP Catchment Area Model) and the efforts being made to integrate it with the treatment plant's advanced supervisory control and data acquisition (SCADA) system to give the operators a valuable tool to deal with loading variability.

THE RYA WASTEWATER TREATMENT PLANT

The Göteborg Regional Sewage Works (GRYAAB) is a public utility owned by seven municipalities in the Göteborg Region on Sweden's west coast. GRYAAB owns and operates the Rya WWTP and an extensive tunnel system (total length approx. 90 km) which transports the wastewater to the plant (Lumley *et al.*, 1988). The Rya WWTP serves about 550 000 persons and has about 793 000 population equivalents connected including industry. The design flow is 328 000 m^3/d (3.8 m^3/s) and the maximum flow which can receive full secondary treatment is 518 000 m^3/d (6 m^3/s). Flows exceeding that level receive only primary treatment. The average daily flow was about 321 400 m^3/d in 1991 and ranged from 179 700 m^3/d to 813 900 m^3/d. The effluent permit allows a maximum BOD_7 of 15 g O_2/m^3 and total phosphorus of 0.5 g P/m^3 in the treated effluent, calculated as quarterly averages.

The water treatment scheme consists of a pumping station with bar screening, primary settling, biological treatment (activated sludge/contact stabilization), chemical treatment (simultaneous precipitation with $FeSO_4$) and secondary settling. Sludge treatment consists of gravity thickening, anaerobic digestion, dewatering with filter belt presses and disposal in a rock cavern or reuse in land reclamation.

The tunnel system, illustrated in Figure 1, consists of about 90 km of tunnels made in granite bedrock with cross sectional areas varying from 6 to 18 m^2 and with bottom slopes of 0.07 to 0.1%. There are 27 access tunnels into the system used for construction and maintenance purposes. The transport time at dry weather flow is 16 to 20 h and the time of concentration at dry weather flow is 0 to 30 h. The tunnel system consists of two main branches, the north branch which leads directly to the Rya WWTP and the south branch which is connected to the north branch just upstream of the plant via two inverted siphons under the Göta River.

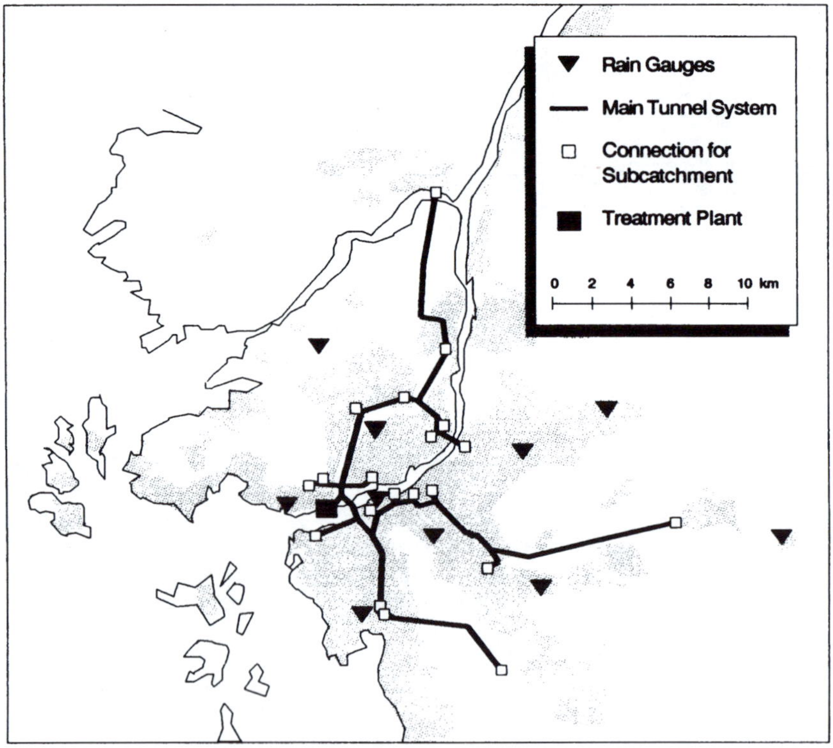

Fig. 1. The Rya WWTP catchment.

The plant's SCADA system has a central process computer and 14 local substation computers. The current system has approximately 1 700 analogue I/O and 4 000 digital I/O signals and 2 800 alarms. A data analysis system is used for long term data storage and advanced statistical analysis. The SCADA system is designed for part-time unmanned operation of the plant (Lumley *et al*, 1992) and provides fully automated remote control of the pumping station and the water and sludge treatment facilities.

RAOM – A HYDROLOGICAL AND HYDRAULIC MODELLING TOOL FOR THE TUNNEL SYSTEM

The RAOM catchment model has been formulated and verified for the 200 km^2 catchment to the Rya WWTP using a deterministic model concept, MOUSE (**M**odelling **Of** **U**rban **SE**wers). The MOUSE package is one of the most widely used tools for the analysis of urban drainage problems in Europe. Different modules are available within the MOUSE package for describing the hydrological effects and the hydraulic performance of sewer systems (Lindberg *et al.*, 1986).

The hydrological processes are described with a general hydrological model, MouseNAM. This model takes into account both the fast response component (FRC) from impervious areas and the slow response component (SRC) caused by infiltration into the sewer system from the surrounding soil. The latter strongly depends on the hydrological history, i.e. preceding events (Gustafsson *et al.*, 1991), and snow, surface, root zone and groundwater storage are taken into account. The hydraulics of the tunnel system, e.g. overflows, storage, controllable structures, etc. are described using the St Venant's equations in the hydrodynamic model MousePIPE.

As a result of an increased concern for the environment, real time control (RTC) of combined sewer systems is becoming more and more common (Lindberg and Carr, 1992). For the assessment of long term effects, continuous modelling is required. However, simulating long term periods (several years) with a dynamic model and small time steps (30 s) is still not practical. The continuous model used here is therefore a combination of a new steady state module and a well proven dynamic model. Based on actual flow conditions the continuous model automatically switches back and forth between steady state and dynamic resolution. The continuous model, including extended flow control possibilities, is an efficient tool for evaluating the potential of RTC. Furthermore, it can be used in assessing optimal operating strategies and as a training tool by operators of both sewage collection systems and WWTPs.

MODEL FORMULATION

The RAOM model includes the whole catchment area of the Rya wastewater treatment plant. Of the total area of about 200 km^2, approximately 20 km^2 are impervious surfaces. The catchment has been divided into 20 subcatchments, each defined by the physical connection points to the tunnel system found in the subcatchment. The classification of the subcatchments has been made taking into consideration the geometrical distribution and size, as well as the flow capacity of the connection points. The general features of the catchment including the subcatchment connection points are shown in Figure 1.

When simulating such a large area, the spatial distribution of the precipitation becomes an important factor. Therefore, precipitation data from 10 rain gauges has been distributed to each subcatchment using Thiessen polygons. The placement of the rain gauges in the catchment is indicated in Figure 1. About 65 km of the total of the 90 km of tunnels in the catchment are described in the model. The 65 km of tunnels are divided into 50 tunnel sections and correspond to a volume of about 560 000 m^3. Existing overflow outfalls are described, as well as the flow capacity for each subcatchment connection to the tunnel system.

The model also describes the large pumping station at the Rya WWTP which pumps up wastewater from the tunnel system for treatment. The operating strategy of this pumping station is quite complex and in RAOM a simplified relationship of the level and volume in the tunnel at the pumping station and the pumping rate is used. Measures taken to control the transport of sediment material to the bar screens are ignored.

VERIFICATION OF THE RAOM CATCHMENT MODEL

Verification of the model with measured data was made in three steps. In the first step the hydrology for the whole catchment area was verified, i.e. verification of a single MouseNAM catchment for the whole catchment. This was done in two parts: first calibration of the model against measured inflow to the treatment plant during a four year period, 1986 to 1989 and then validation against independent data from 1979 to 1984 which did not effect the final values of the model parameters. An example of the verification results is given in Figure 2.

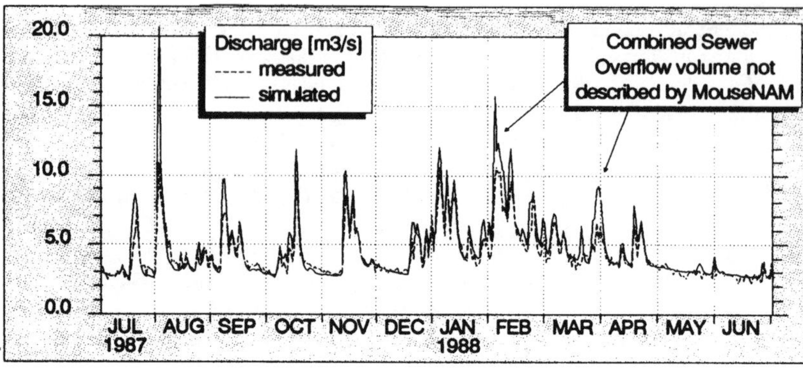

Fig. 2. Verification of the one cell MouseNAM catchment model.

The second step was to verify the distributed hydrological model using the results from the first step, i.e. distribute the hydrological parameters to the 20 subcatchments. The largest and most significant subcatchments were then verified using several years of measured data at the tunnel connection points.

The last step was the verification of the hydraulic processes with emphasis on overflow discharges and friction losses. This was performed using on-line flow and level measurements from different locations in the tunnel system, e.g. data from the treatment plant pumping station. Figure 3 shows two examples of the agreement between observed and simulated inflow to the treatment plant. Comparison with measured flows indicates that the RAOM catchment model can produce accurate flow predictions 4–6 h forward in time with a high degree of accuracy and that 12 h predictions are useful for indicating trends over a longer time horizon.

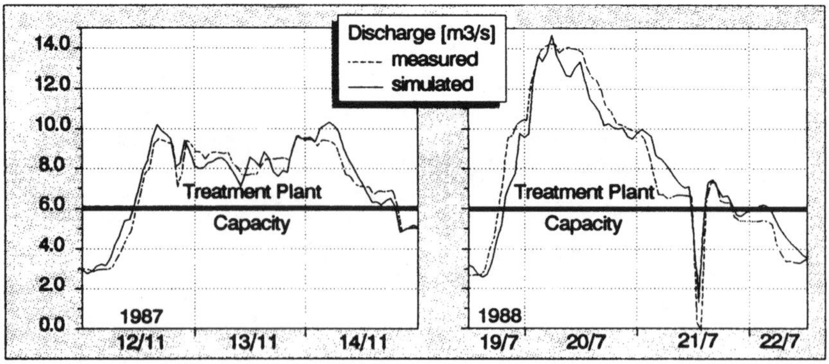

Fig. 3. Observed and simulated flows in the tunnel system using 20 subcatchments.

The differences between calculated and measured flows for the hydraulic modelling can partly be explained by uncertainties in the model parameters, e.g. friction losses in the tunnel, response functions for the subcatchments, etc. The description of the spatial distribution of the rainfall pattern is, however, still quite uncertain (20 km^2 per rain gauge on the average). The disagreement between observed and simulated values can probably, to a large extent, be explained by this factor.

APPLYING RAOM TO THE DEVELOPMENT OF REAL TIME CONTROL STRATEGIES

The modelling tool RAOM described above, has been used to evaluate the potential for real time control of the tunnel system and to study better control strategies for the pumping station at the Rya WWTP. The effects have primarily been evaluated considering CSO volumes from the sewer system and from the treatment plant. The risk for flooding in the tunnel system has also been considered.

Three different operating strategies have been evaluated to examine methods of operating the sewage collection system in cooperation with the treatment plant. These are:

The present system with alternative biological treatment capacities at the treatment plant.

Control of the tunnel system with the present biological treatment capacity (6.0 m^3/s).

Control of the tunnel system with an increased biological treatment capacity (8.0 m^3/s).

In the simulations, control of the tunnel system was made using three existing controllable structures:

The pumping station at the treatment plant.

One of the local pumping stations in the sewage collection system, serving about one third of the total catchment area.

Controllable sluice gates in the inverted siphons under the Göta River.

In addition, three simulated flow attenuation appurtenances (adjustable gates) were placed further up in the tunnel system. A sketch of the control system is given in Figure 4. The simulated adjustable gates are the most northerly control structure and the two most southerly of those indicated in Figure 4. During the simulations the control structures were regulated by five level sensors. Two were existing level sensors located at the WWTP pumping station and in the tunnel system just upstream of the inverted siphons under the Göta River and the other three were simulated level sensors associated with the simulated adjustable gates as indicated in Figure 4. The shaded areas in Figure 4 indicate the relative storage volume available in the tunnel system for each control structure.

The evaluation of long term effects of the different operating strategies has been based on continuous modelling of several years of meteorological data. Figure 5 shows an example of the simulated inflow to the treatment plant over a one year period.

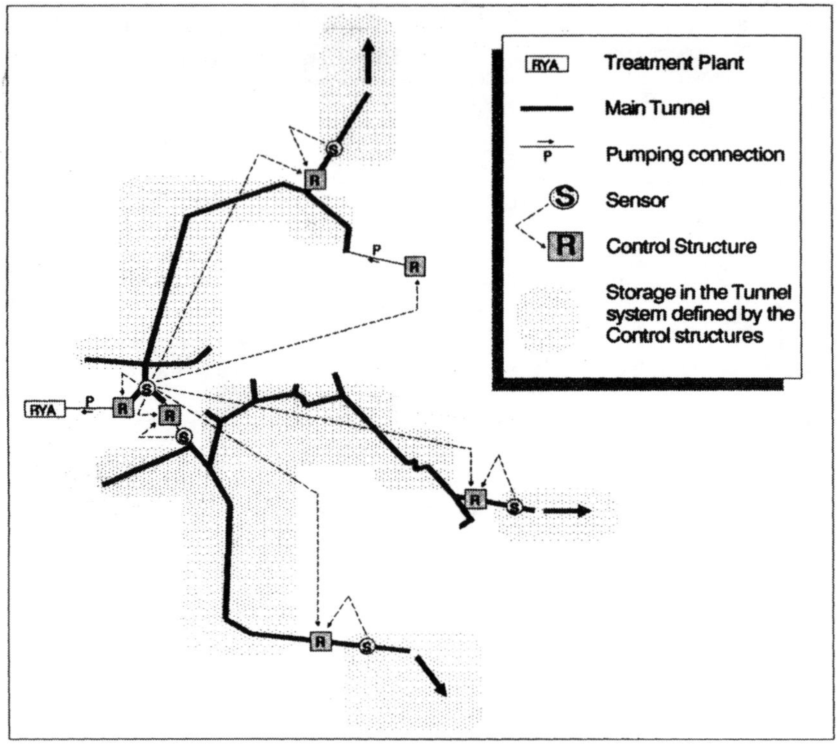

Fig. 4. Sketch of the tunnel system and the control structures used in the simulations.

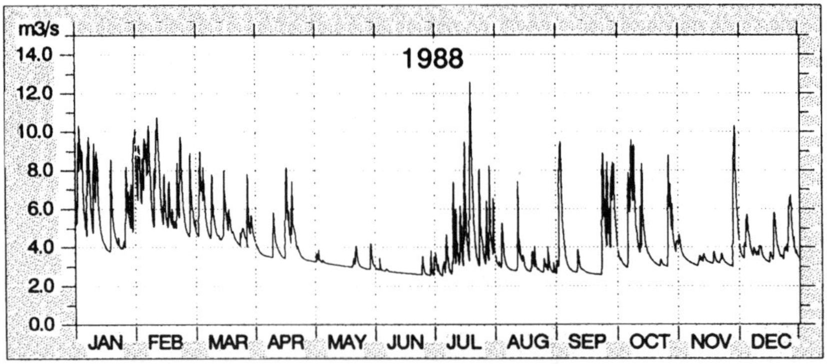

Fig. 5. Simulated inflow to the wastewater treatment plant during 1988.

Simulations for the different control strategies suggest that the yearly overflow volume in the sewage collection system can be reduced by up to 50% through regulation and control of the tunnel connections and by utilizing the storage capacity of the tunnel system. Furthermore, the yearly overflow volume at the treatment plant can be reduced by up to 55%. If the operation strategy also includes increasing the biological treatment capacity at the treatment plant from 6.0 to 8.0 m^3/s, the yearly overflow volume can be reduced by up to 90% at the treatment plant as indicated in Table 1. It is interesting to note in Table 1 that the percentage reduction is about the same for the year with extreme precipitation and the year with average precipitation.

TABLE 1. Overflow Volume in Percent of Present Conditions for Precipitation from an Average and an Extreme Year

	Without control in tunnel		With control in tunnel	
	average year	extreme year	average year	extreme year
Sewage collection system	100%	100%	48%	55%
Biol. treatment cap. 6 m^3/s	100%	100%	46%	--
Biol. treatment cap. 8 m^3/s	21%	23%	8%	10%

Figure 6 shows an example of the effects on the pumping rate to the treatment plant when controlling the tunnel system. The example indicates that the pumping rate can be maintained at the full biological treatment capacity of the plant much longer than what is currently possible yielding significant reductions in overflows in the collection system and at the plant. This effect can also be seen in Figure 7, where duration curves for the pumping rates for different strategies are plotted for an average year. The ability to be able to test different strategies is very valuable for evaluating the total effects of the CSO's in the collection system and the treatment processes on the recipients.

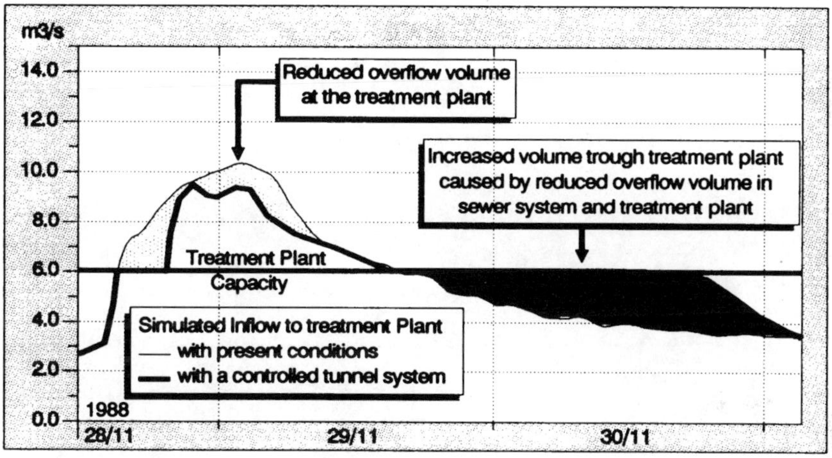

Fig. 6. Effect of control system on CSO's and overflows at the treatment plant.

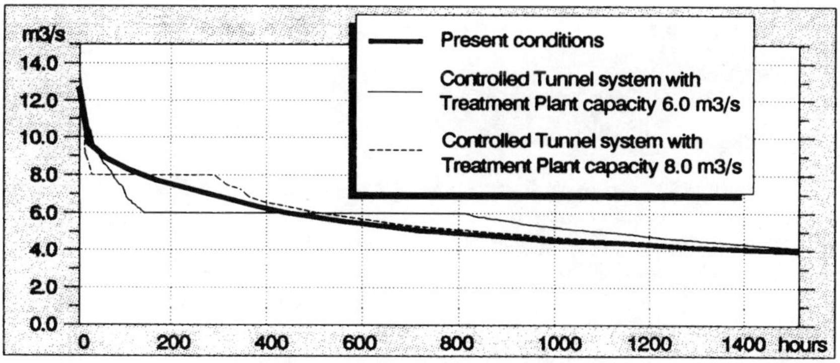

Fig. 7. Flow duration curves for control strategies.

PUMPING STRATEGY AT THE TREATMENT PLANT'S PUMPING STATION

The control strategies currently being used at the Rya WWTP pumping station are being examined to determine what modifications are necessary and possible to best use the information available from the RAOM catchment simulator. The present control strategy is based on an empirical relationship between the pumping rate and the level and volume of the wastewater at the end of the tunnel system. The tunnel level thus determines the pumping rate setpoint and a hysteresis band, tuneable by the operator, is utilized to smooth out moderate fluctuations in level. Pumping rates are maintained to avoid excessive levels in the tunnel system which could lead to flooding at the plant. Changes in pumping rates are ramp changes to make the least impact on the treatment processes. Part of the strategy is also to mobilize the sediment in the tunnels by lowering the tunnel levels once a day for a short period to create high water velocities, which aid sediment transport. This helps to avoid large influxes of sediment during periods of high flows. Manual override of the automatic controls is used for exceptional operating conditions.

Unlike other process industries where the inlet flow of the raw material is a controlled function of a preset production rate, a WWTP must handle ever changing hydraulic and mass loadings. Figure 8 illustrates a period of very high pumping rates to the Rya WWTP. Also indicated in the figure is the maximum capacity for the biological treatment stage (6 m^3/s) following primary settling.

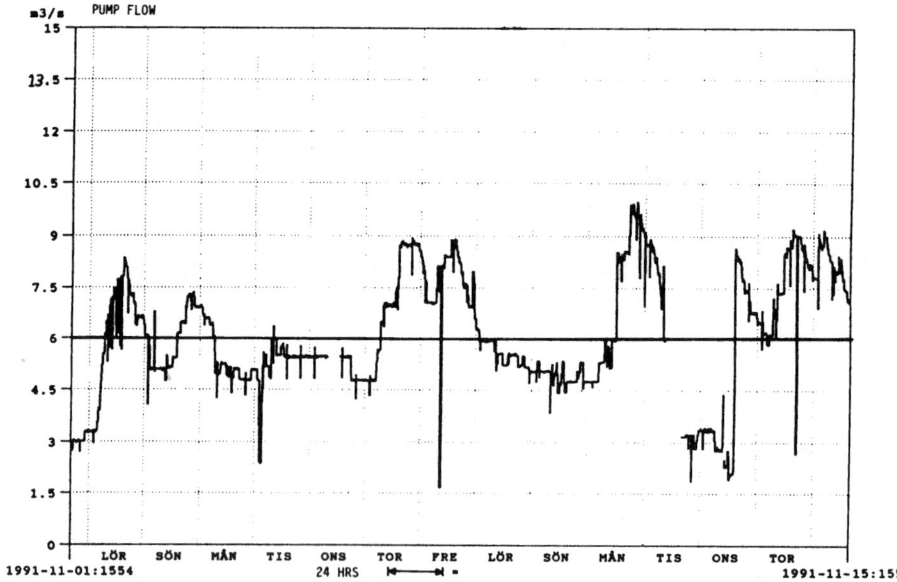

Fig. 8. Pumping rate during a period with high precipitation.

By predicting the flows in the tunnels and by the efficient use of the available storage volumes, the periods with excess flows that receive little treatment can be reduced as described above. This is in keeping with the ultimate objective of minimizing the pollutant load on the recipient. It is of course a very far reaching and difficult task to design and operate such a system. Besides keeping the overflows to a minimum, it is believed that it is advantageous to control the loading on the biological stage, e.g. to maintain a constant substrate flow. This requires a control system that keeps track of the remaining suspended solids, the substrate concentration of the wastewater and the impact of precipitation on the raw wastewater volumes and concentrations. Figure 9 shows the structure of such a proposed system. The blocks represent models of the different process units.

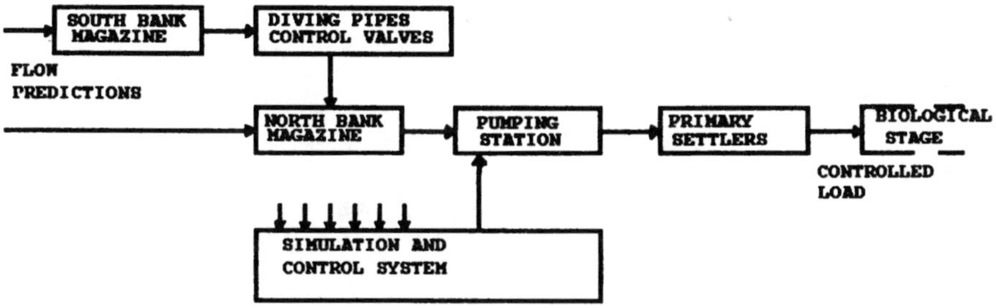

Fig. 9. Proposed simulation and control system.

PUMPING STATION MODEL

The objective for modelling the pumping station was to relate the flows from the two branches of the tunnel system (north and south of the river) to the storage volumes and levels in the tunnel system near the plant. The pump wet well is located at the end of the north tunnel and the two inverted siphons connecting the south branch to the north branch have controllable valves. Figure 10 shows the application of the model, where the two tunnel flows, Q_N (north) and Q_S (south) are estimated using existing on-line sensors.

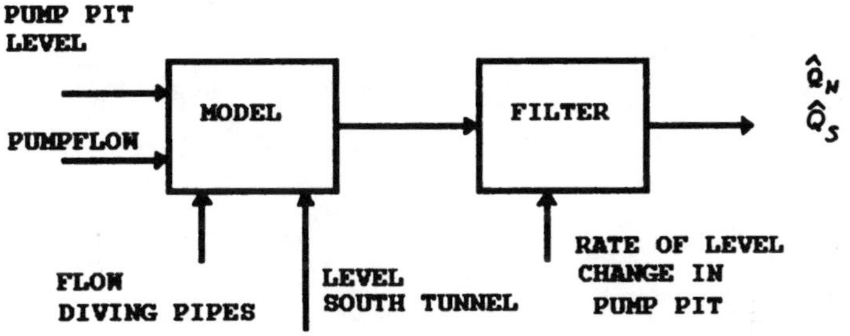

Fig. 10. Flow estimation scheme.

One of the problems encountered when developing the model was the uncertainty of the flow estimate Q when operating at low pump wet well levels. It was therefore necessary to add to the existing volume/level relationship a detailed geometrical description of the last 13 tunnel segments. Another problem was inaccuracies in the tunnel level measurement when large changes in the pumping rate were made. The level sensor is located at the end of the tunnel, not far from the pump wet well. The sensor reading is transient during an estimated 10–15 min period after a rapid change in pumping rate and the level reading is not representative of the liquid volume in the tunnel due to the changing backwater curve. The problem was solved by creating a filter sensitive to large changes in pumping rates.

SUMMARY

The RAOM catchment model has been found to give flow and CSO predictions suitable for use in the operation of the Rya WWTP. The predictions can be used for controlling the pumping station at the plant and reducing overflow volumes. The RAOM model has also been shown to be useful for evaluating different control appurtenances in the tunnel systems which would allow greater utilization of the potential storage capacity. Modelling of the pumping station has shown that corrections must be applied to the on-line level measurements to account for transients in the backwater curve when rapid changes in the pumping rate are made.

FUTURE DEVELOPMENT

The RAOM catchment model is being used to evaluate the design and location of flow attenuation appurtenances in the tunnel system. These will be necessary to utilize the storage capacity of the tunnel system. A further improvement for optimizing operation strategies will be the introduction of an on-line modelling tool MouseONLINE (Lindberg and Carr, 1992; Nielson *et al*, 1993) in mid 1993. This will allow the RAOM catchment model to make on-line flow and CSO predictions based on input from the plant's SCADA system (flows, levels, meteorological measurements). MouseONLINE will be able to examine several pumping strategies on-line and predict the consequences of each strategy. The use of on-line radar precipitation measurements is also being evaluated. The detailed information about the spatial distribution of precipitation that radar measurement can produce should reduce the error in flow predictions in the subcatchments. Also under development are tools such as MouseTRAP which models complex sediment transport and water quality processes. A model to predict the performance of the primary settlers is also being developed as part of the efforts to model the hydraulic and mass loadings to the biological treatment stage.

REFERENCES

Durchschlag, A., Härtel, P., Hartwig, P., Kaselow, M., Kollatsch, D., Otterpohl, R. and Schwentner, G. (1991). Total Emissions from Combined Sewer Overflow and Wastewater Treatment Plants. *European Water Pollution Control*, **1**(6), 13-23.

Gustafsson, L.-G., Lindberg, S. and Olsson, R. (1991). Modelling of the Indirect Runoff Component in Urban Areas. *International Conference on Urban Drainage and New Technologies*, Dubrovnik, June 17-21, 1991.

Lindberg, S., Jorgensen, S. and Willemöes, T. (1986). Mouse – Modelling of Urban Storm Sewer Systems. *International Symposium on Comparison of Urban Drainage Models with Real Catchment Data*. Dubrovnik, April 8-11, 1986.

Lindberg, S. and Carr, R. (1992). MOUSE – Towards an ON-LINE modelling system. *International Symposium on Urban Stormwater Management*, Sydney. February 4-7, 1992.

Lumley, D., Balmér. P. and Adamsson, J. (1988). Investigations of Secondary Settling at a Large Treatment Plant. *Wat. Sci. Tech.*, **20**(4/5), 133-142.

Lumley, D., Haraldsson, J. and Sahlberg, K-Å. (1992). Considerations for Part-time Unmanned Operation of Large Wastewater Treatment Plants. *Proceedings Water Environment Federation 65th Annual Conference and Exposition*, New Orleans. September 20-24, 1992.

Nielson, J., Lindberg, S. and Harremoes, P. (1993). Model Based On-line Control of Sewer Systems. Poster presented at 6th IAWQ Workshop on Instrumentation, Control and Automation of Water and Wastewater Treatment and Transport Systems, Hamilton, Canada. June 21-25, 1993.

Pergamon

Wat. Sci. Tech. Vol. 28, No. 11-12, pp. 55–65, 1993.

0273–1223/93 $7·00 + 0·00

FLOW AND SEDIMENT MOVEMENT MONITORING IN LARGE SEWERS

R. M. Ashley*, I. M. Longair**, D. J. J. Wotherspoon*** and K. Kirby†

** Department of Civil Engineering, Surveying and Building, Dundee Institute of Technology, Bell Street, Dundee, DD1 1HG, UK*
*** Department of Electronic and Electrical Engineering, Dundee Institute of Technology, Bell Street, Dundee, DD1 1HG, UK*
**** Water Services Department, Tayside Regional Council, Bullion House, Dundee, UK*
† Montec International, Systems House, Edge Nook Road, Shadsworth Industrial Park, Blackburn, BB1 2QB, UK

ABSTRACT

The development of more precise computer models for the simulation of the performance of sewerage systems and for management in real-time necessitates the acquisition of more precise data. Existing portable systems for the in-situ measurement of velocity in sewers as commonly used for model development have limited applicability to larger sized sewers (>1200mm). When considered together with the emerging need to monitor the behaviour of sediments in sewers, this lack of instrumentation available to observe detailed velocity gradients in the larger sewers, where sedimentation is most likely to occur, is a severe constraint. A programme of research studying sediment movements in the sewers in Dundee has led to the development of ultrasonic systems for the measurement of three dimensional velocity distributions in large sewers, and sonar systems for the monitoring of bed erosion at points within the sewer network. Data obtained from these systems has contributed to the development of a new cohesive sediment erosion model for sewers.

KEYWORDS

Sewers, velocity measurement, ultrasonic, Doppler-shift, sewer sediments, rheology, erosion, sonar bed monitor.

INTRODUCTION

With the development of increasingly sophisticated (computer based) techniques for the design and operation of sewerage systems, there is concurrently a demand for good quality data relating to the performance of such systems. Traditionally even measurement of the hydraulic performance has been difficult and, because of the problems of monitoring the movement of the surprisingly variable solid-liquid sewage medium, has not normally been very accurate unless as a result of considerable capital and operating investment in the installation. In the UK, a variety of instrumentation systems have been tried for the monitoring of discharge in sewerage systems (Wotherspoon et al., 1990) and it is now normal to measure discharge into a sewage treatment works using a fixed structure with a critical flow device and level measurement system.

For in-sewer monitoring, however, very few fixed devices have been used because of the high capital cost of their installation (Green, 1991 cites an example). If real-time control systems are to become more accepted in the UK then it will be essential for these fixed in-sewer systems to become more widespread as they are in other countries such as Germany.

For the foreseeable future the greatest demand for in-sewer monitoring in the UK and most other countries will be for intrusive portable devices which can be installed inside a sewer at a particular location to collect data over a period of normally one to three months (transient studies). Such information is currently used principally to calibrate and verify computer simulation models of the hydraulic performance of sewer systems; this being deemed essential for most models, although in some countries such as Denmark, the need to check the model performance (MOUSE from the Danish Hydraulic Institute) is not as yet perceived as being necessary. A new generation of models are now emerging which incorporate the significant effects of the sediments which are often ubiquitous, particularly in older combined sewer networks. These models require data which are more precise to ensure that the hydraulic performance is simulated with an accuracy commensurate with the requirements of the theories used to predict sediment transport, deposition and erosion. This necessitates a better measurement of velocity profiles than has hitherto been possible, together with information about the spatial and temporal changes in bed deposits in response to non-uniform and unsteady flows.

The most extensive UK field-based programme of research investigating all aspects of sediments in sewers has been undertaken over the past six years in Dundee. This research has been directed primarily by the UK water industry's Water Research centre (WRc) and has contributed to the development of a new sewer flow quality simulation model known as MOSQITO. This model is currently undergoing trials and pilot applications prior to general release by Hydraulics Research (HRL). The research in Dundee has also been supported by the UK Science and Engineering Research Council (Ashley et al, 1992) and by Tayside Regional Council, whose sewerage system has been studied. A primary aim of these studies has been to assess available instrumentation for use in sewerage systems for sediment related monitoring, and where necessary to develop new equipment. To this end, a relationship was established with Montec International , who are the biggest manufacturers of in-sewer (ultrasonic based) velocity measurement systems in the UK. Systems have now been developed from these studies which can be used to monitor the three dimensional velocity profile in large sewers with sediment deposits, and simultaneously the changing depth of the bed deposits themselves.

VELOCITY MEASUREMENT IN LARGE DIAMETER SEWERS

Existing situation

The options and equipment available for the monitoring of velocity in sewers have been reviewed elsewhere (Wotherspoon et al, 1990). For transient (intrusive) studies many engineers use electromagnetic systems which record the velocity within a small zone (typically 25mm) close to the sensor. This system can be very accurate provided the extrapolation of the velocity profile from this point measurement is possible. Unfortunately a number of manufacturers make unjustified claims about the assumptions in this extrapolation and the consequent resulting accuracy of the estimate of discharge. Obviously the larger the sewer and the more variable the hydraulic conditions, the more inaccurate such a system is likely to be.

In the UK the ultrasonic doppler-shift system has become the standard used for in-sewer velocity measurement. The use of a time of flight ultrasonic system has been investigated (e.g. Guillon et al, 1991) and has been shown to be promising even for flows with concentrations of particles as high as 8000 mg/l. This approach however, has not been developed commercially for in-sewer use. The principal limitations of the Doppler shift system are that the region in the flow field in which the velocity is measured is not precisely known as the ultrasound is emitted in a 'lozenge' and will penetrate into the flow an unknown distance depending upon the turbidity of the fluid. Also, because the system relies upon the receipt of reflections of the ultrasound beam from particles in the flow, if these are not moving as a wash-load then the fluid velocity will be underestimated. Notwithstanding these limitations, the WRc recommend that such a system will be acceptably accurate (5-10%) for estimating the discharge in sewers up to about 1000-1200mm in diameter, provided the hydraulic conditions are stable and as near uniform as possible. The presence of a sediment bed also complicates the measurement problem as the ultrasonic sensor should ideally be located on the invert of the sewer in order to 'look' upward at the vertical central velocity profile in the sewer. In practice under such circumstances the sensor has to be located on the wall of the sewer above the likely sediment bed.

In order to ensure acceptable measurement accuracy for flow modelling even in the smaller sewers, it is normal to calibrate the site when a sensor is installed by visiting it periodically with a hand-held velocity probe and taking point measurements of velocity for comparison. Ironically, an electromagnetic unit is frequently used for this. Unfortunately it is not normally feasible to undertake such calibration measurements when they are most important, during high flows, because of personnel safety requirements. As a consequence calibration data need to be extrapolated from the lower flow conditions.

The lack of a reliable and portable instrument to measure a representative velocity and hence discharge in large sewers under free surface and pipe-full flow is an important problem. In the UK the National Water Council estimated that in the late 1970s the distribution of sizes of the 234,000 km of sewers in the country was:

Diameter (mm)	% of total Length
at least 1000	5%
300 - 1000	25%
less than 300	70%

This illustrates that for the largest 11,700 km of the sewers in the UK the currently available instruments for monitoring velocity in-situ are inadequate.

A three-dimensional velocity measurement system

A circumferential array of ultrasonic transducers (supplied by Montec International) were installed in a 1500mm diameter sewer in Dundee in which sediment movements were being studied in order to enhance the accuracy of velocity measurement. The transducers were initially operated in an autonomous mode, i.e. each transducer emitted an ultrasound beam and also received the reflected signals from particles in the flow from the same beam. In this way a picture of the velocity distribution around the circumference of the sewer could be built up. Unfortunately this revealed only limited information about the velocity profile, particularly along the centre line of the sewer. Figure 1 shows a typical example of velocity profiles measured in this way using eight transducers during storm flows. The velocity is shown plotted against transducer height.

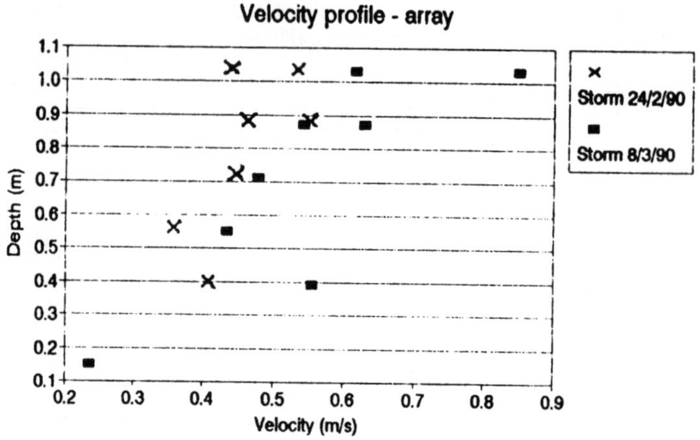

FIGURE 1 - EXAMPLES OF VELOCITY PROFILE MEASURED IN SEWER

Despite the deficiencies, this system can be used to give a better estimation of discharge than for the single transducer as shown in Figure 2.

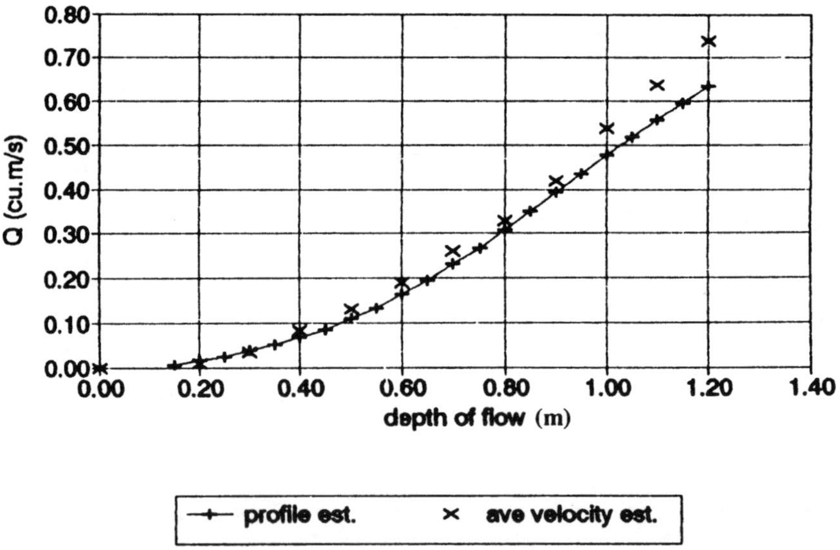

FIGURE 2 - DISCHARGE CALCULATED USING ARRAY DATA
COMPARED WITH STANDARD SINGLE SENSOR - 1.5M SEWER

The discharge determined using the multi-array system for different depths of flow is shown compared with estimating the discharge using a single (conventionally installed) transducer. This figure clearly illustrates the increasing discrepancy with flow depth between the array computed discharge from the multi-velocity measurements and that calculated by assuming that the single transducer has measured an average velocity across the depth of flow.

Further development work has been undertaken to enhance the measurement of velocity using this system. Doppler-shift ultrasonic principles are used in biomedicine (e.g. Cousins, 1978) for velocity measurements in vascular systems.
The basic relationship between the frequency shift ω of the transmitted beam ω_t and the received frequency ω_r is :

$$\omega = \omega_t - \omega_r = \frac{2 \omega_t v \cos \theta}{c}$$

where v is the particle velocity (transported in the fluid)
c the celerity of sound propagation
θ the angle between either the transmitted *or* received ultrasound beam and the flow axis

This type of velocity measurement system thus 'observes' the velocity only where the transmitted and received ultrasound beams intersect (McShane, 1974). The autonomous transducer system is believed to measure velocity in a conical pyramid encompassing the envelope of the transmitted ultrasound beam. The received beam is assumed to be emitted from particles and bubbles throughout the transmitted region.

Each of the eight transducers installed in the sewer had a unique linear calibration of frequency shift with velocity. These were initially calibrated in the laboratory to relate the frequency shifts $\omega_{1..n}$ observed to the range of velocities. Because the transducers observe the reflected ultrasound signal only from the region in which they transmit it is possible to monitor the reflection of the transmitted ultrasound from one transducer by another transducer located elsewhere in the sewer, provided the conical pyramids intersect as shown in Figure 3.

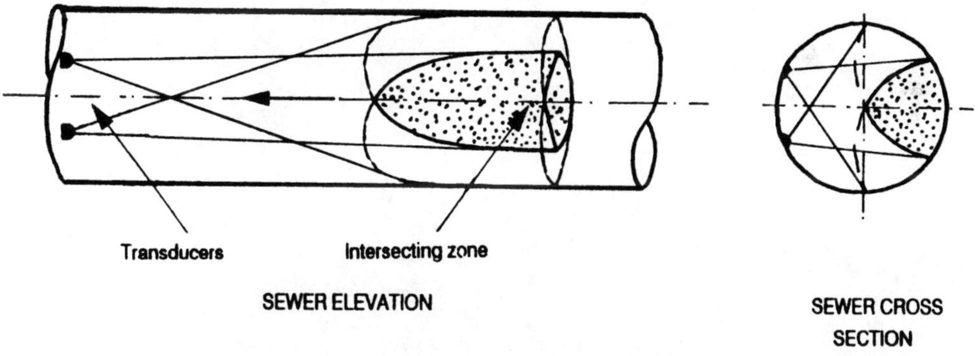

FIGURE 3 - TYPICAL ULTRASOUND ENVELOPE FOR 2 TRANSDUCERS

By locating the transducers at different positions and using each one in turn as a transmitter with all the others receiving, it is possible to observe the velocity in a number of different areas each defined by their coordinates $(x\pm\Delta x, y\pm\Delta y, z\pm\Delta z)$ within the flow. A computer based system is necessary in order to cycle through the transmitters and receivers fast enough to ensure that any temporal changes in the velocity field are on a longer timescale. For n transducers there will be n^2 velocity measurement areas for each cycle of measurements, but because the velocity is measured in the region where each 'pair' overlaps, the number of different areas actually 'seen' in the fluid will in fact be $(n^2-n)/2$. In general the further apart the transducers are located, then the larger the area from which the velocity is being sensed will be. Problems of data averaging and interpretation will occur where the beams intersect on stationary surfaces such as the sewer wall, or where the signal can be reflected from the liquid/air interface.

This type of operation of the array system has been tested by installing six of the transducers in a 700mm wide rectangular inlet channel to the sewage works in Perth. Initial tests in a laboratory flume of the same width had shown that the system was viable even in fairly clean water flows. A typical result of this three dimensional velocity measurement is shown in Figure 4. The longitudinal coordinate of the velocity measurement 'zone' is not shown in the diagram as the flow field was steady and uniform and the measured velocity for $z > 0$ was assumed to be the same as for $z = 0$.

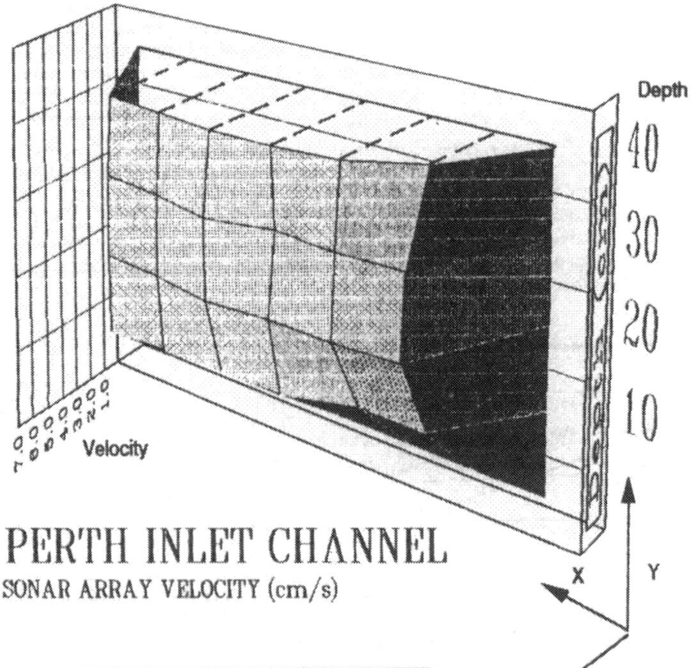

FIGURE 4 - VELOCITY PROFILE

Even for the simple rectangular cross-sectional shape illustrated in Figure 4, the trigonometry required to determine the zones of ultrasound intersection was extensive, and analysis was carried out using a mathematics package (MathCad). The situation when a number of transducers are installed in a circular sewer, or other shape with curving sides, is much more complex and software is currently under development to enable the system to be used routinely where only measures of discharge rate and average velocity are required in large sized sewers. A major difficulty of developing this system further for general use is the difficulty of the user determining within the awkward working environment of a sewer, the precise location of the transducers relative to each other and the angles at which the ultrasound cones are transmitted and received. Subsequently this geometric information will need to be input to a software package which will define for the user exactly where the velocity measurements are being taken.

Questions still remain to be answered about the precise areas in which the velocity is being measured. It is known that the penetration of the ultrasound signal is a function of the concentration of the solids in the flow, but the effect of varying concentrations have not been investigated as yet. As part of the on-going work at Dundee Institute of Technology supported by Montec International , the penetration and precise boundary of the emitted ultrasound beam is being investigated for a range of solids concentrations.

SEDIMENT MOVEMENT MONITORING

The observation of bed depths for coastal and river studies is normally made using sonar systems. This type of system can also be utilised to observe bed deposits in a large sewer. The studies carried out in Dundee have utilised two types of sonar initially developed by WRc: fixed point and scanning (Wotherspoon et al, 1990) to observe respectively the changes in sediment depth at a point and the depth of deposits along a length of sewer. Similar systems have been used in Marseille (Laplace, 1991, fixed point) and in the UK (Anon, 1990, scanning). Recently new instrumentation has been developed in Germany (Lorenzen et al, 1992, Ristenpart et al, 1992) using mechanical and thermal conductivity systems respectively. The former is a scanning system based upon mechanical arms cantilevered from a boat (known as the Gottingen boat), and the latter a fixed device for sediment erosion monitoring at a point. It is not clear for any of the systems what the precise sediment bed/ water interface point is that is being observed. It is known that sediment deposits in sewers frequently comprise a 'weak' surficial layer with a high liquid content overlying a stronger, denser deposit. It is possible that some of the devices described above can only 'see' the denser layer of deposits. The 'Gottingen boat', with the mechanical bed 'follower' will inevitably disturb the weaker deposits and penetrate into the bed. The thermal conductivity system, based on a circumferential quadrant of electrodes attached to the sewer wall, is the device most likely to observe even the weakest sediments, but this system measures the deposit depth at the sewer wall and thus assumes 'a priori' that the sediment bed deposits are planar across the section. Figure 5 shows the results from the scanning sonar used in Dundee which was floated along a 72.5m length of sewer to observe the depth of sediments deposited therein. This clearly shows that there can be important spatial variations in deposit depth both transversely and longitudinally in a sewer.

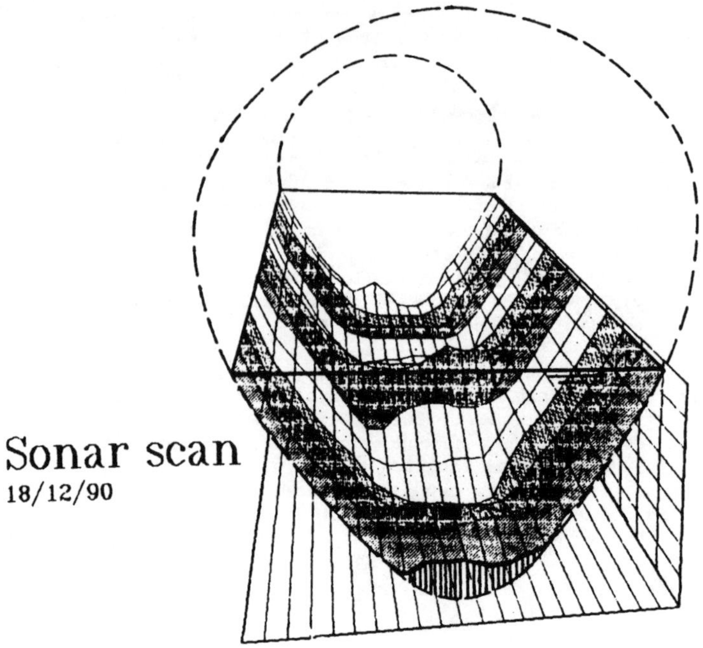

FIGURE 5 - SEDIMENTS OBSERVED IN LENGTH OF SEWER SHOWING TRANSVERSE AND LONGITUDINAL VARIATIONS

The fixed sonar used in Dundee to monitor temporal changes in sediment depth is illustrated in Figure 6. The sonar is incorporated in a floating swinging arm, and thus 'sees' a different location on the bed depending upon the flow depth and the length of the arm. The rotation of the arm is monitored by an inclinometer at the pivot. Problems have occurred due to accumulations of rags on the end of the arm obscuring the sonar sensor, and a longer arm and minimum depth of submergence are preferable if these are to be avoided.

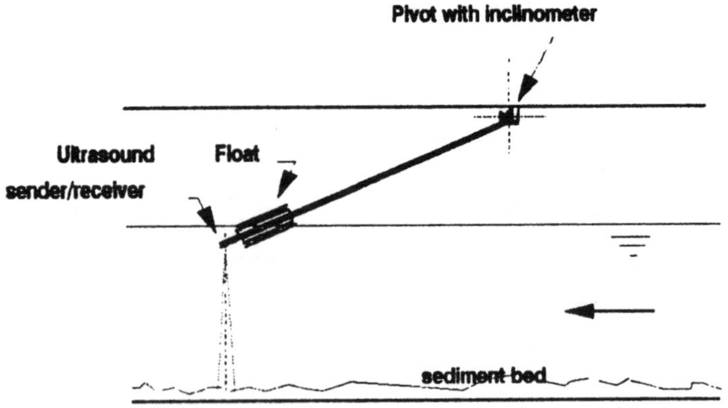

FIGURE 6 - SEDIMENT SONAR MONITOR - DIAGRAMMATIC

The observations of sediment bed deposition and erosion obtained using the fixed sonar system have been used together with the velocity measurements described above to develop a new model for sediment erosion in sewer systems (Wotherspoon, 1992) which includes the cohesive strength of the bed deposits. A typical example of modelled and observed bed erosion is shown in Figure 7.

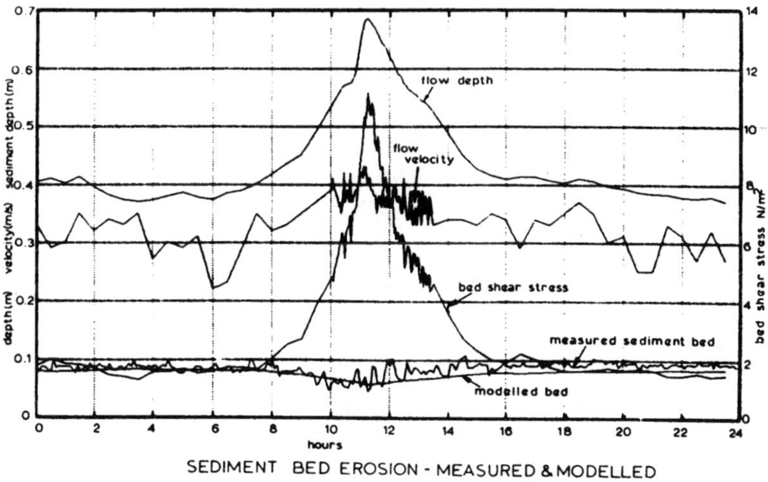

FIGURE 7 - MODELLED AND OBSERVED SEDIMENT EROSION

Attempts were made to measure in-situ the structural strength of the sediment bed at different depths. In-situ rheometrical apparatus based upon the generation and detection of high frequency (100 - 2000 Hz) small amplitude shear waves was developed by researchers from the University College Swansea (Ashley et al, 1992). This allows the vertically distributed rheological properties of viscoelastic cohesive sediment beds to be determined in-situ under a range of conditions. The rheometer is based on a new 'virtual-gap' measuring geometry (Williams & Williams, 1991) utilising small shear wave probes embedded in the sediment. Initial trials at large scale in a laboratory flume showed that the system was viable for determining changes in sediment bed strength when inserted into fine grained homogeneous materials. Unfortunately the technique was found to be unsuitable for use where the sediments are predominantly gritty, coarse grained and inhomogeneous, as found in combined sewer systems of the type being studied. Subsequent rheological investigations of the sewer sediments were carried out using standard laboratory based applied stress rheometry techniques applied to disturbed samples taken from the bed deposits. The new model for bed erosion has utilised these results.

As part of the instrumentation development programme sponsored by Montec International in Dundee, a pre-production prototype of the fixed sonar system is under development. The initial units suffered severely from the harsh conditions experienced by instrumentation left in sewer systems for prolonged periods, and became unreliable after a few months. The experience of Montec International in constructing robust solid state electronic systems for use in sewerage systems is being utilised to ensure greater reliability in the next phase of development of the prototype instruments.

CONCLUSIONS

New and more precise instrumentation is now required for the monitoring of hydraulic and other parameters in sewer systems if the data are to be of an accuracy appropriate for the development of more advanced computer models for the simulation and management of sewerage systems. Whilst instruments are available for the measurement of velocity at a point (electromagnetic) and across a section of flow (ultrasonic Doppler-shift) these are inadequate for large flow depths greater than about 1200mm and where more detailed velocity distributions are required. A system has been developed which utilises the Doppler-shift ultrasonic system with an array of transducers which can collectively observe the velocity within different zones of a flow field. The system is under development to make it more user-friendly for general application in large sized sewers. As part of this work further investigations are underway to define precisely where the ultrasound measurements are observed within a sewage flow of varying solids concentration.

The importance of solids in sewer systems has now been widely acknowledged for the hydraulic and polluting consequences. New models accounting for the solids are under development and it is important that the occurrence and movement of solids can be monitored. A number of systems have been used to carry out observations of sediment bed erosion, and sonar based systems are very effective at monitoring the temporal and spatial variation in sewer sediment beds. Other systems, mechanically and thermally based are also being used, but each system so far developed has limitations.

Attempts to measure in-situ the structural strength of sewer sediment bed deposits using shear wave propagation techniques has only been of limited success. Where fine grained homogeneous deposits are encountered such a system has been shown to be very effective, but because of the highly variable nature of the deposits found in sewers, the technique is not generally viable.

Problems have occurred in the development of the instrumentation described due to the harsh environment in sewer systems. Instruments used in sewers need to be very robust and if possible to have no mechanical parts, as these usually suffer the greatest problems due to clogging. Very sensitive instruments such as the rheometer described can also be susceptible to chemical corrosion unless these are manufactured to take into account the wide range of chemicals likely to be found in sewage flows.

ACKNOWLEDGEMENTS

The instrumentation development continues to be supported by Montec International and the Northumbrian Group PLC. The research studies are supported by the Water Research centre and Tayside Regional Council. The Science and Engineering Research Council funded the original work and supported the third author and also the rheological work outlined, which was the responsibility of Professor D J A Williams, University College, Swansea.

REFERENCES

Anon (1990). Combined sonar and CCTV surveys of sewers in use. Water and sewerage international, p. 24 Aug.

Ashley, R. M., Longair, I. M., Wotherspoon, D. J. J. W., Williams, D. J. A., and Williams, P. R. (1992). The movement of sediment in combined sewers. SERC grant report No. GR/E76308, April.

Cousins, T. (1978). The doppler ultrasonic flowmeter. Flow measurement of fluids. Dijstelbergen, H. H. and Spencer, E. A. (eds). North Holland Publ. Co.

Green, M. (1991). Interim report on storm sewer data. Foundation for Water Research, Report FR 0224, Dec.

Guillon, A., Breuil, B., Herremans, L., Hoc, J. P. and Philippe, J. P. (1991). Performance evaluation of the ultrasonic sensors for the measurement of velocity. Urban Drainage Technologies. Maksimovic, C. (ed), Pergamon, June.

Laplace, D. (1991). Dynamique du Depot en Collecteur D'Assainissement. L'Institut National Polytechnique De Toulouse. PhD Thesis.

Lorenzen, A., Willms, M. and Dinkelaker, A. (1992). The Gottingen boat. Wat. Sci. Tech. **25**(8), 57-62.

McShane, J. L. (1974). Ultrasonic flowmeters. Flow - its measurement and control in science and industry. Dowdell, R. B. (ed). Inst. Soc. of America. 897-913.

Ristenpart, E. and Uhl, M. (1992). Measurement of sediment in sewers. Wat. Sci. Tech. (in press).

Williams, D. J. A. and Williams, P. R. (1991). The determination of dynamic moduli at high frequencies. Non-Newtonian fluid mechanics. (in press).

Wotherspoon, D. J. J., Ashley, R. M. and Woods, S. J. (1990). Imaging in sewer systems. Applications of information technology in construction. Maxwell, J. W. S. et al (eds). Thos Telford (London).

Wotherspoon, D. J. J. (1992). The movement of cohesive sediment in a large combined sewer. Dundee Institute of Technology. PhD Thesis.

Pergamon

Wat. Sci. Tech. Vol. 28, No. 11-12, pp. 67–78, 1993.

0273–1223/93 $7·00 + 0·00

POLLUTION MEASUREMENT ACCURACY USING REAL TIME SENSORS AND WASTEWATER SAMPLES ANALYSIS

G. Ruban*, P. Marchandise* and O. Scrivener**

** Laboratoire Central des Ponts et Chaussées, BP19, 44340 Bouguenais, France*
*** Institut de Mécanique des Fluides, 2, rue Boussingault, 67000 Strasbourg, France*

ABSTRACT

This paper deals with a comparison of the pollution measurement accuracy between real time pollution sensors (especially Suspended Solids – S.S.– determination by turbidimetry) and standard samples analysis on raw wastewaters. The two methods are used in commonly favourable conditions.

Considering separate measurement errors, the S.S. random errors of the two methods are in the same order of magnitude (about ± 25% for a 90% confidence interval), but sampler/analyses lead to a systematic underevaluation of about 17%. These rather high figures for the sampling/analyses method are due to the cumulation of samples compounding and preservation errors.

The determination of the samples collection error when estimating pollution loads is based on a set of 12 rainfall events registered during a 5 month study on a combined sewer overflow. Due to an optimized sampling strategy, the use of a 24 bottle sampler shows an average S.S. load underevaluation of only 5%, compared with the continuous estimation of the pollution sensor considered as a reference.

Real time pollution sensors can, therefore, be considered as convenient systems for field experiments where particulate pollution is of special interest, which is the case for runoff waters. Besides which, they are particularly suitable for the control and automation of sanitary transport and treatment equipment, being easily integrated into computerized systems.

KEYWORDS

Instrumentation; measurement accuracy; runoff; automation; wastewater sampling; pollution sensors.

INTRODUCTION

The increasing use of sensors for the management of wastewater transport and treatment systems requires the consideration of the advantages and drawbacks of these new methods, especially about pollution measurement accuracy. This paper deals with a comparison of the pollution measurement accuracy with on the one hand real time sensors and on the other hand samples analysis on raw wastewater, these two methods being used in common practical conditions. We consider sensors for which a good relation between the measured variable, optical in this case, and standard pollution measurement parameters (Suspended Matters, Chemical Oxygen Demand, etc.) can be established, not automatons reproducing in the field a standard or approximated physico-chemical method. Separate measurement errors are estimated, as are sample collection errors when estimating the pollution loads of a set of real rainfall events (phenomena with fast pollution load variations).

COMPARING CONDITIONS OF THE POLLUTION MEASUREMENT METHODS

So as to make a realistic comparison, the two methods are assumed to be used in common conditions, not particularly unfavourable to one or the other:

– pollution sensor: optical sensor directly immersed in the water investigated, automatic self- cleaning equipment;

– sampling/analyses: sampler in accordance with the recommendations (sample hose diameter larger than 0.95 cm, intake speed within 0.6 and 0.9 m. s^{-1}), 24 h sample preservation period.

Besides which, the sensor and the sampler strainer are supposed to be located at the same measurement point. The comparison can thus be made without taking into account the representativeness error due to the measurement point location in the controlled cross-section.

The purpose of this paper was not to deal with this important type of error, upon which specific investigations are made elsewhere.

SEPARATE MEASUREMENT ERRORS

Sampling Error

This is the compounding error of the sample to be analyzed in comparison with the real compounding at the measurement point in the controlled cross-section. No sampling error is assumed for the pollution sensor as the measurement is made directly in the investigated wastewater. The sampling error concerns primarily the suspended matter measurement with a standard sampler, even if it complies with the above mentioned recommendations because:

– the sampling is never perfectly iso-kinetic (intake speed = water current speed in the considered cross-section) as the current speed fluctuates widely;

– an unavoidable segregation according to the granulometry of the particles occurs during the intake in the sample hose.

An investigation was made by Debarbat (1990) using a sampler testing channel, with a 2 m intake height, current speeds of 0.25, 0.6 and 1 m. s^{-1} (counter-current directed strainer). The materials used to represent the different types of suspended matters found in sewers were the following:

– mineral matter: slag filler with a granulometry from 0 to 80 µm (density 3.1);

– middle-sized organic particles: polystyrene with a granulometry from 200 to 300 µm (density 1.06)

– large organic particles: polystyrene with a granulometry from 800 to 1100 µm (density 1.06).

The experimental results with regard to the different materials appear in Table 1 (initial concentration 300 mg. l^{-1}):

TABLE 1 Suspended Solids Sampling Results

	Samples number	Average	Standard deviation	Under evaluation
Slag filler	20	270 mg. l^{-1}	40.8 mg. l^{-1}	- 10%
Polystyrene 200-300 µm	16	195 mg. l^{-1}	54.2 mg. l^{-1}	- 35%
Polystyrene 800-1100 µm	16	70 mg. l^{-1}	15.7 mg. l^{-1}	- 77%

It may be noted that the suspended matter underevaluation increases greatly with the granulometry. Assuming that raw wastewater carries 80% fine particles, 15% middle-sized and 5% large particles (Chebbo *et al.*, 1990; Verbanck *et al.*, 1990), the resulting S.S. concentration for such a compounded average sample is 249 mg. l^{-1} (i.e. a 17% underevaluation) while the standard deviation is 34 mg. l^{-1} (11%).

Analysis error

Sensor measurement. The measures we made at the laboratory using a field turbidimeter showed a fidelity error (measurement reproducibility on a short time period) ranging from 0.4 to 0.2% on standard Kieselgur solutions of 3 and 15 g. l^{-1} respectively. Besides which, the drift of this device during the 52 days of experiment was very low (+1% in clear water and -1.75% in a 15 g. l^{-1} Kieselgur standard solution). As fidelity errors are negligible and integrated into the sensor calibration, and drift errors are very low, they will both no longer be separately considered.

Good correlations can be established for urban wastewaters between optical parameters (turbidimetry, infrared, visible or ultraviolet ray absorptions) and standard pollution measurement parameters such as Suspended Solids and Chemical Oxygen Demand (Dobbs *et al.*, 1972, Grange *et al.*, 1987, Marchandise *et al.*, 1990).

Table 2 gives some examples of relations established at three different sites between turbidity T (visible light – 400 to 700 nm – absorption measurement) and the above mentioned parameters. These relations apply only to the considered site. Good relations can be established with Chemical Oxygen Demand when the carbonaceous pollution characteristics are relatively constant, which is generally the case for urban wastewaters.

TABLE 2. Relations SS and COD = f (Turbidity) at Three Sites

Location	Parameter	Range mg. l^{-1}	Samples number	Realtion	Correlation coefficient	Standard deviation mg. l^{-1}
Nantes-Sud	SS	81-832	39	0.30 T-84.3	0.94	48.7
	COD	274-2093		0.88 T-176.3	0.93	144.1
Bouguenais	SS	37-147	18	0.24 T-32.4	0.92	14.6
	COD	131-456		0.65 T-28.6	0.93	35.8
La Roche sur Yon	SS	140-960	34	0.87 T-80.2	0.94	63.2
	COD	447-1512		1.18 T+346.9	0.91	110.8

Good correlation coefficients are obtained for S.S. and C.O.D.

For the estimation of the determination accuracy, the single use of the mean standard deviation (S.D.) is not correct as it may vary with the concentrations measured. In fact, Figs 1, 2 and 3 show an increase in the scattering of the points for higher concentrations. A complete investigation of the S.D. variation would require more calibration points.

In order to make a simple evaluation of this variation, linear regressions were performed on each half series of experimental points (lower and above average series).

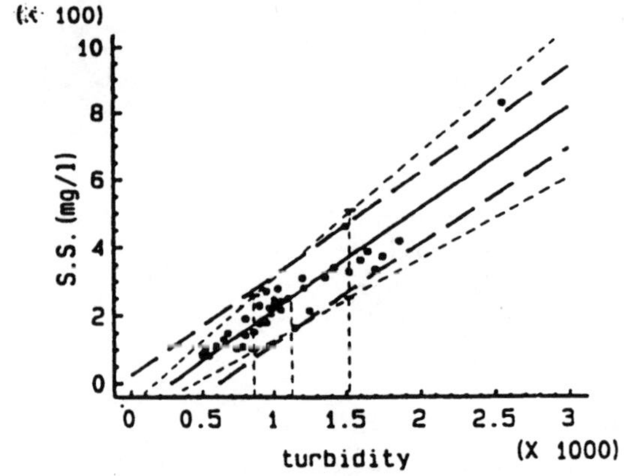

Fig. 1. S.S.-turbidity relation at Nantes-Sud.

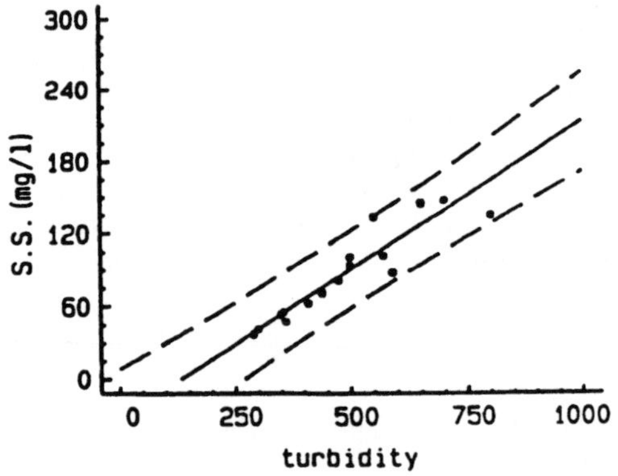

Fig. 2. S.S.-turbidity relation at Bouguenais.

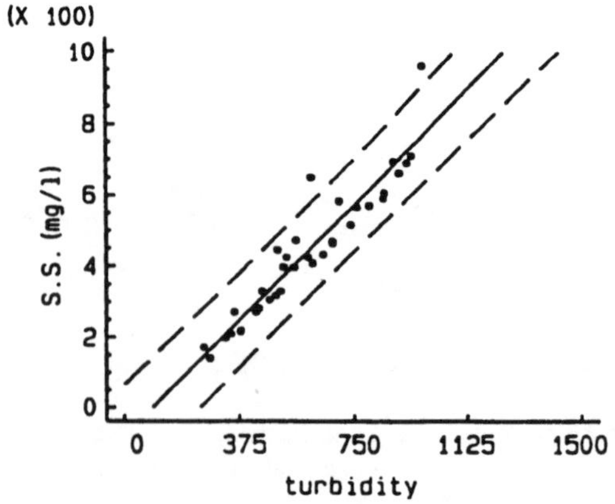

Fig. 3. S.S.-turbidity relation at La Roche sur Yon.

Table 3 shows that:

– the variation coefficient (S.D./average ratio of the corresponding half series) increases for S.S. and decreases for C.O.D. as higher values are considered.

– the regression equations are not statistically different for S.S., but statistically different for C.O.D. when taking into account the standard errors on slopes and intercepts for each couple of series (except for C.O.D. at Nantes-Sud).

TABLE 3 Fluctuation of the Variation Coefficient According to Concentrations Range

Location	Parameter mg. l^{-1}	Slope	Intercept	Variation coefficient
Nantes-Sud	SS*	0.31 ± 0.03	-83.4 ± 27	15%
	SS**	0.39 ± 0.05	-184.7 ± 80	18%
	DCO*	0.80 ± 0.13	-96.7 ± 116	20%
	DCO**	1.02 ± 0.14	-409.6 ± 223	16%
Bouguenais	SS*	0.22 ± 0.02	-27.5 ± 6	5%
	SS**	0.17 ± 0.07	17.1 ± 44	17%
	DCO*	0.66 ± 0.12	-45.2 ± 47	11%
	DCO**	0.35 ± 0.12	165.0 ± 76	9%
La Roche sur Yon	SS*	0.94 ± 0.09	-107.6 ± 42	12%
	SS**	0.89 ± 0.18	-97.1 ± 147	14%
	DCO*	1.97 ± 0.26	7.8 ± 118	12%
	DCO**	0.88 ± 0.17	576.3 ± 132	6%

* lower average series
** above average series

A reasonable mean variation coefficient of 15% can therefore be considered for S.S. determination using this type of optic sensors (the Bouguenais results are less representative as the number of calibration points is lower). For a confidence level of 90%, this corresponds to a relative error of 25% (1.65 S.D.: see Fig.1).

As a decrease of the S.D. for lower values is not visible on the graphs for C.O.D. relations (linear regressions), a definite variation coefficient cannot yet be proposed (Fig.4).

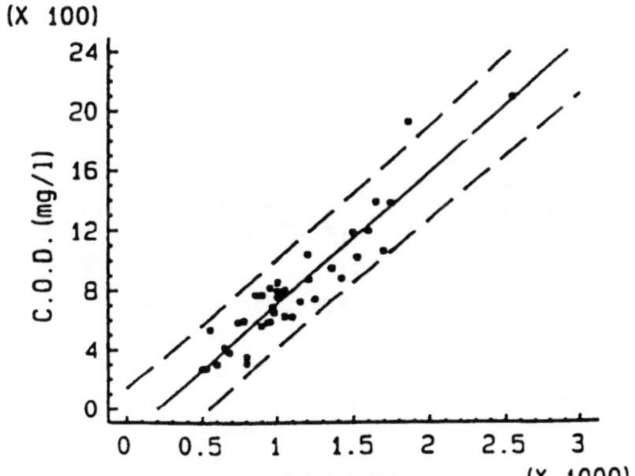

Fig. 4. C.O.D.-turbidity relation at Nantes-Sud.

Samples physico-chemical analysis. The main error causes in this domain are:

– the type and duration of sample conservation,

– the initial sample fractionation into several samples for specific analysis and the assay sampling (especially for particulate pollution),

– the physico-chemical analysis by itself.

Initial sample fractionation and assay sampling may introduce underevaluation of the S.S. concentrations if the samples are not constantly kept homogenized during these operations. A mean underevaluation of 10% in S.S. is mentioned by Paitry and Doubreres (1986) for the sharing of an initial 1 l sample into two 500 ml samples. This error may reach 25% as settling tends to increase for higher concentrations.

The AFNOR standard NF T 90-105 for the determination of S.S. specifies that the whole sample has to be analyzed. We have observed that 50 to 100 ml assay samples may lead to a mean underevaluation of about 10%.

For the following accuracy estimation, it will be assumed that analyses are made strictly so that this type of error is kept negligible.

Ranchet *et al.* (1979) have studied the influence of the type and duration of sample preservation on the determination of S.S., C.O.D. and B.O.D.5. Analyses were made on 25 samples of raw wastewater from 21 locations near the laboratory (100-400 mg. l^{-1} S.S.; 244-1181 mg. l^{-1} C.O.D.). Analyses were performed within 2 to 3 h after sampling.

The 10 l samples were filtered through a 5 mm sieve as soon as they were brought to the laboratory, then carefully homogenized and divided into 1 l samples: 1 was immediately analyzed, 3 were kept in refrigerator (4°C) and analyzed 24 h, 48 h and 8 days later, 2 were kept in freezer (-20°C) and analyzed 8 days and 1 month later.

S.S. and C.O.D. were analyzed using the NF T 90 101 and NF T 90 103 standard methods.

A large S.S. overestimation is observed by freezer conservation (+28% on average). Figures 5 and 6 show the evolution of S.S. and C.O.D. by refrigerator conservation during 24 h, 48 h and 8 days. There is no significant average shifting for S.S., but a slight increase in data scattering. However a decrease in average as well as an increase in data scattering appears for C.O.D.

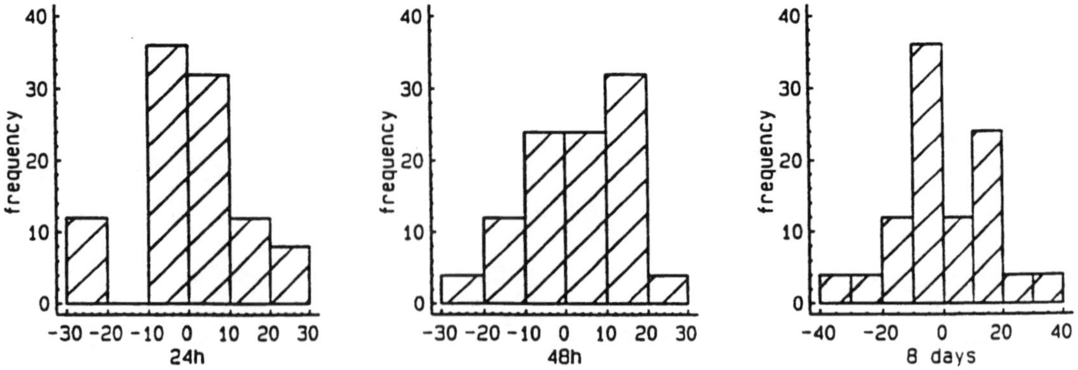

Fig. 5. Suspended Solids evolution by refrigerator conservation (4°C).

Fig.5. Suspended Solids evolution by refrigerator conservation (4°C)

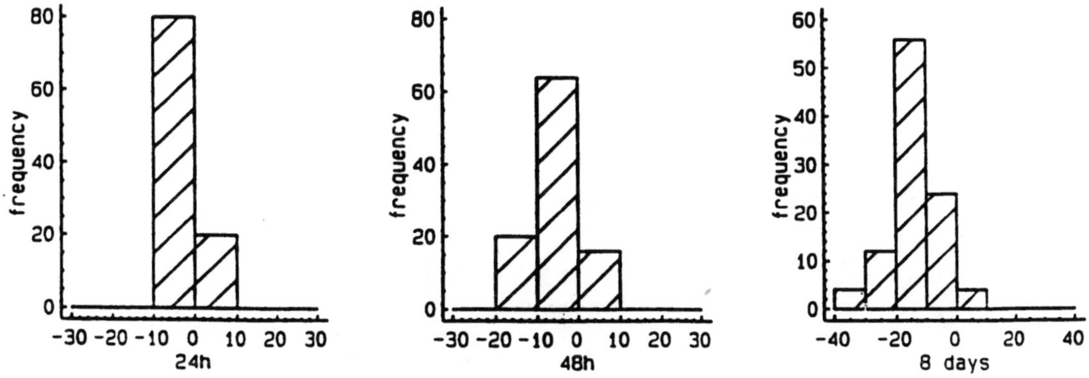

Fig. 6. Chemical Oxygen Demand evolution by refrigerator conservation (4°C).

As we consider a rather favourable sample conservation period of 24 h, the mean errors for a confidence level of 90% (assuming a normal distribution) are:

$$\Delta C_{SS}(90) = \pm 21\%$$

$$\Delta C_{COD}(90) = -3\% \pm 7\%$$

Error Combination

The bibliographic elements available are insufficient to investigate the evolution of sampling and analysing errors with increasing concentrations. However, relative average shifts and relative standard deviations may reasonably be assumed as relatively constant within the concentrations range considered. Given two errors, the characteristics of which are respectively:

- μ_S and σ_S (average shift and standard deviation for sampling)

- μ_A and σ_A (average shift, standard deviation for analysis), the global error characteristics are then:

$$\mu_G = \mu_S + \mu_A; \ \sigma_G{}^2 = \sigma_S{}^2 + \sigma_A{}^2 + \text{covariance (S,A)}$$

As already mentioned above, there are no bibliographic elements available to estimate interactions between sampling and analysing errors, although such interactions could exist especially as concentrations increase. These experimental phases will then be considered as independent, covariance (S,A) = 0. Taking into account the previously determined relative values for S.S.:

$$\mu_S = -17\%; \ \sigma_S = 11\%; \ \mu_A = 0\%; \ \sigma_A = 13\%$$

thus $\mu_G = -17\%$; $\sigma_G = 17\%$. Considering a 90% confidence level (1.65 σ_G):

$$\Delta C_{SS}(90) = -17\% \pm 28\%$$

This result compares with the corresponding accuracy determined for S.S. measurement using an optic sensor, i.e. a negligible systematic error (drift) and a random error of 25%.

It can then be concluded that, given the comparison conditions and assumptions, the random errors of the two measurement methods are in the same order of magnitude, but sampling leads to a systematic underevaluation of about 17%.

SAMPLES COLLECTION ERROR WHEN ESTIMATING POLLUTION LOADS

We have so far dealt only with separate sample errors. Estimating pollution loads requires sample sequences to follow the concentration evolution. The number of samples analyzed with physico-chemical methods is limited for practical and economic reasons. On the contrary, given the present characteristics of the data aquisition systems (high signal treatment accuracy, large data storage capacity), the user can easily choose the most adapted acquisition and storage frequency for the output signal of the pollution sensors (for the experiment used below as an example, measurements were made every 30 s, the mean of 4 measurements being stored every 2 min).

In order to show more clearly the inaccuracy resulting from the samples collection error when estimating pollution loads, the separate sample error will no longer be considered, nor will the flow measurement error. The determination of the pollution loads inaccuracy would in any case be difficult as pollution and flow errors are probably not independent (a better knowledge of the possible interaction is then necessary).
With these hypotheses, the pollution loads estimation using real time sensors may be considered as a reference, as the data aquisition frequency can practically be chosen as high as necessary to follow exactly the concentration evolution.

Comparing conditions of the pollution measurement methods

The comparison will be made on a set of 12 rainfall events registered during a 5 month study on a combined sewer overflow.

As already mentioned above, the methods are assumed to be used in common conditions, not particularly unfavourable to one or the other:

– sampler/analyses: one of the most common programs consists in 24 individual samples with flow proportional sampling (this sampling method is well adapted to pollution peaks measurement). It is, however, difficult to set the water volume run off between 2 samples as the maximum event volume cannot be known in advance (in case of event volume > 24 preset water volumes, the end of the event is not sampled). We adopted the favourable hypothesis of an optimum sampling of the hydrologically larger event, i.e. preset water volume between 2 samples:

$$\frac{\text{maximum event rate}}{24} = 170\text{m}^3$$

Moreover the sampling is supposed to start exactly at the beginning of the event (start switch controlled by water level for example). Simultaneous water level starting and flow proportional sampling is nevertheless not always available on samplers.

– pollution sensor: given the hypotheses mentioned above and the acquisition characteristics (measurements every 30 s, mean of 4 measurements stored every 2 min), the data registered will be considered as a reference for the samples collection error calculation using the sampler/analyses method.

Therefore, the pollution and flow data registered being C_i and Q_i (i ranging from 1 to n between the beginning and the end of the rainfall event), the sampler/analyses data will be the corresponding values when the sampling occurs at t_j (C_j, Q_j, j ranging from 1 to N, N 24). See Fig.7.

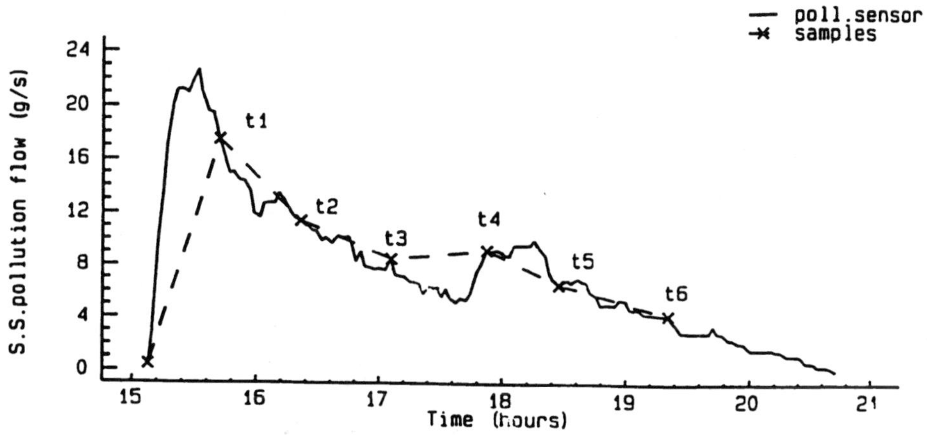

Fig. 7. S.S. flow variation for event no 3.

<u>Samples collection error when estimating pollution loads</u>

The reference pollution load can then be calculated as:

$$\sum_{i=1}^{n} F_i \,\Delta t \qquad (1)$$

F_i being the instant pollution flow ($C_i.Q_i$) and Δt the period between 2 data storage (2 mins)

Considering the sampler/analyses method, the pollution load run off between 2 samplings can be calculated as:

$$m_j = 1/2\ (F_{j-1} + F_j) \cdot (t_j - t_{j-1}) \qquad 1 \leq j \leq N \qquad (2)$$

at the beginning of the event t_0, the pollution flow is considered as negligible $F_0 = 0$,

Therefore, the event pollution load using the sampler/analyses method:

$$MSA = \sum_{j=1}^{N} m_j \qquad 1 \leq N \leq 24 \qquad (3)$$

Then the difference Δm = MSA - MR represents the samples collection error for every rainfall event. The results for the 12 rainfall events considered are shown in Table 4.

It may be noted that the error is larger for small events, about - 10%, except for the event no 4 where the lack of samples is balanced by a load overestimation at the end of the event (Fig.7). This global pollution load balance exists also for event no 3 (Fig.8). The number of samples is of course also smaller than for strong events.

TABLE 4 Samples collection errors for a set of rainfall events

Event Number	MR kg	Δm kg	Error %	Samples Number
1	160	-15	-9	5
2	135	-14	-11	4
3	157	-18	-12	6
4	105	6	5	6
5	901	-6	-1	16
6	272	-32	-12	4
7	855	-25	-3	24
8	432	-5	-1	12
9	222	-28	-13	4
10	525	4	1	17
11	941	-25	-3	21
12	646	-21	-3	16
average	446	-15	-5	11

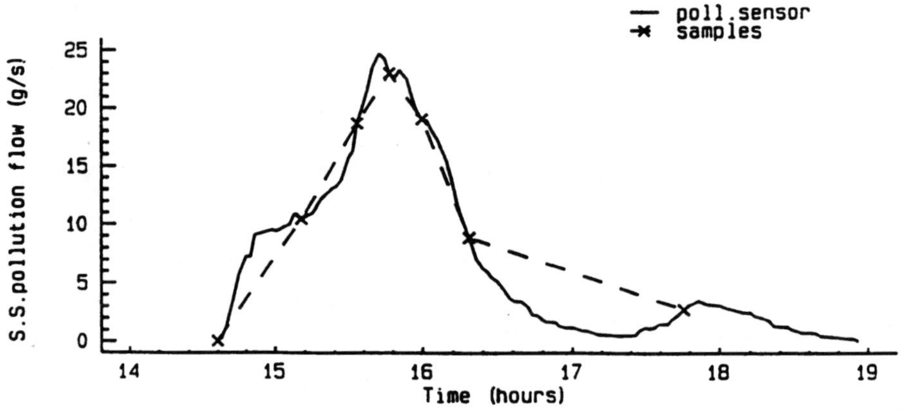

Fig. 8. S.S. flow variation for event no 4

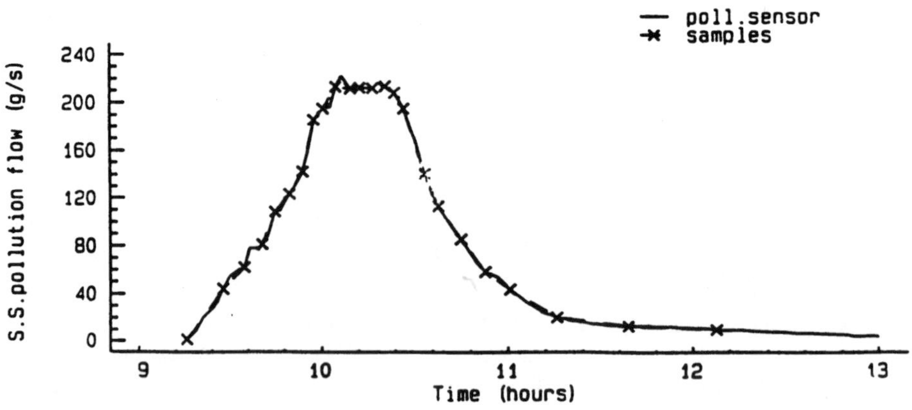

Fig. 9. S.S. flow variation for event no 7

The pollution load underestimation for strong events, as well as the average one (- 5%), is rather small, which means that the sampling is satisfactory, given the rather simple aspect of the pollutograms (few pollution peaks; Fig.9). This small error is mainly due to the very favourable hypothesis considered in the comparison (water volume run off between 2 samples = maximum event volume/24):

– if this volume is too small, the end of strong events is not sampled,

– if this volume is too large, the sampling inaccuracy leads to larger underestimation (as noted above for small events).

Nevertheless, it may be observed that for the set of events investigated, the pollution load estimation is correct starting from half the total possible number of samples (12), and that most sophisticated samplers allow optimized sampling programs (Hydrologic Event Stage Mode from U.S. Geological Survey for example).

CONCLUSION

The accuracy comparison of 2 types of pollution measurement methods (real time sensors and samples analysis) used in common conditions shows that:

- for separate measurements of suspended solids (S.S.): a random error of the same order of magnitude (± 25%), but the sampling/analyses method leads to a systematic underevaluation of about 17%. These figures which seem important are due for the sampling/analyses method, to the cumulation of samples compounding error (hardly avoidable for particulate pollution) and samples preservation error (though in representative conditions: 4°C preservation for 24 h). Further bibliographical elements would be necessary to draw conclusions concerning other pollution parameters such as the Chemical Oxygen Demand.

– samples collection error when estimating S.S. pollution loads: a flow proportional sampling with a 24 bottles sampler leads to an average underevaluation of only 5% for the pollution load estimation of a set of 12 rainfall events. That shows the possibility to get good estimations with well set samplers according to the characteristics of the considered events (maximum water volume in this case).

The available experimental and bibliographical elements allowed us to make an accuracy estimation of S.S. measurement on raw wastewaters. Using complementary elements, this investigation could extend to other parameters, and take into account the increasing accuracy of the measurement methods expected with the technological improvement of equipment (samplers, pollution sensors).

Real time pollution sensors can be considered as convenient systems for field experiments where particulate pollution is of special interest (this is the case for runoff waters pollution, the major part of which is in particulate form). Besides which, they are particularly suitable for the control and automation of sanitary transport and treatment equipments, being easily integrated in computerised systems.

REFERENCES

Chebbo, G., Musquere, P. and Bachoc A. (1990). Solids transferred into sewers. In: *Proceedings of the Fifth International Conference on Urban Storm Drainage*, Osaka (Japan), 23-27/07/1990, pp.891-896.

Debarbat, M. (1990). Recherche d'un prototype de banc d'essais pour préleveurs automatiques en réseau d'assainissement. Mémoire de 3ème année, Ecole Nationale des Ingénieurs des Travaux Ruraux et des Techniques Sanitaires, Strasbourg (France).

Dobbs, R. A., Wise, R. H. and Dean, R. B. (1972). The use of ultraviolet absorbance for monitoring the total organic carbon content of water and wastewater. *Water Research*, **6**, 1173-1180.

Grange, D., Marchandise, P., Paitry, A. and Randon, G. (1987). Mesure instantanée et continue de la qualité de l'eau. *La Houille Blanche*, **4/5**, 339-345.

Marchandise, P., Ruban, G., Cathelain M. and Wacheux H. (1990). Effectiveness of the treatment and retention or rainwater in networks: evaluation with continuous measurement of pollution. In: *Proceedings of the 5th IAWPRC workshop on Instrumentation, Control and Automation of Water and Wastewater Treatment and Transport Systems (Adv. Wat. Pollut. Control no.10)* ed. R. Briggs, pp.265-274.

Paitry, A. and Doubreres, S. (1986). Echantillonnages des Matières en Suspension en réseau d'assainissement. Quelques réflexions... Service Départemental d'Assainissement, Rosny-sous-Bois (France), pp.38

Ranchet, J., Pescheux, F. and Menissier, F. (1979). Influence du temps et du mode de conservation des échantillons d'eau sur les déterminations de DBO5, DCO et MES. Laboratoire de l'Equipement de Trappes (France).

Verbanck, M., Vanderborght, J. P. and Wollast, P. (1990). Size distribution of suspended particles in combined sewers during dry- and wet-weather. In: *Proceedings of the Fifth International Conference on Urban Storm Drainage,* Osaka (Japan), 23-27/07/1990, pp.891-896

Pergamon

Wat. Sci. Tech. Vol. 28, No. 11-12, pp. 79–85, 1993.

0273–1223/93 $7·00 + 0·00

THE PREDICTION OF RAINFALL AND THE DIRECTION OF RAINFALL BY A NARROW AREA RADAR RAIN GAUGE

Shinichi Kondo

Toshiba Corporation, Plant System Application Department, Fuchu Works, Toshiba-cho, Fuchu-shi, Tokyo, Japan

ABSTRACT

Narrow area radar rain gauges are currently used for measuring rainfall. These radar gauges can measure rainfall accurately in a small area. In sewage plants it is important to predict stormwater. To calculate predicted stormwater the results of rainfall and a prediction of the near future are necessary. Recently urbanization has made the arrival time of flooding to the sewage plant much shorter. This paper deals with system technologies for the near future prediction of radar rain gauge rainfall. The method of prediction of rainfall, calculation of results and other considerations are described.

KEYWORDS

Radar rain gauge; prediction of rain movement; sewage plant operation support; urban rainfall.

INTRODUCTION

To prevent houses from being flooded when it rains heavily and continuously in metropolitan areas, the demand for sewage systems that can cope rapidly with the situation has become greater. Consequently, it is becoming an urgent necessity to accurately predict the amount of water flowing into the sewers. To predict the amount of flowing water, it is necessary to measure rainfall conditions in a wide range immediately. The conventional ground gauges can only measure rainfall value at a point. Therefore narrow area radar rain gauges are being introduced into several sewerage systems.

In radar, the directivity of electromagnetic waves and their reflection back are used to locate an object. Radar rain gauges measure the energy reflected from rainfall and calculate the intensity and distribution of the rainfall. The narrow radar rain gauges which have been installed recently at our sewage treatment plants system can measure rainfall in a smaller area. This size is smaller than radar gauges for meteorological observation. The cycle of narrow radar rain gauges is short.

For urban runoff analysis and inflow prediction, it is necessary to predict rainfall in the near future, because recent urbanization has increased the speed at which runoff accumulates.

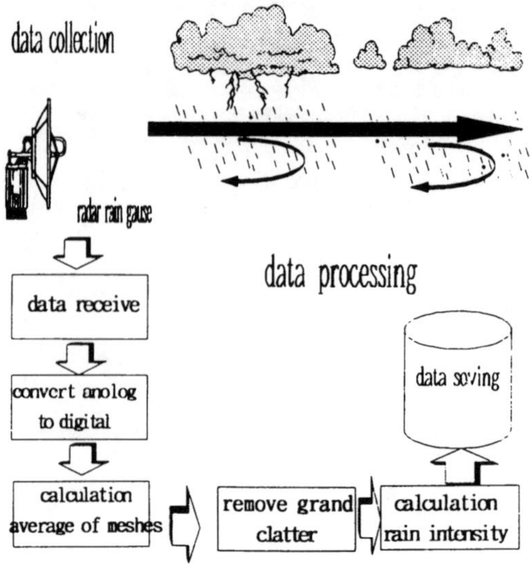

Fig. 1. Block diagram of the radar rain gauge system.

FEATURES OF RADAR RAIN GAUGE

The principle of radar rain gauges that were installed recently at our sewage plants system for operational support is similar to that of radar gauges used in meteorological observation or the military. There are, however, some differences between the two on the purpose of use and the function. Digital computers make it possible to use radar as a rain gauge, and lately engineering workstations make changes in the user interface and techniques for getting information.

The five features of the narrow radar rain gauge are as follows.

(1) The size of meshes measuring rainfall by the narrow radar rain gauge is smaller than that of the meteorological observation. The former's mesh size is about 250 m to 500 m. The latter's size is more than 3 km. The 250 m mesh radar is 144 times more detailed than the 3 km mesh radar.

(2) The observation scope by the narrow radar gauge is small.

(3) The cycle of observation by the narrow radar gauge is short. It is about 2.5 mins where the gauge for meteorological observation is more than 15 mins.

(4) The determination of pump operation at the sewage plants is immediate, therefore the data processing is on-line.

(5) In the sewage plant, the prediction of the rainfall and rain direction can be calculated at high speed and automatically by the computers. In the weather forecast, the specialists predict rainfall in consideration of many weather factors.

RELATION OF THE MOVEMENTS OF RAINFALL TO THE RADAR RAIN GAUGE SYSTEM

The radar rain gauge can determine two-dimensional rainfall which the ground rain gauges cannot determine. The purpose of using these data is to support the pump operation at the sewage plant in case of a shower or typhoon. By continuation of rainfall patterns, prediction of future rainfall is very useful to pump operation.

The amount of pavement in urban areas has risen, and one sewage treatment plant does not charge wide area. The inlet time of rainfall to the sewage plant or pumping station has become short, about several minutes. In support of the pump operation, prediction of flood discharge after 20 to 30 mins is very useful. This calculation of prediction, however, is impossible by only the actual rainfall which was gauged by radar rain gauge. It is necessary to predict the rainfall in the future for calculation of influx. To support the pump operation, the influx after 10 to 20 mins is necessary also.

FEATURES OF RAINFALL PREDICTION FOR THE SEWERAGE SYSTEM

The scope of the ordinary rainfall prediction is several hours in wide area at a meteorological observatory. It considers atmospheric phenomena and geographical elements. The prediction of the sewage system is different from the weather forecast system in the purpose and the function.

The five features of rainfall prediction for the sewerage system are as follows:

(1) It is important to predict in 5 to 60 mins after rainfall in the sewerage system.

(2) The purpose of rainfall prediction is calculation of influx in a sewage treatment plant or pumping station. It is necessary to predict rainfall amount more accurately here than the case of weather forecast.

(3) For the quick operation of pumps, it is necessary to transmit prediction data on-line to each sewage treatment plant or each pumping station from the radar rain gauge station as soon as the calculations are available.

(4) There are different styles of rain like a shower, a typhoon and rain in the wet season. In each case the movement of rainfall is different, but the prediction of movement of rainfall must be calculated promptly and automatically by the computers.

(5) It rains in a small area and strongly in a summer shower. These rains are most dangerous in a sewage treatment plant, but formerly it was impossible to grasp rainfall accurately because the size of the rainfall area was small. It is important to predict shower movement in narrow areas accurately.

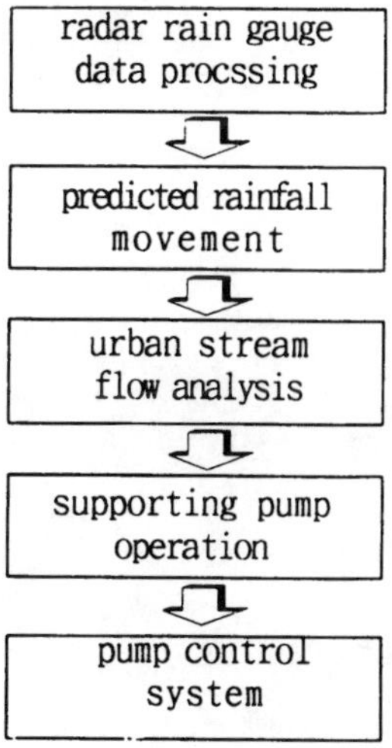

Fig. 2. Flowchart of sewage supporting system using the radar rain gauge.

There are some differences between the prediction rainfall for the sewerage system and that of the usual wether forecast. So for the sewerage system, new methods are necessary to calculate predicted rainfall of a small area.

FEATURES OF RAINFALL PREDICTION FOR SEWAGE SYSTEMS

Preconditions of the Calculation Method

We have developed a calculation model for the prediction of rainfall which satisfies the conditions of the section. The actual data have proved that this calculation model is accurate and useful for supporting the sewage treatment plant operation.

There were some preconditions required to develop the calculation model. The four preconditions are as follows:

(1) The rainfall has continuous movement.

(2) The strength of rain is more effective to the movement of rainfall.

(3) Within a short time, 10 to 20 mins, the shape of rainfall does not change.

(4) Within a small area, the direction of the movement of rainfall stays the same.

Outline of the Calculation Method

According to the above preconditions, the method is defined and programs are made in C language on an engineering workstation. The workstation uses UNIX OS.

The calculation method has the following three parts:

(1) The vector of rain movement is calculated by comparison with rainfall pattern.

(2) The tendency of the entire amount of rain in the future is calculated by interpolation of the amount of rain.

(3) The amount of the calculation of pattern # is quite large. Usually the time of this calculation would be large. To decrease the time calculation, the calculation pattern of recognition is limited by the effect of the former movement factor, so the calculation is able to act on a usual workstation within several seconds. It is unnecessary to use super computers to solve the calculation.

Coefficient of Correlation

It is necessary for a pattern recognition to calculate the coefficient of correlation. There are many coefficients of correlation, for example the following coefficient of correlation is adopted, because time of calculation is shorter and the result of calculation is correct.

$$f = (a_1 - b_1)_2 + (a_2 - b_2)_2 + (a_3 - b_3)_2 = (a_n - b_n)_2$$

a_n: meshes data of the origin (DATA1)
b_n: meshes data of next time (DATA2)

Brief Sample of Calculation for Pattern Recognition

Figure 3 shows the flow of calculation for pattern recognition briefly between DATA1 and DATA2.

case of right shift: $f = (1-1)^2 + (2-2)^2 + (3-3)^2 + (4-3)^2 = 1$
case of left shift: $f = (1-2)^2 + (2-2)^2 + (3-1)^2 + (4-1)^2 = 14$

case of under shift: $f = (1-1)^2 + (2-3)^2 + (3-1)^2 + (4-1)^2 = 14$
case of up shift: $f = (1-1)^2 + (2-1)^2 + (3-2)^2 + (4-1)^2 = 11$

In the sample, it is defined that rainfall moves right, because the coefficient of correlation is smallest in the case of right shift.

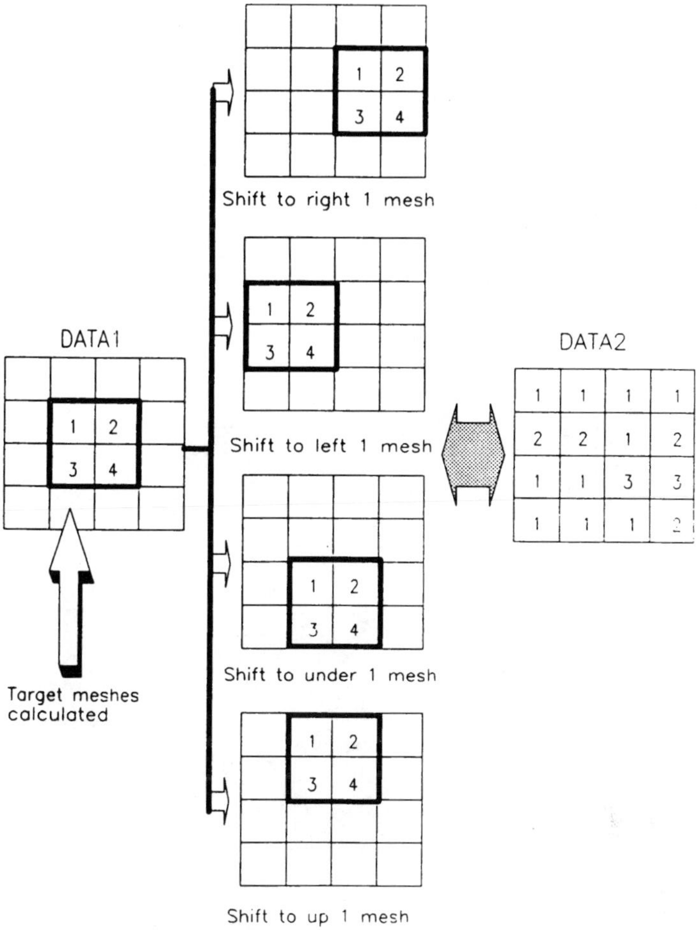

Fig. 3. Flowchart of calculation for pattern recognition.

RESULTS USING RADAR RAIN GAUGE DATA

Figure 4 shows actual rainfall by a narrow radar rain gauge and the rainfall movement vector of the shower. The size of area is 80km by 80km. The mesh size is 0.5km by 0.5km. There is a very strong rainfall area under Yokohama City. The size of the shower is 10km from north to south and 50km from east to west. The strongest rainfall is 50mm/h in the centre of the shower. In the middle of the shower area, it rains heavily, out of the shower it is very fine. It is a peculiar pattern of a shower. It is most dangerous for the sewage treatment plants and the pumping stations because run-off may come into these places suddenly even if it does not rain there, when it rains in the upper area of the sewerage.

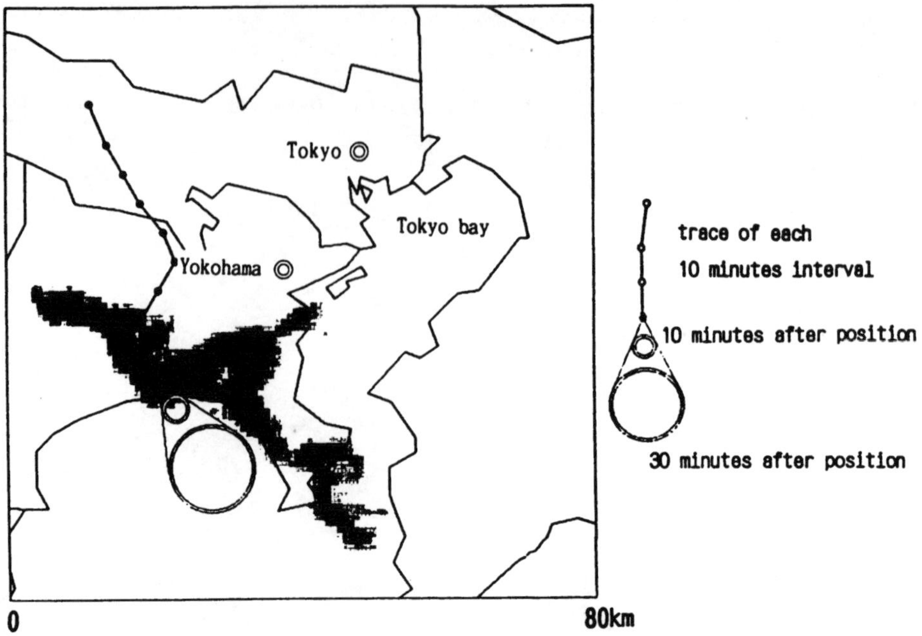

Fig. 4. Actual rainfall and the rainfall movement vector in the shower.

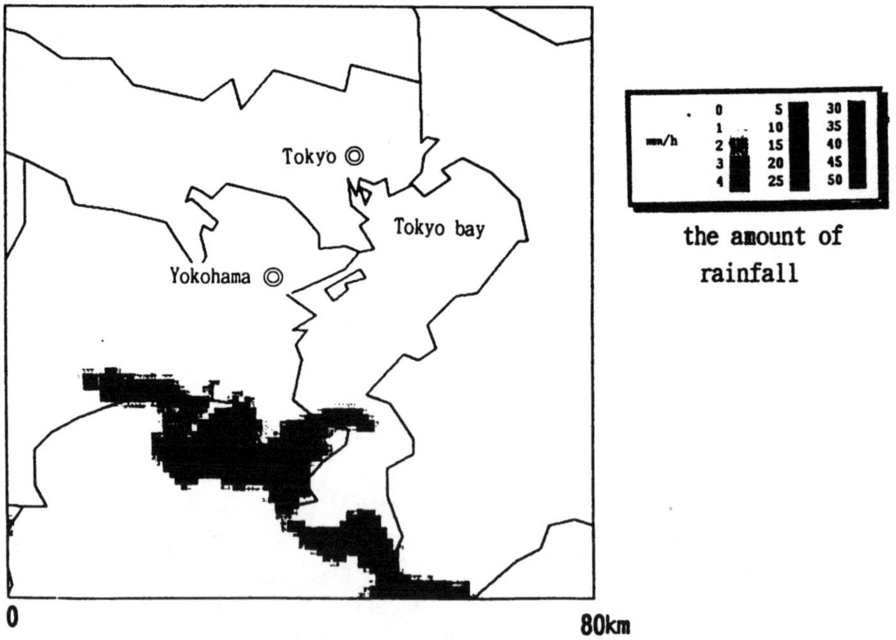

Fig. 5. Actual rainfall 30 minutes after Fig. 4.

In figure 4 the line which has black circles shows the vector of the rainfall movement. The black circles each show a 10 min spot. It is calculated that the rainfall is moving to the south-south-east at a velocity of about 40km per hour. The line of vector curved because the strong point of rainfall in the shower moved to the west on the way.

The direction of rainfall can be calculated. The middle circle shows the point of rainfall centre after 10 mins and the large circle after 30 mins. The rainfall will move continually; after 30 mins it is predicted that the centre of the rainfall will reach west of the peninsula. Figure 5 shows actual rainfall after 30 mins of figure 4. It shows that the rainfall centre reached the west of the peninsula. According to these pictures, the prediction agrees with the actual result. The shape of the rainfall did not change much after 30 mins.

CONCLUSION

In case of a shower and a typhoon, we proved that the vector of the rainfall movement calculated agrees with the actual result, and degree of accuracy is high. The time of calculation by computer is shortened enough, owing to decrease of the calculation count of pattern recognition. By the normal engineering work station, the prediction of the rainfall becomes possible. In the case of showers, the future rainfall within about 30 mins can be predicted. It will be used in the urban stream flow analysis for supporting pump operation.

There are some problems. In the case of a typhoon, the vector of rainfall movement can be calculated, but it is difficult to predict the rainfall area in the near future accurately because the shape of rainfall changes quickly. It is necessary to predict the shape in a different way. In the case of rain in a cold front it is difficult to predict the vector of rainfall movement because the area of rainfall is larger than the area of the narrow radar rain gauge.

BIBLIOGRAPHY

Kanamori, T., Kondo, S. and Kobayashi, S. (1992). System technologies for expanded scope of drainage and sewage plants. *Toshiba Review*, **47**(5).

Takeo Kinosita and Kiyoshi Izumi. (1990). Operation urban disaster prediction system in Tokyo concerning urban storm flooding. *Proceedings of Fifth International Conference on Urban storm Drainage*, Y. Iwasa and T. Sueishi (eds.), Suita, Osaka, Japan, 23-27 July, pp. 203-208.

Eiji Toyokuni and Tsutomu Tanaka (1990). Rainfall runoff characteristics in a small residential basin. *Proceedings of Fifth International Conference on Urban storm Drainage*, Y. Iwasa and T. Sueishi (eds.), Suita, Osaka, Japan, 23-27 July, pp. 91-96.

Pergamon

Wat. Sci. Tech. Vol. 28, No. 11-12, pp. 87–98, 1993.

0273–1223/93 $7·00 + 0·00

MODEL BASED ON-LINE CONTROL OF SEWER SYSTEMS

Jörgen Bo Nielsen*, Sten Lindberg* and Poul Harremöes**

** Danish Hydraulic Institute, Agern Allé 5, DK-2970 Hørsholm, Denmark*
*** Department of Environmental Engineering, Technical University of Denmark, DK-2800 Lyngby, Denmark*

ABSTRACT

A new software package has been developed for use in real-time control of the flow in combined sewer systems (CSS). A central feature of the new package is forecast modelling of the flows and volumes in the sewer system. Based upon the forecast model predictions, an expert system or an optimisation module determines the set-points for all regulators in the system.

The new software package includes a comprehensive tool-set for use in the design of on-line control systems, including a model-based real-time simulator, which can be applied in the testing of control strategies.

The first practical application of the software package for a sewer system in the city of Aalborg is described.

KEYWORDS

Real-time control, combined sewer systems, on-line model, forecast model, design tools, optimisation, rule-based control.

BACKGROUND

On-line monitoring and real-time control (RTC) techniques are gradually becoming accepted as valuable methods for optimizing the operation of urban drainage systems and for ensuring the best possible cost/performance ratio when investing in improved systems.

The basic principles of applying RTC technology to urban drainage systems have been described by several authors. Some of the most comprehensive introductions to the subject have been written by Schilling (1989, 1991).

Schilling points to several types of problems associated with the introduction of RTC technology.

1 Legal problems
Liability problems, global improvement versus local problems.

2 Organizational problems
Traditionally, sewer systems and treatment plants are operated by different groups. Collaboration may initially prove difficult.

3 Technical problems
The lack of software tools dealing with the design and operation of RTC systems is one out of a number of technical problems mentioned by Schilling.

This paper focuses entirely on the technical aspects. It describes a software package which has been developed over the last four years specifically as a set of tools to be used in all phases of the design, implementation, and operation of RTC systems for urban drainage systems.

The software package introduces a new concept in RTC systems for urban drainage systems: the use of numerical on-line and forecast modelling as an integrated part of the RTC system.

MODEL BASED CONTROL

Decisions rely on information. Improvements in the quantity and quality of the available information will lead to improved control decisions, if used properly. The information-theoretical aspects of introducing a forecast capability in the RTC system are illustrated graphically in Fig. 1.

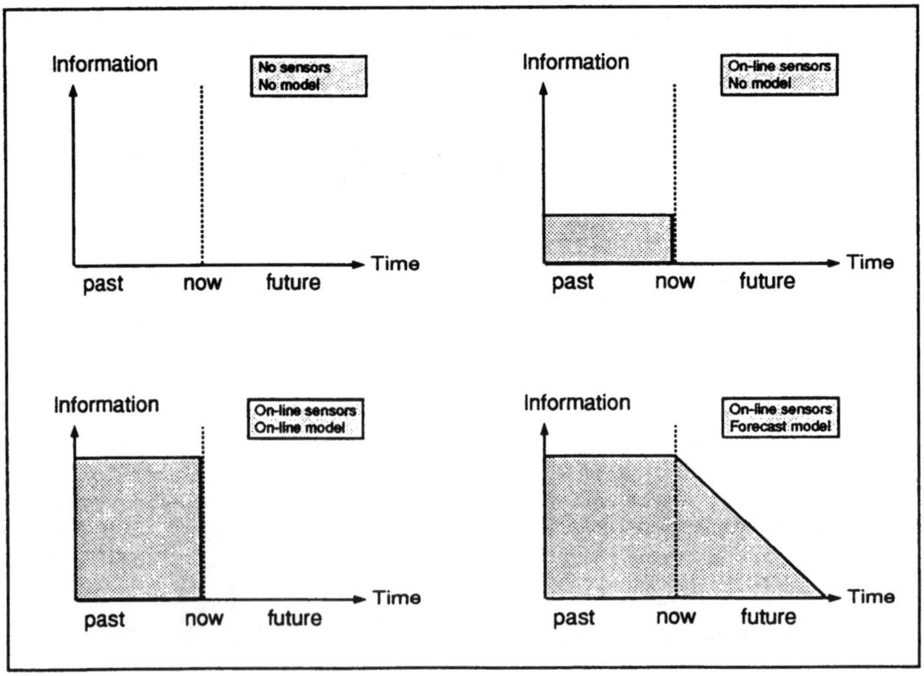

Fig. 1. Models in control. Information-theoretical aspects.

- o Sensors provide information about some of the system now and in the past.
- o An on-line model provides information about the whole system now and in the past.
- o A forecast model provides information about the whole system now, in the past, and (with decreasing accuracy) some time into the future.

Models are usually categorized as either black-box models or white-box models. These two model types use two entirely different methods for encapsulation of knowledge regarding the process which they are both built to model.

A black-box model is a representation of the domain which does not pretend to use or contain any real knowledge about the process itself. Rather, the model contains a number of parameters which can be tuned to represent a range of different processes.

Black-box models can be based on various techniques, which are all basically variations on a common theme. Well-known examples are unit-hydrograph models and models based on Kalman filtering. The most recent example of a general black-box model type is the artificial neural network.

White-box models, on the other hand, encapsulate information about the process to be modelled in a more direct manner. The process is in general described in terms of a set of more or less complex equations which are solved - often numerically - in order to provide an answer.

The main advantages of black-box models are that they can be established without any deeper insight into the process to be modelled. They are also in general relatively efficient in use.

The main disadvantages are that they are limited in scope. They do not (cannot) lead to any new knowledge about the process. They have to be tuned - based on measured data - to give reliable results. If they are used outside the range for which they have been tuned, they may give entirely misleading results, and the results cannot be checked in any simple way, as no additional information is available from the model.

Accordingly, the main reason for using a black-box model is the lack of a sufficiently efficient white-box model.

THE NEW SOFTWARE TOOLS

Three new software tools have been developed for the design, implementation and operation of RTC systems for urban drainage systems. All three are based on a white-box model which simulates the processes in the sewer system. This white-box model is based on the fully dynamical pipeflow model of the MOUSE package (Lindberg *et al.,* 1986; Nielsen *et al.,* 1987). The three new tools are:

MOUSE PILOT — for assessment of the potential for applying RTC to a specific catchment and for evaluation of control strategies based on deterministic long-term simulations.

MOUSE ONLINE — for use as the operational central control unit in the RTC operated urban drainage system.

MOUSE SIMULATOR — for use in the planning and design of the RTC system and in the testing of the final software configuration.

MOUSE PILOT

The MOUSE PILOT tool is the "first leg" of the new software package. The purpose of this tool is to enable the designer of the RTC system to test the long term effects of specific control strategies using deterministic methods.

When introducing RTC in sewer systems, the operation of the control devices will often have dramatic impacts on the hydraulic conditions - indeed that is the purpose of the control. These impacts can only be reliably modelled with a fully deterministic model.

The use of fully dynamic models is normally limited to analysis on a single event basis due to the heavy computations involved. Single event analysis mode is unsuited for the evaluation of the overall potential benefits of applying RTC to new catchments, since this evaluation must be based on the long-term effects.

The MOUSE PILOT module has been designed to assist the engineer in the feasibility phase, combining the dynamic pipe flow model with a less advanced approach for the simulation of dry weather conditions. The module switches automatically between the two modes. The dynamic mode is typically applied during wet weather conditions, eg when combined sewer overflow (CSO) structures are diverting flows or when the flows are actively controlled. Built-in descriptions of movable weirs, gates, valves, and pumps together with a user configurable rule-base allow for testing of a wide range of alternative system layouts and control strategies.

MOUSE ONLINE

MOUSE ONLINE is a modular software package which is installed on top of a traditional SCADA system providing instructions in terms of setpoints to this system, see Fig. 2.

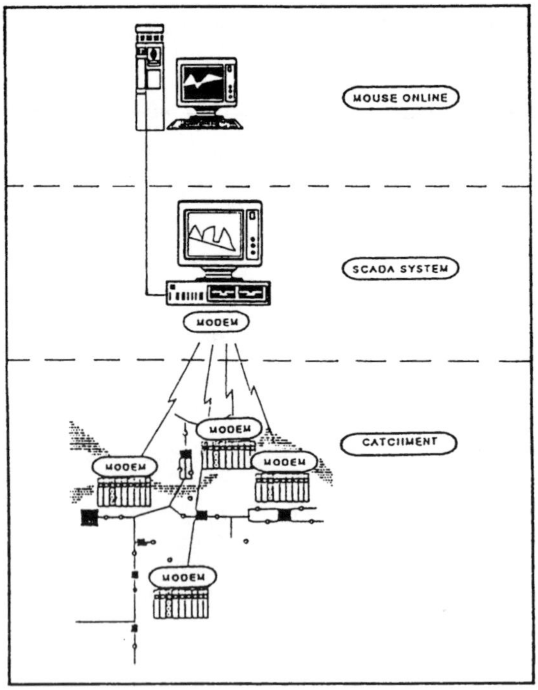

Fig. 2. MOUSE ONLINE as the central control unit.

MOUSE ONLINE derives these setpoints on the basis of all available information:

- o Sensor data which are transferred from the SCADA system
- o Model data from the forecast pipeflow model which is part of the ONLINE package.

MOUSE ONLINE uses a double blackboard architecture. The different modules share and exchange information by writing to and reading from a common data area, the internal blackboard. Another blackboard - the external blackboard - is used for exchanging data between MOUSE ONLINE and the SCADA system. The double blackboard architecture is illustrated in Fig. 3.

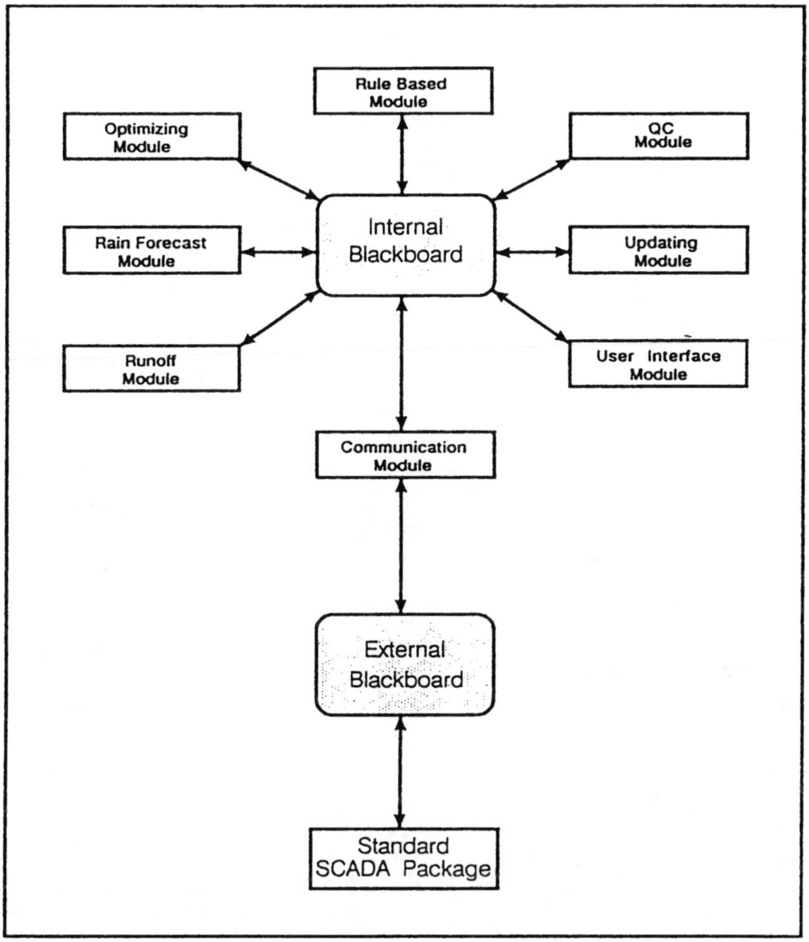

Fig. 3. The double blackboard architecture.

The double blackboard architecture makes the package extremely flexible and open. Modules can be exchanged and added very easily. They simply have to be able to read from and write to the blackboard according to the common specifications.

The double blackboard architecture also makes it very easy to exchange data with external programs. As an example, it is possible to receive and apply information regarding the variations in capacity of a treatment plant, if such information is available. In the other direction, forecasts of the expected flows to the treatment plant can be transferred to the control system for the plant. This information can then be applied, when optimizing the plant operation.

The architecture includes several different options for the assessment of the sequence of control setpoints. One option is to use a rule-based approach. MOUSE ONLINE includes an expert system module (based on NEXPERT OBJECT, a system provided by Neuron Data). By means of this module the system can be configured with a purely rule-based mechanism for assessment of the preferred control sequence - using any combination of forecast data and on-line measurements as input to the rule evaluating mechanism.

The main advantage of such a rule-based approach is the transparency of the process leading to the control actions. Every step is easily tracked and explained in terms of the rules. This can be of major importance in the process leading to general acceptance of the new system within large, diversified organisations.

Another option involves the use of optimizing techniques. A module has been developed which uses the forecast results as input to a computation of the optimum control sequence based on linear optimization methods as described e.g. by Petersen (1987) and Lindberg *et al.* (1990).

A hybrid optimization/rule-based approach can also be implemented with the existing tools. Such a system could eg assess the control sequence in two steps: An optimization step followed by a check (and possible modification) based on a set of rules.

MOUSE ONLINE has been designed to run without continuous supervision by an operator. The system continuously monitors its own performance, and if it discovers that the performance is unsatisfactory, then it shuts down gently - ie changes to default setpoints and hands over the control to the standard SCADA system.

An implication of the lack of continuous operator supervision is the need for the control system to be designed to be fail-safe at the physical level - including eg carefully designed by-pass structures. Even a prolonged total power failure should not lead to system performance which is substantially below the performance of the un-controlled system.

The package includes tools for design of very user-friendly graphical user interfaces for the operators of the control systems. The user interfaces provide an overview of the most relevant sensor and regulator data. The user interfaces are intentionally designed to include only a minimum of options, as the user already has access to a number of more powerful features through the SCADA system. Fig. 4 illustrates this "operator user interface".

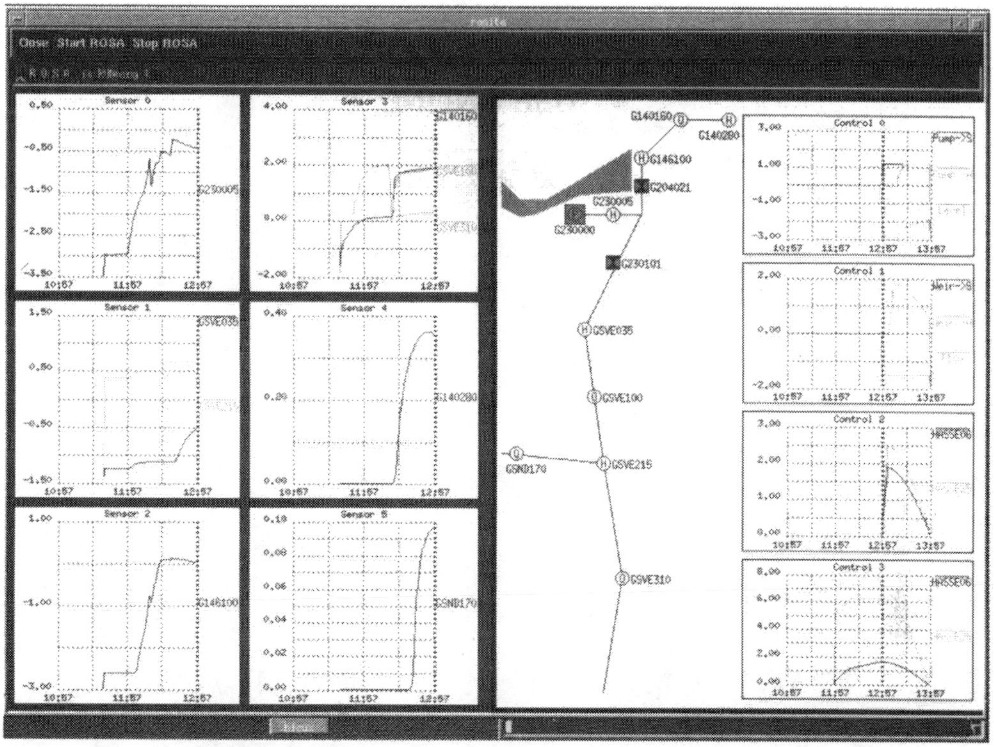

Fig. 4. Example, MOUSE ONLINE operator user interface.

MOUSE SIMULATOR

The basic idea behind the MOUSE SIMULATOR tool is illustrated in Fig. 5. It simply plays the part of the catchment, complete with sensors, regulators, and SCADA software.

The MOUSE SIMULATOR can be instructed to run in either real-time or in fast forward mode. In fast forward mode it simulates as fast as the hardware permits (typically one or two orders of magnitude faster than real-time). In both modes it models the run-off processes in the catchment, including:

Rain	Based on historical rain events selected by the user from a built-in database
Surface runoff	Based on MOUSE surface runoff model
Flow in sewers	Based on MOUSE pipeflow model
Effect of regulators	Receiving setpoints from MOUSE ONLINE or from a human operator. Using realistic characteristics (eg response times) of the individual regulators
Sensors	Collecting "sensor data" from the model results and transmitting these to MOUSE ONLINE or the human operator

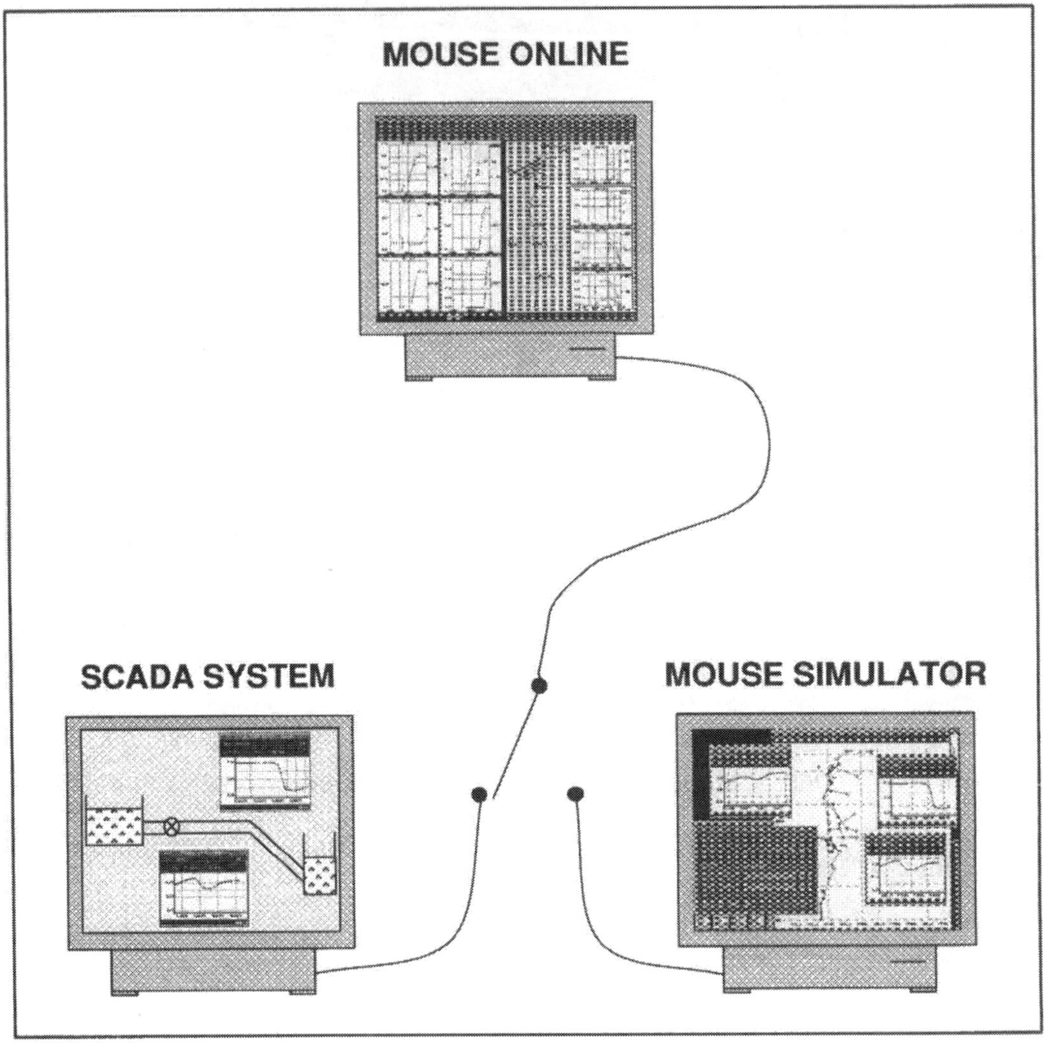

Fig. 5. MOUSE SIMULATOR as "stand-in" for the real catchment.

The MOUSE SIMULATOR can be controlled by any control software (including software based on MOUSE ONLINE) or by a human operator. Accordingly, it can be used in testing such software packages as well as in training of the operators.

Two main applications of this tool are envisaged. The first of these is in the planning and design of the RTC system for a new catchment. The simulator enables the designer to optimize the complete RTC system by simulating any number of RTC system layyouts, including:

o Different locations and physical designs of the regulators

o Different combinations of sensor equipment

o Different overall control strategies/cost functions

The second main application is in the testing of the final RTC software. The design engineer can test a wide range of scenarios, including:

o System behaviour under extreme rain conditions

o Effects of malfunctions in the sensors and/or regulators

o Effects of malfunctions in the sewer network itself (eg blocked pipes)

The MOUSE SIMULATOR is equipped with a very powerful windows-based graphical user interface. This user interface has been designed for use by the design engineer or consultant who has substantial background knowledge concerning the use of advanced numerical models, as well as knowledge on the relevant physical processes. Fig. 6 shows an example from a multiple window session with MOUSE SIMULATOR.

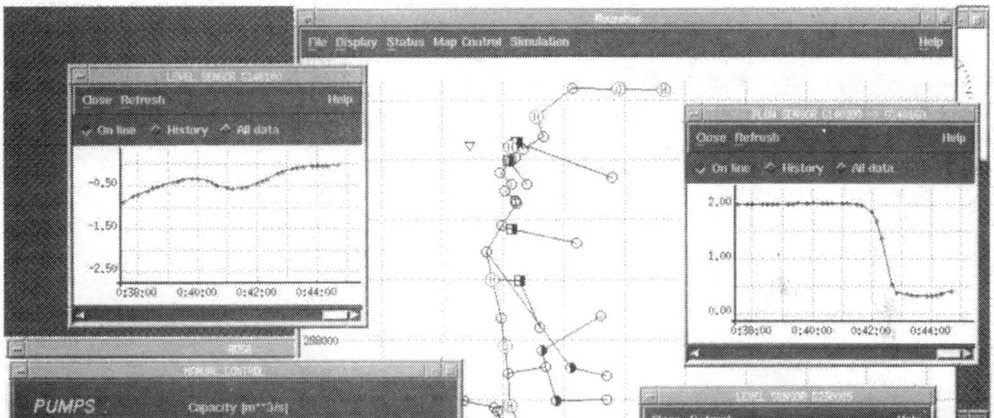

Fig. 6. Typical session with MOUSE SIMULATOR

PRACTICAL APPLICATIONS

The new software packages are being applied in the design of a number of RTC systems in Europe. The first of these systems to go into operation is the RTC system in the city of Aalborg in Denmark (Aalborg West catchment). This system is interesting not just because it is the first system established on the basis of the new tools, but also because it involves two similar, advanced software packages:

STAR For on-line tuning of the control system for the sewage treatment plant

TIPS For management of information related to the sewer system and treatment plant

The two control systems for Aalborg will be fully integrated. The STAR based system will provide on-line information to the MOUSE ONLINE based system regarding the current and projected capacity of the treatment plant. This information will be used as a parameter in the control of the sewer system.

Information will also flow in the other direction: The sewer control system will provide forecasts of the amounts of water which are expected to reach the treatment plant over the coming hour or more. This information can be used in the plant operation, which can be tuned to higher performance over a certain period, provided that sufficient prior warning is received.

Fig. 7 shows an outline of the Aalborg West catchment. Initially, the flow in the main South to North going trunk sewer, which has an average slope of 0.1 %, will be controlled by means of two movable gates. Each gate controls a pipe volume of approximately 4,000 m^3. A third gate will control a similar volume in a branch pipe to the main sewer.

The main problem in the catchment is CSOs at the treatment plant during wet weather. It is expected that the total reductions in CSO due to the control system will correspond to the reductions which can be obtained through the construction of a traditional retention basin of 6,000 - 10,000 m^3 next to the sewage treatment plant. The costs of the RTC based system will be less than half the costs of the traditional system.

Substantially larger effects will be obtained - at relatively even lower costs - if and when the control system is extended to other parts of the Aalborg West catchment. This is expected to take place over the next two to four years.

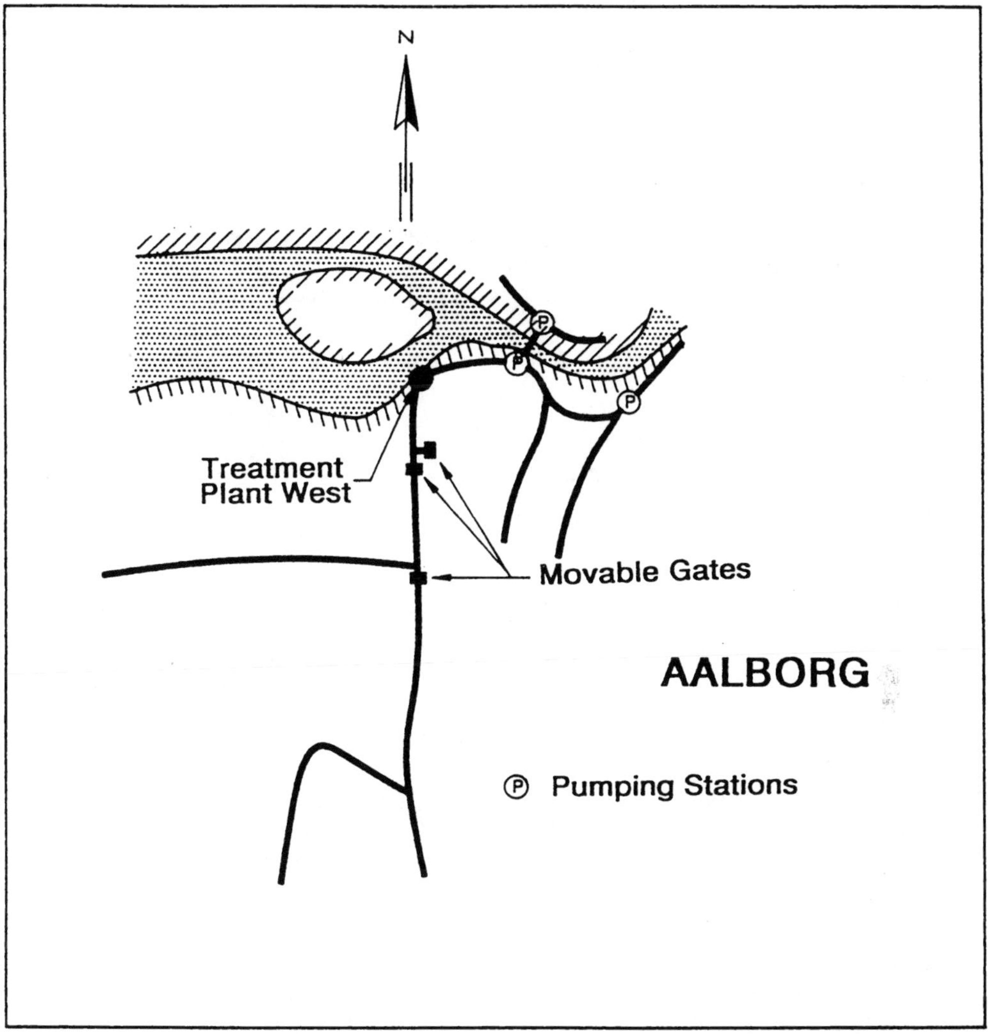

Fig. 7. Aalborg West catchment with control structures and the sewage treatment plant.

The Aalborg system with the three controllable gates has been completed and will be in operation after a brief off-line test period by the end of November 1992.

The new software packages for design and implementation of on-line control of sewer systems will be disseminated in Europe as part of an EEC supported SPRINT project. Pilot projects have been defined for four cities in England, Italy, Spain, and Denmark. A project supervisory committee with the participation of independent experts and end users will supervise the progress of the SPRINT project.

REFERENCES

Lindberg, S., Jørgensen, Willemöes, T. and Emolet, ApS. (1986). MOUSE – Modelling of Urban Storm Sewer Systems. International Symposium on Comparison of Urban Drainage Models with Real Catchments Data, Dubrovnik, April 8-11.

Lindberg, S., Nielsen, J. B., Tetersen, S. O., Amdisen, L. K. and Harremöes, P. (1990). Real Time Optimization of a Sewer System as Part of a Larger Municipal Prototype Project Including Treatment Plant Operation. 5th Int. Conf. on Urban Storm Drainage, Osaka, July 23-27.

Nielsen, J. B., Lindberg, S., Amdisen, L. K. and Jensen, N. E. (1987). Towards the Fifth Generation of Numerical Modelling Systems. Proc. Fourth International Conference on Urban Storm Drainage, Lausanne.

Petersen, S. O. (1987). Real Time Control of Urban Drainage Systems. Department of Environmental Engineering, Technical University of Denmark, ISBN 87-89220-04-08.

Schilling, W. (ed.) (1987). Real Time Control of Urban Drainage Systems – The State of the Art. Report of the IAWPRC Task Group on Real Time Control of Urban Drainage Systems, London.

Schilling, W. (1991). Real Time Control of Urban Drainage Systems, from Suspicious Attention to Widespread Application. In: Advances in Water Resources Technology (ed. G. Tsakiris), Rotterdam.

WATER DISTRIBUTION CONTROL SYSTEMS

Pergamon

Wat. Sci. Tech. Vol. 28, No. 11-12, pp. 99–109, 1993.

0273–1223/93 $7·00 + 0·00

AN OVERVIEW OF THE TRUNK SCHEDULING SYSTEM FOR THE LONDON RING MAIN

David Burnell, Julia Race and Phil Evans

Corporate Modelling Group, Thames Water plc, Nugent House, Vastern Road, Reading RG1 8DB, UK

ABSTRACT

The London Water Ring Main is one of UK's largest water investments. It will eventually carry half of London's water and greatly increase the flexibility of London's trunk distribution network. The Trunk Scheduling System is a core part of decision-support for the enhanced network. It has been written to help Operational Controllers make best use of that extra flexibility.

Cost-effective operation makes best use of the cheapest treatment works, low-energy-loss routes and off-peak tariffs. These need to be combined to meet projected diurnal demands in over 50 zones and leave each of some 20 service reservoirs at a target end volume, whilst taking no more than the declared "reliable output" from each of 10 works. The schedule proposed must be hydraulically viable as well as cheap. Most route constraints arise from pressure limits (so as not to burst mains or leave customers without water). Hydraulic constraints are non-linear and require on-line hydraulic intelligence as the schedule is developed.

The key idea is to regard a schedule as a series of "operating regimes", each of which is applied for a specified duration. Each regime applies within a particular timeslice. It comprises settings for each pump in the network, together with its implications: a unit cost and the effect of those settings on each source and each reservoir under hydraulic equilibrium. Trunk Scheduling provides a series of modules to answer the questions:

- which regimes (of the billions possible) should make up the schedule?
- how long should each selected regime be applied for?
- in which sequence should the chosen regimes be applied?

KEYWORDS

Water, distribution, optimisation, hydraulic equilibrium

CONTEXT

The London Water Ring Main is the largest investment by a UK Water Undertaking. It will significantly increase the flexibility of the trunk distribution network but impose a much greater requirement for integrated operation across London.

Thames Water is developing many new control systems in conjunction with the Ring Main, focused at the London Control Centre at Hampton. These include SCADA systems to monitor the current state of the network, and an Operational Management System (OMS) to bring the data together as the basis for managerial control. Coordination will be achieved from the Centre via Local Area Control.

Feeding off, and back to, OMS is a series of decision-support models. These help make best use of the flexibility which the Ring Main will offer and provide a basis for proactive management:

- **demand prediction** uses time-series data on past demand in each zone plus local judgement, to project demand for each zone;

- **trunk scheduling** starts with demand prediction and projected water treatment works outputs, and generate a schedule for all the major trunk distribution routes;

- **area control modelling** allows local areas to verify the hydraulic viability of the proposed schedule and fill in the local details.

The development of the Trunk Scheduling System - the core decision-support for the Ring Main - has been carried out in-house by Thames Water's Corporate Modelling Group, working closely with London Control staff. Ten man-years' modelling effort has been expended on the development. Much of the System is now in place and in operational use: the rest is undergoing calibration and operational testing.

Scheduling a hydraulically constrained network is an algorithmic task best tackled by well proven procedural methods. Rule-based systems may flounder when confronted with unexpected situations: this System needs to produce a solution whatever the circumstances. Algorithmic methods can still be used in an imaginative manner and in innovative combinations. Trunk Scheduling contains many imaginative leaps: this paper aims to explain these (in simple terms) and set them in the context of what the System has to do.

REQUIREMENTS FOR TRUNK SCHEDULING

Cost-effective operation

The Trunk Scheduling System's task is to help the London Control Centre develop daily schedules for the trunk distribution network. It has to find a cost-effective way of using the Ring Main and other major pumps, to deliver over 2000 Mld from some 10 water treatment works (or other sources) via 20 service reservoirs to 50 pressure zones.

Boundary conditions and non-standard circumstances

The schedule has to meet the day's expected "boundary conditions". These comprise the declared "maximum reliable output" for each works, the starting and target end volumes for each reservoir, and the demand (which varies significantly over the 24 hours) for each zone. All of these factors vary from day to day, so a specific schedule is required for each day.

Another reason for new schedules to be generated daily is that there are variations in the network itself. "Exceptional" circumstances include pumps being unavailable (e.g. for maintenance), stand-by pumps being deployed to a different duty/destination, infusions (i.e. flows achieved by adjusting valves between neighbouring zones with a pressure differential) and reservoir cells being taken out of service or having maximum or minimum levels reset.

Hydraulic constraints

The schedule must not only meet the boundary conditions and take account of the network configuration, it must work within operational constraints on how much water can be delivered along each route at each time. The schedule must use pumps on their operating curves. It must not lead to pressures which are too high (burst pipes) or too low (unmet demand in higher parts of a zone).

The schedule must also make the best of variations in electricity tariff over the 24 hours. It costs much less to pump at night, so this is the best time to replenish reservoirs. However, the network constraints and the numbers of alternative routes mean it is not easy to work out the best way to do this.

KEY SCHEDULING IDEAS

Challenges for Trunk Scheduling

The requirements for Trunk Scheduling are not easily satisfied. The basic difficulty is that:

- the "constraints" on the flows which are possible depend on the hydraulic implications of the plan of action;

whilst

- the hydraulics (i.e. the hydraulic equilibrium of pressures and flows) cannot be computed until a plan of action has been proposed.

A further difficulty is the size of the network to be scheduled. Even when natural boundaries such as strategic reservoirs on the edge of the control area are taken as boundaries for Trunk Scheduling, there remains a vast area to be considered, which cannot easily be subdivided. This gives rise to a vast number of possible pump settings, far too many to be dealt with by (even enlightened) trial and error.

Various steps have been taken to overcome the difficulties and make the problem tractable.

Building in hydraulic intelligence

As a design principle, where constraints arise from hydraulic viability, it is better to make those calculations directly and as required, rather than rely on surrogate rules about what is possible. A hydraulic model offers the best form of hydraulic intelligence, if it can be incorporated within the system. Expert or rule-based systems are a poor substitute for direct calculation where the method of calculation is specifiable. Look-up tables (e.g. relating pressures to pump settings) may also seem attractive, but present formidable problems of maintenance as the network changes.

One innovative feature of Trunk Scheduling has been the development of a hydraulic model using optimisation methods. Because optimisation occurs elsewhere in the System, this has made for economical and compact coding. The approach has turned out to be better at sensitivity analysis than conventional hydraulic methods, allowing pump setting variations to be rapidly evaluated.

Development of a schematic network

To apply a hydraulic model, a schematic network is needed for the Trunk Distribution network. Another design principle is that the same network be used for all stages of scheduling. Doing this minimises on-going maintenance of the System and makes it easier to use. Considerable effort has gone into developing a schematic which captures the key control and transfer elements of the London network.

Time slicing

An important idea to help break down the problem was to schematise time by introducing "time slices" and not allowing boundary and tariff conditions to vary within a timeslice. Currently seven timeslices are assumed, namely night, rising (both customers and demand!), peak, morning, afternoon, teatime and evening. These allow the diurnal variations in demand and in tariffs to be handled with an acceptable level of accuracy.

Basic schedule building blocks: regimes

Given a hydraulic evaluator, a schematised network and a day divided into homogeneous timeslices, the key idea of Trunk Scheduling comes into play. This **is to regard a schedule as being composed of a series of "operating regimes", applied in a particular order and each lasting for a specified duration.**

Each "regime" comprises:

- a list of on/off and speed "settings" for each pump in the network;
- its overall energy use, taking account of the efficiency with which each pump will perform when this regime is applied; and
- details of what "impact" that regime will have on the network.

Each regime operates in a particular timeslice, for which the boundary conditions (supply and demand) are known. When the pump settings are applied to the network, the resulting equilibrium will include an effect (either filling or emptying) on each reservoir and a usage figure for each works. These consequential impacts are an essential part of the regime.

For Trunk Scheduling, regimes are the key building blocks for a schedule. Once the regimes themselves have been specified, a schedule can be specified very compactly as a list of "regimes" and "durations".

A key virtue of dealing with regimes is that their hydraulics can be accurately evaluated and costed (and indeed must be, so the reservoir fill rates can be assessed). Hence, one can avoid considering any regime for the schedule which would not be hydraulically acceptable. Furthermore, if a schedule is made up of regimes which are individually viable, it will in turn be acceptable provided it has the right cumulative effect on each reservoir.

Once regimes are specified together with their reservoir implications, it is a matter of arithmetic to compute the reservoir profiles that will result from a particular schedule. If the cost of each regime is known, it turns out to be a straightforward linear programming (LP) problem to find the best "blend" of a given set of regimes to bring reservoirs to their target end volumes at minimum overall cost.

The concept of "regime" is crucial to Trunk Scheduling. Essentially, London is regarded as being controlled by a series of pump settings, each of which is operationally viable, and which have the right effect in combination. With 100 pumps in the schematic network, however, there are several billion billion possible regimes. Identifying the best of these and showing how they should be combined together is the central task for Trunk Scheduling.

COMPONENTS OF THE SCHEDULING SYSTEM

The Trunk Scheduling System which has been developed to meet these needs comprises a number of separate modules, which can be combined in various ways. The models are depicted in Figure 1, and are discussed in turn below.

The System is written in Turbo Pascal. The system comprises some 120,000 lines of code and comment and the executable code occupies some 900K.

Setting boundary conditions and overall allocations: NETPLAN

The first task of the Scheduling System is to check that the overall boundary conditions for the day's schedule are consistent and achievable, and to gain

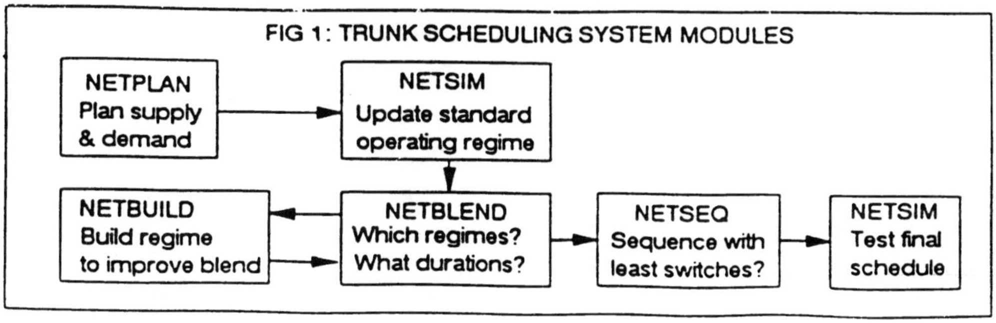

an overview of the way the flows will be allocated.

The boundary conditions comprise:

- "supply" - each water treatment works' declared reliable output;
- "demand" - the projected diurnal demand for each Zone, based on recent trends and local judgement;
- reservoir starting levels and "target end levels" for the schedule.

The "exceptional" features of the network also need to be established so they can be taken into account in this and all subsequent modules.

Clearly, one must ensure works availability exceeds demand plus reservoir refill. It is also be necessary to take account of flow constraints within the network and variations over the day. There are many problems where there is an overall excess but local shortfalls - for example, excess supply in northeast London cannot necessarily be transferred to a zone in the southeast.

The module to check the day's boundary conditions is called NETPLAN and is always called first in Trunk Scheduling. It does not have access to the hydraulic assessor, but relies on user-specified flow constraints on each route in each timeslice. Pump capacity constraints are expressed as an estimate of the lift required of each pump in each timeslice, which are then translated (via pump characteristic curves) into flow constraints.

NETPLAN has to take account not only of the flows along routes but also the scope for using the service reservoirs to smooth out the variations in demand over the day and to maximise cheap pumping. To do this, the flows for the whole day need to be considered together. This is done by considering a multi-tiered version of the schematic network, with one "tier" for each timeslice. The operational flexibility offered by the reservoirs is then incorporated by adding notional arcs between each reservoir in successive timeslices, the capacities of these "routes" being determined by each reservoir's storage limits. A stylised version of the enhanced network is depicted in Figure 2.

When the 300-arc schematic network is replicated seven times (for each timeslice), notional arcs are added between each of 20 reservoirs in each timeslice and boundary links are added in for each source and each zone, a network of some 2,500 route "arcs" and 1,000 location "nodes" is obtained.

Unit costs on each of these arcs arise from the marginal costs of treatment at a works or from the energy use of each pump in achieving its assumed lift (given its operating curve). NETPLAN applies network optimisation to find the cheapest set of flows to meet the specified demand over the day, keeping within route capacities and making best use of the service reservoirs.
The problem is a special case of linear programming. A fast method of solution is used, called Network Simplex. This uses a sophisticated labelling scheme to rapidly improve an initial "solution". The NETPLAN code solves the problem specified, on a 486 PC and starting with zero flows, in under 10 seconds.

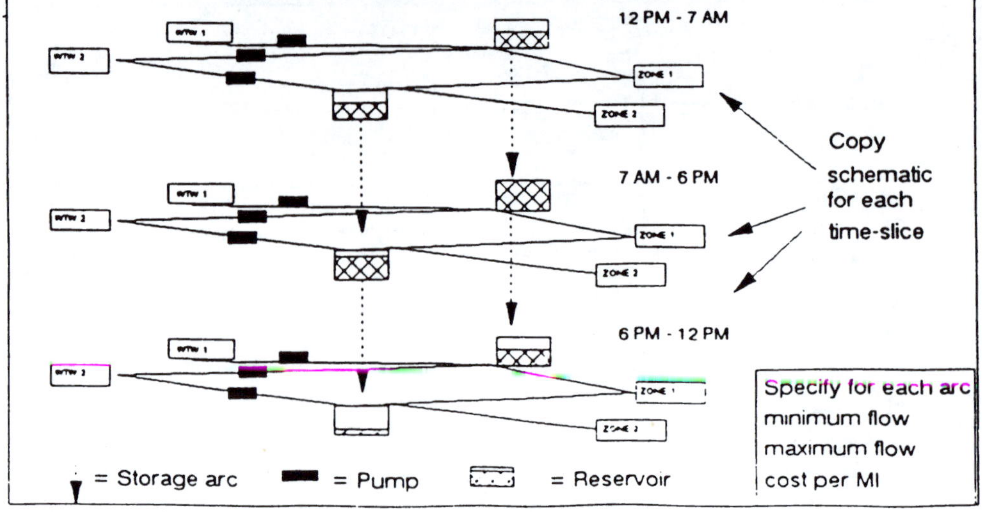

FIG 2 - FLOW BALANCING BY TIME-SLICING

The NETPLAN solution flows are optimal in several ways:

- they make **best use of the sources** available - all other factors being equal, the solution will minimise use of the least efficient works;
- they make **best use of the pumps**, taking account of their efficiencies and of the costs of alternative routes into zones;
- they make **best use of reservoirs** to time-shift as much pumping as possible to the low-tariff periods.

Because NETPLAN can solve the problem so quickly, it can be used as a "what-if" tool to investigate the effects of alternative boundary conditions. This is particularly useful on days where demand is high or supply restricted, and some adjustment to the initial boundary conditions is required (e.g. upping a works output, or reducing an end-of-schedule reservoir target).

"Infeasibilities" arise when supply and demand are out of balance. They manifest themselves in hard-to-reach zones which may not be the real source of the difficulty. NETPLAN handles these problems in an intelligent manner. This is done by building in additional notional arcs when setting up the multi-tier network, with a high unit cost, at points where constraints might need to be relaxed. By graduating these penalty costs, NETPLAN is made to break its constraints (where necessary) in a controlled manner. Hence even "infeasible" solutions give sensible guidance to the user on the best way to modify the day's boundary conditions so the scheduling problem can be solved.

Calibrating standard regimes to the day's boundary conditions: NETSIM

The NETPLAN solution, for all its usefulness, falls short of being an achievable pump schedule. This is for the following reasons.

- It recommends **average** flows by route and timeslice. The way that pumps should be deployed to achieve these flows is yet to be determined.

- NETPLAN uses **flow** constraints as surrogates for **pressure** limitations. This is acceptable for checking overall flow allocations, but means one cannot be sure that a particular set of pump settings will not burst mains or leave customers with inadequate pressure.

In order to build a schedule to meet the agreed boundary conditions, it is necessary to start, not with the NETPLAN flows, but with sets of possible operating regimes as the basic building blocks for the schedule.

Trunk Scheduling exploits the fact that successive days are likely to be fairly similar from a scheduling viewpoint. A reasonable starting point for a schedule is to take yesterday's schedule and consider how it needs modifying to meet today's requirements. More specifically, breaking yesterday's schedule down into a succession of pumping regimes, one can ask:

- what will be the effect of these regimes taking account of today's new boundary conditions and plant availability (and zonal infusions)?

- given an updated implication for each regime on reservoir fill rates,

  how should the regimes be deployed so as to leave each reservoir at or above its target volume by the end of the schedule period?

The first of these questions is dealt with by the NETSIM module: the second is covered by the NETBLEND and NETBUILD modules.

NETSIM's task is to hydraulically simulate a possible set of solution regimes on the latest boundary conditions (for supply and demand), taking account of plant availability and other exceptional circumstances. For calibration purposes, each equilibrium is computed at a standard mid-timeslice volume for each reservoir: this enables regimes to be compared on a like-for-like basis (when NETSIM is re-visited at the end, cumulative volumes are computed).

Finding a hydraulic equilibrium is a standard problem for which long-standing solutions exist. However, Trunk Scheduling needs to find the equilibrium heads and flows for each regime within an integrated system: it would be unwieldy to do this by recourse to an external package. Instead, it uses a method for solving hydraulics via optimisation, suggested only in the late 1970s.

The node "dual costs" used by the Network Simplex algorithm when optimising turn out to correspond in this case to the head at each node. It is surprising to find an internal working variable with real physical significance! This means that the approach can be extended to allow rapid inspection of heads at nodes at equilibrium, and hence of hydraulic viability.

As with NETPLAN, it is possible to make the program react sensibly to unacceptable solutions. This is done by stating, for each critical-head node, a responsive manifold to deal with any violation. Where a head constraint is infringed, the relevant pump settings are then "trimmed" to deal with the infringement and a new hydraulic equilibrium sought.

A set of regimes can be run through NETSIM and hydraulically evaluated on the relevant boundary conditions. Where necessary, pumps are automatically fine--tuned to avoid hydraulic violations. The total pump usage and consequential refill rate of each reservoir passes through to the next stage of calculation.

Finding the best blend of regimes: NETBLEND

Given a set of valid regimes, Trunk Scheduling next asks **what is the best way to combine these so as to achieve the target reservoir volumes by the end of the schedule, keeping each reservoir in bounds at the end of each timeslice?**

Because the effect of each regime on the reservoirs is known and each timeslice is homogenous, this can be answered using linear programming:

- the **variables** to be found are the durations spent operating each regime (which could be zero for some of the candidate regimes);
- the **constraints** are the upper and lower bounds on the cumulative volumes for each reservoir at the end of every timeslice, plus the requirement that enough regimes are used to totally account for the hours in each timeslice;

• the **objective function** to be minimised is the total cost of operating with the selected regimes and durations.

With 20 reservoirs and 7 timeslices, this leads to a fair-sized problem with nearly 300 constraints. However, nearly all of the constraints occur in pairs, so one can use "bounded LP". The number of constraints is halved and extra information carried about whether non-basic variables are at their upper or lower bound. Halving the number of constraints gives an acceptable solution time (4 minutes on the Control Room 486). This type of short-cut makes it sensible to use purposely written code (which can be linked to other modules) rather than accessing a commercial package.

The formulation aims for reservoir end-volumes by imposing penalty costs where these targets are missed, rather than absolute constraints. This gives end targets the status of "soft constraints" - a realistic view, given the variations between predicted and actual demand which will occur before those targets are reached. NETBLEND can then suggest schedules which fall below the end targets if it does not have the regimes needed to hit them all exactly.

NETBLEND is involved in much the same optimisation as NETPLAN. It too knows about variations in tariff between timeslices, in treatment costs and in aggregate pumping costs for alternative regimes. Hence it is no surprise that NETBLEND gives rise to similar reservoir profiles to NETPLAN.

Generating additional regimes: NETBUILD

The limitation of NETBLEND lies in the pre-selection on the regimes it can choose between. To overcome this, another module, NETBUILD, allows "switching" in of new candidate regimes, once an initial solution has been found.

There are two key observations.

• If a regime is not used in a NETBLEND solution, it can be replaced in the LP by another without affecting the current solution, after which the search for a better solution can resume. This will lead to a schedule making use of the new regime if it is cost-effective to do so.

• Marginal cost information for a NETBLEND solution gives insight into the most costly aspects of making up a schedule using the given regimes. Marginal costs show the benefit that would accrue if each constraint could be relaxed. Because constraints apply for each reservoir, one can derive a marginal cost for filling each reservoir in each timeslice, given the current solution.

The task of the NETBUILD module is to complement this process by **generating a suitable regime to add to the current schedule in the light of the marginal fill cost for each reservoir**. (A special case of this requirement is where the original regimes are insufficient to generate a feasible schedule at all.) The relationship between the NETBLEND and NETBUILD modules is depicted in Figure 3.

The extra candidate regime has to be operationally viable and hydraulically evaluated: hence it needs to be derived using the hydraulic assessor. The Trunk Scheduling System needs to be able to alternate between a substantial linear program and a hydraulic model for the whole of London as:

• NETBLEND tries out extra regimes and moves to a new solution; and
• NETBUILD finds another regime to improve on the current solution.

Some complex overlaying is needed for this to be achieved, only possible because source code for all aspects of the System is available.

The NETBUILD module needs to identify which pumps to switch on or off in the

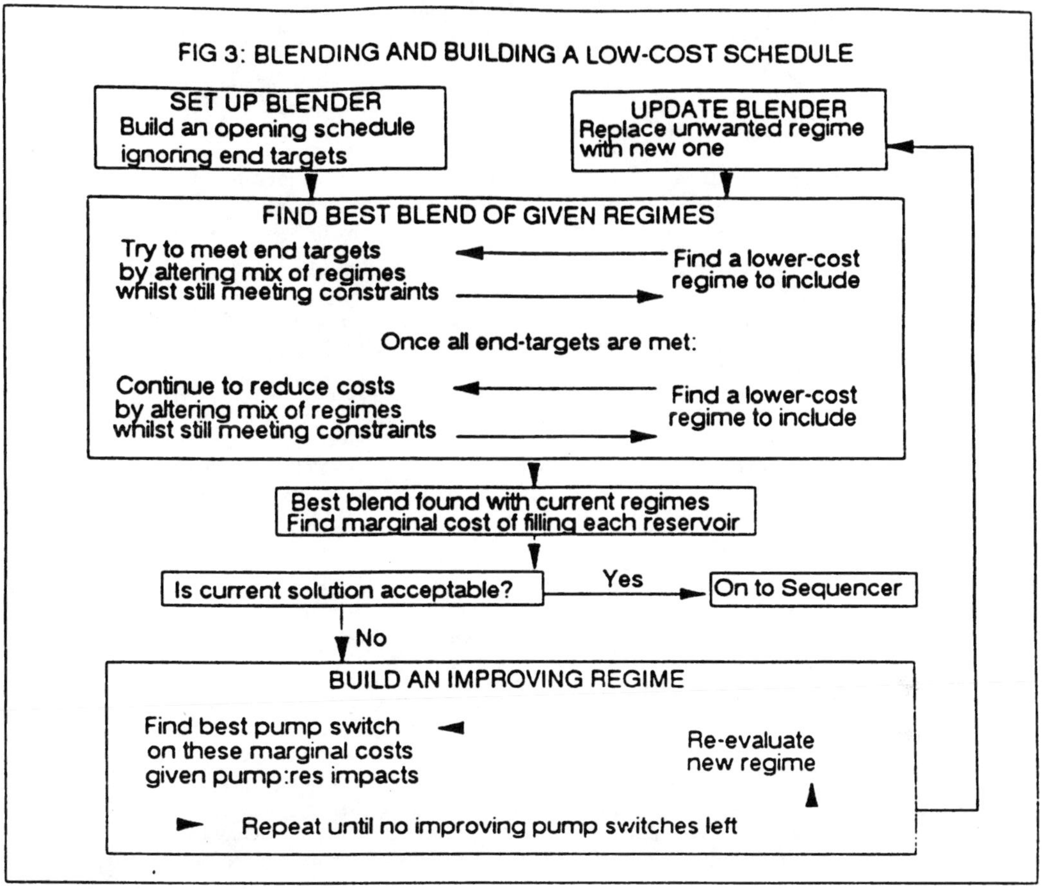

light of marginal reservoir costs, which convey information about which reservoirs it would be profitable to replenish and which to empty, in the context of the current solution.

If there were a simple link between reservoirs and pumps, with each reservoir controlled by a single set of pumps, this task would be easy. However, there are many more pumps than reservoirs, and (via the hydraulics) many links between the two types of element. In particular, changing the pumping at a works manifold can change the fill rate of several reservoirs. The Ring Main itself, as a low-resistance route connecting different parts of the network, increases these interactions.

To identify a suitable regime requires information on the effect of each pump on each reservoir. This is done using "Impact Tables". Impact Tables state the marginal effect of turning on each pump on the instantaneous fill rate for each reservoir and source. This "gradient information" is used to guide the search for a new regime.

Converting a NETBLEND solution to an operational sequence: NETSEQ

The interchange between NETBLEND and NETBUILD continues until the user is satisfied. At this point the "solution" consists of a series of regimes, which if run for specified durations will ensure that the reservoirs are in bounds at the end of every timeslice and meet their end-of-schedule target.

Some work is still needed to move from this to a viable schedule because:

- the order in which the regimes appear in the solution takes no special account of pump switchings or variations in Ring Main flows or water treatment works outputs, and it is desirable for these to be minimised;
- whilst all reservoirs are in bounds at the end of every timeslice there is no guarantee they will stay in bounds within the timeslice.

The task of turning the NETBLEND solution into an operationally viable schedule is carried out by the NETSEQ module. NETSEQ has no control over the regimes chosen or their total duration, but it can adjust the order in which they appear and whether the regime is applied once, for its full duration, or split into a number of shorter spells.

Splitting a regime adds to the pump switches incurred, but makes it possible to keep all the reservoirs in bounds throughout a timeslice. It may be needed, for example, where two regimes have opposing effects on some of the reservoirs, and need to be operated alternately to keep all of them in bounds.

The order in which regimes appear in the schedule needs to be adjusted so as to minimise pump switching and key flow variations. Since the pump settings for each regime are known, it is easy to find the number of pump switches and flow jumps incurred for any specified sequence.

The task of finding the sequence which minimises overall pump switching is analogous to the "Travelling Salesman Problem". In this, the salesman has to find the best tour of a number of cities so as to visit each city once and minimise the total miles travelled. The cities to be visited correspond to regimes. The mileage incurred is the pump and flow switches needed to move from one regime to the other.

A standard way to solve the Travelling Salesman Problem is to try random swaps and inversions of part of the "tour", and keep track of the best sequence found. Ideas from "simulated annealing" are used to avoid getting stuck in a local minimum. Trunk Scheduling applies a similar technique to move to a acceptable sequence or regimes. One benefit of the heuristic approach is that it can ensure reservoirs stay in bounds. Another benefit is that it lends itself to manual intervention if required.

Illustrative results from a call to NETSEQ are shown in Figure 4. The top panel shows profiles for 18 reservoirs and the middle one the variations expected in works outputs, with the schedule proposed. The pumps to be switched during the schedule are shown in the bottom panel. The schedule is made up of some 21 pumping regimes: each vertical line marks the start of a new regime.

Testing and tuning the final schedule: NETSIM revisited

Once NETSEQ has been run, the NETBLEND solution has been converted to potential schedule; this is sent back to NETSIM for final verification.

The schedule comprises regimes chosen to make best use of lower-cost sources, low-energy-loss routes and cheap tariffs, and with durations and an order which ensures all reservoirs will stay in bounds and meet their end targets.

Since only operationally viable regimes are offered to NETBLEND, this schedule should be operationally viable. It is necessary to simulate the schedule taking account of cumulative reservoir volumes when regimes are run in schedule order. Extensive and powerful editing facilities are provided for the Controller to fine-tune the schedule as required.

CONCLUDING OBSERVATIONS

NETPLAN was in the London Control Room for its Royal Opening in January 1992.

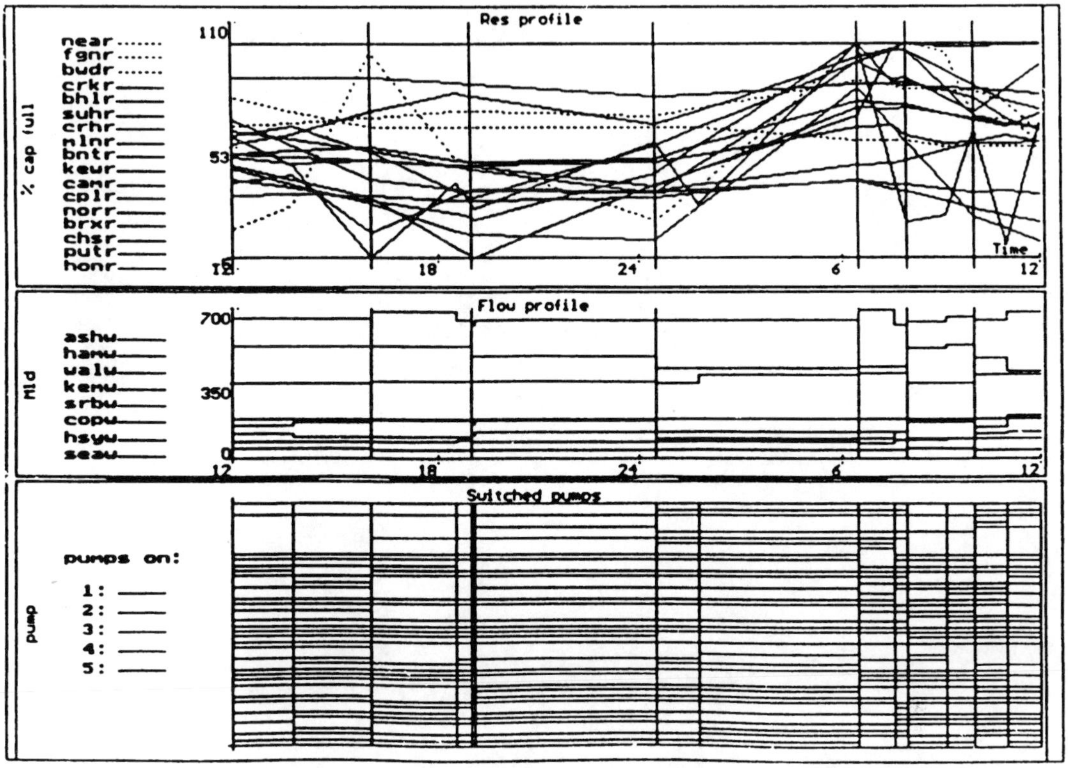

Fig. 4.

A major calibration exercise was undertaken by Operations to check the NETPLAN-recommended flows are in line with those achieved in practice. The interface has been refined following users' suggestions.

NETPLAN has encouraged Operations to formalise procedures relating to the data needed to run the model. Control staff have had one-on-one support from a Corporate Modeller based at the Control Centre. Confidence in NETPLAN is growing and its status is moving from background advisor to a core Control Room System.

The remaining modules were delivered for operational testing in November 1992. An extended handover is now underway, including formal training and calibration.

The London trunk distribution network is a complex facility, undergoing rapid change. The history of the project has been one of intellectual and organisational challenge. Operating the Trunk Scheduling System is likely to throw up many new questions for investigation. However, the solid basis for the System described in this paper is the best guarantee available that the future challenges will also prove surmountable.

ACKNOWLEDGEMENTS

The Authors would like to thank Mike Ribbins (Management Systems Director) and Tony Edwards (Water Supply Manager London) for their support on this project. The views expressed are those of the authors and not necessarily those of Thames Water plc.

Pergamon

Wat. Sci. Tech. Vol. 28, No. 11-12, pp. 111–121, 1993.

0273–1223/93 $7·00 + 0·00

AN OPERATIONAL INFORMATION DATABASE BASED ON THE AUTOMATED PROCESSING OF TIME SERIES DATA AVAILABLE FROM TELEMETRY CONTROL SYSTEMS

R. J. Casey and S. A. Grinney

Thames Water Utilities Limited, Operational Control Systems Department, London Water Control Centre, Lower Sunbury Road, Hampton, Middlesex, UK

ABSTRACT

The paper describes the development and subsequent management of a central database of operational information based largely on the automated processing of time series data retrieved from the Company's telemetry control systems.

Following the implementation of the system, both the quality and quantity of processed statistics available to operational managers has been greatly enhanced above the level of information originally recorded by manual methods. Also significant process modelling systems have been introduced utilising this information to improve operational efficiency.

Key to the success of the system has been the comprehensive provision of facilities to quickly and effectively validate the data, together with the facilities provided to ensure that the configuration of the database accurately matches and tracks configuration changes to the field network. The paper examines these functions in some detail.

KEYWORDS

Water; operational information; automated data processing; process statistics; telemetry control systems.

INTRODUCTION

Background

Thames Water Utilities Limited (TWUL) supplies a population of over 7 million people with a total of 2738 Ml/d of potable water via a supply network consisting of over 30 000 km of mains and over 500 sites. A part of the Company's massive capital programme of more than £2 billion over the next 5 years includes the planned de-manning and subsequent remote control or automation of processes at these sites to cut operating costs and improve levels of service.

A consequence of the development of these control systems will be the availability of vast quantities of process measurement data which, with suitable collation and processing, can be used to drive information and modelling computer systems that further improve operational efficiency, management reporting and future capital planning.

This paper will describe the development and subsequent management of a central database of operational information based largely on the automated processing of time series data retrieved from the Company's telemetry control systems. The database and the applications that run from it, together form the basis of the Company's Operational Management System (OMS), an important part of the current Information Systems Strategy for developing information technology systems within Thames.

Development of Operational Management Systems

The automated transfer of time series data from a Supervisory Control and Data Acquisition (SCADA) system to a separate operational database driving information and modelling systems was first successfully piloted within Thames at the Bourne End Control Centre in 1987 (Parker and Casey, 1989). Its primary function was to drive a pump optimisation modelling package.

The results of this work were then incorporated into the Operations Systems Strategy for the Company in 1988, and application plans for OMS were developed in 1989. The first implementation phase of OMS has been targeted at the London supply network to drive information and decision support/modelling packages such as demand prediction, pump
scheduling and area network analysis systems.

SYSTEM DEFINITION

Operational Requirements

The agreed objectives for the Phase 1 implementation of OMS in London were as follows:-

(A) To retrieve and integrate data from all levels of telemetry, i.e. from continuously polled SCADA systems to data logger systems (where the information is retrieved twice daily via dial-up phone lines).

(B) To validate data retrieved from these systems.

(C) To provide water demand, supply and process statistics.

(D) To establish an historic database of these statistics.

(E) To provide essential data for developing decision support/modelling software to assist operational decisions.

OMS Architecture

The hardware architecture for the OMS is illustrated in Figure 1. Information from field instrumentation is retrieved via telemetry systems based at Area Control Centres and then passed to the operational database (OMS) at the Central Control and Co-ordination Centre for London.

The database and associated data processing software reside on a pair of Digital Equipment Corporation (DEC) VAX 6000-410 computers, each with eight RA70 hard disk drives and one TA90 cartridge tape system for backups and archiving. The system has the capacity to process up to 4800 analogue and 8400 digital signals. The processors are designed to work in tandem rather than having one working and one stand-by machine. In normal use one computer is used to input data into the database and the other to retrieve data for remote users. If one fails then both services are carried out by the single remaining machine.

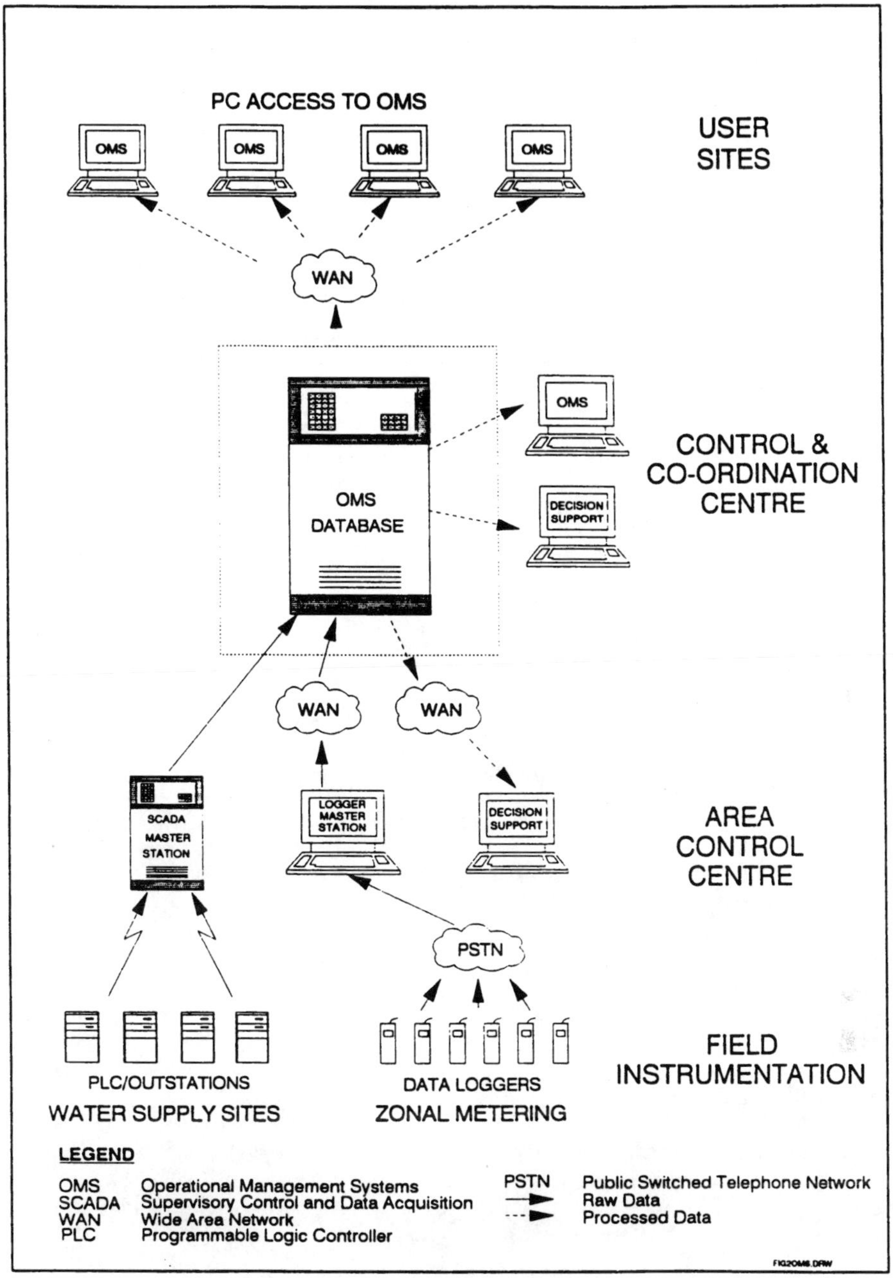

Fig. 1. OMS architecture.

Users, most of whom are situated at sites remote from the database, may access information on the database using PC workstations attached via the Company's Wide Area Network (WAN). Data is requested and then transferred to the PC where local PC software is used to display the data either graphically or as a table on the screen.

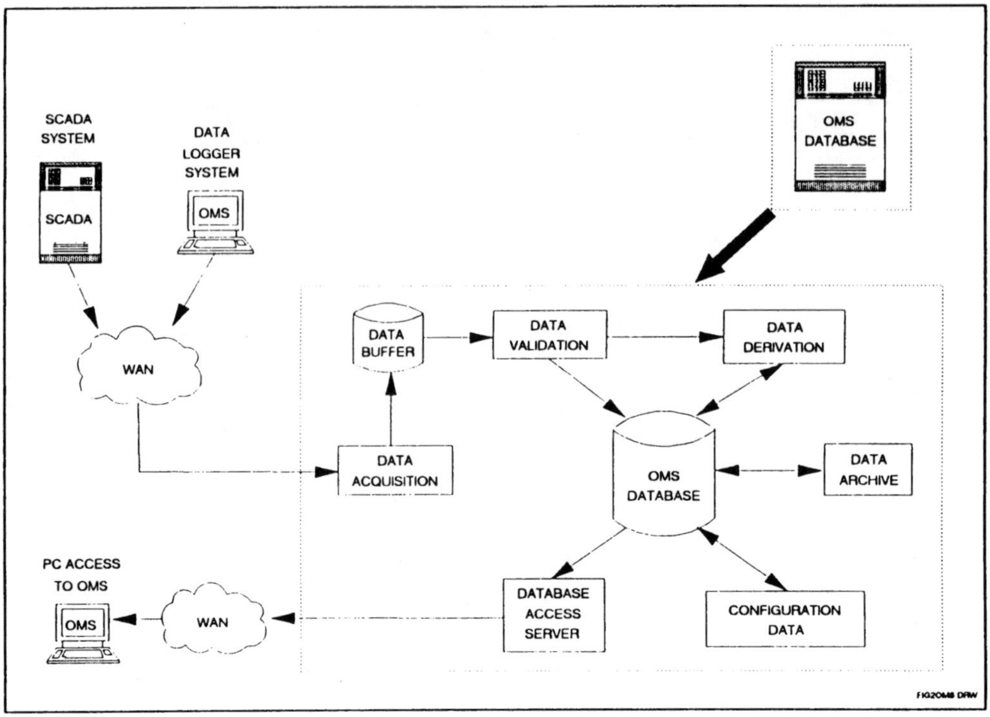

Fig. 2. OMS process overview.

Process Overview

The main processes of the OMS system to meet the operational requirements detailed above are set out in a logical sequence of occurrence as shown on Figure 2. The main functions/facilities are therefore:-

Data acquisition. The OMS system can acquire data from three external sources.

(A) From SCADA via a common data transfer software package used across TWUL, called the General Purpose Data Interface (GPDI). Although the data on a continuously polling SCADA system may be updated at the master station every few seconds, the minimum update frequency for OMS has been set at 15 min to meet the processed data requirements of all applications on OMS.

(B) From data logger systems where the information is recorded every 15 min in the logger but only retrieved by the central master station once or twice daily using standard voice telephone equipment and auto-dial modems. As soon as the data is retrieved by the master station it is transferred to OMS for processing.

(C) Data entered manually via an OMS PC workstation. This function is covered by the Manual Validation process (see Manual Validation below).

Data validation. Once data has been retrieved by OMS it can then be validated by the following processes.

(A) **Individual Validation** is performed by OMS software as the data is retrieved. It involves checking time series individually to highlight any unexpected data values which are marked as suspect.

(B) **Manual Validation** involves the checking and editing of the time series data and is performed manually by the Control Centre operators. This function is covered by the Manual Validation process (see Manual Validation below).

Derivation. This process enables OMS to calculate derived time series from the measured time series retrieved from the various types of telemetry systems or manually entered values. For example, zonal demands can be derived from inter-zonal flows. The derived time series may be computed at the same interval as the measured time series, or less frequently, to produce summary statistics.

Data storage. Both the acquired (raw) data values and validated (best) data values for the measured time series are stored in separate tables on the OMS database. Derived time series are stored solely as best values. Configuration data is also required by OMS to produce process statistics. This is also stored in tables on the database.

Manual validation. This process will be performed by the control centre shift operators after the automatic data acquisition, data validation, data derivation and data storage processes of OMS have run. This manual entry of data may cause the derivation process to be re-run.

Reporting. Facilities exist for the OMS software to automatically generate routine reports for users to access. Configuration data (see Configuration below) is also required to run the reports.

User access. This process covers the access by users from an OMS workstation to OMS data using the data server mechanism. The process is initiated by users to read data in the database to generate reports, graphical displays of data or the copying of data values to the OMS workstation PC to generate *ad-hoc* reports, etc.

Data archiving. Facilities have been provided to automatically archive data from the OMS database to optical disc/magnetic tape. Selected data can then be retrieved as ASCII files when required by users.

Decision support interface. Phase I of the OMS development provides for a data interface to Decision Support Systems. Data will be routinely exported to and imported from packages such as demand prediction, pump scheduling and area network analysis modelling.

Configuration. Those processes that run directly under the control of OMS software such as data acquisition, validation, derivation, data storage, reporting, archiving and the decision support interface will require the initial manual entry of configuration data to ensure that these processes can then function routinely and correctly. Configuration data holds details of measurement series, items and areas and their relationship to one another.

System management. This describes a process by which the overall performance of the OMS software and hardware can be monitored and unexpected occurrences and errors logged. It also provides the mechanism for starting and stopping, or restricting various OMS processes.

IMPLEMENTATION

Timescales

The implementation of the first phase of OMS for London Water commenced in August 1992 with the installation and commissioning of the OMS database on a 6000 series VAX computer at the London Water Control Centre, Hampton. The system will be commissioned across the entire London Water supply network in four stages, each stage covering a geographic area in London. The first stage covering the S.W. area of London became operational in November 1992 with the last stage, S.E. London, due to be completed by June 1993.

From the operational viewpoint, the implementation phase of the project has involved the collection and input of a vast amount of configuration data (see the section on configuration below), and the comprehensive checking of all input time series data, both from SCADA and data loggers, against values displayed on OMS. The derivation of summary statistics and their inclusion on OMS reports has also been checked.

OPERATING THE SYSTEM

Maintaining the OMS Database

For the system to remain operational there are various tasks required to maintain the data on the system. These tasks can be split into two broad categories.

(A) Routine tasks that need to be carried out every day such as checking system error logs, manual validation and the routine entry of non-telemetered time series data.

(B) Non-routine tasks which are event driven processes and generally fall into one of the following areas:-

- (i) Those driven by events in the field, i.e. physical changes outside the computer system but which the software must model internally.
- (ii) Those driven by events within the system, i.e. software errors, hardware or communications failures.
- (iii) Those driven by user request, i.e. system change requests, and software enhancements.

Routine Tasks – Typical Daily Sequence

Generally, all the data values for the previous day should have been retrieved by OMS from the SCADA and Data Logging systems by 3.00 a.m. to 4.00 a.m. The automatic validation and derivation processes on OMS then run and completion is flagged to a summary statistics file which can be inspected by Area Control Centre staff.

The Control Centre staff can then proceed to perform the first phase manual data validation/data entry tasks (see below), reporting to the central Co-ordination and Control Centre when these tasks are complete so that Decision Support applications such as Demand Prediction can be run.

As soon as the peak morning (7.00 a.m.) reservoir level data has been retrieved from SCADA, preliminary standard reports summarising yesterday's water supply statistics are available for operational managers to inspect.

By early afternoon the OMS Database Administrator usually completes any final editing of the data using manual validation so that the final standard reports are available to users.

Data Validation (see Figure 3)

Where configured for a raw data time series, individual valuation is performed automatically by the Data Acquisition Sub-system (DAS) as the raw data is acquired by OMS. DAS automatically generates replacement values for suspect or missing data which is placed along with satisfactory raw data into the best data tables on the database. Both the acquired (raw) data values and the validated (best) data values for each measured time series are stored in separate tables within the OMS database.

Derived time series (including summary statistics) are calculated automatically by the derivation process once all the data to perform the defined calculations has been acquired by OMS and any specified calculation time lag has elapsed. Where configured, further individual validation checks may then be performed on the derived data. The derived time series is then stored in the best data tables within the OMS database.

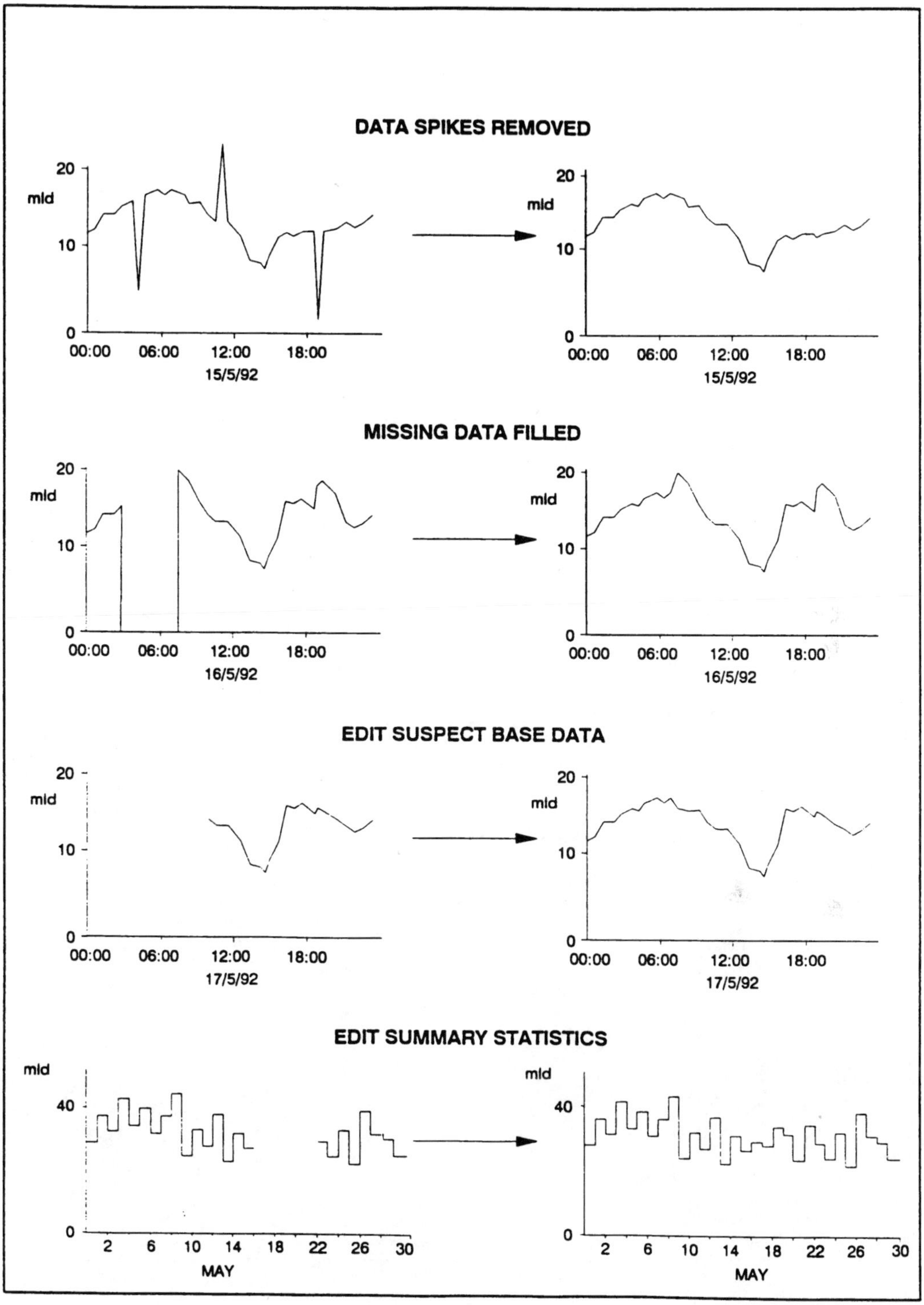

Fig. 3. OMS data validation.

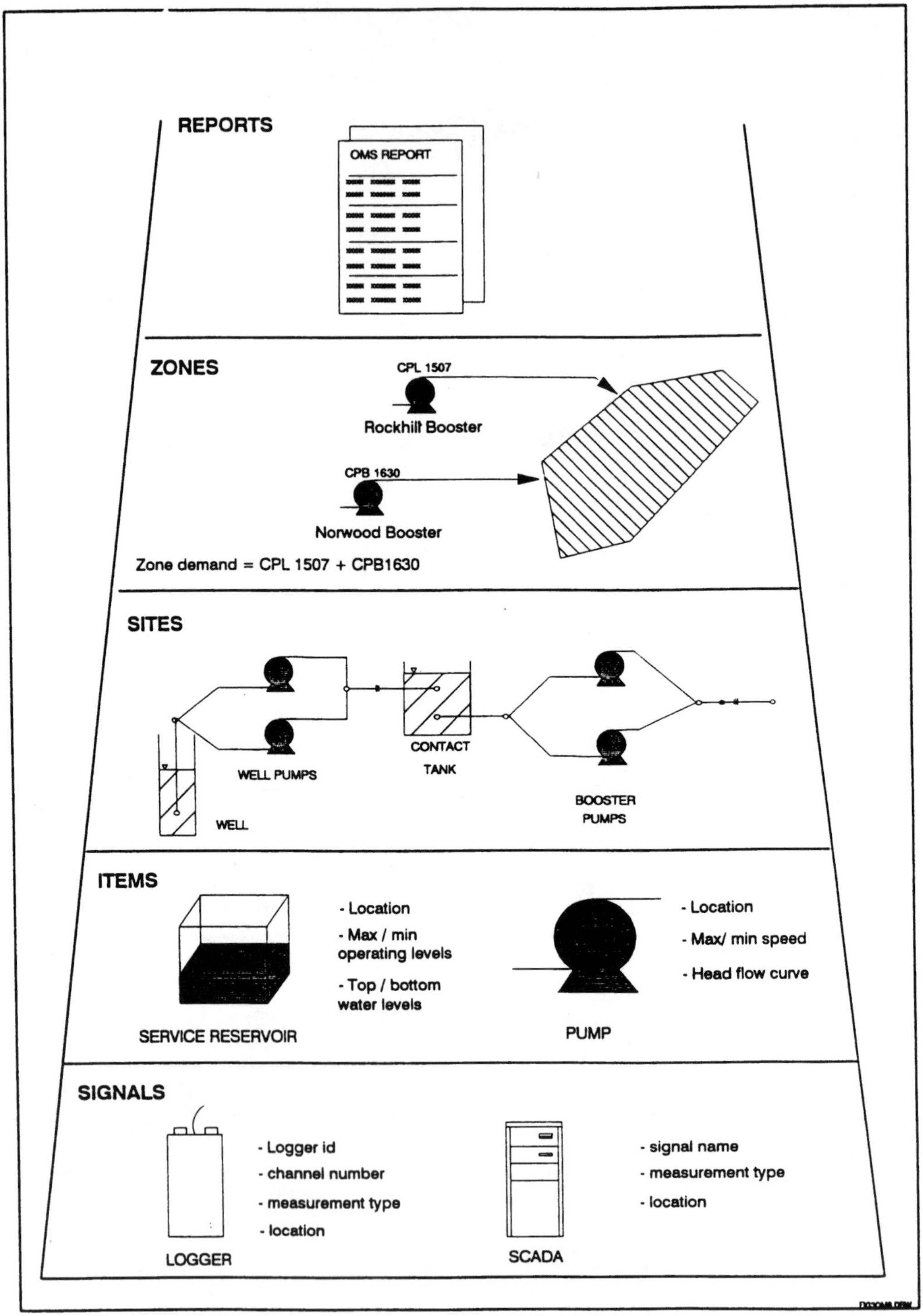

Fig. 4. OMS configuration.

All replacement values generated by the validation processes are marked as suspect. Details of suspect and missing data are reported by OMS as Action Lists within the Manual Validation software. The Primary Action Lists contain details of those time series with the worst quality data, where the number of suspect/missing values are above configured absolute or consecutive thresholds. The Secondary Action List contains details of other time series with missing/suspect values. The lists help direct the operators to suspect/missing data quickly to maximise the efficiency of the whole manual validation process.

Various facilities are available for manually editing the data, including replacement, percent/absolute change, interpolation or curve fitting. The software is designed to ensure consistency of data on the database at the daily summary level.

All manually entered time series data are recorded onto the database using the manual validation application. Once the validation of the data at the daily summary level is complete then the summary reports may be generated and data passed to decision support applications such as demand prediction.

Non-Routine Tasks – Configuration Data Entry

This process, also known as database maintenance, involves either the entry of new configuration data or the amendment of existing configuration data. These configuration tasks are broadly grouped into the following categories (see Figure 4)

(A) Signal Configuration – Maintaining lists of signals to be retrieved from the SCADA and logger based data acquisition systems.

(B) Device Configuration – The input of characteristics about items on the database, e.g. pump head versus flow characteristics, etc.

(C) Site Configuration – The maintenance of site configuration diagrams, and item relationships at a site level.

(D) Derivation Configuration – The maintenance of derivation formulae to calculate zonal statistics, area totals, daily summaries, etc. plus the associated derivation process control data (ie. suspect and missing threshold statistics).

(E) Report Configuration – The maintenance of database tables to ensure links between plant items, sites and organisational areas are correctly maintained. These tables control the composition of standard tabular reports, and assist users to group data for graphical output.

The Configuration Data Entry and Control Centre Application packages have been developed as a part of OMS to carry out the above tasks. A database administrator has been appointed for OMS to ensure that these tasks are performed.

Signal Configuration

Signals are basically terms used to describe measurements stored in the OMS database that vary with time, and details about these time series are held in tables called measurement series. There are four types of measurements held in the database:-

(A) Time series acquired from SCADA.

(B) Time series acquired from logger based acquisition systems.

(C) Manually entered time series.

(D) Time series derived by OMS from other time series.

For signal types (A) and (B) a close liaison has been established with those responsible for changes to signals on the relevant control systems to ensure that OMS reflects these changes.

Device/Site/Derivation Configuration

When new items are entered on the database or existing items amended, physical details about the items (called device configuration details), must also be entered on the database. These details are usually required to perform calculations on measurement series related to these items, or are required by external Decision Support programmes for modelling purposes. For example, a pump may require the input of parameters that define its head versus flow characteristics.

Details about measurements and their relationships with items at the site level must also be entered (site configuration), and these are logged both on the database tables and in the form of site diagrams to aid data interpretation, along with relationships between measurements to derive summary statistics (derivation configuration).

To facilitate the logging of these changes on OMS to exactly match changes occurring in the field, a Control Centre Data Entry application has been developed which includes the following major functions:-

Item availability/restrictions. This facility allows details of the availability or restricted availability of items stored within OMS, such as pumps, reservoir cells, etc. to be recorded.

Unmetered infusions. This facility allows an estimate of any unmeasured flow between zones to be recorded and is used by OMS in the calculation of zonal demands. Both historic and future estimated stop/start times can be entered.

Noticeboard. This facility allows any changes in the field to items, either configured or not configured in OMS, to be recorded descriptively and stored on the database.

Report Configuration

The maintenance of database tables to ensure that links between plant items, sites and organisational areas are correctly established is the responsibility of the Database Administrator. These links are input using the Configuration Data Entry application and control the composition of standard reports, as well as assisting users to group data for graphical output.

SUMMARY

Conclusions

The operation of the system to date has proved the viability of maintaining an operational database of key statistics based on the automated processing of time series data retrieved from telemetry control systems.

Both the quality and quantity of processed statistics available to operational managers has been greatly enhanced above the level of information originally recorded by manual methods. Due to the largely automated processing of data, this has been achieved despite the significant demanning programme associated with the installation of the telemetry control systems.

The enhanced level of information has also produced greater focus on the maintenance and accuracy of the basic instrumentation within the company. The reliability and accuracy of the processed statistics are improved as meter faults are now quickly identified, while inadequate metering is identified and replaced.

Key to the success of the system has been the comprehensive provision of facilities to quickly and effectively validate the data, together with the facilities provided to ensure that the configuration of the database accurately matches and tracks configuration changes to the field network.

ACKNOWLEDGEMENTS

The authors wish to express their thanks to the Management Systems Department, TWUL, for the development of the Operational Management Systems (OMS) software plus Bristol Babcock Ltd and Radcom Technologies Ltd. for the development of the telemetry interfaces.

REFERENCES

Parker, M. J. and Casey R. J. (1989). The effective utilization of automatic data processing software with an advanced multifunctional SCADA system. In: *World Water '89*. Thomas Telford Limited, London.

Pergamon

Wat. Sci. Tech. Vol. 28, No. 11-12, pp. 123–131, 1993.

0273–1223/93 $7·00 + 0·00

UPRATING AND DEVELOPMENT OF A REGIONAL TELEMETRY SCHEME

R. Huntington

Wessex Water Plc, Bristol, BS2 0JQ, UK

ABSTRACT

In 1988 Wessex Water decided that its 10 year old regional telemetry system was in need of major uprating and improvement to meet their changing requirements. This paper explains the philosophy adopted in making the changes and describes the actual improvements carried out. While still based on a centralised system, two important changes included the use of more sophisticated outstations and the provision of a networking facility to VDU's for use by geographically spaced operational staff.

A number of inter-linked software packages developed for the analysis, optimisation and modelling of data are described. Finally conclusions are drawn.

KEYWORDS

Telemetry; automatic control; automatic monitoring; VDU networking; plant control; data base processing; Wesnet; network modelling.

TELEMETRY SYSTEM

Introduction

In the mid-1970s, Wessex Water decided that the operational management of it's business should be centred round a comprehensive telemetry scheme.

In 1978 the first phase was brought on-line and since then it has been expanded and developed to cover all operational aspects of its water supply, sewerage and sewage treatment business. By 1988 the number of sites monitored and controlled had grown to 850 which was 350 more than the originally envisaged maximum; site monitoring and control were also more complex. While reliability of the system as a whole was still good, "back-up" support and product availability was becoming increasingly difficult. The need to present more information on the telemetry displays was leading to loss of clarity on many mimics and causing an element of confusion to control operators. Redrawing of some mimics was adopted as a short term measure.

Data transfer rates and subsequent picture build time over the synchronous data links to regional control were being adversely affected by the high loading placed on each of the computer systems, and speeds were becoming unacceptably slow when computers were required to store historic data. Lack of spares availability was resulting in non-compatible replacements having to be used, appropriate disc systems and core memory were no-longer manufactured. Vulnerability to major failure was becoming unacceptable and therefore a decision was taken to upgrade the master station computer systems.

System Requirements Adopted

The aim was to take advantage of the improvements in computer technology, outstation developments and inter-site data communications, while still maintaining the existing structure of sub-masters, outstations controllers and sensors. Expansion of the multiple VDU-based workstations was required to give access to the system data by operational staff situated at up to 30 locations.

Key criteria for the replacement system were:

- easy to operate;
- high degree of flexibility;
- all existing outstations and sub-systems to be transferred;
- more advanced outstations to be used where appropriate with diagnostic test facilities;
- multiple access availability from up to 30 on-line work stations;
- security to operational facilities by use of password identification;
- integration of existing control facilities to have increased functionability and user features.

During development, importance was placed on increasing the sophistication of the more important outstations to enable them to operate as a network or stand-alone unit in a flexible manner. This was aimed at giving cost effective-control, monitoring, data acquisition and telemetry. The adoption of programmable logic controller outstations at locations such as treatments plants, is a vital step towards moving towards adopting BS 5750 Quality Controls for the telemetry system with treatment plant controls covered by separate Quality Control arrangements. Profiles monitored by telemetry have been developed to provide time variable plant alarms and control.

Dual machine micro VAX 3900s used on a duty-duty basis, were chosen as the heart of the system. This configuration provides a system standby machine but instead of one standing idle, both machines share the load with each capable of running the system at an acceptable response. Each machine keeps the other updated with the work it is doing. These machines will shortly be replaced by the more powerful Micro VAX 3100 model 90s.

Translation of the original system database to the new was not without problems. New standards of presentation of information and increased system flexibility made control and maintenance of a common system approach difficult; a strict disciplined approach was essential. The translated system database required detailed end to end testing using simulated inputs from outstations to final display.

Modification to presentation layouts included placing alarms into prioritized areas of importance according to time of day and associated conditions.

Salient features of the new telemetry system are:

- fully integrated region-wide telemetry coverage;
- system capability – 20,000 analogue values, 80,000 digital states, up to 2000 mimics at any PC workstation;
- main communication links duplicated using resilient digital services with fall back to public switched telephone network;
- integration links with other computer systems for analysis, optimisation and modelling e.g. Wesnet;
- region-wide X25 Packet Switching. Network to provide flexible and secure communications paths for outstations and remote workstations;
- number of sites at present monitored – water supply 599, sewerage 542, sewerage treatment 593.

Workstations: Man-Machine Interface

Operator access to the system is through colour graphic visual display unit (VDUs) workstations. An associated keyboard can be used for system operation but for all normal operational tasks it is not required once an operator has entered his or her password.

Comprehensive indexes are provided listing all sites in tabular and map form. Each site is displayed on one or more mimics and has a status page where all information is listed in tabular form. Each mimic and status page has an associated "help" page on which can be entered free-format text to assist or inform operators. All selection and system operations are performed using pointer boxes on the screen controlled by a "mouse" or "trackerball" depending on user's preference. The pointer or mouse boxes are used to select mimics for display from an index map or list, and to move from one mimic display to another. The use of many linking mouse boxes on a diagram gives flexibility to display the various parts of systems sequentially. New diagrams can be added simply by adding a new mouse box. Clutter is avoided by colour coding names that have interlinking mouse boxes. Historic information over 24 h on a particular variable, can be displayed graphically by operation of the mouse.

All plant operates in automatic control but items of plant can be selected for manual remote control from the associated mimic. Mouse selection presents a statement of options available and current state of the plant. Operation is only permitted by password authorised users.

Alarm Handling

The Wessex region can generate a very large number of alarms, therefore a basic requirement of the system is to separate them into areas and functions so that individual supervisors can receive alarms associated with their own responsibilities without the distraction of inappropriate information. There are 99 areas of availability divided into functional (i.e. water supply, sewerage, etc.) then geographical area of responsibility. Each alarm or, if required, plant state change, can initiate a hard copy report and automatically route to the appropriate depot or other location for action.

Alarms are assigned a priority from "immediate" to "action next day". While any alarm can be displayed at any workstation, they can only be accepted by a specific password controlled operator. An operator can "mouse" onto an alarm message and be routed automatically to the mimic or status page where that alarm and its associated plant is displayed.

Graphical Display

Archive information can be presented in tabular or graphical form. Graphs of a series of parameters can be recalled from a 10 day archive. Longer term records are held in linked systems. Up to eight traces can be displayed simultaneously. "Zoom" and "pan" facilities are provided together with a facility to give measured value and time/date of occurrence of any point on a graph.

Automatic Operation of Plant

Extensive facilities are provided for implementing plant control strategies. Essentially a process variable, such as pressure or level, is compared with a defined operating profile, when outside acceptable variance, adjustments are then automatically made to the controlled item of plant or an alarm is raised, as appropriate.

The telemetry computer can carry out the individual control of plant by comparing the measured parameter with a profile entered by the operator. This can be different for different days of the week etc. There are convenient ways of entering, copying and modifying these profiles in the computer system, which have been available since 1978. Advances in outstations technology have for some time permitted critical profiles to be held in the outstation, and any required change to the plant automatically carried out. Further advances in outstation to control centre computer communications has permitted profile changes to be entered in via a control room workstation and "down-line" loaded to the outstation. This is not conveniently available for display at the outstation and therefore, over the past three years, Wessex has been installing "profilers" on sites to give a graphical display of desired profile and actual measured value with control point and alarm thresholds clearly displayed. The telemetry outstation carries out all the functions of a programmable logic controller. While operational experience has shown reliability to be more than acceptable, there is some concern about the wisdom of placing monitoring, plant control and telemetry all in one unit. Accordingly, policy now provides for the local plant controllers and telemetry outstations to be separate units.

Key areas in the water supply function where automatic profile control is used are:

pump control and associated service reservoir storage
distribution pressure control
distribution zone consumption monitoring

Links with Other Systems

The telemetry system integrates with other computer based systems for analysis of telemetry data with other associated data for optimisation and modelling.

Two important systems are TAP (Telemetry Analysis Package) and Wesnet(Wessex network modelling).

TAP

Tap is a database processing and retrieval program for water supply data. It will accept and process 15 min interval telemetry data to daily, weekly or periodic manual entry, for subsequent archival. Groupings range from single source to zones, areas and region. A full suite of schematics is available to guide the user as and when required. The core database program can be accessed by a demand forecasting module which provides facilities for component analysis and trend projection of historic data at all levels. This is supported by a full graphics capability. The package runs on a PC and can be networked if required.

WESNET

Development

Wesnet has been developed as a PC-based software package for the planning, design and operational management of water supply from source to customers premise's.

Early work commenced in 1988 and since then it has been developed and improved to cover the broad needs of "hands-on" water utilities staff who are charged with the duty of meeting the needs of their customers cost effectively. From the outset Wesnet has been developed by experienced practising engineers for use by practical people who are not well versed in computer programming. Flexibility in its applicability and use have been key elements in its development and in consequence the package has now been sold to 20 water undertakings in 6 countries.

Standard interfaces are available for linkage to telemetry and loggers.

Capabilities

Wesnet has been developed to provide a wide range of capabilities to meet the needs of water supply operational staff. It can:

provide traditional network analysis of water supply distribution systems;
simulate normal and emergency situations for planning and verification of control and emergency procedures;

accept live data from a telemetry system and provide on-line analysis;
provide interactive calibration of network models by comparing simulated results to live telemetry data;
help design new control systems;
produce levels of service reports, e.g. pressures;
calculate costs of pump scheduling and tariff choice options and calculate energy costs using built-in tariffs;
simulate some water quality changes in networks;
enable users to select location of chlorine booster points to maintain chlorine residual at set value.

Features

Important features built into Wesnet packages include:

interactive menu-driven user interface making extensive use of pop-up menus with comprehensive error trapping;
interfaces with mapping systems to use existing files and provide results and live data onto existing maps;
interfaces to real time telemetry and data loggers to provide basic data for analysis or comparison with simulation results;
automatic production of levels of service reports to meet national Regulators' requirements;
ability to accommodate complex pumping arrangements and various control valves in the modelling;
provision of a regional database covering operational plant such as pumps, pipes, valves and reservoirs;
full range of electricity tariffs incorporated for energy cost modelling;
on-line simulation display;
extensive graphical analysis package which allows results to be outputted to an existing standard;
ability to construct a hierarchy of maps and schematics;
ability to model chlorine residuals throughout distribution pipework;
able to model networks on a PC with up to 1000 pipes and 800 nodes;
English, French and German language versions with an open interface for other languages.

Interface with Live Data

While Wesnet can be used for network analysis in the traditional way as a "stand-alone" package, it has been specifically developed for direct interface with telemetry systems and/or data loggers. This arrangement provides two important benefits:

simplifies calibration;
allows on-line analysis of telemetry and logger data.

Wesnet will interface with any telemetry system provided files are set up in the correct format. Its in-built archive and graphics package then provides the flexibility needed to meet the day to day needs of water supply operators.

In addition to having the ability to compare live and simulated data and superimpose the two on a simple display, anomalies can be highlighted. Anomalies can be due to a variety of reasons including:

defective telemetry sensors;
major leaks;
valve openings not in accordance with operating protocol;
pump inefficiencies;
malfunctioning of local loop control.

Good calibration of network analysis models is crucial to give operators confidence in the results they obtain from using the packages. Linkage to telemetry data gives the ability to check accuracy and reliability of models and tune them to the required accuracy. This then provides models which duplicate the behaviour of the real water supply system. Operators, when faced with a genuine emergency, then have the confidence to transfer to a simulation model the real data relating to when the emergency occurred and for the period leading up to it. Alternative options for resolving the emergency can be simulated and a favoured option chosen. After initiating the required action, live telemetry data can be superimposed on the simulation at suitable intervals of time to confirm (or otherwise) that the action taken is meeting requirements. If shown necessary, the effects of the operational action taken can be further simulated and any necessary fine-tuning initiated.

Simulation

Wesnet has been developed to cope with the very wide range of control devices incorporated in modern water supply systems, including the typical range of combinations available through the use of programmable logic controllers (PLCs). These are often brought into play when operators respond to emergencies and have to modify operational protocols such as re-zoning.

An almost infinite number of "what-if" permutations can be considered during a simulation by drawing on separate and linked control and operations files.

Communication with Wesnet is via colour coded menus with options and data chosen by mouse or by key-board entry. An error checking system warns of errors and advises how corrections can be made. During modelling, associated data may be displayed on request without interrupting ongoing procedure. During the progress of a modelling run a wide range of event messages are provided e.g. reservoir only x'% full, locations and times where pressures are below set standard, etc. Operators can interrupt a simulation at any time to examine in detail the results up to that point.

Results can be made available either in tabular form or interactive graphics. Hard copy facilities are available . The graphs used show live and simulated changes over time for one or up to six variables. Variables such as reservoir level, flow, water usage, pressure and energy are typical. The graphs can be modified to meet a user's specific requirements by optimising the graphical capability of a PC. Schematics are used to show the state of a system at a given point in time. The schematics used can range from a comprehensive digital mapping system to ones developed by a user using a standard graphics package. An open package is provided thereby enabling import of a range of computer files and images and allowing presentation of live and simulated data as an overlay onto imported files.

Service Compliance

Management of water utilities is aimed at providing customers with water within set standards at least cost. The need to demonstrate actual achievement against the standards is becoming increasingly important. Water supply systems are a complex mixture of treatment- plants, pipelines, pumps, reservoirs and various associated fitments, and quality of service is a reflection on how they are managed.

Within Wesnet there is a package which allows models to be run to demonstrate when identified parameters such as pressures fall outside set targets. Results may be provided in tabular form or graphical – such as colour contouring of minimum pressures. Service reports produced in this way provide an excellent starting base for the analysis and planning of improvements. The use of Wesnet as a predictive tool enables improvement works to be timed to achieve a realistic balance between vulnerability to failure and premature or late construction.

Management of Energy

Energy is a major cost centre for most water supply undertakers with pumping playing an important role in transferring supplies to customers.

A multiplicity of pumps with various characteristics and the availability of a range of electricity tariffs, provide the opportunity for substantial cost savings provided the inter-relationships involved can be analysed in a manner which is meaningful to the water supply operator. The analysis usually involves a risk analysis between pump capacity availability and assessed demand, or energy savings against proportions of service reservoir storages used to balance flows (remainder retained to meet emergency situations such as bursts).

A Wesnet package enables a supply manager to quantify the savings and risks in a way which identifies variations in movement of reservoirs storage and pressures against a series of pump control arrangements, together with the energy costs arising for a range of tariffs. A module containing all available UK electricity tariffs and other tariffs is easily included. Pump characteristics of all pumps included in networks are readily added to the basic pipeline and reservoir characteristics e.g. volume, depth, top water level.

The outcome is a powerful tool which can not only solve pump optimisation problems but can advise when pumps are operating inefficiently (e.g. throttled) and produce an estimate of financial losses arising therefrom.

Completion of one energy audit on a none Wesnet "best judgements" mode of operation, identified £100,000 per year savings available by switching to a Wesnet based operational regime.

Leakage Investigation and General Management

Linking models to telemetry has enabled network modules to be finely balanced and, after being maintained for a period of time, they are being used to identify major leaks. Numerous leaks have been found, the largest amounting to about 1.5 Ml/d. During the past few years a major drought has resulted in frequent re-zoning of parts of the distribution system and without the use of the networks models it would have been likely major leaks would have been assumed to be simply demand changes.

Modelling of high pressure trunk mains is proving profitable as relatively small leaks under high pressure produce quite large losses.

Models can be used to carry out a check calculation on large telemetry meters by balancing meters in a group against each other. This is an important pre-requisite before carrying out leakage predictive work. A number of important metering errors have been found. The error for one source meter measuring flows up to 20 Ml/d was found to be 20%! Errors in metering result in management not being aware of major leaks or resources being wasted looking for fictitious leaks or incorrect assessments of abstraction licence compliance.

An interesting outcome of developing the use of Wesnet throughout the company has been the ability to confirm (or otherwise) local operators assumptions on how systems actually operate. Unfortunately too often for comfort, the "local knowledge" has been shown to be incorrect. Correcting these has led to between 5 and 10% improvements in efficiency.

Some examples of poor management include:

pumps operating against each other;
local control loops producing imbalance in source usage;
incorrect opening of valves.

Strategic operational modules linked to telemetry are proving invaluable planning aids with substantial savings being obtained from reducing leakage, improved energy efficiency and better resources management.

Water Quality

Water supply operators in the UK and many other countries operate with the objective of maintaining a small chlorine residual throughout their distribution system as a protection against contamination before delivery to customers.

Wesnet provides the facility to predict chlorine residuals at node points throughout a network provided the model has been properly calibrated and degradation coefficients assessed. The model can then be used to predict problem areas such as low chlorine residuals and pin-point suitable locations for chlorine booster installations.

In large distribution systems it is often difficult to maintain a chlorine residual in the extremities of the pipelines without having to provide an excessive residual at sources. In these situations the Wesnet module can be used to design a chlorine booster arrangement which will maintain minimum residuals without excessive chlorine values close to sources. This results in lower `taste' complaints from customers.

CONCLUSIONS

The degree of success obtained when implementing a telemetry system or a major refurbishment of an existing system will depend on the care and expertise brought to bear during the planning stage. Important matters that should be taken into account include:

careful defining of user needs;

careful defining of system requirements;

growth in user requirements and computing power should not be under-estimated;
provision of a software system that is capable of being transported to other hardware solutions;

if equipment is meeting current needs with good reliability and low maintenance, do not assume it has to be replaced because it has been declared redundant;

introduce standards at an early stage, then trying to stick to them – changes can have long term costs in "back tracking" and rectification;

need for a good simple man-machine interface is essential for infrequent users, - with "fast track" routes added for regular users;

simple, clear displays aid correct assimilation in emergencies; sophisticated displays increase "build" times to users and as the system develops can result in unacceptable performance;

a region based system potentially has too much data for assimilation by central operators for effective response during emergencies – targeting geographical and functional areas of interest aids response to important events;

off loading some routine aspects of statistical reports and archiving to secondary systems is beneficial;

continuous monitoring associated with plant control using diurnal operating profiles can introduce significant savings by reducing extremes of operation (peaks and troughs) and highlight problems when variations occur outside controlled patterns;

failure frequency is often over-estimated and complexity of duplication can outweigh advantages – two computers on a duty/duty basis can be beneficial;

as costs of computers and their maintenance continue to fall, it may be cheaper to replace say, in 5 years time, central computing power, rather than buying larger at the beginning or trying to expand in later years on an "older technology" base;

outstation systems should be selected to give wide flexibility in choice of systems linking to them;

modelling and predictive work traditionally has been considered outside the telemetry main machine activity but with a networked solution of computing power each user is provided with requirements to meet their optimum functional needs;

the ability of Wesnet, interlinked with telemetry to provide water supply operators with a set of tools which enables water supply systems to be managed and developed cost effectively.

REFERENCE

Huntington, R. (1990). Wessex Water's Integrated Water Distribution Management System. *Instrumentation, Control and Automation of Water and Wastewater Treatment and Transport Systems. (Adv. Wat. Pollut. Control no. 10)* ed. R. Briggs, Oxford, Pergamon, pp. 9-16.

Pergamon

Wat. Sci. Tech. Vol. 28, No. 11-12, pp. 133–140, 1993.

0273–1223/93 $7·00 + 0·00

WATER DEMAND FORECASTING BY MEMORY BASED LEARNING

Takashi Tamada*, Minoru Maruyama*, Yasuaki Nakamura*, Shigeru Abe* and Kazuo Maeda**

* *Central Research Laboratory, Mitsubishi Electric Corporation, 1-1, Tsukaguchi-honmachi 8-chome, Amagasaki, Hyogo, Japan*
** *Power and Industrial System Center, Mitsubishi Electric Corporation, 1-2, Wadasaki-cho 1-chome, Kobe, Hyogo, Japan*

ABSTRACT

Recently, a "memory based" approach towards various kinds of problems has been proposed. The underlying principles of the memory based approach are : (1) storing past examples in a memory. (2) searching "near" examples to a given input in a memory. In this paper, we apply the memory based approach to water demand forecasting, and present a hybrid method which consists of MBL (Memory Based Learning) and the conventional multiregression. In the memory based method, the distance metric is crucial. In our method, the local distance metric is defined from examples when an input data is given ,then the neighborhood of the input is determined based on the distance metric. If there exist examples within the "neighborhood" of the input data, then the forecast is given by MBL. Otherwise, local multiregression is used. We applied this method to the daily water demand forecasting. The forecasting results by our method are approximately 5% better than those by the multiregression method. Especially, when there exist past examples in the neighborhood, namely in the case where MBL is applicable, the results are approximately 10-30% better than those by the conventional multiregression.

KEYWORDS

Daily water forecasting, Memory-Based Learning, Multiregression, Learning technique, Artificial neural network

INTRODUCTION

For the purpose of stable water supply and efficient management of water-treatment plant and water-distribution systems, highly accurate water demand forecasting is needed [1]. Water demand forecasting can be classified into two categories : long-term forecasting and short-term forecasting. The demand forecasting period may be month or year for the long-term forecasting, and day or hour for the short-term forecasting. The long-term forecasting is used to make an extension plan of facilities according to the change of relationship between the water supply and the water demand. The short-term forecasting is mainly used for the planning of optimal water control. In this paper, we describe the daily water demand forecasting.

The conventional approach to water demand forecasting is to build a statistical model such as a multiregression model [2]. But, in many cases, it is difficult to build an appropriate model from the obtained data [3]. Recently, learning techniques which build a model from training data are often applied to forecasting problems, such as electric load forecasting. Especially, much research has been done on artificial neural networks (ANN), since ANN is capable of non-linear modeling [4]. On the other hand, we have attempted to apply memory-based learning (MBL) to daily water demand forecasting. MBL is a nearest-neighbor method which operates by comparing each example in the training set, retrieving the nearest examples, and using a weighted combination of them as its output [5, 6]. MBL is mainly applied to time series prediction and robotics [3, 7].

In this paper, we apply MBL to daily water demand forecasting, and propose a hybrid method which consists of MBL and multiregression. The forecasting results by our method are about 5% better than local multiregression.

DAILY WATER DEMAND FORECASTING

Fig. 1 shows the daily water demand at a water treatment plant. The figure shows that daily water demand has random variations. Consequently, to build a forecasting model, we have to examine the causes of the variations and select the proper input variables.

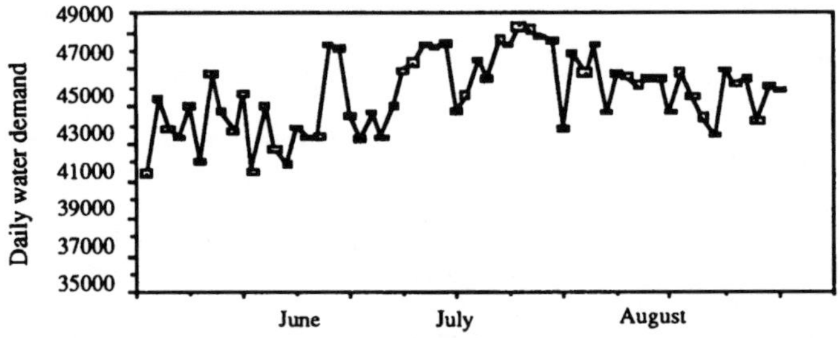

Fig. 1. Daily water demand curve over three months

Fig. 2 shows the relationship between daily water demand and candidate input variables (highest temperature, weather and humidity).

The figure implies

- the relationship between highest temperature and daily water demand is almost linear.
- daily water demand is affected by weather conditions.
- there is no correlation between daily water demand and humidity.

From these facts, we can say that highest temperature and weather conditions are proper inputs. We select three inputs, highest temperature, weather conditions and the latest water demand on the day with the same weather. They were determined based on the analysis by AIC (Akaike's information criterion) [8].

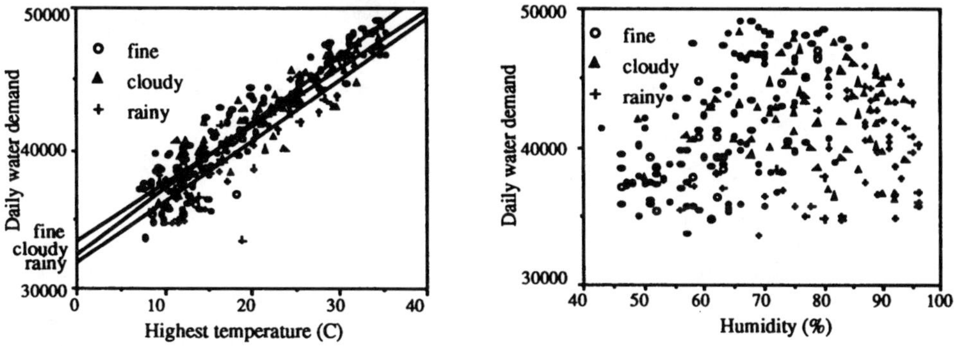

Fig. 2. Relationship between input variables and daily water demand

MEMORY BASED LEARNING FOR DAILY WATER DEMAND FORECASTING

Memory Based Learning

In principle, MBL is a nearest-neighbor method which operates by defining distance metric on the database (past examples), retrieving the nearest data, and by calculating a weighted sum of the nearest data. Consequently, in MBL, definition of the distance metric is crucial. In general, if the database is large enough, then it is likely that the output by MBL is close enough to the desired output. However, if the size of the database gets large, it is time consuming to search for the nearest data. Since the range search on the database is frequently required, we manage the database by tree structure such as an MD tree [9] for efficient management and rapid retrieving of multidimensional data.

In MBL, adding new examples and managing the database structurally is equivalent to learning. Fig. 3 shows the example of MBL in the case that the input variable is two dimensional. The output y for the input O is calculated by using X_1 and X_2 within an oblique area as $y = f(X_1, X_2)$.

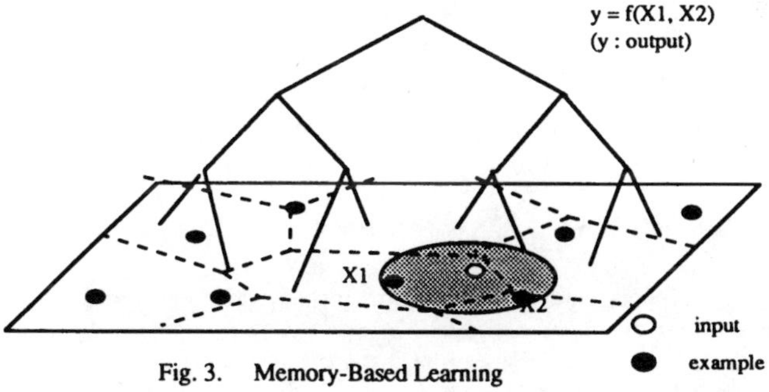

Fig. 3. Memory-Based Learning

If there are enough data which are near to the input data, we can expect that MBL gives accurate forecast. If not, it is difficult for MBL to forecast with high accuracy. We define the neighborhood for each input data so that MBL can give accurate forecast. In our method, if there are examples in the neighborhood then forecasting is performed by MBL, otherwise local regression is used. In the next section, we describe the method for determining the neighborhood for each input.

Definition of the Neighborhood

In our method, a variable distance metric concept is introduced, which determines the proximity by the relationship between the inputs and the data in the database. To determine the neighborhood for each input, we need a computational measure of similarity. In general, the sensitivity of each input variable to the output is different with each other. Consequently, we must adapt the weights of the inputs according to the sensitivity to decide the distance metric. We define the neighborhood U(p) for the input data p such that the following error measure (i) is less than a given maximum error rate E for almost all examples within U(p) considering the sensitivity of each input

$$\text{error} = \frac{|y^{\lambda} - y|}{y} \qquad \text{(i)}$$

where

y^{λ} = daily water demand of examples within U(p),
y = actual daily water demand.

Let $X_1, X_2, \cdots, X_n$ be input variables, and Y be output, $p = (x_1, x_2, \cdots, x_n)$, $x_i \in X_i$ be input data, $\overline{x^j} = (x^j_1, x^j_2, \cdots, x^j_n)$, $x^j_i \in X_i$ be the value of input variables on the day j days before, $\overline{y^j}$ be daily water demand on the day j days before and $d_j = (\overline{x^j}, \overline{y^j})$.

Decision of U(p) Let E be the error bound of forecasting model. We adapt the weights of the input variables according to the value of E to decide the neighborhood of the input data. For a given constant m, we obtain the minimum value of $\overline{\beta^j_i}$ which satisfies the following inequality

$$|j - k| < m, \frac{|y^j - y^k|}{y^j} < E \Rightarrow |x^j_i - x^k_i| < \overline{\beta^j_i} \qquad \text{(ii)}$$

for every $d_j (1 \le j \le n)$. Then let β^j_i be the minimum value of $\overline{\beta^j_i}$. If there are no data which satisfy the left side of (ii), then let β^j_i be ∞. We assume that β^j_i is the sensitivity of X_i on the day j days before and $\tilde{\beta}_i = \min_{(1 \le j \le n)} \beta^j_i$ is the sensitivity of X_i on the forecast day.

This method is based on the fact that if the relationship between input and output is linear, the converse proposition of (ii) is achieved. We can assume that the local relationship between input and output is linear for a closely distributed data set. When the value of E is small enough, inequality (ii) shows the local relationship between input and output.

Consequently, if the value of E is small enough, we can assume that the converse proposition of (ii) is achieved. Then we define the neighborhood U(p) of p as shown in (iii).

$$U(p) = (x_1 - \tilde{\beta}_1, x_1 + \tilde{\beta}_1) \times \cdots \times (x_n - \tilde{\beta}_n, x_n + \tilde{\beta}_n) \qquad \text{(iii)}$$

However if i ($i \leq i \leq n$) exists such that $\tilde{\beta}_i = \infty$, then let U(p) be an empty set. Fig. 4 illustrates determination mechanism of the neighborhood for the two dimensional case.

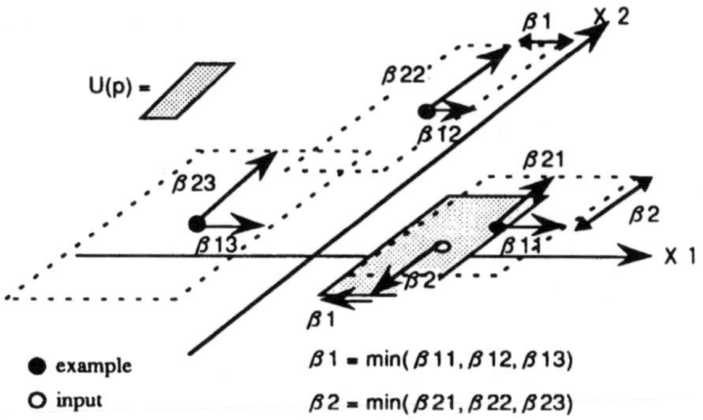

Fig. 4. Definition of neighborhood

<u>The exceptional removal</u> To apply the above method to the practical data set, we have to consider the existence of exceptional data among examples. We propose the following algorithm to remove exceptional data from examples within U(p).

Let $\overline{UD(P)}$ be the set of examples (the past n days) in U(p). For every dj ($1 \leq j \leq n$), that is the element of $\overline{UD(P)}$, if $\beta_i^j = \infty$ ($1 \leq \forall i \leq m$), we assume that dj is exceptional data. The set of neighborhood data UD(p) is obtained by removing exceptional data from $\overline{UD(P)}$. The output y is given by using UD(p), namely, y = f(UD(p)).

Forecasting Algorithm of MBL Model

Let E be the error bound of the forecasting model. Let m and D be the numbers of observations required for MBL and multiregression respectively. If there is no data within the neighborhood, we use multiregression since we can see a linear relationship between input variables and daily water demand. The following is an algorithm for forecasting the daily water demand.

A1. Decide the neighborhood U(p) for given input p.

A2. Obtain UD(p) by removing exceptional data from $\overline{UD(P)}$.

A3. If UD(p) ≠ ϕ, output the average of daily water demand within UD(p). Otherwise, forecast using multiregression from the last D days.

EXPERIMENTS AND DISCUSSIONS

We compared the MBL model (MBL + multiregression) with the multiregression model. We used demand data at a water-treatment plant (April, 1, 1991 - December, 31, 1991) to forecast weekday daily water demand(June, 1 - December, 31) .

We selected five input variables. They are high temperature, weather conditions, last day's weather conditions, latest daily water demand on the day with the same weather and day range. The first four variables are chosen by considering the analysis given in section 2. The last variable (day range) is chosen by the fact that the demand shapes indicate a seasonable influence. And let E = 1.0,1.5(%), D = 40, m = 10,15,20.

Table 1 shows the number of days (DD) MBL was applied and the average error(%) of the daily water demand forecasting on those days. Table 2 shows the average error by our method and the multiregression.

TABLE 1 Forecasting Error(%) on the Day MBL is Applied

(a) Forecasting error for E = 1.0 (%)

	MBL(%)	Multiregression(%)	DD*
m=10	1.45	2.31	14
m=15	1.03	1.51	11
m=20	1.31	1.41	14

(b) Forecasting error for E = 2.0 (%)

	MBL(%)	Multiregression(%)	DD*
m=10	1.34	1.53	33
m=15	1.24	1.46	40
m=20	1.45	1.45	45

DD*: The number of days MBL is applied

TABLE 2 Error(%) for All Forecast Days

	MBL model	
	E = 1.0 (%)	E = 2.0 (%)
n=10	1.69	1.66
n=15	1.68	1.70
n=20	1.74	1.73

Multiregression model	D*
1.74	40

D*: The number of sumples for multiregression

As shown in Table 1, when we use MBL, the forecasting results almost satisfied the error bound E. The forecasting results are improved about 10 - 30% in comparison with the conventional multiregression. However as E is smaller, the number of cases to which we can apply MBL becomes fewer and the accuracy varies according to the value of m. Table 2 shows that the forecasting results by MBL model are better than those by multiregression model. To examine the relationship between MBL and the error bound E, we show the average error by MBL for E = 1.0,1.5,2.0(%), and m= 15 in Table 3.

TABLE 3 Relationship between Error Bound (E) and Forecasting Error of the MBL

Error bound E(%)	Forecasting error (%)	DD*
2.0	1.24	40
1.5	1.13	25
1.0	1.03	11

DD*: The number of days MBL is applied

The error decreases gradually as the value of E becomes smaller, but the number of days MBL was applied decreases. This is caused by the fact that U(p) becomes smaller as E becomes smaller. In all the cases, the forecasting result satisfies the error bound. This shows our definition method of the neighborhood works efficiently. From the above mentioned, since the accuracy of MBL changes as the value of E and m, we have to set the value of E according to the required accuracy of the forecasting model.

CONCLUSION

We have presented a new forecasting model which combines MBL and multiregression. We have applied the method to the daily water demand forecasting. This method gives more accurate forecasting than that of the conventional multiregression. This shows the efficiency of MBL method.

In the future, we will apply MBL model to other water-treatment plants, examine the hybrid model which combines MBL and another global model(such as artificial neural network) and apply this model to other forecasting problems.

REFERENCES

[1] S. Sakazaki and H. Nagakura,"Integrated waterworks management system current status and future tasks"(Adv. Wat. Pollut. Control no 10) ed. R. Briggs, pp. 1-8.

[2] K. Jinno, A. Kawamura and T. Ueda,"On the Dynamic Properties and Predictions of Daily water Deliveries of the Purification Stations in Fukuoka City", Technology Reports of the Kyushu University Vol.59, No.4 (1986).

[3] J. D. Farmer and F. J. Sidorowich, "Prediction Chaotic Time Series", Phys. Rev. Lett., Vol. 59, No.8, pp.845-848 (1987).

[4] D. C. Park, "Electric Load Forecasting Using An Artificial Neural Network", IEEE trans. Power System, Vol.6, No.2, pp442-449 (1991).

[5] C. Stanfill and D. Walts,"Toward memory-based reasoning", CACM, Vol.29, No.12, pp.1213-1228 (1986).

[6] S. M. Omohundro, "Geometric Learning Algorithms", Physica D, Vol. 42, pp.307-321 (1990).

[7] C. Atkeson and D. Reinkensmeyer, "Using associative content-addressable memories to control robots", Artificial Intelligence at MIT, Vol.2, pp.102-127, MIT Press (1990).

[8] H. Akaike, "Information theory and an extension of the maximum likelihood principle, " Proc. 2nd Int.Symp.Information Theory, Supp. to problems of Control and Information Theory, pp.267-281 (1972)

[9] Y. Nakamura, S. Abe, Y. Ohsawa and M. Sakauchi, "MD-tree:A balanced hierarchical data structure for multi-dimensional data with highly efficient dynamic characteristics",Proc 9th ICPR Vol.1, pp.375-378 (1988).

DYNAMIC MODELLING AND SIMULATION

Pergamon

Wat. Sci. Tech. Vol. 28, No. 11-12, pp. 141–150, 1993.

0273–1223/93 $7·00 + 0·00

MODELING AND SIMULATION OF WASTEWATER TREATMENT PROCESSES

John F. Andrews*

Rice University, Houston, Texas, USA

ABSTRACT

Several different types of models which can be useful in describing the dynamic behavior of wastewater treatment plants are discussed. Included among these are (1) visual, (2) linguistic, (3) mental, (4) physical, (5) mathematical, and (6) fuzzy models. Some of the basic concepts of modeling with emphasis on the relationship of these to engineering research and practice are then presented, these being (1) differences in the use of, and expectations from, models by researchers and practitioners, (2) sources of information useful for model development, (3) the amount and type of testing needed for model validation, and (4) the required accuracy of models. Computer simulation is necessary for the solution of most dynamic models. Recent advances in both hardware and software for personal computers have resulted in inexpensive, user-friendly systems suitable for use in both large and small organizations as well as by individual engineers.

KEYWORDS

Dynamic modeling; computer simulation; wastewater treatment; model classification; model development; model testing; model accuracy; mathematical modeling.

INTRODUCTION

Neither modeling nor simulation are new and both have a long history of use in engineering and science. Scaled-down physical models have long been used for simulation in such diverse areas as astronomy (planetariums), hydraulic engineering (river models), architecture (building models), and chemical engineering (pilot plants). Biologists, chemists, and sociologists have studied model organisms (*E. coli* for example), model compounds, and model communities, respectively. Even the hypothesis which is formulated in applying the scientific method can be considered as a verbal model and simulation may be used to test the validity of a hypothesis when it is not feasible to test the real system.

This presentation will discuss modeling and simulation of wastewater treatment plants with emphasis on mathematical dynamic modeling and computer simulation. The literature on these topics is much too voluminous to review herein. For a more detailed description as well as some specific examples the reader is referred to Andrews (1992) and Patry and Chapman (1989). Discussion will be limited to some of the basic definitions and concepts with emphasis being placed on how these relate to both engineering research and practice.

The author strongly feels that progress in developing and applying dynamic models to wastewater treatment can be accelerated by practicing engineers learning how to develop their own dynamic models and conduct

* Address correspondence to John F. Andrews, 1719 Rayview, Fayetteville, AR 72703, USA

their own computer simulations. It is hoped that this paper will contribute to these goals. Rapid advances in both hardware and software for personal computers have resulted in inexpensive, user-friendly systems suitable for use in both large and small organizations as well as by individual engineers.

CLASSIFICATION OF MODELS

There are many types of models with the type selected being primarily dependent upon the purpose for which it is to be used. One possible classification is given below:

Classification of Models
Visual
Linguistic
Mental
Physical
Mathematical
Fuzzy

Each type of model can be further broken down into different categories. For example, visual models may consist of schematic diagrams such as flow diagrams for treatment plants. These diagrams are adequate for illustrating system components and qualitative descriptions of some of the interactions between components. However, they are not adequate for quantitative descriptions and a following step for design engineers is the preparation of plans (another type of pictorial model) to give dimensions of the components and distances between them.

There are also several different categories of linguistic models. An example is a procedural model consisting of a written and ordered list of tasks to be accomplished for a construction project. The addition of time to a procedural model results in a schedule (dynamic model) for completing the project. More recent examples of linguistic models are expert systems. These are a class of artificial intelligence (AI) computer programs usually intended to serve as computer assistants for decision making or for direct use in automatic control. In the context of this paper, they are usually written in the form of If-Then rules for describing dynamic behavior and specifying control actions. They overlap with mathematical models since they are based on the laws of logic. An example of a rule for a pumping station might be "**IF** the water level in the wet well is above 10 ft. **THEN** turn on pump No. 2." A starting reference for expert systems is the book by Bielawski and Lewand (1988). A discussion of the application of expert systems to the activated sludge process has been given by Barnett *et al.* (1992)

A type of model which must precede either visual or linguistic models is a mental model. A common example of the interaction between mental, linguistic, and visual models in engineering is when a young engineer converts his mental model of an idea into a linguistic model in discussions with his coworkers. During the discussions, the key points of the model are often illustrated by sketches or visual models. A more recent example is the expert system computer program in which the mental models of skilled operating engineers are converted to If-Then rules. The emerging field of Knowledge Engineering encompasses the job of gathering and codifying (quantifying) the knowledge of human experts.

Physical models are well known and often used by engineers and scientists. In wastewater treatment plant engineering, three types of physical models are used in taking an idea (mental model) to application in the field. These are bench, pilot, and prototype. Bench scale experiments are designed to give "Yes or No" (qualitative) answers while pilot scale experiments give more quantitative answers. However, the ultimate test is whether the idea will work under field conditions so prototype testing is also an important step in physical modeling.

Mathematical models are used for quantitative descriptions and consist of one or more equations relating the important inputs, outputs, and characteristics of a system. To those not familiar with systems engineering terminology, the term mathematical model is sometimes frightening since it may bring to mind large sets of complex equations and the use of sophisticated mathematical techniques. However, this need not be the case since most common engineering design formulae may also be called mathematical models. For example, as simple an expression as **y = mx + b** can be considered as a mathematical model where, in systems engineering terms, the system output **(y)** is related to the system input **(x)** by the system parameters **(m, b)**.

Mathematical models can also be classified in many different ways, for example as steady state, dynamic, mechanistic, empirical, deterministic, or stochastic. For a more detailed discussion of these classifications, the reader is referred to Andrews (1992). A newer type of empirical model with which the reader may not be familiar is the neural network model. A good description of these models oriented toward their use on personal computers is given in the book by McCord-Nelson and Illingworth (1991)

Fuzzy models are a compromise between the vague statements which humans often use and the strict logic of rule based expert systems (Yes or No or True or False) and quantitative answers provided by other mathematical models. They permit improved communication between the computer and humans by converting (translating) human statements such as high, normal, and low into numbers which the computer can understand (defuzzification) and numbers generated by the computer into statements for humans (fuzzification). A common example of the use of a mental fuzzy model is the ordinary traffic signal in which a green light means **Go** and a red light means **Stop**. However, what does a yellow light mean? If we are very close to an intersection when the light turns from green to yellow we may **Go** on yellow since the alternate could be being hit from behind if we brake too rapidly. However, if we are some distance from the intersection when the light turns from green to yellow we will most likely slow down and then Stop. In other words, we have exerted fuzzy control over our automobile based on vague information (very close or some distance from the intersection).

The author is of the opinion that all of the different types of models are useful under the appropriate circumstances and that practical models for application will ultimately make use of all of the different types. An example of the combination of expert system (linguistic) and neural network (mathematical) models is that of Caudell (1991) which she calls "expert networks". Another example is the work of Barnett and Andrews (1992) on a combination of a mathematical model (differential equations) with an expert system and a fuzzy model for studying the dynamics and control of the anaerobic digester.

SOME BASIC CONCEPTS

Modeling is widely used by many different disciplines as well as for many different purposes. It is therefore quite understandable that there would be disagreements between the different users and perhaps nowhere are these differences more obvious than those between engineering researchers and engineering practitioners. Some of the basic concepts of modeling, as listed below, will be discussed with emphasis on these differences. It is hoped that this discussion will lead to improved understanding of the different viewpoints and more joint efforts by engineering researchers and practitioners in applying modeling to both the design and operation of wastewater treatment plants. The discussion is heavily oriented toward mathematical modeling; however many of the issues addressed are applicable to other types of modeling.

Some Basic Concepts of Modeling

Use of models
Information for model development
Testing of models
Required accuracy

Use of Models

The primary objective of most engineering researchers is to extend engineering knowledge and they see modeling as a tool for guiding research with possibly application of the model in practice at some future date. The researcher's initial models may be considered as mathematical versions of the verbal hypotheses used in the scientific method in which the possibility of being wrong is implicitly recognized. The researcher thus expects and is prepared for the fact that the model may give inaccurate or completely false predictions. This recognition of initial models as hypotheses makes it clear that the development of models is an iterative process (Fig.1) in which models are formulated, predictions compared with observed results, and the model revised if necessary. This procedure is repeated until adequate agreement is obtained between predictions and known facts and can be a time consuming process since years are sometimes required before a research model is ready for practical application.

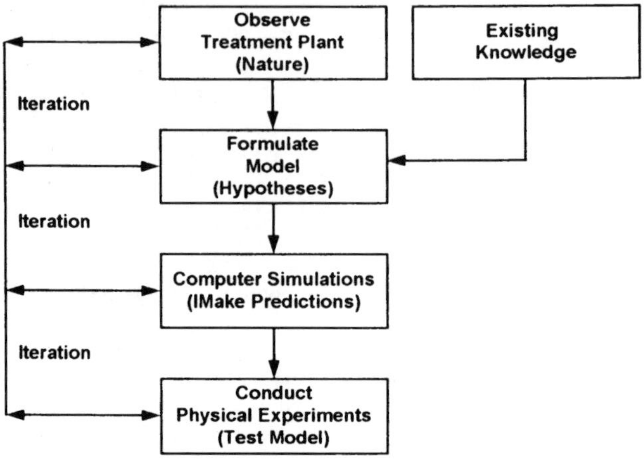

Fig. 1. Steps in the development of mathematical models.

The practitioner must usually operate on a much shorter time scale and is more concerned with immediate application of models to assist in design, construction, or operation of real systems with advancement of engineering knowledge being secondary. Practitioners usually recognize the need for collecting data to obtain numerical values for some of the parameters and expect that there will be some uncertainty in model predictions; however, they normally expect the basic structure of the model to be such that it will give satisfactory predictions without being modified. The practitioner cannot afford highly inaccurate predictions since the consequences (loss of life, reputation, money) can be severe. The practitioner may also not be able to take the time to go through several iterations of the modeling process. These differences in the use of and expectations from models by engineering researchers and practitioners can lead to disagreements as to the value of models unless their intended use is clearly defined.

Information for Model Development

Other points of possible misunderstanding between engineering researchers and practitioners, as well as between scientific and engineering researchers, are the sources of information used in developing mathematical models. Figure 1, which might also be called a block diagram of the scientific method, illustrates the use of four sources of information, these being; (1) existing knowledge, (2) observation of plant behavior, (3) model predictions, and (4) the results of planned experiments. The emphasis placed on each of these different sources of information is very much dependent upon the basic discipline and background of the developer or user of the model. For example, design engineers naturally tend to stress sources of design information whereas operating engineers will emphasize their experience, and that of others, in observing plant behavior. Some scientific researchers go by the maxim that "the experiment is everything" and thus emphasize the conduct of physical experiments while some engineering researchers emphasize modeling and simulation techniques and may be saying that "the model is everything".

The author is of the opinion that many different sources of information have potential value for the development of models. This is especially so for wastewater treatment plant models which require knowledge from a wide variety of scientific and engineering disciplines as well as practical knowledge from plant designers and operators. It is the synthesis of existing knowledge from other disciplines into models that may be one of the distinguishing characteristics between engineering and scientific research. Engineering researchers do consider new applications of existing knowledge from other branches of science or engineering as original research whereas only the discovery of new knowledge is considered as original research by some scientists.

Testing of Models

The amount of testing needed will be a function of the purpose for which the model is to be used and the blend of existing knowledge and new hypotheses incorporated in the model. If the model is entirely based on well established principles from the sciences or other engineering disciplines, the structure of the model should be adequate and all that may be required is determination of numerical values for the parameters in the model. Moreover, first estimates of some of these may either be available or calculable from data available in the literature. An example might be a coupling of hydraulic models for individual processes for simulation of the dynamics of fluid flow in an entire plant as reported by Olsson and Stephenson (1986). Such models could well have been established in previous years since most of the basic knowledge (fluid mechanics) needed to set up the equations comprising the model has been available for some time. However, they may not exist for a reason previously given, this being that most engineers are practical people and inclined to say "Why set up equations for which it is not currently possible to obtain solutions?" In other words, advances in solution techniques may result in a renewed interest in old problems.

In the above case it might be more appropriate to say that the structure of the model requires no testing and the purpose of testing is to determine numerical values for the parameters. Such models are no longer in the realm of engineering research but are instead ready for engineering application.

At the other extreme are research models containing primarily new hypotheses for which only qualitative observations are available to support the hypotheses. Such models need extensive testing to prove or disprove the hypotheses and, if these are proven correct, to structure the model so that it can be applied in practice. Both modeling and simulation can play a significant role in this testing. Simulations using the model can be used to design physical experiments with first experiments usually being at bench scale to prove or disprove hypotheses and obtain rough estimates of parameters. The model is then modified to incorporate these increases in knowledge and then used in simulations to design experiments for pilot and/or full-scale testing.

Figure 1 can also be used to illustrate the difference in attitude toward the testing of models by engineering researchers working on "natural" systems" (streams, lakes, estuaries, etc.) and those working on "man-made" systems such as treatment plants. Experimentation is frequently not possible for natural systems and the usual model testing approach for these systems is to collect data over a long period of time and then analyze the data to examine the "fit" between model predictions and the data. However, what if the dynamic phenomena of interest (a "spill" of a toxic chemical, for example) does not occur during this period of time? Or if the frequency of the measurements is inadequate to quantify the dynamic response?

With man-made systems such as treatment plants, experimentation is often possible and an integral part of model testing should be the design and conduct of carefully planned experiments including appropriate input disturbances and frequency of measurements so that data on the phenomena of interest can be more rapidly and accurately obtained. This type of information may be difficult to obtain from normal operating records since the control exerted in normal operation can remove much of the dynamic information of interest. In other words, the signal to noise ratio is too low. During times of crisis, such as process upsets or pending process failure, significant dynamic phenomena are occurring. However, during such times operating engineers are naturally more concerned with restoring good performance than they are with collecting data for the testing of models! This points out the need for the conduct of carefully planned experiments to evaluate such phenomena.

The testing of models also brings up the issue of measurements (observations) and thus illustrates another difference between the development and use of models by engineering researchers and practitioners. Models for research do tend to be more complex and include more terms than those used for design, some of which may be difficult to measure. This can be an acceptable situation in research since the investigators usually have a good command of the basic theory involved and the ability to use the more sophisticated instruments needed for measurements. However, models for practical applications should be as simple as possible and focus on the inclusion of parameters or variables which are relatively easy to either measure or specify. A model which is too complex to understand or requires complicated measurements for implementation is subject to either "misuse" or "disuse".

Required Accuracy

How accurate must be the prediction of a model? Is a "Yes" or "No" sufficient or must the answer be more quantitative? If it must be more quantitative, how closely must the predicted and observed behavior agree? The answers to these questions depend upon the purpose for which the model is to be used.

In the early stages of model development, a "Yes" or "No" answer may be quite adequate. For example, in the author's early work on the dynamics of anaerobic digestion (Andrews, 1968) it became apparent that the model must be able to predict that process failure can be caused by organic overloading since this phenomena was known to occur in the field and existing models could only predict failure by hydraulic overloading. Once the occurrence of the phenomena could be predicted on a "Yes" or "No" basis, the next step was to expand the model (Andrews, 1969; Andrews and Graef, 1971) so that it could predict the time-dependent behavior of the key variables normally monitored for evaluating process condition. The term "semi-quantitative", which is deliberately vague, was coined to indicate that predictions were in the appropriate direction (up or down) and of the right order of magnitude. Confirmation of these semi-quantitative predictions (model testing) could be obtained from the literature and discussions with experienced operating engineers. However, more quantitative model testing necessitates physical experimentation.

The differences in agreement between model predictions and observed results is due to uncertainty in the inputs to the model, the model itself, and the information available for testing the model. In design, uncertainty is taken into account by the use of safety factors which often result in increases in size. For example, there is usually considerable uncertainty involved in predicting the population of a city at some time in the future (the design period) so the design engineer must use his best judgment to make this prediction. Increases in size to accommodate the uncertainty in these predictions should not be called overdesign, as they are sometimes labeled, but are instead attempts to protect against uncertainty. Since there is always uncertainty in attempts to predict the future, the possibility exists that the load on the plant at the end of the design period can be either larger or smaller than that predicted.

Process control based on model predictions is also subject to uncertainty. However, unlike design, it is possible in many cases to correct for this uncertainty by using automatic feedback control in which the amount of control exerted depends upon the difference between the desired and observed values of performance (the error signal). This means that a reasonable amount of uncertainty can be tolerated in a dynamic model for process control and models with considerable error in the predictions (semi- quantitative) can also be useful. It should be noted that feedback is also involved in process design in that when the design engineer becomes aware of deficiencies in his design, he uses these to correct future designs. However, the feedback is not automatic and is on a much longer time scale (years and sometimes decades) than feedback in process control. It may therefore not be very useful to the operating engineer who has to operate in a much shorter time scale with existing facilities.

COMPUTER SIMULATION

After a dynamic mathematical model has been developed for a process, the equations which comprise the model must be solved in order to predict process behavior with respect to time. This is known as simulation and can be defined as the use of a model to explore the effects of changing conditions on the real system. Obviously, the model must be a reasonable representation of the real system in order for the results to be meaningful since simulation results can be no better than the model and data on which they are based.

Prior to the advent of computers, a computational bottleneck (Fig.2) existed and efforts at mathematical modeling were frequently of little practical value since the equations comprising the model could not be solved. Engineers, being practical people would usually say, "Why set up the equations when it is impossible to solve them?" The ready availability of computers and simulation languages has largely eliminated this bottleneck with the current bottleneck being primarily the development of realistic models.

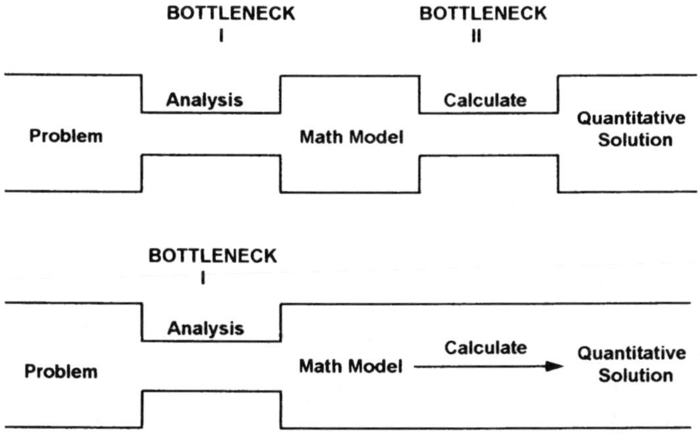

Fig. 2. Difficulties in developing mathematical models before and after the availability of computer simulation (after Franks, 1967).

History of Computer Simulation.

Early computers used for simulation were analog in nature and involved substituting relationships between voltage, current and resistance for those of other physical variables. Some analog computers were general purpose while others were designed to simulate specific systems. The general purpose analog computer could be used to simulate any system for which a deterministic mathematical model could be developed. An example of the use of an analog computer for simulation of a specific system was the McIlroy analyzer in which vacuum tubes with special characteristics were used to simulate the steady state relationships between fluid flow (current) and pressure drop (voltage) in pipe networks for water distribution systems, with the characteristics of the pipes (coefficient of friction, length, diameter) being proportional to the resistance of the tubes.

Although useful in their time, analog computers were not easily programmed by many practicing engineers since they required "scaling" of electrical variables (voltage, current, resistance) to represent numerical values of the variables of interest and "patching" of a computer board to electrically connect the different analog elements. They have been largely replaced by simulation languages for use on digital computers. In these simulation languages, the differential equations comprising the mathematical model are solved (simulated) using numerical techniques to approximate continuous functions by discrete functions. Neither "scaling" nor "patching" is required since the digital computer can handle a wide range of numerical values and can automatically sort the equations to be solved so that variables are calculated in the appropriate order. However, users of simulation languages should always keep in mind that numerical solution of differential equations is an approximation and sometimes may give incorrect answers. It is especially important to be aware of this when studying process stability and one must always ask the question "Is the instability really due to process instability or is it due to instability of the numerical technique used for approximate integration?" Users who plan to make extensive use of computer simulation are urged to learn more about numerical analysis so that they will be aware of the strengths and weaknesses of the different techniques available.

The early use of digital computers for simulation was primarily restricted to specialists since a detailed knowledge of numerical techniques was required and a considerable amount of time was needed to learn programming languages such as FORTRAN. These problems were overcome by the development of simulation languages which were heavily user oriented ("user friendly") thus permitting the engineer or scientist to concentrate on model development and interpretation of simulation results instead of programming details. Most simulation languages could also be learned in a relatively short period of time (hours instead of days or weeks) thus satisfying the engineering desire to "get on with the solution of the problem." Franks (1967) has presented a history of the early simulation languages.

The early simulation languages were batch oriented since most digital computers of that era (1960–1975) were operated as batch systems using relatively expensive computers. This type of operation impeded a more widespread use of computer simulation since the user needed access to a computer center and was not able to interact directly with the computer. These obstacles were removed by low-cost personal computers (PCs) which became available in the early 1980s. A wide variety of simulation languages are now available for use on PCs. The Society for Computer Simulation (1991) publishes an annual directory of simulation software. The 1991 edition of this directory lists 116 general and special purpose simulation software packages with 77 of these being suitable for use on PCs.

The author has used computer simulation since the mid-1960s and during this time has used general purpose analog computers, simulation languages (PACTOLUS and CSMP) for main frame computers and now uses SIMNON (1993) on a personal computer. The PC version of SIMNON can handle the simultaneous solution of up to 300 differential equations. More details, with some simple examples, on the use of SIMNON can be found in Andrews (1992). An example of a more comprehensive simulation system is the general purpose simulator (GPS) which has been described by Patry and Takacs (1992). Their paper also describes the use of GPS in a study of the Hamilton-Wentworth plant in Canada. This simulator, which is designed for use on a Sun workstation, contains a variety of primarily deterministic mathematical models for wastewater treatment processes which can be easily linked together to form a plant model using object oriented concepts and an interactive screen oriented interface. Another simulation package oriented toward the dynamic modeling and control of wastewater treatment plants is SACCESS (Statistical analysis and control of environmental systems) which has been described by Hiraoka and Fujiwara (1992). This package, available in both workstation and PC versions, is based on time series analysis and thus represents modeling using both deterministic and stochastic components. The application of SACCESS to modeling and control of the Kawamata plant in Japan has been described by Hiroaka and Tsumura (1992).

Although space does not permit a detailed review, it should be noted that there are many general purpose PC software packages available for simulation using neural networks, expert systems, and fuzzy logic. The term "general purpose" is used herein to indicate that these packages are not specific to any one application, such as wastewater treatment, but can instead be used for many different types of applications. Special computer chips are also available for neural networks and fuzzy logic so these may be implemented in hardware as well as software. Mention should also be made of the use of commercially available spreadsheets (Excel, Lotus 1-2-3, etc). Most readers will be familiar with the use of spreadsheets for solving explicit algebraic equations and perhaps also their use for iterative ("trial and error") solution of implicit equations. However, they may not know that spreadsheets can also be used for dynamic modeling (solution of differential equations) as discussed by Julian (1989) and El Shayal (1990).

<u>Advantages and Disadvantages of Computer Simulation</u>

Computer simulation has many of the same advantages, and disadvantages, as physical simulation.

Considerable knowledge can be gained about a process through the development of a mathematical model and subsequent computer simulations using the model. Sensitivity analysis, or the response of the model to changes in specific variables, can be used for model improvement by indicating those variables which are most significant. Sensitivity analysis can also be used for model simplification by indicating those variables which have little effect on system performance. Simulation permits examination of large systems, such as river basins, where physical experimentation on the full scale system may not be possible. Time can be compressed on the computer with simulations being conducted in seconds or minutes. This is especially important for the biological processes used in wastewater treatment where rates are relatively slow and physical experimentation may require weeks and even months.

There are, of course, disadvantages to computer simulation. The results of the simulations are no better than the mathematical model and data on which they are based. Poor models and/or poor data can give inaccurate or even completely misleading results. Also, some of the advantages of computer simulation can be disadvantages. Computer simulation can be easier, cheaper, and faster than physical experimentation. In the case of wastewater treatment, it can also be a much cleaner job! These factors can lead to one becoming overly enamored with the tools (modeling and computer simulation) and consequently neglecting physical experimentation and/or the collection of data.

SUMMARY

Several different types of models can be useful in describing the dynamic behavior of wastewater treatment plants. Included among these are; (1), visual, (2), linguistic, (3) mental, (4) physical, (5) mathematical, and (6) fuzzy models.

A discussion of some of the basic concepts of modeling with emphasis on the relationship of these to engineering research and practice is presented. These concepts are:

differences in the use of, and expectations from, models by researchers and practitioners
sources of information useful for model development
the amount and type of testing needed for model validation
the required accuracy of models

Dynamic mathematical models are solved using computer simulation to predict plant behavior with respect to time. The early use of digital computers for simulation was primarily restricted to those with a knowledge of numerical techniques and computer languages such as FORTRAN. These problems were overcome by "user friendly" simulation languages which permitted concentration on model development instead of programming details. However, the early simulation languages were batch oriented for use on relatively expensive computers controlled by computer centers. The availability of interactive simulation languages for use on low-cost personal computers (PCs) has removed these obstacles.

REFERENCES

Andrews, J. F. (1992). *Dynamics and Control of the Activated Sludge Process*, Technomics Publishing Co., Lancaster, PA.

Andrews, J. F. and Graef, S. P. (1971). Dynamic modeling and simulation of the anaerobic digestion process, *Anaerobic Biological Treatment Processes, Advances in Chemistry Series* **105**, 126-162, American Chemical Society, Washington.

Andrews, J. F. (1969). Dynamic model of the anaerobic digestion process, *Journal Sanitary Engineering Division, American Society of Civil Engineers*, **95**, 95-116.

Andrews, J. F. (1968). Mathematical model for the continuous culture of microorganisms utilizing inhibitory substrates, *Biotechnology and Bioengineering*, **10**, 707-723.

Barnett, M. W. and Andrews, J. F. (1992). Expert System for Anaerobic Digester Process Operation, *Journal Environmental Division, American Society of Civil Engineers*, **118**, 949-963.

Barnett, J. W., Patry, G. G. and Hiraoka, M. (1992). Knowledge-based (expert) systems for the activated sludge process, *Dynamics and Control of the Activated Sludge Process* (ed. Andrews, J. F.), Technomics Publishing Co., Lancaster, PA.

Bielawski, L. and Lewand, R. (1988). *Expert Systems Development: Building PC Based Applicatons*, QED Information Sciences, Wellesley, MA.

Caudill, M. (1991). Expert Networks, *Byte*, 108-116.

El Shayal, I. (1990). Solving differential equations on a spreadsheet: Part I: The Euler method, *Chemical Engineering*, 149-150.

El. Shayal, I. (1990). Solving differential equations on a spreadsheet. Part II: The Runge-Kutta method, *Chemical Engineering*, 153-154.

Elmqvist, H., Astrom, K. J., Schonthal, T. and Wittenmark, B. (1993). *SIMNON: User's Guide for M.S.-DOS Computers*, V 3.2, SSPA Systems, Box 24001, S-400 Goteborg, Sweden.

Franks, R. G. E. (1967). *Mathematical Modeling in Chemical Engineering*, John Wiley & Sons, New York.

Hiraoka, M. and Fujiwara, T. (1992). The use of time series analysis in hierarchical control systems, *Dynamics and Control of the Activated Sludge Process* (ed. Andrews, J. F.), Technomics Publishing Co., Lancaster, PA.

Hiraoka, M. and Tsumura, K. (1992). Computer assisted operation of the Kawamata treatment plant, Osaka prefecture, Japan, *Dynamics and Control of the Activated Sludge Process* (ed. Andrews, J. F.), Technomics Publishing Co., Lancaster, PA.

Julian, F. M. (1989). Process modeling on spreadsheets, *Chemical Engineering Progress*, 33-40.

McCord-Nelson, M. and Illingworth, W. T. (1991). *A Practical Guide to Neural Networks*, Addison-Wesley Publishing Co., Reading, MA.

Olsson, G. and Stephenson, J. (1986). The propagation of hydraulic disturbances and flow rate reconstruction in activated sludge, *Environmental Technology Letters*, **6**, 536-545.

Patry, G. G and Takacs, I. (1992). Dynamic modeling and simulation of large scale wastewater treatment plants using the general purpose simulator, *Dynamics and Control of the Activated Sludge Process* (ed. Andrews, J. F.), Technomics Publishing Co., Lancaster, PA.

Patry, G. G. and Chapman, D. (1989). *Dynamic Modeling and Expert Systems in Wastewater Engineering*, Lewis Publishers, Chelsea, MI.

Society for Computer Simulation (1991). *Directory of Simulation Software*, Vol. 2, Society for Computer Simulation, San Diego, CA

Pergamon

Wat. Sci. Tech. Vol. 28, No. 11-12, pp. 151–161, 1993.

0273–1223/93 $7·00 + 0·00

DYNAMIC MODELLING OF AN ASP SEWAGE WORKS – A CASE STUDY

L. Stokes*, I. Takács**, B. Watson** and J. B. Watts***

** Severn Trent Water Ltd, Finham STW, St. Martins Road, Coventry, CV3 6PR, UK*
*** Hydromantis Inc, 1685 Main St West, Suite 302, Hamilton, Ontario, L85 1G5, Canada*
**** Minworth Systems Ltd, Kingsbury Road, Sutton Coldfield, Birmingham, B76 9BL, UK*

ABSTRACT

The application of the GPS-X computer simulation of sewage works to a 55 Ml/d dry weather flow (dwf) nitrifying activated sludge plant (ASP) is described. The role of simulation and the features of a simulation package as required by the end user are outlined. The plant processes, operation, performance and the instrumentation are described. The on- line measurements are collected and displayed by the GFX monitoring system which acts as input/output stage for GPS-X. The simulation package uses the IAWPRC model with modifications based on Dold (1990). The features of the model, its calibration and successful verification are described.

KEYWORDS

Activated sludge; dynamic modelling; simulation; calibration; activated sludge model; intelligent instrumentation; specific oxygen utilisation rate; respirometry.

INTRODUCTION

Severn Trent Water Ltd (STW) operates about 70 ASPs and in the future some will have to meet COD, total N, and total P consent levels as well as BOD, SS and $N\text{-}NH_4$ values. STW sees computer simulation as one way of helping to design and operate plants that have to meet tighter consents and use more complex processes.

A successful model has four roles:

a) Prediction i.e. using on-line instrumentation as inputs to predict effluent quality over the next few hours or over longer times scales so that management have time to react and prevent excursion outside of consent.
b) Trouble shooting and exploration of what-if scenarios.
c) Training supervisors and management in understanding the processes and cause and effects in ASPs. These three roles will involve mainly supervisors and operational management.
d) Design. i.e. use by engineering staff and process experts to design, more efficient processes and more economical plant.

STW has had experience with user unfriendly ASP model simulation and in its view what is required is a very user friendly front end allied to an invisible (to the user) but accurate mathematical model. Until recently no-one has offered this combination in the field of sewage treatment.

Amongst the features that a user requires are the following.

Accuracy in simulation both during normal conditions but also in more extreme but common events such as storms.

Dynamic simulation and easy connection to on-line instrumentation.

Input and output primarily presented graphically.

Plant layout to be on screen with ability to adjust parameters/unit operation on screen.

Icon-type approach to enhance user-friendliness. This is a key concept bearing in mind the potential users (operational management, etc.).

Ability to replace/extend the mathematical model as modelling developments occur.

The hardware is not important. A PC based package would be most attractive but realistically a workstation will probably be necessary to provide the speed required and handle the complexity of a typical ASP.

The GPS-X package from Hydromantis offers these features and STW decided to use it to model Coleshill ASP, employ Hydromantis to customise it for the works, and to develop the software to link with the site monitoring system (GFX – by Minworth Systems). Coleshill was chosen mainly because it had very good on-line monitoring, but also because it will require extensions in the near future.

Coleshill is a 55 Ml/d (dwf) conventional diffused-air nitrifying ASP meeting a 25/45/20 (BOD/SS/N-NH_4) consent (95%ile performance based on 3 samples per week). It is a plant that suffers, more than most, from process instability. Plant data is summarised in Table 1.

TABLE 1. Plant Data

	Crude	Settled Sewage	Final Effluent
BOD_5 (ATU) mg/l	205	117	9
COD mg/l	480	293	59
SS mg/l	323	85	10
N-NH_4 mg/l	28	31	7
TON mg/l	1	0	11
P (total) mg/l	-	8	6
pH	7.3	7.2	7.2
Average flow Ml/d	Sewage 65.1	RAS 48	SAS 1.0
Max. flow to full treatment		130 Ml/d (2.3 dwf)	
MLSS mg/l 2550 (2100-3000)		SSVI (ml/g) 130 (54-200)	
Sludge age days 12			
F/M kg BOD/kg MLSS	0.12 (0.07-0.15)		
Retention hours.	10.5 ASP (nom. dwf)	3.3 Primary (peak)	
Secondary clarifier loading		6.8 kg/sq.m.h (max)	

(Three year averages - 1989/1991)

The plant was commissioned in 1970, the ASP was modified in 1988-90 and instrumentation has been gradually added over recent years. It treats a population equivalent of 248,000. The site has suffered from occasional trade effluent problems although the trade waste component is low (8%). The site is manned daytime only.

All sewage is screened and degritted before storm separation, with 3 storm tanks (9 h retention at dwf). Any storm overflows to the river have to meet a 150 mg/l solids consent. Peak lopping of flows using the storm tanks has been practised in the past during periods of poor MLSS settlement.

The rest of the plant is now operated as two halves - east and west with a combined outfall to the river; each half has 4 primary settlers, 4 aeration tanks, and 4 secondary clarifiers. The flow to each primary settler is measured (flume). Six tanks are used in dry weather with the extra ones manually started in wet weather. The surplus activated sludge (SAS) is returned to the inlet sewer and is cosettled with the primary sludge. The settled sewage is weak and experience in STW is that weak settled sewages are associated with process control problems in ASPs. The aeration bays are plug flow (8 tanks in series) with the first 8% anoxic. The tanks are aerated by fine bubble diffusers in a tapered arrangement (3.4:1 taper). There are 2 control zones using DO probes and feed back to variable speed blowers. The system is good at maintaining set DO levels. The SAS is wasted continuously from the returned activated sludge (RAS). SAS and RAS rates are manually adjusted.

MLSS is the principal variable used by STW to control the ASP process. Custom and practice has shown that an F/M ratio of 0.10 will permit consistent nitrification in the UK with the normal ammonia levels found in the settled sewage. The main intractable problem with the works is the variable settling quality - often it has been poor with SSVIs up to 190, and usually over 110. With poor settlement it is very difficult to take peak flows and keep sludge in the final clarifiers.

ON-LINE INSTRUMENTATION

Coleshill is one of the better instrumented sites in STW with over 150 analogue inputs to the site monitoring system. Regulation is by analysis of samples in the regulator's laboratories, but instrumentation is necessary for plant control. However the operators will only use the instruments if they believe in their accuracy. For modelling purposes it is equally necessary that the instruments are accurate.

To this end it is necessary that instruments are reliable and accurate in this very difficult environment and where possible have intelligent self-checking. Experience at STW is that relatively few instruments meet our requirements. The more important instruments used as inputs to the model are described in Table 2; only the respirometers were installed specifically for the model. The instruments are described in more detail in Snaddon *et al.*, 1980; Watts *et al.*, 1990; and Farrimond and Watts, 1991.

TABLE 2. More Important Instruments

Measurement	Make/type	Number	Self-cleaning	Self-calibrating
Sewage flow	Ultrasonic level	8		
DO	MSL	16	yes	yes
MLSS	Euro control infra-red	4	yes	
Sludge blanket	Markland infra-red	8	yes	
Respirometer	MSL	4	yes	yes
SS	STW design infra-red	3	yes	
$N-NH_4$	ion selective electrode	3		yes
pH	electrode	3	yes	
Temperature	electrode	3	yes	

The last 4 parameters are in 3 water quality monitors on the settled sewage (east and west) and the final effluent.

Four respirometers have just been installed (after delivery of the model). Two are at the beginning of aeration zones and another two at the RAS (east and west). They have MLSS measurement and measure specific oxygen utilisation rates – peak and minimum. These SOUR rates will be input into the model.

Recent work (Watts and Garber, submitted) suggests that the respirometers can be used to establish readily biodegradable substrate and we propose to include this in the model.

For some of the period in which data was used for model development continuous COD, BOD and TOC monitors were in use on the settled sewage and the final effluent. They were not used as inputs to the model but did help in understanding both inputs and outputs.

Several of the instruments are checked remotely using the telemetry to ensure that they were still self calibrating correctly. This serves to reduce the amount of resources required to attend the monitors as the service intervals can be extended without risk. STW are sure that this maintenance technology will become vital in the future as more instruments are introduced into sewage works and there continues to be a scarcity of instrument maintenance engineers.

The signals from the instruments are marshalled onto outstations and displayed on a Unix based monitoring system - GFX supplied by Minworth Systems (Farrimond and Watts, 1991). This system is used on STW's larger works. It is a very friendly and powerful graphical system and is used at Coleshill by operators to monitor plant performance and operation.

MODEL BUILDING

The General Purpose Simulator (GPS-X) is an object-oriented, model- independent, interactive programme which provides an easy-to-use, menu and graphics driven interface for the user to develop a mathematical model of a sewage works.

Some of the most important features of GSP-X are: 1) there is practically no limit on the model size (number of influents, reactors, settlers, bypass streams, etc.) or treatment plant configuration; 2) several user editable libraries (carbon-nitrogen, carbon-phosphorus, user customised industrial pollutants) for optimal modelling efficiency; 3) powerful numerical solver (7 algorithms); 4) user defined interactive runtime controls; 5) multiparameter optimisation routine; 6) sensitivity analysis; 7) 2 steady state solvers (fast, robust); 8) SRT, F/M ratio, BOD load definable for all configurations; 9) modelling of single sludge or different sludge types within one plant; and 10) graphical, digital or report form output.

Coleshill Water Reclamation works is a complex, large-scale plant as described above. Even though in GPS-X it is possible to model all individual unit processes, the data requirement for such a detailed analysis can be quite extensive. Usually it is more appropriate to aggregate some of the parallel unit processes (e.g. settling tanks, aeration tanks) together, assuming that the flow split between these units is not too disproportionate. The calibration can then be done on a simplified layout of the plant such as shown in Fig. 1. The calibrated model parameters can subsequently be imported into the full layout for more detailed analyses.

The following description of the simplified model refers to Fig. 1. The simplified GPS-X model of the Coleshill plant is fed by two influents [1,2](square brackets refer to stream numbers in Fig. 1.), which correspond to the Meriden and Chelmsley inlets. Return from the storm overflow tanks and second ary wastage are also accounted for in the model (streams [70] & [77] respectively). The storm tank was modelled as a pure hydraulic tank in which no reactions take place, i.e. only mass balance and equalization effect of the mixing is simulated. Plant flow, above a specified maximum flow, is automatically diverted to the tank [byp], then returned to the plant [70] when the crude flow is below a preset level. Pump capacities and flow setpoints are defined interactively by the user and the pumps can be operated on automatic or manual control in the model.

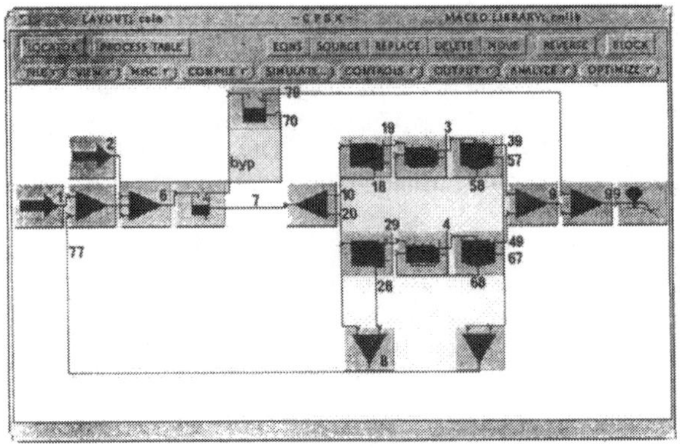

Fig. 1. Simplified plant layout.

The flow to primary settling tanks [7] is divided into east [10] and west [20] flows. The primary settling tanks are modelled using the layered flux model (Takács *et al.*, 1991; Patry and Takács, 1992).

The effluents from both the east [19] and west [29] primary settlers flow to east and west aeration tanks. These tanks are modelled as plug flow by using 8 equal volume Continuous Flow Stirred Tank Reactors (CFSTRs) in series. A modification of the Activated Sludge Model No.1. (Henze *et al.*, 1987) was used. The modifications include a process to account for nitrate uptake when there is a lack of ammonium as nutrient (Dold, 1990), the effect of temperature on kinetic constants (Wentzel *et al.*, 1992), and the inclusion of stoichiometric constants so that BOD, VSS, and TKN can be calculated based on the IAWPRC state variables. Selecting a relatively more complex model for this plant was well supported by results published by Watts and Garber (submitted), which show at least three different organic fractions in the influent. Lumped parameter models (e.g. based on BOD) are not able to simulate correct oxygen requirement patterns in the aeration tank if the influent contains several organic fractions. In the authors' experience such a model (e.g. Chambers and Jones, 1988) produces very conservative results and over-emphasises the taper required in plug-flow systems.

The effluent from the east aeration tank [3] is fed into a secondary clarifier with an overflow [39] and an underflow [38]. Likewise, the west aeration tank has an effluent [4] that is fed to a secondary clarifier with an overflow [49] and an underflow [48]. Both east and west reactor/settler combinations in the model correspond to four reactor/settler combinations at the plant. The final clarifiers are also modelled using the layered flux model. The model calculates a dynamic suspended solids profile along the settler depth. In an extended version of the settling model, the activated sludge model described above is integrated into each of the layers of the settler, allowing the calculation of biological reactions. Among other parameters this model can predict the DO profile in the settler, as well as potential nitrogen- gas generation leading to elevated effluent suspended solids concentration.

The overflows from the two secondary settlers are combined [9], then together with the storm tank overflow [79] discharged to the outfall [99]. The underflow from the east secondary clarifiers [38] is split into a recycle stream [58] and a wastage stream [57]. Similarly, the underflow from the west secondary clarifiers

[48] is split into a recycle stream [68] and a wastage stream [67]. Both the east and west wastages are combined into one stream [77] and recycled to the head works.

CALIBRATION AND VERIFICATION

For calibration purposes data from three sources was used:

1) dynamic data from on-line GFX database, logged every 15 minutes (the signals included: influent sewer flows, primary clarifier effluents solids, MLSS, DO, effluent SS and effluent N-NH_4);
2) routinely logged plant daily records, usually composite samples (influent BOD_5(ATU), SS, N-NH_4 and effluent BOD_5(ATU));
3) data from additional test performed to establish necessary stoichiometric fractions (BOD_5/BOD_{20}, COD/VSS, VSS/TSS, N-NH_4/TKN, and filtered BOD_5/BOD_5).

The plant model was calibrated to the period January-February 1991. During this period the plant was mainly subject to dry weather flow conditions and one aeration tank on the east side was shut down. First a dry weather (steady state) calibration was performed for the average values of the selected two months. The emphasis during parameter identification was placed on primary and secondary settling, carbonaceous BOD removal and nitrification kinetics. After the first steady state optimization, a two day run was performed and the calibrated parameters were fine-tuned to reflect the dynamics of February 24-25, 1991. Then the steady state calibration was checked again for any possible deviation. This process was repeated several times until an optimum set of model parameters was found for the two (average and dynamic) data sets. This procedure allowed for a more robust calibration, as well as a single model parameter set applicable to both steady-state and diurnal conditions. Table 3 summarizes the results of this calibration. The model performs quite well for both settling and biological reactions.

TABLE 3. Results of Steady-State Calibration

	Data	Model
Influent:		
Flow rate (m^3/d)	77,470	77,470
Suspended Solids (mg/l)	320	320
BOD_5 (mg/l)	192	192
Ammonia-N (mg/l)	25	27
Temperature (°C)	12	12
Primary Effluent (avg.):		
Suspended Solids (mg/l)	115	115
BOD_5 (mg/l)	110	108
Ammonia-N (mg/l)	27	27
Aeration Tanks (avg.):		
MLSS (mg/l)	2,550	2,540
DO (mg/l)—last pass	3.7	3.7
Total Recycle Flow rate (m^3/d)	37,000	37,000
SRT (days)	~ 8	7.5
Final Effluent:		
Suspended solids (m/gl)	3.5	3.8
Ammonia-N (mg/l)	9.6	9.5
BOD_5 (mg/l)	7.5	6.7

Some conflict was found between measured stoichiometry (e.g. COD/VSS ratio of the sludge) and the plant wastage data required to maintain the necessary MLSS and SRT values. Since the validity of the stoichiometric data was decreased by several factors (e.g. small number of samples, sampling completed more recently than the calibration data), it was decided that plant operational records would be given more credibility, thus the measured sludge COD/VSS ratio was modified from 1.22 to 1.90 (Table 4). Plant wastage was set to 1.025 Ml/d, which agrees well with normal wastage rates from the plant. The influent

TKN and N-NH_4 compositions were calibrated based on the primary tank ammonia data (monthly averaged data) and the influent stoichiometry.

TABLE 4. Stoichiometric and Kinetic Model Parameters.

Parameter	Value	Source
Stoichiometric Constants		
Heterotrophs		
Yield	0.666	default
N content of active mass (g/ N/g COD)	0.068	default
N content of endogenous mass (g N/g COD)	0.068	default
endogenous fraction	0.080	default
Autotrophs		
Yield	0.150	default
N content of active mass (g/ N/g COD)	0.068	default
N content of endogenous mass (g N/g COD)	0.068	default
endogenous fraction	0.080	default
Sludge		
MLSS BOD5/BOD20	0.66	calibration
MLSS COD/VSS	1.90	calibration
MLSS VSS/TSS	0.86	data
Influent		
influent BOD5/BOD20	0.66	calibration
influent COD/VSS	1.73	data
influent VSS/TSS	0.85	data
influent N-NH4/TKN	0.54	data
influent filtered BOD5/BOD5	0.34	calibration
Kinetic Constants		
Heterotrophs		
max. specific growth rate (l/d)	2.50	calibration
half saturation coefficient (mg COD/l)	11.0	calibration
organism decay rate (l/d)	0.62	default
anoxic hydrolysis factor	0.37	default
max. specific hydrolysis rate (l/d)	2.81	default
hydrolysis half saturation	0.15	default
ammonification rate (l/mg COD d)	0.016	default
Autotrophs		
max. specific growth rate (l/d)	0.43	calibration
half saturation decay rate (mg COD/l)	1.0	default
organism decay rate (l/d)	0.04	default
aerobic/anoxic growth (mg O2/l)	0.002	default
ammonia limit (mg N/l)	0.05	default
nitrate limit (mg N/l)	1.0	default

DYNAMIC CALIBRATION

After optimising the model parameters during the two day calibration period, an extended simulation was conducted in order to validate the model. Results for the four days are plotted in Figs. 2-5. It was found that in this period the model was slightly less sensitive than the recorded data. Three reasons can account for this:

1) the model lumps together the primary settlers, East and West aeration tanks and secondary clarifiers, while the data were more discrete;
2) the influent composition was modelled as a smooth sinusoidal function rather than the real diurnal pattern (which was unfortunately not available);
3) the model did not account for any operational changes (if they occurred)

Even the simple model tracks an impressive number of parameters (211 state variables, and 150 composite variables, e.g. BOD, MLSS, at several locations), but space only allows a few to be shown. Figure 2 shows sewage flow with its diurnal peak to low ratio of about 2. There is a sharp increase in flow starting around 9-10 a.m., presumably the morning loading appearing on the plant after a few hours delay.

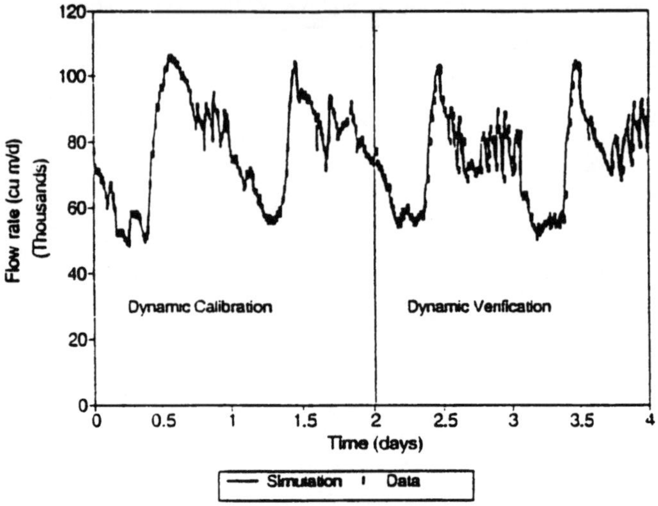

Fig. 2. Influent flow rate.

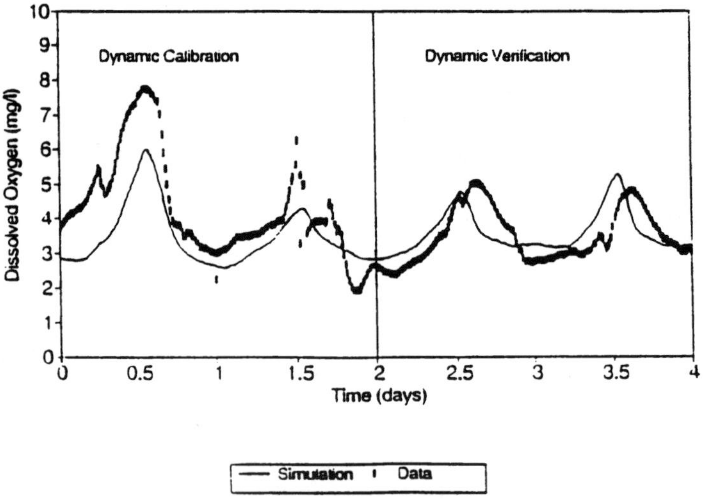

Fig. 3. East DO.

Figure 3 shows the DO values in the last pass of the east aeration tanks. The 1-2 mg/l difference in the first 12 h of calibration could not be reproduced by the model and is suspected to be a DO probe problem which was corrected after 12 h. The simulated DO results show a small time shift compared to the real data, which is attributable to the currently used simplified one-dimensional settler model.

The final effluent SS is shown in Fig.4. It seems the model is able to calibrate to the changes during the first two days, but went out of phase during the verification period. No explanation for this is offered; however the very low concentration range (3-5 mg/l) is simulated properly and the discrepancy is only 1-2 mg/l, which is within the range of noise usually encountered in the case of final effluent solids. Table 5 gives model parameters for primary and secondary settlement.

TABLE 5. Settling velocity model parameters.

Model parameters	Primary	Secondary	Source
max. settling velocity (m/d)	110	300	default
Veslind settling velocity (m/d)	300	500	default
hindered settling coefficient (10^4xm^3/g)	2.6	50	calibrated
flocculent settleable coefficient (10^4xm^3/g)	4.7	100	calibrated
min. attainable solids (mg/l)	50		calibrated
non-settleable fraction (10^3)		1.0	calibrated

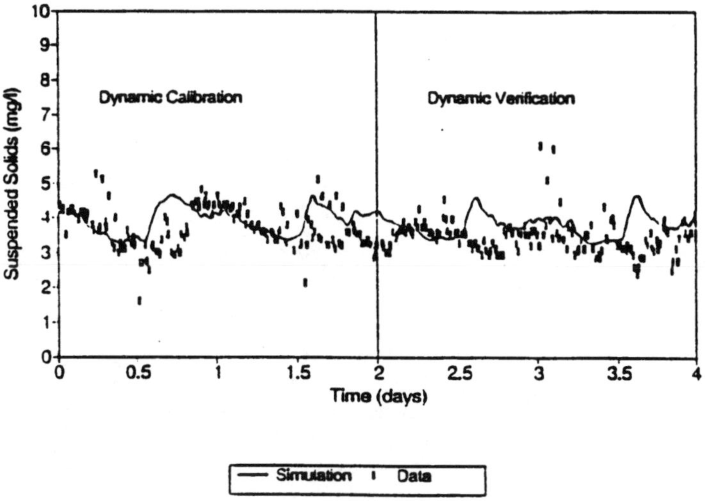

Fig. 4. Effluent suspended solids.

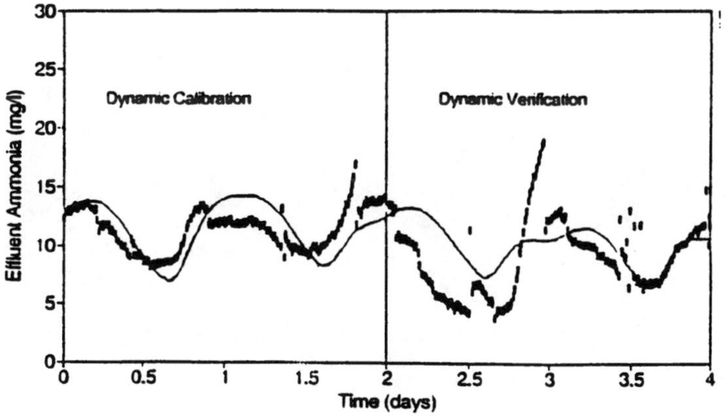

Fig. 5. Effluent ammonia.

Figure 5 shows the final effluent ammonia. This probe can be seen to recalibrate every 12 hours, offering a very interesting possible development of the use of simulation models - ie., using model predictions to initiate self-calibration of the probes. The probe was obviously drifting around 1.8 and 2.9 days which the model was able to detect. Also, it seems that during the period from 2.1 days to 2.5 days (a period of almost 10 hours) the probe might have been off by 3- 5 mg/l, then it recalibrated itself to the correct value. An on-line model could have given warning signals for the maintenance personnel to check if the deviation is caused by the probe or the model's shifting predictions.

CASE STUDIES

Two separate case studies were undertaken with the calibrated model. The first was set-up to study the effect of a SSVI rise. A 4-d simulation was run, during which the SSVI changed instantaneously from 60 to 150 ml/g after 1 d. As the SSVI increased there was a small corresponding increase in effluent SS (about 2 mg/l) and a larger increase in the sludge blanket height. This indicates that the clarification efficiency was still acceptable at the higher SSVI, although the sludge does not compact so well and the sludge blanket level increased to around 1 m from the bottom.

The second case simulated a RAS pump failure and corresponding shut- down of a clarifier. Again a 4-d simulation was made with the pump failure occurring after 1 d. The results indicate that the plant has the capacity to operate for extended periods of time with 3 settling tanks on one side. There was only a small increase in effluent SS (2 mg/l), and the clarifier sludge blanket remained stable despite the increased load. Recently the plant operation has been changed to use only 6 clarifiers (for the first time in 22 years) without problems.

CONCLUSIONS

Models and hardware have developed to a stage where a complex ASP sewage works can be simulated reliably and accurately. The simulation is very fast as a 1 d simulation only takes about 15 s on a 56 MIPS workstation, and at this speed even complex mechanistic models can be optimised.

To work well, reliable autocalibrating, on-line instrumentation is a must to gain maximum benefits.

User friendly inputs, outputs and model building/amending features are seen as essential; doubly so as the end users will often be the operational management.
A large amount of time is spent on parameter optimisation, and current study continues to develop automated methods to reduce time involved and simplify the procedures. Specific Oxygen Uptake Rate (SOUR) from on-line respirometers is to be used to improve modelling.

The simulation can serve as a tool for:

plant optimisation, prediction, on-line warnings and flagging monitoring failures;
investigate what-if scenarios;
operator training; and
design of new plant and upgrading existing ones.

REFERENCES

Chambers, B. and Jones, G. L. (1988). Optimising and updating of activated sludge plants by efficient process design. *Wat. Sci. Tech.*, **20**(4/5), 121-132.

Dold, P. L. (1990). Incorporation of biological excess phosphorus removal in a general activated sludge model, *Proc. 13th Int. Symposium on Wastewater Treatment*, Montreal, Canada, pp. 83-113.

Farrimond, M. S. and Watts, J. B. (1991). A computer monitoring system which integrates intelligent sensors with process knowledge, so allowing the support of site management by remotely located experts, *Proc.64th-Ann.Conf. W.P.C.F.*

Henze, M., Grady, C. P. L., Gujer, W., Marais, G. v. R. and Matsuo, T. (1987). Activated sludge Model No. 1. *IAWPRC Sci. and Tech. Report No.1*, IAWPRC, London.

Patry, G. G. and Takács, I. (1992). Settling of flocculent suspensions in secondary clarifiers. *Wat. Res.*, **26**(4), 473-479.

Snaddon, X. V. M., Dee, J., Lax, F., Augustine, F. and Smith, H. (1980). Continuous monitoring at Minworth Reclamation Works. *J. Inst. W. Eng. & Sci.*, **31**(1), 77-92.

Takács, I., Patry, G. G. and Nolasco, D. (1991). A dynamic model of the clarification-thickening process. *Wat. Res.*, **25**(10), 1263-1271.

Watts, J. B., Evans, K. and Molloy, A. (1990). A wastewater/water process monitoring system allowing operational control – developed by an operational manager. *Wat. Sci. Tech.*, 7/8, 43-154.

Watts, J. B. and Garber, W. S. On-line Respirometry: A Powerful for Activated Sludge Plant Operation and Design. *6th IAWPRC workshop on ICA of Water and Wastewater Treatment and Transport Systems.*

Wentzel, M. C., Ekama, G. A., Marais and G. v. R. (1992). Processes and modelling of Nitrification Denitrification Biological Phosphorus Removal Systems – A Review. *Wat. Sci. Tech.* **25**(6), 59-82.

Pergamon

Wat. Sci. Tech. Vol. 28, No. 11-12, pp. 163–171, 1993.

0273–1223/93 $7·00 + 0·00

MODEL CALIBRATION FOR THE HIGH-PURITY OXYGEN ACTIVATED SLUDGE PROCESS – ALGORITHM DEVELOPMENT AND EVALUATION

Weibo (Weber) Yuan*, David Okrent** and Michael K. Stenstrom*

* *Civil and Environmental Engineering Department, University of California, Los Angeles, CA 90024, USA*
** *Mechanical, Aerospace and Nuclear Engineering Department, University of California, Los Angeles, CA 90024, USA*

ABSTRACT

A model calibration algorithm is developed for the high-purity oxygen activated sludge process (HPO-ASP). The algorithm is evaluated under different conditions to determine the effect of the following factors on the performance of the algorithm: data quality, number of observations, and number of parameters to be estimated. The process model used in this investigation is the first HPO-ASP model based upon the IAWQ (formerly IAWPRC) Activated Sludge Model No. 1. The objective function is formulated as a relative least-squares function and the non-linear, constrained minimization problem is solved by the Complex method. The stoichiometric and kinetic coefficients of the IAWQ activated sludge model are the parameters focused on in this investigation. Observations used are generated numerically but are made close to the observations from a full-scale high-purity oxygen treatment plant. The calibration algorithm is capable of correctly estimating model parameters even if the observations are severely noise-corrupted. The accuracy of estimation deteriorates gradually with the increase of observation errors. The accuracy of calibration improves when the number of observations (n) increases, but the improvement becomes insignificant when n>96. It is also found that there exists an optimal number of parameters that can be rigorously estimated from a given set of information/data. A sensitivity analysis is conducted to determine what parameters to estimate and to evaluate the potential benefits resulted from collecting additional measurements.

KEYWORDS

Activated sludge; model; calibration; complex; oxygen; parameter estimation.

INTRODUCTION

The high-purity oxygen activated sludge process (HPO-ASP) was invented in the late 1960s, and has been widely used since then in large, mature cities with developed infrastructures and where land available for expansion is difficult to find. Many advantages of the HPO-ASP have been claimed over the air systems, such as more active biomass, better settling characteristics of sludge, etc. From the perspective of oxygen transfer, the most important advantage is the tremendous increase of oxygen transfer rate (OTR). The typical average oxygen transfer driving force for a HPO-ASP system operating at a mixed liquor D.O. level of 6 mg/l is approximately 25 mg/l as compared with 7 mg/l for an air system operating at a D.O. level of 2 mg/l.

This means that on an overall basis the high-purity oxygen activated sludge process will transfer about 3.6 times as much oxygen as an air system.

The HPO-ASP is more complicated than an air system. The complexity mainly results from the use of high purity oxygen (purity over 97%) and multi-stage (usually 3 to 4 stages), closed aeration tanks. Figure 1 shows the flow diagram of a typical HPO-ASP.

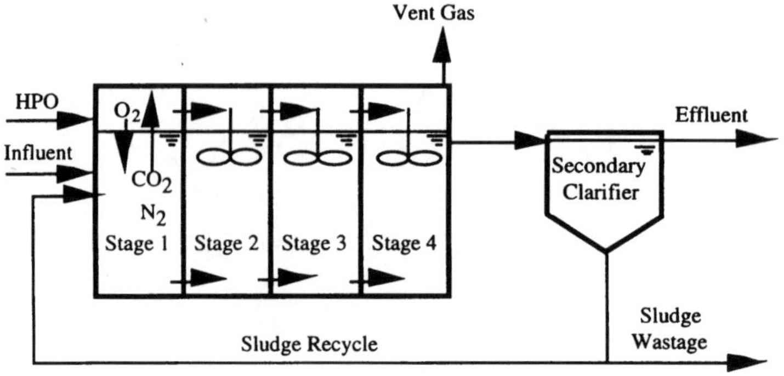

Fig. 1. High-purity oxygen activated sludge process.

The Sacramento Regional Wastewater Treatment Plant, which has a treatment capacity of approximately 6.57 m^3/sec, was used as a model for this investigation.

Process Model. Process models have been developed and applied to better understand and control the sophisticated process. An incomplete list of recent publications on this topic is Clifft and Andrews (1986), Stenstrom *et al.* (1989), Stenstrom (1990), Clifft and Barnett (1990), Clifft (1992), and Tzeng (1992).

The HPO-ASP model used in this study is a new process model which was developed based on the IAWQ Activated Sludge Model No. 1 (Henze, *et al.*, 1987; Yuan, 1993). None of the above publications used Activated Sludge Model No. 1.

The process model was applied to static conditions in this research. Sinusoidal excitation was applied to the process model to mimic the diurnal change in influent flow and substrate.

Model Calibration. Model calibration represents the process of assigning values to the parameters used in the model such that the error between the model predictions and the observations from the real plant is minimal. Model calibration has been done by hand in many cases in the past, but is difficult for process models that have a large number of parameters. Activated Sludge Model No. 1, after being simplified for the HPO-ASP, has 11 biokinetic parameters - 4 stoichiometric coefficients and 7 kinetic coefficients. The total number of parameters of the process model is well over 30. The necessity for developing a model calibration algorithm is obvious. Tzeng (1992) performed the first model calibration that used observations from a full-scale oxygen activated sludge treatment plant. The particular process model that Tzeng calibrated uses the Clifft and Andrews structured activated sludge model (Clifft and Andrews, 1986).

Many questions need to be answered before a model calibration algorithm is developed and the corresponding calibration is implemented, such as: What observations are to be used in the calibration? How many data points are adequate for the calibration? How many parameters can be rigorously estimated from a given set of observations? The answers to these questions are crucial to the success of the calibration.

The objective of this study is to present the development and evaluation of a model calibration algorithm for the high-purity oxygen activated sludge process. More specifically, we intend to find out the effects of the following factors on the performance of the calibration algorithm: data quality, number of data points, and the number of measured variables. The investigation is accomplished through numerical experiments in which the observations are generated numerically by corrupting model predictions with Gaussian noise. By doing so, we guarantee that the model is not the cause if a good fit is not achieved. We also have the advantage of knowing the 'true' values of the parameters and have the flexibility of adjusting the data quality (e.g. greater or lesser variability in data).

MODEL CALIBRATION METHOD

The basic approach in all calibrations is usually the same: An objective function (or merit function) is designed that measures the agreement between the data and the model with a particular choice of parameters. The objective function is conventionally arranged so that small values represent close agreement. The parameters of the model are then adjusted to achieve a minimum in the objective function. The adjustment process is thus a problem in minimization in many dimensions.

Objective function. Least-squares is the most common form of objective function used. Three kinds of least-squares objective functions can be used: absolute least-squares (ALS) (Eq. 1), weighted absolute least-squares (WALS) (Eq. 2), and relative least-squares (RLS) (Eq. 3). ALS is not appropriate for the activated sludge process because the magnitudes of observations differ significantly. For example, the ranges of some observations in a full-scale HPO-ASP are: DO 5-12 mg/l; vent gas oxygen purity 0.40-0.80 (oxygen mole fraction); MLSS 1000-2000 mg/l. One possible solution to the uneven-weight problem associated with the ALS objective function is WALS which solves the problem by assigning an appropriate set of weights to different variables measured. Another possible solution is to normalize the function by dividing the difference between the model-predicted and observed values by the model-predicted value. This is RLS (Eq. 3). Eq. 4 represents the relative least-squares objective function that uses observations of 1st stage DO, 4th stage DO, vent gas oxygen purity, and MLSS.

$$f=\sum_{j=1}^{m}\left(\sum_{i=1}^{n}\varepsilon_{a_{ij}}^{2}\right)=\sum_{j=1}^{m}\left(\sum_{i=1}^{n}(S_{ij}-S_{ij}^{*})^{2}\right) \qquad \text{(Eq. 1)}$$

$$f=\sum_{j=1}^{m}\left(\sum_{i=1}^{n}\gamma_{ij}\varepsilon_{a_{i}}^{2}\right)=\sum_{j=1}^{m}\left(\sum_{i=1}^{n}\gamma_{ij}(S_{ij}-S_{ij}^{*})^{2}\right) \qquad \text{(Eq. 2)}$$

$$f=\sum_{j=1}^{m}\left(\sum_{i=1}^{n}\varepsilon_{r_{ij}}^{2}\right)=\sum_{j=1}^{m}\left(\sum_{i=1}^{n}\left(\frac{S_{ij}-S_{ij}^{*}}{S_{ij}^{*}}\right)^{2}\right) \qquad \text{(Eq. 3)}$$

$$f=\sum_{i=1}^{n}\left(\left(\frac{p_{O_{2_4}}^{*}-p_{O_{2_4}}}{p_{O_{2_4}}^{*}}\right)_i^2+\left(\frac{S_{O_{2_1}}^{*}-S_{O_{2_1}}}{S_{O_{2_1}}^{*}}\right)_i^2+\left(\frac{S_{O_{2_4}}^{*}-S_{O_{2_4}}}{S_{O_{2_4}}^{*}}\right)_i^2+\left(\frac{MLSS^{*}-MLSS}{MLSS^{*}}\right)_i^2\right) \qquad \text{(Eq. 4)}$$

where S - the observed variable; i - as subscript, denotes the ith observation; j - as subscript, denotes the jth observed variable; * - as superscript, denotes predicted values; m - total number of observed variables; n - total number of observations; γ - weight; ε_a - absolute error; ε_r - relative error. The model includes a mass balance on the gas phase so that the oxygen uptake rate (OUR), which is frequently used by other investigators for parameter estimation, is not linearly independent, and its inclusion would not add further information, except to identify errors in other data. Only limited OUR data are available from the plant, and is of questionable value since it is infrequently collected.

Most measurements in biological systems have normally or log-normally distributed relative errors instead of absolute errors. Saez and Rittmann (1992) demonstrated that the relative least-squares objective function (Eq. 3) is superior to the absolute least-square criterion (Eq. 1) where observations (state variables) differ significantly in magnitude, such as in the activated sludge process.

Constraints. The parameters in the HPO-ASP model have clear physical implications. For example, the heterotrophic yield Y_H must have a value between 0 and 1. In reality, Y_H has an even narrower bound, such as between 0.5 and 0.8. This is true for all parameters in the process model. To ensure that the parameter values selected by the calibration program are physically feasible, it is necessary to impose constraints on them. In the calibration program, the constraints are given in the form of upper and lower bounds. With well defined parameter bounds, it takes less iterations for the objective function to converge to its minimum. However, if the calibration algorithm is robust, the constraints should not be too stringent, that is, the range between the upper and lower bound should not be too narrow.

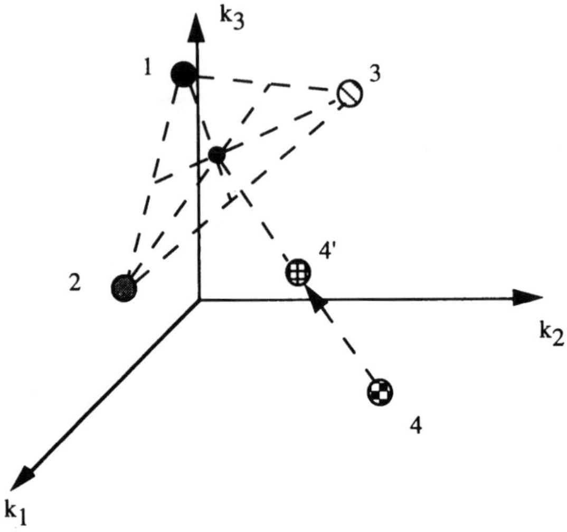

Fig. 2. Illustration of the complex method.

Observations. The observations used in this study are generated mathematically as follows. For a state variable S at time t, the relative error of the observation, S_i, to the exact solution S_i^*, is defined as:

thus $S_i = S_i^* + \varepsilon_r S_i^*$. If ε_r is normally distributed, with mean $\mu = 0$ and variance $= \tau$, then the observation S_i is also normally distributed, with mean $\mu = S_i^*$, and variance $= \tau\, S_i^*$. If $\tau = 0.10$, this means the variance of the observation equals to 10% of the exact solution S_i^*.

Optimization method. After the objective function and constraints of parameters are defined, the calibration problem is transformed into a nonlinear, constrained optimization problem. Many optimization techniques are available for this purpose, such as the Gauss-Newton method. The particular algorithm used in the research is the Complex Method. The Complex method was developed in 1965 by Box (Box, 1965). Its derivation is not rigorously based on mathematical theory. It is an empirical method. It, however, has been proved to be a very robust optimization algorithm for many engineering problems (Little and Williams, 1992). The Complex method is derivative-free, and thus remarkably robust to messy problems such as discontinuities and difficult-to-evaluate or nonexistent derivatives.

Figure 2 illustrates the basic idea of the Complex method for a 3-parameter system. Each point (k_1,k_2,k_3) in 3-D space represents a set of parameters to be estimated. The method requires at least four sets of parameters to be initially selected, and scatters them randomly in the domain. For this 3-parameter system, four sets of parameters are chosen. They are denoted as points 1, 2, 3, 4 in Figure 2. Mathematically, this is implemented as:

$$\theta_{i,j} = L_i + r_{i,j}\,(U_i - L_i)$$

where

i = as subscript, denotes the ith parameter ($i=1,2, ..., N$),
j = as subscript, denotes the jth set of parameter ($j=1, 2, 3, ..., K$),
N = number of parameters,
K = number of parameter sets,
$r_{i,j}$ = a random real number between 0 and 1,
$\theta_{i,j}$ = value of the ith parameter in the jth set of parameters,
U_i = upper bound of the ith parameter, and
L_i = lower bound of the ith parameter

Each point (each set of parameters) is used to make predictions and the objective function is calculated from model predictions and observations. The point (set of parameters) which causes the largest objective function (largest error) is identified, say, as point 4. Point 4 is then moved half way to the middle point between the centroid of points 1, 2, and 3 and point 4. Now point 4 becomes point 4'. The parameter values of point 4' are used to make predictions, the objective function is calculated and its value is compared to those of points of points 1, 2, and 3. A highest value is found, and that point is adjusted through the same procedure as described. This is iterated again and again, until all sets of parameters give the same function value (Kuester and Mize, 1973).

Other. The initial conditions and inputs used for observation generation and prediction are the same. Sinusoidal inputs are used to represent the diurnal changes in influent quantity and quality. Tabulated in Table 1 are the values of the stoichiometric and kinetic coefficients used for observation generation.

ALGORITHM EVALUATION

The effects of the following factors on the performances of the calibration algorithm - quality of observations, number of observations, number of measured variables, and number of parameters estimated- are evaluated in the following case studies.

Quality of observations

The quality of observations is indicated by the values of the variances of the Gaussian noises. Hourly observations of 1st stage DO, 4th stage DO, vent gas oxygen purity, and MLSS are used for calibration. A total of 60 hours of hourly observations is used. The variance of the Gaussian noises is set to 0.10, 0.20, 0.30, and 0.40 respectively.

Number of observations

The observations used are the same as in A except the number of observations are changed from 24, 48, 72, to 96. This analysis is intended to find the optimum sampling period which meets the requirement of correctly calibrating the model but avoids wasting resources by not taking too many observations. The variance of the Gaussian noise is fixed as 0.20 in this analysis.

TABLE 1. Parameters in the simplified IAWPRC Activated Sludge Model Symbol.

Symbol	Name	Unit	Value[a]
		Stoichiometric parameters	
Y_H	yield for heterotrophic biomass	g Cell COD formed/g substrate COD oxidized	0.67
f_p	fraction of biomass yielding particulate products	dimensionless	0.08
i_{XB}	mass N/Mass COD in biomass	g N(g COD in biomass)-1	0.086
i_{XP}	Mass N/Mass COD in products from biomass decay	g N (g COD)-1 in endogenous mass	0.06
		Kinetic Parameters	
μ_H	Maximum specific growth rate of heterotrophs	day-1	6.0
K_S	Half-velocity coefficient of heterotrophs	g COD m^{-3}	20.0
K_{OH}	oxygen half-saturation coefficient for heterotrophs	g O_2 m^{-3}	0.20
b_H	decay coefficient of heterotrophic biomass	day^{-1}	0.62
k_a	ammonification rate	$m^3 \cdot (g\ COD.day)^{-1}$	0.08
k_h	maximum specific hydrolysis rate	g slowly biodegradable COD (g cell COD · day)$^{-1}$	3.0
K_X	half-saturation coefficient for hydrolysis of slow biodegradable substrate	g slowly biodegradable COD (g cell COD)$^{-1}$	0.03

[a] typical value at 20°C

Number of measured variables

The observations used here are the same as above except that the measurements of ammonia nitrogen are added. The results obtained are compared with those in which ammonia nitrogen is not considered in the objective function.

RESULTS AND DISCUSSION

A. The relationship between the variances of the relative errors and the quality of data may be explained as follows: the larger the values of the variances are, the worse the quality of the data. The results of this study are shown in Figure 3. The line denoted as 'exact' represents the exact solution of process model. The line denoted as 'predicted' represents the solutions of the process model using the estimated parameter values. For a given set of observations of reasonable quality, with a reliable calibration, the 'estimated' should be very close to the 'exact'. As we can see from Figure 3, the 'estimated' fits the 'exact' quite well in all cases, though it deteriorates gradually and slowly with increasing τ.

B. The results of this analysis are shown in Figure 4 and Table 2. There is no significant difference between n=24 and n=96 so far as the fit between the 'predicted' and the 'exact' solutions is concerned, but it is true the fit improves as the number of observations increases. The estimated parameters show the same pattern: the relative error between the estimated value and the 'true' value decreases when the number of observations increases (Table 2).

Theoretically, for a longer observation period, a more accurate calibration should be obtained. In the real world, however, the observation period is always limited to a finite time interval due to the limitation in resources. This analysis provides useful information when a decision needs to be made on a possible tradeoff between accuracy and resources.

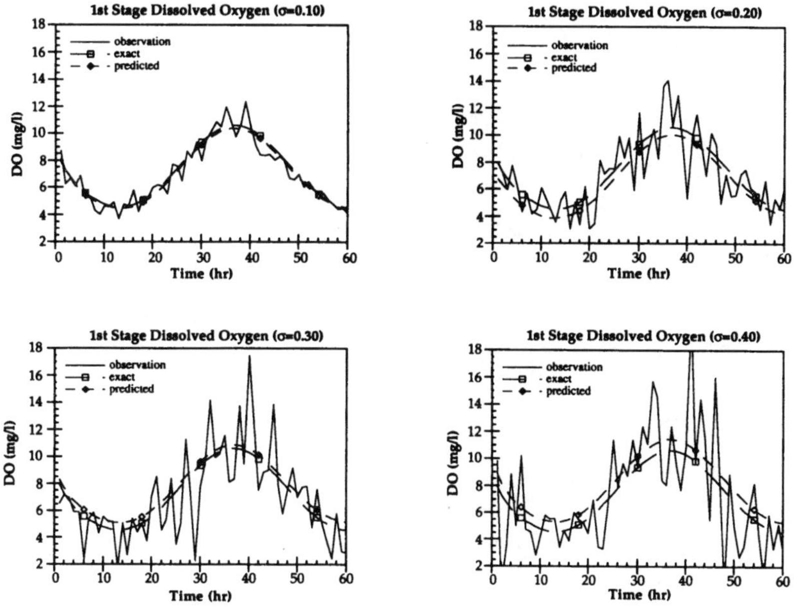

Fig. 3. Effect of on 1st stage DO calibration.

TABLE 2. Relative Error of Parameter Estimation for Different n (Number of Observations).

parameter	n=24	n=48	n=72	n=96
Y_H	5.76%	2.05%	1.34%	0.36%
μ_H	8.50%	4.34%	2.61%	0.79%

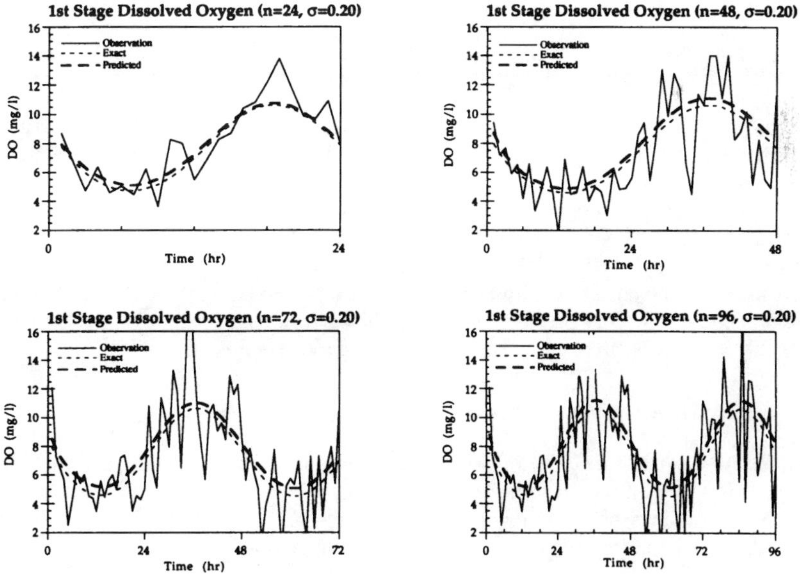

Fig. 4. First stage dissolved oxygen for different n.

C. The results of A and B reveal that some parameters, such as i_{XB} and i_{XP}, cannot be accurately estimated if the objective function only includes 1st stage DO, 4th stage DO, vent gas oxygen purity, and MLSS (Table 3).

TABLE 3. Relative errors between the 'true' and 'estimated' values of parameters

parameter	s=0.10	s=0.20	s=0.30	s=0.40
Y_H	0.60%	1.34%	0.10%	2.30%
i_{XB}	25.47%	35.58%	49.88%	49.19%
i_{XP}	2.83%	12.30%	49.83%	16.00%

A steady-state sensitivity analysis is then conducted to systematically identify the sensitivities of observations to the changes of parameters. The sensitivity of the *ith* state variable S_i with respect to the j_{th} parameter τj is defined by the following formula:

$$\Sigma(S_i,\theta_j) = \frac{\frac{\partial S_i}{S_i}}{\frac{\partial \theta_j}{\theta_j}} \cong \frac{\frac{\Delta S_i}{S_i}}{\frac{\Delta \theta_j}{\theta_j}}$$

From the definitions of i_{XB} and i_{XP}, it is natural to conclude that i_{XB} and i_{XP} have little effect on 1st stage DO, 4th stage DO, and 4th stage oxygen purity (vent gas oxygen purity), but strongly affect the nitrogen balance in the system. Table 4 confirms this point. In contrast, DO_1 and DO_4 are much more sensitive to Y_H and μ_H. Vent gas oxygen purity is not sensitive to any parameter, thus its measurements are of limited value for model calibration.

TABLE 4. Sensitivities

$\Sigma(DO_1,i_{XB}) = 5.8306\times10^{-4}$	$\Sigma(DO_4,i_{XB}) = 6.1926\times10^{-4}$	$\Sigma(p_{O_{24}},i_{XB}) = 1.6484\times10^{-4}$
$\Sigma(DO_1,i_{XP}) = -1.9889\times10^{-4}$	$\Sigma(DO_4,i_{XP}) = -4.3348\times10^{-4}$	$\Sigma(p_{O_{24}},i_{XP}) = 3.5604\times10^{-3}$
$\Sigma(DO_1,Y_H) = 2.4264\times10^{-2}$	$\Sigma(DO_4,Y_H) = 6.0553\times10^{-2}$	$\Sigma(p_{O_{24}},Y_H) = 4.6264\times10^{-3}$
$\Sigma(DO_1,\mu_H) = 7.6667\times10^{-3}$	$\Sigma(DO_4,\mu_H) = -3.5814\times10^{-3}$	$\Sigma(p_{O_{24}},\mu_H) = 5.0330\times10^{-3}$

It is obvious that for a given set of information (observations) only a certain number of parameters can be rigorously estimated. If the observations are not sensitive to a parameter, then the parameter cannot be rigorously estimated from the given observations. On the other hand, if a parameter is to be estimated, only observations sensitive to the parameter should be used. If we intend to estimate i_{XB} and i_{XP}, additional observations, such as measurements of ammonia nitrogen, should be used.

This analysis is useful to the design and implementation of a model calibration. It provides insights into the algorithm, and lets one know what parameters can be reliably estimated from given observations, and what possible improvements may be attained by adding additional measurements.

SUMMARY AND CONCLUSIONS

A model calibration algorithm is developed for the high-purity oxygen activated sludge process. The objective function is formulated as a relative least-squares function and the optimization method is the Complex Method. The effects of the following factors on the performance of the algorithm are evaluated: data quality, number of observations, and number of estimated parameters.

The following conclusions are drawn from this study.

1. The model calibration algorithm is robust to the quality of observations. As long as the errors are random (no systematic errors), the algorithm is capable of estimating the parameters to an acceptable accuracy.

2. The accuracy of calibration improves as the number of observations increases. However, the improvement slows down as the number of observations increases. No significant improvement is observed beyond n = 96 (hourly observations of 4 days). In this study n = 24 (hourly data of 1 day) was able to estimate Y_H and μ_H within 10% of their true values. For applications in which 10% is an acceptable estimation error, hourly sampling for 24 hours should be an adequate sampling scheme.

3. For a given set of observations, only a certain number of parameters can be reliably estimated. A sensitivity analysis should be performed before the calibration is performed to determine what parameters are to be estimated. If additional variables are to be measured and used for calibration, the potential benefits resulted from the addition should be quantitatively evaluated in advance.

REFERENCES

Box, M. J. (1965). A New Method of Constrained optimization and a comparison with other methods, *Computer J.*, **8**, 42-52

Clifft, R. C. (1992). Gas Phase Control for Oxygen-Activated Sludge, *J. Envir. Engrg.*, ASCE, **118**(2), 390-401

Clifft, R. C. and Barnett, M. W. (1988). Gas Transfer kinetics in Oxygen Activated Sludge, *J. Envir. Engrg.*, ASCE, **114**(2), 415-432

Henze, M., Grady, Jr., C. P. L., Gujer, W., Marais, G. v. R. and Matsuo, T. (1987). Activated Sludge Model No. 1, Scientific and Technical Reports No. 1, IAWPRC (now IAWQ).

Kuester, J. L. and Mize, J. H. (1973). *Optimization Techniques with FORTRAN*, McGraw-Hill Book Company.

Little, K. W. and Williams, R. E. (1992). Least-square calibration of QUAL2E, Water Environ. Res., **64**(2), 179-185

McWhirter, J. R. (1978). Oxygen and the activated sludge process, In: *The use of high-purity oxygen in the activated sludge process*, J. R. McWhirter, ed., Vol. 1, CRC Press.

Saez, P. B. and Rittmann, B. E. (1992). Model parameter estimation using least squares, *Wat. Res.*, **26**(6), 789-796

Stenstrom, M. K. (1990). Westpoint Treatment Plant Oxygen Process Modeling, UCLA ENG 90-17, School of Engineering and Applied Science, University of California, Los Angeles

Stenstrom, M. K., Kido, W. Shanks, R. F. and Mulkerin, M. (1989). Estimating oxygen transfer capacity of a full-scale pure oxygen activated sludge plant, *J. Water Pollut. Control Fed.*, **61**(2), 208-220

Tzeng, Chwen-Jeng (1992). Advanced Dynamic Modeling of the High-purity Oxygen Activated Sludge Process, Ph.D. dissertation, University of California, Los Angeles

Yuan, Weibo (Weber) (1993). Dynamic Models and Expert Systems for the Activated Sludge Process, Ph.D. dissertation, University of California, Los Angeles (in preparation)

Pergamon

Wat. Sci. Tech. Vol. 28, No. 11-12, pp. 173–183, 1993.

0273–1223/93 $7·00 + 0·00

REDUCED ORDER MODELS FOR ON-LINE PARAMETER IDENTIFICATION OF THE ACTIVATED SLUDGE PROCESS

U. Jeppsson and G. Olsson

Department of Industrial Electrical Engineering and Automation (IEA), Lund Institute of Technology (LTH), P.O. Box 118, S-221 00 Lund, Sweden

ABSTRACT

A reduced order dynamic model for an activated sludge process performing carbonaceous removal, nitrification, and denitrification is proposed. Based on directly measurable real time data – by methods available today – all model parameters are identified on-line. A simplified extended Kalman filter is used for the actual identification. Verification of the results is based on computer simulations of the *IAWQ Activated Sludge Model No. 1*. The reduced order model presented herein may serve as a tool for predicting the dynamic behaviour of a biological wastewater treatment plant since the parameters under varying operating conditions can be tracked on-line. The model parameters are effectively estimated even when the measurements are affected by a significant level of noise. The model is aimed for operation and control purposes as an integral part of a hierarchical control structure.

KEYWORDS

Activated sludge; dynamic modelling; simulation; model calibration; model reduction; parameter estimation; extended Kalman filtering.

INTRODUCTION

The complexity of wastewater treatment processes has increased dramatically over the last twenty years. In order to determine the most suitable design, optimize the control and predict the behaviour of full scale plants under varying operating conditions, computer simulations are today increasingly used (Patry-Chapman, 1989, Andrews, 1992). During the same period dynamic models of enhanced complexity have been developed in order to incorporate a more detailed description of the reactions and phenomena taking place, particularly in the biological reactors (Dold *et al.*, 1980, Henze *et al.*, 1987, Dold, 1992). In comparison there is a lack of structure in many thickener/clarifier models.

The high complexity of the models causes a severe problem of identifiability and verifiability (Godfrey-DiStefano, 1985; Jeppsson, 1993). Often the models are derived from simpler unit operations and later combined to large plant models. The parameter values consequently may not be the same. Thus, the parameter identification problem becomes an awkward task. Several parameter combinations can often explain the same dynamical behaviour. This is further accentuated when the influent wastewater composition is taken into consideration; a change in its characteristics can quite often be explained by kinetic parameter changes.

Even if a major problem of the models has to do with the complex structure and the too large number of states and parameters to be uniquely identified, instrumentation problems amplify the difficulties. Simply stated, available on-line instrumentation and laboratory procedures are usually not adequate to verify the details of such a complex model. Furthermore, for a reliable identification result the true process has to be perturbed (or purposefully disturbed) in such a way that all the interesting dynamical modes are excited. This creates a demand not only in amplitude but also in the time frame of the disturbances. Another problem is that results from experiments on full scale plants are seldom reproducible and the process inputs (incoming flow rate and wastewater characteristics) cannot be manipulated and controlled in the way that would be needed.

Though most available models are highly complex they are still greatly simplifying the representation of many species of organisms. As the microbial population changes this would be reflected in changing kinetic parameters and new added state variables (Cech *et al.*, 1985). For example, filamentous organisms ought to be represented during many operating conditions. Consequently, parameters of an accurate model have to be readily updated on-line, since they are in fact varying with time (the adaptive behaviour of the organisms) and with changing operating conditions. Their values, determined in bench or pilot scale experiments should be regarded as suitable 'initial values'.

The purpose of our research is to develop simple dynamical models for the activated sludge system, which are suited for operation and control. Such a model has to be sufficiently complex to describe major phenomena but still so limited that its parameters can be updated during full scale plant operations. Still we recognize the need for the highly complex models for design purposes which imply that the operational model has to be considered as a special case, either for certain operational levels or for particular time scales. In an operational model it is not always necessary to know the absolute values of certain parameters but more of their relative change.

This paper presents some results of using a simplified model for simulating the activated sludge process, capable of modelling carbonaceous and nitrogeneous activities. Other researchers (e.g. Ayesa *et al.*, 1991, Larrea *et al.*, 1991), have also approached the on-line model identification problem. They have considered another simplified form of the IAWQ Activated Sludge Model No. 1 (Henze *et al.*, 1987), excluding denitrification. The difficulties of complex model representations are illustrated by a simple example. Physical reasoning is used to motivate the structure of the reduced order model and its transient behaviour is compared with the IAWQ model The on-line model calibration is illustrated with an extended Kalman filter (Náhlík-Burianec, 1988, Söderström-Stoica, 1989) using the IAWQ model as the basis for comparison. Finally, the operational implications of the reduced model are considered.

A SIMPLE EXAMPLE

Most of the microbiological kinetics is traditionally modelled using a Monod type structure (Monod, 1942) for the description of the organism growth rate, e.g.

$$\mu_H(S_S) = \frac{\hat{\mu}_H S_S}{K_S + S_S} \tag{1}$$

Let us consider the very basic single-substrate/single-organism batch reactor model,

$$\begin{cases} \dfrac{dX_H}{dt} = \mu_H(S_S)\, X_H - b_H\, X_H \\ \dfrac{dS_S}{dt} = -\dfrac{1}{Y_H}\, \mu_H(S_S)\, X_H \end{cases} \tag{2}$$

without any other growth limitations. The parameters are defined in the nomenclature. *Theoretically*, with perfect measurements of S_S and $X_{B,H}$ all four parameters are simultaneously identifiable (Holmberg,1982). If more realistic noisy measurements are assumed or if measurements are scarce in the sensitive regions, the situation is different.

Figure 1 illustrates the necessary sensitivity of the identification procedure. By simulating the model (2) with apparently different sets of model parameters almost identical responses have been achieved. The results are compared to simulations of the same model, but with white noise added to the measurements (standard deviation 10% of typical state variable values) and yet another set of parameters.

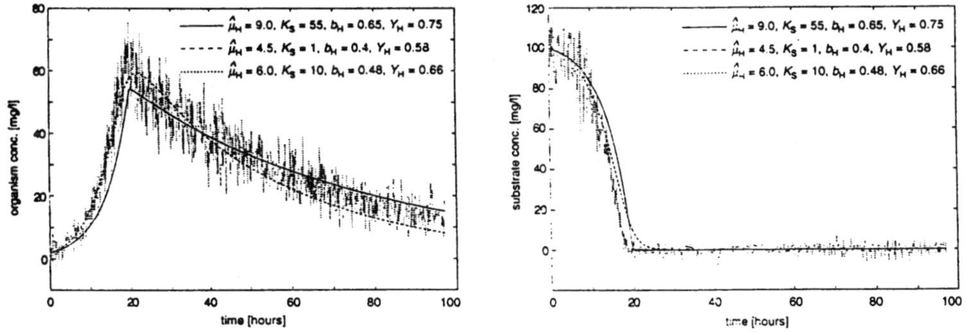

Fig. 1. Simulations of system (2) with very different sets of model parameters.

By applying a 'state-of-the-art' off-line low pass filter with no time lag (Little-Shure, 1988) to the noisy measurements, the data is filtered according to Fig. 2. A robust Nelder-Mead (NM) simplex optimization algorithm (Fletcher, 1987) is then used to determine the 'optimum' parameter set, based on the filtered data. As Fig. 2 shows, the resulting parameters deviate considerably from the true values, even though the optimization is performed under very favourable conditions. As the time instances between the measurements increase, the results deteriorate quickly (Jeppsson, 1993).

The situation demonstrates the basic *practical* identifiability problem when noise is present. One possible way of attacking this is to apply more measuring instruments (i.e. monitor the oxygen uptake rate, OUR) another is to use a simplified Monod formulation, such as:

$$\mu_H(S_S) = \begin{cases} r_H\, S_S & \text{if } S_S < S_{S_{sat}} \\ r_H\, S_{S_{sat}} & \text{if } S_S \geq S_{S_{sat}} \end{cases} \qquad (3)$$

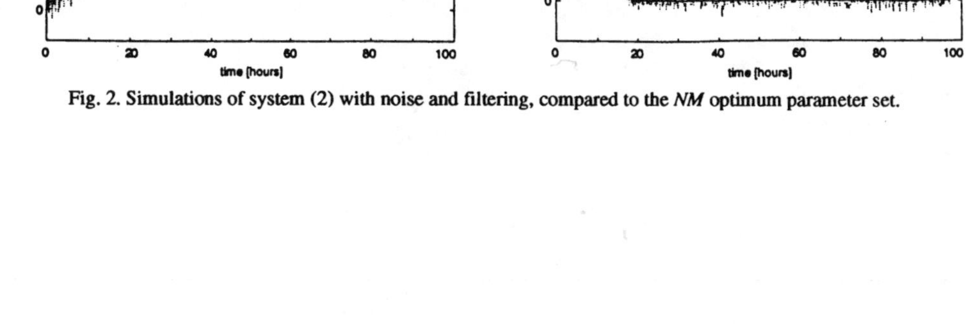

Fig. 2. Simulations of system (2) with noise and filtering, compared to the *NM* optimum parameter set.

This modification is applied to model (2). All four model parameters are again optimized from noisy data and in Fig. 3 the new system output is compared to the original system (with and without noise) for the organism concentration. The convergence is both faster and more reliable which imply that the optimization method can not compensate for an inadequate model structure. Moreover, at most WWT plants receiving municipal wastewater the substrate concentration is often sufficiently low to motivate the use of a first order reaction only. This further simplifies the model.

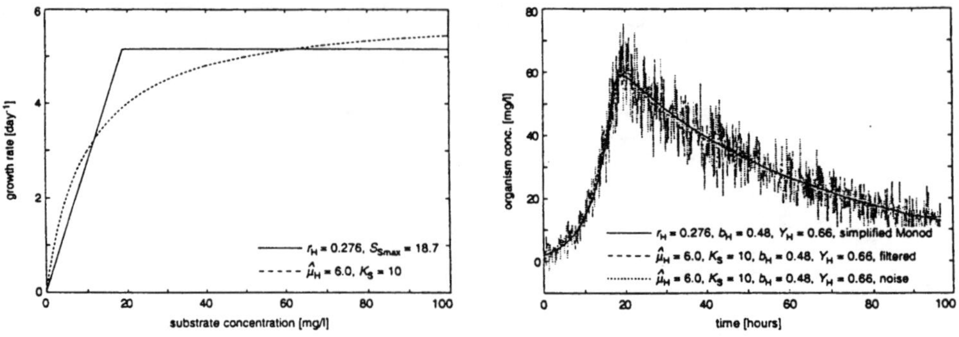

Fig. 3. Simulations of the standard and simplified Monod equations, applied to system (2).

MODEL REDUCTION BY PHYSICAL REASONING

The IAWQ model is the basis for the reduction procedure. The goal of the reduced model is to adequately represent both the carbonaceous and nitrogeneous activities. Some of the problems that exist in the IAWQ model are focused on in the development of the reduced model, for example:

- lack of identifiability (the model may be adjusted to measurement data using different combinations of parameter values and influent wastewater characteristics);
- troublesome non-linearities, such as the Monod and the switching functions;
- difficult estimation and updating of time varying parameters;
- complicated characterisation of the incoming wastewater;
- limited usefulness as a basis for control.

Instead of applying general methods for model reduction we think it is important to make use of physical insight into the process behaviour. This means that major phenomena still should be described in a physically reasonable manner.

Dissolved oxygen (DO) is excluded as a state variable. We assume that the DO is controlled on a routine basis, so that corresponding growth expressions become independent of the DO. Still the DO mass balance contains plenty of useful information. It is the basis for the estimation of the oxygen uptake rate or respiration, which is recognized as fundamental information for further control. The respiration can be estimated using either the whole aerator as a respirometer (Holmberg, 1991) or using separate instrumentation of various construction and philosophy (Spanjers-Olsson, 1992; Vanrolleghem-Verstraete, 1993). Furthermore, the substrate concentration can be estimated from respirometry data (Spanjers *et al.*, 1993). This makes up an important module which needs to be synchronized with an overall model of the reactions and plant behaviour.

The time scale is a crucial factor. The reduced model does not take rapid uptake phenomena into consideration. Rather it makes some averaging of biosorption and growth by combining soluble and stored substrate. Consequently fast dynamics (of the order less than an hour) is neglected. As already remarked, the fast DO dynamics is taken care of separately. By these simplifications the model becomes less stiff, i.e. the ratio between the smallest and the largest dynamical time constant is reduced.

The following quality variables are assumed to be measurable on-line:

biodegradable organic substrate;
ammonia nitrogen;
nitrate nitrogen;
flow rates.

The assumed measurements of the chemical oxygen demand (COD) cannot distinguish between soluble and particulate organic substrate. Since only two nitrogen fractions are considered measurable some of the mechanisms are combined as summary reactions.

In the IAWQ model the hydrolysis of entrapped organic nitrogen is described. This reaction continues via the ammonification process to form ammonia nitrogen. In the reduced model, ammonia nitrogen is assumed measured and its formation mechanism is not considered. The hydrolysis of entrapped organics in the IAWQ model is simplified. Therefore the decay material (of Heterotrophs and Autotrophs) is formed into organic substrate and ammonia directly. Another reason for this simplification is that the true nature of the hydrolysis mechanism is not well known.

As illustrated by the simple example in the previous section there are good reasons to reconsider the parametrization of the Monod and the switching functions. The basic idea is the old approach of approximating the functions by straight lines (Fig. 3), i.e. a first order reaction followed by a zero order reaction. It is useful to note that during most normal operations there is only need for first order reactions. Due to the existing DO control, the DO influence on the switching functions is constant. The switching functions are regarded solely as functions describing growth limitation due to DO or various kinds of substrate. Therefore the parameter in the first order reaction includes both maximum growth rate and possible limitation by DO. Precautions have been taken in the practical implementation of the model to guarantee that no concentration is allowed to become negative.

From the eight basic processes represented in the IAWQ model only four remain in the reduced model. The components and processes of the reduced model are summarized in Fig. 4. The parameters shown in bold in the figure indicate those that are assumed to be variable and consequently should be identified. The parameter *isdo*4(XB) is considered known and constant.

	Anoxic environment						Aerobic environment					
Component → / Process ↓	X_{COD}	S_{NH}	S_{NO}	$X_{B,H}$	$X_{B,A}$	Process rate	X_{COD}	S_{NH}	S_{NO}	$X_{B,H}$	$X_{B,A}$	Process rate
Growth of heterotrophs	$-\frac{1}{Y_H}$	$-i_{XB}$	$-\frac{1-Y_H}{2.86\,Y_H}$	1		$\mathbf{r_H}\, X_{COD}\, X_{B,H}$	$-\frac{1}{Y_H}$	$-i_{XB}$		1		$\mathbf{r_H}\, X_{COD}\, X_{B,H}$
Growth of autotrophs								$-i_{XB}-\frac{1}{Y_A}$	$\frac{1}{Y_A}$		1	$\mathbf{r_A}\, S_{NH}\, X_{B,A}$
Decay of heterotrophs	$1-i_{XB}$	i_{XB}		-1		$\mathbf{b_H}\, X_{B,H}$	$1-i_{XB}$	i_{XB}		-1		$\mathbf{b_H}\, X_{B,H}$
Decay of autotrophs	$1-i_{XB}$	i_{XB}			-1	$\mathbf{b_A}\, X_{B,A}$	$1-i_{XB}$	i_{XB}			-1	$\mathbf{b_A}\, X_{B,A}$

Fig. 4. Matrix representation of the reduced order model (Jeppsson, 1993).

The thickener is modelled as a constant compaction ratio (Olsson-Andrews, 1978). This ratio is in turn computed from the sludge retention time. All COD is considered as part of the floc in the thickener. The settler retention time is taken into account by a subsequent time lag. The clarifier behaviour can be excluded from this part of the model verification.

MODEL IDENTIFICATION

In order to test the reduced order model, the IAWQ model was used as the 'true plant'. Gaussian white noise (with a standard deviation of 10% of typical state variable amplitudes) was added to both measurement data and to the influent wastewater concentrations and flow rate in order to make the identification more realistic. The IAWQ model has been simulated using the simulation platforms Simnon (SSPA Systems, 1991) and Simulink (MathWorks, 1992).

The simulator generates data series from which the parameters of the reduced model are updated. There are two principal ways of doing this. In an *off-line* approach all parameters are adjusted according to some suitable optimization criterion to fit the data in the best possible way, as was done in the first simple example. The other method, an *on-line* parameter estimation, is applied here using an extended Kalman filter.

The dynamical model of the plant is described in the traditional format,

$$\frac{d\mathbf{x}}{dt} = \mathbf{f}(\mathbf{x}, t) + \mathbf{g}(\mathbf{u}, t) \tag{4}$$

where $\mathbf{x}(t)$ is a vector which represents all the concentrations indicated in Fig. 4, i.e. five for each completely mixed sub-reactor. The term $\mathbf{g}(\mathbf{u}, t)$ indicates the input from the influent flow to each reactor, respectively.

For the extended Kalman filter not only are the internal states **x** updated according to the measurements but also the unknown parameters. The Kalman filter is divided into two phases, *prediction* and *correction*. The structure of the identification procedure is illustrated in Fig. 5.

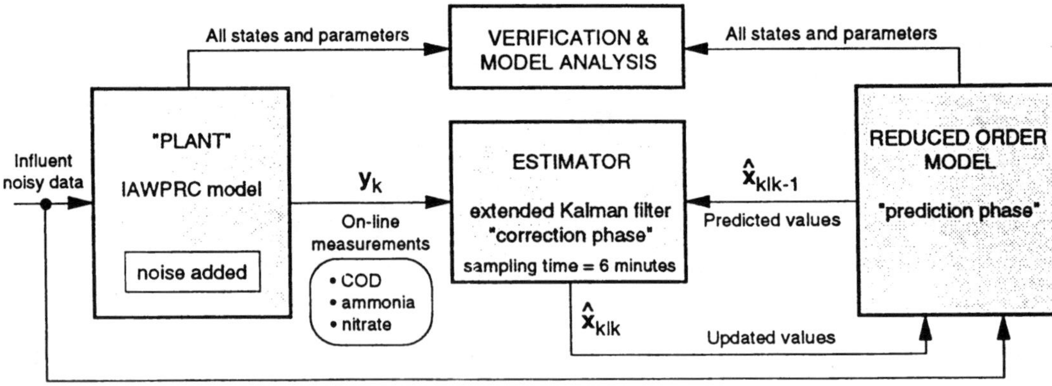

Fig. 5. Structure of the extended Kalman filter identification procedure.

In the prediction phase the dynamical equations are integrated between two measurements, from time t_{k-1} to time t_k. The states (i.e. the concentration variables) and the parameters are now based on measurements up until time t_{k-1}. The generalized state vector (including the unknown parameters) is denoted o(**x**;sup3(^))($t_k|t_{k-1}$). As new measurements are acquired at time t_k they are used to update both the concentration variables and the parameters (see Fig. 4). The latter is called the *correction* phase and the updated vector is denoted o(**x**;sup3(^))($t_k|t_k$). It is based on the calculation,

$$\hat{\mathbf{x}}(k|k) = \hat{\mathbf{x}}(k|k-1) + \mathbf{K}\left[\mathbf{y}(k) - \mathbf{C}\cdot\hat{\mathbf{x}}(k|k-1)\right] \tag{5}$$

The last term corresponds to the assumed measurements at time t_k while **y** symbolizes the real measurements.

A Kalman filter is based on the assumption that the dynamics are linear, which is not the case here. In order to calculate the gain matrix **K** the dynamical equations are linearized around the existing operating point for each measurement instance. At the correction time, **K** is calculated from the linearized equations at time t_k and depends not only on the linearized state but also on the character of the noise that affects both states and measurements. In this work the gain matrix **K** has been kept constant in order to simplify the computations. The value of **K** will influence the convergence rate of the parameters towards their final values. The purpose of this study has primarily been to verify that there is a convergence at all. The chosen value of **K** has been calculated as the ultimate value that is obtained in the Kalman filter calculation. The steady state value from the IAWQ model under the applied conditions were selected as the operating point.

The property of identifiability can be defined in a strict mathematical sense (Söderström-Stoica, 1989). It is related to the problem, if the model parameters can be uniquely calculated from dynamical data. For the Kalman filter implementation this means, that both the states and the parameters have to converge towards the same values, independently of the initial estimates.

In order for a model to be practically identifiable other factors have to be considered too. The convergence has to be acceptable even for noisy measurements. Furthermore, numerical problems may appear. Such a problem is detected already in the small example with the yield factor Y_H. In the organism mass balance (2), the reaction rate factor r_H (3) is identified. The corresponding substrate balance equation contains a term r_H/Y_H. These two parameters can be identified separately. A better method, however, is to rename the term r_H/Y_H as a new parameter. Such a new parametrization is better from a numerical point of view.

The different concentrations in the model have significantly different values, expressed in mg/l. In order to obtain reasonably accurate identification results, it is mandatory to *normalize* or *scale* the equations in such units that the concentration values are expressed in the same order of magnitude. This is even more important when the parameter values are considered. If scaling is not properly performed the **K** matrix will contain such different elements that the estimation becomes infeasible.

Certain states, particularly those directly related to the measurements, are updated much quicker than other states and parameters. This means that the size of the correction may be reduced towards zero too quickly. Consequently the correction term in (5) may decrease to a very small value too early and some parameters be updated too slowly. The fact that **K** is chosen constant may amplify this problem.

It should be noted, that the Kalman approach is *not* the only method to identify the model. Off-line optimizations can be modified to regularly update the model. The difference between the real measurements **y** and the model outputs o(**y**;sup3(^)) can be used to formulate an error criterion J, which is a function of all the unknown model parameters. They are adjusted until J has obtained a minimum. If there is only one unique minimum for J then we define the system as globally identifiable.

In our particular application we have considered the plant configuration illustrated in Fig. 6. The plant is assumed to consist of a predenitrification-nitrification system, having an anoxic reactor followed by an aerobic reactor. Both are considered to be completely mixed. The applied values for the IAWQ model states and parameters as well as the operational variables are also shown in the figure.

In the reduced model (Fig. 4) each reactor is represented by five state variables. In addition there are four parameters to be estimated for the anoxic reactor and six for the aerobic one, i.e. 20 states and parameters to be updated altogether. There are also ways of further reducing the size of the model. Some parameters can be considered identical for the two reactors, for example the decay rate factors. Such a simplification enhances the possibilities of global identifiability without deteriorating the behavior of the model (Jeppsson, 1993).

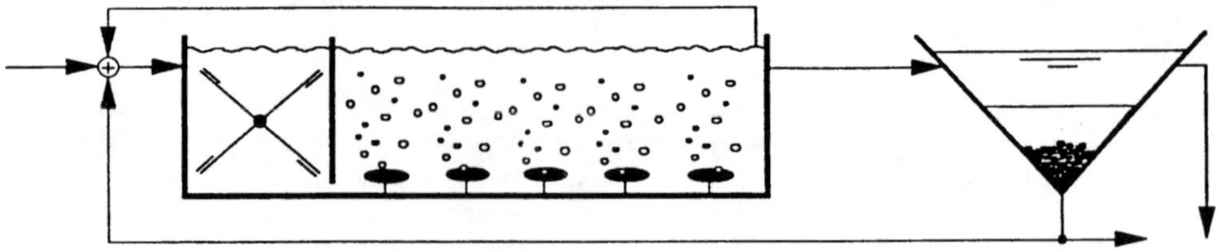

Operational variables		
Influent flow rate (Q_{in})	=	3000 m^3/day
Recycle flow rate	=	0.5*Q_{in}
Internal recycle flow rate	=	3*Q_{in}
Anoxic tank volume	=	250 m^3
Aerobic tank volume	=	1000 m^3
Settler volume	=	1250 m^3
Sludge age	=	10 days
Hydraulic retention time	=	10 hours

Influent wastewater characteristics		
S_S	=	30 mg COD/l
X_S	=	70 mg COD/l
X_I	=	10 mg COD/l
S_{NO}	=	2 mg N/l
S_{NH}	=	10 mg N/l
S_{ND}	=	1 mg N/l
X_{ND}	=	0 mg N/l
S_O	=	0 mg (-COD)/l
$X_{B,H}$	=	0 mg COD/l
$X_{B,A}$	=	0 mg COD/l

Model parameters [Henze et al.,1987]		
$\hat{\mu}_H$	=	6.0 day^{-1}
K_S	=	20 mg COD/l
$K_{O,H}$	=	0.2 mg O_2/l
K_{NO}	=	0.5 mg NO_3-N/l
b_H	=	0.62 day^{-1}
$\hat{\mu}_A$	=	0.8 day^{-1}
K_{NH}	=	1.0 mg NH_3-N/l
$K_{O,A}$	=	0.4 mg O_2/l
b_A	=	0.2 day^{-1}
k_a	=	0.08 mg/(mg COD day)
k_h	=	3.0 mg COD/(mg COD day)
K_X	=	0.03 mg COD/(mg COD)
η_g	=	0.8
η_h	=	0.4
Y_H	=	0.67 mg COD/(mg COD)
Y_A	=	0.24 mg COD/(mg COD)
f_P	=	0.08
i_{XB}	=	0.086 mg N/(mg COD)
i_{XP}	=	0.06 mg N/(mg COD)

Fig. 6. Configuration of the modelled plant.

REDUCED MODEL VERIFICATION BY SIMULATION

The reduced order model has been verified against the IAWQ model to investigate whether it incorporates the dynamical phenomena of importance in the actual time scales or not. Notice, that the simulation data are now noise free. The parameters of the reduced order model have first been identified under perturbed steady state conditions, using the extended Kalman filter. The model is then simulated using the obtained parameter values and directly compared with simulations of the IAWQ model, exposed to the same input data.

In the comparisons between the models, pulse disturbances in the influent flow have been considered. The reduced order model has a different behaviour for fast transients and its accuracy in the rapid responses is not so good. A main reason is that the hydrolysis is neglected. Thus, the pulse transients appear to be the most decisive model test.

In Fig. 7 the transient effects of a combined pulse disturbance (50% increase) of the flow rate (starting at t=3), biodegradable organic substrate concentration (starting at t=3.5), and ammonia concentration (starting at t=4) in the influent wastewater, are illustrated for some key variables in the two models. Each pulse is assumed to last for a period of one day.

It is clear that the qualitative behaviour of the models is very similar. Some differences are apparent for the organic substrate concentration. The influent disturbance consists of both soluble and particulate substrate, i.e. $S_S + X_S$ in the IAWQ model and combined as X_{COD} in the reduced model. This means that the reduced model receives a larger quantity of substrate which is considered directly available for growth (no hydrolysis) and the corresponding COD transient has a lower amplitude than that of the IAWQ model. The differences for the ammonia concentration appears mainly because the Monod kinetics for the Autotrophs in the IAWQ model is in its most non-linear region. Therefore the reduced model (with a linear Monod approximation) will predict a higher nitrification rate as the influent ammonia concentration increases. Through the recirculation this also leads to a somewhat higher concentration of nitrate in the anoxic reactor.

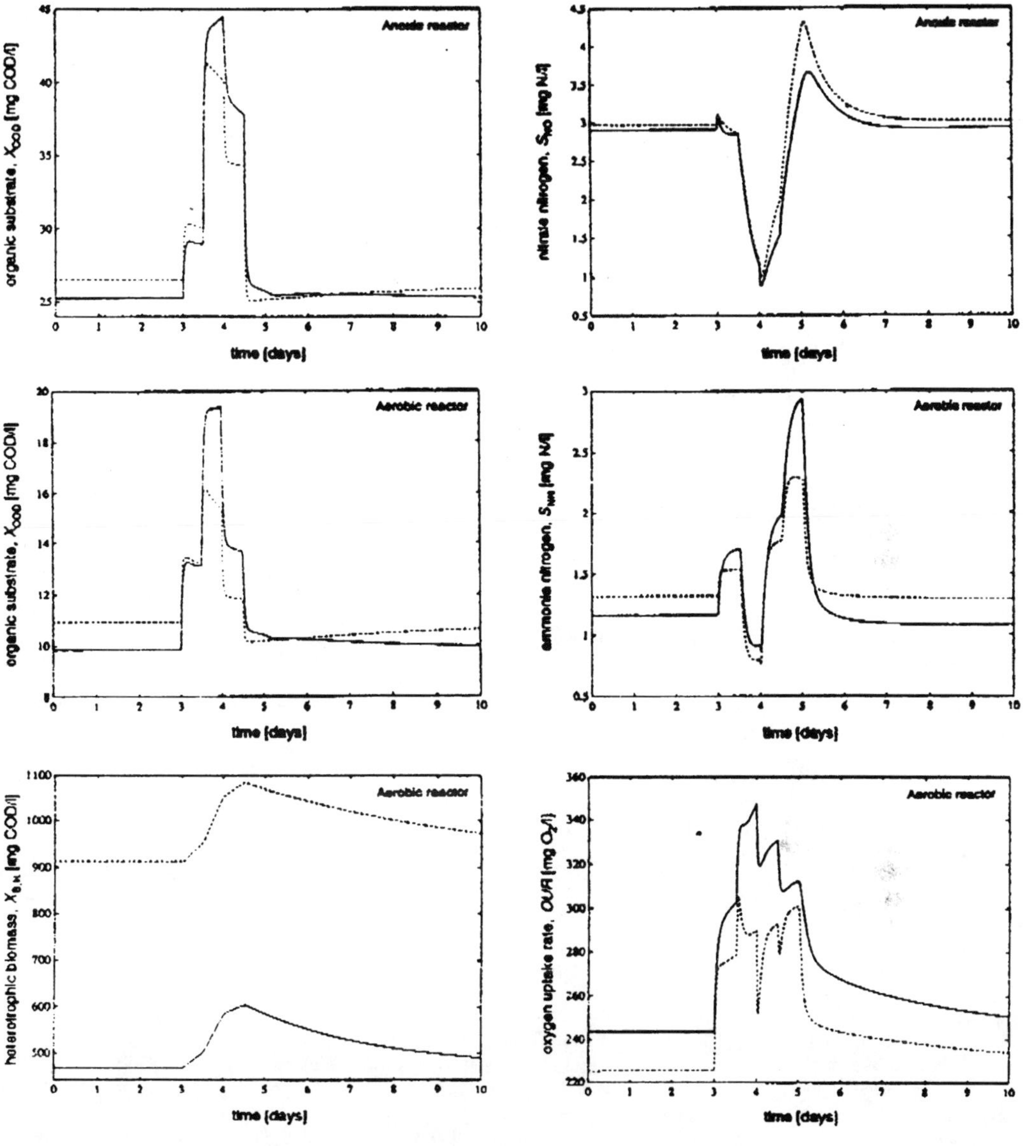

Fig. 7. Transient responses for a combined flow rate, organic substrate, and ammonia disturbance.

The biomass concentration and the *OUR* is not used by the identification algorithm. Still the predictions of the *OUR* are similar for the two models and the relative changes of the organism concentrations appear the same.

CONCLUSIONS

The problem of over-parametrization in complex models has been approached. Existing complex models of the activated sludge system dynamics do not have a unique set of parameters which can explain a certain behaviour. An attempt has been made to derive a reduced order model with less number of states and parameters. Furthermore, available instrumentation emphasizes the need for a realistic complexity of models for operational purposes.

It has been shown that it is possible to find the reduced model parameters on-line, even under noisy conditions. The identification method may be improved for better convergence and alternative methods have to be tried out. Simulation comparisons between the IAWQ model and the reduced model indicate that the main features of the dynamics have been retained.

The model is primarily aimed as an on-line tool for supervision and control. Since its parameters can be gradually updated from on-line measurements, any deviation between the real plant and the model predictions can be used as an early warning system for process diagnosis purposes.

ACKNOWLEDGEMENTS

This work was financially supported by the Swedish National Board for Industrial and Technical Development (NUTEK) and performed within the Swedish STAMP (Swedish abbreviation for *Control of Wastewater Treatment Plants Methods and Processes*) research group (Olsson, 1991).

REFERENCES

Andrews, J. F. (1992). *Dynamics and Control of the Activated Sludge Process*, Technomic Publishing Company, Inc., Lancaster, Pennsylvania, USA.

Ayesa, E., Florez, J., García-Heras, J. L. and Larrea, L. (1991). State and coefficients estimation for the activated sludge process using a modified kalman filter algorithm. *Wat. Sci. Tech.*, **24**(6), 235-247.

Cech, J. S., Chudoba, J. and Grau, P. (1985). Determination of Kinetic Constants of Activated Sludge Microorganisms. *Wat. Sci. Tech.*, **17**(2/3), 259-272.

Dold, P. L., Ekama, G. A. and Marais, G. v. R. (1980). A General Model for the Activated Sludge Process. *Prog. Wat. Tech.* **12**(6), 47-77.

Dold, P. L. (1992). Activated Sludge System Model Incorporating Biological Nutrient (N & P) Removal, Dept. of Civil Eng. & Eng. Mech., McMaster University, Hamilton, Ontario, Canada.

Fletcher, R. (1987). *Practical Methods of Optimization*, John Wiley & Sons Ltd., Chichester, UK.

Godfrey, K. R. and DiStefano, J. J. (1985). Identifiability of Model Parameters. *Proc. IFAC Identification and System Parameter Estimation*, York, UK.

Henze, M., Grady Jr., C. P. L., Gujer, W., Marais, G. v. .R. and Matsuo, T. (1987). Activated Sludge Model No. 1, IAWQ Scientific and Technical Report No. 1, IAWQ, London, UK.

Holmberg, A. (1982). On the Practical Identifiability of Microbial Growth Models Incorporating Michaelis-Menten Type Nonlinearities, *Math. Bioscience*, **62**, 23-43.

Holmberg, U. (1991). Relay Feedback of Simple Systems, Ph.D. Dissertation, CODEN:LUTFD2/(TFRT-1034), pp. 130-144, Dept. of Automatic Control, Lund Institute of Technology, Lund, Sweden.

Jeppsson, U. (1993). On the Verifiability of the Activated Sludge System Dynamics, Licentiat thesis, Dept. of Ind. Elec. Eng. & Automation, Lund Institute of Technology, Lund, Sweden.

Larrea, L., García-Heras, J. L., Ayesa, E. and Florez, J. (1991). Designing Experiments to Determine the Coefficients of Activated Sludge Models by Identification Algorithms, *IAWQ Seminar on Interactions of Wastewater, Biomass and Reactor Configurations in Biological Treatment Plants*, Aug. 21-23, 1991, Copenhagen, Denmark.

Little, J. and Shure, L. (1988). Signal Processing Toolbox for use with MATLAB , The MathWorks, Inc., Natick, MA, USA.

MathWorks (1992). *Simulink , A Program for Simulating Dynamic Systems*, The MathWorks, Inc., Natick, MA, USA.

Monod, J. (1942). The Growth of Bacterial Cultures, *Ann. Rev. Microb.*, **3**, 371-394.

Náhlík, J. and Burianec, Z. (1988). On-line Parameter and State Estimation of Continuous Cultivation by Extended Kalman Filter, *Appl. Microb. and Biotech.*, **28**, 128-134.

Olsson, G. and Andrews, J. F. (1978). The Dissolved Oxygen Profile a Valuable Tool for Control of the Activated Sludge Process, *Wat. Res.*, **12**, 985-1004.

Olsson, G. (1991). Operations and Control in Wastewater Treatment - some Swedish Experiences, *Wat. Sci. Tech.*, **24**(6), 193-200.

Patry, G. G. and Chapman, D. (1989). *Dynamic Modeling and Expert Systems in Wastewater Engineering*, Lewis Publishers, Inc., Chelsea, Michigan, USA.

Spanjers, H. and Olsson, G. (1992). Modelling of the Dissolved Oxygen Probe Response in the Improvement of the Performance of a Continuous Respiration Meter, *Wat. Res.*, **26**(7), 945-954.

Spanjers, H., Olsson, G. and Klapwijk, A. (1993). Estimation of Short-term Biochemical Oxygen Demand and Respiration Rate in the Aeration Tank, Paper submitted to the *6th IAWQ Workshop on Instrumentation, Control and Automation of Water and Wastewater Treatment and Transport Systems*, June 17-25, Banff-Hamilton, Canada.

SSPA Systems (1991). *Simnon, User's guide for UNIX Systems*, SSPA Systems, Gothenburg, Sweden.

Söderström, T. and Stoica, P. (1989). *System Identification*, Prentice Hall, Inc., Hertfordshire, UK.

Vanrolleghem, P. A. and Verstraete, W. (1993). Simultaneous Biokinetic Characterization of Heterotrophic and Nitrifying Populations of Activated Sludge with an On-line Respirographic Biosensor, Paper submitted to the *6th IAWQ Workshop on Instrumentation, Control and Automation of Water and Wastewater Treatment and Transport Systems*, June 17-25, Banff-Hamilton, Canada.

Pergamon

Wat. Sci. Tech. Vol. 28, No. 11-12, pp. 185–195, 1993.

0273–1223/93 $7·00 + 0·00

SIMULATION OF A STRATEGY TO START UP NITRIFICATION AT BROMMA SEWAGE PLANT USING A MODEL BASED ON THE IAWPRC MODEL NO. 1

Anders Finnson

Stockholm Water Ltd, Box 6407, S-113 82 Stockholm, Sweden

ABSTRACT

At the Bromma sewage plant in Stockholm a project is being carried out where the aim is to elaborate different operational strategies for the plant. One of the strategies, the seeding of nitrificating bacteria from one line to other lines in the activated sludge process, was simulated with the assistance of the Danish computer program EFOR. The simulations showed that it was possible to use EFOR for this kind of simulation and that the seeding effect which was achieved in the sewage plant was of relatively small importance for reducing the NH_4-N in the effluent water.

KEYWORDS

IAWPRC-model no 1; EFOR; activated sludge process; wastewater treatment; nitrification; simulation; mathematical modelling; seeding; operational strategies.

INTRODUCTION

The new requirements for nitrogen removal in Sweden place high demands on the operation of sewage plants. At many sewage plants, as at the Bromma plant in Stockholm, upgradings together with efficient operational strategies will be necessary to fulfil the new effluent standards. The complexity of the activated sludge process makes it difficult for a process engineer at a sewage plant to predict all the effects of a new operational strategy. Therefore, computer-aided simulation programs will be very useful tools for process engineers in the near future. This paper reports on a study of an operational strategy realized at the Bromma sewage plant and simulated on a computer simulation program called EFOR based on the IAWPRC model no.1 (Henze *et al.*, 1987).

PRESENTATION OF BROMMA SEWAGE PLANT

Introduction

The Bromma sewage plant serves the north-western parts of Stockholm. The plant is at present dimensioned for a flow of 160 000m^3/d and a BOD_7- load of 30 000 kg/d.

Table 1 lists the present and future effluent standards.

TABLE 1. Effluent Standards – Present and in the Future

Parameter	Unit	Present Effluent standards	Probable Future Effluent standards	Remarks
BOD_7	mgO_2/l	15	10	
Total-P	mgP/l	0,5	0,3	
NH_4-N	mgN/l	4	-	July-Oct
Total-N	mgN/l	-	15	

The effluent standards for BOD_7 and total phosphorous are based on 3-months average values but the total nitrogen requirement will be based on a yearly average. The NH_4-N effluent standard is based on the average value for the period of July - October.

The Bromma sewage plant consists of two plants, the Åkeshov plant and the Nockeby plant. At the Åkeshov plant the mechanical and chemical treatment (pre-precipitation with ferrous sulphate) is situated together with the sludge treatment. At the Nockeby plant the wastewater is treated biologically. The biological step is a conventional activated sludge process upgraded for pre-denitrification, six parallel lines with an anoxic zone at the beginning of the basin (1/8 of the total volume), an aerobic zone and final settlers. Some of the ferrous sulphate which is added in the chemical treatment reacts with phosphorous in the activated sludge tank i.e. simultaneous precipitation occurs.

The treated wastewater is then led to the Baltic sea via a tunnel and a heat exchange plant.

Operational Strategies for the Bromma Sewage Plant

An operational strategy involves various steps being taken to maintain as even and good purification as possible, under changing external and internal conditions. The measures taken will differ from sewage plant to sewage plant. Improved knowledge of the sub-processes taking place in the plant may make it possible to distribute the load between the parts of the plant in various ways according to the load and the current capacities.

The introduction of biological nitrogen reduction places high demands on the operation of the plant. The risk of loss of nitrifiers during the cold months of the year makes it considerably more difficult to achieve a high degree of nitrogen removal, even when the requirement is based on annual averages. Strategies for keeping the sludge in the plant, alternatively for quickly starting up nitrification again by seeding are of considerable importance.

The mechanical and the chemical pretreatment at the Bromma sewage plant is not as sensitive as the biological step which means that the strategies which are used today have as an aim to secure the function of the activated sludge process under various conditions. Some of the present and future operational strategies are listed below:

– Equalization of inflow peaks in the inlet tunnels
– By-pass of the biological step during very high inflow peaks
– Maintaining different values of MLSS in the activated sludge tanks depending on the season to maximize the yearly nitrogen reduction.
– Keeping nitrification in one of the six lines in the biological step to make it possible to seed when starting up the nitrification process.
– Improvement of precipitation of organic material, phosphorous and nitrogen in the preprecipitation process
– Treatment of the NH_4-N concentration in the digester supernatant

Downflow sandfilters are under construction at the plant to enable it to fulfil the future effluent standards. The sandfilters will make it possible to keep a higher value of MLSS in the activated sludge tank without having the risk of high SS concentration in the effluent.

SIMULATION OF AN OPERATIONAL STRATEGY WITH THE HELP OF EFOR

The Aim of the Simulation

One of the above-mentioned operational strategies was simulated in the Danish computer program EFOR.

The aim of the computer-assisted simulations described in this article was to:

1. Investigate if it was possible to simulate the reduction of NH_4-N in effluent water during a period of approximately two weeks after the kind of nitrification start-up carried out in June 1992.
2. Simulate a comparison between the reduction of NH_4-N in the strategy used in June 1992 for starting up nitrification and the reduction of NH_4-N in the strategy used in previous years.

Background to the EFOR Program

The Danish computer program EFOR (Edb modell FOr Renseanlaeg) is a program which allows the user to simulate the activated sludge process at a municipal sewage plant (Aspegren *et al.*, 1992; Pedersen, 1992). The calculation function of the program is constructed around a further development of IAWPRC's model no. 1 which was originally first presented by Henze et al. (1987) and around a sedimentation model based on the so- called solid flux theory presented by Dupont and Henze (1992). The program is relatively user-friendly and is a program of the generation intended for use not only by university researchers but also by operating engineers and operators at sewage plants. The EFOR program's usefulness for operating engineers stems to a large part from the so- called conversion function, a function in the program which converts Tot-COD (or Tot-BOD_7), Tot-N (or K_j-N) and Tot-P to most of the variables included in the IAWPRC model. This conversion function may, however, require extensive preparatory work to equip the function with the necessary conversion constants. What need to be analysed above all are the relationships between the particulate and dissolved fractions of COD, BOD_7, K_j-N and Tot-P.

Background to the Experiment with Seeding in 1992

Nitrification is only a requirement at the Bromma Sewage Treatment Plant during the period 1 July-31 October, and the plant is accordingly operated without nitrification during the remainder of the year. Various kinds of methods for starting up nitrification have been discussed over the years. It was decided that nitrification in the 1992 season would be maintained throughout the winter season on one (line L1) of the six parallel activated sludge lines (Fig. 1).

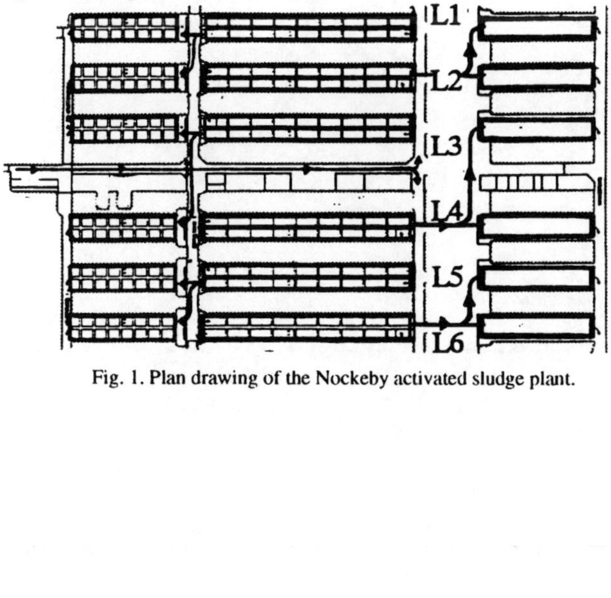

Fig. 1. Plan drawing of the Nockeby activated sludge plant.

Inflow to line L1 was reduced by approximately 35 per cent to make possible a sufficiently high sludge content in the activated sludge tank so that extensive nitrification could be maintained during the winter.

Nitrification was maintained in line L1 during the winter season at an average content of 4.5 mg NH_4-N/l in effluent water (January-April 1992). Unlike the previous year's start-up of nitrification, there was now a possibility of seeding nitrificating sludge from line L1 to lines L2-L6.

The seeding process took place as follows:

Line L1 was completely emptied through sludge being pumped to the inlet of the biological stage where the sludge was evenly distributed to lines L2-L6 respectively. Line L1 was then started up by being filled in the course of a day with excess sludge from lines L2-L6. The procedure took about 36 h. The pumping of excess sludge was then sharply reduced on all lines in order to quickly build up a high sludge content in the basins. The point of time for the start up of nitrification has been estimated from the time when line L1 was again filled with sludge
(4 June 1992).

The Sampling Program

Four different kinds of sampling periods were carried out to get the necessary data for the computer simulation.

– Three 30-h periods of intensive sampling were carried out in spring 1992 to estimate the constants included in the conversion function (an intensive sampling period is fifteen 2-h accumulated samples taken during the 30-h period). The tests were immediately filtered and BOD_7, COD and K_j-N were then analysed for the filtered and unfiltered samples (Table 2) and (Fig. 2).

TABLE 2. Conversion Constants Determined from the Intensive Sampling Periods

	Intensive sampling periods			
	Jan.13-14	Feb.10-11	May 11-12	Average
Soluble COD / Total COD				
Average value	0.50	0.41	0.42	0.44
90% percentile	0.62	0.48	0.52	
10% percentile	0.45	0.34	0.34	
Soluble BOD7 / Total BOD7				
Average value	0.42	0.30	0.31	0.34
90% percentile	0.48	0.45	0.36	
10% percentile	0.33	0.18	0.23	
Soluble Kj-N / Total Kj-N				
Average value	0.76	0.82	0.74	0.77
90% percentile	0.83	0.83	0.77	
10% percentile	0.71	0.80	0.70	
NH4-N / Soluble Kj-N				
Average	0.82	1.0	0.90	0.91
90% percentile	1.0	1.0	0.92	
10% percentile	0.71	0.87	0.88	

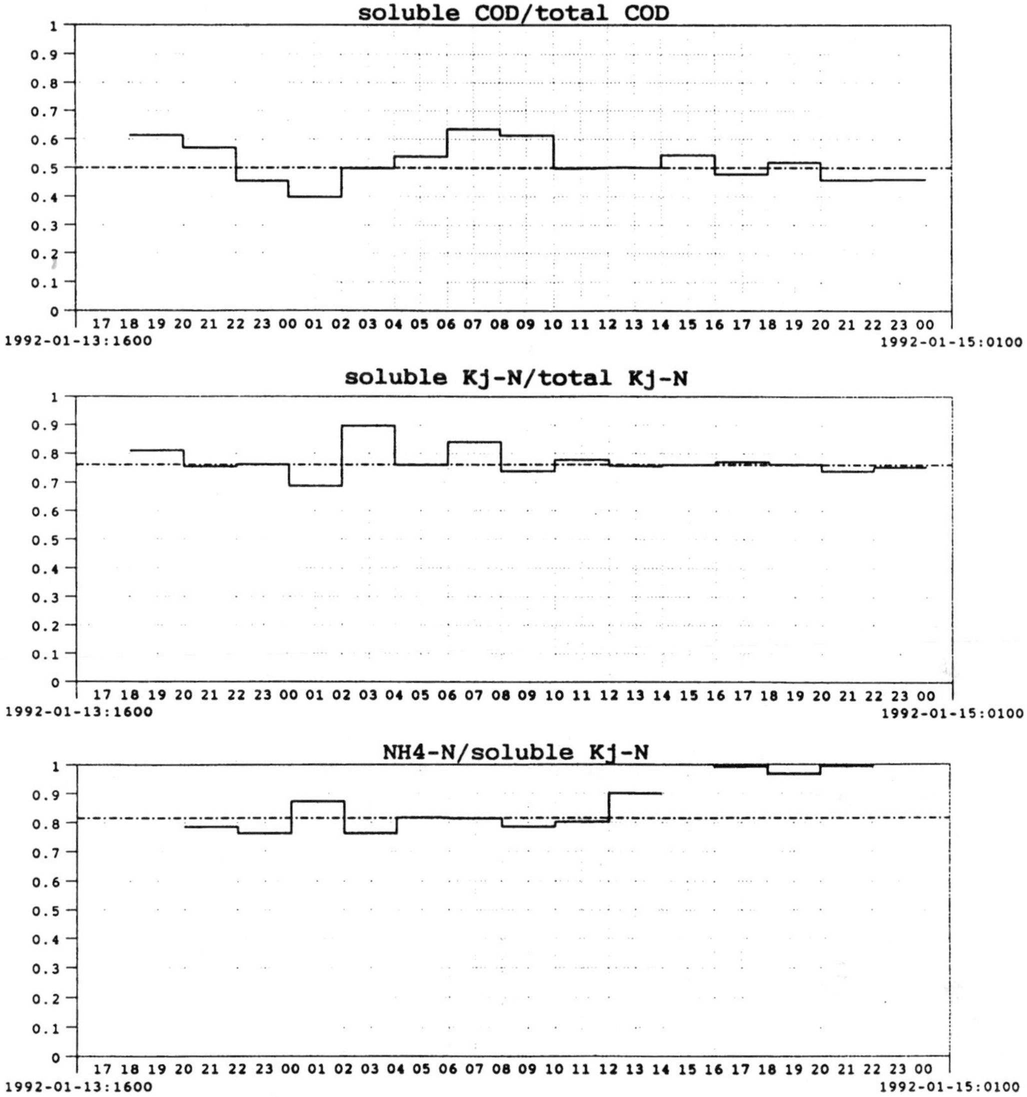

Fig. 2. Presettled wastewater: soluble COD/total COD, soluble K_j-N/total K_j-N and NH_4-N/soluble K_j-N from the intensive sampling period of January 13-14 1992. The averages are marked with a dotted line.

- The average values of the accumulated weekly samples for the period 18 May - 4 June (Table 3) were used to generate a series of measurements for the period before nitrification started. The measurements were used for steady-state simulations.

TABLE 3. Average Values from the Period of Measurement 18 May - 4 June

	Flow m_3/h	Excess sludge m_3/h L1	m_3/h L2-L6		COD mg/l	Tot-P mg/l	Kj-N mg/l		Alk $mgHCO_3/l$	temp oC
18/5-4/6 Presettled water	5042	7	185		250	3.1	34		255	16.3
	NH_4-N mg/l L1	NH_4-N mg/l L2-L6	NO_3-N	NO_3-N mg/l L1	SS mg/l L2-L6	MLSS	MLSS mg/l L1	mg/l L2-L6	mg/l	
18/5-4/6 Effluent water	1.5	21		10	2		9 tank	4300	1900	

When nitrification was started up, the process was monitored by a 24 h sampling program during the period 4 June-17 June in which COD, NH_4-N, NO_3-N, SS and alkilinity were analysed daily.

- During the period 18 June-30 June accumulated weekly samples for COD, NH_4-N, NO_3-N, SS were analysed again.

While collecting data as above, information has also been collected regarding flow, temperature, MLSS and MLVSS.

Simulation Strategy

After the physical design of the sewage plant had been fed into the model the current return sludge flows, excess sludge flows, control and regulatory functions, set values, water distribution and any chemical dosage, etc. were also given in the model.

The simulations were then divided into two main types:

- Steady-state simulation

- Dynamic simulation

Steady-state simulation. Steady-state simulation provides the basis for the later dynamic simulation. Steady-state simulation was carried out to be able to adjust, under stable conditions, the constants which need to be changed so that the effluent content according to the model agrees with the actual effluent concentration of NH_4-N. Simulation is made over a sufficiently long time so that no further changes take place in the activated sludge tank with regard to the content of nitrification bacteria, X_{ba}, heterotrophic bacteria X_{bh} or inert particulate material X_i.

Steady-state simulation is carried out in this case on the basis of the following measurements:

– Measurements influent: average values from the period 18 May-4 June 1992
– Measurements effluent: average values from the period 18 May-4 June 1992
– Conversion constants: processed material from the three intensive 30-h periods
– Process constants: only the default values in the model are used except for the growth rate of nitrification bacteria (μ_a) and the number of nitrification bacteria in influent water (X_{ba}). These two constants are adjusted so that the correct content of NH_4-N in effluent water from L1 and L2-L6 in the model is obtained at the same time.

The dynamic simulation. The dynamic simulation was started after the steady-state simulation was carried out. In the dynamic simulation, the process constants were used which were adjusted for the steady-state simulation. The dynamic simulations can be categorized according to two different main principles:

Forecasts

Answer the question: What result will the process give?
Forecasts simulate the time period which is immediately to follow. The series of averages for 18 May-4 June have also been used as measurements for the period 4 June-30 June since no better estimates were considered to exist for how the flows, temperature and contents in the presettled water would be during the period 4 June-30 June. Pumping of excess sludge in the model is changed as planned to increase the age of the sludge.

Retrospective simulation

Answers the question: What measures could have been taken to have achieved another result of the process?
Here a simulation is made of what has already taken place in reality.
The real 24 h values for the period 4 June-30 June are used as indata in this case. Excess sludge pumping in the model is changed to the actual flows during the period.

RESULTS

For the steady-state case the constants μ_a and X_{ba} were adjusted so that the effluent values of NH_4-N in the model corresponded to the real effluent values of line L1 and lines L2-L6 respectively.

In this particular case, Bromma sewage plant, there were two processes running at the same time at different sludge retention times (SRT) which consequently led to different NH_4-N content in the effluent from the two processes in L1 and in L2-L6. Line L1 was running with nitrification and at an SRT of approximately 10 d and lines L2-L6 were running without nitrification and at an SRT of 3 d.

Under these circumstances it seems that there is only one combination of μ_a and X_{ba} which can simulate the NH_4-N in the effluent of both L1 and L2-L6.

The combination of the μ_a and X_{ba} which agreed best with the effluent values of NH_4-N from L1 and L2-L6 in the steady-state simulation was $\mu_a = 0.40$ d-1 and $X_{ba} = 0.018$ g COD/m^3.

When the total NH_4-N content (L1 plus L2-L6) in the effluent from the plant was used for steady-state simulation, another result was obtained. In that case several different combinations of μ_a and X_{ba} could simulate the same content of NH_4-N in the effluent (Table 4).

TABLE 4. Steady-state Simulated NH_4-N in the Effluent for Different Combinations of μ_a and X_{ba}.

Influent		Steady-state simulated effluent NH_4-N (mg/l)		
μ_a (d^{-1})	X_{ba} (mgCOD/l)	L1	L2-L6	L1+L2-L6
0.80	0.005	0.2	21	19
0.60	0.009	0.5	21	19
0.40	0.018	1.5	21	19

However, every combination of μ_a and X_{ba} gave different values of NH_4-N in the effluent when the dynamic simulations were run (Fig. 3).

It seems that if an activated sludge process is run only at one SRT (in time or space), there is a need to make laboratory tests to determine at least one of the constants μ_a and X_{ba} to have sufficient data for simulations of NH_4-N in the effluent.

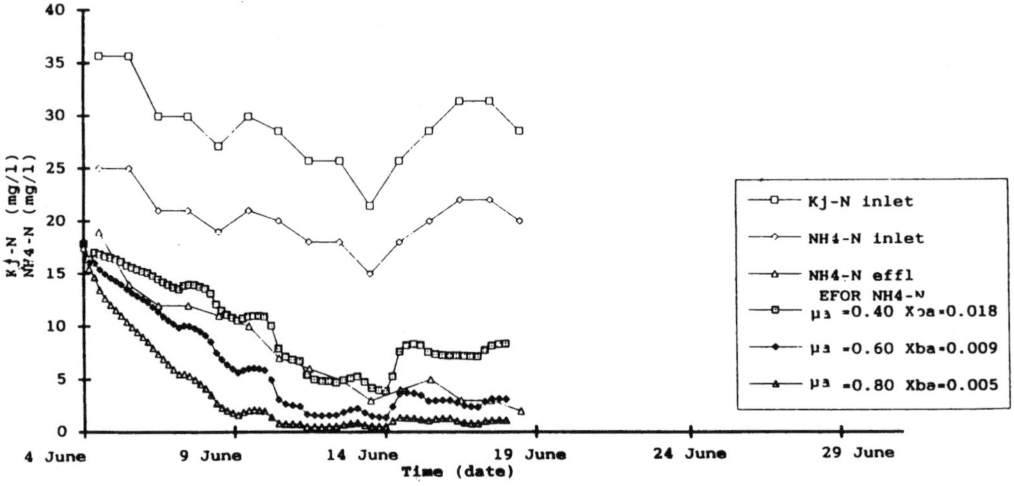

Fig. 3. Dynamic simulated NH_4-N in the effluent for different combinations of μ_a and X_{ba}.

The dynamic simulations are initiated on the basis of the results from the steady-state simulation.

The agreement of the dynamic forecasts with reality is not especially good (Fig. 4). It may be apposite here to point out that the averages from the measurements between 18 May-4 June were used for the period 4 June-30 June since no better estimation for this period could be made for the incoming flows, temperature and concentrations.

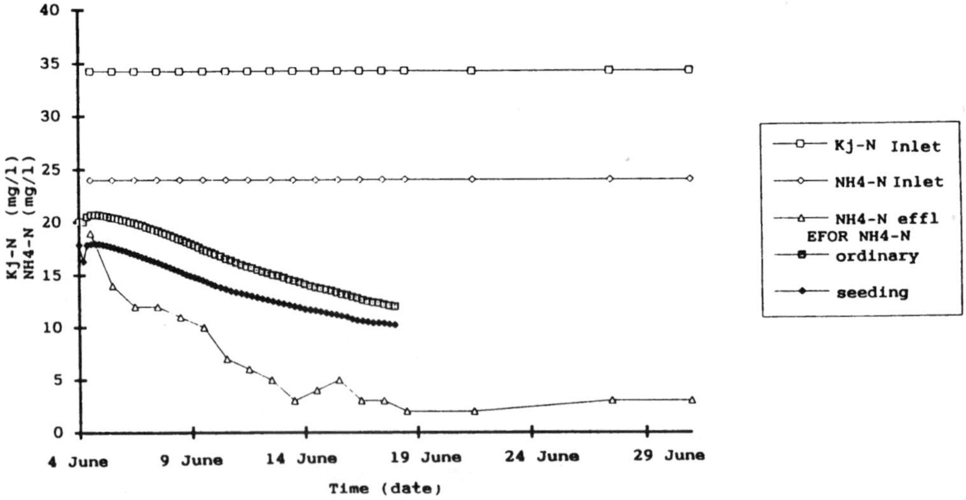

Fig. 4. Forecasts with estimated averages as indata series

If the averages for the flows, temperature and concentrations in the presettled water could be better predicted for the period in question 4 June-30 June, a considerably better agreement would be achieved between the real and simulated NH_4-N content in the effluent water.

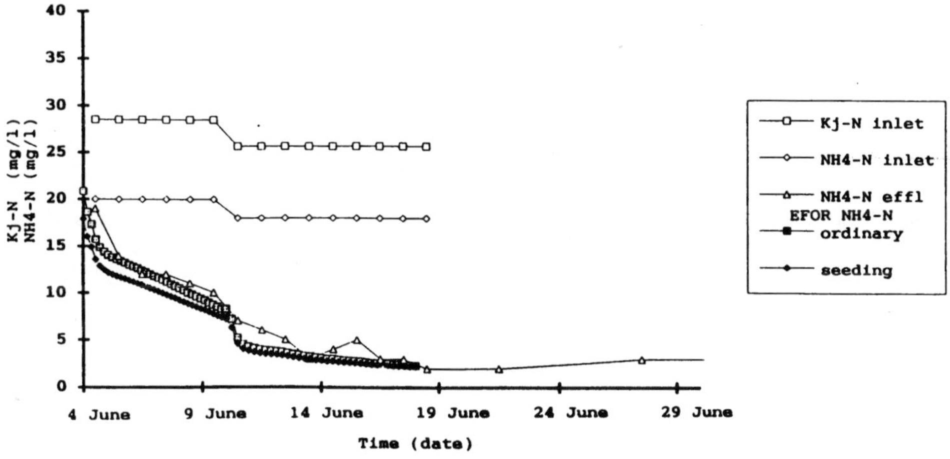

Fig. 5. Forecasts with real average values divided into two periods as in data series.

In Figure 5 the real average values for the flows, temperature and concentrations from the period 4 June-18 June are divided into two periods. The real average values have been used as indata for simulation (i.e. the correct average values could be "predicted" here) and a considerably better agreement was achieved between real and simulated content of NH_4-N in the effluent water.

The forecast which was obtained from the simulation does not indicate any dramatic difference in effluent NH_4-N content between the real case when LI was seeded in L2-L6 and the strategy which was previously used for starting off nitrification.

In this case the ability to predict the data in a series of measurements has great importance for how good a forecast can be made.

In a retrospective simulation, the problem of needing to predict flows, temperature and influent values naturally disappears. All the relevant data have been daily collected and are available for the period 4 June-30 June.

In this case too, constants are used which have been obtained during the steady-state simulation ($\mu_a = 0.40$ and $X_{ba} = 0.018$).

The result of the dynamic retrospective simulation shows a good agreement between the model and reality during the first ten simulated days (Fig.6).

Only a minor difference can be observed in effluent NH_4-N content between the real strategy where L1 is seeded into L2-L6 and the ordinary strategy without seeding which was used in earlier years for starting up nitrification.

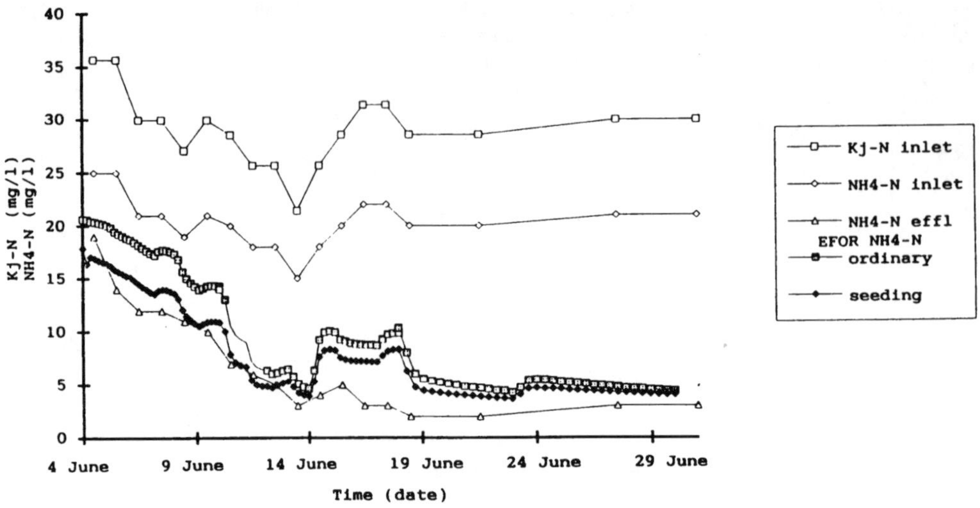

Fig. 6. Retrospective simulation with real dynamic daily series as in data

DISCUSSION

The Constants in the Model

The IAWPRC model no. 1 contains around 20 different constants which are to be determined. Already when the IAWPRC model no. 1 was presented (Henze *et al.*, 1987) a full set of constants was proposed for the model. A set of default values similar to the values in the IAWPRC model no. 1 was also proposed for the EFOR program.

An enormous effort has been put into the determination of these constants. Nevertheless, all constants do not have the same impact on the result of a simulation. A study of the sensitivity and variability of the model was presented by Henze (1988) which showed the differences between the impacts of different constants on the model.

A strategy for the constant adjustment can be to concentrate solely on those constants which the simulated process normally can be very sensitive. In this study where the main aim was to simulate the nitrification process the strategy was only to adjust two of the constants μ_a and X_{ba}, to get the model to simulate the NH_4-N in the effluent.

The Simulation of the Seeding

In previous years when the nitrification starting-up process at the Bromma sewage plant ran without seeding, the time to reach low NH_4-N content in the effluent used to be much longer than during 1992 when the seeding strategy was used. Therefore the staff at the plant believed strongly in the seeding strategy which at first seemed to be the reason for the extremely fast start-up of nitrification in June 1992.

The simulations which were done, showed however, that the seeding strategy had only a minor effect on the NH_4-N in the effluent and probably only during the first week after the start of the starting-up process (Fig. 6). According to the simulations the reason for the fast reduction of NH_4-N in the effluent seems instead to have been the extremely advantageous conditions which prevailed during the starting-up period of 1992.

Comparison can be made with the forecast simulation (Fig. 4), where the average inflow values approximately reflect ordinary conditions during previous years of the period of starting-up and the speed of the NH_4-N reduction is much slower than during the period of 1992.

The situation was dramatically different in June 1992, the inflow, the high temperature and the decreasing K_j-N content in the influent water seemed together to be the conditions which made the fast reduction of NH_4-N possible.

The EFOR program seems to be applicable for simulating the strategy which was presented in this paper. Worth mentioning in the discussion even though it can be seen as a trivial problem is the great importance of the accuracy of the indata values. Without an accurate indata measurement the result of a simulation will not be very reliable regardless of how carefully the constants in the model are determined. When the indata values are of such a great importance for the simulation it means also that a result of a simulation based on forecasted indata values will not be better than the accuracy of the used indata values (which in many cases can be difficult to forecast).

However, with accurate indata values and with the constants most sensitive for the simulated process reasonably well adjusted, it is possible to simulate a nitrification starting up process in EFOR.

CONCLUSIONS

– It is possible to simulate a start-up of nitrification retrospectively with good agreement with reality.

– The result of a forecast simulation is to a very great extent dependent on how the flows, temperature and influent concentrations of K_j-N can be forecasted.

– Seeding seems to be of relatively small importance for how quickly NH_4-N diminshes in effluent water.

–The EFOR model seems to be useful in this particular context but it cannot be clearly foreseen whether other strategies can be simulated.

ACKNOWLEDGEMENTS

The help, opinions and encouragement from the staff at the Bromma sewage plant and from the laboratory at Stockholm Water Ltd are gratefully appreciated.

This project is a part of a joint project between Stockholm Water Ltd and the Royal Institute of Technology (Department of Water Resources Engineering and Department of Biochemistry and Biotechnology) and has been supported by the Swedish National Board for Industrial and Technical Development (NUTEK) and Stockholm Water Ltd. The author is a Ph D student at the Department of Water Resources Engineering.

REFERENCES

Aspegren, H., Andersson, B., Nyberg, U. and Jansen, J. C. (1992). Model and sensor based optimization of nitrogen removal at the Klagshamn WWTP. *Wat. Sci. & Tech.*, **26**(5/6), 1315-1323.

Dupont, R. and Henze, M. (1992). Modelling of the secondary clarifier combined with the activated sludge model no. 1. *Wat. Sci. & Tech.*, **25**(6), 285-300.

Henze, M., Grady, C.P.L., Gujer, W., Marais, G.v.R. and Matsuo, T. (1987). Activated sludge model no. 1. *Scientific and technical reports no.1 (1)*, IAWPRC.

Henze, M. (1988). Constants in the Mathematical Models. Material from the IAWPRC technology transfer seminar in Rome 4-6 May 1988 on mathematical modelling of biological wastewater treatment processes.

Pedersen, J. (1992). Controlling activated sludge process using EFOR. *Wat. Sci. & Tech.* **26**(3/4), 783-790.

Pergamon

Wat. Sci. Tech. Vol. 28, No. 11-12, pp. 197–207, 1993.

0273–1223/93 $7·00 + 0·00

DYNAMIC MODEL SIMULATION AND VERIFICATION OF A TWO-STAGE HIGH-RATE ANAEROBIC TREATMENT PROCESS WITH RECYCLE

J. Keller, M. Romli, P. L. Lee and P. F. Greenfield

Department of Chemical Engineering, University of Queensland, Queensland 4072, Australia

ABSTRACT

The effect of overload on a two-phase high rate anaerobic wastewater treatment system with recycle was determined experimentally and simulated dynamically using a structured model. The experimental system consisted of a well mixed continuous reactor, controlled at pH 6 by sodium hydroxide addition, as the acidification stage and a fluidized bed reactor for the methanogenic stage, with an additional recycle connection from the second to the first reactor. Step changes in the feed concentration as well as in the feed flow rate were investigated and compared to simulation predictions. Operation without recycle was modelled accurately, with the simulation data of most process variables matching the experimental results quantitatively. The application of the same model to recycle operation showed significant discrepancy between the experimental and simulated data for the sodium hydroxide consumption rate and the gas flow rate in the acidification reactor. Although the actual values were different, the qualitative responses of the experimental and simulated step changes were similar. Other simulated variables showed good agreement with the experimental measurements. The model provides a useful tool for design and control studies in systems without recycle and, with some modification, for operation with recycle.

KEYWORDS

anaerobic, wastewater, modelling, recycle, overload, step-changes, carbon dioxide

INTRODUCTION

The use of two separate reactors for anaerobic degradation processes has several advantages over the single reactor concept, especially at high dilution and load rates (Bull *et al.*, 1984; Dinopoulou and Lester, 1989; Zhang and Noike, 1991). The separation of the two major reaction steps in anaerobic degradation, acidogenesis and methanogenesis, into two different reaction vessels is inherent in such systems. It can be further enforced by using specific operating parameters for the two reactors, usually different liquid and solid residence times as well as pH. The improved operating conditions for the two major bacteria groups can lead to better performance of the reactor system, in particular with high organic loadings and during overload periods.

In high rate wastewater treatment systems it is often necessary to control the pH by addition of base in order to neutralise the high levels of organic acids produced by the acidogenic bacteria. In two-phase systems the base consumption is increased further since the acids produced in the first reactor are only consumed in the second stage by the methanogenic bacteria. The addition of base is therefore often a major component in the overall operational cost, and any reduction of this consumption rate improves the economic value of the process. One possible way to reduce the base consumption involves the introduction of a recycle stream from the second to the first reactor as suggested by Costello (1989) based on simulation results from an anaerobic degradation model. Experimental studies in this field have recently been carried out and steady state results have shown the expected behaviour (Romli *et al.*, submitted). The dynamic response of such systems with and without recycle to major changes in the feed concentration and feed flow rates has also been studied and the results are presented in this paper.

A structured dynamic model of the anaerobic process has been successfully applied to a number of single stage systems (Costello *et al.*, 1991). Further development of this model has allowed a single set of model parameters to be used to describe single-stage as well as two-stage processes (Keller *et al.*, in preparation). The same model was used in the present work to investigate its applicability to two-stage processes with recycle.

REACTOR SYSTEM AND EXPERIMENTS

The laboratory scale two-stage anaerobic wastewater treatment system used in this work consists of a continuous stirred tank reactor (CSTR) as the acidification reactor and a fluidized sand bed reactor as the methanogenic reactor (Fig. 1). The liquid volume in the first stage was 1.64 l and in the second

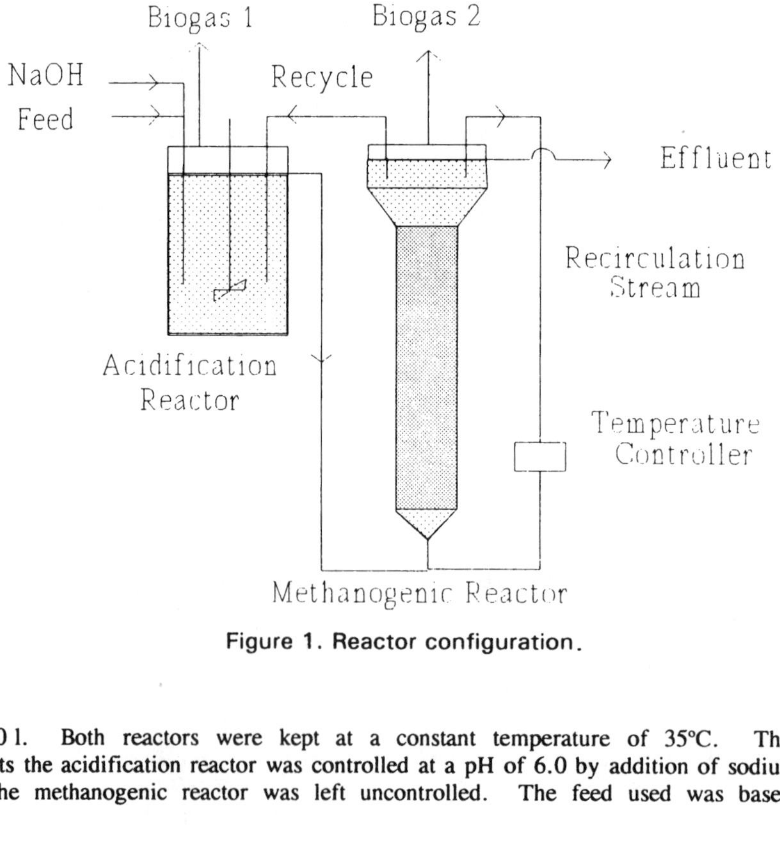

Figure 1. Reactor configuration.

stage 3.30 l. Both reactors were kept at a constant temperature of 35°C. Throughout the experiments the acidification reactor was controlled at a pH of 6.0 by addition of sodium hydroxide, whereas the methanogenic reactor was left uncontrolled. The feed used was based on diluted molasses with the addition of nitrogen and phosphorous as well as trace elements to ensure good bacterial growth. The details of the reactor system, feed composition and analytical methods are described elsewhere (Romli *et al.*, submitted)

All the experiments in this study are step changes in the total loading of the reactor system based on steady state operation before the disturbances. This is achieved by either increasing the molasses concentration in the feed by 100 % (concentration step) or by doubling the feed flow rate (hydraulic step). The duration of these disturbances was 12 hours in the case of the concentration steps and 8 hours for the hydraulic steps. In these periods a pseudo steady state in terms of degradation rates is reached without major changes to the biomass in the reactors. At the end of the disturbances the feed concentration or flow rate was set back to its original value. For a total time of at least 24 hours the following measurements were recorded at hourly intervals: pH, gas flow rate, gas composition, volatile fatty acid (VFA) concentration and total organic carbon (TOC) concentration in both reactors, plus the base flow rate into the acidification reactor. Both step change experiments were done with and without a recycle stream; the recycle flow rate was equal to the feed flow rate (i.e. a recycle ratio of 1).

STRUCTURED MODEL

An improved version of a dynamic structured model of the anaerobic degradation process which was developed by Costello *et al.* (1991) is used for the simulations. The model is implemented in a recently developed simulation package NIMBUS (CAPE Centre, The University of Queensland, Australia, 1991). In NIMBUS the model is set up in a modular form according to the treatment

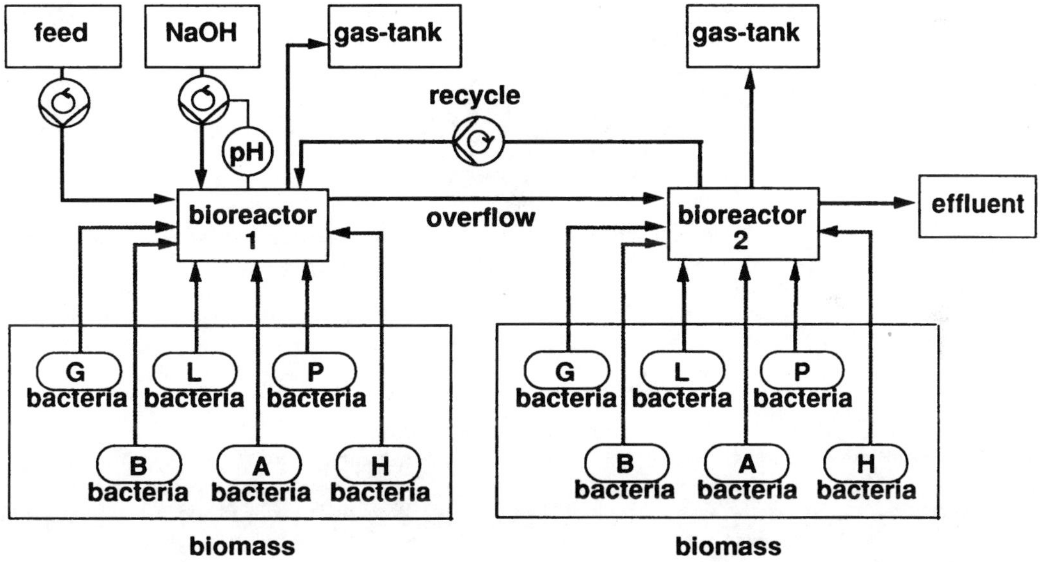

Figure 2. Model schematic of a two-stage anaerobic treatment process.

plant flowsheet (Fig. 2). To simplify the equations for the bioreactor modules, the biomass calculations are done in special units, one for each group of bacteria involved in the degradation process. The advantages of the modular design are the better understanding of the model concepts and the reduction of code to be typed since the same module can be used several times.

The model calculations are based on carbohydrate degradation, with glucose as the carbon source, and include lactic, propionic, butyric and acetic acids as the acid components. Six different bacterial groups based on substrate requirements are incorporated, while gas phase calculations are carried out for methane, carbon dioxide and hydrogen. The pH value in each reactor is calculated on an overall ion balance. The model equations also account for pH and hydrogen inhibition of the biomass growth. As in the experimental set-up the pH in the acidification reactor is controlled by the addition of 1 M sodium hydroxide solution to the first reactor. The complete two-stage model with recycle consists of 49 differential and 308 algebraic equations. Further details may be found elsewhere (Keller *et al.*, in preparation).

The set of parameters used in the original model was based mainly on literature values and steady state studies for single-stage processes. During subsequent development of the model, some changes to the parameters as well as minor structural improvements to the model itself have been made to improve its performance and make it suitable for two-stage systems. The changes were verified by experimental studies with extensive analysis of different process variables. These studies were done with two-stage operation without recycle and with molasses as the carbohydrate feed. The experiments described in this work, however, were carried out with sugar cane syrup feed, and included operation of the two-stage system with recycle. The calculations of the model glucose feed concentrations used in this study were based on COD (chemical oxygen demand) measurements of the experimental feed. No other changes or adaptations within the model were done to obtain the results presented in this work.

The whole periods of the experiments were simulated in a single run. During this simulation run any input variable could be altered at a specified time in order to achieve the same step change profile as in the experimental study described above. Therefore, even small variations in the biomass concentrations due to the disturbances or the change in the operation mode were taken into account for the subsequent step changes.

INCREASED LOAD WITHOUT RECYCLE

To illustrate the observed and simulated behaviour of the two-stage reactor system, six different process variables are plotted in Figures 3 to 6. They are obtained from two classes of experiment. Figures 3 and 5 show the effect of a 100% step increase in the feed concentration for a period of 12 hours, whereas the results of a 100% step increase in the feed flow rate for a period of 8 hours are displayed in Figures 4 and 6. For comparison of effects of the different step changes, the same variables (sodium hydroxide consumption rate and pH value in the second reactor) are plotted in Figures 3 and 4.

In both types of experiments there was an increase in the loading rate of 100%. Therefore the results are not expected to be significantly different. Nevertheless due to the decrease in liquid residence time in the system by the feed flow step change, some differences in the reactor behaviour should occur. These effects are shown in Figures 3 and 4. The NaOH consumption rate in both cases is similar but reaches a slightly lower maximum value during the disturbance in the case of the flow rate increase. This is the case for both the experimental as well as the simulated results although the simulation gives a lower overall estimation of the consumption rate. The behaviour of the pH in the second reactor shows some qualitative differences between the two experiments. A decrease in pH value can be observed in both cases mainly due to a higher feed concentration in the overflow from the first into the second reactor and subsequently increased production of VFA's in the second reactor. The actual drop in pH during the feed flow step change is much larger than with

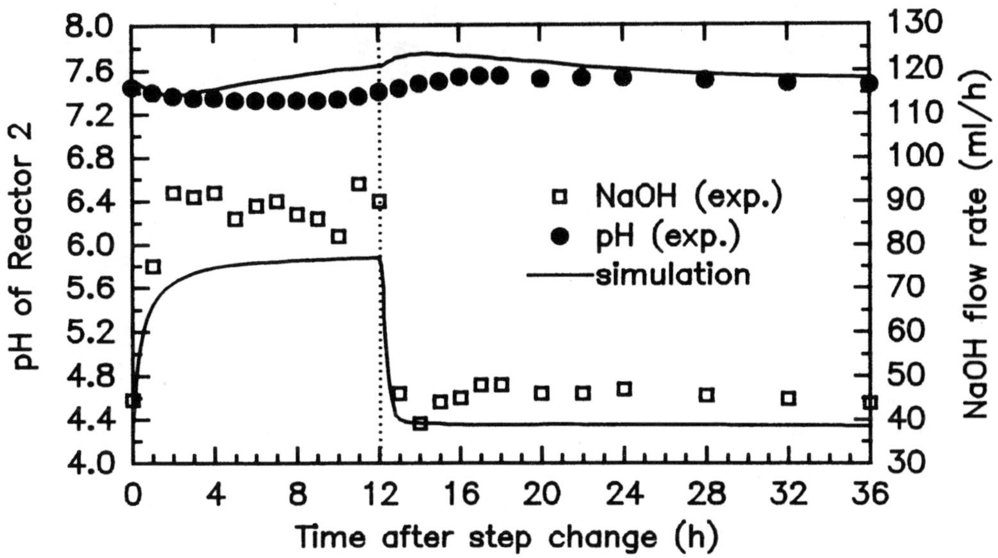

Figure 3. Effect of 100% step increase in feed concentration on pH and NaOH consumption rates.

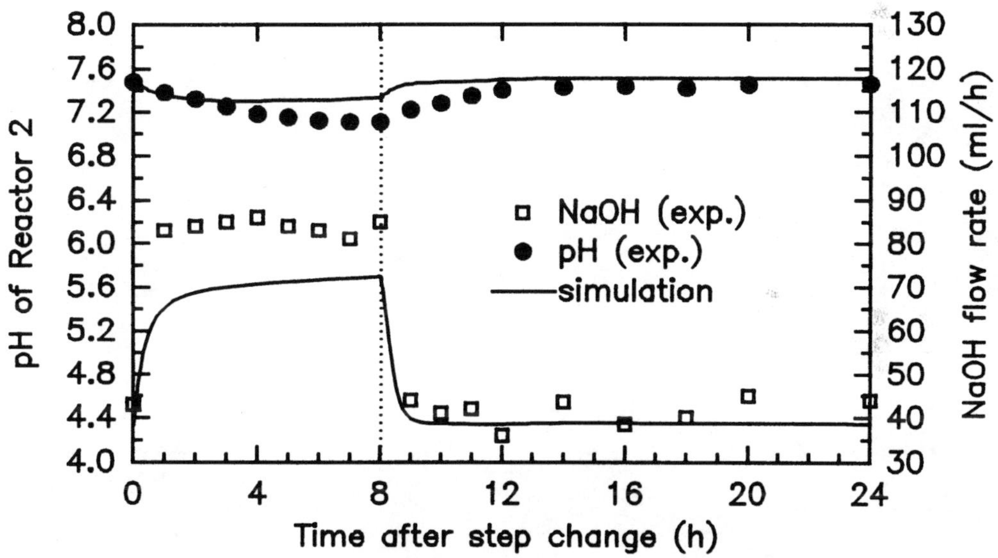

Figure 4. Effect of 100% step increase in feed flow rate on pH and NaOH consumption rates.

the feed concentration step where an increase in the pH value towards the end of the disturbance can be observed. The simulation results show the same qualitative effects although the experimental results are not perfectly matched. The pH increase during the feed concentration step starts earlier in the simulation and reaches a higher value than in the experiment. On the other hand during the feed flow step, no increase in pH is observed for the whole period of increased load.

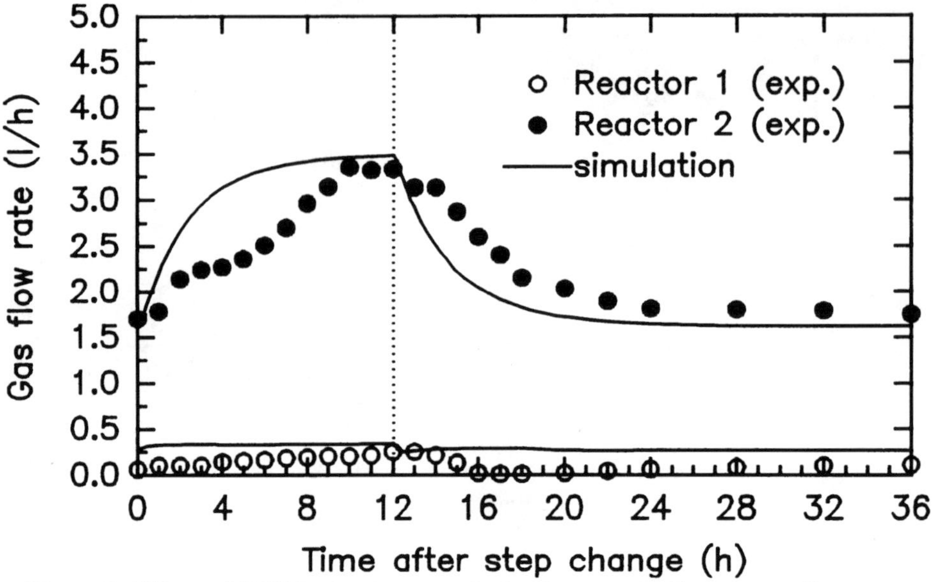

Figure 5. Effect of 100% step increase in feed concentration on gas flow rate.

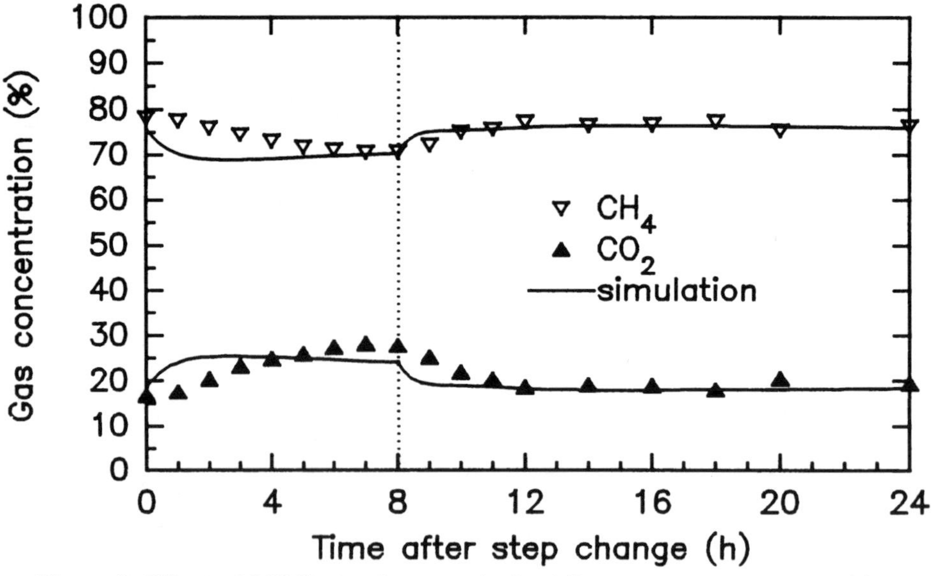

Figure 6. Effect of 100% step increase in feed flow rate on gas composition.

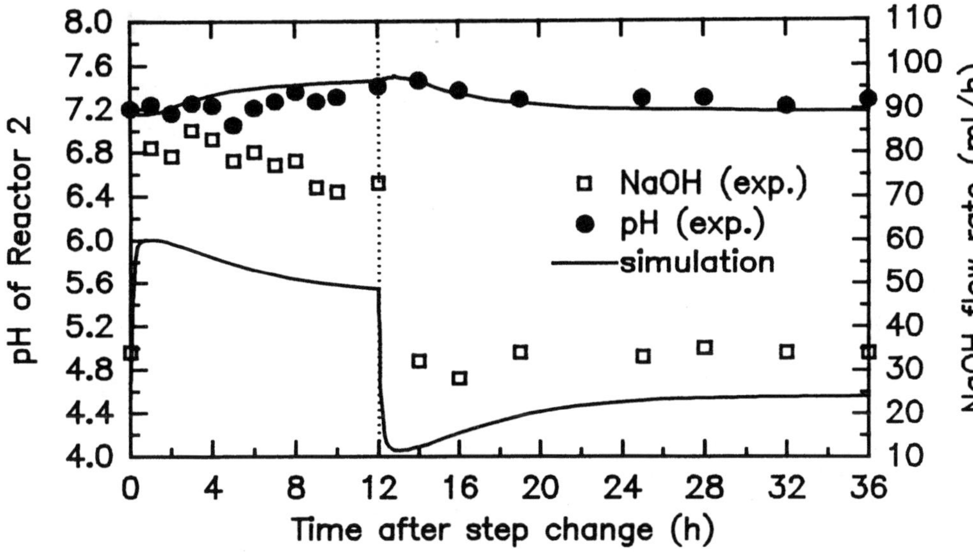

Figure 7. Effect of 100% step increase in feed concentration on pH and NaOH consumption rates (with recycle).

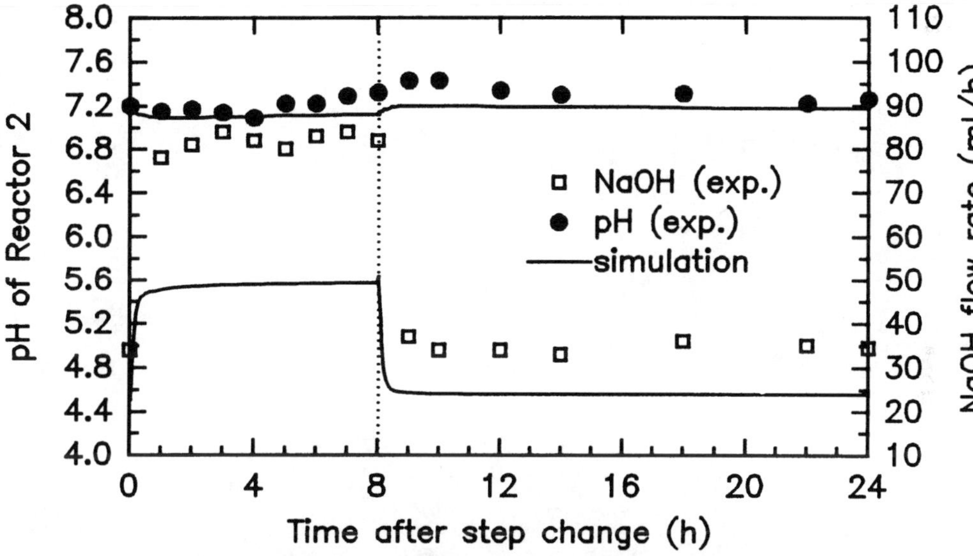

Figure 8. Effect of 100% step increase in feed flow rate on pH and NaOH consumption rates (with recycle).

notably. In addition to these effects the recycle of wastewater of neutral pH and high inorganic carbon concentration into the first reactor, which is controlled at pH 6, reduces the sodium hydroxide consumption rate due to stripping of CO_2 in the first reactor. The experimental and simulated results

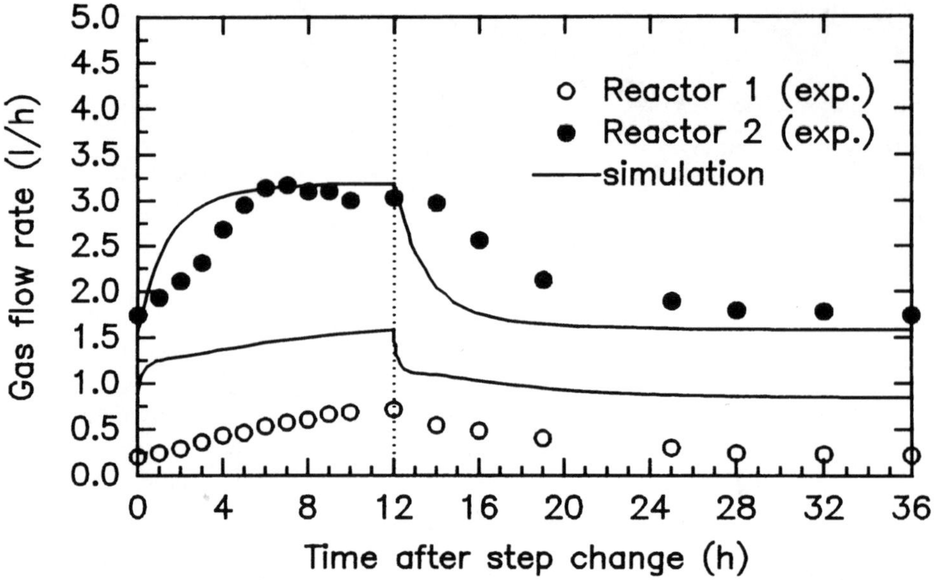

Figure 9. Effect of 100% step increase in feed concentration on gas flow rate (with recycle).

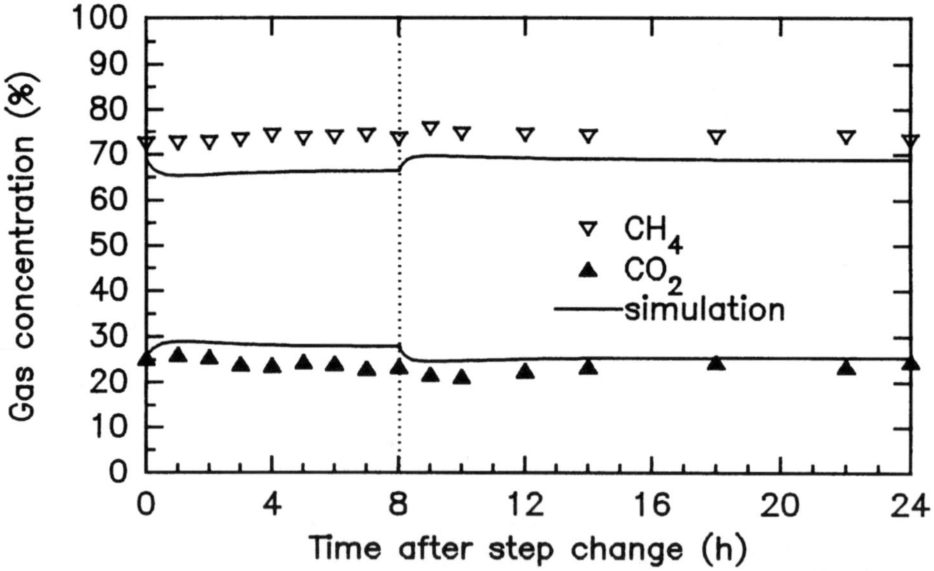

Figure 10. Effect of 100% step increase in feed flow rate on gas composition (with recycle).

for the same process variables as for operation without recycle are shown in the following graphs. Figures 7 and 9 are outcomes of a step increase in the feed concentration whereas Figure 8 and 10 are obtained from feed flow rate step changes.

The effects of the different step changes on other measured process variables are similar in terms of absolute values and trends to the ones described above but are not shown in detail in this work. To demonstrate the applicability of the model, the experimental data gas flow rates and concentrations are displayed in Figures 5 and 6. Since the first reactor is used as an acidification reactor only very low gas flow rates are measured (Fig. 5). Even during the disturbances only a slight increase can be observed. The simulation results are very similar although the predicted steady state value is somewhat higher than in the experiment. The main gas production is obtained in the second reactor where the VFA's produced in the first reactor are degraded to methane and carbon dioxide. In this reactor a major increase in gas flow rate during the high loading period can be observed; the production rate at the end of the disturbance reaches about twice the steady state value. This indicates that the overall degradation rate in the system remains relatively constant even under these severe operational changes. The VFA measurement in the effluent is proof of this observation since the total VFA concentration increases initially but decreases over the duration of the disturbance (data not shown). The simulation results are in good agreement with the measured flow rates. The steady state as well as the maximum gas production rate is quantitatively close to the corresponding measured values. The rate increase in the experimental system is initially slower and shows a second order delay function behaviour, whereas the simulation would be better fitted by a first order delay function. This discrepancy is probably due to some unknown experimental effects since the measurements from different experiments do not give precisely the same pattern. One possible explanation could be the observed build-up of foam on the liquid surface in the reactor during the disturbances. This could influence the gas transfer rate from the liquid to the gas phase.

The gas concentrations in the second reactor (Fig. 6) show a similar pattern as the gas flow rate in the same reactor. The simulated steady state and final values during the step change are in good agreement with the experimental results although the dynamic response of the model is slightly faster than that of the actual reactors.

Apart from the six process variables shown in these figures, many others were monitored during the disturbances and can be compared to the results of the simulation runs. In particular, the concentrations of different VFA's in both reactors are useful for these comparisons since they are also included in the model. Their dynamic behaviour represents directly the biological activity in the reactor and, therefore, they provide important information during model development.

The overall accuracy of the dynamic simulation in the operation of this two-stage reactor system is very high. Although the model parameters were developed for a different waste stream, the obtained simulation data are in good agreement with the experimental results. This indicates that the calculation of the feed concentration in the model based on the COD measurement of the reactor inflow is satisfactory, presumably because both the reference waste and that used in this study were predominantly simple carbohydrates.

EFFECT OF RECYCLE

After completion of the above experiments a recycle stream from the second to the first reactor was introduced. The recycle to feed rate ratio was set to 1, i.e. the same volumetric flow rate was used for the recycle stream as for the feed flow. The same step changes as for the operation without recycle were applied to the system and the same process variables were monitored. The only change in the model was the introduction of an additional stream from the second to the first reactor; no other parameter or structural changes were made.

The results obtained are significantly different from those without recycle. This can be expected since the liquid residence time in both reactors is reduced by a factor 2 and the introduction of methanogenic bacteria from the second into the first reactor changes the degradation structure

The sodium hydroxide consumption rate during steady state operation is reduced by approximately 25 % due to the introduction of the recycle stream. This is in good agreement with earlier results in steady state operation (Romli *et al.*, submitted) and can be explained by the mechanisms described in the previous paragraph. The consumption rate during the loading experiments is reduced as well although the decrease is less than expected. For the concentration step change a reduction in the range of 10-20% is recorded but for the feed flow rate change the savings in base addition are insignificant. This behaviour is probably related to unknown liquid-gas interface effects or washout problems in the first reactor due to the very low liquid residence time (0.85 h) during the feed flow step change. The observation is in accordance with the trends in the gas flow rate in the acidification reactor. During the concentration step increase, the NaOH consumption rate reaches a maximum in the initial phase of the disturbance and decreases thereafter. This corresponds with a steady increase in the gas flow rate in the same reactor as indicated in Fig. 9. In case of the feed flow rate step, no reduction in the base consumption rate can be observed (Fig. 8), which again corresponds with the measurements of the gas flow rate since a constant flow rate at a low level is observed during the disturbance (data not shown). Further studies in this area are currently being conducted to resolve the reasons for this behaviour.

The outcome of the simulation runs is a further indication that unknown and therefore unmodelled processes play a significant role in the operation with recycle. The steady state as well as the dynamic simulation results show consistently lower sodium hydroxide consumption rates than in the experiments. Although the quantitative results show significant discrepancy the qualitative behaviour of the simulation is remarkably consistent with the measurements (Fig. 7 and 8). This indicates that the model is applicable to the recycle operation but that some additional development work is required. The gas flow rate in the first reactor is most likely related to the base consumption rate as discussed previously. Therefore the differences between simulation and experiments (Fig. 9) are probably caused by the same processes as for the NaOH consumption rate. Again the qualitative behaviour is good but the actual values are significantly different from the experiments.

Other variables monitored are in good agreement with the simulation results. The pH values in the second reactor show different response functions during the two overload experiments (Fig. 7 and 8) but the simulation runs give very similar results. Also, the predicted gas flow rate in the second reactor matches the experimental measurements closely with similar differences in the short term dynamic behaviour as found in the experiments without recycle. The gas concentrations in the same reactor show little change during the feed flow disturbance although the gas flow rate increases by about 75%. The simulation results are also approximately constant but show a slightly lower methane concentration than actually measured. Nevertheless, the overall fit of the predicted to the experimental data is good with the few exceptions discussed in the previous paragraph.

CONCLUSIONS

The applicability of a dynamic structured model to a two-stage anaerobic wastewater treatment system has been shown for operation with and without recycle. In case of operation without recycle the predicted results for concentration and hydraulic step changes are in good agreement with experimental data. Both the predicted and the observed dynamic behaviour of most process variables is similar for both types of disturbance.

The use of the same model for the operation with recycle yields results which correspond well with most monitored process variables, although significant discrepancies exist with the predicted sodium hydroxide consumption rate and the gas flow rate in the first reactor. These differences are likely to be related to each other and a further investigation is being conducted to determine the reasons for these deviations.

The model is sufficiently reliable to be used as a tool for design and control projects in two-stage anaerobic treatment systems without recycle. The correct dynamic behaviour of this model is of major importance in these studies. Additional experimental investigations and possibly some modification of the model will be necessary to improve its accuracy for operation with recycle.

ACKNOWLEDGMENT

Financial support from the Swiss National Science Foundation (to J.K.) is gratefully acknowledged.

REFERENCES

Bull, M.A., Sterritt, R.M. and Lester, J.N. (1984). An Evaluation of Single- and Separated-Phase Anaerobic Industrial Wastewater Treatment in Fluidized Bed Reactors, *Biotechnol. Bioeng.* **26**, 1054-1065

Costello, D.J. (1989). Modelling, Optimisation and Control of High-Rate Anaerobic Reactors, PhD dissertation, University of Queensland

Costello, D.J., Greenfield, P.F., and Lee, P.L. (1991). Dynamic Modelling of a Single-Stage High-Rate Anaerobic Reactor - I. Model Derivation, *Water Res.* **25**, 847-858

Dinopoulou, G. and Lester, J.N. (1989). Optimization of a Two-Phase Anaerobic Digestion System Treating a Complex Wastewater, *Environ. Technol. Letters* **10**, 799-814

Keller, J., Romli, M., Greenfield, P.F. and Lee, P.L. (in preparation). Modelling and Simulation of A Two-Stage High-Rate Anaerobic Treatment System Using Simulation Package NIMBUS.

Romli, M. Greenfield, P.F. and Lee, P.L. (submitted). Effect of Recycle on a Two-Phase High-Rate Anaerobic Wastewater Treatment System, submitted to *Water Res.*

Zhang, T. C. and Noike, T. (1991). Comparison of One-Phase and Two-Phase Anaerobic Digestion Processes in Characteristics of Substrate Degradation and Bacterial Population Levels, *Wat. Sci. Tech.* **23**(7-9), 1157-1166

Pergamon

Wat. Sci. Tech. Vol. 28, No. 11-12, pp. 209–218, 1993.

0273–1223/93 $7·00 + 0·00

EVALUATION OF SENSITIVITY AND OBSERVABILITY OF THE STATE VECTOR FOR SYSTEM IDENTIFICATION AND EXPERIMENTAL DESIGN

E. Ayesa*, J. Flórez**, L. Larrea* and J. L. García-Heras*

** Department of Environmental Engineering, CEIT, Apdo. 1555, 20080 San Sebastián, Spain*
*** Elecricity Department, Escuela Superior de Ingenieros Industriales, University of Navarra, Apdo. 1674, 20080 San Sebastián, Spain*

ABSTRACT

This paper presents two new algorithms to evaluate numerically the state and coefficient's observability and sensitivity in dynamic models. The observability algorithm computes the numerical condition of observability matrix as an index of the ability to identify the unknown states or coefficients. The sensitivity algorithm quantifies dynamically the influence of unknown variables on measurable variables. Both algorithms are useful tools to predict the identification accuracy and to design the best operational conditions to estimate each one of the unknown parameters. The algorithms have been applied to discuss the identification of kinetic coefficients in the IAWPRC biodegradation model, usinganexperimental procedure based on changing the values of reactor volumes.

KEYWORDS

Observability, sensitivity, identification, activated sludge, modeling, kinetic coefficients.

INTRODUCTION

Identification of high dimension dynamic systems has been frequently limited by the overparametrized structure of model equations. In these cases, it is a useless effort to develop better identification algorithms improving their accuracy. The first logical step must be the a priori identifiability analysis of the involved kinetics (Beck, 1989).

A high sensitivity does not guarantee the identifiability of a coefficient, specially when non-linear kinetics appear in the model structure. Some errors tend to be self-canceling among the coefficients, in terms of the net model response, and therefore not detectable. But this self-cancellation in non-linear structures is generally a function of the numerical range of variables involved in the kinetics, and it can disappear when the amount of separate information for each one increases.

The observability matrix (with different structures) provides information about the ability to identify the unknown variables using the measurable variables (Bastin, 1990, Chen and Bastin, 1991). This characteristic can be evaluated in an instantaneous sense and also storing the cumulative capability during a specific experiment (Brammer and Siffling, 1989).

For a practical utilization of these concepts, it is important to develop an algorithm that computes in a numerical way the degree of system observability. This algorithm must also be flexible to define the unknown and measurable variables considered in different experiments.

The observability results will provide a very interesting tool to design experimental conditions from the point of view of system identification and also additional information, more realistic to validate the identification results than the error covariance matrix used in recursive parameter estimation and filtering.

In this paper, a new type of algorithm is presented that is based on observability and sensitivity concepts. It has been applied to an experimental procedure presented in Larrea *et al.*, (1992), analyzing from the new point of view the operational conditions used during the identification.

DEVELOPED ALGORITHMS

Observability algorithm

The observability in a linear system is determined by the ability of every state to influence the output of the system. In a linear discrete free system, where $\mathbf{A}(k)$, $\mathbf{B}(k)$, and $\mathbf{C}(k)$ are the plant matrix, input matrix and measurements matrix.

$$\mathbf{x}(k+1) = \mathbf{A}(k)\cdot \mathbf{x}(k) + \mathbf{B}(k)\cdot \mathbf{u}(k) \tag{1}$$

$$\mathbf{y}(k) = \mathbf{C}(k)\cdot \mathbf{x}(k) \tag{2}$$

The state $\mathbf{x}(k_1)$ of system (1) is said to be observable in the interval $[t_0,t_1]$ if it can be uniquely determined on the basis of knowing y(k) in the interval $[t_0,t_1]$. It can be demonstrated (Brammer and Siffling, 1989)that this condition is equivalent to the non-singularity of the observability matrix $\mathbf{M}(k_1,k_0)$

$$\mathbf{M}(k_1,k_0) = \sum_{\kappa=k_0}^{k_1} \mathbf{\Phi}^{T}(\kappa,k_1)\cdot \mathbf{C}^{T}(\kappa)\cdot \mathbf{C}(\kappa)\cdot \mathbf{\Phi}(\kappa,k_1) \tag{3}$$

where $\mathbf{\Phi}(\kappa,k_1)$ is the state transition matrix from discrete period k_1 to κ. The equation (3) admits a recursive expression,

$$\mathbf{M}(k+1,k_0) = \mathbf{A}^{-T}(k)\cdot \mathbf{M}(k,k_0)\cdot \mathbf{A}^{-1}(k) + \mathbf{C}^{T}(k+1)\cdot \mathbf{C}(k+1) \tag{4}$$

$$\mathbf{M}(k_0,k_0) = \mathbf{C}^{T}(k_0)\cdot \mathbf{C}(k_0) \tag{5}$$

and considering that,

$$\mathbf{M}(k_0,k+1) = \mathbf{\Phi}^{T}(k+1,k_0)\cdot \mathbf{M}\cdot (k+1,k_0)\cdot \mathbf{\Phi}(k+1,k_0) \tag{6}$$

a recursive law can be obtained to actualize in a simple manner the matrix $\mathbf{M}(k_0, k+1)$ (Acuña, 1992),

$$\mathbf{M}(k_0,k+1) = \mathbf{M}(k_0,k) + \mathbf{\Phi}^{T}(k+1,k_0)\cdot \mathbf{C}^{T}(k+1)\cdot \mathbf{C}(k+1)\cdot \mathbf{\Phi}(k+1,k_0) \tag{7}$$

$$\mathbf{M}(k_0,k_0) = \mathbf{C}^{T}(k_0)\cdot \mathbf{C}(k_0) \tag{8}$$

The ability to invert the matrix $\mathbf{M}(k_0, k_1)$ shows the capability to observe the initial state $\mathbf{x}(k_0)$ during the period $[t_0, t_1]$. The final state $\mathbf{x}(k_1)$ will be related with the initial by the state transition matrix $\mathbf{\Phi}(k_1, k_0)$.

The numerical condition has been chosen to evaluate in a numerical manner the degree of singularity of a square matrix. It is defined by the next expression (Forsythe *et al.*, *1977*):

$$cond(\mathbf{A}) = \max\|a_j\| \cdot \max\|a'_j\| \qquad j = 1, n \tag{9}$$

where,

$$\|\mathbf{a}_j\| = \sum_{i=1}^{n} |\mathbf{a}_{ij}| \tag{10}$$

$$\|\mathbf{a}'_j\| = \sum_{i=1}^{n} |\mathbf{a}'_{ij}| \tag{11}$$

a_{ij} = element of matrix **A**

a'_{ij} = element of matrix $\mathbf{A}^{-1}$

The values obtained by the algorithm can be compared with the maximum condition number computed up to the machine epsilon of a non-singular matrix (Forsythe *et al.*, *1977*). For example, in the μVAX-3500 working with double precision used in the identification, the maximum condition for a non-singular matrix is :

$$Maximum\ condition \cong 2.0 \times 10^{14}$$

Sensitivity algorithm

The numerical computation of the sensitivity for every state or coefficient is a very interesting tool to discuss the results obtained in the system identification. The sensitivity has been computed as the influence from a change in the numerical value of a state or coefficient to the measurable variables. Hence, it is also an intrinsic property of the plant and measurements models.

The different sensitivity values have been grouped in a sensitivity matrix $\mathbf{S}(k)$, where a generic term $s_{ij}(k)$ shows the evolution during the discrete period of the measurable variable $y_i(k)$ respect to the numerical change in the state or coefficient $x_j(k)$. Hence,

$$s_{ij}(k) = \frac{\Delta y_i}{\Delta x_j} \cong \frac{\partial y_i}{\partial x_j} \tag{12}$$

$$\mathbf{S}(k) = \mathbf{C}(k) \cdot \mathbf{A}(k) \tag{13}$$

For a non-linear continuous system, every generic element of matrix $\mathbf{S}(k)$ can be evaluated as,

$$s_{ij}(k) = \frac{\partial y_i}{\partial x_j} = \frac{\partial h_i}{\partial x_j} + \sum_{l=1}^{n} \frac{\partial h_i}{\partial x_l} \cdot \frac{\partial^2 f_l}{\partial t\, \partial x_j} \cdot T +$$

$$+ \sum_{l=1}^{n} \sum_{m=1}^{n} \frac{\partial h_i}{\partial x_l} \cdot \frac{\partial^2 f_l}{\partial t\, \partial x_m} \cdot \frac{\partial^2 f_m}{\partial t\, \partial x_j} \cdot \frac{T^2}{2!} + \cdots \tag{14}$$

where,

$$\dot{\mathbf{x}}(t) = \mathbf{f}\{\mathbf{x}(t)\} \tag{15}$$

$$\mathbf{y}(t) = \mathbf{h}\{\mathbf{x}(t)\} \tag{16}$$

The high dimension of matrix **S**(k) recommends the utilization of a non-dimensional vector that stores the sensitivity of every state or coefficient to the whole of the measurable variables. The structure of this vector is (Ayesa, 1992),

$$sens_i(k) = \sum_{j=1}^{m} abs(s_{ij} \cdot \frac{\overline{x}_i}{\overline{y}_j}) \qquad (17)$$

where,

$\overline{x}_i$ = mean value of x_i during the experiment

$\overline{y}_j$ = mean value of y_j during the experiment

In this manner, every element of the sensitivity vector shows normalized information on the influence from the state or coefficient to the whole of the measurable variables.

EXPERIMENTAL PROCEDURE

The experimental procedure used to verify the behaviour of the observability and sensitivity algorithm was presented in a previous paper (Larrea *et al.*, 1992) and it is funded on devising operational conditions in which the coefficients have considerable influence on the measurable variables.

Activated Sludge Mathematical Model

The model proposed by the IAWPRC Task Group on Modeling (Henze *et al.*, 1986) has been simplified for the oxic reactions. Thus denitrification has not been considered and it is assumed that there are no oxygen limitations. It has also been assumed that there is neither soluble (SI) nor particulate (XI) inert organic matter in the influent wastewater and the same for the associated nitrogen. A steady state equation was used for thickening in the secondary settler and no suspended solids content in the effluent was assumed. With these assumptions the unknown states of the model are:

SS: Readily biodegradable substrate (g $CODm^{-3}$)
XS: Slowly biodegradable substrate (g $CODm^{-3}$)
XBH: Active heterotrophic biomass (g $CODm^{-3}$)
XBA: Active autotrophic biomass (g $CODm^{-3}$)
XP: Particulate products arising from biomass decay (g $CODm^{-3}$)
SNO: Nitrate and nitrite nitrogen (gNm^{-3})
SNH: NH_4 nitrogen (gNm^{-3})
SND: Soluble biodegradable organic nitrogen (gNm^{-3})
XND: Particulate biodegradable organic nitrogen (gNm^{-3})

It is considered that the following variables can be measured on-line:

COD_T = Total Chemical Oxygen Demand (g $CODm^{-3}$)
COD_F = Filtered Chemical Oxygen Demand (g $CODm^{-3}$)
NH_4-N = Ammonia Nitrogen (g Nm^{-3})
NO_3-N = Nitrates Nitrogen (g Nm^{-3})
TKN_T = Total Kjeldahl Nitrogen (g Nm^{-3})
TKN_F = Filtered total Kjeldahl Nitrogen (g Nm^{-3})
OUR = Oxygen Uptake Rate ($gO_2m^{-3}day^{-1}$)

The stoichiometric and kinetic coefficients considered unknowns are:

Y_H: Yield for heterotrophic biomass (g cell COD formed (gCOD oxidized)$^{-1}$)
Y_A: Yield for autotrophic biomass (g cell COD formed (gN oxidized)$^{-1}$)
μ_H: Maximum specific growth rate for heterotrophic biomass (day^{-1})
K_S: Half-saturation coefficient for heterotrophic biomass (gCODm^{-3})
μ_A: Maximum specific growth rate for autotrophic biomass (day^{-1})
K_{NH}: Ammonia half-saturation coefficient for autotrophic biomass (gNm^{-3})
K_H: Maximum specific hydrolysis rate (g slowly biodegradable COD(g cell COD day)$^{-1}$)
K_X: Half-saturation coefficient for hydrolysis of slowly biodegradable substrate (g slowly biodegradable COD(g cell COD)$^{-1}$)
K_A: Ammonification rate (m^3(gCOD day)$^{-1}$)
b_H: Decay coefficient for heterotrophic biomass (day^{-1})
b_A: Decay coefficient for autotrophic biomass (day^{-1})

The relationship used between measurable variables and state variables in the mixed liquor are:

$COD_F = SS$ (18)
$NH_4\text{-}N = SNH$ (19)
$NO_3\text{-}N = SNO$ (20)
$COD_{VSS} = COD_T - COD_F = XBH + XBA + XP + XS$ (21)
$TKN_F - (NH_4\text{-}N) = SND$ (22)
$TKN_T - TKN_F = i_{XB} \cdot (XBH + XBA) + i_{XP} \cdot XP + XND$ (23)

$$OUR = \frac{1 - Y_H}{Y_H} \cdot \frac{\mu_H \cdot XBH \cdot SS}{K_S + SS} + \frac{4.57 - Y_A}{Y_A} \frac{\mu_A \cdot XBA \cdot SNH}{K_{NH} + SNH} \qquad (24)$$

The unknown coefficients and states in the two in-series biodegradation reactors at time kT are grouped in the vector $\mathbf{x}(k)$, and the on-line measurements available from the plant at time kT in the vector $\mathbf{y}(k)$

Experimental test

The experiment is proposed to be carried out in a continuous AS pilot or lab-scale plant that has two in-series biodegradation reactors and a settler with sludge recycle to the first reactor. The latter has a volume smaller than the second reactor and it acts as an oxic selector. In this work a total volume of 30 litres for the reactors has been chosen.

Once the starting steady state has been achieved the test goes on by reducing in steps the first reactor volume and increasing the volume of the second one to maintain the overall hydraulic retention time. The used steps are listed in the Table 1.

The presumed real values for the model coefficients (Table 1) have been taken from those proposed by the literature (Henze *et al.*, 1986, Henze *et al.*, 1987, Bisdrup and Grady, Jr., 1988). The characteristics used for the influent wastewater can also be found in Table 1. They correspond to a synthetic wastewater. Both the wastewater characteristics and the plant dimensions have been chosen according to the experimental work presently being carried out at CEIT. A more complete description of the experimental procedure can be found in Larrea *et al.*, 1992.

TABLE 1.

Operational conditions of the proposed experiment			Influent wastewater	Presumed real coefficients
$V_1 + V_2$	= Overall reactors volume	= 30 liters		
q	= Overall Hydraulic retention time	= 7.6 hours		
q_c	= Sludge retention time	= 8.2 days		μ_H = 6 day^{-1}
	Volume changes			K_S = 20 gCOD·m^{-3}
Time	V_1	V_2		μ_A = 0.8 day^{-1}
(days)	(liters)	(liters)		K_{NH} = 1 gNH$_4$-N·m^{-3}
0	10	20	SS_1 = 200 gCOD·m^{-3}	K_H = 3 gCOD·(gCOD·day)$^{-1}$
0.1	8	22	XS_1 = 300 gCOD·m^{-3}	K_X = 0.1 gCOD·gCOD^{-1}
0.2	6	24	SI_1 = 0 gCOD·m^{-3}	K_A = 0.08 m^3·(gCOD·day)$^{-1}$
0.3	5	25	XI_1 = 0 gCOD·m^{-3}	b_H = 0.62 day^{-1}
0.4	4	26	SNH_1 = 30 gNH$_4$-N·m^{-3}	Y_H = 0.67 gCOD·gCOD^{-1}
0.5	2.5	27.5	SND_1 = 15 gN·m^{-3}	Y_A = 0.24 gCOD·gN^{-1}
0.6	1.5	28.5	SNO_1 = 0 gN·m^{-3}	b_A = 0.1 day^{-1}
0.7	1	29	XND_1 = 6 gN·m^{-3}	fp = 0.08
0.8	0.8	29.2	XBH_1 = 0 gCOD·m^{-3}	i_{XB} = 0.086 gN·gCOD^{-1}
0.9	0.6	29.4	XBA_1 = 0 gCOD·m^{-3}	i_{XP} = 0.06 gN·gCOD^{-1}

Identification procedure

The identification algorithm used to estimate the values of states and coefficients of the plant is based on an extended Kalman filter (EKF) modified to improve their numerical behaviour in high dimension systems, and to estimate simultaneously the states and coefficients (Ayesa *et al.*, 1991, Ayesa, 1992).

To assess the validity of the identification algorithm, simulated results of measurable variables, generated from presumed real coefficients, have been employed as real results. It is assumed that these results arrive as on-line measurements every 0.002 days (approximately 3 minutes) and are affected by a Gaussian white noise whose standard deviation laws are linear approaches from routine analysis at CEIT laboratories.

Several initial estimates for the coefficients have been tested to verify the behaviour of the identification algorithm. The initial erroneous estimates have been elaborated considering that a proper one should produce, by computer simulation of the experiment, measurable variables values not too distant from the real experiment results obtained (Larrea *et al.*, 1992).

RESULTS AND DISCUSSION

The biggest difficulty to identify the dynamic behaviour of the plant is the estimation of coefficients related with non-linear (Monod) kinetics. The rest of the unknown coefficients are identified rapidly and accurately (Ayesa, 1992). In the model presented here, there are three non-linear kinetics: growth of heterotrophic and autotrophic biomass and hydrolysis.

Figures 1 to 6 show the convergence curves for the coefficients μ_H, K_S, μ_A, K_{NH}, K_H and K_X, related with the three Monod kinetics. The lines A, B, C and D represent the evolution of the estimation from different erroneous initial points (Larrea *et al.*, 1992). The trajectory of every estimation can be analyzed using the information of the observability and sensibility algorithms, and discussing separately each one of the Monod kinetics.

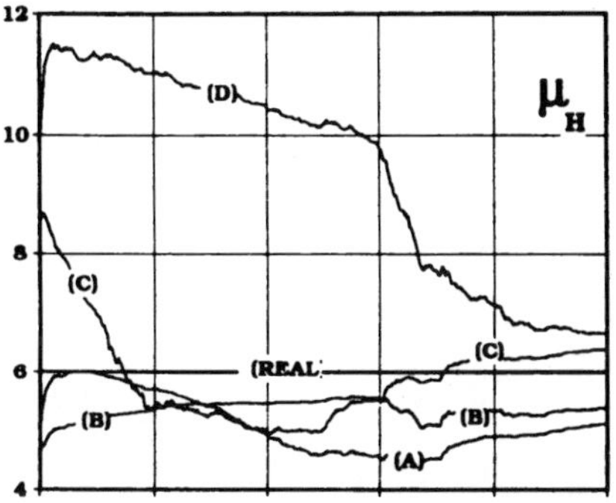

(Fig. 1)

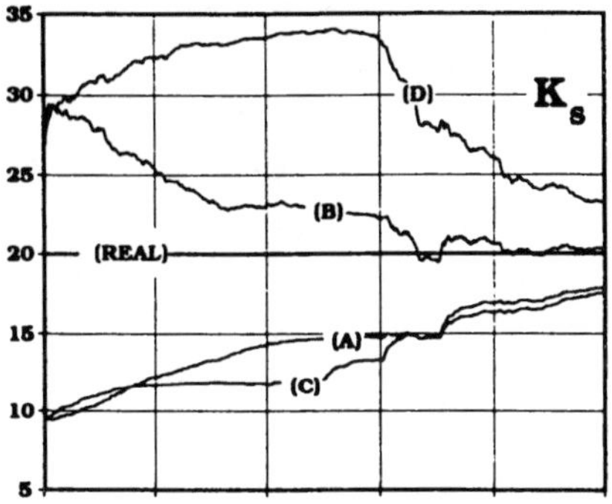

(Fig. 2)

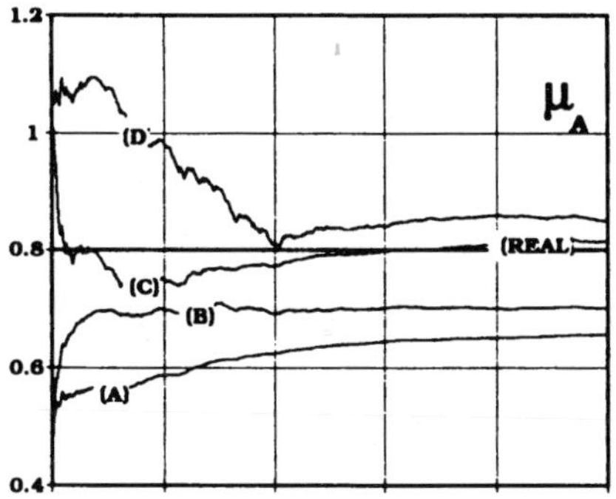

(Fig. 3)

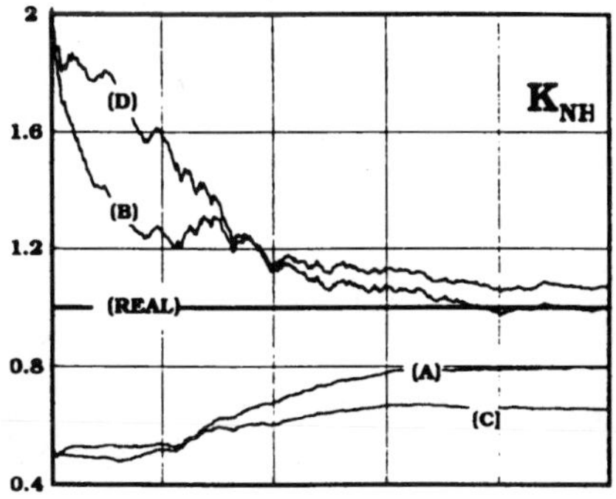

(Fig. 4)

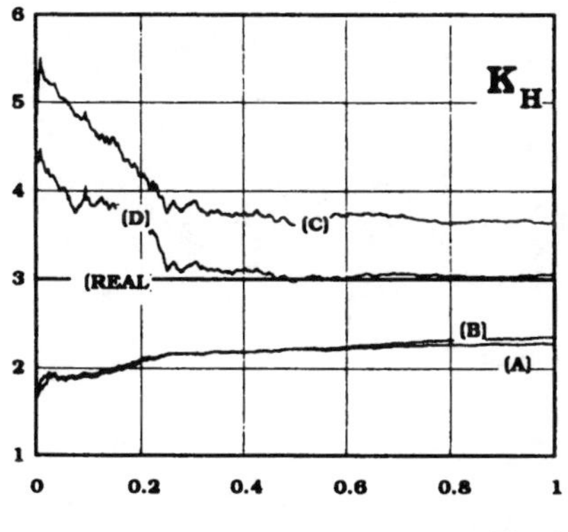

(Fig. 5)

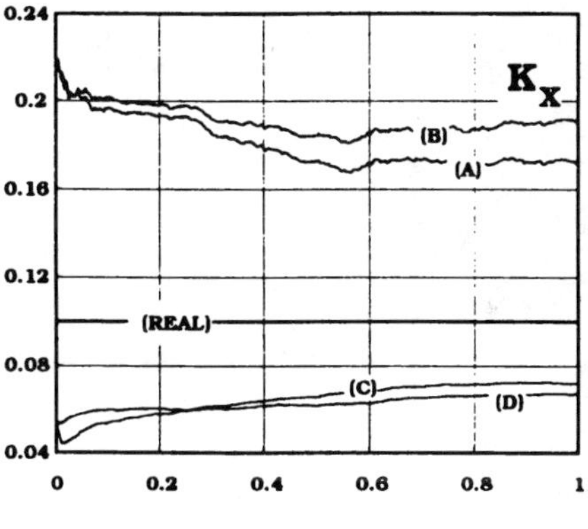

(Fig. 6)

Time (Days)

Figures 1-6: Estimation of Monod kinetic coefficients by the identification algorithm.

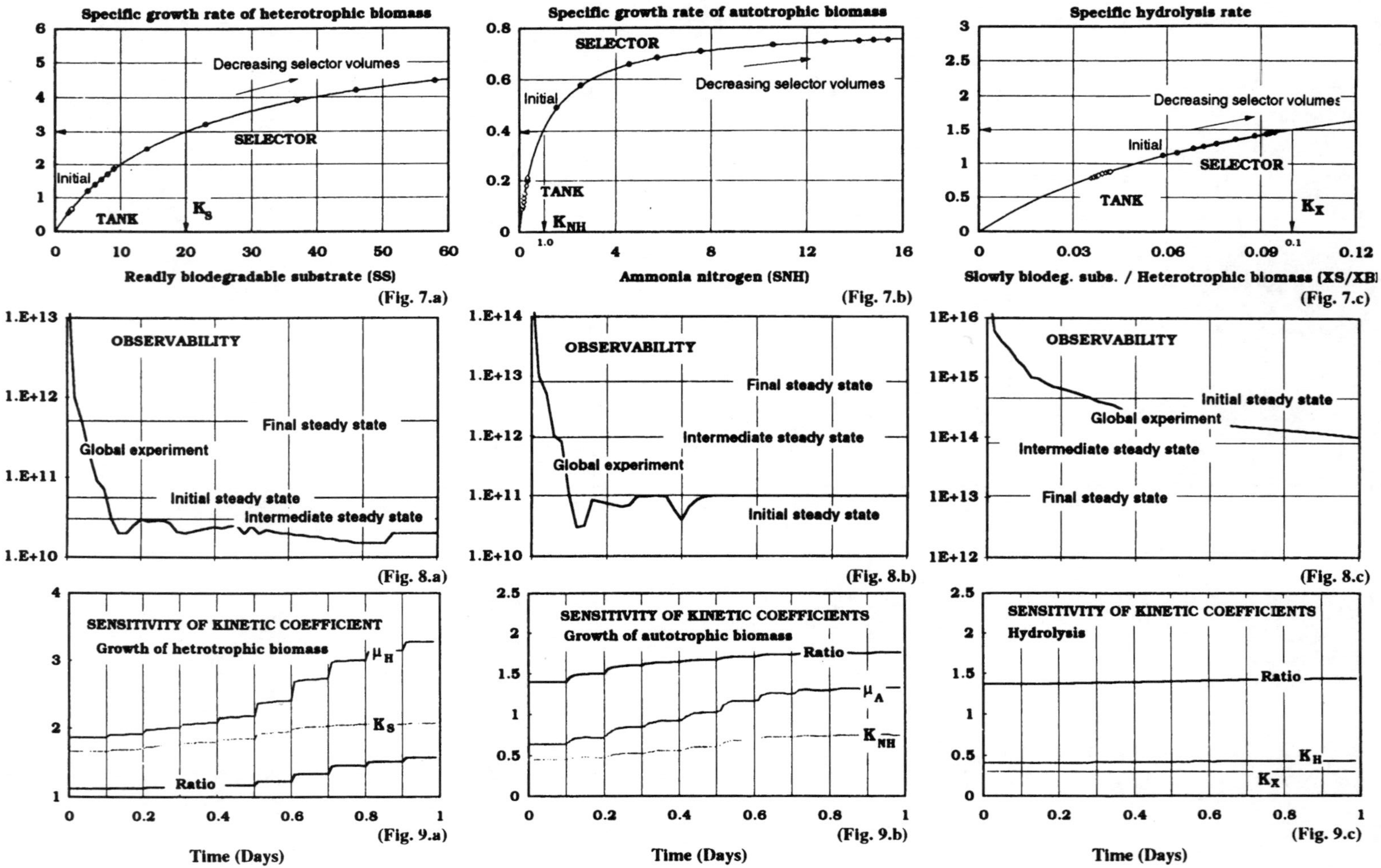

Figures 7(a,b,c), 8(a,b,c) and 9(a,b,c): Operational points on the kinetic curves, numerical conditions of observability matrices and sensitivity trajectories for the Monod kinetics in the IAWPRC (simplified) biodegradation model.

Figures 7.a to 9.a represent the growth of heterotrophic biomass. The first figure shows the shape of the kinetic curve and several superposed points that correspond with the operating points traversed by the states in selector and aeration tank during the experiment. The possibility to identify the maximum rate (μ_H) and the saturation parameter (K_S) separately is clearly related with the ability to obtain operating points that go over a good big portion of the curve,especially near the saturation parameter. Hence, one can observe the very important function of selector tank, whose points traverse the bigger part of the operating points of the plant. Also, it is easy to observe that the operating points nearest the saturation parameter are traversed during the intermediate steps, in the second half of the experiment.

Figure 8.a shows the evolution of the numerical condition of the observability matrix $\mathbf{M}(k_0,k_1)$ when the variables considered unknown are all the states of the plant and only the two coefficient related with the analyzed kinetic (μ_H, K_S). The horizontal lines represent the condition of the observability matrix corresponding to simulate the steady states of initial, intermediate and final steps of experiment. The complete experiment obtains better observability (less condition) but it does not represent a very significant improvement from the intermediate step.

Finally, figure 9.a shows the evolution of normalized sensitivity for the coefficients μ_H and K_S and also the ratio of sensitivities. The possible compensation between the maximum specific rate and the saturation parameter is due to the lack of some separate information from each one of these coefficients. Operating conditions that produce constant proportionality in their sensitivities can eliminate the ability to discriminate between the numerical values of the maximum specific rate and the saturation parameter. In other sense, the use of plant measurements generated under transient proportionality between these coefficients, breaks their possible compensation and improves the system observability. In the case analyzed here the proportionality changes especially during the second half of the experiment.

The study of operating conditions points over the kinetic (fig. 7.a), the numerical condition of observability matrix (fig. 8.a) and sensitivity trajectories (fig. 9.a) point out that the better conditions to identify the coefficients correspond to the second half of the experiment. This conclusion is clearly corroborated by the identification results (fig. 1 to 2).

The next figures (fig. 7.b to 9.b) show the same analysis referred to the growth of autotrophic biomass. The operating points nearest the saturation parameter, the steady state with minimum numerical condition of observability matrix and the biggest change in the sensibility ratio correspond to the initial part of the experiment. This conclusion is also corroborated by the convergence in the identification (fig. 2 to 3).

Finally, the last Monod kinetic corresponds to the hydrolysis kinetic (fig. 7.c to 9.c). In this case there is an important lack of operating points near the saturation parameter (fig. 7.c), the condition of the observability matrix is very high for all the steps (fig. 8.c) and the ratio of sensitivities remains constant (fig. 9.c). This situation indicates a typical case of low observability, that is corroborated by the low convergence in the estimation (fig. 5 to 6).

In summary, a general analysis of the results shows that the observability and sensitivity algorithms provide very interesting information about the ability to identifiy the coefficients of non-linear kinetics under different operational conditions. These algorithms evaluate in a numeric way an intrinsic characteristic of the plant and measurement models, independently of the identification procedure and the initial estimations.

SUMMARY AND CONCLUSIONS

This paper presents two new algorithms to evaluate the observability of dynamic systems under different operational conditions. The observability algorithm computes the numerical condition of the observability matrix and providing information about the best conditions to identifiy each kinetic coefficient. The join of these operational conditions, simultaneously or consecutively, will determine the optimal experimental design to identify the unknown values.

The sensitivity algorithm shows the operational conditions that provide more differentiated information about the maximum specific rate and the saturation parameter in a non-linear kinetic. Hence, it also determines the best conditions to identify separately the coefficients.

It is very important to note that the most significant information about the kinetic coefficients is stored in a narrow band of operational conditions, near the saturation parameter. When the objective of an experiment is the identification, and it is impossible to traverse all the points on the kinetic curve, it would be desirable to operate in these privileged zones.

Further research of this group aims to improve the computation of numerical conditions of observability matrices, developing simultaneously other criteria to evaluate the degree of singularity of a square matrix. Other interesting objectives are focused on the systematic analysis of the steady states observability under several operational conditions and different unknown coefficients, to establish in a more rigorous way the best strategy to identify each kinetic.

ACKNOWLEDGMENTS

The authors wish to express their gratitude to the Basque Government for its financial support.

REFERENCES

Ayesa E. (1992). *Diseño y desarrollo de un algoritmo para la identificación on-line del proceso de fangos activos*. PhD thesis. Universidad de Navarra.

Ayesa E., Flórez J., García-Heras J. and Larrea L. (1991). State and coefficients estimation for the activated sludge process using a modified Kalman filter algorithm. *Wat. Sci. Tech.* Vol. 24, No. 6, pp. 235-247.

Acuña A. (1992). Estudio de Observabilidad y Sensibilidad de un Modelo Dinámico de Biodegradación por Fangos Activos. *Proyecto fin de carrera E.T.S. de Ingenieros Industriales de San Sebastián.* Universidad de Navarra, San Sebastián.

Bastin G. and Dochain D. (1990). *On-line Estimation and Adaptive Control of Bioreactors*. Elsevier Science Publishers B.V. Amsterdam.

Beck (1989). System Identification and Control. In: *Dynamic Modeling and Expert Systems in Wastewater Engineering*. Lewis Publishers, Inc., Chelsea, Michigan.

Brammer K. and Siffling G. (1989). *Kalman-Bucy Filters*. Artech House, Inc, Norwood.

Chen L. and Bastin G. (1991). *On the Model Identifiability of Stirred Tank Bioreactors*. Internal Report, Université Catholique de Louvain, Belgium.

Henze M., Grady Jr. C.P.L., Gujer W., Marais G.vR. and Matsuo T. (1986). *Activated Sludge Model No.1*. Technical Report by IAWPRC Task Group on Mathematical Modelling for Design and Operation of Biological Wastewater Treatment.

Larrea L., García-Heras J., Ayesa E. and Flórez J. (1992) Designing experiments to determine the coefficients of activated sludge models by identification algorithms. *Wat. Sci. Tech.* Vol. 25, No. 6, pp. 149-165.

Pergamon

Wat. Sci. Tech. Vol. 28, No. 11-12, pp. 219–229, 1993.

0273–1223/93 $7·00 + 0·00

PARAMETER ESTIMATION AND SENSITIVITY ANALYSIS OF AN ACTIVATED SLUDGE MODEL USING MONTE CARLO SIMULATION AND THE ANALYST'S INVOLVEMENT

Marcos von Sperling

Department of Sanitary and Environmental Engineering, Escola de Engenharia - UFMG, Av. Contorno 842 - 7° andar, 30110-060 - Belo Horizonte, Brazil

ABSTRACT

The present work describes an adaptation of the regionalized sensitivity analysis based on Monte Carlo simulations for the parameter estimation and sensitivity analysis of an activated sludge model. The procedure described should be used when observed data are available for the model calibration, which is nevertheless still limited by the problems inherent to activated sludge models (uncertainty and lack of identifiability). The selection between *good* and *bad* performance of the model is judged based on the Coefficient of Determination. The application of the procedure to an 11-parameter 4-state dynamic activated sludge model used for operational control was considered satisfactory. The method is simple and yet robust, and the analyst's involvement in the interpretation of the results and decision upon the next steps to be taken increases its controllability.

KEYWORDS

Modelling; parameter estimation; sensitivity analysis; Monte Carlo simulation; activated sludge.

INTRODUCTION

Recently a number of control strategies for the activated sludge process have been developed, based on the utilization of dynamic models of the process. Together with the identification of the model structure, the estimation of the parameter values is the core of the model performance and, as a result, of the control strategy and ultimately of the effluent quality. An additional tool in the whole procedure of model structure identification and parameter estimation is represented by the sensitivity analysis, which gives an indication of the relative sensitivity of the model predictions to changes in the values of the parameters.

The methods developed for the estimation of the parameter values of well defined systems (i.e. "those for which a valid model structure is known and for which parameter values can be accurately specified") should not be applied to poorly defined systems, such as biological wastewater treatment processes (Whitehead and O'Connell, 1984). Activated sludge systems can be considered ill-defined systems because of (a) difficulty in the quantification of the various complex biochemical reactions involved, (b) difficulty in the representation of the reactor hydraulics in full-scale systems, (c) difficulty in gathering all the required field data, (d) the usual large number of parameter and state variables included in the current models, (e) the high non-linearities in the model equations, and (f) lack of identifiability in some equations.

The most common method for block parameter estimation (in which the parameters are kept constant during the estimation period) is the minimization of an objective function representing the sum of the squares of the errors (an error, or residual, represents the difference between the observed and the predicted values). However, this procedure requires the compliance to certain assumptions about the distribution of the errors, such as that they should be normally distributed, with zero mean and constant variance. Violation of these assumptions would still permit the estimation of the parameters, but their interpretation would be incorrect (Clarke, 1983). Unfortunately, complete fulfilment of these criteria in the case of wastewater treatment modelling is probably more an exception than a rule.

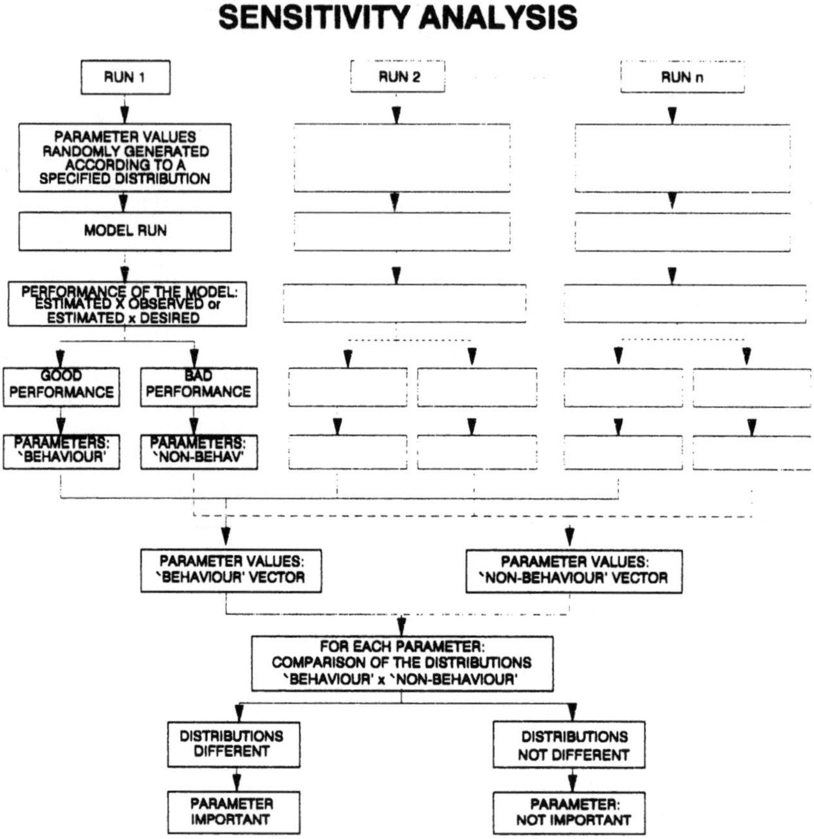

Fig. 1 Flow chart of the procedure used for the sensitivity analysis

An alternative approach, first applied to the study of environmental systems by Spear and Hornberger (1980), is the use of the regionalized sensitivity analysis based on Monte Carlo simulations (see Figure 1). The concept behind this method is the utilization of a simulation model together with a classification algorithm, which translates whether the predicted variables are representative of the observed (or expected) behaviour of the system (Whitehead and O'Connell, 1984). Because of the recognition of the ill-definition of the system, uncertainty is introduced by specifying the parameters via probability density functions covering a particular range of values (and not point estimates). A large number of model runs (Monte Carlo simulations) is carried out, each time with a different set of parameter values randomly selected from the specified distributions. The result of each run is then interpreted by the classification algorithm, which incorporates the parameter values in the class *behaviour* if the result was good (good representation of the observed or expected behaviour) and in the class *non-behaviour* if the result was `not good'. After all the runs have been carried out statistical analysis is conducted, comparing the distributions of each of the parameter values under *behaviour* against those under *non-behaviour.* If the two distributions are significantly different, the parameter is considered to be important for the performance of the model, that is, the model predictions are sensitive to the parameter values. If the distributions are not significantly different, then the parameter is taken as non-important (under the specified range and distribution).

The use of the generalized sensitivity analysis has been reported by many authors, including Spear and Hornberger (1980), Hornberger and Spear (1983), Buckley (1985), Tomlin (1986), Lumbers (1987) and Jaffe *et al* (1988). In most of these works the classification of the parameters into the category of *behaviour* was based on boundaries established for the expected behaviour of the model. This expected behaviour was generally set according to what could be supposed as a typical output from the model, under the given conditions, and not as a function of experimental data collected from the system. Therefore, the model response is judged on the basis of an expected rather than observed behaviour. This approach is undoubtedly important and is especially indicated when exploring different scenarios under a substantial amount of uncertainty, or when there are incomplete or unavailable data. Indeed, these where the original motivations for the development of the approach (Beck, 1987). A typical application is that of projective forecasts, in which future conditions are to be evaluated and for which there cannot obviously be any observed data (van Straten and Keesman, 1991). A high degree of subjectivity is sometimes unavoidable in the specification of the criteria for defining the boundaries of the expected model response, which will define the classification between *behaviour* and *non-behaviour.*

The situation analysed in this paper is a different but only apparently a simpler one. The objective here is to present the stepwise procedure adopted for the simultaneous parameter estimation and sensitivity analysis of a dynamic activated sludge model, using Monte Carlo simulation. In this case, observed data were available, and the assessment of *behaviour* or otherwise was done based on the goodness-of-fit between the estimated and the observed data. Even with the availability of the observed data, uncertainty still remains, together with the lack of identifiability in the Monod-based equations. The model comprises the aeration tank and the final clarifier of an activated sludge system, and is described elsewhere (von Sperling and Lumbers, 1989; von Sperling, 1990). The observed data available includes two-hourly values of the variables, collected during a period of 35 days at a full scale oxidation ditch in UK (50000 population inhabitants). In order to concentrate on the aspect of parameter estimation, the 11 model parameters will not be defined here, being simply designated as a_1, a_2, ..., a_{11}. Even though the sensitivity analysis and the parameter estimation were conducted simultaneously, for the clarity of explanation they are described here in separate sections.

PROCEDURE FOR THE SENSITIVITY ANALYSIS

The sensitivity analysis was based on Monte Carlo simulations, according to the flow chart presented in Figure 1. An explanation of the main items is given below.

Generation of the parameter values

At each run of the Monte Carlo simulation, the values of the parameters were generated randomly, according to a rectangular (uniform) distribution. Initially, the range of each parameter was selected in such a way as to be wide enough to cover values normally quoted in the literature, as well as the values previously found by trial and error on initial runs of the model. To achieve the concomitant objective of parameter estimation, each range was progressively narrowed down, as explained in the section related to the parameter estimation.

Criteria for *behaviour* and assessment of the performance of the model

Different criteria for defining and assessing *behaviour* are reported in the literature. Van Straten and Keesman (1991) defined the *behaviour* space by symmetrical ranges around the smoothed values of the time series of the observed state variables. The predicted states were then required to be within this "corridor" of constraints (Beck, 1987) in order to lead to a *behaviour*. A similar concept of a corridor around the observed values was adopted by Rose *et al* (1991), but in this case the classification of *behaviour* was a function of the number of days in which the predicted states were within the acceptable range. Klepper *et al* (1991) adopted a different approach, with the performance of the model being judged by the sum of the absolute values of the errors. Successful runs were considered to be those which led to a sum of errors situated between the minimum sum found (C_{min}) and 1.1 C_{min}.

In the present work, the classification of each run was a function of the relative performance of the model in the estimation of the four main state variables. These are mixed liquor suspended solids (MLSS), return activated sludge suspended solids (RASS), dissolved oxygen in the reactor (DO) and effluent ammonia. The evaluation of the degree of fit between estimated and observed values for each state variable was computed by the *Coefficient of Determination* (COD), given by (Mendenhall and Sincich, 1988):

$$COD = 1 - \frac{\Sigma (Y_{obs} - Y_{est})^2}{\Sigma (Y_{obs} - Y_{av})^2} \qquad (1)$$

where

Y_{obs} = observed value of the variable
Y_{est} = estimated value of the variable
Y_{av} = average of the observed values

In this context, COD values can vary between $-\infty$ and 1. For positive CODs ($0 \leq COD \leq 1$), the value represents the fraction of the total variance of the observed variable that is explained by the model. COD equal to 1 indicates perfect agreement between estimated and observed variables and COD equal to 0 indicates that the model is doing no better than a constant line passing through the average of the time series of the observed points. Unlike models based on regression analysis, COD values can be here also negative, implying that the performance of the model is worse than that obtained by a constant value equal to the average of the observed values.

Judgement based purely on the COD statistics can be sometimes misleading, because COD values are very influenced by the stationarity of the observed series. When the observed series is approximately stationary, with a small variance, the denominator in Equation 1 is small, and thus it is more difficult to obtain high COD values. This can be further complicated if this relatively stable series is significantly influenced by noise, which of course is not expected to be reproduced by the model. If however the time series of the observed variable presents trends, then it is easier to obtain a higher COD, provided that the model is able to approximately follow the main trends. In spite of these limitations, the COD statistics are still probably the best single statistical method of analysing the performance of a model. COD is directly related to the `residuals sum of squares' (numerator of Equation 1) and, thus a parameter estimation procedure that seeks to maximize COD is the same as that which minimizes the residuals sum of squares. If residuals sum of squares is an absolute measure of adherence, COD is a relative one, and can be compared with the results from other simulations or other variables (provided that the limitations above described are taken into account). Based on all these considerations, it was decided to adopt COD as the criterion for defining *good* and *bad* performances, or *behaviour* and *non-behaviour*.

For the initial Monte Carlo simulation, the criterion was that the COD of each of the four main variables had to be greater than or equal to zero for defining a *behaviour*. This criteria had to be simultaneously fulfilled for the four variables and, if at least one failed, the run resulted in a *non-behaviour*. In the stages of parameter estimation, the criterion for COD was successively made more stringent, in parallel with the improvement in the parameter values and model performance. Because this is a multi-variable model, it would be difficult to judge the model performance based on a single goodness-of-fit function. This is the reason why each state was analysed individually, thus requiring the analyst's participation.

Statistical analysis for comparison of the distributions of parameter values under *behaviour* and *non-behaviour*

As already mentioned, the sensitivity analysis of a parameter is based on the comparison of the distribution of the parameter values under ***behaviour*** with that under ***non-behaviour***. Since the resulting distributions are not known, the two-sample test has to be based on non-parametric methods, due to their property of being distribution free. The method adopted in this work was the Kolmogorov-Smirnov two-sample test, which was also used by various of the above named workers in their applications of the regionalized sensitivity analysis.

The Kolmogorov-Smirnov test calculates the maximum separation between the cumulative distribution functions of the two samples. The higher the separation, the more distinct are the two samples. The test is sensitive not only to differences in central tendency but to any difference in the distribution functions. The null hypothesis H_0 is that the two distributions are equal, and if the resulting significance level of the test (α) is less than a certain specified value, the null hypothesis is rejected (Massey, 1951; Gibbons, 1971). Under these conditions, the interpretation is that the model is sensitive to this parameter, since it is responsible for a significant separation between ***behaviour*** and ***non-behaviour***. An example is shown in Figure 2, which shows (a) separation – parameter important and (b) no separation – parameter not important.

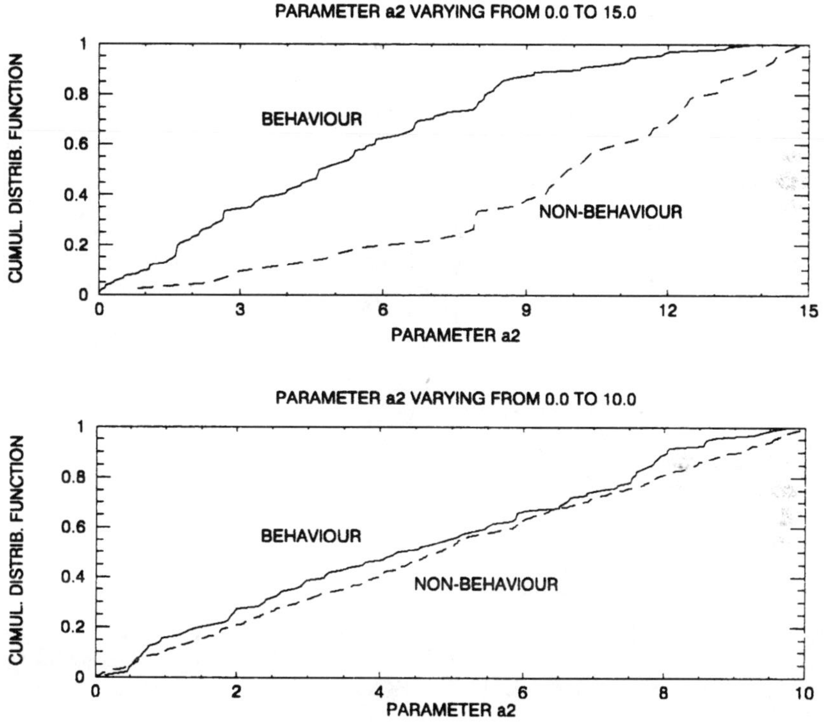

Fig. 2. Cumulative distribution function of the *behaviour* and *non-behaviour* parameter vector. (a) Large separation: parameter important. (b) Small separation: parameter not important

The degree of sensitivity of the model can be classified in terms of the significance level α, under the concept that, the lower its value, the more distinct are the two distributions. The classification system adopted in this work is presented in Table 1.

TABLE 1. Classification System for the Sensitivity Analysis

Class	Significance level	Interpretation
1	$0.00 < \alpha \leq 0.01$	critical
2	$0.01 < \alpha \leq 0.10$	important
3	$0.10 < \alpha \leq 1.00$	unimportant

Based on this classification scheme, the relative importance of each parameter can be assessed. It should be remembered, however, that "the conclusions about each parameter are confined to the specified model structure and multidimensional region of parameter space defined by the limits of the `a priori' distribution of the values of the parameters" (Hornberger and Spear, 1983).

Correlation between parameters

It is possible that a single parameter might exhibit no separation (Class 3, for instance), and yet it could be important, due to a strong correlation with other parameters under *behaviour*. In this case, it is not the single parameter, but rather the interaction between parameters which is important. Therefore, inspection of the correlation matrix of the parameters is essential, and when substantial correlation exists, the univariate methods of analysis cannot be applied, and multivariate techniques must be employed (Whitehead and O'Connell (1984).

In the Monte Carlo simulations carried out, only the correlation matrix of the parameters under *behaviour* was calculated, as suggested by Whitehead and O'Connell. When large correlations were identified, no sensitivity analysis of the parameters was conducted. However, the interpretation of their distributions and COD values was still performed, in order to investigate possibilities of reduction of the range of parameter values in the process of parameter estimation.

PROCEDURE FOR PARAMETER ESTIMATION

The procedure developed for parameter estimation using Monte Carlo simulations differs in its structure from the others mentioned in this paper. The flow chart is presented in Figure 3, which also shows the integration with the stage of sensitivity analysis. The procedure is repeated until the performance of the model is judged satisfactory, or when no further significant improvements are expected to occur.

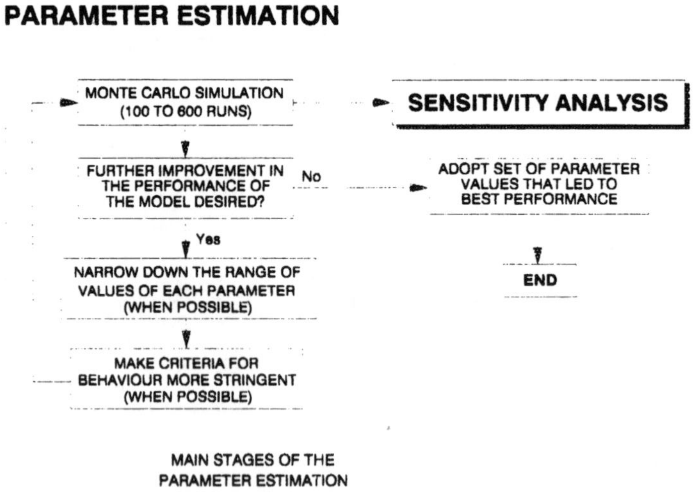

Fig. 3. Flow chart of the procedure for the parameter estimation based on Monte Carlo simulations

Reduction of the ranges of parameter values

After each Monte Carlo simulation (one simulation comprising 100 to 600 runs, each with a different randomly generated parameter set), the results were analysed, a sensitivity analysis was conducted, and a new range of parameter values was established. The objective of reducing the ranges was to approximate the region of best values for the parameters, and also to increase the likelihood of achieving a better performance of the model (as measured by the CODs) on the next Monte Carlo simulation.

For each parameter, the new range was established by the simultaneous analysis of the following criteria:

(1) Visual inspection of the frequency histogram of the parameter values under *behaviour*, and identification of the range where the frequency of *behaviours* was more predominant. In the example given on Figure 4, it can be seen that values of a_2 greater than 9.0 have less chance of producing *behaviours*, and could be excluded from the next simulation.

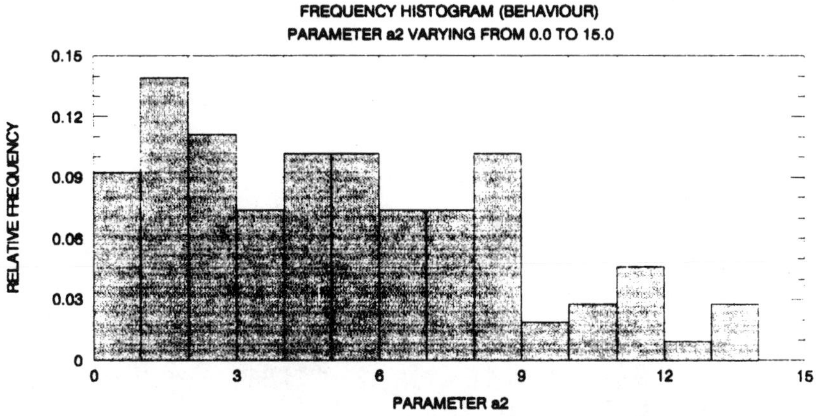

Fig. 4. Example of a frequency histogram of the parameter values under behaviour.

(2) Interpretation of the skewness coefficient of the parameters under *behaviour*, which is complementary to the analysis of the histogram. In the example given, the skewness is positive, thus indicating that the frequency is lower on the right hand side of the X-axis. Therefore, the range can be reduced by excluding a region on the right side (higher values), as already concluded from the histogram.

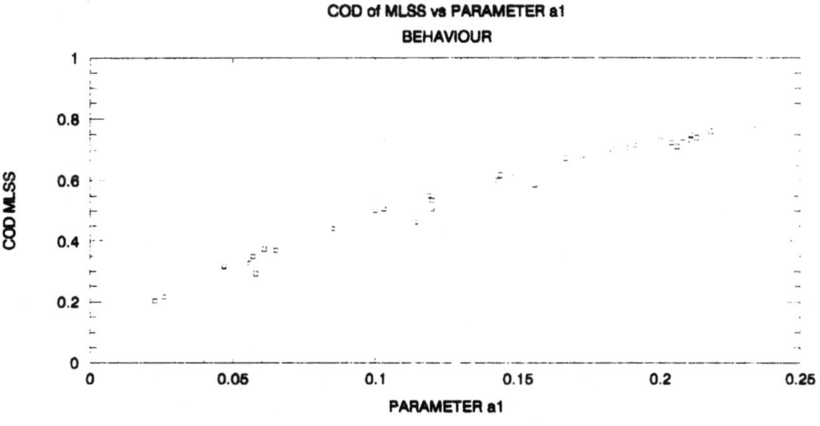

Fig. 5. Example of a scatter-plot of COD values vs parameter values.

(3) Visual inspection of the plot of CODs (*behaviour*) versus the parameter values, and adoption of the range of the parameter values where the CODs are highest. From another example, given in Figure 5, it can be seen that the highest COD values for MLSS are obtained when a_1 is greater than 0.10, suggesting that this could be adopted as the new lower limit for the range of a_1 in the next Monte Carlo simulation. A similar graph has to be made with the COD from the other variables, in order to avoid contradictions.

Naturally these criteria are not strictly formal and have a degree of subjectivity. There was no priority of one criterion over the other, and it was always the objective to define the new parameter range to satisfy as many criteria as possible.

Increase in the precision required to satisfy the criteria for *behaviour*

After the range of parameter values had been narrowed, the possibility of making the criteria for defining *behaviour* more stringent was analysed (increasing the minimum COD for each variable). The reason for this was that, due to a better selection of the parameter ranges, an increase in the COD values was expected. Therefore, in order to keep a balance between the number of *behaviours* and *non-behaviours* for the next sensitivity analysis, the minimum acceptable COD values for each variable should be raised. There was no formal criterion for increasing the values, since the outcome of the next Monte Carlo simulation could not be anticipated. A separate analysis for each variable (MLSS, RASS, DO, ammonia) was carried out, aiming at producing realistic propositions for the increase (or not) of each COD.

End of the Monte Carlo simulations

After the range of parameter values had been reduced and the criteria for *behaviour* made more stringent, another Monte Carlo simulation and sensitivity analysis was carried out, comprising another large number of model runs. When the performance of the model showed scope for further improvement, the whole sequence described was repeated. When the performance of the model was judged to be satisfactory, or when no further improvements were found, the procedure was stopped. Since the ranges had been narrowed down to their practical minimum width, any set of parameter values situated inside the last ranges were considered capable of producing a good performance. Alternatively, also from the last simulation, the set of parameters leading to the highest total COD (sum of the CODs for the four variables) could be adopted as the *best* parameter set.

TABLE 2. Distribution between *Behaviour* and *Non-behaviour* in the Monte Carlo Simulations

SIMULATION	NUMBER OF RUNS	BEHAVIOUR		NON-BEHAVIOUR	
		NUMBER	%	NUMBER	%
1	150	34	22.7	116	77.3
2	150	108	72.0	42	28.0
3	600	18	3.0	582	97.0
4	500	96	19.2	404	80.8
5	500	188	37.6	312	62.4
6	100	93	93.0	7	7.0
7	100	58	58.0	42	42.0
8	100	12	12.0	88	88.0
9	100	45	45.0	55	55.0
10	100	89	89.0	11	11.0
TOTAL	2400				

DESCRIPTION, RESULTS AND STATISTICS OF THE SIMULATIONS

A sequence of ten Monte Carlo simulations was carried out, each of them with either different variables involved, number of runs, criteria for *behaviour* and/or range of parameter values. Table 2 presents some statistics of the simulations, including the number of runs and absolute and relative numbers of *behaviours* and *non-behaviours* obtained. Eventual decreases in the percentage of *behaviours* from one simulation to the other can be attributed to a more stringent criterion for *behaviour* (higher minimum COD values) or to the narrowing of the ranges of the parameter values.

The average COD values (mean of the runs under *behaviour*) of each variable in each of the 10 simulations are presented in Figure 6. The progressive increase can be clearly seen, indicating the successful stepwise selection of the ranges of parameter values. There are no indications that further increases should be expected after Simulation 10, suggesting that the procedure of parameter estimation could be stopped.

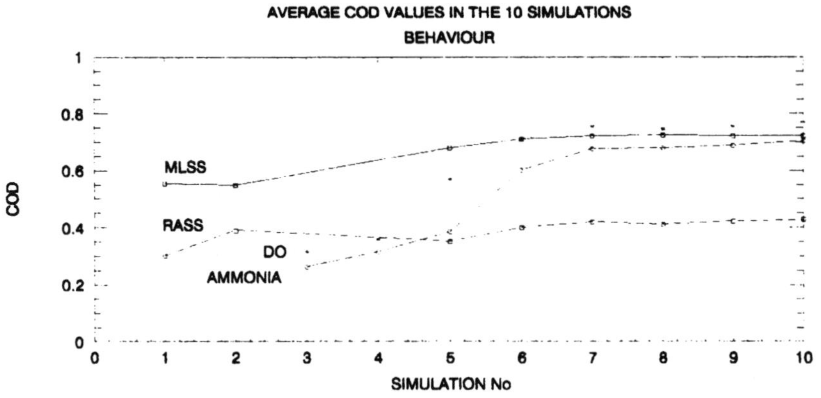

Fig. 6. Average COD values of the state variables (behaviour) in each of the 10 simulations

An example of the progressive reduction in the ranges of a parameter is shown on Figure 7. Parameter a_2 represents a fixed term for the biomass growth (gMLSS/m^3.h), and the low values finally obtained are indicative of the small net growth rate occurring in extended aeration systems. In this particular example, all the reductions in the parameter range occurred only in the upper limit.

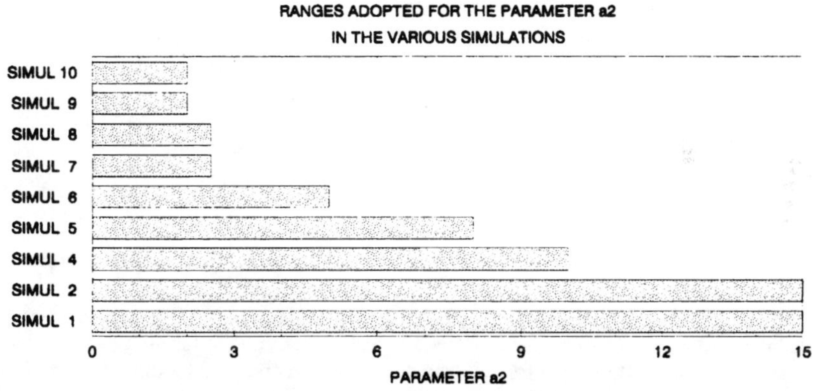

Fig. 7. Range of values adopted for the parameter a2 in each of the 10 simulations.

The interpretation of the main statistics for each of the 10 simulations reveals the following points (the statistics are presented in von Sperling, 1990):

(1) There was a general tendency of the *skewness coefficient* of the parameters under *behaviour* to come close to zero, reflecting the improved selection of the parameter ranges and the resulting more symmetrical distribution of the parameter values around the mean.

(2) In the last simulation (Simulation 10) the *kurtosis coefficient* for most of the parameters was close to 1.8, which is the theoretical value for a rectangular distribution. The interpretation of this is that the parameter values were uniformly distributed inside the *behaviour* range, without any significant modal tendency, thus indicating that the frequency of *behaviours* was approximately the same inside the range. This is another indication that the ranges had been narrowed down sufficiently in this process of parameter estimation.

(3) From Simulation 1 to 10, the *Kolmogorov-Smirnov test* showed the parameters to be progressively changing from critical (Class 1) or important (Class 2) to unimportant (Class 3). The results from Simulation 10 are a confirmation of this, with none of the parameters being critical, and most of them being unimportant. This is still a further indication that the ranges of the parameters had been reduced to a sufficient degree, and variations of the parameters inside their respective ranges were not likely to significantly affect the model performance. At this stage, it could be concluded that no further simulations would be necessary, and that a satisfactory convergence had been achieved.

(4) For comparative purposes, use of the *Mann-Whitney test* led to virtually the same results (distribution into the three classes) obtained with the Kolmogorov-Smirnov test.

DISCUSSION

Based on the results obtained, there is evidence that the Monte Carlo simulation was a satisfactory method for parameter estimation and also for the assessment of the relative contribution of each parameter to the performance of the model (sensitivity analysis).

In the whole process, a certain degree of subjectivity existed mainly in the following steps: (a) selection of the range of parameter values, (b) selection of the criteria for *behaviour* and *non-behaviour* and (c) selection of the criteria for the classification of the degree of importance of each parameter.

The process of parameter estimation requires a substantial input from the modeller in various stages, which makes its automation difficult. However, by involving the modeller and by using different criteria for the definition of the ranges of parameter values, the method provides a higher chance that the reduction in the ranges is moving in the right direction. There is no formal guarantee that the actual *best* set of parameters is the one obtained by the Monte Carlo simulations, but only that the ranges obtained are likely to have a higher probability of satisfying the imposed criteria for *behaviour*, and also a higher probability of producing better COD values. However, a similar and possibly stronger limitation can be found in methods for parameter estimation based on conventional single optimization techniques, in which the optimal solution found cannot be guaranteed to represent the global optimum. This would be especially the case for this model, with its 11 parameters to be estimated.

In the sensitivity analysis, the judgement regarding the relative importance of each parameter is influenced by the criteria for defining *behaviour* and by the selected range of parameter values at each stage. The interpretation of the sensitivity analysis has to be made under this perspective. Therefore it is difficult to establish in absolute terms the degree of importance of each parameter in terms of the performance of the model. However, a good insight into the model structure could be obtained by the adopted method, which is also another justification for the closer involvement of the modeller.

CONCLUSIONS

For the purposes of the modelling studies undertaken, the final range of parameter values was considered satisfactory, since it was believed that further increases in COD values would be only marginal, not justifying additional simulations. Since (a) the final parameter ranges were reasonably narrow, (b) the distributions of the parameters under *behaviour* were approximately symmetrical and uniform, and (c) none of the parameters was considered critical, the proposed set of parameter values could be considered satisfactory. The usefulness of the procedure can be seen through the progressive increase in the COD values from simulation to simulation. As a result, it is judged that the objectives of the Monte Carlo simulation were accomplished.

An overall evaluation suggests that this methodology of parameter estimation and sensitivity analysis employing Monte Carlo simulations is a simple, useful and robust one, and can be applied to other models in which observed data are available. The analyst's involvement helps to maintain the controllability of the procedure and also to increase the level of understanding of the processes represented by the model.

ACKNOWLEDGEMENTS

The author wishes to thank Dr Jeremy Lumbers for the supervision of the work, on part of which this paper is based.

REFERENCES

Beck, M. B. (1987). Water quality modelling: a review of the analysis of uncertainty. *Water Resources Research*, **23**(8), pp. 1393-1442.

Buckley, J. M. (1985). *A generalized sensitivity analysis of the kinetics of the activated sludge process (with particular reference to dissolved oxygen control).* M.Sc. Dissertation, Imperial College, University of London.

Clarke, R. T. (1973). A review of some mathematical models used in hydrology, with observations on their calibration and use. *Journal of Hydrology*, **19**, pp. 1-20.

Gibbons, J. D. (1971). *Nonparametric statistical inference.* McGraw-Hill. 306 pp.

Hornberger, G. M. and Spear, R. C. (1983). An approach to the analysis of behaviour and sensitivity in environmental systems. In: *Uncertainty and forecasting of water quality.* Beck, M. B. and Van Straten, G. (ed). Springer-Verlag, pp 101-116.

Jaffe, P. R., Paniconi, C. and Wood, E. F. (1988). Model calibration based on random environmental fluctuation. *Journal of the Environmental Engineering Division. ASCE*, **114** (EE5), pp. 1136-1145.

Klepper, O., Scholten, H. and Van De Kamer, J. P. G. (1991). Prediction uncertainty in an ecological model of the Oosterschelde estuary. *Journal of Forecasting*, **10**, pp. 191-209.

Lumbers, J. P. (1987). *Rotating biological contactors: mechanisms, modelling and design.* Ph.D. Thesis, Imperial College, University of London, London.

Massey, F. J. (1951). The Kolmogorov-Smirnov test for goodness of fit. *American Statistical Association Journal*, **46**, pp. 67-78.

Mendenhall, W. and Sincich, T. (1988). *Statistics for the engineering and the computer sciences.* Dellen Publishing Company, San Francisco. 1036 pp.

Rose, K. A., Smith, E. P., Gardner, R. H., Brenkert, A. L. and Bartell, S. M. (1991). Parameter sensitivities, Monte Carlo filtering, and model forecasting under uncertainty. *Journal of Forecasting*, **10**, pp 117-133.

Spear, R. C. and Hornberger, G. M. (1980). Eutrophication in Peel Inlet: II. Identification of critical uncertainties via generalized sensitivity analysis. *Water Research*, **14**, pp. 43-49.

Tomlin, S. (1986). *The development of a dynamic model for an activated sludge sewage treatment process.* M.Sc. Dissertation, Imperial College, University of London.

Van Straten, G. and Keesman, K. J. (1991). Uncertainty propagation and speculation in projective forecasts of environmental change: a lake-eutrophication example. *Journal of Forecasting*, **10**, pp. 163-190.

Von Sperling, M. (1990). *Optimal management of the oxidation ditch process.* Ph.D. Thesis, Imperial College, University of London, London.

Von Sperling, M. and Lumbers, J. P. (1989). Control objectives and the modelling of MLSS in oxidation ditches. *Wat. Sci. Tech.*, **21**(10/11), pp. 1173-1183.

Whitehead, P. G. and O'Connell, P. E. (1984). *Water Quality Modelling, Forecasting and Control.* Proceedings of an International Workshop, Institute of Hydrology, Wallingford. Report No. 88. 123 pp.

PROCESS CONTROL STRATEGY DEVELOPMENT

Pergamon

Wat. Sci. Tech. Vol. 28, No. 11-12, pp. 231–237, 1993.

0273–1223/93 $7·00 + 0·00

DIAGNOSIS EXPERT SYSTEM FOR THE ACTIVATED SLUDGE PROCESS USING BIOTA OBSERVED BY MICROSCOPIC EXAMINATION

M. Ichikawa*, K. Simizu* and K. Iwahori**

** R&D Center, Meidensha Corporation, 2-1-17, Osaki, Shinagawaku, Tokyo 141, Japan*
*** Department of Environmental Engineering, Osaka University, 2-1, Yamadaoka, Suita, Osaka 565, Japan*

ABSTRACT

A diagnosis expert system for the activated sludge process was developed in conformity to our experience and reported studies. Using the biota observed by microscopic and data measurement easily obtained, i.e. pH or DO etc., the process treatability or the activated sludge condition was diagnosed and evaluated by this system. In this paper, the effectiveness of the diagnosis expert system was investigated using the microscopic and measurement results from both the actual sewage treatment plant and the bench-scale experiment. As a result, the low loading and the progress of nitrification were presumed in the case of the actual plant and the variations of treatment condition depending on DO set values were evaluated satisfactorily in the case of the bench-scale experiment. Therefore, it was shown that this system was applicable to the diagnosis of the activated sludge process.

KEYWORDS

Activated sludge process; biota; protozoa; metazoa; diagnosis expert system; knowledge base; certainty factor; treatment conditions.

INTRODUCTION

We have learned that the efficiency of biological treatment in the activated sludge process largely depends on the kinds and biological activities of microorganisms and also that the treatment condition can be readily observable by microscopic examination. It has been proposed that the appearance of the microorganism can be utilized to control the activated sludge process (Sudo and Inamori, 1983). Accordingly, we input the data on protozoa and micro-metazoa into a computer database, conducted a microscopic examination in order to diagnose the treatment condition, and developed a biota diagnosis expert system to achieve improved operation and maintenance of an activated-sludge processing plant (Ichikawa *et al.*, 1989). In this study, we applied data from microscopic observations, conducted an actual sewage treatment plant and a bench-scale experiment to estimate the treatment condition in the activated sludge process.

OUTLINE OF BIOTA DIAGNOSIS EXPERT SYSTEM

Figure 1 shows the schematic diagram of the biota diagnosis expert system.

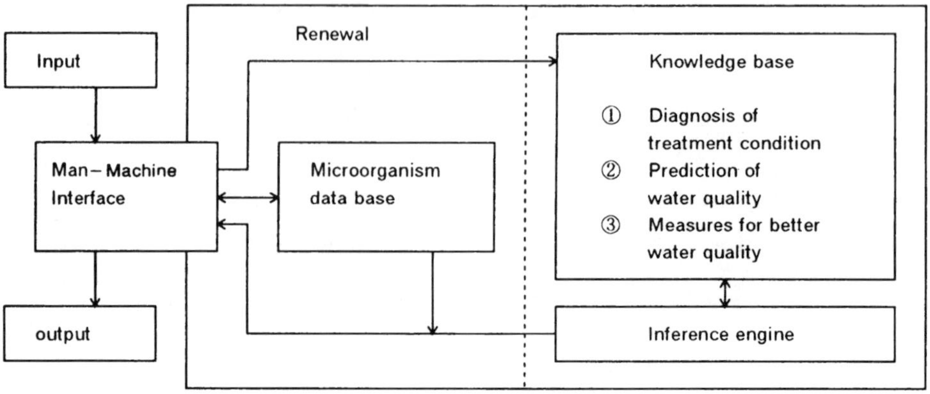

Fig. 1. Schematic diagram of the biota diagnosis expert system.

In the figure, the knowledge base consists of three types of knowledge: 1) a diagnosis of the treatment condition based on microscopic observation of microorganisms, 2) prediction of treated water quality, and 3) measures to obtain better water quality inferred from the first and second types of knowledge. The microorganism database includes data on the shapes, preferred living environments, and other characteristics of those microorganisms which are useful indices for evaluating the treatment condition. A selection of data can be displayed on CRT display to help identify microorganisms. The inference engine serves to diagnose the treatment condition and search for appropriate measures while referring to the database with respect to the input data on an appearing microorganism.

Classification of Input Data

We classified the input data into survey data and microorganism data. The survey data included measured values of pH, DO and other properties, color and odor of activated sludge in the aeration tank, and the quantity of scum and the transparency of the water in the final sedimentation pond. The microorganism data included 28 groups of protozoa and metazoa, according to a re-classification of 48 kinds of frequently appearing protozoa and metazoa. There were five input categories for the number of appearing microorganisms: (1) was the first category and assigned for 1 to 99 microorganisms per ml, (2) for 100 through 399 microorganisms per ml, (3) for 400 through 999 microorganisms per ml, (4) for 1000 through 1999 microorganisms per ml and (5) for 2000 microorganisms or more per ml. As for the number of filamentous microorganisms, we used two input categories, 'many' and 'few', determined intuitively during microscope observation.

Flow of System

Figure 2 shows general flow chart of the system.

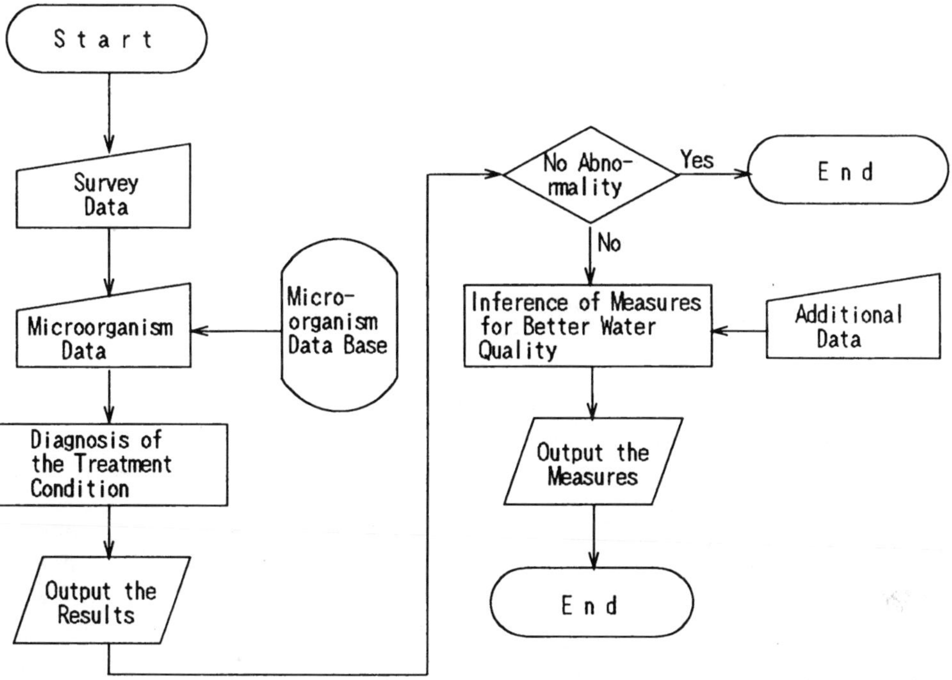

Fig. 2. General flow chart of the system.

First the survey data and categories for the number of each microorganism are input to the system. Next the treatment conditions are inferred based on the data so that the results are given by the system. After this inference, if there are no abnormalities, the job is terminated. But if there are any abnormalities or the symptoms of them, the job goes on the inference of measures for better water quality. The measures are searched out from the rules if "then" format in the knowledge base. During the search, the system sometimes requires the input of additional data for more accurate inference.

Inference Method

The biota diagnosis expert system inferred the treatment condition as one of six types: (1) high loading and white turbidity in effluent, (2) low DO in the aeration tank, (3) low loading and progress of nitrification, (4) decomposition of sludge, (5) influence of toxic materials, and (6) filamentous bulking.

We define the certainty factor CF_{nm} as the likelihood that microorganisms of the kind n appearing at a rank Wn are in the m-th processing state. Then, we have the following rule,

$$\text{THEN } CF_{nm} = Wn \times \alpha nm \tag{1}$$

where αnm is the weight coefficient, ranging from -0.2 to +0.2 and preset for each microorganism. Some examples of the weight coefficients are shown in Table 1. According to equation (1), CF_{nm} can take values between -1.0 and +1.0 because Wn is an integer between one and five. A negative value indicates the possibility that the treatment condition may never occur. When two kinds of microorganisms in the same group, n and K, appear simultaneously, the certainty factor may be evaluated as follows:

$$CF_{nkm} = CF_{nm} + CF_{km} - CF_{nm} \times CF_{km} \tag{2}$$

TABLE 1. Weight Coefficients of Index Microorganism and Certainty Factor of Survey Data

		Treatment condition No.					
		1	2	3	4	5	6
Micro-organ-isms	A	−.05	−.10	+.10	−	−	−
	B	−.10	−.10	+.10	−.05	−	−
	:	:	:	:	:	:	:
	:	:	:	:	:	:	:
	P	−	+.20	−.15	−	−	+.10
Survey data	pH 6.5 or less	−	−	+.60	−	+.25	−
	pH 8.0 or more	−	−	−.60	−	+.25	−
	DO 2.0 or less	+.50	+.30	−.30	−	−	−
	:	:	:	:	:	:	:
	:	:	:	:	:	:	:
	transparency low	+.40	+.20	−.20	+.20	+.10	−
	〃 high	−.40	−.20	+.20	−.20	−.20	−

When three kinds of microorganisms or more in the same group appear simultaneously, equation (2) applies first of two kinds, then the thus obtained value of CF_{nkm} and the value of the third microorganism go on the right side of equation (2), and this evaluation is repeated through the last kind of microorganism. If CF is negative, the following equation applies:

$$CF_{nkm} = CF_{nm} + CF_{km} + CF_{nm} \times CF_{km} \tag{3}$$

The survey data gives a certainty factor value to each of the treatment conditions for each property item. Some certainty factors are shown in Table 1. Hence, equations (2) and (3) equally apply. If a certainty factor is positive, we call it an affirmative-diagnosis certainty factor; if negative, a negative-diagnosis certainty factor. The final result of diagnosis will be a summation of the two diagnosis certainty factors.

APPLICATIONS OF THE BIOTA DIAGNOSIS EXPERT SYSTEM

Input Data and Diagnosis Results

We continued microscopic examination of the activated sludge mixed liquor at the outlet of the aeration tank in the actual sewage treatment plant (about 30000 m^3/d for 1 year. We extracted 13 interesting samples from the data obtained in the examination and applied them to the present expert system. We also utilized the five samples of measurement given in the results of the experiments using a bench-scale experiment (with DO concentrations of 0.2 to 5.0 mg/l in the aeration tank) reported by Handa *et al.* (1980). Table 2 lists both results of these measurements and presents the results of diagnosis based on the microorganism data and the survey data on DO, MLSS and pH. Samples 1 through 13 are the measurements taken from the actual plant; samples 14 through 18 are the measurements from the bench-scale experiment. In this table, blanks in the microorganism data indicate that corresponding kinds of microorganisms did not appear. We set the pH values of samples 14 through 18 at 7.0 because pH values were not given in the report. For a similar reason we set the transparency of samples 14 and 18 as "low" and all the other properties in the survey data of samples 14 through 18 as "no abnormality".

TABLE 2 Input Data and Results of Diagnosis

			SAMPLE NUMBER																	
			NO.1	NO.2	NO.3	NO.4	NO.5	NO.6	NO.7	NO.8	NO.9	NO.10	NO.11	NO.12	NO.13	NO.14	NO.15	NO.16	NO.17	NO.18
MICROORGANISM DATA	Rotatoria	*Philodina, Rotaria*	1	1	1	1		2	2	1		2		1			1	1	1	1
		Colurella	2	3	2	2	3	2	2	3	3	3	2	4	4		1	1	1	1
		Lekcane, Monostyla	1	1	2	1		1					2							
	Nematoda	*Nematoda*		2								1	1			1	1	1		1
	Oligochaeta	*Aelosoma, Nais*	1	1	2										1					
CILIATA	Peritrichida	*Vorticella*	4	4	4	4	5	5	5	3	4	4	4	5	4	1	3	3	3	3
		Carchesium, Zoothamnium	2		2	5		5	4	5	5	5	5	4	5		2	2	3	2
		Epistylis, Opercularia	2	2	2	4		4	3	4	4	4	2	4		2	3	5	4	3
	Suctrida	*Acineta, Tokohrya*				1							2	3			1	1	1	1
	Hypotrichida	*Aspidisca*	1	5	2		2	2	4	4		3	5	2		3	5	4	4	5
	Gymnostomatida	*Litonotus*	2	3	2	1	2	2	2			1	2	2		1	1	2	1	1
		Trachelophyllum	3	2	2	3		3	2	1	3		3	2	3	2	2	2	2	2
		Amphileptus														2	3	2	2	1
		Chilodonella			1				1			1	2							
	Trichostomatida	*Paramecium*														1				1
	Hymenostomatida	*Uronema, Colpidium*														5	3	1	3	5
		Glaucoma, Tetrahymena	1	3		2	2	2	3	2		1	4	2						
	Sarcodina	*Arcella, Euglypha*	3	5	5	3	5	3	5	4	4	5	5	5	5	2	4	4	4	3
		Amoeba, Vahikampfia	4	3	3	2	4	4	4	3	5	4	5	5	5	5	5	5	5	5
	Mastigophora	*Oicomonas, Monas, Bodo*	3	2	2	1	3	1	2	2	1	2	2	2		5	4	3	3	4
		Peranema	2	3	1	2	1	2	1		1		1	2	2		1	1		1
		Entosiphon														2	1	1	2	2
	Filamentous	*Beggiatoa*														few				
Survey data		pH (−)	6.2	6.2	5.7	6.3	6.6	6.1	7.0	6.5	6.4	6.2	6.5	6.6	6.6	7.0	7.0	7.0	7.0	7.0
		DO (mg/l)	6.6	3.0	3.2	3.1	3.7	5.7	5.9	3.5	3.1	3.0	2.6	2.6	3.5	0.2	1.0	2.0	3.0	5.0
		MLSS (mg/l)	2210	2420	2600	1710	1340	1920	1840	1890	1720	2080	2020	1830	1790	2870	2420	2376	2250	2515
RESULTS OF DIAGNOSIS	The Final Result of Diagnosis	Category 1														0.53				0.29
		Category 2														0.57				
		Category 3	0.47	0.41	0.60	0.46	0.34	0.54	0.41	0.44	0.65	0.54	0.34	0.43	0.72			0.16	0.20	
		Category 4														0.18				0.18
	Affirmative Diagnosis Certainty Factor	Category 1	0.37	0.44	0.20	0.28	0.61	0.28	0.44	0.36	0.10	0.28	0.52	0.36	0.00	0.93	0.79	0.69	0.51	0.82
		Category 2	0.15	0.10	0.28	0.05	0.15	0.05	0.10	0.10	0.05	0.10	0.10	0.10	0.00	0.97	0.78	0.46	0.41	0.77
		Category 3	0.96	0.96	0.96	0.89	0.89	0.96	0.96	0.93	0.93	0.96	0.95	0.92	0.92	0.47	0.81	0.81	0.81	0.87
		Category 4	0.72	0.45	0.45	0.30	0.60	0.72	0.72	0.45	0.75	0.60	0.75	0.75	0.75	0.80	0.75	0.75	0.75	0.86
	Negative Diagnosis Certainty Factor	Category 1	0.73	0.80	0.80	0.68	0.80	0.73	0.81	0.75	0.77	0.82	0.80	0.83	0.83	0.40	0.76	0.76	0.76	0.53
		Category 2	0.80	0.81	0.80	0.64	0.78	0.80	0.86	0.74	0.75	0.82	0.76	0.84	0.84	0.40	0.71	0.71	0.71	0.70
		Category 3	0.50	0.55	0.36	0.42	0.55	0.42	0.55	0.49	0.28	0.42	0.62	0.49	0.20	0.93	0.76	0.65	0.61	0.82
		Category 4	0.68	0.78	0.73	0.66	0.73	0.69	0.76	0.74	0.68	0.75	0.78	0.73	0.70	0.62	0.76	0.74	0.74	0.68
Real Data		BOD or COD loading* (1/d)	0.10	0.21	0.10	0.17	0.34	0.20	0.13	0.23	0.25	0.18	0.20	0.21	0.27	0.11	0.09	0.09	0.12	0.10
		COD treated water (mg/l)	7.2	11.0	12.0	19.0	6.4	6.2	6.8	11.0	9.2	6.6	10.0	12.0	15.0	33.0	10.9	10.4	10.1	13.1
		SS treated water (mg/l)	7.8	3.9	11.4	14.4	3.6	3.6	3.9	17.4	7.6	7.4	6.9	8.1	16.4	−	−	−	−	−
		Nitrification rate (%)	−	100	100	91.8	90.8	96.6	−	100	87.5	−	−	−	−	0	8.3	15	41.2	71.3

Category 1 : high loading and white turbidity in effluent　　Category 2 : low DO in the aeration tank
Category 3 : low loading and progress of nitrification　　Category 4 : decomposition of sludge
* : from NO.1 to NO.13 BOD loading, from NO.14 COD loading

Applicability of Diagnosis Expert System

Applicability in actual sewage treatment plant. The facilities are characterized by nitrification and a high nitrification in winter. The final diagnosis results were "low loading and progress of nitrification", with a certainty factor ranging from 0.34 to 0.72. The average nitrification rate was a very high 87.5% for the seven samples. There was no correlation between the final results of diagnosis and the measured nitrification rate, although this may indicate progress of nitrification. Accordingly, the present expert system did not make a quantitative evaluation.

In the activated sludge process, bacteria work directly to purify the sewage and alter the environmental conditions so that protozoa and micro-metazoa can come into existence. As a result, there is a time-lag between initial purification and the appearance of the indexed microorganisms. Although we did not take it into account in the present study, this problem will be of future importance along with the method of inferring weight coefficients and other empirical factors.

The affirmative-diagnosis certainty factor for "low loading and progress of nitrification" ranged from 0.89 to 0.95, and the negative-diagnosis certainty factor from 0.20 to 0.62. Sample 11 had a high appearance rank of *Glaucoma* and its negative-diagnosis certainty factor was the largest (0.62), thereby lowering the final diagnosis result. According to the other observations, *Glaucoma* often appeared when the loading was high, and we gave a negative value to its weight coefficient for a "low loading and progress of nitrification" in the present study.

On the other hand, for the treatment condition of a "high loading and white turbidity in effluent", the affirmative-diagnosis certainty factor ranged from 0.2 to 0.61, and the negative-diagnosis factor from 0.68 to 0.83. The final diagnosis results were negative for all samples. The results were similar for "low DO in the aeration tank". For "decomposition of sludge", when large quantities of *Amoeba* and *Vahlkampfia* appeared, the affirmative-diagnosis certainty factor was approximately the same. Therefore, the final results were between -0.36 and 0.13, and, in our judgement, decomposition of sludge was not very probable.

Applicability in bench-scale experiment. It is noted that despite the appearance of many kinds of microorganism, the affirmative- and negative-diagnosis certainty factors were high. As the set value of DO became larger, the nitrification rate increased, and the final diagnosis result for "low loading and progress of nitrification" improved, if not as much as with the actual plant. For sample 14, in which the DO concentration was set at 0.2 mg/l during operation, the final results for "low DO in the aeration tank" and "high loading and whiter turbidity effluent" were high indicating that no nitrification took place and hence that the present system was applicable.

For sample 18, in which the DO concentration was set at 5.0 mg/l, the final results for "decomposition of sludge" and "high loading and white turbidity effluent" were affirmative, but those for "low loading and progress of nitrification" were negative. However, nitrification actually took place. This disagreement was due to the high values of the affirmative- and negative-diagnosis certainty factors, because *Colpidium* appeared at a high probability, namely with a rank of 5. This is similar to the case of *Glaucoma* in sample 11.

CONCLUSIONS

We obtained data both by microscopic examination and by visual inspection at an actual sewage treatment plant and a bench-scale experiment. We input this data in our biota diagnosis expert system to determine whether or not the system was applicable to the activated sludge process. The results indicated that the system was useful for a making a qualitative evaluation of the treatment condition. Under particular environmental conditions differing from normal operations, diagnosis of the treatment condition yielded high certainty values, and the system seemed applicable to the diagnosis of abnormalities in the activated sludge process. However, since activated sludge mixed liquor had a significant buffer action, a large variation in biota did not directly produce a rapid effect on the treatment condition. Therefore, the evaluation of the treatment condition was quantitatively insignificant. We feel that to take this unfavorable effect into consideration in order to improve the diagnosis accuracy, it will be necessary to adjust the weight factors, increase the number of indexed microorganisms, improve the inference method by further classifying the treatment condition, improve the man-machine interface and introduce other decision criteria.

ACKNOWLEDGEMENT

We would like to thank Prof. Dr Ryuichi Sudo of Tohoku University and Dr Yuhei Inamori of the National Institute for Environmental Studies for their assistance in constructing the present system.

REFERENCES

Handa, T., Furuya, N., Fujimoto, E. and Inamori, Y. (1980). Effect of DO on the activities of organisms in activate sludge. *Japanese Journal of Water Treatment Biology*, Vol. **16**, 1-10. (In Japanese).

Ichikawa, M., Ide, S., Shimizu, K., Sudo, R. and Inamori, Y. (1989). Biota diagnosis expert system in the activated sludge process. *Proceedings of the 26th J.S.W.A. Annual Technical Conference*, 693-695. (In Japanese).

Sudo, R. and Inamori, Y. (1983). Illustrated diagnosis of processing functions in view of biota. *SANGYO YOSUI CHOSAKAI*. (In Japanese).

Pergamon

Wat. Sci. Tech. Vol. 28, No. 11-12, pp. 239–248, 1993.

0273–1223/93 $7·00 + 0·00

POTENTIAL USE OF STEP FEED CONTROL USING THE BIOMASS IN THE SETTLER

M. T. Sorour*, G. Olsson** and L. Somlyody***

** Sanitary Engineering Department, Faculty of Engineering, Alexandria University, Alexandria, Egypt*
*** Industrial Automation, Lund Institute of Technology, Box 118, S-22100 Lund, Sweden*
**** International Institute for Applied System Analysis (IIASA), A-2361 Laxenburg, Austria*

ABSTRACT

In spite of the widespread availability of step feed configurations in activated sludge systems, little use has been made of their control potential. The main aim of this study is to provide a better understanding of applying step feed control considering the role of biomass activity in the settler. The control strategy was evaluated by simulating the dynamics of an activated sludge system. The model contains both biomass activity in the settler and step feed capabilities. The model and the strategy then have been evaluated against pilot scale and full scale experimental data.

KEYWORDS

Step feed; dynamic modeling; final settler; biological activity; operational control.

INTRODUCTION

The activated sludge process is recognized as one of the most commonly used and the major unit process for treatment of wastewaters. It has been well known by operators that activated sludge plants are seldom in steady-state. Despite this fact traditional design and operational methods are based on static criteria. Consequently, the plant will behave poorly if the variations in influent flow rate or concentration are too large (Olsson, 1977).

Disturbances related to the influent flow rate have to be met by operational control techniques. As traditionally operated, the process offers only three possible control variables, i.e. the sludge recycle rate, the sludge wasting rate, and the aeration rate (Busby-Andrews, 1975). A fourth control variable is the wastewater feed pattern, which can be manipulated in systems with tanks in-series and step feed capability. This has been recognized for a long time (Torpey, 1948). In spite of the widespread availability of the step feed configuration, surprisingly little use has been made of its control potential.

If the feed point of influent is moved towards the outlet, the suspended solids concentration in the inlet to the final settler becomes smaller. However, the reactor volume available for treatment of soluble organics also decreases. Therefore, high concentrations of soluble organics can be expected in the final settler. Consequently the role of the settler in utilizing part of these soluble organics ought to be established.

The aim of this study was to provide a better understanding of using step feed strategy considering the role of the biomass in the settler. The step feed strategy was evaluated using simulation of the dynamics of the system. The model dynamics is based on previously established biological models with modifications to incorporate the soluble organics utilization in the final settler.

Only a very limited knowledge concerning biological activity in settlers is found in the literature. Therefore the experimental verification of the model was crucial.

MODEL DEVELOPMENT

In order to increase the understanding of how step feed control would be influenced by possible biomass activity in the settler, a dynamical model of the aerator-settler system has been developed. Two important goals of the process model are to:

Predict realistic estimates of the effluent suspended solids and soluble substrate concentrations;

Include relevant variables to develop adequate control strategies, such as mixed liquor suspended solids concentration, sludge blanket level, dissolved oxygen concentration etc.

The reactor model employed is the one proposed by the IAWPRC Task Group (Henze *et al.*, 1987). In this study only the organic carbon was considered. The aerator model is complemented with a settler model that also contains the mechanisms of soluble organics removal. The settler model includes both clarification and thickening. The thickener model employed is that developed by Vitasovic (1986). The clarification function of the model which was based on empirical relationships in Vitasovic's model will be replaced by the modification of Takacs *et al.* (1991) to provide estimates of effluent suspended solids. (Appendix 1). The model has been simulated by using ten layers. The feed point is assumed to be in layer 6.

In order to incorporate soluble COD utilization in the settler model it was assumed that each layer acts as a single continuous stirred tank reactor (CSTR). Each CSTR is modeled according to the IAWPRC group model (Henze *et al.*, 1987). Aerobic and anoxic growth of Heterotrophs results in soluble COD removal, while hydrolysis of particulate COD produces soluble COD. The difference is the soluble COD utilized in each layer. Aerobic growth of biomass also depends on the dissolved oxygen concentration, which is assumed to act as a complementary limiting nutrient through adoption of a double Monod expression.
The material balance for soluble substrate, particulate substrate, and dissolved oxygen have been included for each horizontal layer in the settler. The material balance equations are shown in detail in Sorour (1991) and are not repeated here due to limited space. The model has been written in the simulation language SIMNON (Elmqvist *et al.*, 1990).

EXPERIMENTAL PROCEDURE

The experiments were performed to support the hypothesis that the final settler is able to remove part of the soluble substrate which escapes from the reactor in the step feed system. The experimental work was conducted on a small pilot-scale plant consisting of an activated sludge unit, using raw sewage from the Southern Budapest Treatment plant.

The pilot plant consists of three 15 l completely mixed reactors in series followed by a 0.18 m diameter and 0.025 m^3 volume settler as shown in Figure 1. The wastewater is screened where approximately 15% of the solids are removed. Influent total COD varied within a range of 288 to 450 mg/l. Because of the size of the pilot plant and the large number of experiments, it was not possible to use model sewage (synthetic sewage). To reduce the effect of this wide range, influent COD concentration of 350 mg/l ± 10% range was used for data analysis. Any data obtained with influent COD concentration outside this range was excluded from the analysis.

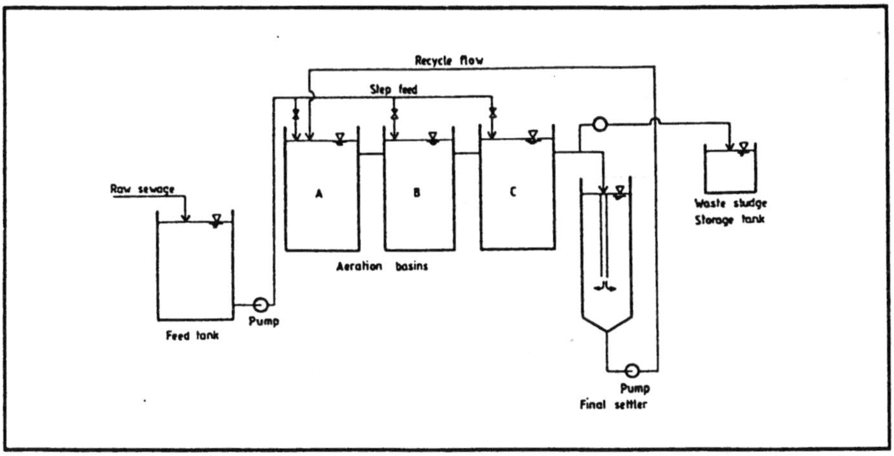

Fig. 1. Layout of activated sludge pilot plant.

Wastewater was fed at a rate of 200 l/d, the feed point was directed to the selected reactor according to the experimental plan, resulting in hydraulic retention times of 5.4, 3.6 and 1.8 hours respectively. The mixed liquor suspended solids, then passed through the settler from which the sludge was recycled at three different ratios, 70, 100 and 150%.

The mean cell residence time (MCRT) was controlled using the waste sludge flow rate. In the experiments the MCRT was held constant at 5, 10 and 15 d respectively. The dissolved oxygen (DO) in the reactors varied but never fell below 3.0 mg/l. The settler had a lower DO concentration-measured at 10 cm from the water surface-ranging between 2.0 and 4.0 mg/l. All data were collected under steady-state conditions. The system was kept in steady state for at least three time constants of the adjusted variable before samples were collected for analysis.

Suspended solids were measured with 0.45 mm filters using the techniques described in the Standard Methods (1976). Any material passing through those filters were considered to be soluble. COD measurements were also performed using the Standard Method procedure. The filtrate contained organic matter attributable to colloidal and dissolved compounds but is referred to here as soluble COD.

Data and analysis depicting the performance of the pilot plant for each of the operational conditions are found elsewhere (Sorour, 1991). Here, only the data of runs No. 13 and 14 are presented (Table 1). The values of all the dependent variables (DO, pH, effluent COD, effluent suspended solids, and the concentrations of MLSS and MLVSS) were averaged during their best steady-state conditions. These averages are summarized in Table 1. The table shows that the soluble COD concentration in the effluent is lower than its concentration from the last reactor. Such an observation supports the hypothesis that the final settler is able to remove part of the soluble COD. Comparing runs 13 and 14, a higher soluble COD concentration escaped from the reactor in run 14 because of the short hydraulic retention time (1.8 h). Logically, both the high levels of DO concentration (2.0-4.0 mg/l), and the sufficient detention time in the settler, encourage the utilization of a large part of this high concentration of soluble COD.

TABLE 1. Results of Run No 13 and 14.

Run No.	Experiment conditions		DO conc. in reactor mg/ℓ	DO conc. in settler mg/ℓ	pH in reactor	pH in settler	Temp. °C		MLSS conc. mg/ℓ	MLVSS % VSS	Effluent SS* conc. mg/ℓ	Last reactor soluble COD conc. mg/ℓ	Effluent soluble COD conc. mg/ℓ	Effluent NH-N mg/ℓ	Effluent PO_4-P mg/ℓ
13	SRT, d	5	A 5.0	3.7	A 7.1	7.8	Reactor u	24.0	A 1400	70.3	12.6	30.3	23.9	3.0	5.6
	Feed to reactor i.e.HRT,h	A 5.4	B 5.0		B 7.9				B 1380						
	R %	100	C 5.3		C 7.5		Settler	24.7	C 1383						
14	SRT, d	5	A 4.9	2.1	A 7.7	7.5	Reactor u	23.5	A 1639	71.0	9.5	64.8	24.9	17.10	3.5
	Feed to reactor i.e.HRT,h	C 1.8	B 4.9		B 7.7				B 1622						
	R %	100	C 3.0		C 7.2		Settler	23.0	C 847						

The periods used for data averaging are shown in Figure 2.

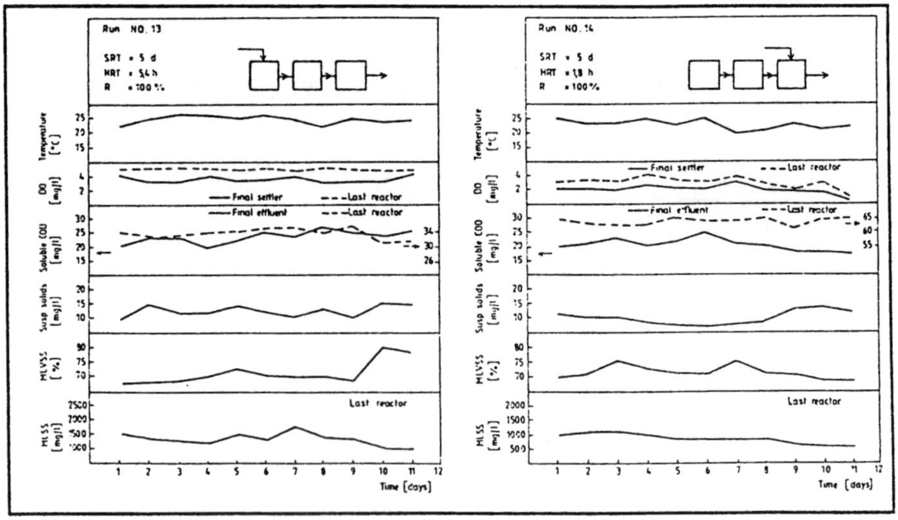

Fig. 2. Operational data for run no. 13 and 14 respectively.

MODEL EVALUATION

In order to indicate the qualitative validity of the proposed model, it has been tested against full scale data from the Hamilton Water Pollution Control Plant (Takacs *et al.*, 1990). The Hamilton WPCP is an activated sludge plant designed to treat an average combined sewer flow of 410.000 m^3/d. The plant is also able to cope with a short term storm water flow of approximately 600.000 m^3/d, which occurs quite frequently.

The data from three dynamic events were selected. The second one will be used to identify the model, while first and third events will serve as model verification tests:

Event # 1:(August 4, 1989) a small storm on with peak load of 409.000 m^3/d;
Event # 2: (August 10, 1989) peak flow rate of 600.000 m^3/d;
Event # 3:(September 22, 1989) hurricane "Hugo" downgraded to a tropical storm that hit Hamilton. The hydraulic load of the plant increased to 727.000 m^3/d and bypassing had to be activated in order to prevent flooding of the plant. The maximum influent load to the plant was kept around 600.000 m^3/d.

Calibration

Calibration of the model requires that the stoichiometric, kinetic and settling parameters are quantified. However, Dealing with a model describing biological activities, and therefore being confronted with the problem of limited and uncertain data, confidence in sophisticated numerical calibration techniques may not be too well placed. The model was calibrated by varying some of the parameters to give the best fit with the data of event # 2 which was selected as a database for dynamic calibration.

Default stoichiometric, kinetic and settling parameters available in the literature were incorporated. Table 2 presents the final model key parameters. Because the measurements available from the Hamilton treatment plant were in terms of BOD, suspended solids and TKN, more constants were needed in order to convert these terms to the IAWPRC state variables (Appindex 2). After having identified the parameters all simulations were performed without changing any of them for the same plant.

TABLE 2. Calibrated Parameters Values of the Proposed Model

Parameter		Value	
Stochiometric constants			
heterotrophic yield	Y_h	0.67	
autotrophic yield	Y_a	0.24	
unbiodegradable fraction of biomass	f_u	0.08	
COD/VSS ratio	icv	1.48	
VSS/TSS ratio	ivt	0.68	
Kinetic constants			
max. heterotrophic growth rate	μ_h	3.10	d^{-1}
saturation coefficient for S_s	K_s	20.00	$gCOD/m^3$
heterotrophic oxygen half saturation	K_{oh}	0.20	gO_2/m^3
nitrate half saturation	K_{no}	0.50	gN/m^3
anoxic growth factor	η_g	0.80	——
max. autotrophic growth rate	μ_a	0.43	d^{-1}
autotrophic s_{nh} half saturation	K_{oa}	0.40	gO_2/m^3
heterotrophic decay	b_h	0.62	d^{-1}
autotrophic decay	b_a	0.12	d^{-1}
ammonification rate	K_a	0.08	$m^3(gCOD\ d)^{-1}$
max. hydrolysis rate	K_h	3.00	d^{-1}
hydrolysis half saturation	K_x	0.03	
anoxic hydrolysis factor	η_h	0.40	
Settler parameters			
sludge blanket conc.	x_t	3000	g/m^3
floc settling parameter	r_{floc}	0.00035	m^3/g
colloids settling parameter	r_{coll}	0.0024	m^3/g
non-settleable fraction of susp. solids	f_{ns}	0.0015	
max. settling velocity	v_{max}	256	$m\ d^{-1}$

Figure 3a shows the influent flow during event #2 and Figure 3b the actual data points with the calibrated simulations of influent and effluent BOD. The model fit to the effluent BOD concentration is not accurate. This is natural, since the measurement inaccuracy is related to the absolute value of the concentration

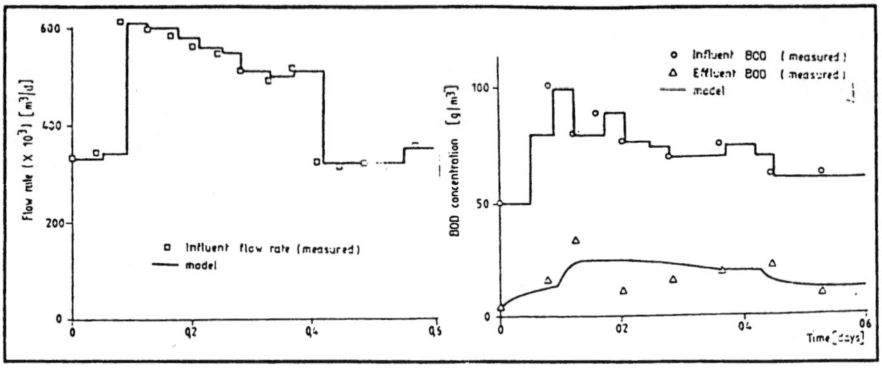

Fig. 3. Observed and simulated influent flow rate and BOD concentrations (calibration).

Model Verification

The proposed model was verified against the data of events #1 and #3. The analysis was focused on the effluent BOD concentration both *before* the final settler (IAWPRC+Vitasovic model) and *after* the settler (our proposed model). Results from event #1 are shown in Figure 4 and compared with the recorded observations. This event was a small and steady storm with peak flow of 409.000 m^3/day.

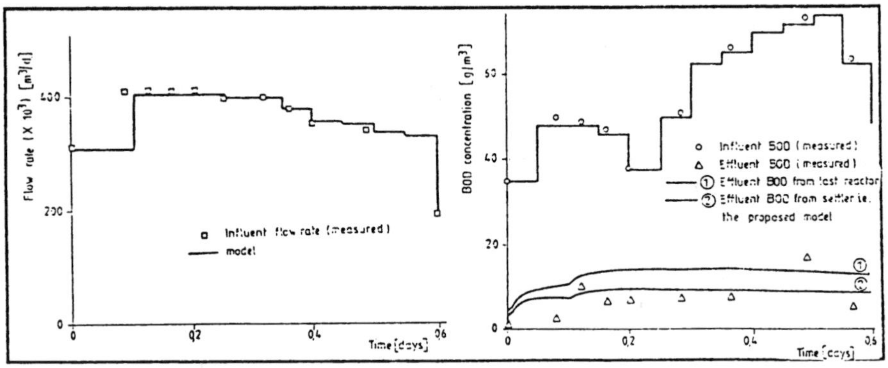

Fig. 4. Event # 1. Observed and simulated influent flow rate and BOD concentrations.

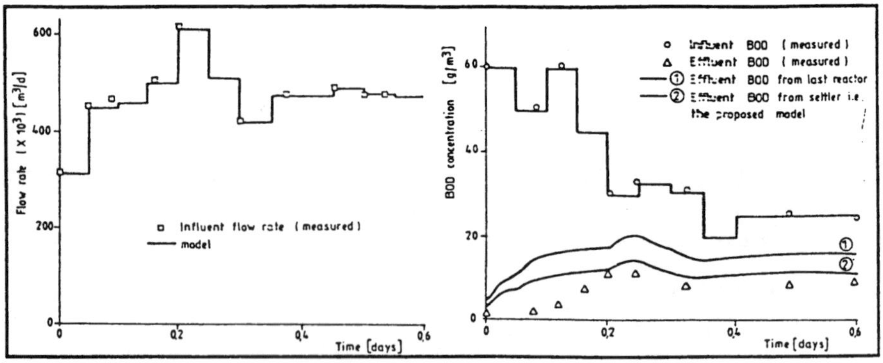

Fig. 5. Event # 3. Observed and simulated influent flow rate and BOD concentrations.

Simulated and measured results of events #3 (a tropical storm) are shown in Figure 5. Concerning the effluent BOD concentration a distinction must be made between the original and modified models. Figures

4b and 5b show the measured and predicted values of BOD in the effluent of the two studied cases. Comparing the observed and predicted responses it is evident for both cases that the proposed model gives a better agreement with the full-scale data than the original model. The difference between the two models is the soluble substrate utilized in the final settler. This result strongly supports the settler biological activity hypothesis.

The model predictions are sufficiently close to the observed data to support the model assumptions.

In spite of the widespread availability of step feed configurations in activated sludge systems, little use has been made of its control potential. The main aim of this study is to provide a better understanding of applying step feed control considering the role of biomass activity in the settler. The control strategy was evaluated by simulating the dynamics of an activated sludge system. The model contains both biomass activity in the settler and step feed capabilities The model and the strategy then have been evaluated against pilot scale and full scale experimental data.

STEP FEED CONTROL CONSIDERING SETTLER BIOLOGICAL ACTIVITY

Step feed control makes the sludge move between different parts of the aerator. This makes it possible to attenuate large hydraulic or organic shock loads by control (Olsson, 1985). Step feed control becomes even more interesting if biomass activity in the settler is considered. The step feed control will be evaluated using the proposed model applied to a pilot scale plant.

Pilot Plant Experiments

The pilot activated sludge plant located at the Wastewater Technology Center, Burlington, Ontario, Canada, was chosen as the test bed for the step feed control and the proposed model. This plant consists of three $22m^3$ aeration tanks in series followed by a final settler with the depth 2.7 m. Settled sewage enters the plant at a normal flow rate of 288 m^3/day and can be fed into any of the three reactors. The characteristics of the plant are summarized in detail by Thompson (1988). To examine the step feed control on this plant the following steps have been carried out:

an adequate model representation of the pilot plant;

calibration of the model to reflect more closely the actual dynamical behaviour;

computer simulations to evaluate the control strategy.

To prevent a significant loss of solids during storm flow conditions, Thompson (1988) experimented with the step feed of sewage to the second reactor. The data of this experiment was selected to serve as a base for model calibration, i.e. comparisons were made between simulation results and the data of these experiments to achieve the model parameters. Strom flow conditions were simulated by an increase in the flow rate from 288 to 720 m^3/day, while the recycle flow rate was kept constant at 86.5 m^3/day. There were no major changes needed for the aerator parameters, while parameters of the settler model Takacs *et al* (1991) were adjusted to fit more closely with the actual data of the plant. The calibrated model parameters were (cf Table 2):

$$v_{max} = 200 \text{ m/d};\ r_{floc} = 0.006 m^3/g;\ r_{coll} = 0.01\ m^3/g;\ f_{ns} = 0.005$$

Figure 6 shows the actual and simulated results. As indicated in Figure 6a, the biomass concentration increased in the first reactor, while solids from the third reactor were washed out from the system, decreasing the solids loading to the settler. Simulated and measured effluent suspended solids are also in good agreement (Figure 6b).

After quantifying the parameters, simulations were performed to evaluate the control strategies. All simulations were performed without channging any of the calibrated parameters.

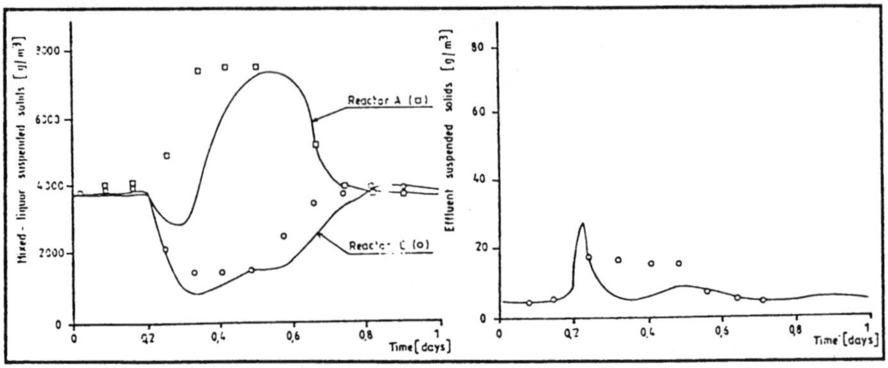

Fig. 6. Thompson's experiment. Storm flow with step feed reactor B (calibration).

Step Feed Control to Prevent Solids Washout

Because conventional activated sludge plants have limited hydraulic dampening capacity, high flow rates transfer additional solids from the aeration tank to the final settler and increase the sludge blanket depth (Olsson *et al.*, 1986). Solids washout occurs when the top of the blanket reaches the effluent weirs. Large fluctuations were simulated using sinusoidal pattern as flow rate with an average of 750 m^3/d, and an amplitude of 150 m^3/d during a period of 24 h.

To evaluate the effect of step feed control strategy on the effluent quality simulations were performed using the sinusoidal flow and the pilot plant layout. The flow was directed either to reactor B or to reactor C. The responses for the latter case is shown in Figure 7. In neither case there was any other change made in operational variables of the plant.

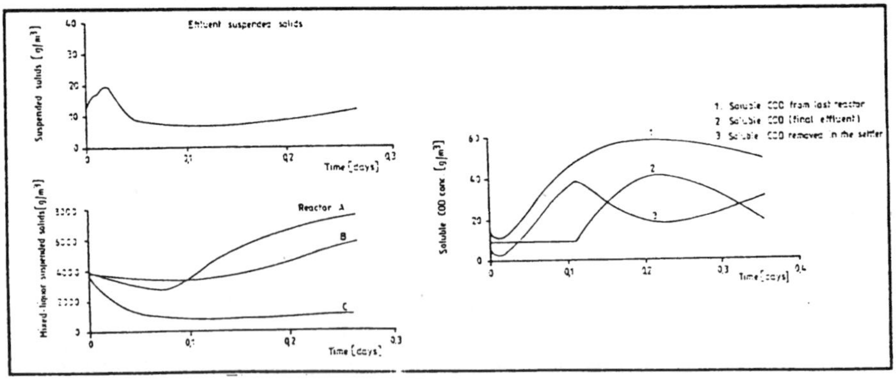

Fig. 7. Simulation of step feed to reactor C.

A step feed to reactor C caused an increase in the MLSS concentrations in reactors A and B, while solids from the reactor C were washed out from the system, decreasing the solids loading to the settler. It is noted that step feed to reactor C on one hand caused decrease in the effluent suspended solids and on the other hand caused an increase in the effluent soluble COD. Thus it should be concluded that the step feed control requires that the advantage of reduced settler solids loading be balanced with the potential disadvantages of poorer soluble organics removal.

For activated sludge plants which are frequently subject to large fluctuations in flow rate or rainstorms, the steady-state formula of sludge age is not adequate to predict the waste sludge flow rate. An alternative control law can be used to calculate the actual waste sludge flow rate due to the dynamical changes of the total sludge mass. This is further development in Sorour (1991).

CONCLUSIONS

Ignoring the final settler biological activity in modeling the activated sludge system, with step feed capability, causes error in predicting the final effluent soluble substrate, especially during peak flow events. Using the proposed model, experimental data and computer simulation, it was possible to justify the incorporation of this process. Simulation results showed that the propoesd model gave a good agreement with the full scale data: such a finding strongly supports our hypothesis.

For an activated sludge system with tanks in series, and step feed capability, increasing the detention time in the settler while providing adequate DO levels, would allow the utilization of a large part of the soluble organics that escape from the last reactor during peak flow events.

The step feed contrrol strategy is a critical strategy. Particular attention should be paied to balancing the advantage of reducing solids loading to settler with the disadvantage of less removal of soluble organics in the aerator. Since the dynamics are slow it is realistic to implement a rule base as an operator guidance.

REFERENCES

Busby, J. B. and Andrews, J. F. (1975). Dynamic modeling and control strategies for the activated sludge process. *Water Pollut. Control Fed.*, **47**, 1055-1080.

Elmqvist, H., Astrom, K. J. and Schonthal, T. (1990). SIMNON-User's Guide for Ms-Dos Computers. Dept. of Automatic Control, Lund Institute of Technology, Lund, Sweden.

Henze, M., Grady, Jr., C. P. L., Gujer, W., Marais, G. V. R. and Matsuo, T. (1987). Activated Sludge Model No. 1, IAWPRC, London, UK.

Olsson, G. (1977). State of the art in sewage treatment plant control. *AIChE Symp Ser.*, **72**, 52-76.

Olsson, G. (1985). Control strategies for the activated sludge process. Comprehensive Biotechnology, Pergamon Press, New York, USA, 107-1119.

Olsson. G., Stephenson, J. and Chapmaan, D. (1984). Computer detection of the impact of hydraulic shocks on plant performance. *J. Water Pollut. Control Fed.*, **58**, 954-959.

Sorour, M. T. (1991). Activated sludge modeling for operational control. An approach to step feed system. Ph. D. Disseration, The Hungarian Academy of Sciences, Budapest, Hungary.

Standard Methods for the Examination of Water and Wastewater (1976). 14th ed., American Public Health Association, Washington D.C., USA.

Takacs, I., Patry, G. G. and Snodgrass, W. J. (1990). Hamilton water pollution control plant, analysis of plant performance. Dept. of Civil Engineering and Engineering Mechanics, McMaster University, Hamilton, Ontario.

Takacs, I., Patry, G. G. and Nolasco, D. (1991). A generalized dynamic model of the clarification/thickening process. *Instrumentation, Control and Automation of Water and Wasterwater Treatmenent and Transport Systems* (ed. R. Briggs), pp. 487-94, Pergamon Press.

Thompson, D. (1988). Activated sludge: step feed control to minimize solids loss during strom flow. M. Eng. Thesis, McMaster University, Dept. of Chem. Engineering, Hamilton, Ontario, 25 pp.

Torpey, W. N. (1948). Practical results of step aeration. *Sewage Works J.*, **20**, 781-788.

Vitasovic, Z. Z. (1986). An intergated control strategy for the activated sludge process. Ph. D. Diseration, Rice Universsity, Houston, Texas, 288 pp.

APPENDIX 1. MODIFICATION OF THE CLARIFIER DYNAMICS

The settling velocity is suggested by Takacs *et al.* (1991) as

$$v_{s,i} = v_{max} \cdot \exp\left(-r_{floc} \cdot X^*_i - r_{coll} \cdot X^*_i \right)$$

where

v_{max} = maximun settling velocity (L/T)
r_{coll} = collids settling parameter (L^3/M)
r_{floc} = floc settling parametre (L^3/M)
$v_{s,i}$ = settling velocity of the solids particles in layer (m/d)
$X^*_i = X_i - X_{min}$
X_i = supsended solids concentration in layer i (g/m^3)
X_{min} = minimum attainable suspended solids concentration (g/m^3)
$X_{min} = f_{ns} \cdot X_{in}$
f_{ns} = non-settlebale fraction of suspended soilds
X_{in} = mixed liquor suspended solids enetring the settler (g/m^3)

Pergamon

Wat. Sci. Tech. Vol. 28, No. 11-12, pp. 249–256, 1993.

0273–1223/93 $7·00 + 0·00

MODELLING, CONTROL AND ON-LINE ESTIMATION OF ACTIVATED SLUDGE BULKING

J. Chen and M. B. Beck

Department of Civil Engineering, Imperial College, London SW7 2BU, UK

ABSTRACT

Sludge bulking is a problem widespread in the operation of activated sludge processes. Over the past fifteen years much research has been applied to identifying the nature of the microbial species responsible for bulking and to developing an understanding of their population dynamics. While some isolated attempts have been made at developing a unit process model capable of simulating the behaviour of bulking, none have been extensively evaluated against field data from full scale plants. This paper describes the development of a multiple-species model of the activated sludge process (for both the reactor and settler), its application to the assessment of various operational strategies for the control of bulking, and its simplification for incorporation into an on-line estimation scheme using a Kalman filter. Routine operating data from the Davyhulme Wastewater Treatment Works in Manchester are used for identification (calibration) of the model. Simulation studies of control strategies are based on the same Works. By using the Kalman filter to reconstruct real-time estimates of the "unmeasurable" states of the process model, the paper also explores the extent to which this filter-based control can bring improvements over similar control based entirely upon conventionally measured operating variables.

KEYWORDS

Dynamic modelling; operational control; activated sludge process; on-line estimator; sludge bulking; Kalman filtering.

INTRODUCTION

Sludge bulking is one of the most common and widespread problems in the operation of activated sludge processes (Tomlinson, 1976; Wagner, 1984; Blackbeard *et al.*, 1986; Pujol and Boutin, 1989). The prevailing view is that bulking is caused by the excessive growth of filamentous microbial organisms (Sezgin *et al.*, 1978; Strom and Jenkins, 1984; Gujer and Kappeler, 1992). However, the observed causes of excessive filamentous biomass are numerous and often contradictory (Farquhar and Boyle, 1972; Chudoba *et al.*, 1973a, 1973b, 1974; Simpson *et al.*, 1991; Jenkins, 1992). Yet two such causes have been found to be especially important in most cases: these are either a low organic loading, or an organic overloading coupled with a low DO condition (Chiesa and Irvine, 1985). The paper will thus focus on bulking as a function of these two factors alone.

Despite the proposed use of various parameters and operational variables as early warning indicators of the onset of bulking (Tomlinson, 1976; Eikelboom, 1981; Lee *et al.*, 1982; Jones and Franklin, 1985; Linne and Chiesa, 1987), none of these indicators seems to offer any greater reliability than the potential use of a mathematical model. Hitherto, however, no such application of a model has been reported and it is known that these models have suffered from various deficiencies (Smouse, 1980; Lau, *et al.*, 1984; Takamatsu, *et al* 1984; Hermanowicz, 1987). So great did these deficiencies seem some ten years ago that there was a strong tendency to move away from the use of the more conventional forms of model towards techniques such as expert systems and qualitative simulation as possible means to control the occurrence of bulking (Beck, 1986; Beck *et al.*, 1990). In the following a dynamic multiple species model is developed and evaluated against field data, from which appropriate strategies for the control of sludge bulking might be derived.

Much of the earlier work has been notable for its concern primarily with the design (or re-design) of a works for the prevention of bulking, as opposed to the use for such purposes as operational control manipulations of existing designs. Consequently the dominant techniques for bulking control are either the operational use of chemicals (van Leeuwen and van Rossum, 1990) or the construction of selectors in the aeration tank (Patoczka and Eckenfelder, 1990; Albertson, 1991), which latter are relatively costly and of debatable effectiveness. It is apparent, then, that the most desirable and economic means of controlling bulking might be through operational manipulation of the more routine control variables and indeed some associated successes have already been reported (Rensink *et al.*, 1982; Chiesa and Irvine, 1985). This paper likewise examines the use of routine control variables for preventing bulking.

IDENTIFICATION OF THE MODEL

The Model Structure

Our multiple species model contains three interdependent parts, comprising the reactor and the settler, the latter having two functions of thickening and clarification.

The reactor sub-model. The reactor sub-model is essentially a modification of the IAWPRC Task Group model (Henze, *et al.*, 1987), the heterotrophic biomass being divided into floc-forming and filamentous biomasses. In addition, the behaviour of organic nitrogen is excluded. In other words, the following processes are simulated in our study: aerobic and anoxic growth of the floc-forming and filamentous biomasses; the aerobic growth of autotrophic bacteria; the decay of heterotrophic and autotrophic organisms; and the hydrolysis of particulate organics by heterotrophs. A detailed description of the model can be found in Chen (1993). The hydraulic characteristics of the model are assumed to approximate those of a sequence of constant-volume, continuously stirred tank reactors (CSTRs). Provision has also been made for the influent flow, recycle flow and air input to be distributed differently to each such CSTR element. This allows simulation of operational control through step-feed, step-sludge and tapered aeration schemes.

The thickener sub-model. The thickener is represented by two conceptual zones: the sludge blanket and compression zones. As is usual, the model is based on the flux theory, with an exponential function chosen to describe the gravitational settling velocity of the suspended solids (SS) (Lessard, 1989; see also Daigger and Roper, 1985; Wahlberg and Keinath, 1988). However, instead of assigning a constant value to this parameter, the initial settling velocity is regarded as a function of the floc-forming and filamentous biomasses, so that the effects of bulking on the performance of the settler can be incorporated. The detailed reasoning for this assumption can be found elsewhere (Chen, 1993). Flow through the zone of the sludge blanket is similar to that of Lessard (1989), and is represented by a (vertical) series of one to several variable-volume layers, or CSTR elements.

The clarifier sub-model. Clarification is represented in a semi-empirical fashion. The major assumption is that when the mixed liquor of the biological reactor enters the settler, all the SS will settle into the thickening zone, so that any SS in the clarification zone results from "resuspension". Empirical expressions are used to characterize the flux of solids mass resuspended from the thickening zone and introduced into the clarification zone, largely as a function of hydraulic factors (Pflanz, 1969) and the occurrence of bulking (Beck, 1984; Takamatsu, *et al.*, 1984; Chen, 1993). The movement of SS through the clarification zone is then governed by its resettling and hydraulic dispersion characteristics. The volume of the clarification zone is variable and equal to the volume of the settler above the sludge blanket.

Identification of the Model Against Field Data

The field data. The field data for this study were obtained from the Davyhulme Wastewater Treatment Works of North West Water Ltd. Bulking occurs from time to time at this Works. However, there is no clear evidence of the causes of this bulking. During a period of several weeks in the late summer/early autumn of 1990, the dominant filamentous species was *S. natans*, often reported in association with a low DO in the aerator (Lee, *et al* 1982). Yet DO has been kept continuously at a high level in the Works (3-6 mg/l), implying thus that a low DO should not be identified as the cause of the bulking. Further observation of the influent characteristics suggests that a prolonged period of low influent organic loading, i.e. "the single organic limit" (Pavlou *et al.*, 1990), might have led to the occurrence of bulking at this plant.

In the present study, the simulation period is set at 50 days (more than six times the sludge age), during the last 20 days of which bulking is found to occur. All data are taken from the routine operating records of the Davyhulme Works and include daily influent flow, COD, ammonia-N, nitrite-N and nitrate-N concentrations. Recycle sludge flow was set by the plant operators at 1.5 times the influent flow and the sludge wastage rate was adjusted daily as a function of the sludge age.

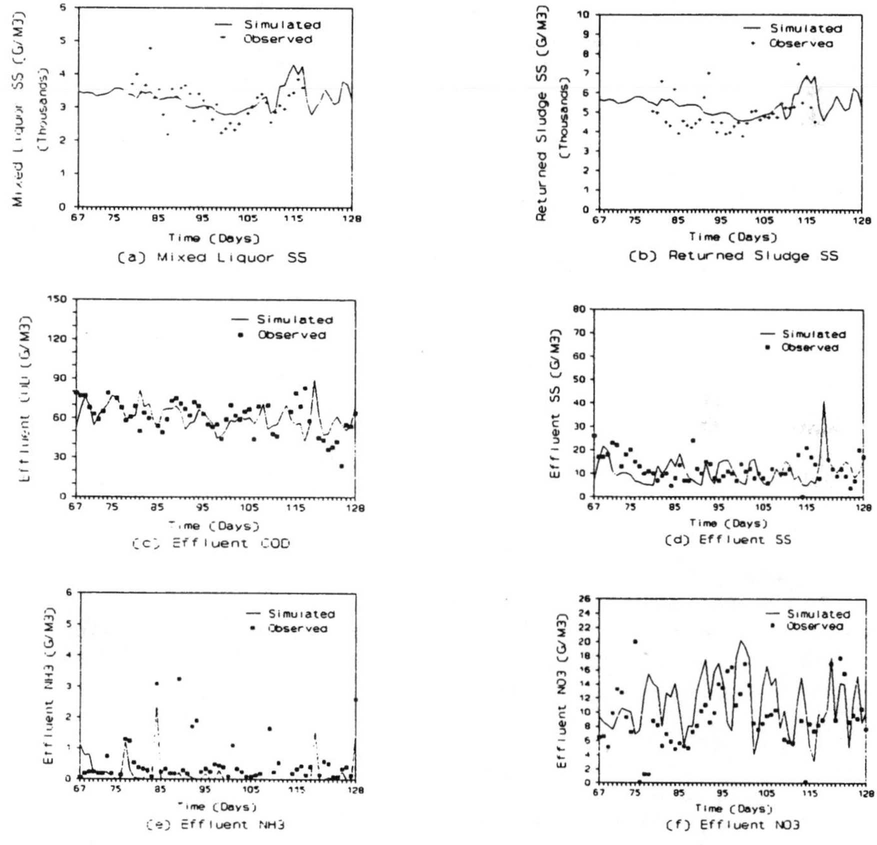

Fig. 1. The evaluation of the multiple-species model against field data.

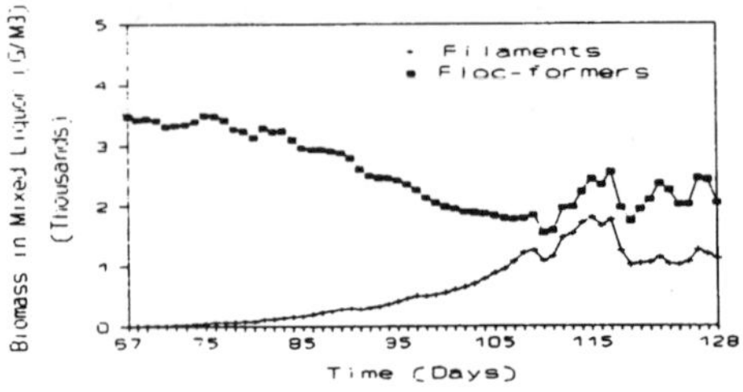

Fig. 2. Simulated floc-forming and filamentous biomasses.

Identification (calibration) results. The basic structure of the model has been evaluated by reference to a second set of field data from the Norwich Sewage Works (Chen, 1993). Following these preliminary results, the model parameters have been estimated simply by trial and error for the Davyhulme Works; the results are presented in Figure 1. From this figure it is apparent that the model performs satisfactorily in simulating the overall trend of the system's behaviour for most of the time. Figure 2 shows the accompanying varying pattern of the composition of filamentous and floc-forming biomasses in the mixed liquors. It is evident that filamentous biomass increases steadily from the beginning of the simulation. It takes over one month to develop a significant amount of filamentous biomass (10% of the total biomass); during the last 20 days this part of the biomass is persistently (relatively) high and in fact bulking occurred just after the end of the plotted record. Analysis of the net growth rates of both biomasses confirms that a low COD in the aeration tank is the major factor promoting the growth of the filamentous bacteria.

OPERATIONAL CONTROL OF SLUDGE BULKING

Using the model thus identified, a comprehensive assessment of conventional and advanced control strategies has been undertaken in order to assess the potential to suppress bulking by operational means (Chen, 1993). The priorities for such control were specified as follows:

(a) avoid gross failure of the plant, i.e. avoid bulking;

(b) maintain performance complying with specified effluent standards for total COD, SS and NH_3-N, the values of which are 100, 30 and 3 g/m^3 respectively;

(c) maintain the ratio of filamentous to floc-forming biomasses as close to a stable point as possible, so that there is spare capacity for the biomass to endure possible future influent perturbations promoting the occurrence of bulking; and

(d) maximise effluent quality as much as possible.

The first pair of the above points is clearly aimed at avoiding long-term and short-term failures of the process, while the second pair is geared towards pursuing the best performance of the plant in terms of both its internal (biomass) state and its external output response. Some 34 control strategies of various kinds have been analysed, including feedforward proportional control of recycle flow, sludge age control, feedback control, optimal control and the combined optimal control of sludge recycle and wastage (Chen, 1993). Consideration has also been given to the frequency of applying control action (hourly or daily), spatially distributed feed-points of the sewage and sludge recycle, and the role of uncertainty in schemes of both controllers and on-line load forecasting for predictive control.

Three important conclusions can be drawn.

(a) Conventional control strategies, such as sludge age control and a fixed invariant proportional control of the recycle rate do not prevent the sludge from bulking. Neither are the feedback controllers of sludge recycle and wastage effective due to the long time-lag of the activated sludge system. However, it is found that the manipulation of recycle rate has a considerable effect on the behaviour of the filamentous biomass (as opposed to the floc-forming biomass).

(b) All of the optimal controllers of recycle rate, step-feed and step-sludge are very effective in controlling bulking, in spite of significant uncertainties in forecast information of the influent characteristics and the model predictions of the controllers. Figure 3 illustrates thus the successful result of an optimal manipulation of the recycle rate. The success of the strategy results from manipulation of the internal state of the system (the biomass composition) such that its propensity to bulk is no longer in phase with the timing of "adverse" conditions in the (external) sewage input. Corresponding optimal controllers of the wastage rate are far less effective than this.

(c) Some sludge bulking control strategies can cause a significant deterioration in effluent quality, so that there may be conflicts between the control of bulking and satisfaction/improvement of the effluent standards/quality.

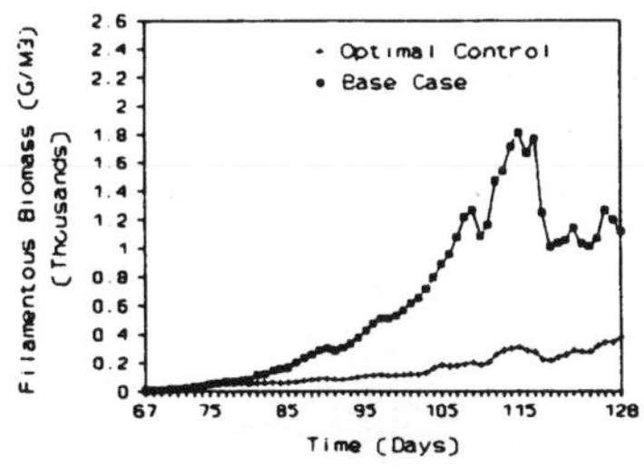

Fig. 3. Example of operational control of sludge bulking: optimal implementation of recycle rate.

In particular, it is found that effective control of sludge bulking requires early action. Generally speaking, the earlier the action can be taken, so the more effective is the control. In the light of this, and given that a better control might be exercised on the basis of information about the state of the system that is not directly measurable, such as parameters associated with the filamentous biomass, the utility of an on-line estimation algorithm has also been investigated. Given too that there may be conflict between the intention to control bulking and the need to improve effluent performance, there is also considerable interest in whether such an estimator can be used to forecast the likelihood of bulking occurring in the future.

ON-LINE ESTIMATION

The on-line estimator used in this study has been based on an extended Kalman Filter (as in Beck, 1981). However, since the model discussed above involves a large number of state variables and parameters, with but few observed input and output variables, it is necessary first to simplify this model.

The Simplified Model

Development of the simplified model has been guided by the results of a sensitivity analysis of the full model (Chen, 1993). It contains the following assumptions, which differ from those of the full model.

(a) Only the biological reactor (aeration tank) is included in the simplified model, although the observed sludge recycle SS and its flow rate are used as input variables that can account for interactions between the secondary settler and the reactor.

(b) The simplified model is directed only at the description of sludge bulking dynamics, so that the dynamics of nitrification are excluded.

(c) The configuration of the aeration tank is considered to comprise two CSTRs with different volumes, so that step-feed and step-sludge control strategies can be taken into account in the overall estimation/control algorithm.

(d) Only those parameters identified as critical in the composition of the full model are estimated on- line; all other parameters retained in the simplified model are considered as constants and assigned the same values as in the full model.

Detailed information of the simplified model can be found in Chen (1993).

The On-line Estimator

The estimator has been applied to the same set of data as presented earlier. However, a simulation period of 21 d less than previously has been used; sludge concentrations in the recycle flow and mixed liquor were not available for the first 11 and last 10 d of the original period.

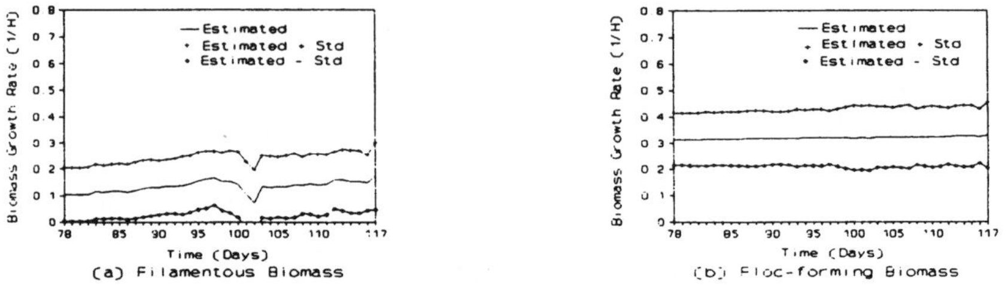

Fig. 4. On-line estimated maximum specific growth rates of biomass

An example of the performance of the estimator is given in Figure 4, which shows the estimated parameters (and their uncertainties) for both the maximum specific growth rates of the filamentous and floc-forming biomasses. It is apparent that the maximum growth rate of the floc-forming biomass is quite invariant with time, fluctuating but little around a constant value. However, that of the filamentous biomass shows a steady upward trend. This may well indicate that in the activated sludge process the physiological behaviour (expressed by the aggregated parameter of maximum growth rate) of floc-forming biomass is quite stable, but that of the filamentous biomass is not. In fact the reason for this is readily apparent, since the amount of floc-forming biomass is so large in the system that it is (in aggregate) well buffered against perturbations induced by the influent. In contrast, the amount of filamentous biomass in the system is so small that it possesses little resistance to the influence of outside changes, thus its growth rate is therefore less stable. The same reason can also explain why a sudden decrease is estimated in the growth rate of the filamentous biomass during the time period of days 96 to 101, over which the influent organic flow and COD have been found not to favour the growth of filamentous biomass. In other words, the problem of preventing the sludge from bulking could become one of how to predict the unstable filamentous biomass behaviour. It is evident, then, that the slowly changing characteristics of the growth rate of the filamentous biomass may be used as a good early warning, or supporting confirmation, of the likely occurrence of bulking.

Figure 5 shows thus one of the situations when both the estimated filamentous biomass and its maximum growth rate are relative high on days 112 to 115. From this prediction, it is found that the filamentous biomass is quite constant for the first 9 d of the prediction (from day 115 to 123), but then suddenly increases to a level capable of inducing bulking due to the increase of the estimated maximum growth rate of the filamentous biomass. Such a prediction is consistent with both what actually happened in the Davyhulme Works and the prediction of the full model.

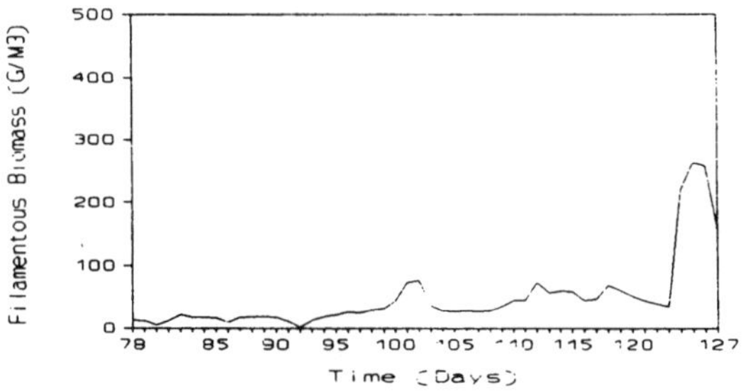

Fig. 5. Predicted filamentous biomass in the mixed liquor by EKF.

CONCLUSION

A multiple species model of the activated sludge process has been developed for the purposes of examining the use of conventional control variables (such as sludge recycle and wastage) for the avoidance of sludge bulking. The reactor sub-model is essentially a modification of the IAWPRC Task Group model, with the heterotrophic biomass separated into floc-forming and filamentous bacterial species. The sub-model of the thickening function of the secondary clarifier incorporates an exponential form of settling velocity dependent upon the relative proportions of these two species of bacteria. The sub-model of clarification allows for the upward advection, dispersion and downward settling of any solids resuspended and entrained into the clarification zone above the sludge blanket.

Evaluation of the model against routine operating data from the Davyhulme Works in Manchester shows that its performance is adequate. The simulated ratio of floc-forming and filamentous biomasses is quite consistent with the recorded development and occurrence of bulking. It is further shown that had an alternative strategy of manipulating the sludge recycle been implemented over this period, bulking may have been prevented. The success of the strategy depends upon altering the relative timings of adverse influent conditions and the propensity of the sludge to bulk. Such control action must be exercised well in advance of the expected occurrence of bulking. It may arguably be more effective if based on reconstructed estimates of the state of the biomass, as opposed to conventional, readily observable process indicators. A simplified model of the system incorporated within a Kalman filter can fulfil both of these tasks: of on-line forecasting, for predicting the likelihood of bulking occurring; and on-line estimation of the composition of the biomass and the relative growth rates of the floc-forming and filamentous bacteria. The stability of the filamentous population is a key factor linked with such occurrences.

Further work – enabling a better identification of the model against specially monitored data on process performance and allowing assessment of the extent to which control strategies are robust in the face of model uncertainty – is now needed.

ACKNOWLEDGEMENT

The authors are indebted to North West Water Ltd for their permission to the use of the field data presented in this paper.

REFERENCES

Albertson, O. E. (1991). Bulking sludge control – progress, practice and problems. *Wat. Sci. Tech.*, **23**(4-6), 835-846.

Beck, M. B. (1981). Operational estimation and prediction of nitrification dynamics in the activated sludge process. *Wat. Res.*, **15**, 1313-1330.

Beck, M. B. (1984). Modelling and control studies of the activated sludge process at Norwich sewage works. *Trans. Inst. M. C.*, **6**(3), 117-131.

Beck, M. B. (1986). Identification, estimation and control of biological wastewater treatment Processes. *IEE Proceedings*, **133**(5), 254-264.

Beck, M. B., Lumbers, J. P., Mackenzie, H. E. C. and Jowitt, P. W. (1990). Un prototype de système expert pour le contrôle d'un procédé de boues activées. *Sciences et Techniques de l'eau*, **23**(2), 161-167.

Blackbeard, J. R., Ekama, G. A. and Marais, G. V. R. (1986). A survey of filamentous bulking and foaming in activated sludge plants in South Africa. *Wat. Pollut. Control*, **85**(1), 90-100.

Chen, J. (1993). Modelling and control of the activated sludge process: towards a systematic framework. *PhD thesis*, Imperial College, University of London, UK.

Chiesa, S. C. and Irvine, R. L. (1985). Growth and control of filamentous microbes in activated sludge: an integrated hypothesis. *Wat. Res.*, **19**(4), 471-479.

Chudoba, J., Ottová, V. and Mad ra, V. (1973a). Control of activated sludge filamentous bulking – I. Effect of the hydraulic regime or degree of mixing in an aeration tank. *Wat. Res.*, **7**, 1163-1182.

Chudoba, J., Grau, P. and Ottová, V. (1973b). Control of activated sludge filamentous bulking – II. Selection of microorganisms by means of a selector. *Wat. Res.*, **7**, 1389-1406.

Chudoba, J., Bláha, J. and Mad ra, V. (1974). Control of activated sludge filamentous bulking – III. Effect of sludge loading. *Wat. Res.*, **8**, 231-237.

Daigger, G. T. and Roper, R. E. (1985). The relationship between SVI and activated sludge settling characteristics. *Journal of Wat. Pollut. Control Fed.*, **57**, 859-866.

Eikelboom, D. H. (1981). Biosorption and prevention of bulking sludge by means of high floc loading. *Proc. Wat. Res. Centre Conf.*, Cambridge, England.

Farquhar, G. J. and Boyle, W. C. (1972). Control of *Thiothrix* in activated sludge. *Journal of Wat. Pollut. Control Fed.*, **44**(1), 14-24.

Gujer, W. and Kappeler, J. (1992). Modelling population dynamics in activated sludge systems. *Wat. Sci. Tech.*, **25**(6), 93-103.

Henze, M., Grady, C. P. L. Jr., Gujer, W., Marais, G. V. R. and Matsuo, T. (1987). Abbreviated report: a general model for single-sludge wastewater treatment systems. *Wat. Res.*, **21**(5), 505-515.

Hermanowicz, S. W. (1987). Dynamic changes in populations of the activated sludge community: effects of dissolved oxygen variations. *Wat. Sci. Tech.*, **19**(5-6), 889-895.

Jenkins, D. (1992). Towards a comprehensive model of activated sludge bulking and foaming. *Wat. Sci. Tech.*, **25**(6), 215-230.

Jones, G. A. and Franklin, B. C. (1985). The prevention of filamentous bulking of activated sludge by operation means at Halifax sewage treatment works. *Wat. Pollut. Control*, **84**, 337-356.

Lau, A. O., Strom, P. F. and Jenkins, D. (1984). The competitive growth of floc-forming and filamentous bacteria: a model for activated sludge bulking. *Journal of Wat. Pollut. Control Fed.*, **56**(1), 52-61.

Lee, S-E. Koopman, B. L., Jenkins, D. and Lewis, R. F. (1982). The effect of aeration basin configuration on activated sludge bulking at low organic loading. *Wat. Sci. Tech.*, **14**(6-7), 407-427.

Lessard, P. (1989). Operational water quality management: control of storm discharges. *PhD thesis*, Imperial College, University of London, UK.

Linne, S. R. and Chiesa, S. C. (1987). Operational variables affecting performance of the selector-complete mix activated sludge process. *Journal of Wat. Pollut. control Fed.*, **59**(7), 716-721.

Patoczka, J. and Eckenfelder, W. W. (1990). Performance and design of a selector for bulking control. *Res. Journal of Wat. Pollut. Control Fed.*, **62**(2), 151-159.

Pavlou, S., Kevrekidis, I. G. and Lybertos, G. (1990). On the coexistence of competing microbial species in chemostat under cycling. *Biotech. Bioeng.*, **35**(3), 224-232.

Pflanz, P. (1969). Performance of (activated sludge) secondary sedimentation basins. In: *Advances in Water Pollution Research.*, S. H. Jenkins (Ed)., Pergamon Press, 569-581.

Pujol, R. and Boutin, P. (1989). Control of activated sludge bulking: from the lab to the plant. *Wat. Sci. Tech.*, **21**(6/7), 717-726.

Rensink, J. H. *et al.* (1982). The influence of feed pattern on sludge bulking. In: *Bulking of Activated Sludge: Preventative and Remedial Methods*, B. Chambers and E. J. Tomlinson (Ed).

Sezgin, M., Jenkins, D. and Parker, D. S. (1978). A unified theory of filamentous activated sludge bulking. *Journal of Wat. Pollut. Control Fed.*, **50**, 362-381.

Simpson, J. R., List, E. and Dunbar, J. M. (1991). Bulking sludge: a theory and successful case histories. *Journal of the Institution of Wat. and Environ. Management*, **5**(3), 302-311.

Smouse, P. E. (1980). Mathematical models for continuous culture growth dynamics of populations subsisting on a heterogeneous resource base – I. simple competition. *Theoretical Population Biol.*, **17**.

Strom, P. F. and Jenkins, D. (1984). Identification and significance of filamentous microorganisms in activated sludge. *Journal of Wat. Pollut. Control Fed.*, **56**(5), 449-459.

Takamatsu, T., Suteaki, S. and Hirokazu, K. (1984). Dynamics and control of an activated sludge process as a mixed culture system. *AIChE Journal*, **30**(3), 368-376.

Tomlinson, E. J. (1976). Bulking – a survey of activated sludge plants. *Wat. Res. Centre, Technical Report TR35.*

van Leeuwen, J. and van Rossum, P. G. (1990). Trihalomethane formation during sludge bulking control with chlorine. *Journal of the Institution of Wat. and Environ. Management*, **4**(6), 530-534.

Wagner, F. (1984). Studies on the causes and prevention of bulking sludge in germany. *Wat. Sci. Tech.*, **16**(10/11), 1-14.

Wahlberg, E. J. and Keinath, T. M. (1988). Development of settling flux curves using SVI. *Journal of Wat. Pollut. Control Fed.*, **60**(12), 2095-2100.

Pergamon

Wat. Sci. Tech. Vol. 28, No. 11-12, pp. 257–261, 1993.

0273–1223/93 $7·00 + 0·00

EXPERIENCE WITH THE RELAY PROCEDURE FOR TUNING CONTROLLERS IN AUTOMATIC CONTROL OF CHLORINATION

M. Truett Garrett, Jr., Zaki Ahmad and Shelly Young

City of Houston, Department of Public Utilities, Houston, TX 77251, USA

ABSTRACT

The recent requirements by U.S.E.P.A. for dechlorination and biomonitoring have increased the importance of automatic control of effluent chlorination in wastewater treatment plants. Difficulties with the Ziegler-Nichols controller tuning procedure were reported at the Kyoto Workshop, 1990. Problems are caused by the noise of incomplete mixing, a long time constant, and the disturbances of changing flow and chlorine demand. The Astrom-Hagglund relay feedback procedure provides acceptable control while data is logged to determine the controller constants. Experiences in using the procedure in existing facilities (not redesigning the mixing point) and the quality of control are presented.

INTRODUCTION

Dechlorination of chlorinated effluents was required in Houston beginning in 1990 with seven wastewater treatment plants required to dechlorinate that year and the remainder of the thiry four plants scheduled one year after the permit renewal dates. It was subsequently required that biomonitoring tests be performed without dechlorination at the laboratory. This has meant that failure to remove all chlorine causes a failure of a $2000 biomonitoring test instead of failure of a $20 residual chlorine test. Such failure can even trigger a "Toxicity Reduction Evaluation" (TRE) which can cost $100,000. This greatly increases the importance of automatic control of chlorination and dechlorination.

The city had determined to use sodium hypochlorite for chlorination at the larger plants for safety, so sodium bisulfite was chosen for dechlorination at all plants for safety and uniformity of equipment. It was found that dechlorination of a known concentration of total residual chlorine could be accomplished with feedforward control based on the effluent flow and chlorine concentration. Two on-line residual chlorine analyzers were used, one at the end of the chlorine detention tank to control the bisulfite feed and the other after dechlorination to monitor the result. Many plants are manned only eight or less hours per day. For these plants, automatic chlorine control is desirable to avoid having to dechlorinate excessive chlorine residuals during the early morning hours. For chlorination control an additional on-line analyzer was added just after the mixing point; and a PLC was provided to control both chlorination and dechlorination, and having capacity to control and monitor other functions as needed.

EQUIPMENT FOR CHLORINATION/DECHLORINATION

A schematic diagram of the equipment used for chlorination and dechlorination is shown in Figure 1.

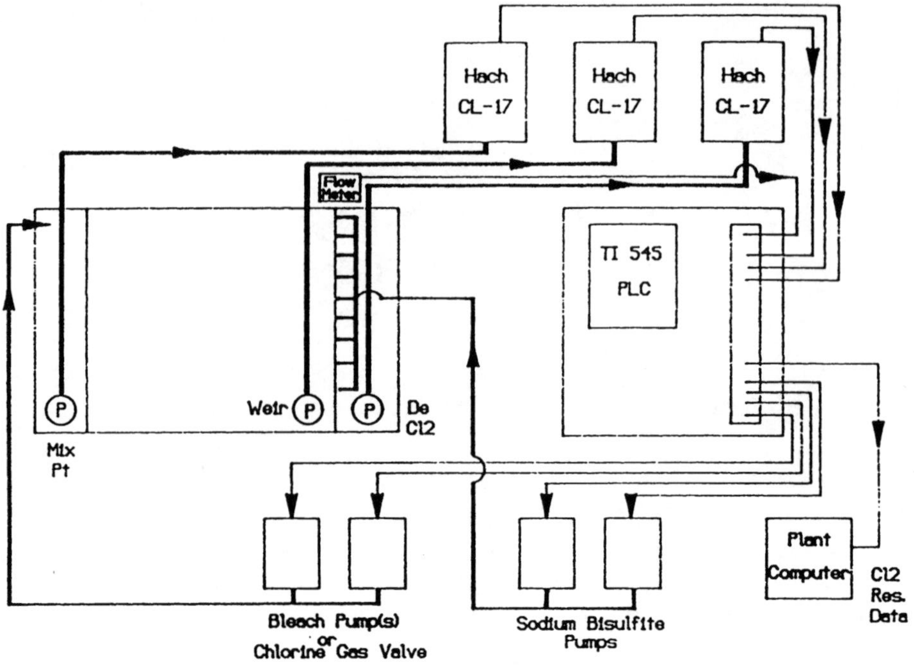

Fig. 1. Schematic diagram of chlorination and dechlorination components.

PLANT DATA

Table 1 shows the typical flow and chlorine detention tank size at each of the plants in this study.

TABLE 1. Flow and Chlorine Detention Tank at Study Plants

Plant	DW Des. Flow, L/s	Aug.'92 Flow, L/s	Mixing Vol., m^3	Est. Mixing Time, min.
Gulf Meadows	61.3	20.5	13.1	11
Turkey Creek	525.7	163.1	107.2	11
Northwest	788.6	417.0	1109.8	44
White Oak	175.2	55.8	50.8	15
Beltway	584.4	235.1	86.5	6
Keegan's Bayou	740.4	464.3	305.8	11

GULF MEADOWS PLANT

The Gulf Meadows Plant was the first plant required to have dechlorination. Equipment was also installed to control chlorination through an electrically controlled valve on the gas chlorinator. The Hach total residual chlorine analyzer produced an analysis of a discrete sample at 2.5 min intervals.

The control was designed for the loop to control the chlorine dose, mg/l, with residual chlorine after the mixing point as the process variable. A PLC special function was used to multiply the dose by the flow and appropriate constant to give the signal for the valve position for chlorine feed. The graph of the Ziegler-Nichols open loop tuning procedure for a step decrease in dose is shown in Figure 2 as reported at the 1990 Workshop (Garrett *et al.*, 1990). The erratic shape of the curve was considered to be the result of inadequate mixing intensity. Therefore the point at 25 min was ignored and the "maximum" slope drawn as shown. The controller gain, K_p, was determined as follows for a proportional-integral controller.

$$K_p = 0.9/(0.12/0.5) = 3.75 \text{ mg/l/mg/l}$$

The actual controller gain in %/% depends upon the span of the process variable and output.

The integral time, T_i, is as follows,

$$T_i = 3.3 \times l = 3.3 \times 13.3 = 44 \text{ min.}$$

These settings produced excessive swings in the chlorine residual, and the gain was reduced to give better control. However before satisfactory results were obtained, the flow signal to the PLC was lost which gave zero chlorine feed. The isolator on the signal line to the PLC had failed. Therefore it was decided to reconfigure the controller without the flow. The output was set up to control the chlorine valve directly.

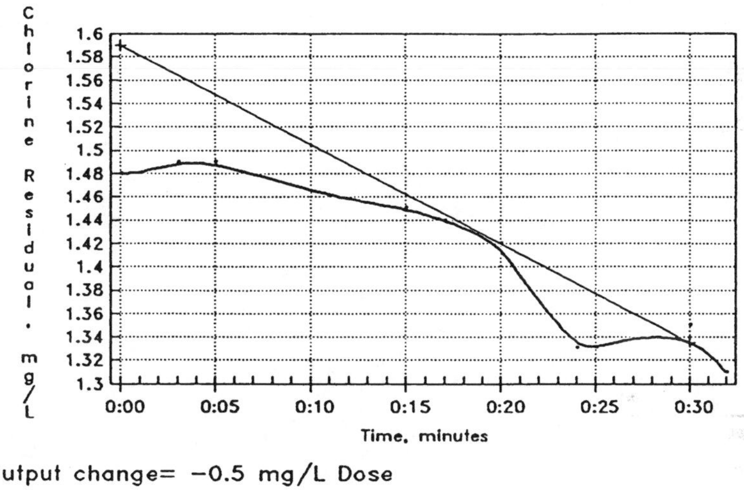

Fig. 2. Chlorine controller tuning. Z-N step response method. Gulf Meadows WWTP, Houston.

The Astrom-Hagglund (1988) relay feedback procedure was used because the procedure provided control while data was recorded. The relay was written as a special function on the PLC to switch the feed to 9 lb/day at a chlorine residual of 2.1 mg/l and to switch the feed to 65 lb/day at a chlorine residual of 1.3 mg/l. Data was recorded in the PLC and retrieved daily for 2 days. The calculated K_p values ranged from 0.93 to 1.51, and the T_i values ranged from 26 to 149 min. After some trials, a gain of K_p = 1.2 was used. However the plant operators preferred an integral time of T_i = 18 min. These values have provided satisfactory control of the chlorination for the past 10 months. With a setpoint of 1.8 mg/l after mixing, the chlorine residual at the tank outlet has remained above 1.0 mg/l.

TURKEY CREEK PLANT

The controller for the Turkey Creek plant was set up for direct control of the chlorine feed rate, and tuning values determined by the Astrom-Hagglund procedure. Figure 3 shows a short time of flow and chlorine residual at the mixing point. In order to overcome the apparent effect of flow on the chlorine residual, the controller was reconfigured to control the dose, and the tuning values were rerun. Figure 4 shows the variations of flow and chlorine residual with dose control. Although this gives flow proportional control with automatic control of the dose, there is still some variation in the chlorine residual that is caused by dead time and process lag and other unknown disturbances. Furthermore the flow meter and the hypochlorite feed pumps were not calibrated by the people tuning the controller. In spite of the variation at the mixing point, the residual chlorine at the outlet has been satisfactory.

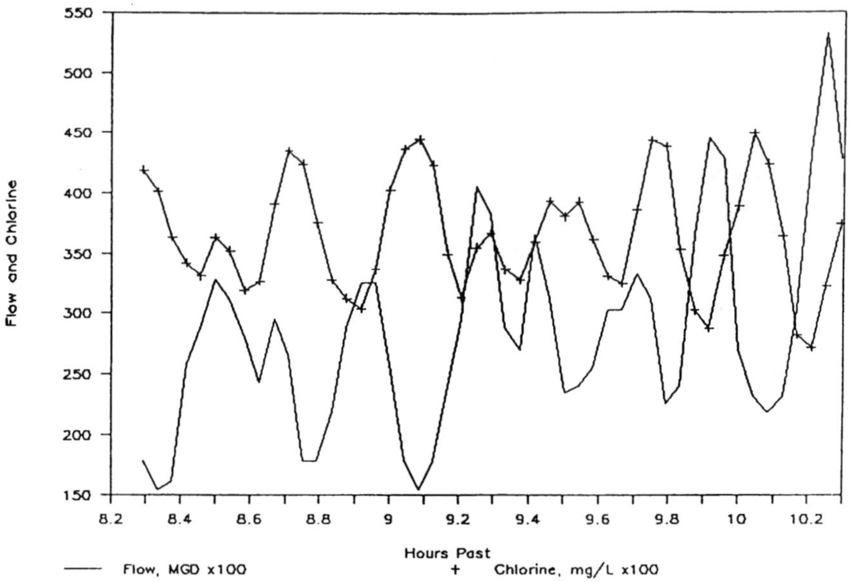

Fig. 3. Turkey Creek; control of chlorine feed.

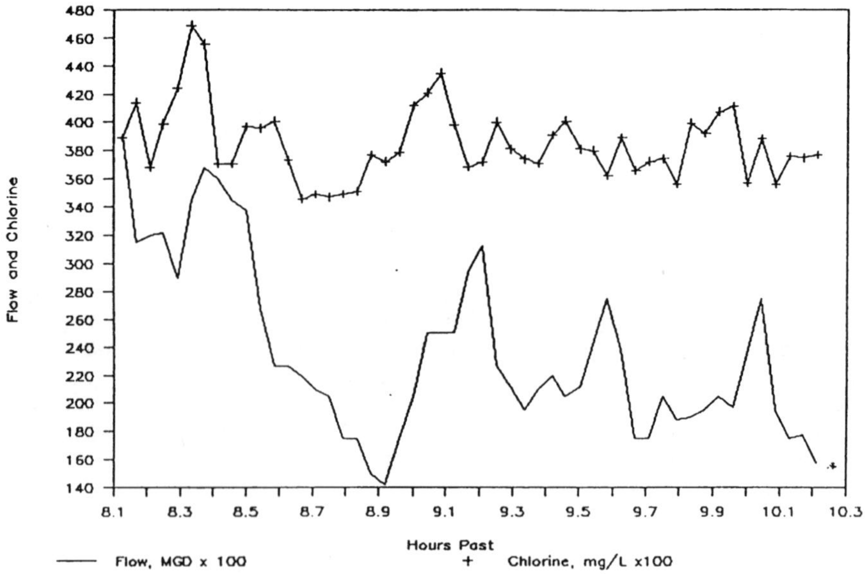

Fig. 4. Turkey Creek; control of chlorine dose.

NORTHWEST PLANT

The Northwest plant controller was tuned with the Astrom-Hagglund procedure to control the chlorine dose. At one time the controller reduced the output to zero while the residual chlorine remained near the set point. This raised the question of an interference in the chlorine analyzer, such as nitrite. A nitrite test proved negative and an effluent sample before chlorination did not show any color with DPD. Next the piping was searched for any leaks and none were found. Last, the discharge of the hypochlorite pumps was checked. The pumps have variable speed motor drives and the output showed a great bias at 10% speed. A pump calibration was then included in PLC special function. In spite of misinformation, the PI controller can adjust to give reasonable control.

WHITE OAK PLANT

At the White Oak Plant a gas chlorinator with control valve was found to be off in its calibration. A calibration curve was run and included in the PLC. At this plant control of the chlorine feed has been very satisfactory based on the Astrom-Hagglund tuning procedure.

BELTWAY AND KEEGAN'S BAYOU

These plants were tuned using the Astrom-Hagglund procedure to control the dose. The controller has given satisfactory results without further attention.

DISCUSSION

The Astrom-Hagglund procedure has worked well with PI loops in a PLC. The procedure provided adequate control during tuning tests. Data retrieved from the PLC was down loaded to a laptop PC Lotus 123 program with software from TI. This data was easy to analyze and store.

The control engineer that controls the loops is NOT the instrument technician that installed and calibrated the equipment. In order for loop values to have meaning as an actual dose or feed rate, the equipment calibration must be within close tolerances. Therefore when results are obtained that appear out of range, the control engineer should check the accuracy of meters and pumps. Corrections may be made in the PLC without adjusting the equipment to improve the information in the controller output.

Another concern is that noise in an input signal passes through to the output signal. We found the signal from flow meters to have considerable noise at a frequency of 2 to 10 Hz from waves on the water surface. Electrically adjusted valves should not receive noise signals. This is a good justification to leave flow out of the control loop with gas chlorinators. On the other hand the variable speed motors and electronically adjusted pulse rate solenoid pumps have not been affected by the noisy output when flow was included in the control. It is recommended that the noise be considered in the specification of devices used in automatic control systems. The manufacturer is in the best position to make changes to control the noise.

REFERENCES

Astrom, K. J. and Hagglund, T. (1988). *Automatic Tuning of PID Controllers*, Instrument Society of America, Research Triangle Park, N. J.

Garrett, M. T., Jr., Olsson, G., Gossett, B., Ahmad, Z., Hill, R. and Hyare, G. (1990). Implementation of computer controlled energy use optimization in a large activated sludge plant. *Instrumentation and Control of Water and Wastewater Treatment and Transport Systems*, (*Adv. Wat. Pollut. Control* no. 10) ed. R. Briggs, Pergamon Press, pp. 359-60

Pergamon

Wat. Sci. Tech. Vol. 28, No. 11-12, pp. 263–272, 1993.

0273–1223/93 $7·00 + 0·00

DEVELOPMENT OF NEW POLYMER DOSAGE CONTROL METHOD FOR SLUDGE DEWATERING PROCESS

Chiaki Igarashi*, Kazunari Tanaka*, Teruaki Kitamura**, Kazuo Fujita**, Kazuaki Arai*** and Kenichi Kameyama***

** Ebara Research Co., Ltd., 4-2-1 Honfujisawa, Fujisawa, Kanagawa, Japan*
*** Ebara Infilco Co., Ltd., 1-6-27 Konan, Minato, Tokyo, Japan*
**** Japan Sewage Works Agency, 5141 Shimosasame, Toda, Saitama, Japan*

ABSTRACT

In a sludge dewatering process using polymer, controlling the polymer dosage is an important factor for upgrading performance and achieving a saving on costs. A new automatic polymer dosage control method has been developed by the authors, which features the use of a streaming current detector or a colloid titrator. These detectors detect the colloidal charge of the filtrate from sludge. The said method was used with actual dehydrators and the following advantages were confirmed.

(1) The said method is applicable for various types of dehydrators.

(2) The said method is effective even when there are changes in sludge characteristics.

(3) The said method is more economical than a conventional solids proportional control method in that less polymer is used.

KEYWORDS

Sludge dewatering; polymer; dosage control; colloidal charge; breakpoint; streaming current detector.

INTRODUCTION

Polymer is being widely used in the sludge dewatering process of sewage treatment plants and facilities. Controlling the polymer dosage in the sludge dewatering processes is an important factor, as this dosage has not only great effect on the dewatering efficiency but also great impact on the running cost of the process. One practical polymer dosage control method is the Solids Proportional Control Method (hereafter referred to as S.P. control method). However, this method has a deficiency that it cannot fully trace changes in properties other than solids concentration in sludge, as in this method a polymer charge is proportional to solids concentration. Another deficiency is that in many cases the polymer dosage is set to an overly conservative value, as the readout of the sludge concentration meter is not sufficiently reliable. For these reasons there is demand for a better method to control the polymer dosage for the sludge dewatering process.

The authors are currently engaged in the research and development of a new method to optimally control polymer dosage by detecting the breakpoint (this will be explained later; hereafter this method is referred to as Breakpoint method). This method involves new sensors which enable a fully automatic measurement of colloidal charges in the filtrate.

Expected Advantages of New Polymer Dosage Control Method

Figure 1 shows the flow sheet of the Breakpoint method. By this method, the colloidal charge of the filtrate or centrate is automatically measured and the polymer dosage is feedback-controlled to hold a target value.

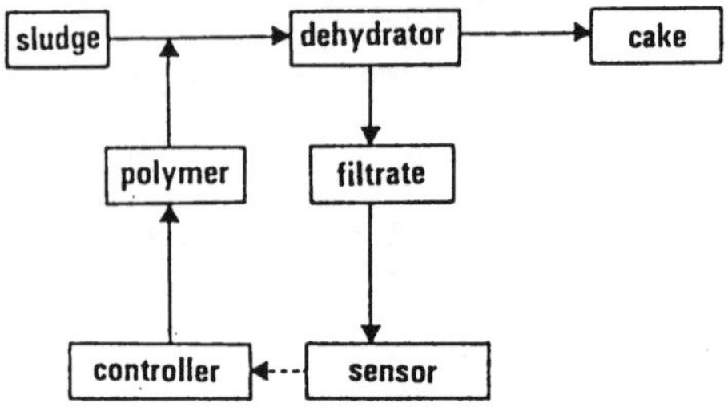

Fig.1. Flow sheet of breakpoint method.

The expected advantages of this control method are as follows.

1) Favorable operation of the dehydrator, even when there occur changes in the properties of the sludge.
2) No need to measure solids concentration or polymer concentration.
3) Cost-saving in use of polymer.
4) Labor-saving in operation and management.

Research Targets

Research was carried out on the following items and factors.

Colloidal charge sensors. This involved research on how to automatically measure the amount of colloidal charges by the colloid titration method and other methods.

Research on actual dehydrator system. Previous research (Igarashi *et al.*, 1986, 1987) which formed the basis of this research, yielded mainly lab-test results on the dehydrator system. This time, tests using an actual-scale beltpress filter and centrifuge were carried out to confirm the reproducibility of these previous results.

Automatic control by colloidal charge sensors. Experiments were carried out to study basic controllability factors in using our sensors to obtain optimal polymer dosage.

Reliability of Breakpoint method and confirmation of effectiveness. Long-term test runs were carried out at actual sewage treatment plants to establish the reliability of our sensors and estimate the effects of this method.

RESEARCH RESULTS

Colloidal Charge Sensors

Principle of colloid titration. Figure 2 shows the principle of colloid titration. It is based on: (1) both the positive and negative charges of the colloidal substance being lost by reacting at one to one ratio, and (2) a use of the reagent which changes color according to the sign of the electric charge in the liquid. The case where the reagent changes from blue to pink is more apparent than a change in the opposite direction. For this reason the inverse titration method, whereby positive colloid is added beforehand, is used for samples with negative charges such as filtrates.

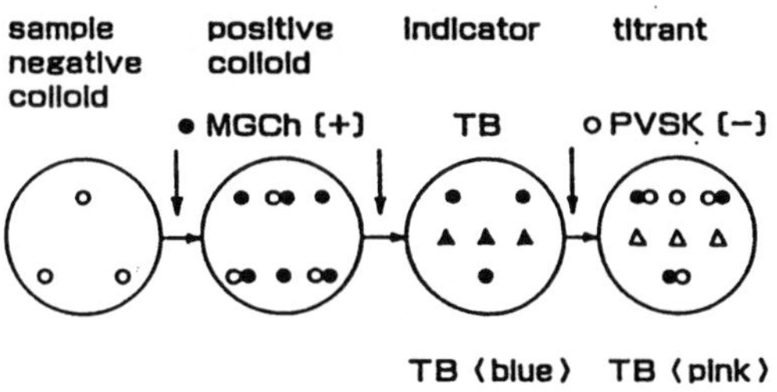

Fig.2. Principle of colloid titration.

Operational procedures of colloid titration. A positive colloid (Methyl-Glycol-Chitosan solution, MGCh) with a given concentration and an indicator (Toluidine-Blue indicator solution, TB) are added to the sample for reaction. The residual amount of positive colloid was titrated with a negative colloid (Potassium-Polyvinyl-Sulfate solution, PVSK). Igarashi (1988) reported that the end point of titration was detected by the color change of a reagent and was electrically determined accurately by utilizing a dual-frequency photometric cell.

The colloidal charge was calculated using the following equation:

$$\text{Colloidal charge (eq/l)} = 1/800 \cdot (a - b) / V$$

where V represents the extracted amount of sample (ml), "a" represents the added volume of titrant (ml) at the titration end point of the sample, "b" the titrant volume (ml) at the titration end point of blank (pure water), and 1/800 represents the concentration (N) of the titrant.

Fully-automated colloid titrator. The aforementioned operational procedures are carried out by this system. It takes about ten minutes to measure the colloidal charge of one sample. The relation between the results of this system and manually analyzed results is extremely favorable.

Streaming current detector. A streaming current detector for measuring colloidal charges was used instead of the colloid titration system. Figure 3 shows details of the detector cell. Samples were put into the space between the piston and the holder. A deviation in the electric charge was produced between the upper and lower sections of the holder, when the piston was made to reciprocate up and down. The phenomenon where an electric current (streaming current) generates between the upper and lower electrodes, corresponding to the sign and amount of the electric charge in the sample, was applied here. No reagents were used for the streaming current detector from which output signals were fed continuously.

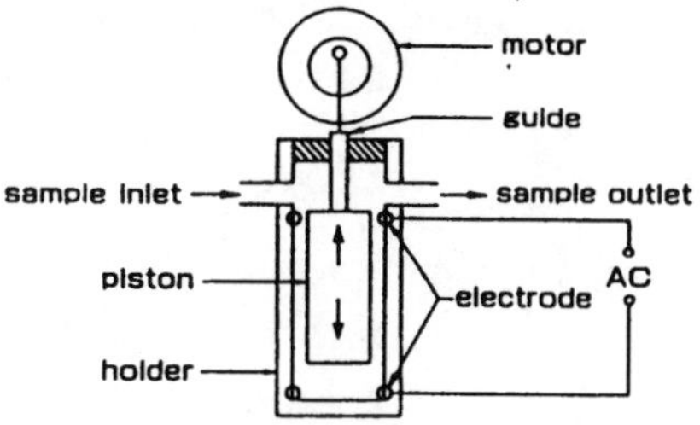

Fig.3. Streaming current detector cell.

Figure 4 shows the relation between the colloidal charge in the filtrate and the streaming current detector output (hereafter referred to as streaming current). It is apparent that the relation between the two was favorable. It was thus confirmed that the streaming current detector could be used to continuously measure the colloidal charge, although in an indirect manner.

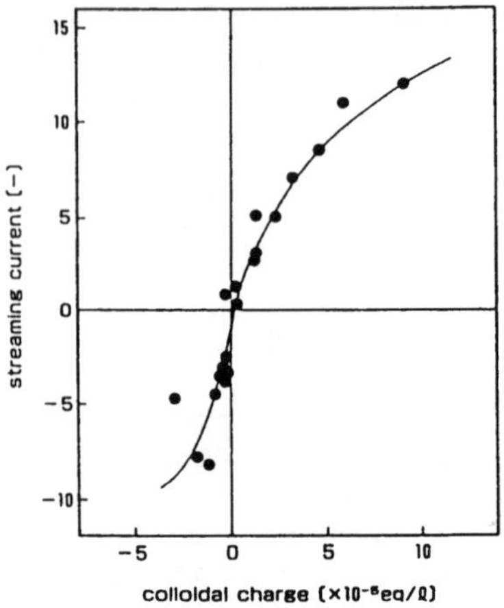

Fig.4. Relation between manually analyzed colloidal charge and streaming current.

Research on Actual Dehydrator System

Research on an actual dehydrator system was carried out at sewage treatment plant N which used a beltpress filter for the dehydrator, and at sewage treatment plant S which used a centrifuge for the same. The conventional activated sludge process was used for sewage treatment at both plants.

Sewage treatment plant N. The sludge dewatered at this plant included sludge deriving from this plant as well as that which was pumped up to this plant from five other treatment plants in the same city. The sludge, first stored in a sludge reception tank, was led into the dehydrator via the comminutor, feed pump, and storage tank. The width of the belt was 3 m. Under normal operation conditions, the sludge flow rate was 19 m^3/h, and the belt speed at the draining section was 2 m/min while that at the compression and shear section was 0.5 m/min.

Figure 5 shows the effects of the polymer dosage on the colloidal charge, streaming current in the filtrate extracted from the draining section, and moisture content of dewatered cake. The colloidal charge, negative at low polymer dosage regions, approached the value of zero with an increase in the polymer dosage. It then gradually increased after reaching a positive value. The same tendency was observed in the streaming current experiments. This was assumed to be because the colloidal charge becomes negative, due to fine residual sludge particles carrying a negative charge at a low polymer dosage and because the colloidal charge becomes positive, due to residual excess polymer in the filtrate at a high dosage. The moisture content of dewatered cake decreased with an increase in the polymer dosage but increased when this dosage was further increased. The moisture content of the dewatered cake reached minimal values when the colloidal charge was zero (hereafter this value is referred to as the breakpoint) or when it was slightly positive.

Analogous tests were carried out with changes made in the sludge flow rate, belt speed, and belt tension and the dewatering efficiency was found to be best at the breakpoint in all cases.

It was confirmed from these results that the beltpress filter operation can be maintained favorably by controlling the polymer dosage so that the colloidal charge would be in the vicinity of zero, i.e. near the breakpoint.

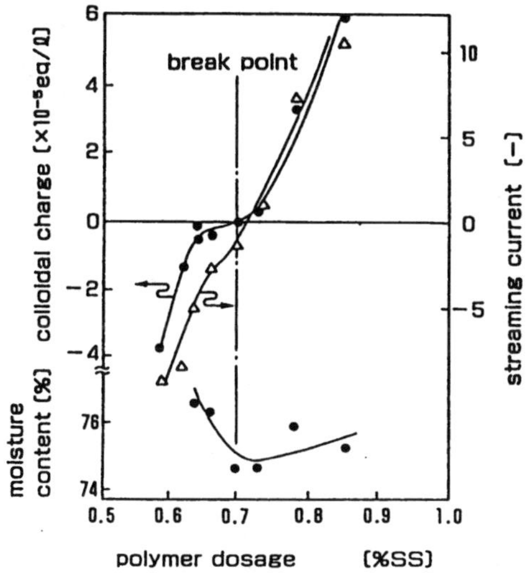

Fig. 5. Effects of polymer dosage on colloidal charge, streaming current, and moisture content of dewatered cake (Plant N;.Q=19m^3/h).

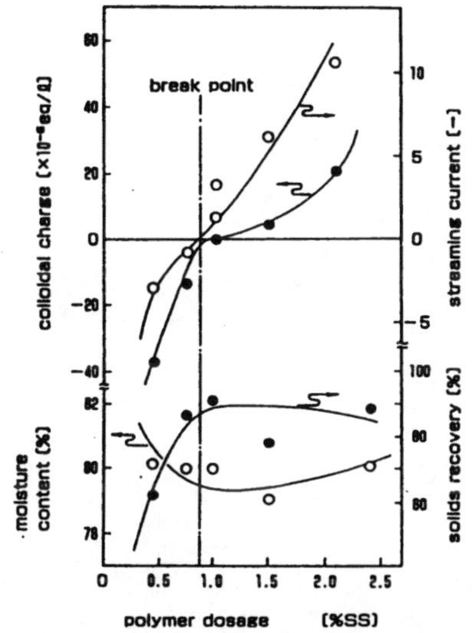

Fig.6. Effects of polymer dosage on colloidal charge, streaming current, and moisture charge, streaming current, moisture content of dewatered cake (Plant S; Q=7 m^3/h).

Sewage treatment plant S. A solid-bowl scroll centrifuge is used for the dehydrator at this plant. A mixture of sludge, made up from centrifugally thickened excess sludge and gravity thickened raw primary sludge, was treated. Under normal operation conditions, the sludge flow rate was 7 m^3/h and the scroll speed differential was 4 rpm.

Figyre 6 shows the effects of the polymer dosage on the colloidal charge, streaming current in the centrate, moisture content of dewatered cake, and solids recovery. The colloidal charge changed from negative to zero and then to positive with the increase in the polymer dosage. The same tendency was observed in the streaming current data. The moisture content of dewatered cake, although there was slight scatter in its value, decreased with an increase in the polymer dosage but showed increase later on. The solids recovery became greater with an increase in the polymer dosage and then maintained an almost constant value. It was confirmed that dewatering could be maintained favorably when the colloidal charge was in the vicinity of zero, i.e. near the breakpoint, even in the case of using a centrifuge for the dehydrator. Analogous tests were carried out with changes made in the sludge flow rate and the scroll speed differential, and similar results were obtained in all cases.

It was also confirmed from tests using different types of sludge and polymers that favorable dewatering is carried out when the colloidal charge is in the vicinity of zero, i.e. near the breakpoint.

Research on Fully Automatic Control

Results of beltpress filter control at sewage treatment plant H. Two types of newly developed colloidal charge sensors were used at this plant to enable a fully automatic operation by the Breakpoint method. Comparisons were made between results of the Breakpoint method by the colloid titrator (hereafter referred as C.T. control method) and that by the streaming current detector (hererafter referred as the S.C. control method) and also results of the conventional solids proportional control method (S.P. control method). At this plant, two beltpresses of same size and same style were used. One day, two beltpresses were operated simaltaneousely, which one was controled by C.T.control method and another one was controled by S.P.control method. Then, in other day, C.T. control method was changed to S.C. control method.

At this plant H, thickened sludge was extracted at given intervals, temporarily stored in the sludge mixing tank, and quantitatively led into the beltpress filter through a feed pump. The width of the belt was 2 m. Under normal operation conditions the sludge flow rate was 14 m^3/h and the belt speed was 2.3 m/min.

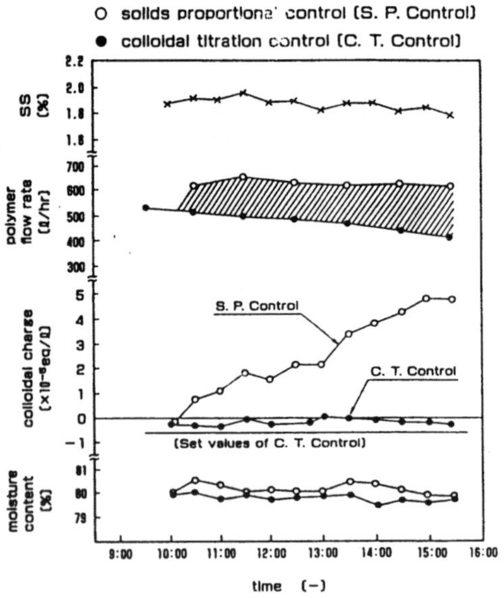

Fig.7. Comparison of solids proportional control method (S.P.control) and Breakpoint method by the colloid titrator (C.T. control method) (Plant H; Q=14 m^3/h).

Figure 7 shows a comparison of control results by the breakpoint method using the colloid titrator, i.e. the C.T. control method, with that by the S.P. control method. The polymer dosage for the S.P. control method was set at 0.6% SS, a standard value at Plant H. The target control value of the colloidal charge was set as -

0.5×10^{-5} eq/l, a value obtained from previous tests. Although the polymer flow rate by the S.P. control method was almost at a constant at 650 l/h, that by the C.T. control method decreased from 530 l/h (measured at the early stage of the operation) to 430 l/h (measured at the final stage of the operation). The daily mean polymer dosage for the S.P. control method was 0.59% SS, while that for the C.T. control method was 0.46% SS. This means that the C.T. control method achieved a 20% saving on polymer as compared with the other.

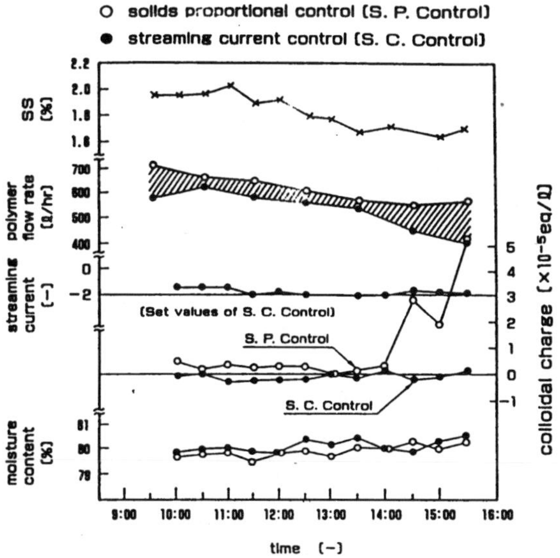

Fig. 8. Comparison of solids proportional control method (S.P.control) and Breakpoint method by streaming current detector (S.C. control) (Plant H; Q=14 m3/h) .

Figure 8 shows a comparison of control results by Breakpoint method using the streaming current detector (the S.C. control method) and that by the S.P. control method. The polymer dosage for the latter method was set at 0.6% SS, as it was for the tests mentioned earlier. The target control value by the former method was set as -2, a value obtained from previous test results.

The solids concentration observed during the morning hours was stable and about 2%; however, this value decreased gradually in the afternoon and dropped to around 1.7% by late afternoon. The polymer flow rate for the S.P. control method decreased with the decrease in solids concentration. The polymer flow rate for the S.C. control method also decreased in a similar manner, and there was a significant drop from 550 l/h to 400 l/h which occurred after 13:30. The daily mean polymer dosage for the S.P. control method was 0.62% SS, while that for the S.C. control method was 0.55% SS. This means that this method achieved a saving of almost 10 %.

Control results of centrifuge at sewage treatment plant S. The S.P. control method was not suitable for Plant S because the operation of the solids concentration meter at its dewatering facility was unstable. It was also difficult to control by the colloid titrator due to the slow output response versus extremely abrupt changes in solids concentration. Tests were therefore carried out using the streaming current detector and by the S.C. control method.

The target control value of the streaming current was determined as 0~+2, a value from previous test results shown in Fig. 6. Figure 9 shows control results by the S.C. control method. Changes in many factors (such as solids concentration, polymer flow rate, streaming current, moisture content in dewatered cake, and solids recovery) with time are indicated. The polymer flow rate increased with a rapid rise in the solids concentration, while the streaming current was maintained at almost a constant value. The moisture content of the dewatered cake was maintained almost constant at 80% but the solids recovery slightly dropped. The latter seemed to have resulted from slightly unfavorable matching of the polymer, due to changes in the mixing ratio of the excess sludge and primary raw sludge.

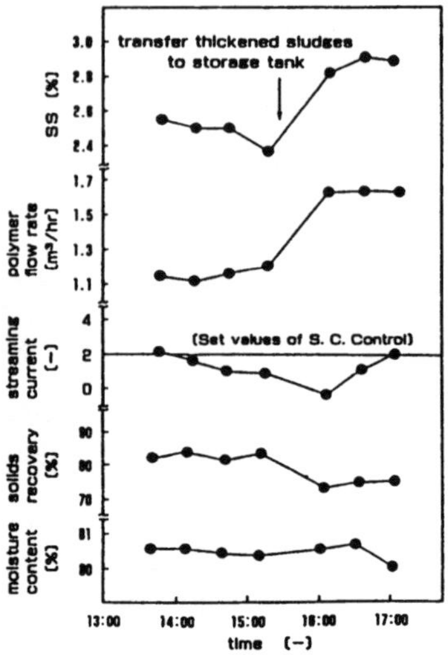

Fig.9. Control results of Breakpoint method by streaming current detector (S.C. control) (Plant S; Q = 7 m^3/h).

Comparison between two types of colloidal charge sensors. As for the practicalization of the Breakpoint method, more simplified sensors have to be developed. Although the physico-chemical significances of its output signal are apparent, this fully-automated titration system is not favorable for automatic control because the output of the titrator is discontinuous. Igarashi (1991) invented a continuous type colloid titrator which features fast response speed but it carries a deficiency that reagents have to be used. In contrast, although the streaming current detector output was relative value, it was at least confirmed in this research that a close relation exists between its output signal and colloidal charge. It can be established that it is a practically promising sensor from the facts that no reagents have to be used in its operation and the output signal is continuous.

Therefore, a streaming current detector was used as the colloidal charge measuring sensor for the research discussed in the following.

Reliability of Breakpoint Method and Confirmation of Effectiveness

The topics discussed so far are according to short-term experimental results. For confirming the reliability of the present method, a fully automatic experimental apparatus with a washing mechanism was installed in wastewater treatment plant T. A long-term measurement test was carried out for approximately 6 months.

Reliability of streaming current detector The probe and piping were washed automatically, using industrial water, following analysis of the filtrate. Fig.10 shows daily changes in the streaming current of industrial water. This measurement was made each day following the completion of the washing. The measured results indicate a stable value of approximately -1. The sensitivity of the streaming current detector was measured using standard fluid but no significant changes were found.

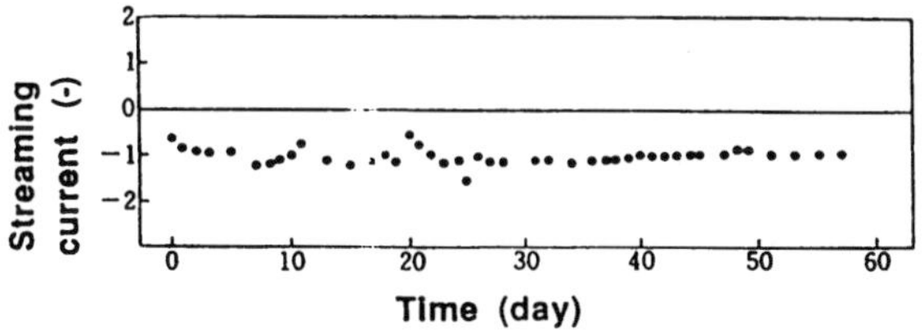

Fig.10. Daily changes in streaming current of industrial water (after completion of washing).

Responsiveness and stability of control system. The responsiveness and stability of the control system were studied. Variations in the sludge concentration and quality were made by altering the mixing ratio of primary and excess sludge. Figure 11 shows results of the dewatering process on two distinct types of sludge (Sludge A: main element being primary sludge, and Sludge B: main element being excess sludge). Disturbance was caused in the dewatering process by alternating the sludge. The set value of the streaming current detector was +4, according to the result from a preliminary test. It is acknowledged that changing the quality of the sludge results in swift regulation of the polymer flow rate, a constant output of the streaming current detector, and a stable moisture content of dewatered cakes. These test results suggest favorable responsiveness and stability of the control system versus variation in the sludge quality.

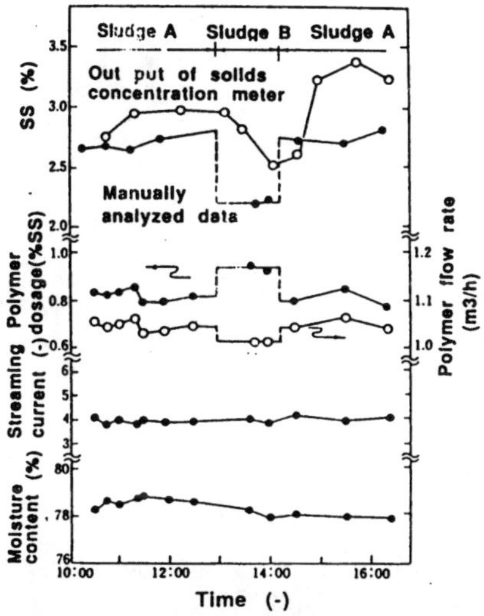

Fig. 11. Results of dewatering versus changes in sludge quality.

Cost-saving achieved by the Breakpoint method. One advantage of the Breakpoint method lies in minimizing the amount of polymer used and this clearly constitutes a cost-saving factor. Table 1 shows an estimate on cost-saving effects achieved by the Breakpoint method in regard to polymer. Assuming a daily influent amount of 100,000 m^3/day, the saving rate would amount to 14,040 ~ 28,080 Yen/d (5~10 million Yen/year). As there are other advantageous effects by the Breakpoint method, i.e. other than savings made on polymer costs, a great improvement of the sludge dewatering process can be looked forward to.

TABLE 1. Estimate on Effects of Cost-saving on Polymer

Influent	m^3/d	100,000
Influent BOD	ppm	200
Sludge generation ratio	kgSS/kgBOD	0.9
Amount of generated sludge	kgSS/d	18,000
Polymer dosage	% SS	0.6
Amount of charged polymer	kg/d	108
Unit price of polymer	Yen/kg	1,300
Cost of polymer	Yen/d	140,400
Polymer saving ratio	%	10~20
Saving rate	Yen/d	14,040~28,080

CONCLUSIONS

The following are conclusive points obtained in the research and development on the polymer dosage control method for sludge dewatering processes.

1. It was confirmed that optimal dewatering is achieved, in cases where an actual-scale beltpress filter or centrifuge was used, when the colloidal charge of the filtrate or centrate is in the vicinity of zero, i.e. at the breakpoint.

2. When using the streaming current detector, it is possible to control the polymer dosage so that the colloidal charge of the filtrate or centrate remains in the vicinity of zero, i.e. near the breakpoint.

3. The streaming current detector was a sensor which featured easy maintenance and reliable output.

4. A stable dewatering operation and a reduction in the polymer dosage were made possible by the Breakpoint method using streaming current detector. Experimental results obtained in this research showed that a 10~20 % reduction in the polymer dosage was achieved by the present method in comparison with that by the solids proportional control method.

The research results mentioned so far confirmed all "expected advantages" of the present method mentioned at the beginning of this paper.

REFERENCES

Igarashi, C., Sato, H. and Suzuki, H. (1986). Polyelectrolyte adsorption and its effect on dewatering characteristics of night soil treatment sludge. *Kagaku Kogaku Ronbunshu*, **12**, 694-700.

Igarashi, C., Gohda, S., Eto, S. and Suzuki, H. (1987). An index for optimal polyelectrolyte dosage for dewatering of municipal sewage sludge by a commercial scale continuous belt press. *Kagaku Kogaku Ronbunshu*, **13**,392-398.

Igarashi, C. (1988). Effect of salts and turbidity on determination of cationic polyelectrolytes by colloid titration. *Japan J.W.P.R.*, **11**, 774-782.

Igarashi, C. (1991). Measurement of cationic polyelectrolyte concentration by continuous flow colloid titration, *Japan J.W.P.R.*, **14**, 31-37.

Pergamon

Wat. Sci. Tech. Vol. 28, No. 11-12, pp. 273–282, 1993.

0273–1223/93 $7·00 + 0·00

REAL-TIME CONTROL OF WASTEWATER TREATMENT SYSTEMS USING ORP

David G. Wareham*, Kenneth J. Hall** and Donald S. Mavinic**

** Department of Civil and Structural Engineering, Hong Kong University of Science and Technology, Clear Water Bay, Kowloon, Hong Kong*
*** Department of Civil Engineering, University of British Columbia Vancouver, British Columbia, V6T 1Z4, Canada*

ABSTRACT

Oxidation-Reduction Potential (ORP) can be used as a process control parameter for real-time automated control of batch sewage treatment systems. Results suggest that ORP has great potential for the control of anoxic sequences, since the distinct breakpoint in the ORP-time curve correlates to the disappearance of nitrates.

KEYWORDS

Oxidation-Reduction Potential (ORP); Real-time control; aerobic-anoxic sludge digestion; biological phosphorus removal, sequencing batch reactors.

INTRODUCTION

The past few years have witnessed renewed interest (Sekine *et al.*, 1985; Charpentier *et al.*, 1987, 1989; Heduit and Thevenot 1989; and de la Menardiere *et al.*, 1991) in the use of Oxidation-Reduction Potential (ORP) as an *in situ* control parameter for wastewater treatment systems. Some (Rimkus *et al.*, 1985; Watanabe *et al.*, 1985), have advocated process control based upon the absolute value of ORP, while others (Peddie *et al.*, 1990; Wareham *et al.*, 1993a, 1993b, 1993c) have favoured wastewater control strategies based upon the relative change in ORP with time.

These latter approaches recognize that ORP-time profiles contain certain distinctive features, which can be correlated with biologically-mediated reactions. Moreover, when computers are used to detect these features, they can provide rapid, real-time information, regarding the biological "state" of the system.

The main ORP-time feature used for control is the "nitrate breakpoint". As Figure 1 indicates, the breakpoint is associated with the disappearance of nitrates; which, from a bacterial perspective, corresponds to a transformation from respiratory (oxygen or nitrates) to non-respiratory (fermentative) processes (Koch and Oldham, 1985).

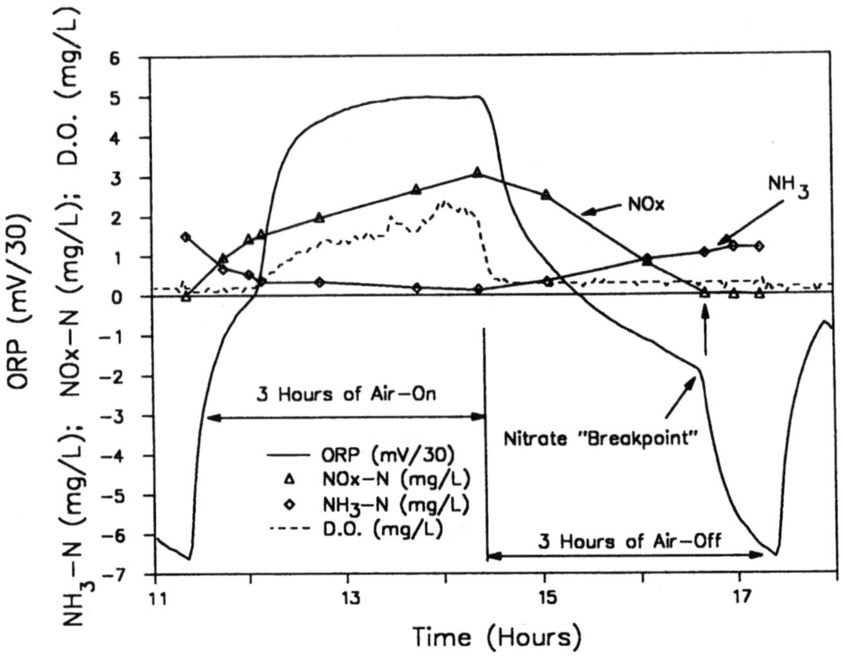

Fig. 1. Nitrate breakpoint occurring in ORP-time profile.

RESEARCH APPROACH AND OBJECTIVES

The main objective of this research was to demonstrate the use of ORP for automated control of Sequencing Batch Reactor (SBR) wastewater treatment systems. As mentioned, control depended upon the unique nitrate breakpoint phenomena; with its presence signifying the termination of anoxic conditions (presence of nitrates, absence of oxygen) and the start of true anaerobic conditions (absence of both nitrates and oxygen). The wastewater treatment strategies studied included Aerobic-Anoxic Sludge Digestion (AASD) and Biological Phosphorus Removal (Bio-P). For both processes, the operating strategies consisted of a Control (or Fixed-Time (FT)) reactor and an Experimental (or Real-Time (RT)) reactor.

During the AASD trials the FT reactor had its aerobic/anoxic sequence, "Fixed" at set time intervals; that is, a set time period for both the aerated and non-aerated portion of the cycle (3 h each). During the Bio-P experiments, the FT reactor received a dose of 30 mg of acetate (volatile fatty acids) per litre of influent at a "Fixed" time interval (1 h and 25 min) into the non-aerated sequence. Under the reactor conditions studied, this seemed to be a reasonable period to allow for complete denitrification.

In contrast, the RT reactor had at least one sequence controlled by computer detection of the nitrate breakpoint. During the first AASD experiment (AASD#1) the computer (upon detection of the breakpoint) re-initiated the air supply to ensure that the bacteria at all times were provided with a source of highly-efficient electron acceptors (be they oxygen or nitrates). Thus, less efficient processes (in terms of solids degradation (i.e. true anaerobic conditions)) were not allowed to proceed. During the second AASD experiment (AASD#2), the RT reactor recorded the length of time for denitrification (as recorded by detection of the breakpoint) and used this time for the length of the subsequent aerobic period. The resulting 50/50 operating strategy could then be directly compared to the FT 50/50 strategy of 3 h air-on and 3 h air-off.

During the Bio-P experiments, the RT reactor had its acetate dose triggered by computer detection of the nitrate breakpoint. This correlated to a **known** time for completion of denitrification, in contrast to the **assumed** time, used in the operation of the FT reactor. In theory, if denitrification was known to be complete, the acetate would always be used exclusively for storage by Bio-P organisms, rather than as a carbon source by denitrifiers.

EXPERIMENTAL DESIGN AND OPERATION

The major components of the sequencing batch reactor set-up are shown in Figure 2. In general, the reactors were made of plexiglass (diameter = 12 cm, volume = 5.4 l) and filled to the 4.8 l mark with activated sludge. The sludge and/or feed wastewater was obtained from a nearby pilot plant operating in biological phosphorus removal mode. Wasting and feeding of the reactors were performed manually during the AASD experiments and automatically during the Bio-P experiments. Air was supplied by an in-house compressor coupled to two pressure regulators connected in series. The air line was split into two, with each line passing through an air solenoid (ON/OFF regulation controlled by computer in the case of the AASD trials) before continuing on through an adjustable air flow meter. From there, the lines looped below the reactor underside to flow through a diffusing stone before entering the reactor.

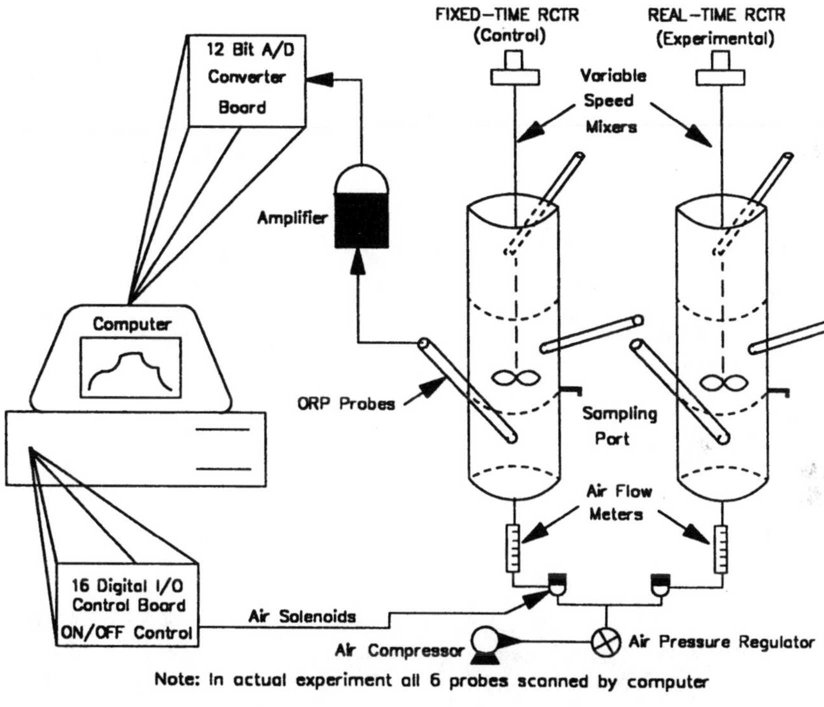

Fig. 2. Schematic of experimental apparatus.

As shown in Figure 2, combination oxidation-reduction potential probes (Broadley-James Corporation (reference electrode Ag/AgCl)) were inserted into each reactor at three strategic locations. These probes were attached to a large input impedance amplifier whose output was connected to a commercially available analog-to-digital (A/D) converter card, mating into a single input/output (I/O) expansion slot in the computer's motherboard. For control purposes, a commercial I/O expansion board was linked to a series of solid state relays; which in turn, were wired to a power bar. Each electrical socket in the power bar had been modified, so that it could be controlled independently by a single solid state relay. Details of the software used to detect the nitrate breakpoint and control the relays have been presented elsewhere by Wareham *et al.*, (1993a). Finally it should be noted that before implementing the ORP-regulated strategies, both reactors were operated in a "Fixed-Time" manner, to ensure that approximately identical conditions existed in each system.

RESULTS AND OBSERVATIONS

ORP vs. Time Profiles

Figure 1 is characteristic of the typical curve shape associated with the FT operating strategy for both AASD experiments. Figure 3 however shows an ORP-time profile indigenous to the RT AASD#1 operating strategy. Clearly evident is the intersection between the ORP breakpoint and the point of zero nitrate concentration. Consistent with the objectives of this strategy, the profile does not proceed into the anaerobic region, but instead; rises rapidly, as the ORP probe responds to the influx of oxygen after the computer detects the breakpoint and re-activates the air supply.

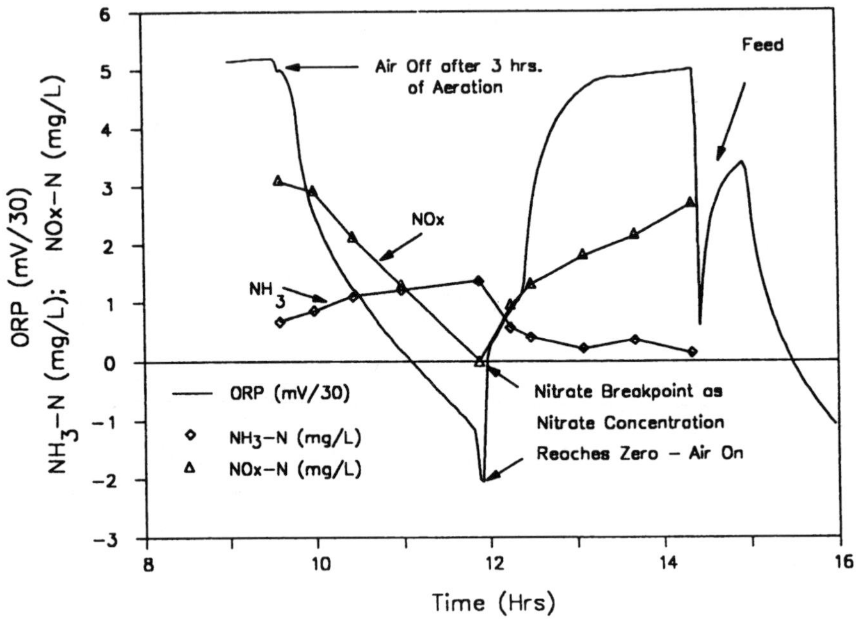

Fig. 3. Breakpoint in real-time AASD operating strategy.

Figure 4 portrays the switch from FT to RT conditions under the AASD#2 operating strategy. As can be seen, a distinctive pattern developed, reflecting the RT reactor's tendency to "collapse" in on itself. (eventually arriving at an equilibrium state), with very short on/off times for both the aerated and non-aerated portions of the cycle.

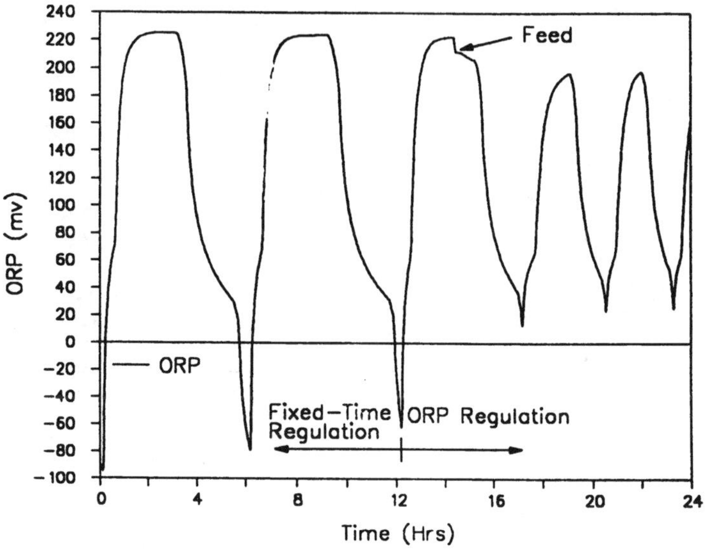

Fig. 4. Switch from fixed-time to real-time operation: AASD#2.

The most distinctive characteristic evident in real-time AASD strategies is the self-adjusting ability of the reactor to reduce the nitrates to zero, regardless of the initial nitrate concentration. This is attributed to the reactor's ability to delay switching on the air until denitrification is fully complete. For example, daily feeding of the reactor produces an increase in the carbon level. This surplus is reflected in relatively short denitrification times during subsequent anoxic periods. Gradually, as the overall carbon level diminishes, longer and longer denitrification times become necessary. The advent of feeding the next day causes the denitrification time to immediately shorten once again.

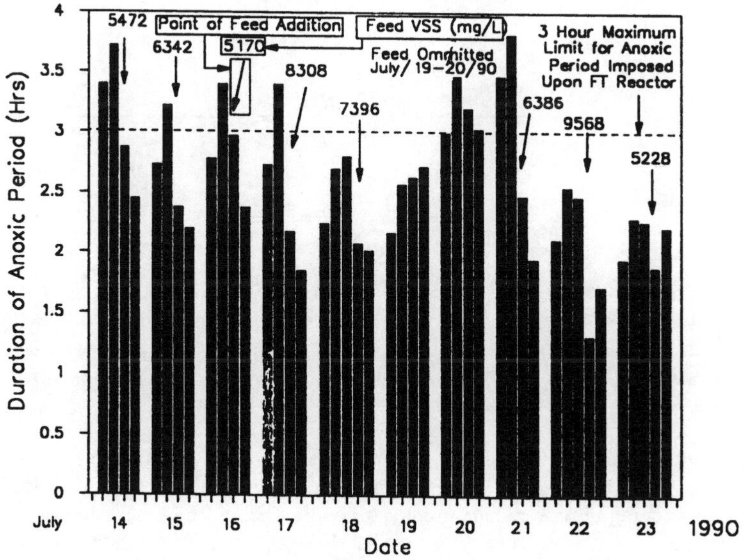

Fig. 5. Cyclical pattern of length of anoxic period.

This is illustrated by the cyclical pattern in Figure 5, (from AASD#1) which reflects the amount of carbon available in the system on a real-time basis. The flexibility associated with a real-time AASD operationalstrategy becomes clearly evident here; since the anoxic periods furthest from the time-of-feeding, occasionally extend beyond the 3 h air-off limit arbitrarily imposed on the FT reactor. Thus, some sequences in the FT reactor would not fully denitrify; whereas, if the reactor had been operating in real-time, the denitrification reaction would have been complete.

With regards to biological phosphorus removal, Figure 6 shows the three distinct zones of the Bio-P process, with a sharp, pronounced nitrate breakpoint separating the non-aerated period into anoxic and anaerobic segments. The breakpoint occurred prior to the addition of acetate, which caused the classical release/uptake phenomena of phosphorus and carbon. In this example however, the reactor failed to take up excess phosphorus and thus the effluent was discharged from the reactor at essentially the same level at which it entered.

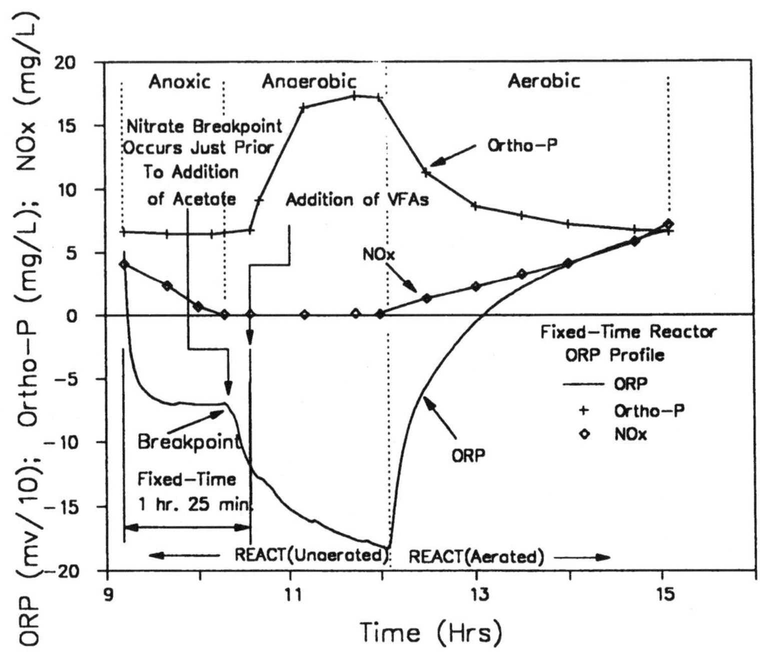

Fig. 6. Three phases of the Bio-P SBR process.

On occasion, the acetate would actually induce the breakpoint as shown in Figure 7. Clearly seen is the sharp drop in the nitrate concentration prior to the breakpoint, but immediately following the addition of acetate to the FT reactor. Even though in this case the phosphorus was removed to a very low level, it appeared that the acetate was being partitioned between being used for nitrate reduction by denitrifiers and being used for carbon storage by Bio-P organisms.

A detailed examination of the time-of-occurrence of the breakpoints will show that if the nitrate breakpoint occurs either simultaneously with, or after (i.e greater than or equal to) the acetate addition; then all or a portion of the acetate is being used to reduce nitrates by denitrifiers. If the breakpoint occurs before (i.e. less than) the acetate addition; the acetate is being primarily used by micro-organisms capable of excess P removal. Categorizing the nitrate breakpoints into different groupings is illustrated in Figure 8. This shows a detailed snapshot over two days, itemizing the breakpoints into those that occurred before the addition of acetate (i.e. acetate used solely for carbon storage); those that were directly attributable to (i.e induced by) the addition of acetate; and those that occurred after VFA addition (i.e. had a portion (or all) of the acetate utilized for denitrification).

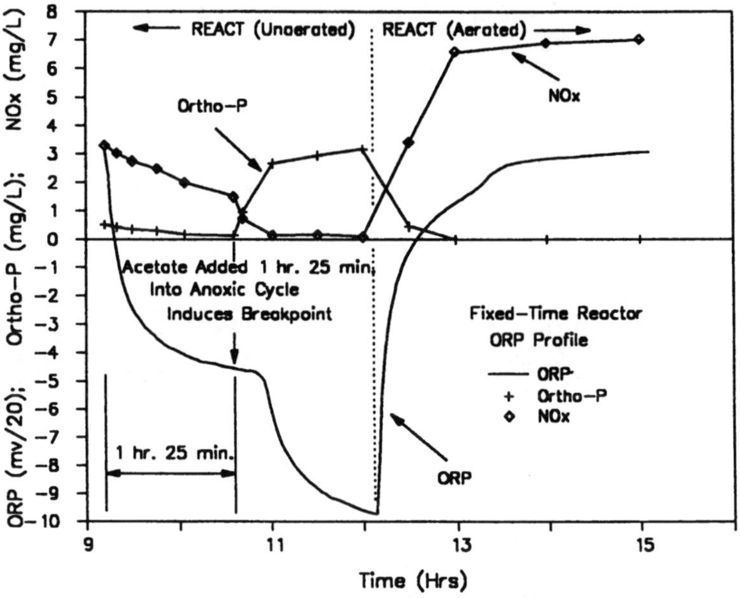

Fig. 7. Acetate addition induces nitrate breakpoint.

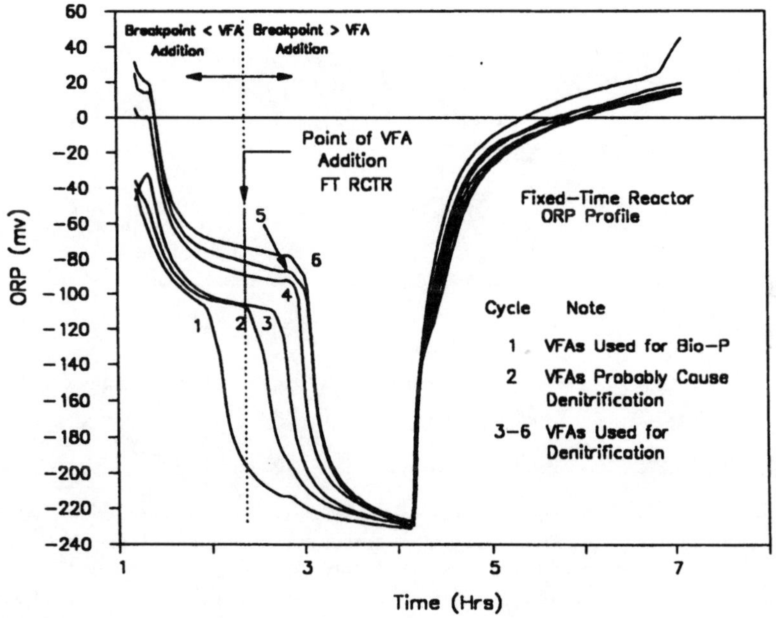

Fig. 8. Categorization of nitrate breakpoints in the Bio-P process.

Evaluation of Chemical Parameters Through Mass Balances

Nitrogen (TKN, NO_x, NH_3), phosphorus (TP, ortho-P) and solids (TSS, VSS) measurements were made on a daily basis (on the digester effluent plus feed) in the AASD experiments. This enabled a mass balance around each reactor to be performed. Any missing solids were presumed to have been degraded by bacterial processes, while any missing nitrogen was attributed to losses in the form of nitrogen gas (since a neutral pH precluded losses through stripping of NH_3 gas). Phosphorus should theoretically have been conserved since there was no biological or physical mechanism for its removal under the operating conditions studied.

Table 1 summarizes the mass balance results, for both AASD experiments. The data presented reveals small differences (up to 6 % depending upon the parameter and experiment) in the performance of the two reactors. Due to the lack of replication, no conclusive statistical statements can be made; however, from a mass balance perspective, the experimental (i.e. ORP-regulated) reactor appeared to remove marginally more than the control (Fixed-Time regulated) reactor for the TSS, VSS and nitrogen. The apparent increase in phosphorus may have been due to the TP digestion technique used; an anomaly, which is being investigated further.

TABLE 1. Percent Removals Obtained from Mass Balances*

Mass Balance Parameter	AAAD#1 Fixed-Time Reactor	AASD#1 Real-Time Reactor	AASD#2 Fixed-Time Reactor	AASD#2 Real-Time Reactor
Total Suspended Solids	15.8 %	15.7 %	14.8 %	19.9 %
Volatile Suspended Solids	17.7 %	18.3 %	16.7 %	21.5 %
Total Nitrogen	17.9 %	21.1 %	19.4 %	25.9 %
Total Phosphorus	-6.2 %	-5.8 %	-9.8 %	-5.3 %

* All values expressed as percent reduced.

Chemical data for the Bio-P experiments is available (Wareham, 1993c); however, difficulties were encountered in achieving steady state conditions, thus the data is not presented here.

Evaluation Through Unsteady Process Input Conditions

The suitability of ORP as a control parameter for AASD systems was demonstrated by observing the probe response to spikes of chemicals (sodium nitrate, ammonium chloride and hydrogen peroxide); all of which influenced the time-of-occurrence of the nitrate breakpoint. For illustrative purposes, Figure 9 shows the response to the sodium nitrate spike, for both reactors, during AASD#1. As can be seen incomplete denitrification occurred when no discernable nitrate breakpoint was observed in the ORP-time curve. The RT occurrences of this are due to a maximum-allowable-denitrification time (arbitrarily set at 4 hours) in the event that the software failed to detect the breakpoint. As can be seen in this example, the RT reactor failed to completely denitrify twice, while the FT reactor failed to completely denitrify 4 times.

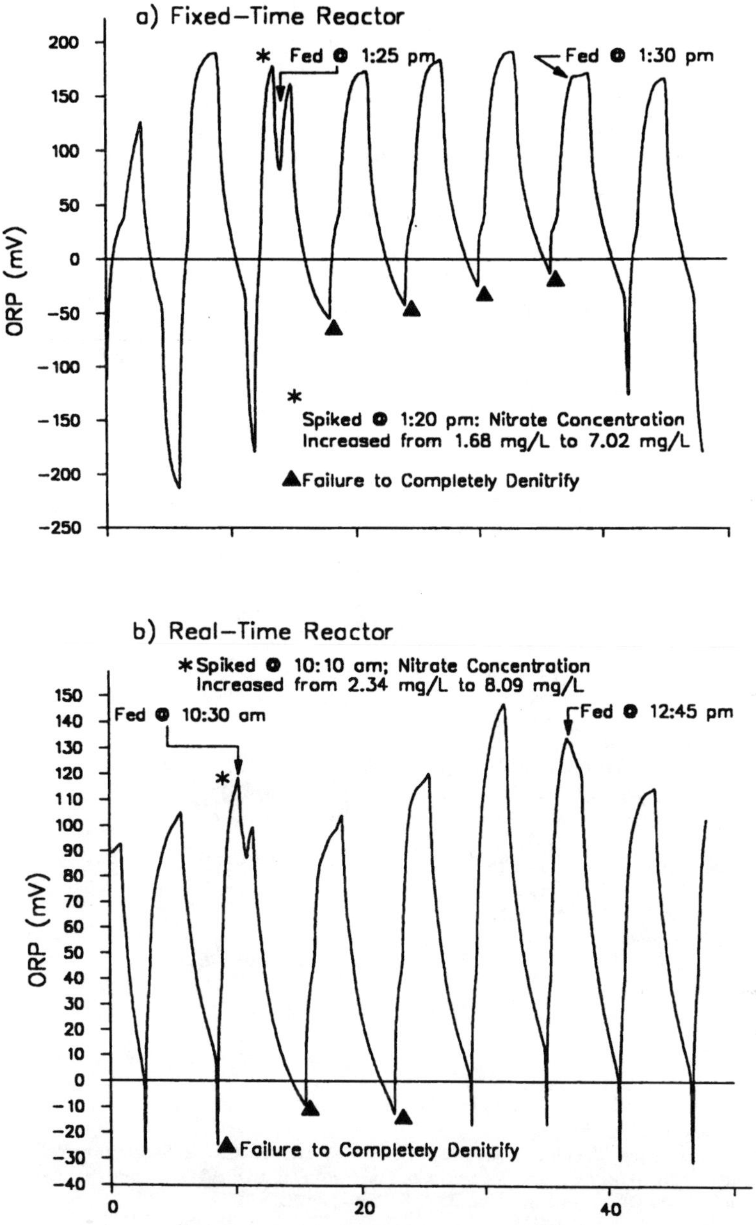

Fig. 9. Effect of sodium nitrate spikes on fixed-time and real-time reactors.

The RT reactor consistently outperformed the FT reactor in its ability to more readily accommodate and recover from the chemical stresses considered in this research. This was true for both sets of AASD experiments. Unfortunately, due to time constraints, spikes of this nature were not performed for the Bio-P experiments.

CONCLUSIONS

This research compared two different operating strategies for aerobic-anoxic sludge digestion and biological phosphorus removal. The first strategy relied on an operator externally "fixing" the length of time associated with a biological sequence in the SBR. In effect, this type of approach operated in a manner insensitive to the anoxic reactions occurring in the system. More specifically, there may have been either insufficient or an excess of time allowed for denitrification. In contrast the real-time strategy made use of the distinctive breakpoint in the ORP-time profile to control the anoxic sequences in the process. Since the breakpoint correlated to nitrate disappearance, the ORP probe demonstrated great potential for automated control of anoxic segments in batch sewage treatment systems.

ACKNOWLEDGEMENTS

Financial support for this work came from the Natural Sciences and Engineering Research Council of Canada (NSERC) in the form of operating grants to the second and third authors. The excellent technical support provided by the staff of the environmental engineering lab and pilot plant, UBC, is acknowledged and appreciated.

REFERENCES

Charpentier, J., Florentz, M. and David, G. (1987). Oxidation-Reduction Potential (ORP) Regulation: A Way to Optimize Pollution Removal and Energy Savings in the Low Load Activated Sludge Process. *Wat. Sci. Tech.*, **19**(3-4), 645-655.

Charpentier, J., Godart, H., Martin, G. and Mogno, Y. (1989). Oxidation-Reduction Potential (ORP) Regulation as a Way to Optimize Aeration and C, N, and P Removal: Experimental Basis and Various Full-Scale Examples. *Wat. Sci. Tech.*, **21**(10-11), 1209-1223.

De la Menardiere, M., Charpentier, J., Vachon, A. and Martin, G. (1991). ORP as a Control Parameter in a Single Sludge Biological Nitrogen and Phosphorus Removal Activated Sludge System. *Water S.A.*, **17**(2), 123-132.

Heduit, A. and Thevenot, D. R. (1989). Relation Between Redox Potential and Oxygen Levels in Activated-Sludge Reactors". *Wat. Sci. Tech.*, **21**(8-9), 947-956.

Koch, F. A. and Oldham, W. K. (1985). Oxidation-Reduction Potential – A Tool for Monitoring, Control and Optimization of Biological Nutrient Removal Systems. *Wat. Sci. Tech.*, **17**(11-12), 259-281.

Peddie, C. C., Mavinic, D. S. and Jenkins, C. J. (1990). Use of ORP for Monitoring and Control of Aerobic Sludge Digestion. *ASCE J. of Env. Eng.*, **116**(3):461-471.

Rimkus, R. R., Ziols, R. and Shaikh, A. (1985). Computer Control of Raw Sewage Chlorination for Odour Control. *Instrumentation and Control of Water and Wastewater Treatment and Transportation Systems, Advances in Wat. Pollut. Control* no. 2, ed. R. A. R. Drake, 297-302.

Sekine, T., Iwahorik, K., Fujimoto, E. and Inamori, Y. (1985). Advanced Control Strategies for the Activated Sludge Process. *Instrumentation and Control of Water and Wastewater Treatment and Transportation Systems, Advances in Wat. Pollut. Control* no. 2, ed. R. A. R. Drake, 269-276.

Wareham, D. G., Hall K. J. and Mavinic, D. S. (In Press). Real-Time Control of Aerobic-Anoxic Sludge Digestion Using ORP *ASCE J. of Env. Eng.*

Wareham, D. G., Mavinic, D. S.,and Hall, K. J. (In Press). Sludge Digestion Using ORP-Regulated Aerobic-Anoxic Cycles" *Wat. Res.*

Wareham, D. G., Hall K. J. and Mavinic, D. S. (submitted). An ORP Screening Protocol for Biological Phosphorus Removal Processes". *ASCE J. of Env. Eng.*

Watanabe, S., Baba, K. and Nogita, S. (1985). Basic Studies of an ORP/External Carbon Source Control System for the Biological Denitrification Process. *Instrumentation and Control of Water and Wastewater Treatment and Transportation Systems, (Advances in Wat. Pollut. Control* no. 2), ed. R. A. R. Drake, pp. 297-302.

Pergamon

Wat. Sci. Tech. Vol. 28, No. 11-12, pp. 283–293, 1993.

0273–1223/93 $7·00 + 0·00

CONTROL DESIGN FOR NONLINEAR WASTEWATER TREATMENT PROCESSES

D. Dochain* and M. Perrier**

** CESAME, UCL, Place du Levant 3, B-1348 Louvain-la-Neuve, Belgium*
*** PAPRICAN, 570 Boulevard St. Jean, Pointe Claire, H9R 3J9, Canada*

Abstract

This paper deals with the design of adaptive nonlinear control algorithms of biological wastewater treatment processes. The control design is based on the dynamical mass balance equations of the process and includes the on-line estimation of uncertain parameters (specific growth rates and yield coefficients). The procedure is illustrated by two examples (activated sludge process, anaerobic digestion).

Keywords

Control algorithms; wastewater treatment; activated sludge; anaerobic digestion; adaptive nonlinear control.

1. Introduction: The Why and How of Controlling Wastewater Treatment Processes?

The treatment of wastes from industrial plants is clearly a key question these days. The situation of the Pulp and Paper industry in a country like Canada is particularly symptomatic and illustrative in that respect. Among the different waste treatment processes, the biological wastewater treatment processes are susceptible to play an increasing role. Two typical biological wastewater treatment processes are usually considered : the activated sludge process and the anaerobic digestion. In this paper, we shall concentrate on the design of control algorithms for both processes.

As for any industrial process, the need for efficient monitoring and automatic control systems of biological wastewater treatment processes in order to maintain the wastewater degradation at a desired level at the output of the process unit and optimize their operation is obvious. The efficiency of any control system depends highly on the design of the control and monitoring techniques and the care taken in its design. Indeed, a control algorithm will prove to be efficient if it is able to incorporate the important well-known information on the process while being able to deal with the missing information (lack of on-line measurements, uncertainty in the dynamics,...) in a "robust" way, i.e. such that these missing informations will not significantly deteriorate the control performance of the process. The present generalized use of computers make possible fairly easy to design and implement controllers which are more sophisticated (and susceptible of better performance) than the classical PID. These controllers may refer to quite complex theory (nonlinear control, adaptive control) but, as it will be shown, their structure and implementation may remain rather simple while including the key features of simple PID's. These controllers will also look simple with respect to the complexity of the processes under consideration. It is also important to draw the attention of the reader that these control systems are not just the object of academic research but are already used in several applications (see e.g. [1], [2]).

The objective of this paper is to show how to incorporate the well-known knowledge of the wastewater treatment process (basically, the reaction network and the material balances) in a control algorithm which is moreover capable of dealing with the process uncertainty (in particular on the reaction kinetics) by introducing an adaptation scheme. Our intention in this paper is to introduce a systematic approach for designing controllers for wastewater treatment processes based on the above concepts. The design of the control algorithm is based on a dynamical model of the process which will be the object of Section 2. The design of the controller will be developed in Section 3. and its performance illustrated in Sections 4 (activated sludge process) and 5 (anaerobic digestion).

2. The Basic Knowledge : the Dynamical Model of the Process

The dynamical model of the process under study is undoubtedly one of the cornerstone for the derivation and evaluation of a control algorithm. Since it serves as a basis for the control scheme derivation, it must fairly well represent the key phenomena of the process. Because of their inherent complexity (in particular due to the involvement of living organisms), it is usually quite difficult to obtain accurate models for biochemical reactors. However, simply speaking, we can emphasize two important different quality levels in the information about bioprocess models. On a first level, we can find the *well-known* information, basically the *important reactions* and the *transport dynamics*, which can be formalized into a dynamical model by considering material balances (referred here as the *basic dynamical model*). The second level concentrates on the different sources of *uncertainties* in the model, among which one of the most important is the *reaction kinetics* but which may also include other (less important) reactions, or possibly some uncertainty in the transport dynamics (e.g. non-perfect mixing).

In this section, we shall introduce the *basic dynamical model* for the activated sludge and anaerobic digestion processes on which the derivation of the control algorithm will be based. Extensions of the basic dynamical model (by considering a particular model for the process kinetics and by including microorganism death and an endogenous respiration reaction) will be used to simulate the process and test the performance and robustness of the controller in realistic conditions.

The *basic dynamical model* of a wastewater treatment process in stirred tank reactors is derived by considering material balances and can be written under the following matrix form :

$$\frac{d\xi}{dt} = -D\xi + K\varphi + F - Q \tag{1}$$

where ξ is the component concentration vector, D is the dilution rate, K is the yield coefficient matrix, φ is the reaction rate vector, F is the feedrate vector and Q is the gaseous outflow rate (see [1] for further details).

Example #1 : activated sludge process

Let us consider an activated sludge process operated in two sequential tanks (aerator + settler) characterized by a simple microbial growth reaction in the aerator (e.g. [3]):

$$k_1S + k_2C \rightarrow X$$

where S, C and X are the substrate, dissolved oxygen, and biomass, respectively, and k_1 and k_2 are yield coefficients. The dynamics of the process can be described by model (1) with the following definitions :

$$\xi = \begin{bmatrix} S \\ C \\ X \\ X_R \end{bmatrix},\ K = \begin{bmatrix} -k_1 \\ -k_2 \\ 1 \\ 0 \end{bmatrix},\ F = \begin{bmatrix} D_{in}S_{in} \\ Q_{in} \\ 0 \\ 0 \end{bmatrix},\ D = \begin{bmatrix} D_1 & 0 & 0 & 0 \\ 0 & D_1 & 0 & 0 \\ 0 & 0 & D_1 & -D_2 \\ 0 & 0 & -D_3 & D_4 \end{bmatrix} \tag{2.a}$$

$$\varphi = \mu X, Q = 0, D_{in} = \frac{F_{in}}{V}, D_2 = \frac{F_R}{V}, D_1 = D_{in} + D_2, D_3 = \frac{F_{in} + F_R}{V_S}, D_4 = \frac{F_R + F_w}{V_S} \quad (2.b)$$

where F_{in}, F_R, F_W, V, V_S, Q_{in} are the influent, recycle and waste flowrates, the aerator and settler volumes, and the oxygen feedrate, respectively. Note that here because of the presence of two tanks, the dilution rate term in (1) is a matrix. The oxygen feedrate in activated sludge is usually set equal to the gas-liquid oxygen transfer rate :

$$Q_{in} = k_L a(C_S - C) \quad (2.c)$$

where $k_L a$ is the oxygen mass transfer coefficient and C_S the saturation constant. In line with [9], we shall consider in the following that $k_L a$ is locally a linear function of the air flow rate W :

$$k_L a = a_0 W \qquad a_0 > 0 \quad (2.d)$$

Example #2 : anaerobic digestion (4 population model)

The five bacterial population model structure proposed by Mosey [4] has been used by several researchers ([5], [6]) to represent adequately the dynamical behaviour of anaerobic digestion reactors. The model presented here includes four bacterial populations where the butyrate-utilizing and propionate-utilizing acetogens are merged into the Obligate Hydrogen Producing Acetogens (OHPA). The process is then characterized by the following reaction network :

Acid Formers Bacteria (X_1) :	$k_1S_1 \rightarrow X_1 + k_3S_2 + k_6S_3 + k_8S_4 + k_{11}S_5$
OHPA (X_2) :	$k_2S_2 \rightarrow X_2 + k_6S_3 + k_9S_4 + k_{12}S_5$
Acetoclastic Methanogenic Bacteria (X_3) :	$k_4S_3 \rightarrow X_3 + k_{13}S_5 + k_{14}P_1$
Hydrogenophilic Methanogenic Bacteria (X_4) :	$k_7S_4 + k_{10}S_5 \rightarrow X_4 + k_{15}P_1$

where S_1, S_2, S_3, S_4, S_5 and P_1 are glucose, propionate, acetate, hydrogen, inorganic carbon, and methane, respectively. k_i (i = 1 to 15) are the yield coefficients. The process dynamics can be represented by model (1) with the following definitions :

$$\xi = \begin{bmatrix} X_1 \\ S_1 \\ X_2 \\ S_2 \\ X_3 \\ S_3 \\ X_4 \\ S_4 \\ S_5 \\ P_1 \end{bmatrix}, K = \begin{bmatrix} 1 & 0 & 0 & 0 \\ -k_1 & 0 & 0 & 0 \\ 0 & 1 & 0 & 0 \\ k_3 & -k_2 & 0 & 0 \\ 0 & 0 & 1 & 0 \\ k_5 & k_6 & -k_4 & 0 \\ 0 & 0 & 0 & 1 \\ k_8 & k_9 & 0 & -k_7 \\ k_{11} & k_{12} & k_{13} & -k_{10} \\ 0 & 0 & k_{14} & k_{15} \end{bmatrix}, \varphi = \begin{bmatrix} \varphi_1 \\ \varphi_2 \\ \varphi_3 \\ \varphi_4 \end{bmatrix}, F = \begin{bmatrix} 0 \\ DS_{in} \\ 0 \\ 0 \\ 0 \\ 0 \\ 0 \\ 0 \\ 0 \\ 0 \end{bmatrix}, Q = \begin{bmatrix} 0 \\ 0 \\ 0 \\ 0 \\ 0 \\ 0 \\ 0 \\ Q_{H_2} \\ Q_{CO_2} \\ Q_{CH_4} \end{bmatrix} \quad (3)$$

φ_i (i = 1 to 4) is the reaction rate of reaction i :

$$\varphi_i = \mu_i X_i$$

with μ_i the specific growth rate of reaction i (d^{-1}).

3. Design of the Adaptive Linearizing Controller

The design of the control algorithm is based on the dynamical model (1) or on a reduced-order form of (1). In the following, we shall present two control design examples : one based on the dynamical model (1) (activated sludge process), and one based on a reduced-order version of (1) (anaerobic digestion). In each case, the control design follows the following line of reasoning. Let us first define the control problem .

The objective is to control the concentration of some reactant components (one in example #2, two in example #1) by acting on flow rates (the air flow rate and the recycle flow rate in example #1, the dilution rate in example #2) under the following practical constraints :

C1. the controlled components are assumed to be measured on-line;

C2. the concentrations of the other components (particularly of the biomass) are not available for on-line measurement;

C3. the kinetics are unknown;

C4. most of the yield coefficients are unknown (only the ratio k_1/k_2 will be assumed to be known, in example #1);

C5. the mass transfer coefficients are known;

C6. the feedrates F, the dilution rate D and the gaseous outflow rates Q are known (either by user's choice or by measurement).

By defining y the controlled component(s), the dynamics of y are simply the equation(s) of y in model (1) and can be rewritten as follows :

$$\frac{dy}{dt} = -Dy + K_y\varphi + F_y - Q_y \tag{4}$$

where the index y holds for the rows of K, F and Q corresponding to the controlled output y. By considering the above control problem and defining u the control input, the output equation (4) can be rewritten as follows :

$$\frac{dy}{dt} = f(F_y, Q_y, K_y\varphi) + g(F_y, y)u \tag{5}$$

Assume now that you wish to have a linear stable closed-loop (process + controller) dynamical behaviour, i.e. :

$$\frac{dy}{dt} = C_1(y^* - y) \tag{6}$$

By combining equations (5)(6), we readily obtain the control law :

$$u = g(F_y, y)^{-1}\{C_1(y^* - y) - f(F_y, Q_y, K_y\varphi)\} \tag{7}$$

Since the kinetics and most of the yield coefficients are assumed to be unknown, they are replaced by on-line estimates of selected parameters :

$$\boxed{u = g(F_y, y)^{-1}\{C_1(y^* - y) - f(F_y, Q_y, \widehat{K_y\varphi})\}} \tag{8}$$

The above controller (8) is also known as the *model reference adaptive linearizing control law* (see e.g. [1]). As we shall in the examples, the unknown parameters appear in the linear way in the equations and will therefore be estimated by using linear regression techniques (example #1) or via a Lyapunov design estimation approach (example #2). Moreover, the unknown components that may appear explicitly in the output equation (5) via the reaction rate φ will be replaced either by an auxiliary variable easy to compute (example #1) or by gaseous outflow rates (example #2)(in the latter instance, the effect will also to make the kinetics termsdisappear).

Comment. Note that the adaptive linearizing controller contains all the important features of a PI controller : the ***proportional*** term $(C_1(y^* - y)$ and the ***integral*** action via the parameter adaptation (see [10] for an explicit analysis). Moreover the other terms can be viewed as *feedforward* action terms.

4. Example #1 : Activated Sludge Process (continued)

The following example is treated in detail in [11]. The control problem we shall consider now can be formulated as follows. Like in any aerobic fermentation processes, proper aeration is crucial to process efficiency, and adequate control of the dissolved oxygen concentration in the aerator is very important. But it is also important to limit load variations and substrate concentration variations by acting on them. As suggested by many authors (e.g. [3], [7]), load variations can be expected to be smoothed by using the recycle flow rate as a control input. We shall consider the above control algorithm under the following conditions :

1) the controlled outputs are the effluent BOD (Biological Oxygen Demand) and the dissolved oxygen concentrations, which are assumed to be measured on-line (see [8] for an on-line measuring device of the BOD in activated sludge processes);
2) the control inputs are the recycle flow rate and the air flow rate.

Therefore the input u and the output y are defined as follows :

$$u = \begin{bmatrix} F_R \\ W \end{bmatrix}, \; y = \begin{bmatrix} S \\ C \end{bmatrix} \tag{9}$$

and the control problem is multivariable (2 inputs, 2 outputs). The expressions of f and g in (5) are readily obtained from model (1)(2) :

$$f = \begin{bmatrix} -k_1 \mu X + D_{in}(S_{in} - S) \\ -k_2 \mu X - D_{in} C \end{bmatrix}, \; g = \begin{bmatrix} \frac{-S}{V} & 0 \\ \frac{-C}{V} & a_0(C_S - C) \end{bmatrix} \tag{10}$$

g is invertible (i.e. g^{-1} exists) as long as S and C_S - C are different from zero. Under these physically realistic conditions, the linearizing controller (7) can be applied. The remaining problem is how to deal with the "unknown" parameters k_1, k_2 and μ, and variable X. The solution proceeds as follows. Let us first define the auxiliary variables z (for a systematic definition of these auxiliary variables, see [1], chapter 1, section 1.7) :

$$z = \begin{bmatrix} z_1 \\ z_2 \\ z_3 \end{bmatrix} \stackrel{\Delta}{=} \begin{bmatrix} k_1 \\ k_2 \\ 0 \end{bmatrix} X + \begin{bmatrix} 1 & 0 & 0 \\ 0 & 1 & 0 \\ 0 & 0 & k_2 \end{bmatrix} \begin{bmatrix} S \\ C \\ X_R \end{bmatrix} \tag{11}$$

The dynamical equation of z is derived from model (1)(2) and definitions (11) :

$$\frac{d}{dt}\begin{bmatrix} z_1 \\ z_2 \\ z_3 \end{bmatrix} = \begin{bmatrix} -D_1 & 0 & D_2k_1/k_2 \\ 0 & -D_1 & D_2 \\ 0 & D_3 & -D_4 \end{bmatrix}\begin{bmatrix} z_1 \\ z_2 \\ z_3 \end{bmatrix} + \begin{bmatrix} D_{in}S_{in} \\ Q_{in} \\ -D_3C \end{bmatrix} \tag{12}$$

One important feature of the above dynamics (12) is that these are independent of the (unknown) kinetics. Let now rewrite the specific growth rate μ as follows :

$$\mu = \alpha SC \tag{13}$$

where α is an (unknown) positive function of the process components : this equation (13) simply implies that there is no growth in absence of one of the limiting substrate. By introducing (11)(13), the function f in (10) becomes :

$$f = \begin{bmatrix} -\alpha SC(z_1 - S) + D_{in}(S_{in} - S) \\ -\alpha SC(z_2 - C) - D_{in}C \end{bmatrix} \tag{14}$$

This expression of f is used in the computation of the control law (8) :

$$F_R = -\frac{V}{S}[C_{11}(S^* - S) - D_{in}(S_{in} - S) + \hat{\alpha}_1 SC(z_1 - S)] \tag{15.a}$$

$$W = \frac{1}{a_0(C_S - C)}\left[\lambda_2(C^* - C) - \frac{\lambda_1(S^* - S)C}{S} + \frac{D_{in}S_{in}C}{S} + \hat{\alpha}_2 C(Sz_2 - Cz_1)\right] \tag{15.b}$$

C_{11} and C_{12} (> 0) are the control design parameters (we have considered a diagonal matrix C_1).The variables z_1 and z_2 are computed via the use of equations (12) (which only requires the knowledge of the ratio k_1/k_2) and the parameter α is estimated on-line by using e.g. a recursive least square (RLS) algorithm. In discrete-time, if α is estimated via the substrate concentration equation S (α_1, to be used in equation (15.a)) or via the dissolved oxygen concentration C (α_2, to be used in equation (15.a)), the RLS algorithm is written as follows :

$$\hat{\theta}_{i,t+1} = \hat{\theta}_{i,t} + g_{i,t}\phi_{i,t}e_{i,t+1} \qquad i = 1, 2 \tag{16.a}$$

$$g_{i,t} = \frac{g_{i,t-1}}{\gamma_i + g_{i,t-1}\phi_{i,t}^2} \qquad 0 < \gamma_i \le 1 \tag{16.b}$$

with :

$$\hat{\theta}_{1,t} = \hat{\alpha}_{1,t}, \; \phi_{1,t} = TS_tC_t(S_t - z_{1,t}) \tag{17.a}$$

$$e_{1,t+1} = S_{t+1} - S_t + TD_{1,t}S_t - TD_{in,t}S_{in,t} - \hat{\alpha}_{1,t}\phi_{1,t} \tag{17.b}$$

$$\hat{\theta}_{2,t} = \hat{\alpha}_{2,t}, \; \phi_{2,t} = TS_tC_t(C_t - z_{2,t}) \tag{17.c}$$

$$e_{2,t+1} = C_{t+1} - C_t + TD_{1,t}C_t - Ta_0W_t(C_S - C_t) - \hat{\alpha}_{2,t}\phi_{2,t} \tag{17.d}$$

and t the time index, and γ_i (i = 1, 2) a forgetting factor. Note that now because of the variables z, the biomass X does not appear explicitly anymore in the controller (15).

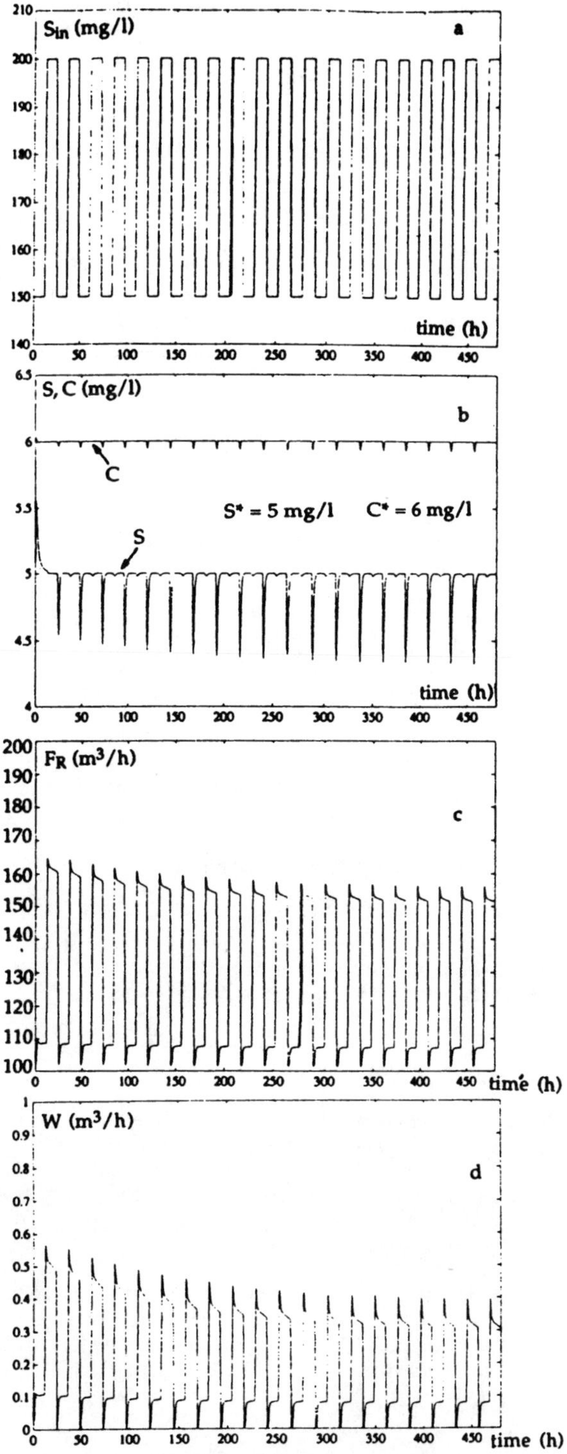

Fig.1. Control of an activated sludge process in presence of an unknown biomass mortality.

A typical simulation result is shown in Fig.1 (see also [11]). The activated sludge process has been simulated by numerical integration of the *basic dynamical model* equations (1)(2) (using a 4th order Runge-Kutta method) with the following (Monod-type) model for the specific growth rate:

$$\mu = \mu_{max}\frac{S}{K_S+S}\frac{C}{K_C+C}$$

and a decay rate ($-k_dX$) which has been added to the biomass equation (in order to simulate biomass mortality). The specific growth rate model and the decay rate term are (obviously) completely ignored by the control algorithm. The decay rate term can be viewed as an unmeasured disturbance. The model parameters have been set to the following values (inspired from literature data, in particular from [5], [7]) :

$$\mu_{max} = 0.2\ h^{-1}, K_S = 75\ mg/L, K_C = 2\ mg/L, k_1 = 1.2, k_2 = 0.565$$
$$a_0 = 0.018\ m^{-3}, C_S = 10\ mg/L, V = 100\ m^3, V_S = 50\ m^3, k_d = 0.001\ h^{-1}$$

In Fig.1, a square wave signal of the influent BOD concentration (disturbance input) S_{in} (from 150 mg/L to 200 mg/L) (in order to simulate the periodical variation of the pollutant load) has been applied over a period of 20 days (480 hours). The sampling period has been set to 3 minutes following the constraints of the commercially available BOD measuring device proposed by [8]. The following initial process conditions have been considered in the simulation :

$$S = 5\ mg/L, C = 6\ mg/L, X = 1225\ mg/L, X_R = 2333\ mg/L$$
$$S_{in} = 150\ mg/L, F_{in} = F_R = 10\ m^3/h, F_W = 0.5\ m^3/h, W = 100\ m^3/h$$

The controller parameters have been set to the following values :

$$S^* = 5\ mg/L, C^* = 6 mg/L, \lambda_1 = 1\ h^{-1}, \lambda_2 = 10\ h^{-1}$$

Auxiliary variables z, the unknown parameters and tuning estimation variables have been initialized as follows :

$$z_1 = 1400\ mg/L, z_2 = 750\ mg/L, z_3 = 1400\ mg/L,$$

$$g_{1,0} = g_{2,0} = 10^{-3}, g_1 = g_2 = 0.9, \hat{\alpha}_{1,0} = \hat{\alpha}_{2,0} = .00025\ L^2.mg^{-2}.h^{-1}$$

Note the ability of the controller to maintain the controlled outputs S and C close to their desired values in spite of the unknown disturbance.

5. Example #2 : Anaerobic Digestion (continued)

In the example #1, we have considered a multivariable control problem. The specificity of the second example is that the design of the controller is based on a "reduced-order" form of the basic dynamical model (1). Let us first motivate the use of a reduced order model. The design of a controller based on a reduced order model will result in a simpler controller structure but will still contain the important process features if the model reduction has been correctly performed.

Two typical phenomena (at least) motivate model order reduction in bioreactor applications ([1]). First of all, the presence of low solubility products, like CH_4. Secondly, the presence of "fast" and "slow" reactions. In both cases, it can be shown that the model order reduction (via the use of "singular perturbation") results in setting the corresponding component concentration ξ_i (the low solubility product or the "fast" component) and its time derivative to zero :

$$\xi_i = d\xi_i/dt = 0 \tag{18}$$

and replace its differential equation by an algebraic equation. For instance, the order reduction applied to the methane (P_1) equation in model (1)(3) results in the usually considered gaseous outflow rate equation for CH_4 :

$$Q_{CH4} = k_{14}\varphi_3 + k_{15}\varphi_4 \tag{19}$$

i.e. the gaseous outflow rate of CH_4 is equal to its production rates in reactions 3 and 4. But what about the "fast" and "slow" reaction case? It is difficult to know a priori which reactions in the anaerobic digestion reaction network will be, in a broad range of operating conditions, fast or slow. On the other hand, one may wish to avoid undesirable effects such as accumulation of propionate or of hydrogen : then, in that situation,the corresponding reaction (reaction 2 or 4) can be considered as the limiting, i.e. "slow", reaction, and the others as "fast" reactions. For instance, assume that your main problem is to avoid accumulation of propionate, then reaction 2 is the slow reaction and reactions 1, 3 and 4 are the fast reactions. Model reduction is carried out by setting S_1, S_3 and S_4 and their time derivative to zero. Including the order reduction on P_1, we obtain a set of four algebraic equations :

$$\begin{bmatrix} k_1 & 0 & 0 & 0 \\ k_5 & k_6 & -k_4 & 0 \\ k_8 & k_9 & 0 & -k_7 \\ 0 & 0 & k_{14} & k_{15} \end{bmatrix} \begin{bmatrix} \varphi_1 \\ \varphi_2 \\ \varphi_3 \\ \varphi_4 \end{bmatrix} = \begin{bmatrix} DS_{in} \\ 0 \\ Q_{H_2} \\ Q_{CH_4} \end{bmatrix} \tag{20}$$

If we invert the above expression and express φ in terms of DS_{in}, Q_{H2} and Q_{CH4}, we can introduce the obtained expressions in the dynamical equation of the propionate concentration S_2 in model (1)(3) which can then be rewritten as follows :

$$\frac{dS_2}{dt} = -DS_2 - \theta_1 Q_{CH_4} + \theta_2 DS_{in} - \theta_3 Q_{H_2} \tag{21}$$

where θ_1, θ_2 and θ_3 are nonlinear combinations of the yield coefficients :

$$\theta_1 = \frac{k_2k_4k_7}{k_4k_9k_{15} + k_6k_7k_{14}},\ \theta_2 = \frac{k_3}{k_1} + \theta_3\left(\frac{k_8}{k_1} + \frac{k_5k_7k_{14}}{k_1k_4k_{15}}\right),\ \theta_3 = \frac{k_2k_4k_{15}}{k_4k_9k_{15} + k_6k_7k_{14}} \tag{22}$$

The above equation is the reduced order model on which the control design (7) is based. Note that because of the order reduction, the unknown reaction rates φ_i have been replaced by known quantities (D, S_{in}, Q_{CH4} and Q_{H2}). One important feature of the above reduction procedure is that it is systematic and based on a rigourous mathematical line of reasoning (for further details, see [1], [12]).

Let us go back to our control problem. If we consider that the control input is the dilution rate D, the expressions of f and g are here equal to :

$$f = -\theta_1 Q_{CH4} - \theta_3 Q_{H2},\ g = \theta_2 S_{in} - S_2 \tag{23}$$

and the controller (8) is written here as follows :

$$D = \frac{C_1(S_2^* - S_2) + \hat{\theta}_1 Q_{CH_4} + \hat{\theta}_3 Q_{H_2}}{\hat{\theta}_2 S_{in} - S_2} \qquad C_1 > 0,\ S_2^* = \text{reference value of } S_2 \tag{24}$$

The unknown parameters are now θ_1, θ_2 and θ_3. They can be estimated on-line by using an appropriate updating, for instance here a "Lyapunov design" estimation equation :

$$\frac{d\hat{\theta}_i}{dt} = C_{2i}(S_2^* - S_2) \qquad C_{2i} > 0,, \quad i = 1, 2, 3 \tag{25}$$

A typical simulation is shown in Fig.2 (see also [10][13]). In this simulation, the hydrogen gaseous outflow rate Q_{H2} has been assumed to be negligible, and the parameter θ_2 has been assumed to be known. Therefore, only θ_1 has been estimated on-line. In Fig.2, a step of influent organic matter concentration has been applied at time t = 2 days. Note the ability of the controller to maintain the propionate concentration S_2 close to its desired value (= 0.28 g/l). Note also the glucose concentration S_1 and the acetate concentration S_3 remain almost constant, while the hydrogen concentration increases from 3 µM to a value a little larger than 4 µM.

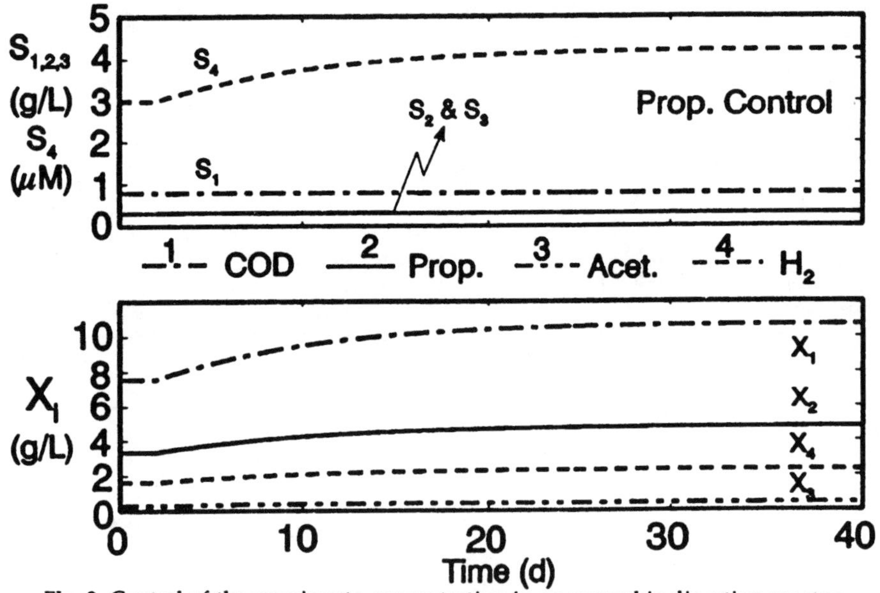

Fig. 2. Control of the propionate concentration in an anaerobic digestion reactor.

Remark : a similar approach has been used to design an adaptive linearizing controller for the hydrogen and the COD (Chemical Oxygen Demand), which resulted in similar controller structures ([1], [13]). This type has also been applied to a pilot-scale anaerobic digestion reactor (e.g. [14][15]).

An important comment : in the above examples, we have presented two possible control solutions. These examples must be viewed as illustrative examples and any combination of these two solutions can be considered, depending on the specific control problem under study. The above procedure can be used for instance to treat other control problems, e.g. the control of the denitrification or dephosphatation processes.

6. Conclusions

This paper has been concerned with the design of control algorithms for biological wastewater treatment processes. Our intention has been to present a systematic approach for the design of controllers based on the dynamical mass balance equations of the process while including the on-line estimation of uncertain parameters (specific growth rates or yield coefficients) and unmeasured components (biomass). The procedure has been illustrated on two control examples, but can be used and extended to other typical control problems arising in wastewater treatment processes, like the control of denitrification or dephosphatation processes.

References

1. BASTIN G. and D. DOCHAIN, *On-line Estimation and Adaptive Control of Bioreactors*. Elsevier, Amsterdam, (1990).
2. CHEN L., G. BASTIN and V. VAN BREUSEGEM, "Adaptive nonlinear regulation of fed-batch biological reactors : an industrial application". *Proc. 30th IEEE CDC*, 2130-2135 (1991).
3. MARSILI-LIBELLI S., "Optimal control of the activated sludge process". *Trans. Inst. Meas. Control*, **6**, 146-152, (1984).
4. MOSEY F.E., "Mathematical modelling of the anaerobic digestion process : regulatory mechanisms for the formation of short-chain volatile acids from glucose". *Wat. Sci. Tech.*, **15**(8/9), 209-232 (1983).
5. JONES R.M. and E.R. HALL, "Assessment of dynamic models for a high rate anaerobic treatment process", *Env. Technol. Lett.*, **10**, 551-566 (1989).
6. COSTELLO D.J., P.F. GREENFIELD and P.L. LEE, "Dynamic modelling of a single stage high rate anaerobic reactor, I. Model derivation, II. Model verification", *Wat. Res.*, **25**(7), 847-871 (1991).
7. HOLMBERG A., "A microprocessor-based estimation and control system for the activated sludge process". In A. Halme (Ed.), *Modelling and Control of Biotechnical Processes*, Pergamon, 111-120 (1983).
8. KOHNE M., "Practical experiences with a new on-line BOD measuring device". *Env. Technol. Letters*, **6**, 546-555 (1985).
9. HOLMBERG A. and J. RANTA, "Procedures for parameter and state estimation of microbial growth process models". *Automatica*, **18**, 181-193 (1982).
10. PERRIER M. and D. DOCHAIN, "Evaluation of control strategies for anaerobic digestion processes". In preparation (1992).
11. DOCHAIN D . and M. PERRIER, "Adaptive linearizing control of activated sludge processes", Proc. Control Systems'92, Whistler, Canada (1992).
12. VAN BREUSEGEM V. and G. BASTIN, "Reduced order modelling of reaction systems : a singular perturbation approach", *Proc. 30th IEEE Conf. Dec. Control*, 2130-2135 (1991).
13. PERRIER M. and D. DOCHAIN, "An analysis of the impact of controlled variables selection on the operation of anaerobic digestion processes", *Proc. IFAC-BIO-2*, Keystone, Colorado (1992).
14. RENARD P., D. DOCHAIN, G. BASTIN , H. NAVEAU and E.J. NYNS. "Adaptive control of anaerobic digestion processes : a pilot-scale application". *Biotech. Bioeng.*, **31**, 287-294 (1988).
15. RENARD P., V. VAN BREUSEGEM, N. NGUYEN, H. NAVEAU and E.J. NYNS, "Implementation of an adaptive controller for the start-up and steady-state running of a biomethanation process operated in the CSTR mode". *Biotech. Bioeng.*, **38**, 805-812 (1991).

Pergamon

Wat. Sci. Tech. Vol. 28, No. 11-12, pp. 295–302, 1993.

0273–1223/93 $7·00 + 0·00

ACTIVATED SLUDGE CONTROL ADVISER

K. R. Stimson

Water Research Centre (WRc), Frankland Road, Blagrove, Swindon, Wiltshire SN5 8YF, UK

ABSTRACT

This paper concerns the development and use of a computer-based expert system (ASCA) to advise operators of activated sludge wastewater treatment plants in the event of an actual or potential effluent problem. The system provides a diagnosis of problems and gives advice on remedial actions. An incident database facility permits the recording and later inspection of problematic events. The system contains a general knowledge base applicable to all such treatment works. This is supplemented by site-specific knowledge when customised to particular plant. Development required close collaboration with users. On-site trials are being performed at two sites. Early results indicate both usability and usefulness, and demonstrate that the system can be readily customised for further sites. Further possible enhancements are discussed.

KEYWORDS

Wastewater, activated sludge, expert systems, decision support, diagnosis.

INTRODUCTION

Activated sludge wastewater treatment plants are in wide use, particularly in urban areas. Plant operation is subject to increasingly stringent effluent standards. Operating utilities must meet these standards and ensure cost-effective operation. Control is exercised by operators of varying levels of expertise and recent developments in automation and working practices are making this expertise scarce. Operators may need to make decisions on incomplete or uncertain information, particularly when the plant experiences problems. Experienced operators may be able to diagnose the causes of these problems and implement corrective action. However, they may find difficulty in assessing the likelihood of alternative diagnoses and in realising possible alternative courses of action. Inxperienced operators may make even more superficial judgements.

The use of expert systems to meet these requirements was first investigated in the late 1970s. Problem diagnosis in particular was well established in other industries and essential aspects of decision support for activated sludge plants were proposed (Maeda, 1985). More recent developments include prototype systems and systems developed for specific works (Schmuller and Morlino, 1990; Beck et al, 1989; Gall and Patry, Williams et al, 1989).

The system (ASCA) described in this paper addresses the problem or potential problem of unacceptable effluent and provides both diagnosis and advice on remedial actions. User involvement was an essential part of the development. Prior to the development, it was identified that users' confidence would be gained if they were able to refer to past incidents recorded in an incident database. To ease future customisations it was decided to partition the knowledge-base into general knowledge and site-specific knowledge. This paper describes the system, its development and use, and discusses possible further enhancements.

OVERVIEW

ASCA is a pc-based system developed using the GoldWorks 3 expert systems tool kit. The system employs colour graphics and user interaction is achieved using both mouse and keyboard. Presentation is in a WINDOWS 3 environment.

The main facilities are:

Diagnosis - presentation of one or more causes of an effluent problem, in response to users input of the plant state.

Remedial Action - presentation of possible actions to be taken for a given diagnosis.

Incident Recording - record of the current consultation and other incidents for later reference.

In addition, the user is able to halt and resume a consultation, change previously entered plant state data, log comments and obtain help both in using the system and explanation of its reasoning.

DIAGNOSTIC SYSTEM

The fundamental premise of the diagnosis is the existence of an effluent problem. The diagnosis proceeds by asking the user for relevant information on the current operating status of the plant. By this process, and reference to rules in the knowledge base the diagnosis is further refined.

In order to help the user assess the merits of alternatives, the system is based on a diagnostic tree (see Figure 1). For example, supporting operational information may indicate that EFFLUENT PROBLEMS are caused by AERATOR PROBLEMS which are caused by LACK OF BACTERIAL ACTIVITY which is caused by ADDITION OF BIOCIDES. The

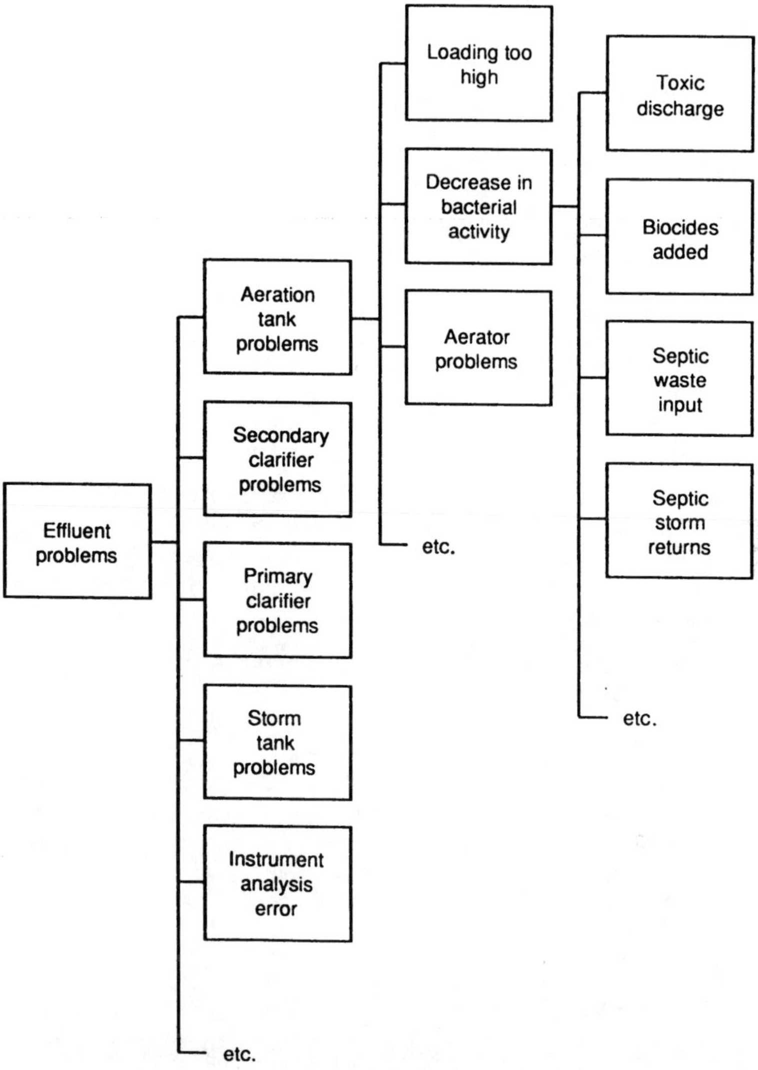

Fig. 1. Extract from diagnosis tree

tree currently contains 150 diagnoses at four levels of depth.

The system determines the likelihood of these causes using the terms CERTAIN, PROBABLE, POSSIBLE, PROBABLY NOT and CERTAINLY NOT. Thus if EFFLUENT PROBLEMS is CERTAIN, the user is requested to input the value of effluent suspended solids. The rule

"If EFFLUENT PROBLEMS is CERTAIN and SUSPENDED SOLIDS are HIGH or VERY HIGH then PROBLEMS IN SETTLEMENT TANK is CERTAIN"

could be used to deduce settlement tank problems. There are over 600 such rules in the knowledge base.

Users may input numeric values, e.g. effluent ammonia = 3mg/l or select a range (LOW, MEDIUM, HIGH etc) tailored to a specific works. When requested data are unavailable, the user may enter "NOT KNOWN". In general the system will still produce a diagnosis even though not all required data are known.

The diagnosis proceeds selectively, choosing lines of enquiry that are more likely. This avoids asking unnecessary questions. The user is also able to ask for an explanation (WHY) from the system about the reason for pursuing a certain line of enquiry. The user may also enter his own comments to be recorded in the consultation log for later reference.

As the diagnosis continues, the parts of the diagnostic tree under investigation are illustrated on the screen in a colour coded fashion which indicates likelihood. At the end of the diagnosis, or when the diagnosis is halted, the user may investigate this tree using the HOW facility. By selecting one of the displayed diagnoses, the user is presented with an explanation of how the diagnosis was made.

The diagnosis finishes when all the most likely causes have been investigated.

REMEDIAL ACTION

The diagnoses recommended by the system are those deemed most likely at the lowest level of the tree and corresponding causes at higher levels of the tree. The user can select any of these recommended diagnoses. This produces a check-list of corresponding remedial actions. Some of these will be general actions applicable to all works, e.g. the diagnosis TOXIC DISCHARGE INTO SEWER would generate the action CALL TRADE EFFLUENT OFFICER. Other actions will be site specific.

INCIDENT RECORDING

ASCA allows users to build up a database of works incidents for later reference. This can be performed both independently of using the system for diagnosis and as a means of recording an incident for which diagnostic advice was sought. Typical key items in the database are date/time, nature of event, plant state (effluent and other operating parameter values), action taken, degree of success and users comments.

The search facility allows retrieval on these key items. Thus, the user experiencing an effluent problem can obtain both diagnostic advice and investigate previous occurrences of the problem.

EXAMPLE CONSULTATION

A typical example concerns a nitrifying plant where the level of effluent ammonia is high or even above the regulated level. The user therefore has an effluent problem and initiates the diagnosis system.

ASCA prompts for (i) level of effluent ammonia (9mg/l), (ii) Effluent Solids (LOW) and (iii) Effluent BOD (LOW). The use of numeric or qualitative values (HIGH, LOW etc) depends upon site preference. A HELP facility will explain the meaning of LOW effluent solids at a particular works if the user is unsure.

Based on the above information ASCA makes a second level diagnosis, ascribing a likelihood of a problem in the major parts of the works. Thus in this example the diagnosis is:

"PROBLEM IN AERATION TANKS is CERTAIN".

Problems in the storm tanks, primary clarifiers or secondary clarifiers are deemed to be PROBABLY NOT or CERTAINLY NOT. This simple second level diagnosis is based on a rule:

"PROBLEMS WITH AERATION TANKS are CERTAIN
if EFFLUENT AMMONIA is >4".

As the diagnosis proceeds the diagnosis tree (see Figure 1) is constructed and displayed as each potential problem is investigated. The diagnoses are colour coded according to likelihood. To obtain the above explanation the user simply selects the diagnosis using the mouse.

As an aeration tank problem is the most likely, ASCA looks for further possible causes (level 3). Typical questions now asked concern the dissolved oxygen (DO) level, the DO set point (if controlled), whether normal or abnormal power is being used by the aerators, sewage flow, waste activated sludge flow and recycle ratio. At any question the user can use the WHY facility to obtain an explanation of why the information is sought. For example for the request for sewage flow the reason given is

"Sewage flow is required to calculate sewage retention time and assess the probability of the diagnosis RETENTION TIME TOO LOW".

The most likely third-level diagnosis is:

"DECREASE IN BACTERIAL ACTIVITY is PROBABLE".

The explanation given is:

"DECREASE IN BACTERIAL ACTIVITY is PROBABLE if DO is controlled and the DO level in the aeration lanes is >1.4".

Further questioning at level 4 elicits the fact that the colour of the settled sewage is unusual and leads to the most likely diagnosis:

"DISCHARGE OF A TOXIC SUBSTANCE is PROBABLE".

This completes the diagnosis stage.

The most likely diagnoses at each level are then displayed. The user can select any of these and obtain a checklist of possible remedial actions. For the above level 4 diagnosis the actions suggested (based on a real example) are:

- Analyse sewage to confirm diagnosis
- Increase aeration
- Divert sewage to storm tanks
- Trace and remove source of toxic substances
- Call in the trade effluent officer.

The user may select one or more of these actions and record the selection, e.g. "Increase aeration". The whole consultation can be logged for future reference. To obtain further confirmation of this decision the user then interrogates the ASCA database searching for instances of "TOXIC SUBSTANCE" events in the past and a record of the action taken.

DEVELOPMENT

The development of the ASCA system began in early 1991. At that stage earlier work had identified the potential for expert systems in this application and an early prototype system was available. The development involved production of the basic system and customisation at two sites operated by two UK Water Companies.

Contact with likely users was an essential component of the development. Users at each site were identified early in the project. The system was developed in a staged manner and users took part in demonstration and feedback sessions at intervals. This allowed both confirmation of the development and provided opportunity to improve the user interface.

The diagnostic knowledge-base was developed by process engineers and based on theory and experience of the operation of activated sludge plants. This knowledge base was applied to both sites and confirmed by local operators. Site specific knowledge, concerning the existence of potential works problems, works characteristics, remedial actions and identification of parameter ranges (HIGH, MEDIUM, LOW etc), was obtained by interviewing local operators. Typically four to six interviewing sessions with two to four operators were needed at each site. The interviewing team consisted of a knowledge engineer and a process engineer.

The customised systems were then subject to on-site trials at each works (Summer 1992). The trials lasted two to three months. Operators were encouraged to use the system for any incidents that arose, for exploring how potential incidents could be handled and for training purposes. Feedback was obtained from questionnaires, the system log and interviewing the staff. Final acceptance of the system was based on the acceptable performance both in usability and for predefined test cases.

CONCLUSIONS

Although the trials are not fully completed (September 1992) there are encouraging conclusions that can be drawn. The operators have found the system easy to use and they have approved both the accuracy and completeness of the diagnosis system. There were some circumstances where the more experienced operators considered that the advice given was obvious. However, the confirmation of the diagnosis was felt to enhance confidence. The presentation of alternative possible diagnoses and an action chech list ensured that possibilities other than the obvious were not overlooked. The system performs as expected for the test cases identified at both sites. Customising the system at two successive sites has confirmed the value of developing ASCA with a site-independent basis which can be completed by the addition of local knowledge. ASCA can therefore be easily customised to further activated sludge plants.

The benefits to plant operation will come from using ASCA to make the best possible expertise and practice available to all operators. This, in turn, provides for more consistent and efficient operation and reduces the likelihood of failure to meet effluent quality standards. In addition, the system can be used for operations planning in order to anticipate and provide for potential problems. Novice operators can use the system for training.

FUTURE ENHANCEMENTS

The increasing use of telemetry and computer based laboratory systems will make many of the ASCA data requirements available in electronic form. An interface to these data acquisition systems will reduce data entry errors and operator workload.

Actions recommended by ASCA corresponding to the different diagnoses aim to remove or alleviate the problem. However, it would be useful to evaluate alternative actions and plan the recovery of the works to an acceptable operating state. The use of a dynamic model would assist in both cases. The current work has demonstrated that linking ASCA to a dynamic model is technically feasible. Further development would be needed to provide a user interface which makes the running of such a model straightforward. In order to speed up the investigation of many alternatives, a simplified qualitative model, embodying the results of carrying out different actions in different circumstances, is a further possibility. In this case the system would search for a suitable match to the current plant status and investigate the results of alternative actions, or the system itself would conduct a search for best solutions.

ACKNOWLEDGEMENTS

The author acknowledges Thames Water Utilities and North West Water Limited and the collaboration of their staff at their respective activated sludge plants.

REFERENCES

Maeda, K. (1985). An intelligent decision support-system for activated sludge wastewater treatment processes. In: *Instrumentation and Control of Water and Wastewater Treatment and Transport Systems*, R.A.R Drake (Ed), Pergamon Press, London, pp 629-632.

Schmuller, J. and Morlino, M.R. (1990). The Activated Sludge Adviser Prototype. In: *Expert systems for Environmental Applications*, J.M. Hushon (Ed), American Chemical Society Symposium Series 431, Washington DC, pp 139-145.

Beck, M.B., Lumbers, J.P., Mackenzie, H.E.C., and Jowitt, P.W. (1989). A Prototype Expert System for Operational Control of the Activated Sludge Process. *Institution of Water and Environmental Management*, Proceedings of Metropolitan Branch Workshop, 12 July 1989.

Gall, R.A.B. and Patry, G.G. (1989). Knowledge-based System for the Diagnosis of an Activated Sludge Plant. In: *Dynamic Modelling and Expert Systems in Wastewater Engineering*, G.G. Patry and D Chapman (Eds), Lewis Publishers Inc, Michigan, pp 193-240.

Williams, R., Knight, B., Watts, P. and Burns, J. (1989). An Expert System for the Control of the Activated Sludge Process. In: *Research and Development in Expert Systems VI*, N. Shadbolt (Ed), Cambridge University Press, pp 276-289.

Pergamon

Wat. Sci. Tech. Vol. 28, No. 11-12, pp. 303–314, 1993.

0273–1223/93 $7·00 + 0·00

CONTROL OF BNR PROCESSES BY REGULATION OF THE VFA/M RATIO

William B. Armiger*, George J. Lee*,
Benedict R. Schwegler, Jr.** and Terry J. Mah***

** BNR Engineers and Consultants, Division of BioChem Technology, Inc., 100 Ross Road, P.O. Box 61366, King of Prussia, PA 19406-0857, USA*
*** Walt Disney Imagineering, Glendale, CA, USA*
**** University of British Columbia, Vancouver, B.C., Canada*

ABSTRACT

Diurnal variation is characteristic of nearly all POTWs currently in operation. Equalizing the organic load to the process can significantly benefit the performance of the biological treatment process. The potential benefits of flow equalization have been recognized for over 20 years, although subsequent work has not been able to demonstrate reliable engineering designs to capitalize on these theoretical benefits. During the past decade, stricter environmental regulations have increased the interest and investment in nutrient removal processes. The successful development of Biological Nutrient Removal (BNR) treatment processes has accelerated this process. The data in this paper demonstrate that the ability to accurately measure and track biological reactions is crucial to the ability to equalize plant loadings. It is important to reiterate, in this regard, that the "bioreactor" which carries out the treatment reactions is a living cell and not a concrete tank. This manuscript presents new technology for the control of BNR treatment plants based upon direct measurements of cell metabolic activity, and presents results of preliminary experiments to control the waste load to biomass ratio independent of hydraulic flow fluctuations.

KEYWORDS

Process control; Biological Nutrient Removal (BNR); load balancing; NADH; diurnal variation; load equalization.

INTRODUCTION

Diurnal variation is characteristic of nearly all POTWs currently in operation. The primary objective of using flow equalization basins has been to dampen the diurnal variation in the flow to achieve nearly constant flow rates through the downstream treatment processes (Downing, 1976; Munksgaard *et al.*, 1980; Young *et al.*, 1978). In addition, equalizing the organic load to the process can significantly benefit the performance of the biological treatment process. However, of the nearly 17,000 articles published on wastewater treatment operation during the last 25 years, less than 25 have had flow and or waste load balancing as their principal topic.

The potential benefits of flow equalization have been recognized for over 20 years (Boon and Burgess, 1972; Dold *et al.*, 1984; EPA, 1974), although subsequent work has not been able to demonstrate reliable engineering designs to capitalize on these theoretical benefits. Foess *et al.* (1977) tested both in-line and side-line flow equalization methods, and found that neither was effective in improving effluent quality, principally because the effect of waste load balancing by blending was swamped by the effects of mixed liquor settleability, current densities and organic loading. However, this evaluation did not make use of any "real-time" process control technology.

To date, improvements in effluent quality due to stabilized mass loading of nutrients to biological systems have not been implemented due to the inability to accurately measure and control plant loads on a real-time basis (Beck, 1986; Edwards and Sherrard, 1982). This situation has now changed, and the technology described in the presentation solves this problem that has plagued the industry for many years.

During the past decade, stricter environmental regulations have increased the interest and investment in nutrient removal processes. The successful development of Biological Nutrient Removal (BNR) treatment processes has accelerated this process. However, most common BNR processes used worldwide (Bardenpho, UCT, modified UCT, A^2/O, VIP) are characterized by a number of requirements including high BOD/P ratios (Daigger *et al.*, 1988; Oldham and Stevens, 1985; Sadick *et al.*, 1992; Stevens and Oldham, 1992; Tetreault *et al.*, 1986; Weston, 1985; Wilson *et al.*, 1992) and some type of process control of the recycle stages. The design practice of constructing separate basins in BNR plants arose in part because of the necessity to control each stage of the bioprocess.

The data in this paper demonstrate that the ability to accurately measure and track biological reactions is crucial to the ability to equalize plant loadings. It is important to reiterate, in this regard, that the "bioreactor" which carries out the treatment reactions is a living cell and not a concrete tank. This may seem self-evident in light of the development of the BNR process, but from the standpoint of control technology, it is essential.

A focus on the cell as "bioreactor" leads to two productive lines of inquiry: (1) the internal biochemistry of the cells in the activated sludge and (2) characterization of that fraction of the influent and recycle waste streams that the cells actually use as a biochemical substrate. This manuscript reviews both of these areas with respect to new technology for the control of BNR treatment plants, and presents results of preliminary experiments to control the waste load to biomass ratio independent of hydraulic flow fluctuations.

CRITICAL ROLE OF NADH IN BNR OPERATIONS

The optimization of such a complex process as BNR requires an assessment of the metabolic activity of the activated sludge in a variety of environments. Unlike dissolved oxygen, NADH metabolism is involved in all environmental zones. Thus, NADH is an excellent indicator of metabolic activity that can be used to control BNR processes (Armiger *et al.*, 1990b; Armiger *et al.*, 1991; Yang *et al.*, 1991).

Theory and operation of BNR treatment plants have been extensively discussed in the literature (Randall *et al.*, 1992; Sedlak, 1991). All are characterized by the coordination of cellular phosphorus release with simultaneous storage of volatile fatty acids (VFA) in the initial anaerobic zone, followed by denitrification in the anoxic zone and BOD oxidation with phosphorus uptake and nitrification in the aerobic zones, respectively.

The dominant organisms and the active biochemical pathways vary with the environmental zone of the bioreactor. However, the one common denominator is the requirement to transfer energy by the oxidation of available energy sources. These reactions are summarized in Figure 1 which shows the critical role of nicotinamide adenine dinucleotide (NAD+, NADH) in energy transfer. Energy for the uptake of VFA is provided by the depolymerization of polyphosphate within the cell. After transport across the cell membrane, the VFAs (acetate and propionate) are synthesized into an organic storage product poly-β-hydroxvalate (PHV) and poly-β-hydroxybutyrate (PHB) with NADH providing the driving force. In the anoxic zone following the anaerobic zone, the recycle of nitrified mixed liquor makes an oxidizing agent (NOx) available for stabilization of BOD and oxidation of NADH, resulting in denitrification. When the BNR process is operated at a high rate of denitrification, phosphorus is reassimilated in the anoxic zone. In the aerobic portion of the process, rapid oxidation of the remaining BOD occurs along with the nitrification, provided sufficient NADH is available. Active aerobic metabolism is necessary to insure complete assimilation of phosphorus in this zone. Phosphorus is then removed from the system by sludge wasting when the plant is operating optimally, or by precipitation with metal salts when necessary.

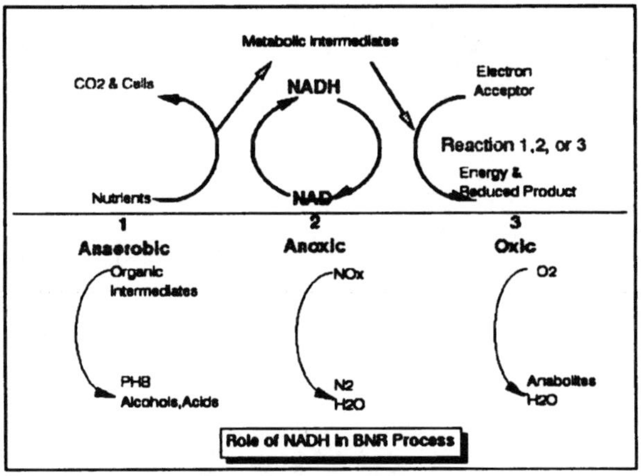

Fig. 1. Role of NADH in BNR process

PRINCIPLES OF NADH MEASUREMENT

The on-line NADH measurement is based upon the principle that NADH, which is found in all living cells, fluoresces at 460 nm when irradiated with light at 340 nm (Armiger *et al.*, 1986). As illustrated in Figure 1, the environmental conditions of the activated sludge determine the metabolic pathways by which NADH is constantly recycled from the oxidized to the reduced form. Specifically, in BNR processes, the reduction state [NADH/(NAD+NADH)] of the activated sludge varies as the mixed culture flows from the anaerobic zone to the anoxic zone to the oxic zone. The Biological Activity (BA) in each zone is defined as the reduction state of the activated sludge times the viable cell population. By constantly measuring the fluorescence from the NADH, it is possible to monitor changes in the BA of the activated sludge resulting from normal operating conditions as well as process disturbances.

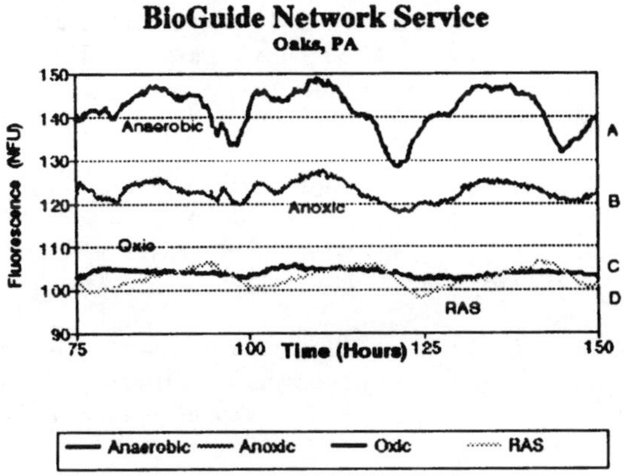

Fig. 2.

Armiger *et al.* (1990a) have documented a typical "fingerprint" for the normal operation of a three-stage BNR plant. The biological activities for the anaerobic, anoxic, and aerobic zones of the bioreactor shown in Figure 2 are highest in the highly reduced environment of the anaerobic zone and lowest in the highly oxidized environment of the aerobic zone. These data from a three-day period illustrate the typical diurnal variation due to an increase in the VFA loading as the flow rate increases during the morning with subsequent VFA loading decreasing toward the end of the day. The large decline in the anaerobic zone results from the increased demand for NADH as the organisms in the anaerobic selector absorb VFA to be stored as PHB and PHV with the simultaneous release of phosphate. A performance profile shown in Figure 3 indicates the TOC, PO_4, NOx, and NH_3 concentrations in each stage confirming the biological measurements from individual sections within each zone.

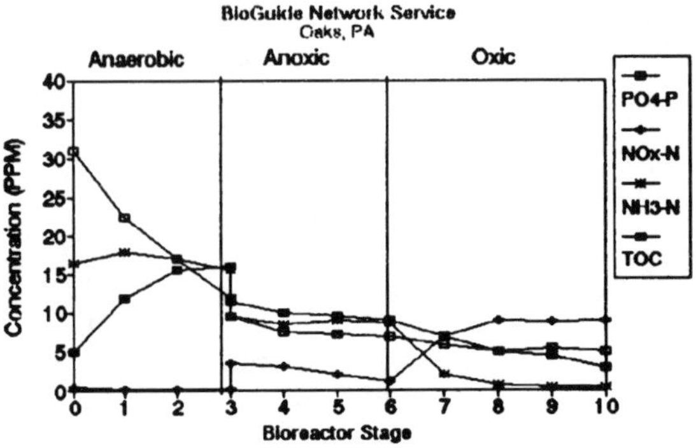

Fig. 3. Typical "snapshot" profile of each section of the bioreactor during normal plant operation.

The multiple goals of nitrogen, phosphorus, BOD and suspended solids removal, make BNR process control even more critical than in conventional wastewater treatment processes. The F/M ratio has normally been used as the operational parameter to regulate BNR activated sludge. However, the inability to provide a rapid assessment of this parameter has impeded process optimization. With the recent development of this technology for measuring biological activity, this barrier to process control and optimization has now been removed.

BNR PROCESS CONTROL

BNR process control is implemented through the BioGuide Network Service provided by BioChem Technology which combines patented NADH detectors, proprietary software, and process engineering expertise. The NADH detectors are immersed directly in the process streams at various locations throughout the plant, where data are transmitted to a local computer workstation. There, proprietary state-of-the-art software manages a local database for processing the data, indicating plant performance, and providing management information to implement process control decisions.

Two of the BNR treatment plants using the BioGuide Network Service for controlling plant loading are presented as case studies in this paper. The first plant is a 4.0 MGD modified UCT design in the City of Penticton, British Columbia, and the second is a 10.0 MGD modified Bardenpho plant located at the Reedy Creek Improvement District (RCID) near Orlando, Florida.

As shown in Figure 4, the modified UCT process utilizes an equalization tank between the primary clarifier and the bioreactor. The plant also incorporates a complete mix fermentor for additional VFA generation. The fermentor discharges into the primary clarifier. The original purpose of the equalization tank was to stabilize the hydraulic load to the process. However, this design is still subject to large variations in VFA loading as are all BNR processes.

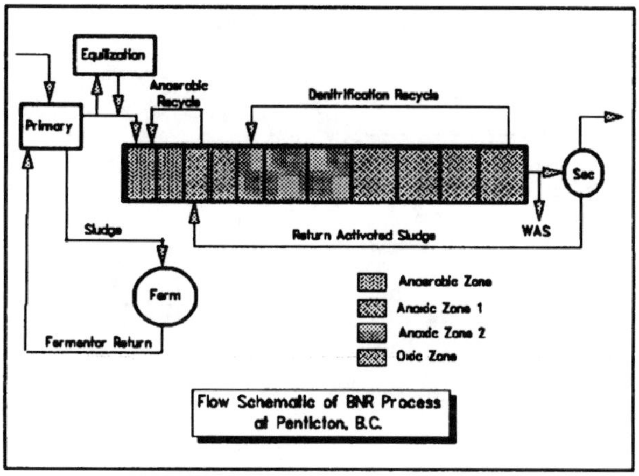

Fig. 4.

Due to the importance of VFA loading on the effective operation of a BNR process, the F/M ratio has been redefined as VFA/M resulting in a new process control parameter that is directly related to nutrient removal biochemistry. The importance of this new ratio is illustrated in Figure 5a which shows the relationship of the COD, VFA, and biological activity (BA) in the primary clarifier effluent. The BA response is inversely related to the influent concentration of VFA and not correlated to COD concentration. Furthermore, VFA's comprise a small, but significant and systematically variable fraction of the COD loading. These data are consistent with the expected fingerprint for BA from the dilute microbial population present in the primary effluent. Following the typical diurnal cycle of the plant load, they show an increase in the NADH concentration as the decrease in the VFA concentration causes a lower demand for NADH, and they show a decrease in the NADH concentration as the increase in the VFA concentration causes a greater demand for NADH.

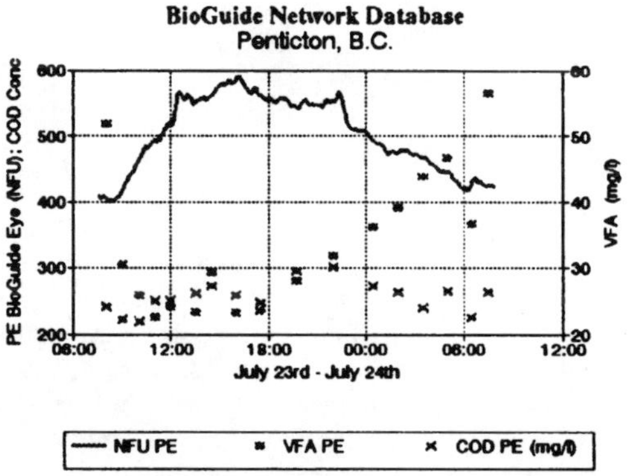

Fig. 5a.

Figure 5b shows a linear correlation between the measured biological activity in the primary effluent and the VFA concentration. Even though the VFA concentration changes by nearly 300% over a typical diurnal cycle, the correlation with the on-line continuous signal is greater than 90%. Using the on-line indication of VFA concentration combined with the on-line flow rate produces a real-time measure of the plant VFA load.

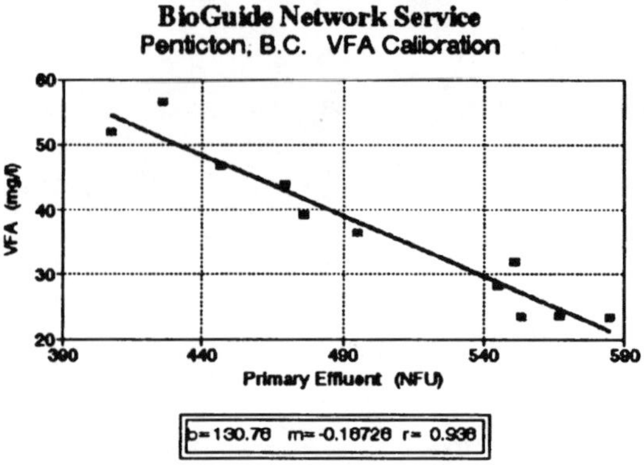

Fig. 5b.

Figure 6 shows the spike in the plant VFA load beginning at about 06:30 when the flow into the plant begins to increase and washes out the high concentration of VFA that has accumulated in the primary clarifier overnight.

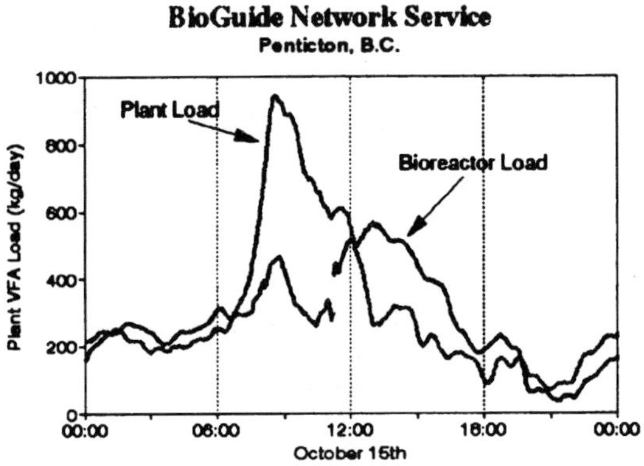

Fig. 6.

The high concentration of VFA in the primary clarifier in the early morning hours is the combined result of 1) increased VFA production in the collection system due to the lower night time flow rate and 2) continuous addition of fermentor- produced VFAs to the primary clarifier. This clarifier has a residence time of 3 times the peak flow residence time. As this initial surge passes through the primary clarifier, the peak load decreases to a more steady load that is maintained until the early afternoon hours when the VFA load from the collection system decreases due to the shorter fermentation time for the water in the collection system. In the evening, the VFA to the plant begins to increase as declining flow in the collection system increases the residence time resulting in an increase in VFA production.

The data for the bioreactor VFA load are also shown in Figure 6. The attenuation of the peak VFA load and the shift in the bioreactor loading compared to the plant loading are the result of controlling the flow rate into and out of the equalization basin in order to regulate the VFA/M value. The strategy for VFA/M regulation is discussed in the next section.

The biological activity measured in the bioreactor is a function of the concentration and the reduction state of the activated sludge. For a BNR process, a typical fingerprint of biological activity has already been shown in Figure 1 which illustrates the diurnal variation of BA. Within each environmental zone, variations in BA are caused by changes in activated sludge concentration and changes in the reduction state of the cells due to fluctuations in nutrient availability. Similar diurnal fluctuations are measured in each zone of the modified UCT process in Penticton. The changes in the reduction state can be determined by normalizing the BA with respect to activated sludge concentration. Thus, within the anaerobic zone, where the VFA concentration regulates the reduction state of the activated sludge, the BA activity is a function of VFA and MLSS.

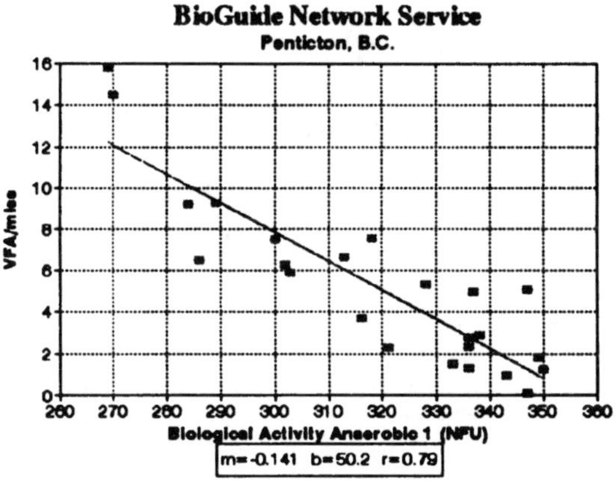

Fig. 7.

Figure 7 shows the correlation of the VFA/M ratio to the BA measurements in the anaerobic zone of the bioreactor. The relationship shows that (1) a decrease in the VFA/M ratio caused by an increase in MLSS concentration results in a higher BA reading and (2) an increase in the VFA/M ratio caused by an increase in the VFA concentration results in a lower BA reading due to the lower reduction state caused by the greater demand for NADH. Thus, it is now possible to regulate the process toward a constant VFA/M ratio by control of the VFA loading and/or the activated sludge concentration.

PROCESS CONTROL RESULTS: REGULATION OF VFA/M CONTROL OF VFA LOADING

By far, the best method for regulating the VFA/M ratio is to balance the organic load while operating the plant under a hydraulic equalization. A control logic diagram for this type of operation is shown in Figure 8 and was implemented through the BioGuide Network Service on the modified UCT process in Penticton, B.C. Data on the plant and bioreactor flow rates, the biological activity in the primary effluent and the mixing zone of the bioreactor were logged at minute intervals. Those data were used to calculate the instantaneous VFA loads exiting the primary clarifier and entering the bioreactor and the instantaneous VFA/M ratio in the anaerobic zone. The flow rates to and from the equalization tank were then adjusted according to the logic diagram. For example, in the early morning hours when the plant flow rate begins to increase as shown in Figure 6, the VFA/M ratio was greater than the set point of 5 and the VFA load coming from the primary clarifier exceeded the load at the mixing zone. To maintain a stable VFA/M ratio, the net VFA load to the equalization tank (defined as the flow times the concentration entering minus the flow times the concentration leaving the equalization tank) was increased. The other control actions taken under different circumstances are evident from the logic diagram.

Figure 9 shows the results of the control scheme over a 24-hour period in which the VFA/M was regulated to a set point of 5.0±1 ms/mg. The figure shows the VFA/M data along with the VFA loading to the equalization tank. The data show that it is indeed possible to regulate the VFA ratio by manipulating the VFA load to the equalization tank. The VFA/M ratio was maintained within its set point operating range except for two relatively brief excursions. The first occurred in the late morning hours when the peak load hit the plant. In this case, control was lost due to insufficient dilution water availability, rather than an intrinsic control algorithm failure. Control was regained as the peak load declined. This situation can now be avoided by making some minor changes in the plant operation. The second time control was lost was due to an operational error. For a brief period, as the VFA/M ratio was increasing, the load to the equalization was decreased instead of increased. The situation was corrected, but it took approximately one hour for the plant to totally recover the VFA/M ratio into the targeted operating range.

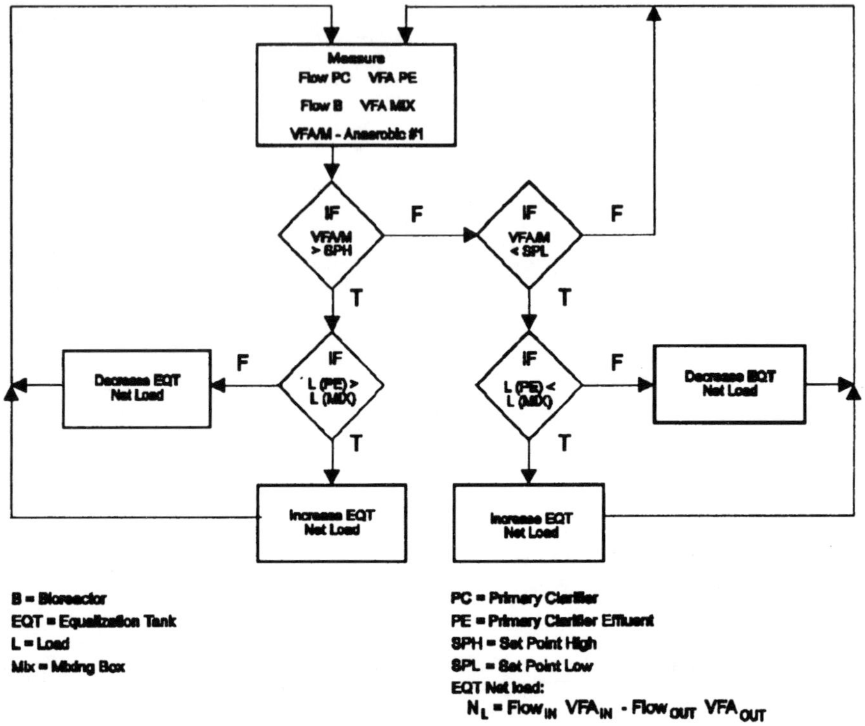

Fig. 8.

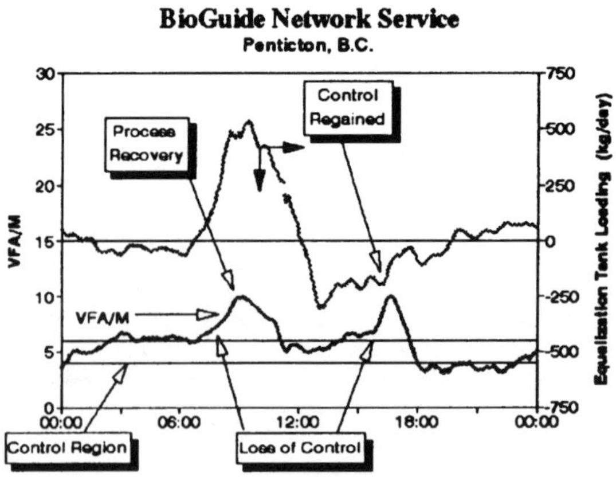

Fig. 9.

Figure 10 compares the results of on-line VFA/M regulation by organic load control (OLC) with the operation of the plant under hydraulic equalization (HDE) without OLC for two different 24-hour periods. From the data it is obvious organic load control is superior to simply hydraulic equalization (HDE). In the

HDE mode, the VFA/M ratio varies by 300% and is within the control region for one half of the day and out of control for the other. These large diurnal fluctuations are detrimental to the smooth performance of the bioreactor. In contrast, the OLC operation maintained the VFA/M ratio at 5.0 + 20% for approximately 85% of the time with two brief excursions of about 80% from set point. The control algorithms continue to be tuned and with additional operating experience, the OLC operation should continue to improve.

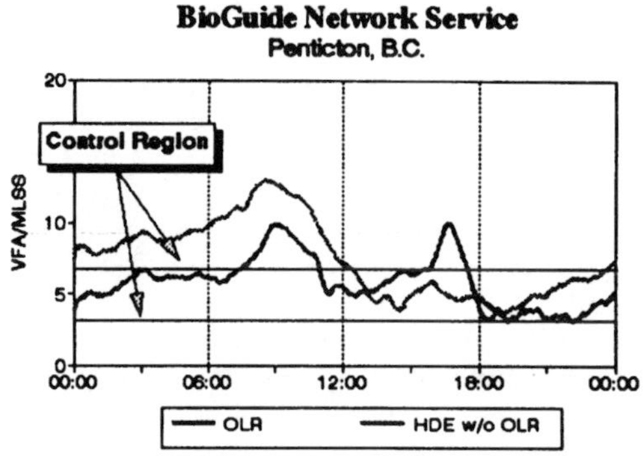

Fig. 10. Comparison of organic load control to hydraulic equalization without organic load control.

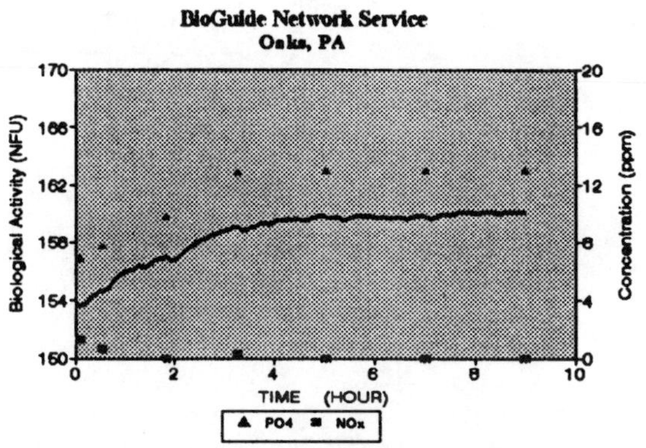

Fig. 11a. Effect of aging on biological activity of RAS.

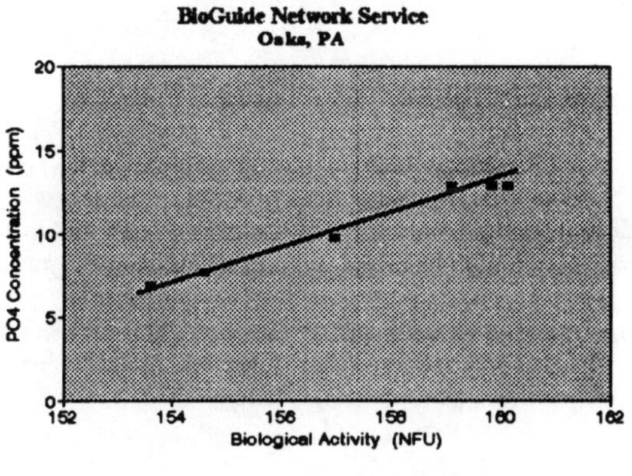

Fig. 11b. Biological activity vs. PO_4 release in RAS.

PROCESS CONTROL RESULTS: CONTROLLING VFA/M BY BIOLOGICAL ACTIVITY OF RAS

It is well known that the release of phosphorus occurs under anaerobic conditions, particularly when the bacterial cells have been induced to take up additional phosphorus, as in the BNR process at the RCID treatment plant. Figures 11a and 11b show laboratory data which correlate the BA response to the levels of P released from return activated sludge over a 24-hour period (Yang *et al.*, 1991).

Chemical analysis and the biological activity as measured by the reduction state showed that the RAS was not in an anaerobic stage, but in an anoxic stage, immediately after it was collected. Nitrate was completely consumed after 3.5 h. Meanwhile, phosphate was released to the medium at a higher rate in the beginning and gradually slowed down afterward.

In BNR operational practice, the age of the sludge blanket in the secondary clarifiers has an important effect on the DO levels and consequently the rate of phosphorus release. Discharge limits notwithstanding, it is also important to recycle sludge with a relatively high internal P level, to provide energy for VFA uptake. Thus, it can be seen that interactive control of the RAS is highly desirable for overall process control.

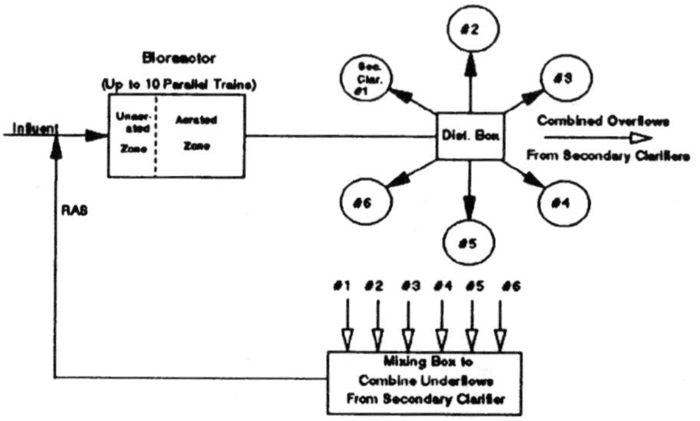

Fig. 12. Flow schematic of process at Ready Creek improvement District.

The aeration bays at the RCID treatment plant are followed by 6 clarifiers which are fed from a common splitter basin (Figure 12). Hydraulic flows to the clarifiers are balanced by controlling valves in the splitter basin. However, it has not been possible to accurately measure the resultant flow differences. Return activated sludge is pumped to a common collection well where it is mixed and pumped to the head of the aeration bays. Pumping cycles are identical for all six clarifiers. Figure 13 shows the results of continuous PO_4 analysis in five of the six clarifiers. Clarifiers 3 and 4, which receive higher flows and consequently longer solids retention, have marked PO_4 peaks.

The instrumentation of the clarifier sludge blanket has proved to be a significant technical challenge. However, our results to date show that BA data track the PO_4 release in Clarifier 3 with an accuracy comparable to that of the laboratory experiments. These preliminary results show that the timing and rate of sludge recycle pumping has the potential to be an important process control variable.

Return activated sludge control has shown early promise as a method to control both phosphorus levels in effluent and the uptake of VFAs in the initial anaerobic stage. Much additional work needs to be performed to take advantage of improved process performance through VFA/M control by regulation of biological activity in the RAS line.

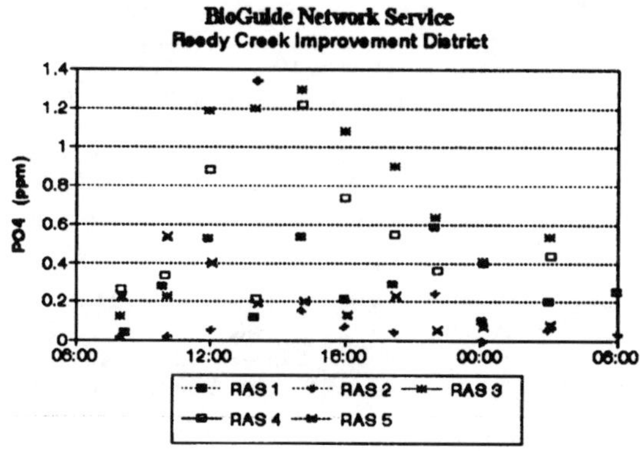

Fig. 13. Diurnal variation in PO_4 in RAS due to sludge aging.

CONCLUSION

Previous work has shown that load balancing using COD has been ineffective at reducing effluent concentrations; these results suggest that, with the high background variability of COD, it may not be an appropriate control signal. Our results from Penticton show that, with respect to the cellular metabolism, VFAs are superior to BOD or COD as a control parameter for the recycle rates, since phosphorus release is and subsequent cell growth is strongly correlated to VFA concentrations. Determination of BOD and COD measurements are slow relative to process operations, and mask important differences between the easily degradable (VFAs) and more refractory components of the influent, e.g. bacterial biomass, cellulose, and vegetable fibers.

REFERENCES

Armiger, W. B., Forro, J. R., Montalvo, L. M., Lee, J. F. and Zabriskie, D. W. (1986). The Interpretation of On-Line Process Measurements of Intracellular NADH in Fermentation Processes, *Chem Eng. Comm. Vol.* **45**, pp. 197-206,

Armiger, W. B., Lee, G. Bordacs, K. and Tracy, K. D. (1990). Automated On-Line Control Strategy for the Optimization of Nutrient Removal Processes, *Instrumentation, Control and Automation of Water and Wastewater Treatment and Transport Systems, (Adv. Wat. Pollut. Control no. 10)* ed. R. Briggs, Pergamon Press, pp. 187-194.

Armiger, W. B., Lee, J. F., Roshong, W. and Doutrich, J. (1990). Keeping a Constant "Eye" on Plant Performance, *Operations Forum, Vol.* **7**, pp. 30-32

Armiger, W. B., Lee, J. F., Maneshin, S. and Yang, X. (1991). Control of Process Performance by On-Line Monitoring of Biological Activity, Presentation at 64th Conf. and Expos. of WPCF, Toronto.

Boon, A. G. and Burgess, D. R. (1972). Effects of Diurnal Variations in Flow of Settled Sewage on the Performance of High-Rate Activated-Sludge Plants, *Water Poll. control,* Vol. **71**.

Beck, M. B. (1986). Identification, Estimation and Control of Biological Wastewater Treatment Processes, *IEEE Proceedings - D Control Theory and Applications. Vol.* **133**(5), pp. 254-264.

Daigger, G. T., Waltrip, G. D., Rom, E. D. and Morales, L. M. (1988). Enhanced Secondary Treatment Incorporating Biological Nutrient Removal, *Journal WPCF.*

Dold, P. L., Buhr, H. O. and Marais G. v. R. (1984). An Equalization Control Strategy for Activated Sludge Process Control, *Wat. Sci. Tech.* **17**(2/3), 221-234

Downing, A. L. (1976). Variability of the Quality of Effluent from Wastewater Treatment Processes and its Control, *Prog. Wat. Tech,* **8**(1), 189-199, Pergamon Press.

Edwards, G. L. and Sherrard, J. H. (1982). Measurement and Validity of Oxygen-Uptake as an Activated-Sludge Process-Control Parameter, *Journal WPCF,* **54**(12), 1546-1552.

EPA Technology Transfer Seminar Publication. (1974). Flow Equalization, Environmental Protection Agency.

Foess, G. W., Meenahan, J. G. and Blough, D. (1977). Evaluation of In-Line and Side-Line Flow Equalization Systems, *Journal WPCF.*

Munksgaard, D. G. and Young, J. C. (1980). Flow and Load Variations at Wastewater Treatment Plants, *Journal WPCF,* **52**(8), 2131-2144.

Oldham, W. K. and Stevens, G. M. (1985). Operating Experiences with the Kelowna Pollution Control Centre, Proceedings of the Seminar on Biological Phosphorus Removal in Municipal Wastewater Treatment, Penticton, B.C., April 17-18.

Randall, C. W., Barnard, J. L. and Stensel, H. D. (1992). Design and Retrofit of Wastewater Treatment Plants for Biological Nutrient Removal, Technomic Publishing Company.

Sadick, T. E., Semon, J. A., Waltrip, G. D., Daigger, G. T. and Newbry, B. W. (1992). Operational Experiences with BNR Facilities in the Chesapeake Bay and Long Island Sound Regions of the USA, Presentation at European Conference on Nutrient Removal from Wastewater, University of Leeds, Leeds, England, September 1-4

Sedlak, R. I. (1991). Phosphorus and Nitrogen Removal from Municipal Wastewater, Lewis Publishers.

Stevens, G. M. and Oldham, W. K. (1992). Biological Nutrient Removal Experience at Kelowna, British Columbia, European Conference on Nutrient Removal from Wastewater, University of Leeds, Leeds, England, September 1-4.

Tetreault, M. J., Benedict, A. H., Kaempfer, C. and Barth, E. F. (1987). Biological Phosphorus Nutrient Removal: A Technology Evaluation, *Journal WPCF*, **58**(7).

Weston, R. F., Inc. (1985). Emerging Technology Assessment of PhoStrip, A/O, and Bardenpho Processes for Biological Phosphorus Removal, USEPA Contract No. 68-03-3055, February.

Wilson, A. W., Stevens, G. M. and Do, P. (1992). Retrofitting Biological Nutrient Removal Processes at Existing Wastewater Treatment Facilities, European Conference on Nutrient Removal from Wastewater, University of Leeds, Leeds, England, September 1-4.

Yang, X., Armiger, W. B., Lee, J. F. and Maneshin, S. (1991). On-Line Measurement of NADH for Monitoring Biological Nutrient Removal Processes, Presentation at 64th Conf. and Expos. of WPCF, Toronto.

Young, J. C., Cleasby, J. L. and Baumann, R. E. (1978). Flow and Load Variations in Treatment Plant Design, *J. Env. Eng. Div. ASCE*, **104**(2), pp. 289-303.

Pergamon

Wat. Sci. Tech. Vol. 28, No. 11-12, pp. 315–323, 1993.

0273–1223/93 $7·00 + 0·00

ON-LINE AND OFF-LINE EXPERT SYSTEMS FOR THE OPERATION OF WASTEWATER TREATMENT PLANTS

Gösta Ladiges and Rolf Kayser

Inst. f. Siedlungswasserwirtschaft, Technische Universitat Braunschweig, P. O. Box 33 29, D-3300 Braunschweig, Germany

ABSTRACT

To provide the operators of wastewater treatment plants with the required knowledge to operate their plant both an on-line and an off-line expert system were developed. The on-line system recognises plant failures from on-line signals and laboratory measurements, immediately starts the diagnosis and gives proper advice displayed on the screen without controlling the plant directly. The off-line system is used for the diagnosis of plant upsets as well as for plant operation and the choice of control parameters. Both expert systems are PC-based and are applied to nutrient removing municipal treatment plants. They contain rules and models for nitrogen- and phosphorus removal, sludge losses, sludge bulking and the on-line control of different plant functions.

KEYWORDS

Activated sludge process; nitrogen removal; phosphorus removal; knowledge based systems; expert systems; operation; control.

INTRODUCTION

The new German requirements for wastewater treatment plants concerning nutrient removal have caused more complications in plant operation. Even the earlier plants that were built for carbon removal (BOD, COD) were not easy to operate, and the newly demanded processes of nitrification, denitrification and biological and chemical phosphorus removal make it much more difficulty to get an overview of a plant and all the relationships between the different processes in all situations. But this is necessary to get a stable plant operation. Many parameters have to be adjusted during the varying circumstances, and also it is necessary to react rapidly and correctly at plant failures, so that the given effluent requirements are not exceeded.

The German effluent standards as listed in Table 1 are very stringent. The values are the minimum requirements. The local water authorities may set stronger limits, e.g. for small rivers. The values have to be kept within the so called "qualified grab sample" or in a two hours composite sample, which means that there is little time to lose in overcoming the effluent requirements.

Especially at smaller treatment plants that are operated by only one or two (non-academic) operators difficulties can arise. The operators gain most of their experience by practical plant operation, and that may be sufficient for most of the normally occurring problems. But in the case of unexpected or unusual failures or when an experienced person is not available, the upsets are recognised and removed very late. As a result the required effluent concentrations are exceeded.

TABLE 1. German Effluent Standards for Nutrient Removal

Ammonia-nitrogen		< 10 mg/l
Inorganic nitrogen (NH_4-N + NO_2-N + NO_3-N)		< 18 mg/l
		or 70% nitrogen removal
	(> 5,000 p.e. at temperatures > 12°C)	
Total phosphorus	(> 20,000 p.e.)	< 2 mg/l
	(> 100,000 p.e.)	< 1 mg/l

But also the large treatment plants are faced with the same problems. An increasing number of plants are operated by engineers who know some of the background of wastewater treatment, but they also have to react fast and correctly, which is not always easy with the complex processes that depend on one another.

Modern process control systems give a helpful overview with the display of measured values and some alarms. Further help by finding reasons for upsets or for their removal or even a support for the choice of control parameters is not given. The operators are dependent on their experience in estimat-ing the right values, on different handbooks and maybe on the use of some formula or programs.

To provide the operators with the missing and difficult available knowledge, the Institute of Sanitary Engineering in Braunschweig started at the end of 1988 on the development of an expert system. In this project a system was developed that recognises plant failures from on-line and laboratory measurements, immediately starts the diagnosis and gives proper advice. This expert system has been in operation on a treatment plants since early 1991 and has been permanently improved since then.

In a second project, two expert systems for off-line use are under development. They will be used for the diagnosis of plant upsets as well as for plant operation and choice of control parameters.

This paper presents both expert system projects. It will discuss the aims within the projects, the way of developing the systems and of course the contents of the knowledge bases. It will be based on both common facts about expert systems and other projects with similar aims. There is a rich amount of literature concerning this field.

ON-LINE SYSTEM

A lot of plant troubles only lead to major upsets because they are recognised too late or the operators do not react in the right way. In order to give support the Institute of Sanitary Engineering in Braunschweig started developing an expert system in 1988. A major aspect was to achieve a very small time delay in reaction. Therefore the system had to be connected to the on-line signals but should be able to work with laboratory values as well.

The project was supported by a grant of the German Research Ministry (BMFT). It had a duration of three years. Within that time the system was developed in co-operation with computer scientists, the Institut fuer Betriebssysteme und Rechnerverbund in Braunschweig, Prof. Langendoerfer (Kayser *et al.*, 1992). After the project was finished the expert system was improved by our institute.

Prerequisites for the expert system were:

to enable the use on smaller treatment plants the hardware should not be too expensive, so the system was developed on a PC;

the user interface should be easy enough to be accepted by operators without computer experience;

the expert system should not control the plant directly but give advice and information to the operators.

The expert system was developed on a municipal treatment plant for 35,000 p.e. representative for German plants of that size.

Plant Description

The treatment plant of the city of Salzgitter-Bad serves a population of about 35,000 p.e. The sewerage system of the 600 hectare watershed consists of about 65% combined sewers and 35% separate sewers. At storm water flows, up to three times the dry weather flow is treated biologically.

In its major parts the treatment plant consists of two biological reactors (V = 2,800 m^3 each) and works without primary settling. Each reactor is equipped with three mammoth rotors (37 kW each) that are controlled according to the on-line signals from the NO_3 and NH_4 monitors. The final settling tanks have a diameter of 32 m. In figure 1 a flow scheme of the plant is shown. The loading data are listed in Table 2.

TABLE 2. Loading Data of the Salzgitter-Bad Wastewater Treatment Plant

Wastewater flow		8,030 m^3/d
COD	2,860 kg/d	356 mg/l
BOD_5	970 kg/d	121 mg/l
Suspended solids	1,520 kg/d	189 mg/l
Total nitrogen	353 kg/d	44 mg/l
Total phosphorus	48 kg/d	6 mg/l

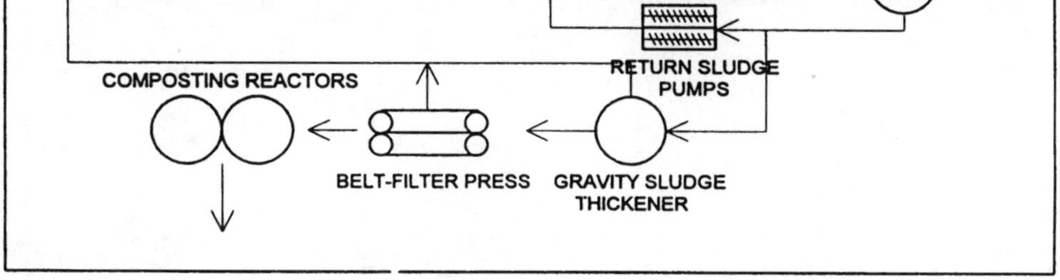

Fig. 1. Flow scheme of the Salzgitter-Bad wastewater treatment plant.

The treatment plant is able to fulfil the nitrogen requirements during the whole year, i.e. that the nitrification and denitrification processes are stable at measured temperatures down to about 8°C. Nevertheless some problems with the effluent concentrations occur as in every treatment plant. Increased and sometimes very high concentrations of ammonia and also smaller ones up to very high sludge losses via the final settling tanks are found. These problems are the main reasons for implementing the knowledge bases.

Data Collection and Operation Mode of the Expert System

For the recognition and diagnosis of these problems the system uses the current and normal plant signals such as the effluent and return sludge water flow, nitrate as the signal for the aeration control, digital signals from different aggregate conditions and of course the laboratory values. To enable the system to react immediately when sludge losses occur a turbidity probe was installed in the outlet of the final clarifiers.

A PC-based data collection and display system gathers the analogous and digital signals and the laboratory measurements and leads them to the expert system computer (Fig. 2). Also the expert system computer is equipped with a data display mode which will be interrupted as soon as a diagnosis is needed (Fig. 3). The expert system checks if the measured values show an unusual situation of the plant, i.e. pre-set values are exceeded or the data curves do not fit with the expected course. In such a case the reasons will be investigated and the diagnosis will be displayed on the screen together with possible remedies.

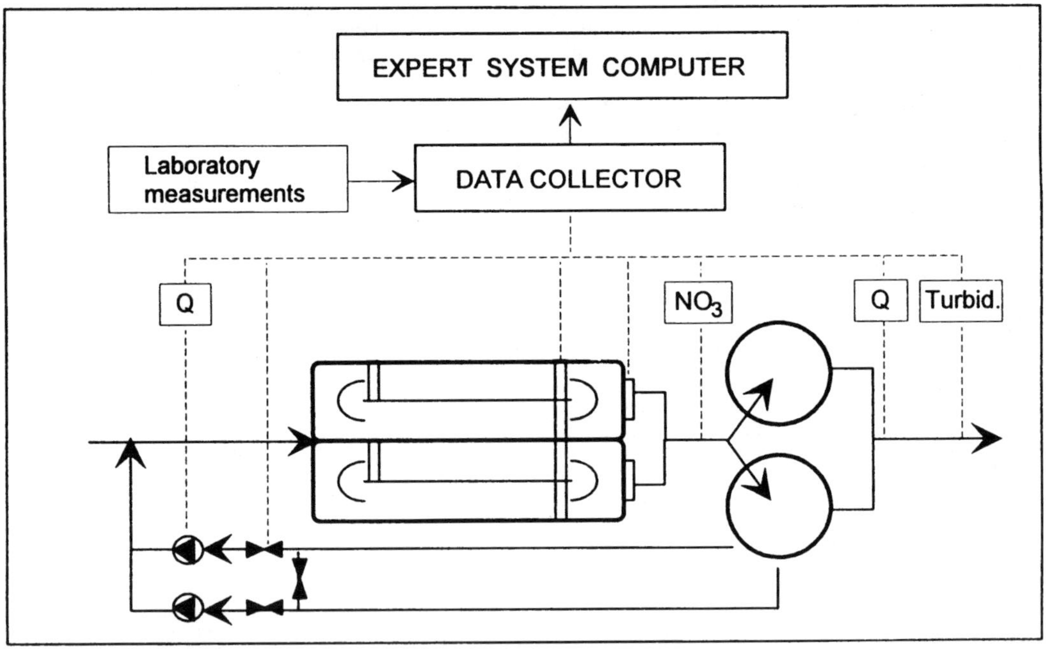

Fig. 2. Configuration of data collection and expert system at the wastewater treatment plant of Salzgitter-Bad.

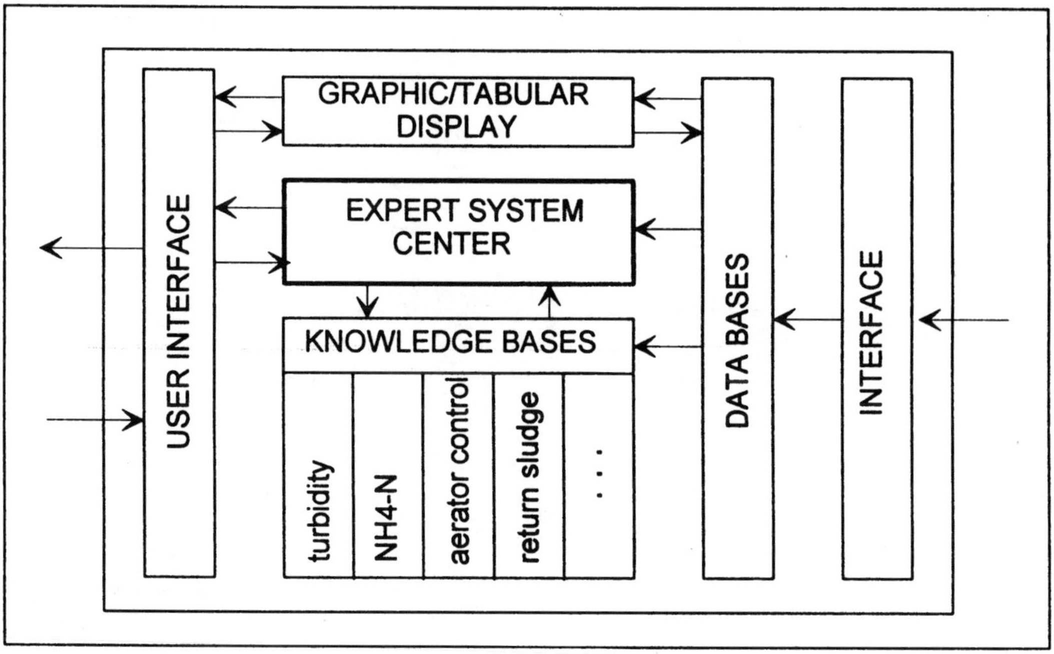

Fig. 3. Structure of the on-line expert system.

Knowledge Bases

To give support on the mentioned problems, four knowledge bases were developed:

1. Diagnosis of nitrification problems according to the laboratory value of NH_4-N.

2. Control of the sludge separation in the final clarifiers according to the on-line signals from the turbidity probe in the effluent.

3. Control of the aeration control according to the NO3 and NH4 monitor signals and the condition sig nals from the switch-able aerators.

4. Control of the return sludge pumps according to the current water flow-, pump condition- and valve signals to keep an eye on the return sludge transport and watch one of the reasons for sludge loss before it occurs.

When the system finds a disturbance it examines the rules in the relevant knowledge base. It checks all the current signals, calculates the needed values and if necessary requests some information at the screen. The relevant diagnosis is given at the screen containing both the possible reasons and the required remedies to get back to normal conditions in the plant.

The ammonia knowledge base is used as an example here. The structure of that knowledge base is shown in Figure 4.

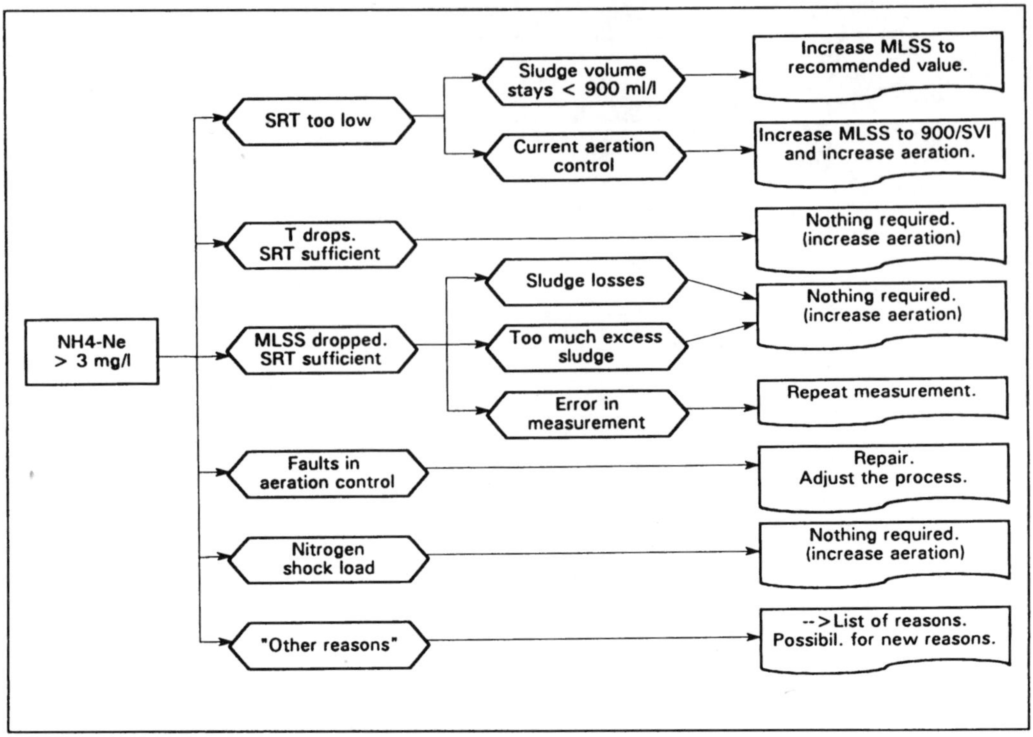

Fig. 4. Structure of the ammonia knowledge base.

The knowledge base reacts on the input of the ammonia laboratory value. Since the NH4-N effluent concentration seldom exceeds 3.0 mg/l the expert system starts working when the ammonia is higher than this.

The estimation of the current and required aerobic solids retention time (SRT) is based on the approach of Kayser (1987). The system works with constant loading data since this is applicable for this plant. This has the main advantage that the SRT can be calculated rapidly, especially without waiting for the BOD_5. When the system finds that the SRT is too low it recommends the calculated value for MLSS. This can only be done up to the point when the MLSS cannot be kept within the plant at storm water flows (plant specific sludge volume of 900 ml/l). In this case the system recommends a smaller MLSS enhancement together with an increased aeration.

Sudden temperature drops and greater sludge losses can also cause nitrification problems although the SRT is sufficient. Aeration faults (pre-set durance of aerator stops) can as well as nitrogen shock loads result in high ammonia effluents.

If the system does not find any reason in the implemented knowledge bases it gives the user the possibility to put in his/her an own relevant reason and its remedies. This piece of information is stored in the system and recalled in the next session at that point. In this way an interface is given to collect new coherence at the plant which later can be modified and implemented as new rules. This possibility is also given within the other knowledge bases.

The structure of the other knowledge bases is similar except for the fact that the other ones are connected to the on-line signals which means that they can react directly without any time delay. Details are not given here, they are available in the final report for this project (Kayser *et al.*, 1992).

System Development and Applied Tools

The expert system has been developed in co-operation between a sanitary engineer as the expert and a computer scientist as the knowledge engineer. The knowledge was developed within several sessions and written down. In the following session it was checked again and changed if necessary. Except for the first sessions this correction took about half of the time. After the knowledge had been collected it was implemented by several programmers.

Within the first of the three project years the first version of the knowledge base was developed. Within this time several software tools were examined and suitable tools were chosen. In the second year the knowledge bases were implemented and improved. At the beginning of the third year (early 1991) the system was installed at the plant and tested there for the first time. The occurring software and hardware errors and problems are removed meanwhile. The development of a test environment was very helpful to examine the system. This enabled the user and the developer to test the system behav-iour in all upset situations independent from the plant operation (Ladiges, 1992).

One of the aims in the project was to examine which tools were the most suitable ones for the develop-ment of an on-line expert system: a procedural programming language (C and C++, especially the on-line flexibility) or a commercial shell (Nexpert Object, partially programmed by the expert and ready made expert system functions). Both methods were successful and required about the same amount of time. Since the development of the version in C was started earlier it was applicable before the shell version. Therefore this was applied to the plant and has been improved after finishing the project.

OFF-LINE SYSTEM

In early 1992 a new project was started to develop an expert system for off-line consulting. This system works without data connection and can be started at occurring plant problems. It works with two main tasks: one is to give support by finding and solving current and simulated plant problems, the other is to find the required control parameters and strategies e.g. the choice of the optimum solids retention time SRT or to give recommendations and required values when some tanks have to be taken out of operation (Fig. 5).

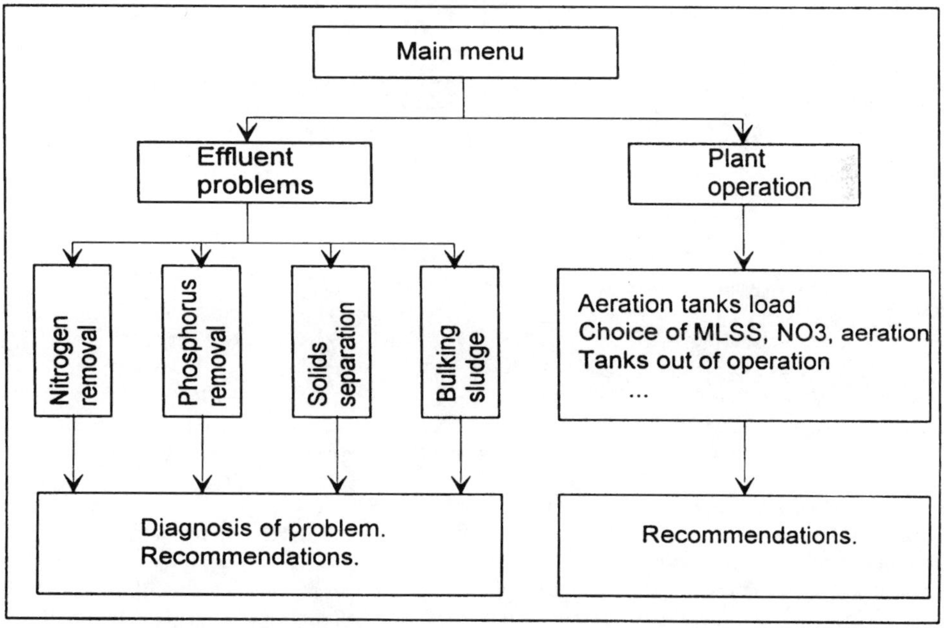

Fig. 5. Structure of the off-line expert system.

The expert system is designed for the main problems of nutrient removing treatment plants. It contains different problem solving parts. In the next step within the project these parts will be connected among one another to form a model of the current plant as a whole so that various important effects caused by a single action can be examined. The system contains knowledge bases for

Nitrification

Denitrification

Phosphorus removal

Sludge separation in the final clarifiers

Bulking sludge

The expert system combines heuristic rules with deterministic models for, e.g. nitrification and denitrification, inhibition or alkalinity balances. Some models are applicable at any treatment plant e.g. the NH_3 or HNO_2 inhibition of the nitrification, others have to be adjusted to the current plant, e.g. models for nitrogen removal. This enables the system not only to give qualitative advice but also required and useful control values for problem solving and plant operation.

The system is developed with the shell Nexpert Object. This method enables the expert to program the first versions of the knowledge bases by himself. This avoids the known problems by forwarding the knowledge to the programmers. The shell allows a "rapid prototyping" to give the user a rapid view of the system and its way of working. This gives the operators the opportunity already in an early stage of progress to give their opinions and new ideas to the contents of the system. The user interface is developed with Windows ToolBook and put on top of the "rapid prototyping"-interface afterwards.

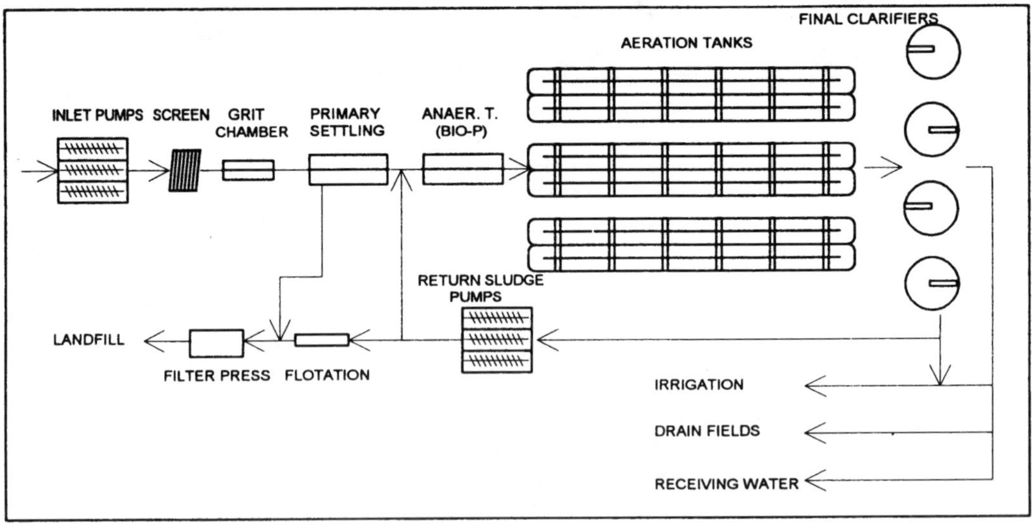

Fig. 6. Flow scheme of the wastewater treatment plant in Braunschweig

The system will be applied at the wastewater treatment plants both in Salzgitter-Bad (the same as for the on-line system) and Braunschweig. The latter serves a population of about 270,000 p.e. and is built for nitrogen and phosphorus removal. It consists of three aeration tank units and works with primary sedimentation and biological phosphorus removal (Fig. 6). After the treatment the wastewater in different and changing amounts is irrigated and pumped to drain fields or into receiving water depending on the effluent requirements. In summer the nutrients are needed for agriculture purposes and in winter they have to be removed. The varying amount of water that cannot be used in agriculture but is discharged into the river has at all times to fulfil the requirements listed in Table 1. This shows that the plant operation has to be changed very often. The expert system will be a very suitable help for the operators for this.

At the plant in Braunschweig the expert system is combined with the operational manual (Betriebsanweisung) that is prescribed for each wastewater treatment plant in Germany (ATV, 1984). This contains possible reasons, recommendations and, if available, alternative operational strategies for troubles at the different parts of the plant and some major upsets, e.g. total electricity breakdowns. Both programs are developed with the same shell and are implemented on the same PC. The aim of both programs is the same, just the field of working is different. Both programs are able to call parts of the other one if needed, so that one is a good supplement for the other.

FUTURE WORK

Many expert systems have been developed and a lot of current systems are under development to provide the operational staff of wastewater treatment plants with the required knowledge for the operation of their plants. The progress is promising. The bottleneck of the investigations is the collection, preparation and combination of the knowledge.

Systems have to be developed that are applied at the plants. They have to be improved and enhanced according to the gained experiences in exchange with the operators, both in the manner of knowledge representation and in the possible fields of application.

Some systems are already very useful for the treatment plants. Nevertheless they are designed especially for one certain plant or they can only solve a few of the different kinds of problems. By further development, the power of the expert system will be increased, e.g. by generalisation of present approaches, the combination with efficient models or by giving the systems a kind of self adaptation to the plant. It is also possible to combine the knowledge based systems with other techniques of artificial intelligence such as fuzzy logic or neural networks.

Consulting systems that can be started at current problems are a great benefit for the operators. They can be included within more powerful systems that judge the situation at the plant on their own according to either laboratory measurements or on-line signals and in this way will give advice for plant operation. As one of the future steps the expert systems can be equipped with own competence, e.g. to modify control strategies or to take measures depending on some situations.

The development of expert system has not reached the end yet, but it is a long way to the often discussed "totally automated wastewater treatment plant". Good staff can only be replaced by good staff.

ACKNOWLEDGEMENT

The author thanks the BMFT and the state of Lower Saxony for financial contribution.

REFERENCES

ATV (1984). Arbeitsblatt A 124, Dienst- und Betriebsanweisungen fur das Betriebspersonal von Klaranlagen, Gesellschaft zur Forderung der Abwassertechnik, St. Augustin.

Kayser, R. (1984). Bemessung von Belebungsanlagen zur Stickstoffelimination, Veroffentlichungen des Institut fur Stadtbauwesen, Heft 42, 67-83, Braunschweig, ISSN 0341-5805.

Kayser, R., Langendorfer,H., Hofmann, M. and Ladiges, G. (1992). Expertensysteme als Hilfsmittel beim Klaranlagenbetrieb, final report of a BMFT project, Forderkennzeichen 02 WA 8828, Institut fur Siedlungswasserwirtschaft, Institut fur Betriebssysteme und Rechnerverbund, Braunschweig.

Ladiges, G. (1992). Entwicklung eines Online – Expertensystems zur Betriebsunterstutzung auf Klaran-lagen, Conference "Einsatz von Expertensystemen zur Unterstutzung des Klaranlagenbetriebes", Institut fur Wassergute und Abfallwirtschaft, TU Wien, together with Symbios and aqua-system, Vienna.

FUZZY CONTROL

Pergamon

Wat. Sci. Tech. Vol. 28, No. 11-12, pp. 325–332, 1993.

0273–1223/93 $7·00 + 0·00

INTELLIGENT OPERATION SUPPORT SYSTEM FOR ACTIVATED SLUDGE PROCESS

Shoji Watanabe*, Kenji Baba*, Mikio Yoda**, Wen-Chih Wu**, Ichiro Enbutsu*, Masakatu Hiraoka*** and Kazushi Tsumura***

** Hitachi Research Laboratory, Hitachi Ltd., Omika-cho 7-1-1, Hitachi-shi, Ibaraki, 319-12, Japan*
*** Omika Works, Hitachi Ltd., Omika-cho 5-2-1, Hitachi-shi, Ibaraki, 319-12, Japan*
**** Faculty of Engineering, Kyoto University, Yoshida-honcho, Sakyo-ku, Kyoto-shi, 606, Japan*

ABSTRACT

We have proposed the Intelligent Operation Support System (IOSS) for sewage plants, which can be applied to bulking prediction and control for the activated sludge process. The IOSS can provide guidance and control messages to operators on actions to take based on various kinds of integrated information including on-line process data, image signals on microorganisms, heuristics on plant control and data analyzed manually. To get on-line microorganism image information, a high resolution submerged microscope was developed and applied in a full scale plant. By combining image data with heuristics, the IOSS allows prediction of the symptoms of abnormal phenomena like bulking occurrence, and then provides ways to control the abnormalities thus decreasing the operators' work loads. Furthermore, diagnosis methods for detecting sensor troubles are proposed. Simulation results indicate the IOSS is effective.

KEYWORDS

Sewage plant; activated sludge; filamentous microorganism; zoolegal microorganism; floc; bulking; bulking index; support system; image processing; submerged microscope; knowledge engineering; fuzzy; neural network; monitoring; diagnosis; control; guidance.

INTRODUCTION

Sewage plant operators are required to observe many items covering water qualities and microorganisms conditions. In particular, observing microorganisms conditions in the activated sludge is very important to keep plant conditions satisfactory. Skilled operators grasp normality of abnormality of treatment conditions based on their experiences, learned data and visual observation results. Their abilities have ensured operators are important to satisfactory plant operation, but the number of skilled operators in Japan will decrease during the next decade, while the number of sewage treatment plants will increase. Then some measures must be developed to cover this gap.

In a previous work, we proposed the Intelligent Operation Support System (IOSS) for water and wastewater treatment processes. It integrates a physical model, image processing, knowledge engineering including fuzzy theory and a neural network. The IOSS was applied to coagulant injection control for a water purification plant. Now it is applied to bulking control for the activated sludge process.

INTELLIGENT OPERATION SUPPORT SYSTEM

Figure 1. is a block diagram of IOSS which is composed of a management system and control systems. These include workstations (WS) and image processing apparatus. The image processing apparatus receives signals from the newly developed high resolution submerged microscope. The guidance software was installed in the WS connected with the information LAN. The WS presents appropriate messages to improve process conditions based on inference, which uses on-line process data including image data of the microorganisms, off-line process data gathered manually, and production rules including rules extracted from historical data using artificial neural networks (ANN).

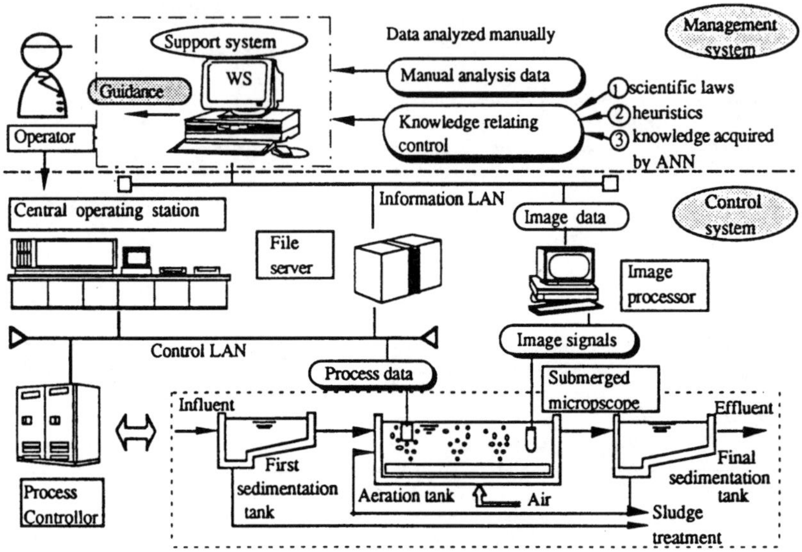

Fig. 1. Intelligent operation support system for activated sludge process.

Figure 2 shows functions of the IOSS. The database has on-line process data on the inflow rate, MLSS, aeration flow rate, microorganism image data and so on, and off-line data on the manually analyzed water qualities and visual observation results. The knowledge base was constructed by gathering scientific laws, heuristics and production rules extracted from historical data using the ANN.

Both the data and knowledge bases were used for inference for monitoring and diagnosing the activated sludge process, in which forward-inference was applied to the diagnosis of the sensor trouble and fuzzy-inference was applied to monitoring of bulking occurrence at pre-determined regular intervals. The type identification of filamentous microorganisms for precise inference and statistical analysis of variable data were done using SACCESS (Statistical Analysis and Computer Control of Environmental SystemS), which was developed at Kyoto University. SACCESS was installed in the IOSS for supplementary support. Guidance messages on bulking occurrence prediction and bulking control strategies were represented on the CRT based on these functions. Operators could select the best control action.

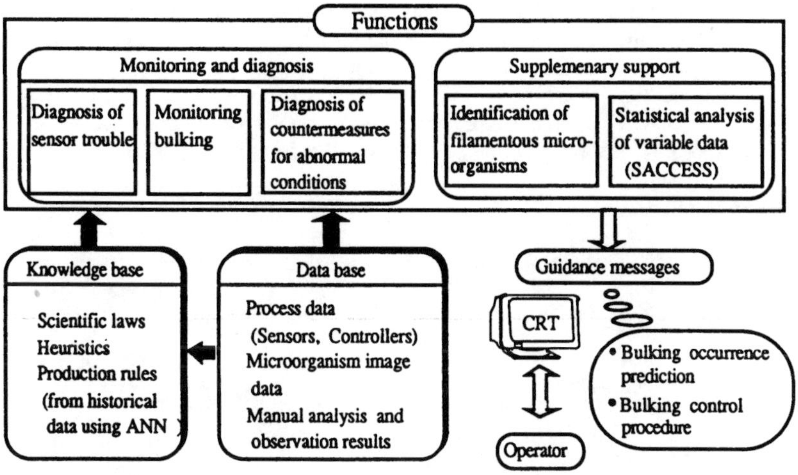

Fig. 2. Functions of intelligent operation support system for bulking control.

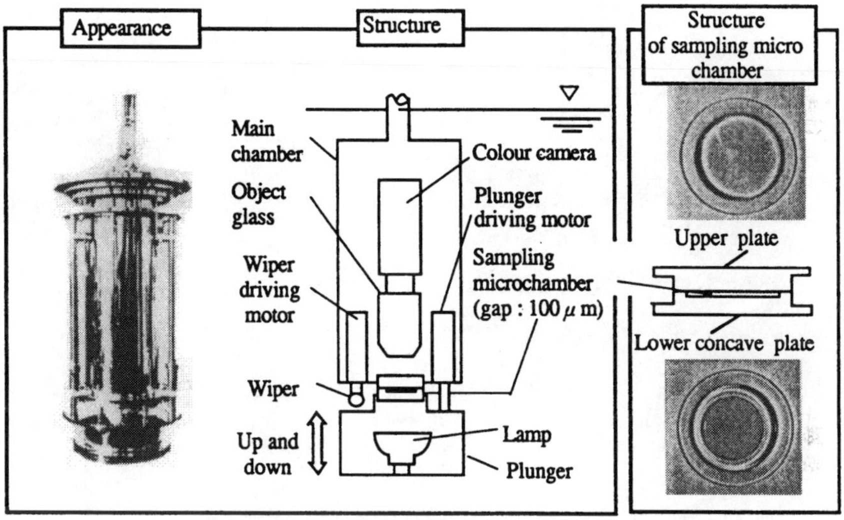

Fig. 3. Structure of high resolution submerged microscope.

ON-LINE IMAGE PROCESSING

High Resolution Submerged Microscope

In the IOSS the high resolution submerged microscope for the on-line image processing system was newly developed to monitor conditions of the activated sludge microorganisms. Figure 3 illustrates in structure, which allows on-line image taking of microorganisms in activated sludge contained in the aeration tank. The submerged main body is made of stainless steel and consists of a main chamber and a plunger, which are mounted vertically. A colour television camera with object glass (extension rate=4), a wiper driving motor and a plunger driving motor are contained inside the main body. The plunger has an illumination light set inside it. The plunger can move up and down in the water to get a very small amount of liquid sample. Its micro sampling chamber is set between the main body and the plunger. Its lower concave plate can be separated from the upper plate by the up-down movements of the plunger. This separating movement introduces liquid sample fro the aeration tank. This cycle time is about 90 s. The gap between the upper and lower plates is 100 μm, which nearly equals the separation between a coverglass and a slideglass when they are set up on a conventional type microscope. The main body has a wiper which washes both the upper plate and the lower concave plate. These mechanisms allow a high resolution image to be kept as with a conventional type microscope without fluctuation of liquid flow in the aeration tank.

Image recognition

Figure 4 shows a gray image obtained by the high resolution submerged microscope and a processed image. The latter was obtained by an image processing algorithm for recognizing flocs and filamentous microorganisms, which was developed previously (Enbutsu *et al.*, 1993).

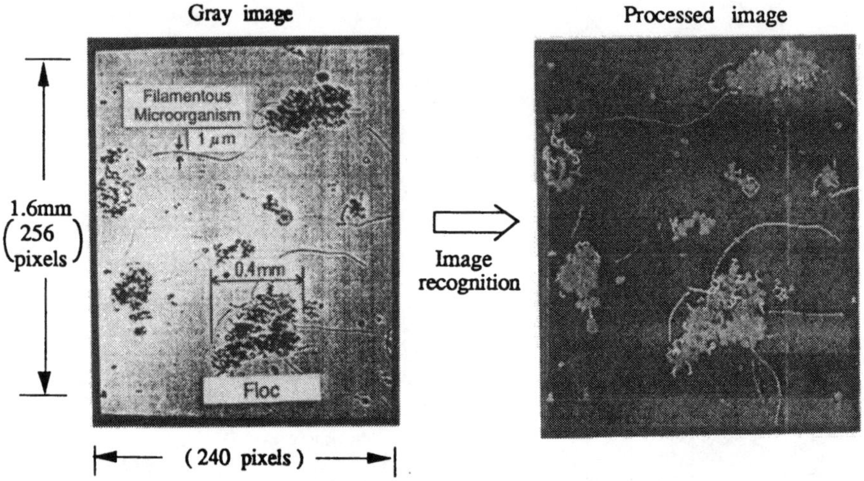

Fig. 4. Example of images from on-line system.

Bulking Index

The bulking index, BI was proposed previously (Watanabe *et al.*, 1990) to predict the occurrence of bulking phenomena on the activated sludge. The BI is defined as follows:

$$BI = \frac{R^2}{F/V} \quad (1)$$

where: BI: Bulking index
R: Floc size (μm)
F: Filamentous microorganisms length (m/mg MLSS)
V: Floc volume

The BI monitoring is effective for diagnosis of the settling ability of activated sludge. The production type rules were extracted from monitoring results of the BI as well as the length of the filamentous microorganisms, organic load, dissolved oxygen concentration and sludge age. These rules were used for inference in the IOSS.

KNOWLEDGE BASE

Knowledge Source

Figure 5 represents the knowledge base construction procedure carried out in this study. Various kinds of information which include heuristics, image measurement data, physical models, reference values, literature, manual analysis data and visual observation results were provided for the basic source knowledge.

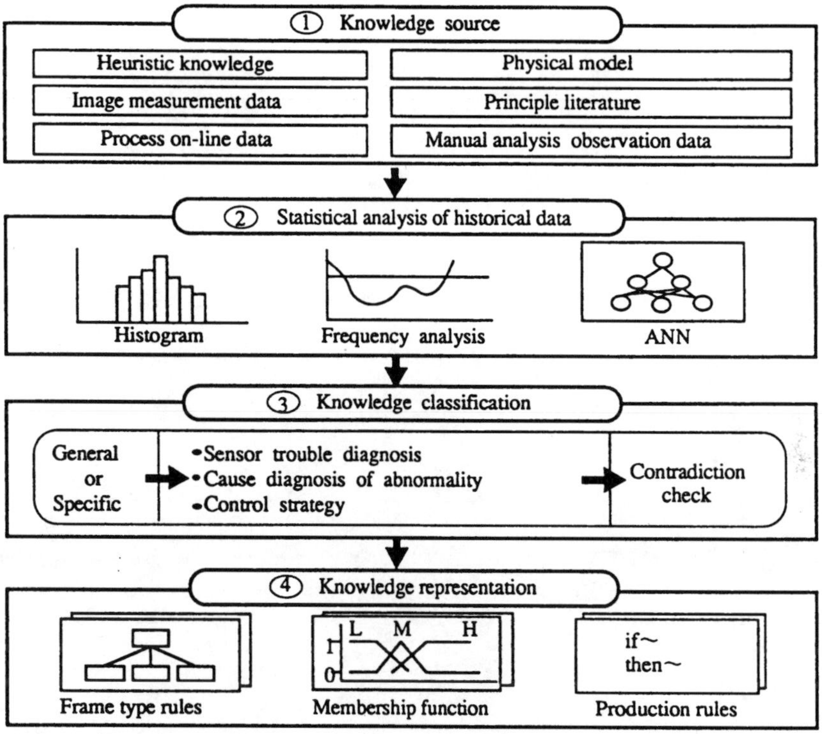

Fig. 5. Knowledge acquisition and representation.

Statistical Analysis of the Historical Data

Daily or seasonal changes in the historical data were examined by statistical analysis such as histograms and frequency analysis. SACCESS was applied to the statistical analysis. Furthermore, production type knowledge describing the relationship between various kinds of data were extracted from historical data using the ANN. The extraction algorithm has already been proposed by the authors (Baba *et al.*, 1990). A new algorithm for fuzzy rule extraction is reported elsewhere (Enbutsu *et al.*, 1993).

Classification of Knowledge

The knowledge base acquired from the knowledge source was divided into two parts based on whether it was general or specific. After this division, knowledge was further classified into several uses such as sensor trouble diagnosis of an abnormality and control strategies. The knowledge was constructed to get some functional modules. Then it was checked for contradictions while comparing it with the experimental results.

Knowledge Representation

The arranged knowledge was represented as production type if~then~ rules. The membership functions used in the fuzzy reasoning were defined through man-machine interface. Knowledge relating criteria values to judge process conditions was represented in frame type rules. Fuzzy reasoning and forward-inference were applied to get operation guidance. In the present system, messages regarding bulking control strategies were represented on the CRT.

FULL SCALE PLANT EXPERIENCE

Microorganism Image Measurement

The optimum location for the submerged microscope was studied experimentally in a full scale sewage plant. Three location points along the aeration tank line were examined. Figure 6 represents the image measurement results, which were acquired every two hours during a 24 h period. Seventy-two images per one sample were processed to get one datum. The variation range decreased according to the flow direction through the aeration tank, while the average values were different at each measuring location. This suggested that the optimum point is the outlet of the aeration tank because the data variation is smallest there. The reliable number of submerged microscope images processed was investigated at the outlet of the aeration tank. It is more than 25.

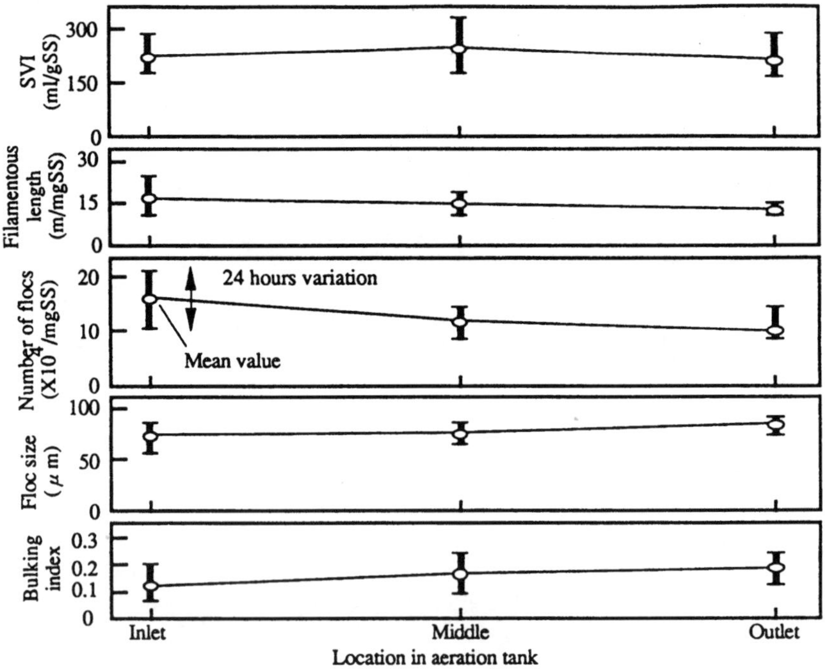

Fig. 6. Variation of image measurement data.

Bulking Reasoning and Control Guidance

Figure 7 shows simulation results for the IOSS and the conventional expert system, using full scale plant data. The certainties of bulking occurrence were inferred in both cases, in which only process data were used for the inference expert system in this study. Figure 7 emphasizes the effectiveness of the IOSS on the bulking certainty prediction.

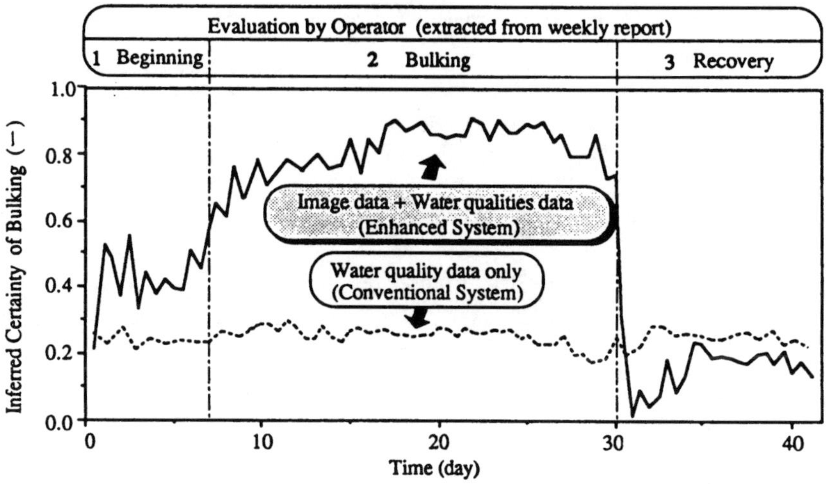

Fig. 7. Simulation results using full-scale plant data.

Table 1 shows guidance and control messages obtained by the IOSS. These messages were considered appropriate by operators in the full scale plant. Inference in the IOSS is effective for predicting and controlling bulking occurrence.

TABLE 1. Guidance and Control Messages (123 indicate the terms shown in Figure 7)

Conditions	1	2	3
Guidance Messages	·Filamentous amount increase ·Bulking symptom	·Large amount of filaments ·Bulking occurrence	·Filaments amount decreases ·Recovery
Control Messages	·Increase air flow rate ·Identification of filaments type	·Increase sludge withdraw rate ·Increase sludge return rate	·Continue present operations

Diagnosis for Sensor Troubles

Figure 8 shows an example diagnosis for sensor troubles. Sensor troubles were diagnosed based on the average values, standard values, upper and lower limit values and changing rates obtained by statistical analysis using historical data. Furthermore, periodical change patterns during a day and the concurrent values in the same type sensors on other lines were considered for the diagnosis. When abnormal sensors were found, data from them were eliminated from that used for inference. This improved the inference accuracy.

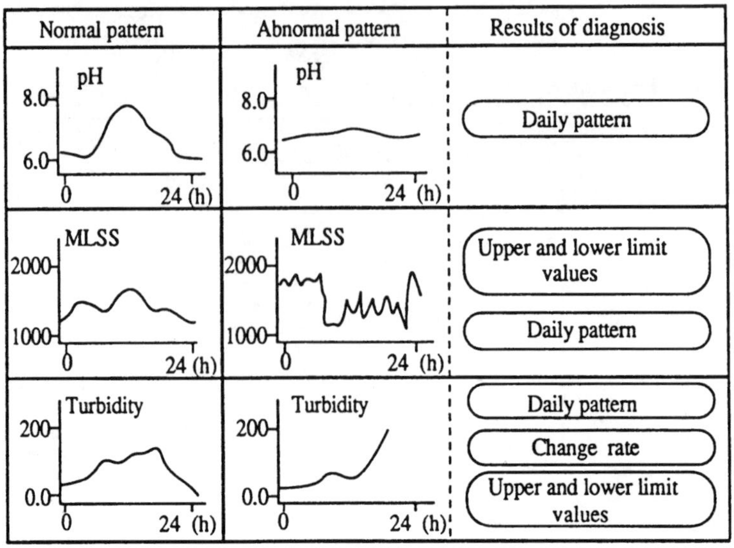

Fig. 8. Example diagnosis for sensor troubles.

SUMMARY

The Intelligent Operation Support System (IOSS) for sewage plants has been proposed, in which the IOSS was applied to bulking prediction and control of the activated sludge process. The IOSS can provide guidance and control messages on what operators should do based on various kinds of information integrating on-line process data, image signals on microorganisms, heuristics on plant control and data analyzed manually. To get on-line microorganism image information, the high resolution submerged microscope was developed and applied in a full scale plant. By combining these image data with heuristics obtained from various kinds of knowledge sources, the IOSS enables prediction of symptoms of abnormal phenomena like bulking occurrence and then provides ways to control them, decreasing the operators' work loads. Furthermore, diagnosis methods for detecting sensor troubles were proposed. Simulation results suggested the effectiveness of the IOSS.

REFERENCES

Baba, K., Enbutsu, I., Matuzaki, H. and Nogita, S. (1990). Intelligent support system for water and sewage treatment plants which includes a past history learning function - coagulant injection guidance system using neuralnet algorithm. *Instrumentation Control and Automation of Water and Wastewater Treatment and Transport System* (*Adv. Wat. Pollut. Control* no. 10) ed. R. Briggs, Oxford, Pergamon, pp. 227-234.

Watanabe, S., Baba, K., Matuzaki, H., Hiraoka, M. and Tsumura, K. (1990). Operating conditions diagnosis system for activated sludge processes based on microorganisms image information. *Instrumentation Control and Automation of Water and Wastewater Treatment and Transport System* (*Adv. Wat. Pollut. Control* no. 10) ed. R. Briggs, Oxford, Pergamon, pp. 683-690.

Enbutsu, I., Baba, K., Hara, N., Waseda, K. and Nogita, S. (1993). Integration of multi AI paradigms for intelligent operation support system – fuzzy rule extraction from a neural network – *Wat. Sci. Tech.* **28**(11/12) (this volume).

Fujiwara, T. (1988). Research on automatic control of municipal incineration plant by use of autoregressive model, Doctoral thesis, Kyoto University (Japan).

Pergamon

Wat. Sci. Tech. Vol. 28, No. 11-12, pp. 333–340, 1993.

0273–1223/93 $7·00 + 0·00

INTEGRATION OF MULTI AI PARADIGMS FOR INTELLIGENT OPERATION SUPPORT SYSTEMS – FUZZY RULE EXTRACTION FROM A NEURAL NETWORK

I. Enbutsu*, K. Baba*, N. Hara**, K. Waseda*** and S. Nogita†

** Hitachi Research Laboratory, Hitachi Ltd., 7-1-1 Omika-cho, Hitachi-shi, Ibaraki-ken, Japan*
*** Omika Works, Hitachi Ltd., Omika-cho, Hitachi-shi, Ibaraki-ken, Japan*
**** Electrical Engineering Dept., Hitachi Ltd., 6, Kanda-Surugadai 4, Chiyoda-ku, Tokyo, Japan*
† Department of Urban System Engineering, Ibaraki University, 4-12, Nakanarusawa-cho, Hitachi-shi, Ibaraki-ken, Japan

ABSTRACT

Some artificial intelligence (AI) paradigms have been applied to water and wastewater treatment systems. An artificial neural network (ANN), which can learn historical data of a plant, provides operational guidance for plant operators, and a fuzzy system (FS) provides a framework to put operators' heuristics into practical use as fuzzy rules in a fuzzy rulebase. In application, however, the practical problems remain that the ANN is a blackbox model which is unfamiliar to plant operators, and the FS usually requires much time-consuming work by system engineers and operators for knowledge acquisition and rulebase maintenance. The authors think that integration of the paradigms can give appropriate solutions to these problems. As one method which realizes such integration, an automatic fuzzy rule extraction method using an ANN is proposed. Simulation results of the proposed method using full-scale plant data demonstrated that an FS whose rulebase was modified automatically with extracted rules had better performance than a conventional FS whose rulebase included only operators' heuristics. This effect is thought to be realized by enhancement of knowledge source with the proposed method.

KEYWORDS

Artificial intelligence; artificial neural network; fuzzy system; rule extraction; knowledge acquisition; operation support; water and wastewater treatment; coagulant injection.

INTRODUCTION

Water and wastewater treatment plants are subject to changes due to unpredictable disturbances from their external environments. Therefore, it is difficult for conventional methods to formulate a general operation strategy which covers a wide range of operational conditions. To get better operating performance, some artificial intelligence (AI) paradigms, such as an artificial neural network (ANN) and fuzzy theory, have been applied to their operation support systems. Early studies (Yagishita *et al.*, 1984; Baba *et al.*, 1989) demonstrated that these AI paradigms were quite promising to our field. The authors proposed the concept which covered several AI paradigms (Intelligent Operation Support System; IOSS) (Baba *et al.*, 1989), but practical problems still remain that an ANN is a blackbox model which is unfamiliar to plant operators, and that a fuzzy system (FS) usually puts great burdens on both system engineers and operators for knowledge acquisition and rulebase maintenance. To get appropriate solutions to these problems, the authors propose a method to integrate an ANN and fuzzy theory. In this paper, the procedure outline and simulation results are described.

INTEGRATION OF MULTI AI PARADIGMS

Knowledge Sources for Operation Support Systems

As knowledge sources for operation support systems in water and wastewater treatment, three main ones are available as shown in Fig. 1. Among these resources, operators' heuristics are used for an FS as if-then type fuzzy rules, and historical data are used for an ANN system as training patterns for learning.

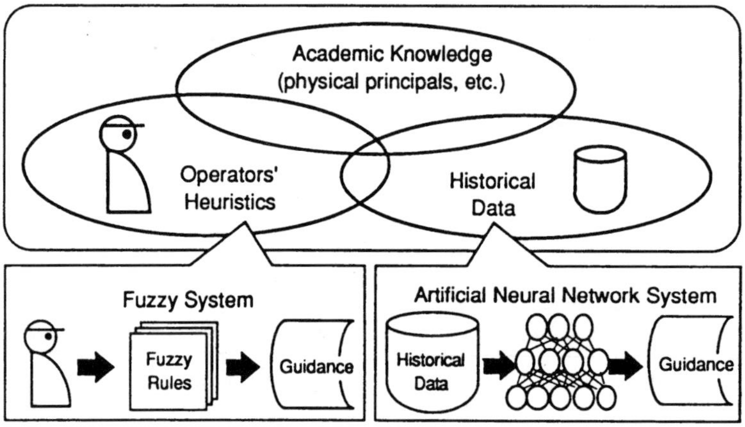

Fig. 1. Knowledge sources for operation support.

However, in conventional operation support systems which are comprised of only one AI paradigm, the knowledge resources are limited to only one, because an FS cannot deal with numerical data which historical data consist of, nor can an ANN system deal with if-then type rules. The authors think that an enhancement of knowledge sources gives the operation support system better performance, and to realize the enhancement, one knowledge representation has to agree with other knowledge. So the authors propose a fuzzy rule extraction method using an ANN, which converts numerical data into if-then type rules. This proposed method can integrate the merits of an ANN and an FS, and solve conventional problems described above.

Fuzzy Rule Extraction from an ANN

Figure 2 shows the procedural outline of fuzzy rule extraction using an ANN. In the conventional knowledge acquisition method, fuzzy rules are acquired by interviews with experienced operators, while the new method acquires fuzzy rules from plant historical data instead of the operators. This acquisition is executed by an ANN which can learn the implicit knowledge included in a large number of historical data. After the learning, the acquired knowledge is represented by a large number of parameters (so-called weights) in an ANN model, then the parameters are evaluated to translate into if-then type rules. Furthermore, by repeating this procedure at predetermined periods, the fuzzy rulebase of an FS can be refined automatically according to changes in the environment outside the plant.

This method features a modified structure of an ANN and an index to evaluate the parameters in an ANN model. The details of these features are described in the next section.

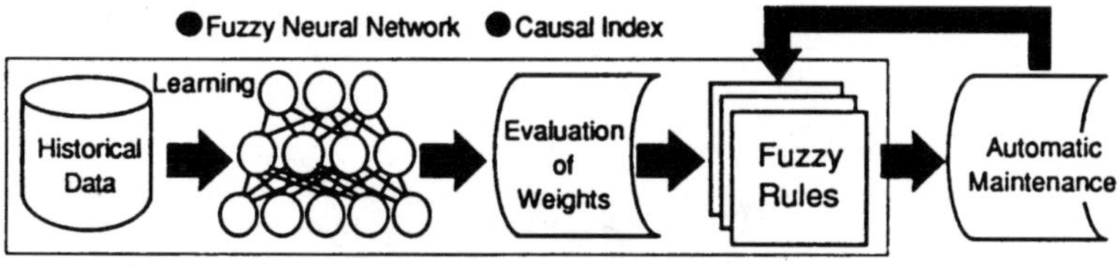

Fig. 2. Procedural outline of the proposed method.

Fuzzy Neural Network

To extract fuzzy rules from an ANN, the authors focus on the causal relationship between the neurons in the input and output layers. Figure 3 shows the structure of a fuzzy neural network, which was proposed by the authors (Enbutsu *et al.*, 1991). This structure is distinguished from a standard ANN by assignment of fuzzy neurons in the input and output layers. Fuzzy neurons represent fuzzy variables (e.g. "Low", "Medium", "High") of each input or output, and have the function of converting input data to the grades of fuzzy variables using given membership functions. By learning the historical data using a fuzzy neural network, causal relationship between operational factor such as coagulant injection rate in a water purification plant and influential factors is acquired in the network. As a learning algorithm of a fuzzy neural network, the Error Backpropagation (BP) Method (Rumelhart *et al.*, 1986), which is one of the most popular algorithms for an ANN, is available, however, the modified BP method (Baba *et al.*, 1990a) which is specific for rule extraction is used for learning in this method.

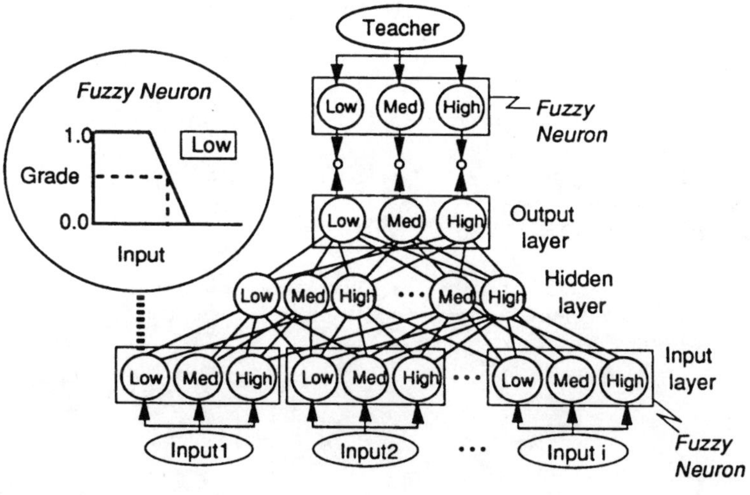

Fig. 3. Structure of a fuzzy neural network.

Causal Index

To get fuzzy rules from a learned ANN which includes knowledge, weights which represent the intensities of the connections between each neuron have to be evaluated to quantify the degrees of influences of the input factors on the output factors. To do this, the Causal Index (CI) was proposed by the authors (Baba *et al.*, 1990b; Enbutsu *et al.*, 1991).

CI is defined on the basis of the differential shown in Equation (1).

$$CI(\gamma_k,\alpha;x_i,\beta) \equiv \left(\frac{\delta\gamma_O(\gamma_k,\alpha)}{\delta x_I(x_i,\beta)}\right) = \sum_j [f'\{u_O(\gamma_k,\alpha)\}\cdot f'\{\mu_T(j)\}\cdot w_O(\gamma_k,\alpha;j)\cdot w_T(j;x_i,\beta)] \quad (1)$$

$$\mu_T(j) = \sum_i \sum_\beta \{w_T(j;x_i,\beta) \cdot x_I(x_i,\beta) \quad (2)$$

$$\gamma_T(j) = f\{\mu_T(j)\} \quad (3)$$

$$u_o(y_k,\alpha) = \sum_j \{W_o(y_k,\alpha;j)\cdot y_T(j)\} \quad (4)$$

$$y_o(y_k,\alpha) = f\{u_o(y_k,\alpha)\} \quad (5)$$

where $CI(y_k, \alpha; x_i, \beta)$: Causal Index for output fuzzy neuron "y_k=α" and input fuzzy neuron "x_i=ß",
$f(u)$: sigmoid function,
$W_T(j;x_i,\beta)$: weight from fuzzy neuron "x_i=β" to j-th hidden neuron,
$w_O(y_k, \alpha;j)$: weight from j-th hidden neuron to fuzzy neuron "y_k=α",
x_i : i-th input in the input layer,
$x_I(x_i,\beta)$: output from fuzzy neuron "x_i=β" (grade of membership function)
y_k : k-th output in the output layer,
α, β : fuzzy variable {"Low", "Medium", "High"}.

Since this index means a correlation between each input and output fuzzy variable, the input-output relationship whose absolute value of CI is greater than the others is considered to be more significant than other relationships. Also the sign of CI represents a positive or negative correlation of the input-output relationship.

Rule Translation

The purpose of rule extraction is to get useful knowledge for plant operation as an explicit form, therefore in the proposed method only significant input-output relationships are translated into fuzzy rules, and relationships whose absolute values of CI are less than others are omitted as trivial.

For a positively signed CI, this index is translated into a fuzzy rule as shown in Equation (6). For a negatively signed CI it is translated into a fuzzy rule which has a negative meaning in the condition clause as shown in Equation (7).

when $CI(y_k, \alpha; x_i, \beta) > 0$, if x_i is β then y_k is α (6)
when $CI(y_k, \alpha; x_i, \beta) < 0$, if x_i is not β then y_k is α (7)

But, since fuzzy rules which have a negative meaning in the condition clauses are unfamiliar to operators, the proposed method provides an additional method (Enbutsu *et al.*, 1991) which can convert a negative meaning to an equivalent meaning that is expressed by another positive form.

SIMULATIONS

Simulation Method

The purpose of the simulation was to demonstrate the knowledge acquisition ability of the proposed method for enhancement of knowledge sources. Figure 4 illustrates the simulation method which the authors prepared for this purpose. In the conventional method, only the fuzzy rules which are acquired from experienced operators through interviews are used for fuzzy reasoning, and the fuzzy rulebase is not modified during the simulated period. On the other hand, in the proposed method, the fuzzy rulebase which included operators' rules is refined automatically with extracted rules using an ANN. In the refinement of the fuzzy rulebase, operators' rules which conflict with the extracted rules are replaced by the extracted rules.

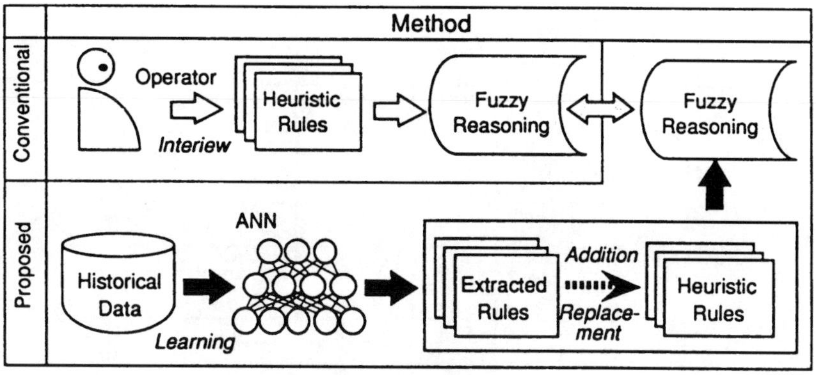

Fig. 4. Outline of simulation method.

Training Data

As training data for an ANN to extract fuzzy rules, historical data of a water purification plant during one year were prepared. The historical data included one operational factor, coagulant (poly aluminium chloride; PAC) injection rate, and ten influential factors on the operational factor:

(1) alkalinity
(2) pH
(3) temperature
(4) flow rate
(5) turbidity
(6) floc density
(7) number of floc
(8) floc volume
(9) floc diameter
(10 illumination intensity.

Influential factors (1)-(5) are water qualities of raw water, factors (6)-(10) are properties of the flocs which were measured in the floc formation tank with the image processing system developed by the authors (Baba *et al.*, 1988). Since the operational strategy of coagulant injection is much influenced by the level of turbidity, normal conditions (turbidity <10 mg/l) and abnormal conditions (turbidity ≥ 10 mg/l) were learned with each ANN. Twenty training data each were selected from the historical data, and membership functions of each factor were given in advance for fuzzy neurons.

RESULTS AND DISCUSSION

Table 1 shows example results of CI values which were calculated from ANNs for normal and abnormal conditions. The values in this table had already been converted with additional calculations to get only positive meaning rules, and normalized in the range of 0-100%

TABLE 1. Example Results of Causal Index (Abnormal Conditions)

Input \ Output			PAC injection rate		
			Low	Medium	High
			1	*13*	*86*
Water quality	alkalinity	Low	2	3	0
		Medium	5	0	1
		High	2	2	0
	temperature	Low	0	1	7
		Medium	0	2	6
		High	0	9	0
	turbidity	Low	2	1	1
		Medium	2	0	1
		High	1	1	21
	⋮	⋮	⋮	⋮	⋮
Floc property	floc density	Low	2	3	1
		Medium	2	2	3
		High	14	1	6
	number of floc	Low	0	3	3
		Medium	5	7	0
		High	0	1	7

CI values (in italic in Table 1), which correspond to each output fuzzy variable, represent the relative degrees of the correlation between each output fuzzy variable and all input fuzzy variables. Under the abnormal conditions, "PAC is High" was found to be more significant as a conclusion clause of a fuzzy rule than other output fuzzy variables, because it had the largest value (86) of all output fuzzy variables. And as a condition clause for "PAC is High", "turbidity is High" (CI(PAC, High; turbidity, High)=21) had the maximum degree of correlation, therefore it was selected for translation into a fuzzy rule. As explained in Equation (6), it was translated into "If turbidity is High, then PAC is High".

In the same way as above, fuzzy rules for abnormal and normal conditions were extracted automatically from ANNs, and through automatic refinement using the extracted rules, some fuzzy rules which could not be acquired from the operators were added to the rulebase, and some of the operator's rules were replaced by other rules as shown in Table 2.

TABLE 2. Results of Refinement of the Fuzzy Rulebase

	Operators' Rules	Extracted Rules from ANN	Results of Refinement
Abnormal Conditions	*If turbidity is High then PAC is High* *If temperature is Low then PAC is High* *If alkalinity is High then PAC is High* ---->$^{}$ ⋮ *If pH is Medium then PAC is Medium* (10 rules)	*If turbidity is High then PAC is High* *If temperature is Low then PAC is High* *If alkalinity is High then PAC is Medium* O ⋮ *If floc density is Medium then PAC is Medium* ● (13 rules)	● Additions 7 O Replacements 3 (17 rules)
Normal Conditions	*If turbidity is Low then PAC is Low* *If pH is Low then PAC is High* *If temperature is Medium then PAC is Medium* ---->$^{}$ ⋮ *If alkalinity is Medium then PAC is Medium* (10 rules)	*If turbidity is Low then PAC is Low* *If floc diameter is large then PAC is Medium* *If temperature is Medium then PAC is Low* O ⋮ *If flow rate is Low then PAC is Medium* ● (7 rules)	● Additions 4 O Replacements 2 (14 rules)

( ) : number of fuzzy rules in rulebase

Figure 5 shows PAC injection rates calculated by fuzzy reasoning with an original rulebase including only operators' rules, and with the refined rulebase by the proposed method. As found in this figure, the prediction was greatly improved by the proposed method, especially under abnormal conditions when operational guidance was more necessary for the operators.

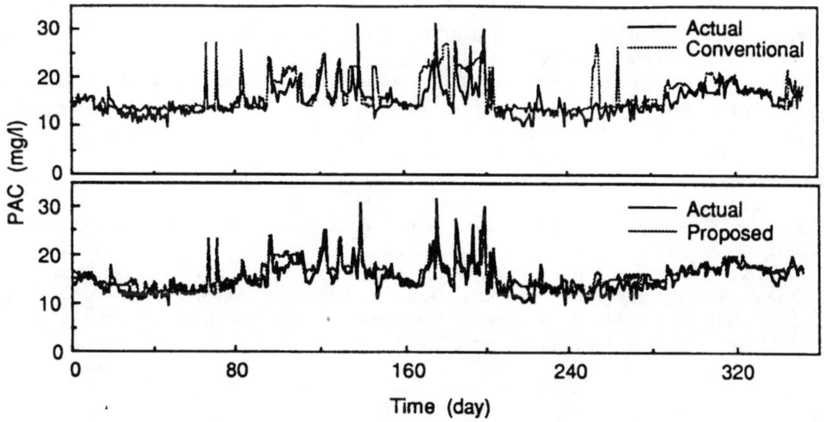

Fig. 5. Comparison of the results of fuzzy reasoning.

Table 3 shows the mean error during the simulated period. About 30% reduction of the mean error (4.88mg/l → 3.40mg/l) in abnormal conditions was realized. This effect was realized by the addition of new rules and replacement of inappropriate rules. In the conventional operation support system, it is difficult to maintain the rulebase without expending a large deal of manpower. However, this simulation demonstrated that the proposed method which combines different knowledge sources (operators' heuristics and historical data) can provide better performance without time consuming tasks. This is thought to be the most significant merit to be given by the integration of AI paradigms such as ANN and FS. This method is also applicable and promising for an operation support system for wastewater treatment (e.g. Watanabe *et al.*, 1993), because the operation of wastewater treatment requires several knowledge sources.

TABLE 3. Comparison of Mean Error

	Conventional		Proposed
Normal Conditions	1.43 (mg/l)	-10.5%	1.28 (mg/l)
Abnormal Conditions	4.88 (mg/l)	-30.3%	3.40 (mg/l)

CONCLUSION

The operation support system which integrates multi AI paradigms, featuring a fuzzy rule acquisition method using an ANN, was proposed, and it demonstrated better performance than the conventional single AI paradigm system, or FS. As the operation support systems for water and wastewater treatment plants where objective data are considered to be more important than subjective know-how of operators, a single AI paradigm system (Baba *et al.*, 1989) which uses only an ANN without fuzzy theory can be applied, however, where operators' know-how is respected as valuable knowledge, the proposed system provides a significant advantage.

REFERENCES

Baba, K., Yoda, M., Ichiki, H. and Osumi, A. (1988). A floc monitoring system with image processing for water purification plants. *Wat. Supply*, Vol.**6**

Baba, K., Enbutsu, I., Matsuzaki, H. and Nogita, S. (1989). Operation support system for water and wastewater treatment plants featuring historical data learning. (Japanese) *Proceeding of the 3rd domestic workshop on instrumentations and control for water and wastewater treatment systems*

Baba, K., Enbutsu, I. and Yoda, M. (1990). Explicit representation of knowledge acquired from plant historical data using neural network. *Proceedings of IJCNN '90 San Diego*

Baba, K., Enbutsu, I., Matsuzaki, H. and Nogita, S. (1990). Intelligent support system for water and sewage treatment plants which includes a past history learning function - coagulation injection guidance system using neuralnet algorithm *Instrumentation, Control and Automation of Water and Wastewater Treatment and Transport Systems* (*Adv. Wat. Pollut. Control* No.10) ed. R. Briggs, Oxford, Pergamon, pp. 227-234

Enbutsu, I., Baba, K. and Hara, N. (1991). Fuzzy rule extraction from a multilayered neural network. *Proceedings of IJCNN '91 Seattle*

Rumelhart, D. E. (1986). Learning internal representations by error propagation. *Parallel Distributed Processing* Vol.**1**, MIT Press

Watanabe, S., Baba, K., Yoda, M., Wu, W., Enbutsu, I., Hiraoka, M. and Tsumura, K. (1993). Intelligent Operation Support System for the Activated Sludge Process. *Wat. Sci. Tech.* **28**(11/12) (this volume).

Yagishita, O., Itoh, O. and Sugeno, M. (1984). Application of fuzzy theory to the control of coagulant injection in a water purification plant. (Japanese) *System and Control*, **28**(10).

Pergamon

Wat. Sci. Tech. Vol. 28, No. 11-12, pp. 341–345, 1993.

0273–1223/93 $7·00 + 0·00

STATISTICAL ANALYSIS OF OPERATION DATA AND AUTOMATIC CONTROL ON FLUIDIZED BED INCINERATION PLANT WITH DRYER

Shigeki Minami*, Hidekazu Nagasawa*, Yoshinori Saito*, Motoharu Yamagishi**, Masakatsu Hiraoka***, Nobuo Takeda*** and Shigenobu Okajima***

** Tsukishima Kikai Co. Ltd, Tsukuda, Chuo-ku, Tokyo 104, Japan*
*** Sewage Works Bureau, City of Yokohama, Naka-ku, Yokohama, Kanagawa 231, Japan*
**** Department of Environmental and Sanitary Engineering, Kyoto University, Sakyo-ku, Kyoto 606-01, Japan*

ABSTRACT

Continuous operation data were obtained on a fluidized bed incineration plant with dryers, and two autoregressive models were then prepared through statistical analysis of the data. Based on the results, an automatic plant control system using fuzzy theory was designed. An incinerator system of this type is characterized by energy efficiency, for which optimum and stable moisture control of the dried sludge is important. The large difference in time constants between incinerator and dryers makes energy saving difficult. Based on these analyses and design, control operations at a commercial plant with a capacity of 150 wet-tons/day were studied. It was confirmed that reduction of auxiliary fuel consumption and reduction of CO and NOx in the exhaust gas were optimized, while the moisture content of dried sludge and the furnace temperature were kept stable.

KEYWORDS

Sludge treatment; indirect heating type dryer; fluidized bed incinerator; energy conservation; autoregressive model; fuzzy control.

INTRODUCTION

Incineration is an effective process for volume reduction and stabilization of sewage sludge. As finding new final disposal sites is difficult, many sludge incineration plants have been constructed in large cities in Japan. Currently, more than 60% of the sludge produced throughout Japan is incinerated. The fluidized bed incineration system with dryers consists of dryers for dewatered sludge, a fluidized bed incinerator, a heat recovery unit, and a gas purification unit. Dewatered sludge treated with organic polymer coagulant is exposed to dryers which use the waste heat of the incinerator, before being fed to the furnace. This process is characterized by its energy efficiency.

The authors carried out continuous automatic measurement of the operation conditions and analyzed the data, which were reported at the ICA workshop in Kyoto in 1990 (Hiraoka *et al.*, 1990). Based on those results, we studied operation control at a commercial plant of Yokohama Nanbu Sludge Treatment Center.

INCINERATION PLANT

The study was carried out in the No 2 incineration plant (capacity: 150 wet tons/day) at the Center. Figure 1 shows the flow diagram and basic specification of the No 2 incineration plant. Sludge, which is transported by pipeline from the sewage treatment plants in the city, is thickened and anaerobically digested. The sludge is then dewatered using organic polymer coagulant and incinerated. The No 2 incineration plant consists of two indirect-heating-type disk dryers, one fluidized bed incinerator, and other devices. The heat in the exhaust gas of the furnace is recovered by an air heater and a waste heat boiler. The exhaust gas is then purified by an electrostatic precipitator (ESP) and a scrubber, before being discharged into the atmosphere.

At this plant, a part of the dewatered sludge (DI-sludge) is dried by disk-type dryers, and the dried sludge is mixed with the remaining dewatered sludge (WI-sludge) just before being fed into the furnace. The temperature in the furnace had conventionally been controlled by a feedback control system, in which the control variables are the moisture content of the mixed sludge (ratio of WI-sludge and DI-sludge), the sludge quantity and the feed quantity of auxiliary fuel. However, the response speeds of the dryer and furnace differ greatly. Therefore, the plant had to be run by skilful operators who had sufficient experience with controlling the dryers in order to maintain optimal furnace temperature.

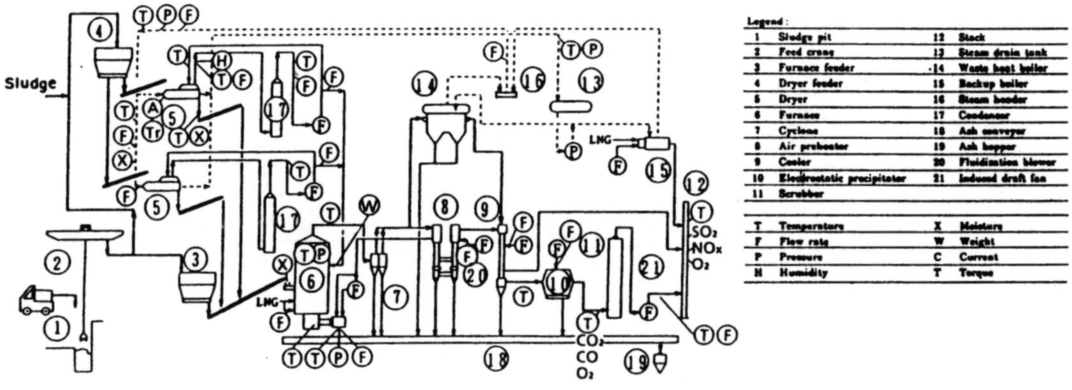

Dryer : Heat source = Steam	Furnace :	Capacity = 150 t/day
Heating area = 100m²		Fuel = LNG
		Size = ϕ 5500×12270H

Fig. 1. Plant flow diagram and basic specification.

DATA ACQUISITION AND PLANT CONTROL SYSTEM

A controller system consisting of a personal computer and a sequencer was constructed and installed at the plant in order to obtain plant operation data and to study control operations.

ANALYSIS OF THE PROCESS

Autoregressive (AR) models for the dryer and the furnace were designed in order to analyze the dynamic characteristics of the processes. An AR model is a dynamic statistical model which expresses a time series of multi-variables as the sum of linear factors in the past and noise factors. The AR model is represented by the following equation:

$$X(s) = \sum_{m=1}^{M} A(m) \cdot X(s-m) + U(s)$$

where

X(s): vector of system variable at time s
A(m): autoregressive coefficient of (k x k)th matrix (m = 1, 2, ... M)
U(s): intrinsic noise vector at time s.

We used SACCESS, a computer program developed by Hiraoka *et al.* for the data analysis, model construction and simulating computation (Hiraoka *et al.*, 1988). Figures 2 and 3 show examples of the step response simulation using the constructed AR model. The results of the analysis showed that the DI-sludge quantity and the weir height of the dryer discharge part influence the moisture content of dried sludge, and we have confirmed that these two factors can be used as the control variables for the dryer. On the other hand, it was found that the WI-sludge quantity can be the control variable for the furnace temperature. Also, NOx and CO in the exhaust gas decrease when the furnace temperature is stably maintained between about 810 and 850°C.

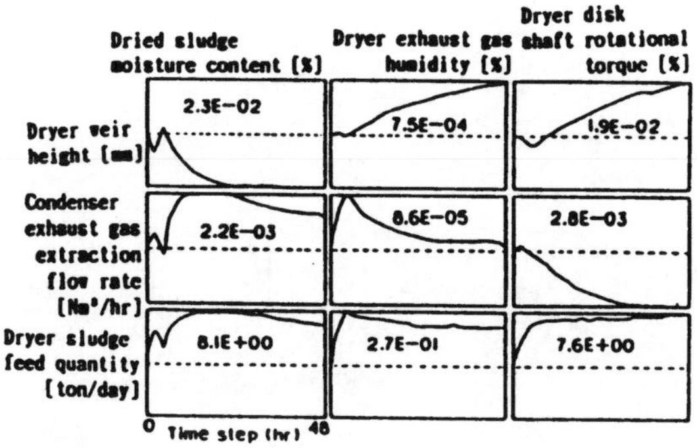

Fig. 2. Step response simulation of dryer model variables.

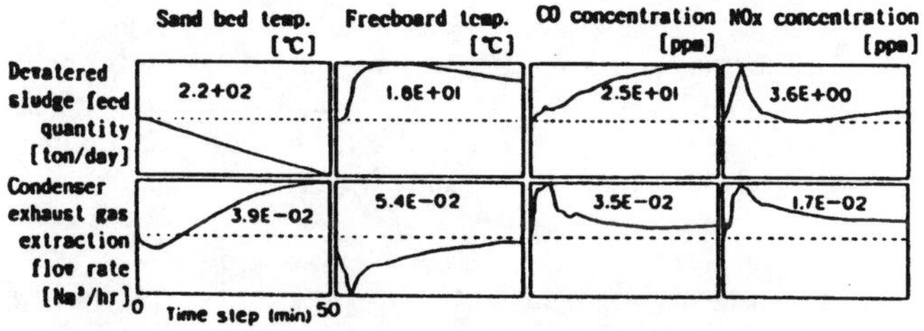

Fig. 3. Step response simulation of furnace model variables.

PLANT OPERATION CONTROL

The combustion characteristics of sludge are changed by various factors, such as seasonal influences, and the conditions of wastewater treatment and sludge treatment processes. A control system appropriate to these factors is desirable. Therefore a control system based on fuzzy theory was designed and used for the study. Fuzzy control systems are built using a combination of simple expressions of phenomena, and can be applied for a process with a slow response speed, such as the dryer.

Figure 4 shows the process control strategy. The initial values of the DI-sludge quantity and the WI-sludge quantity are determined from computed mass-and-heat balance values, etc. Through the control operation, the moisture content of dried sludge and the furnace temperature are changed to their optimum values. The control cycle of the dryers is set at 30 minutes and the incinerator is set at 10 minutes.

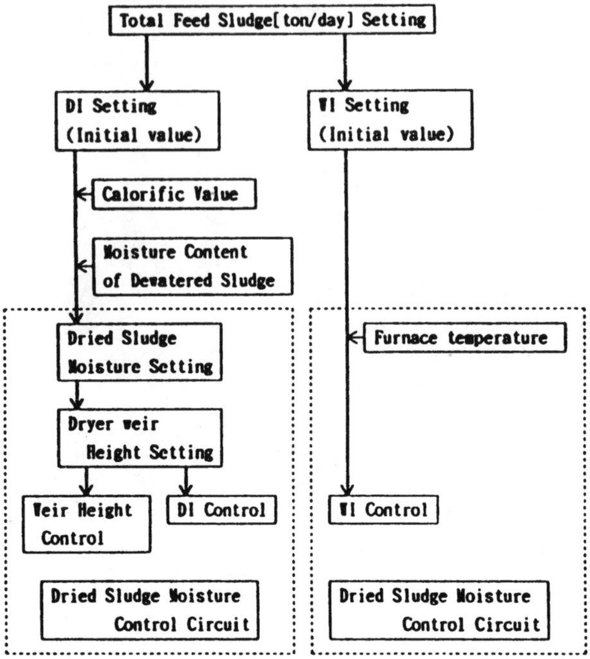

Fig. 4. Control strategy.

RESULTS

Dried Sludge Moisture Control. The moisture content data for dewatered sludge and dried sludge were continuously monitored by an automatic moisture meter installed at the plant. The time series data of the moisture content of dried sludge under conventional control and during this control study, are shown in Figure 5. Since fine control was difficult under the conventional control method, the moisture content of dried sludge had to be kept higher than the optimum value for the sake of safe operation. Using the fuzzy control system, the moisture content of dried sludge can be kept at an optimal, fixed value.

Furnace Temperature Control. Data for auxiliary fuel consumption for the incinerator under the conventional control and during the current study, are shown in Figure 5. The furnace temperature could be kept at an optimal fixed value by controlling the WI-sludge quantity, using the fuzzy control system. As a result, we confirmed that consumption of auxiliary fuel can be cut in half by combining optimum control of the dried sludge moisture content and the furnace temperature.

Exhaust Gas Data. Data for NOx and CO concentrations are shown in Figure 6. Both NOx and CO concentrations remained low and stable by optimizing the furnace temperature.

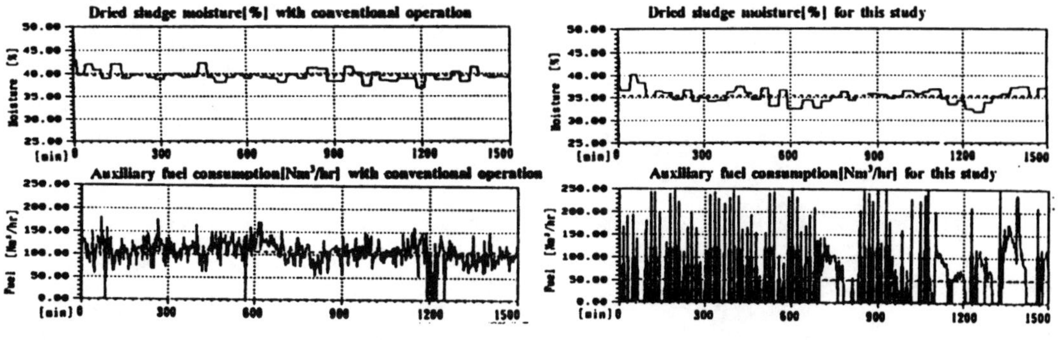

Fig. 5. Time series data of the plant.

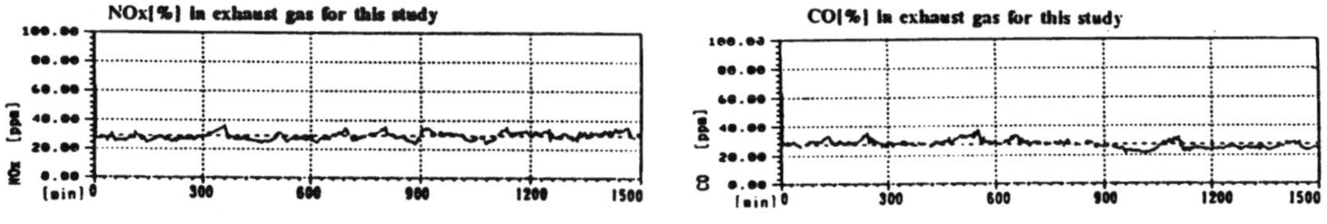

Fig. 6. Time series data of NOx and CO for this study.

SUMMARY

We analyzed the operation data of the fluidized bed incineration plant with dryer, designed a fuzzy control system and applied it to the operation of a commercial plant. The developed control system was able to keep the moisture content of dried sludge at a fixed value, decrease the consumption of the auxiliary fuel, and reduce the NOx and CO concentration of the exhaust gas.

REFERENCES

Hiraoka, M., Takeda, N., Fujisaki, K., Kuwabara, H., Minami, S. and Kimura, K. (1990). Automatic measurement and analysis of operation data on drying fluidized-bed incineration plant. *Instrumentation, Control and Automation of Water and Wastewater Treatment and Transport Systems (Adv. Wat. Pollut. Control* no 10) ed. R. Briggs, Oxford, Pergamon, pp. 129-136.

Hiraoka, M. and Tsumura, K. (1988). System identification and control of the activated sludge process by use of statistic model. *Wat. Sci. Tech.* **21**(10/11), 1161-1172.

Pergamon

Wat. Sci. Tech. Vol. 28, No. 11-12, pp. 347–354, 1993.

0273–1223/93 $7·00 + 0·00

STABLE COMBUSTION IN SLUDGE MELTING FURNACE BY FUZZY LOGIC CONTROL

Shunichi Shiono, Takashi Usui, Morio Iriyama, Kazuyuki Suzuki and Yohta Naka

Ebara-Infilco Co. Ltd., 1-6-27 Kohnan, Minato-ku, Tokyo 108, Japan

ABSTRACT

This paper discusses combustion control using fuzzy logic for a sewage sludge melting furnace. The sewage sludge melting process is gaining attention as an effective method for reducing sludge volume. The slag from molten sludge is stable, causes no pollution, and can be used as a by-product. However, the operation of a sludge melting plant requires expert operators who have to deal with complex phenomena that take place in a melting furnace. In an effort to facilitate the operation, the authors have employed a control system which features the use of fuzzy logic for automatic combustion control in a cyclone type sewage sludge melting furnace. Confirmation was made that this fuzzy logic combustion control system functioned effectively for controlling the furnace temperature, O_2 concentration in the furnace outlet flue gas, and concentration of NOx.

KEYWORDS

Sewage sludge, Melting furnace, Fuzzy logic control, Slag temperature, O_2 , NOx.

INTRODUCTION

Better life-styles, which provide comfort and conveniences, produce abundant waste and sewage sludge whose disposal is posing a serious social problem. Many attempts are made to solve the sewage sludge disposal problem. The development of the sewage sludge melting process was one such attempt and some processes have been practicalized and are currently used in several sewage treatment plants in Japan.

The sewage melting process involves incinerating dewatered and dried sludge within the temperature range of 1,300℃ to 1,400 ℃ .

Inorganic matter contained in the sludge does not turn into ash when incinerated but into slag, which becomes a vitreous solid when cooled by water or air (Naka et al., 1992).
One advantage of the present process is that toxic matter, such as heavy metals, contained in sludge become sealed in the slag after the melting process, thus causing no contamination. Some other advantages would be that the volume of the slag is smaller than that of ash resulting from incineration processes, that there is no problem of scattering ash, and that the slag can be easily handled. Slag can be used as a by-product, in particular as material for construction purposes.

The quality of sludge, i.e. its composition and nature, fed into the melting furnace,greatly changes according to various factors.
Such factors would be results of changes in the operational conditions of sewage and sludge treatment processes, including seasonal factors. Sensors are generally used indirectly for measuring the furnace temperature, which is extremely high. Although doing so protects the sensors, there is a lag in the measured data. Furthermore, it is difficult to achieve optimum control by conventional methods as controlling the process involves multiple factors which interfere with each other; it therefore becomes necessary to rely on the ability of experienced operators, similar to the case of controlling incineration processes (Shiono et al., 1992). The authors had previously studied control methods featuring fuzzy logic for a sewage sludge rotary-kiln (Suzuki and Nakamori., 1991) and a municipal refuse incinerator (Tanaka et al., 1993). The use of fuzzy logic control for the present process was decided based on these experiences.

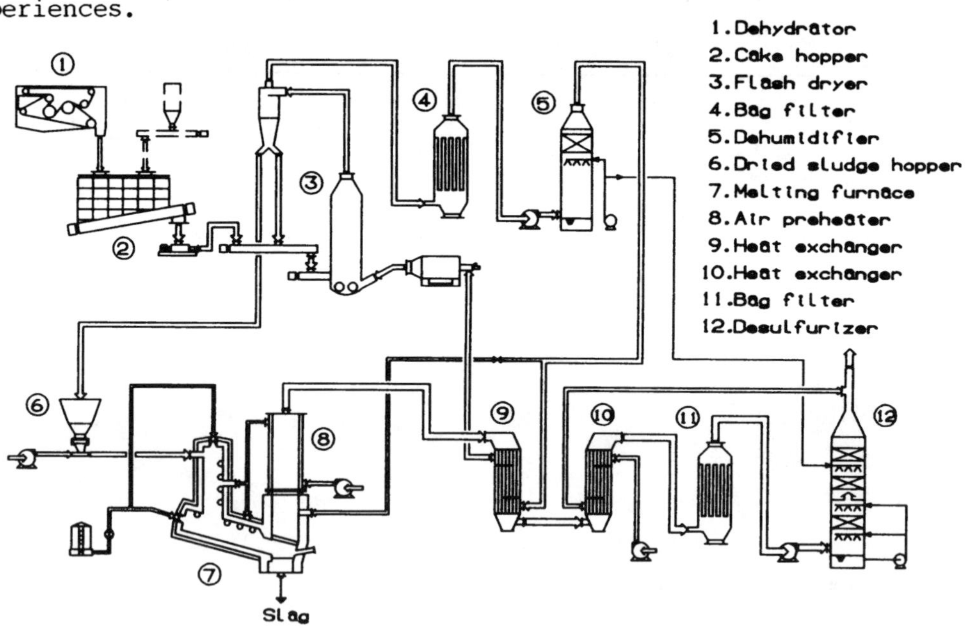

Fig.1 Flow diagram of sewage sludge melting system.

SEWAGE SLUDGE MELTING FURNACE PLANT

Figure 1 shows a flow diagram of an actual sewage sludge melting plant currently in operation. Sludge dewatering, drying, melting furnace, heat recovery, and flue gas treatment processes are featured. The treatment capacity of this plant is 15 tons of dewatered cakes per day. A control system using fuzzy logic is implemented for melting furnace process.

Figure 2 shows the cross-section of the cyclone melting furnace and its control devices. The operation is carried out as follows.

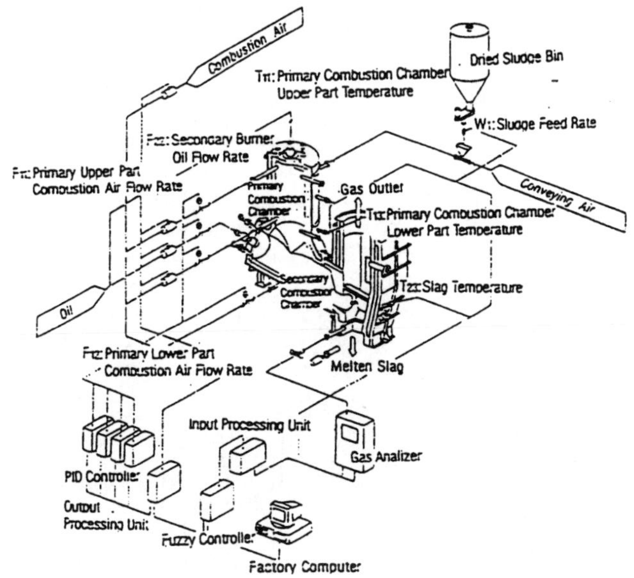

Fig.2. Cross-section of melting furnace and its control devices

Sludge feed rate. A preset amount of sludge is fed into the system based on the treatment plan.

Auxiliary fuel flow rate. A secondary oil burner is used to heat the sludge in case the sludge has insufficient thermal efficiency, a situation which makes it impossible to maintain the flow of slag in the furnace. The oil burner is made to maintain an appropriate temperature by regulating its intake of fuel.

Total combustion air flow rate. The air ratio (excess air number) is maintained at a minimal constant by monitoring the O_2 concentration in the flue gas at the furnace outlet and regulating the total combustion air flow rate into the melting furnace.

Air flow rate in each section. The temperature in each section of the furnace and the concentration of NOx in the flue gas at the furnace outlet is controlled proportionally by the combustion air blown into each section of the furnace.

Pressure regulation. The pressure in the furnace is regulated to be slightly below atmospheric pressure, by adjusting the opening/closing of the flue gas induced draft fan damper, for safety reasons.

DEVELOPMENT OF FUZZY LOGIC COMBUSTION CONTROL SYSTEM

Various measurements and analysis had to be carried out under manual operation in the plant, which extended over a period of ten months, prior to installing the fuzzy logic combustion control for the combustion control. This included measuring the thermodynamics of the process, statistical analysis of operational data, and establishing rules of thumb in terms of operation.

Control Targets & Control System Variables

Combustion control of the melting furnace involved controlling the furnace temperature, slag temperature, combustion air ratio, concentration of NOx at the furnace outlet, and the internal pressure of the furnace. Conventional PID control was used for controlling the internal pressure of the furnace, while automatic control using fuzzy logic was used for the rest. The objective was to facilitate control by the operator and stabilize the process. The input-output variables of the control system are listed below.

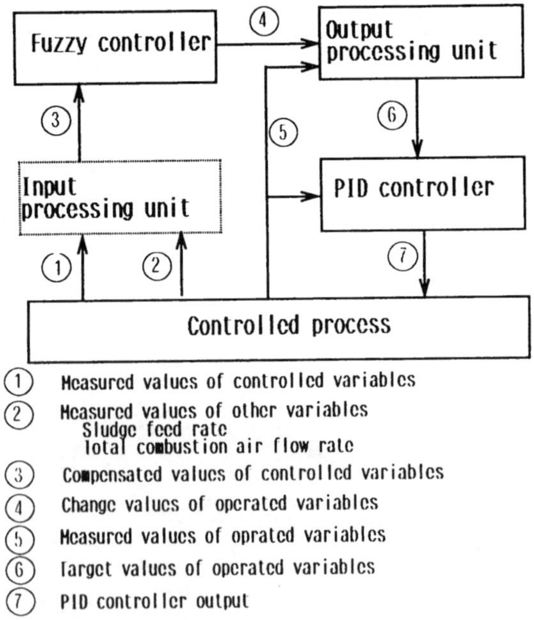

Fig.3. Fuzzy logic control system

Variables Input into Control System

T_{11} : Temp. of primary chamber upper section

T_{13} : Temp. of primary chamber lower section
T_{23} : Slag temp.
O_2 : O_2 concentration in flue gas at furnace outlet
NOx: NOx concentration in flue gas at furnace outlet
(W_1 : Dried sludge feed rate)
(F_0 : Total combustion air flow rate)

Data Output of Control system
F_0 : Total combustion air flow rate
F_{11} : Combustion air flow rate at upper section of primary combustion chamber
F_{12} : Combustion air flow rate at lower section of primary combustion chamber
F_{22} : Fuel oil feed rate to secondary burner

Fuzzy Logic Control System

Figure 3 shows a block diagram of the fuzzy logic control system. The system was configured in three stages, the top of which featured the fuzzy logic control unit, the middle the sequence control unit, and the bottom the PID control unit. The functions of each of these units are discussed below.

Fuzzy logic control unit. Monitored data on control variables (part of values are processed for compensation) are input into unit. The input data are then processed by fuzzy inference, based on fuzzy production rules and membership functions. The processed data is then output as change values, appended to or subtracted from, manipulated variables for optimum control of the combustion. The MAX-MIN and the center of gravity operation were applied for fuzzy inference.

Sequence control unit. Change values of manipulated variables from the fuzzy logic control unit are merged with current values of manipulated variables and these data become output as the target values for the PID control units.
Current and change values of manipulated variables become sampled at 10 ms intervals in the sequence control unit. Mean values from every 90 seconds of sampling become calculated out as the target values of manipulated variables. These data become individually input into the PID control units at an interval of three to thirty minutes. The frequency of output was determined by considering the time lag in the response of the control system and data retrieval. The sequence control unit also functions as a warning system for a fault in the O_2 analyzer and when a control limit is reached in the fuzzy logic control system.

PID control unit. The PID control units control the opening/closing of motor operated valves, which function as final control elements, following up the target value set by the sequence control unit.

Fuzzy Production Rules and Membership Functions

Table 1 shows some fuzzy production rules, while Figure 4 shows examples of membership functions. A preliminary set of control rules and membership functions were first installed based on data from manual operation and previous experiences. These were both tuned during test runs using an actual furnace.

Table 1 Some fuzzy production rules

FUZZY RULES	Premise Part of Fuzzy Inference Rules				Consequent Part of Fuzzy Inference Rules	
	Primary chamber upper part temperature T_{13}	Primary chamber lower part temperature T_{11}	Flue gas NO_x conc. NO_x	Flue gas O_2 content O_2	Primary chamber upper part air flow rate change F_{11}	Primary chamber lower part air flow rate change F_{12}
f 01	–	NL	ZR	–	PS	NS
f 02	–	NL	PS	–	PS	NS
f 03	–	NL	PM	–	PS	NS
f 04	–	NL	PL	–	PS	NL
f 05	–	NS	–	–	PS	NS
f 06	ZR	ZR	ZR	–	ZR	ZR
f 07	PS	ZR	ZR	–	ZR	ZR
f 08	PL	ZR	ZR	–	NS	ZR
f 09	ZR	ZR	PS	–	ZR	NS
f 10	PS	ZR	PS	–	ZR	NS
f 11	PL	ZR	PS	–	NS	ZR
f 12	–	ZR	PM	–	NS	ZR
f 13	–	ZR	PL	–	NS	ZR
f 14	ZR	PS	ZR	–	ZR	ZR
f 15	PS	PS	ZR	–	ZR	ZR
f 16	PL	PS	ZR		NS	PS
	–	PS	PS			ZR
		PS				

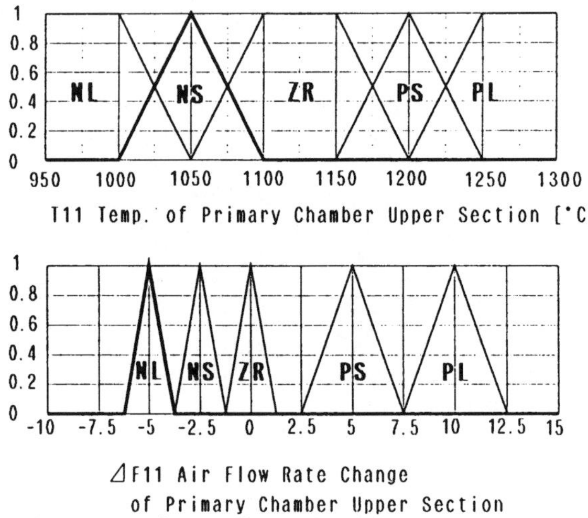

Fig.4 Examples of membership functions.

Compensation Input Values

The melting point (temp.) of sludge fluctuates between 1,150℃ to 1,350℃ during a one year span due to changes in sludge quality. It is desirable that the internal temp. of the furnace (T_{11}) and slag temp. (T_{23}) are not higher than necessary. To achieve this, the temperature must be regulated properly according to the melting point of the sludge. Accordingly, it becomes necessary to constantly detect the melting point of sludge, considering changes in sludge quality, and use the obtained data to set target values which would optimally control the internal temp. of the furnace (T_{11}) and slag temp. (T_{23}). Estimation of the sludge melting point, by the present control system, is carried out using data on the total combustion air flow, including that of air for blowing sludge into the furnace, on O_2 concentration in the flue gas at the furnace outlet, and on the feed rate of dried sludge. The estimated value of the sludge melting point is used for compensation the measured internal temp. of the furnace (T_{11}) and slag temp. (T_{23}). Equation (1) for calculating the estimation value of the sludge melting point (Ts) is shown below.

$$Ts = 1,490 - \left(\frac{21 - O_2}{21} \times \frac{A}{W \times \frac{100-M}{100}} \right) \times 69 \ [^\circ C] \quad (1)$$

where
A: Total combustion air flow rate [Nm^3/h]
O_2 : O_2 percentage at furnace outlet [%]
W: Dried sludge feed rate [kg/h]
M: Moisture content in sludge [%]

The method used in applying the estimated sludge melting value for the control system involved compensating measured values without changing the membership functions of T_{11} and T_{23}. Equations (2) and (3) for calculating the compensating temp. (T) are shown below.

$$T = T_0 + \Delta T \quad [^\circ C] \quad (2)$$

where,
T: Compensating temp. [°C]
T_0 : Measured temp. [°C]
ΔT: Temp. compensation value [°C]
accordingly,

$$\Delta T = a\,(Ts - b) \quad (3)$$

Parameters a and b were set to attain optimum operating conditions under T_{11} and T_{23}.

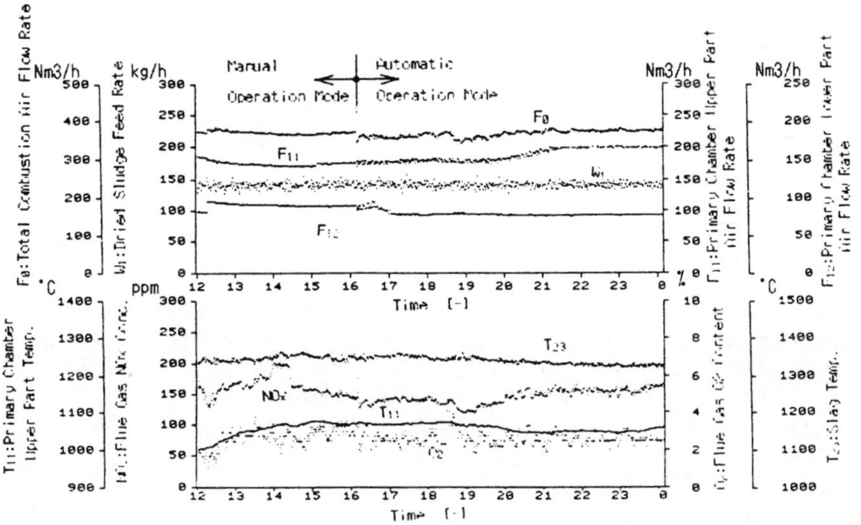

Fig.5 Result of fuzzy logic control.

TEST RESULTS OF COMBUSTION CONTROL USING FUZZY LOGIC CONTROL

Figure 5 shows changes which occurred before and after the operation of present process was shifted from manual to automatic. Manual adjustment was carried out (until 16:00), i.e.for the total combustion air flow (F_0) and the combustion air flow at the upper (F_{11}) and lower (F_{12}) sections of the primary combustion chamber, to control the temp. at the upper section of the primary combustion chamber (T_{11}), the slag temp (T_{23}), and the NOx and O2 percentage in the flue gas. This was all shifted to automatic (fuzzy logic control) after 16:00.

As soon as the operation was shifted to automatic, the fuzzy logic control system began to control the F_0 to compensate for the rather high O_2 concentration. It is acknowledged that an optimum level of approx. 2.5% was attained (see Fig.5). The fuzzy control system detected that the T_{11} was rather low between 20:00 and 21:00, and it increased the F_{11} gradually. As a result, the T_{11} started increasing at approx. 23:00, lagging about two hours behind.

CONCLUSION

A control system comprising a fuzzy logic control unit, PID control units, and sequence control unit was applied to control the combustion in a sewage sludge melting furnace. Operation under this system yielded satisfactory results. Perspectives in using fuzzy logic control include the development of a graphic analysis system for monitoring slag flow and a system for assessing the balance of the entire melting process. Further research is currently underway on these points.

REFERENCES

Naka,Y. and Yokoo, M. (1992), Development of Noble Sludge Melting Process, Proc. of WEF The Future Direction of Municipal Sludge Management, Vol. I, pp.745-758

Shiono, S., Usui., T., Iriyama, M., Suzuki, K. and Naka, Y. (1992), Fuzzy Control of a Sewage Sludge Melting Farnace, Proc. of 2nd WEF Asia/Pacific Rim Conference on Water Pollution Control, pp.551-558

Suzuki, K. and Nakamori, Y. (1991), Model Predictive Control based on Fuzzy Dynamic Models, Proc. of 4th International Symposium on Process Systems Engineering (PSE '91) volII , pp18.1-18.15

Tanaka, K., Sano, M., Suzuki, K. and Watanabe, H. (1993),Self-Organizing Fuzzy Identification of a Municipal Refuse Incinerator, to be presented at International Fuzzy System and Intelligent Control Conference 93 (14-17 March 1993 U.S.A)

Pergamon

Wat. Sci. Tech. Vol. 28, No. 11-12, pp. 355-367, 1993.

0273-1223/93 $7·00 + 0·00

FUZZY CONTROL OF A DYNAMIC ACTIVATED SLUDGE PROCESS FOR THE FORECAST AND CONTROL OF EFFLUENT SUSPENDED SOLID CONCENTRATION

Y. P. Tsai, C. F. Ouyang, M. Y. Wu and W. L. Chiang

Graduate Institute of Civil Engineering, National Central University, Chung-Li, Taiwan

ABSTRACT

The effluent total BOD (or COD) concentration of the activated sludge process (A.S.P.) usually increases with suspended solid concentration. How to reduce effluent S.S. concentration, therefore, is the key issue of treatment efficiency for A.S.P. The varied return sludge and influent flow rate are two major operational factors of those affecting effluent S.S. concentration. However, the wastewater flow rate and substrate concentration in municipal wastewater treatment plant, due to the differences of city scale and life style, are significantly time-varied every day. Based on the above, the purpose of this study is to control in timely fashion return sludge flow rate with the variation of influent flow rate to minimize effluent S.S. concentration and meanwhile decrease the effluent total BOD (or COD) concentration. The fuzzy control theory is utilized in this study to forecast and control effluent S.S. concentration and further predict the MLSS concentration in aeration tank. It reveals that the inferred control strategies not only enable one to decrease effluent S.S. concentration but also make it more stable.

KEYWORDS

Activated sludge process; dynamic control; fuzzy control; real time control; effluent S.S. control.

INTRODUCTION

The A.S.P. has been widely used in wastewater treatments and the removal of organic compounds can be described quite well by the activated sludge model developed by IAWPRC group (Henze *et al.*,1986). On the other hand, solid flux and its related theories are needed to be used to control the effluent S.S. concentration. However, the theories involve a complex graphic process, their feasibilities become slim, especially when used in on-line control. Thus, establishing a simple and practical control methodology can efficiently benefit the operation of full-scale plants. The effluent S.S. concentration is affected by the physical and biochemical behaviors in the system. Since the latter is hard to control, most researchers (Cashion and Keinath, 1983; Chapman, 1983; Parker, 1983; Tuntoolavest *et al.*, 1983; Severin and Poduska, 1985; Tekippe and Bender,1987) studied emphatically the former influence on the secondary clarifier. Most conclusions aimed directly at various influence factors, and Tuntoolavest *et al.*(1983) classified them into two categories as follows:

(A). factors that affect bioflocculation in the aeration basin :
- (1). the solid retention time (SRT) of the activated sludge;
- (2). the mixed liquor suspended solids (MLSS) concentration in the aeration basin;
- (3). the turbulence level in the aeration basin;
- (4). the sludge recycle rate (R).

(B). factors that affect settling of the sludge in final settler :
- (1). the surface overflow rate in the secondary clarifier;
- (2). the settler hydraulic retention time;
- (3). the MLSS concentration in the aeration tank;
- (4). the sludge recycle rate (R).

The above mentioned SRT, MLSS, turbulence level and R are all operating influence factors. Of these SRT calculated by the day cannot be applied in real time control; moreover, turbulence level from aeration is also difficult to control because it depends on D.O. requirement. Consequently, MLSS and return sludge flow rate are chosen as the control factors for effluent S.S. concentration in this study.

In general, the control of effluent S.S. concentration from steady state influent is no longer a problem; nevertheless, the discrepancy of city scale and life style makes flow rate and substrate concentration in municipal wastewater treatment plants time-varied significantly every day. In order to prevent system unsteadiness, inefficiency and even failure, it is necessary to make real time control actions. Generally, there are three dynamic control methods for the secondary clarifier:

(1) based on solid flux and its related theories: this method is too complicated to be suitable for real time control;

(2) based on expert systems: the uncertainty and highly-nonlinear behaviors of systems are very difficult to handle;

(3) based on time series analysis: the theory is quite good but when the number of variables involved is more than one, the mathematic model becomes too complicated to solve.

Fuzzy control theory has some advantages such as (1) quantifying the system uncertainty rationally, (2) handling highly-nonlinear systems easily, (3) being used effectively when suitable class of model is unknown, (4) quantifying the system's qualitative behaviors well. It is feasible to use the fuzzy control theory for complex systems, especially the activated sludge systems which contain the above mentioned characteristics.

FUZZY CONTROL THEORY

Fuzzy control theory was proposed firstly by Mamdani and Assilian (1975) from fuzzy set theory originated from Zadeh (1965). It has become a popular and valuable tool for modelling uncertainty and recently there have been applications of fuzzy control in environmental engineering such as control of A.S.P. (Tong *et al.*, 1980) and anaerobic digester (Boscolo *et al.*, 1993) etc. The fuzzy control rules used in the above two applications were mainly based on human operator's experience and/or control engineer's knowledge. When a process is complex, there might not exist any expert who can control it adequately. It is then necessary to build a fuzzy relational equation based on operational and/or experimental data to derive control rules. Some basic properties of fuzzy relational equations are introduced in the following.

A fuzzy set **A** defined on the universe of discourse **U** can be defined by a membership function m which maps **U** onto the membership space M=[0,1]. $\mu_A(y)$ represents the grade of membership of y, each element of **U**, in **A**. The raw data ought to be fuzzified before fuzzy set theory is adopted because all the calculations involved are through membership functions. The membership function can be chosen by the model designer or through the fuzzy clustering method. Fuzzy systems are composed of control variables **U**, state variables **X** and output variables **Y** and fuzzy relational set **R** correlates them. The fuzzy relational equation can be delineated as Eq. (1).

$$\mathbf{U} \circ \mathbf{X} \circ \mathbf{R} = \mathbf{Y} \tag{1}$$

where **U**, **X**, **Y** are all fuzzy sets represented by membership function, **R** is the fuzzy relational set, and ° is max-min operation. Equation (2) can be obtained from Eq.(1) through membership function representation.

$$\underset{1\leqq k\leqq l}{y'_k} = \bigvee_{\substack{1\leqq i\leqq m\\ 1\leqq j\leqq n}} (u_i \wedge x_j \wedge r_{ijk}) \tag{2}$$

where y, u, x are membership functions of Y, U, X respectively, Λ, V are min and max operators respectively, and l, m, n are cardinity number (level number) of y, u, x respectively. Euclidean distance is usually adopted as performance index as shown in Eq.(3).

$$Q = \sum_{p=1}^{N} \sum_{k=1}^{l} \left[\bigvee_{\substack{1\leqq i\leqq m\\ 1\leqq j\leqq n}} (u^P_i \wedge x^P_j \wedge r_{ijk}) - y_k \right]^2 \tag{3}$$

where Q is performance index and N are data numbers.

In general, two approaches relate fuzzy relational equation (1). They are

(a) **U**, **X**, **Y** are given, find **R** which can minimize Q; and

(b) **X**, **Y**, **R** are given, find **U** which can minimize Q.

The above two optimal problems in (a) and (b) can be solved by Eq.(4) and (5).

$$\partial Q / \partial \mathbf{R} = \mathbf{0} \tag{4}$$

$$\partial Q / \partial \mathbf{U} = \mathbf{0} \tag{5}$$

For the reason of simplification, modified Newton's Method (Pedrycz, 1983) is suggested to solve the partial differential equations of Eq.(4),(5). For example, the Newton's iteration of Eq.(4) is shown as Eq.(6).

$$r^{(c+1)}_{vst} = r^{(c)}_{vst} - \alpha \cdot \partial Q / \partial r_{vst} \mid r_{vst} = r^{(c)}_{vst} \tag{6}$$

where v = 1,2,...,m, s = 1,2,...,n, t = 1,2,...,l, and then

$$\partial Q / \partial r_{vst} = 2\sum_{p=1}^{N} \left[\bigvee_{\substack{1\leqq i\leqq m\\ 1\leqq j\leqq n}} (u^P_i \wedge x^P_j \wedge r_{ijk}) - y_k \right] \mathbf{P}_{pvst} \tag{7}$$

$$P_{pvst} = \begin{cases} 1, & \text{if } u^P_v \wedge x^P_s \wedge r_{vst} \geq \bigvee_{i \neq v,\ j \neq s} (u^P_i \wedge x^P_j \wedge r_{ijt}) \\ & \text{and } r_{vst} \leq u^P_v \wedge x^P_s \\ 0, & \text{otherwise} \end{cases} \tag{8}$$

The parameter α is adjusted due to the concrete form of the equation under consideration in order to obtain a good rate of convergence.

In order to utilize the controller, the solution of Eq.(1) represented by a fuzzy set must be transferred to crisp value. Defuzzification is therefore a necessary step and the gravity center method is usually used as defined in Eq.(9).

$$Y = \frac{\int y \cdot \mu_y(y) \cdot d_Y}{\int \mu_y(y) \cdot d_Y} \tag{9}$$

EXPERIMENTAL MATERIAL AND METHODS

The pilot plant used in this study contains aeration tank and settler and the process is controlled automatically according to initial setting values. Some of the plant and process features are:

Aeration tank volume	: 32 l
Final settler volume	: 15 l
Final settler surface area	: 0.09 square meter
System temperature	: 20 + 1 °C
Final settler scrubber speed	: 0.5 rpm
Inflow pattern	: continuous
Influent substrate	: artificial wastewater
COD on-line analyzer	: ORGANIC POLLUTANT ANALYZER, OPSA-120, HORIBA Ltd.
MLSS on-line analyzer	: MLSS METER, MODEL ML-3, KAKOH SEIKI, TOKYO, JAPAN
S.S. on-line analyzer	: DIGITAL DIRECT-READING TURBIDIMETER, SERIES 965, ORBECO ANALYTICAL SYSTEMS INC.
D.O. on-line analyzer	: DISSOLOVED OXYGEN ANALYZER, μ AI SERIES 4060, WALTRON Ltd. (D.O. concentration in aeration tank is maintained above 1.5 mg/l)
Programmable pump	: MASTERFLEX COMPUTERIZED DRIVE, MODEL 7550-60, COLE-PARMER INSTRUMENT CO.
Adapted card	: RS232
Control program language	: C LANGUAGE

The compositions of artificial organic substrate are listed in Table 1. Figures 1 to 3 show the correlations between electrical signal (mV) and estimated data (analyzed following Standard Methods, APHA, 1985) of total COD, MLSS and S.S. concentration respectively. The electrical signal is obtained from each on-line analyzer and can be used to calculate the estimated value by regression analysis. The major function of the control facility is the control of the pump speed in order to attain the target values of influent flow rate, substrate concentration and sludge concentration in the aeration tank. The time interval of control actions is 1 min so 1440 data sets are collected every day. The raw data is simplified by moving the average approach (per 20 min) for further analysis.

TABLE 1. Composition of Artificial Substrate (mg/l)

Composition	Conc.
Full cream powdered	135.00
Saccharose	13.33
Ammoniumferrosulfate	2.86
Urea	8.57
Acetic acid 99%	29.80
Phosphoric acid	7.20
Iron (III) chloride 6-hydrate	0.10

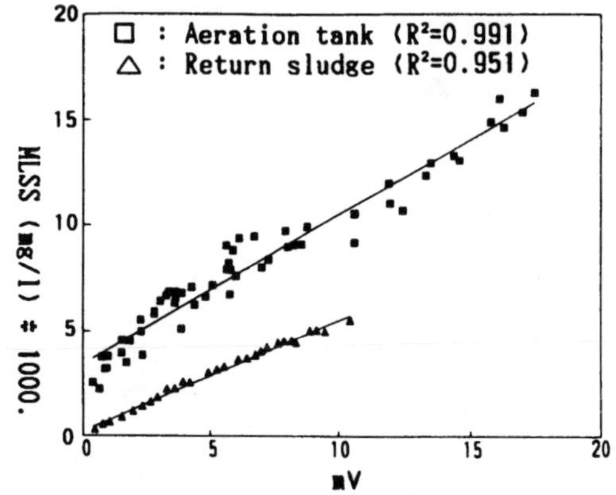

Fig. 1. Relationship between analyzed data and electrical signal of MLSS.

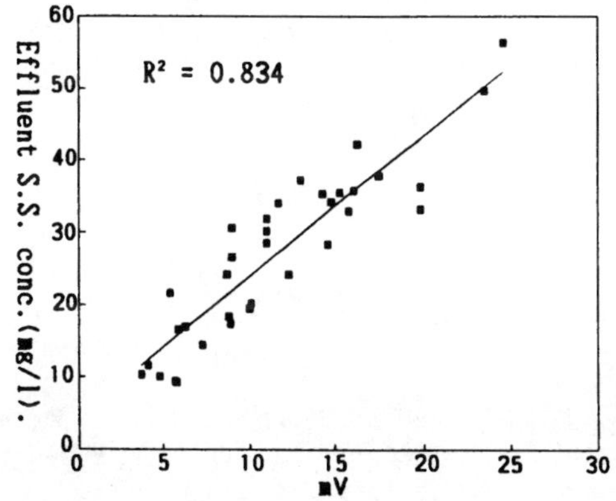

Fig. 2. Relationship between analyzed data and electrical signal of total COD.

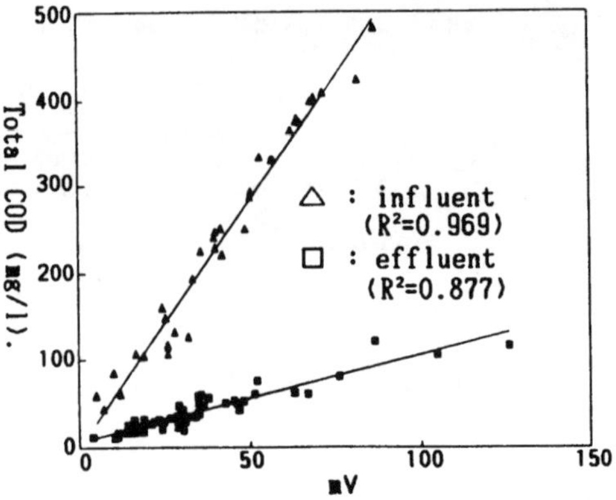

Fig. 3. Relationship between analyzed data and electrical signal of effluent S.S.

RESULTS AND DISCUSSIONS

Experimental Results

The setting pattern of influent flow rate and substrate concentration as shown in Fig.4-a and Fig.4-b are used to simulate the actual inflow pattern for the purpose of studying the effects of dynamic inputs for A.S.P. The real time control of return sludge flow rate is required in order to properly maintain F/M ratio (Fig.4-c). The theoretical return sludge flow rate derived from mass balance of sludge in system is shown in Eq.(10).

$$Qr = \frac{X' * Qin * t + (X'-X) * Va}{(Xr - X') * t} \tag{10}$$

where Q_r is the theoretical return sludge flow rate (ml/min), Q_{in} is the influent flow rate (ml/min), X is the actual MLSS concentration in aeration tank (mg/l), X_r is the return sludge concentration (mg/l), V_a is the volume of the aeration tank (ml), t is the required time for changing X to X' (min), and X' is the required MLSS concentration (mg/l) in the aeration tank, defined by Eq.(11), in order to maintain an adequate F/M ratio.

$$X' = \frac{1.44 * Qin * (Sin - Seff)}{a * Va * F/M'} \tag{11}$$

where S_{in} is the influent BOD concentration (mg/l), S_{eff} is the effluent total BOD concentration (mg/l), F/M' is the target value of F/M ratio (1/d), "a" is the ratio of MLVSS/MLSS, and 1.44 is the conversion factor of the unit.

The theoretical return sludge flow rate and MLSS concentration in the aeration tank are, through the calculation of Eq.(10),(11) by the on-line monitor data, shown in Fig.4d and Fig.4e. When low influent loadings occurred, the pump of return sludge stopped so as to dilute MLSS concentration in the aeration tank with inflow to maintain proper F/M ratio; while on peak inflow time, a large amount of sludge was pumped into aeration tank to raise MLSS concentration. Since the solid flux of the final settler was larger at this moment, the return sludge concentration decreased. In addition, the large return sludge flow rate made the HRT of the settler shorter while the effluent S.S. concentration (Fig.4h, solid line) increased. The results of experimental data analysis indicate that total COD concentration was evidently increased with S.S. concentration as shown in Fig.5a and total BOD concentration also increased with total COD concentration as shown in Fig.5b, i.e., S.S. concentration is the main effecting factor of the increase of effluent total COD and BOD concentration. This phenomenon can be demonstrated by Fig.4h. Therefore, decreasing of effluent S.S. concentration only can reduce effluent total COD and BOD concentration.

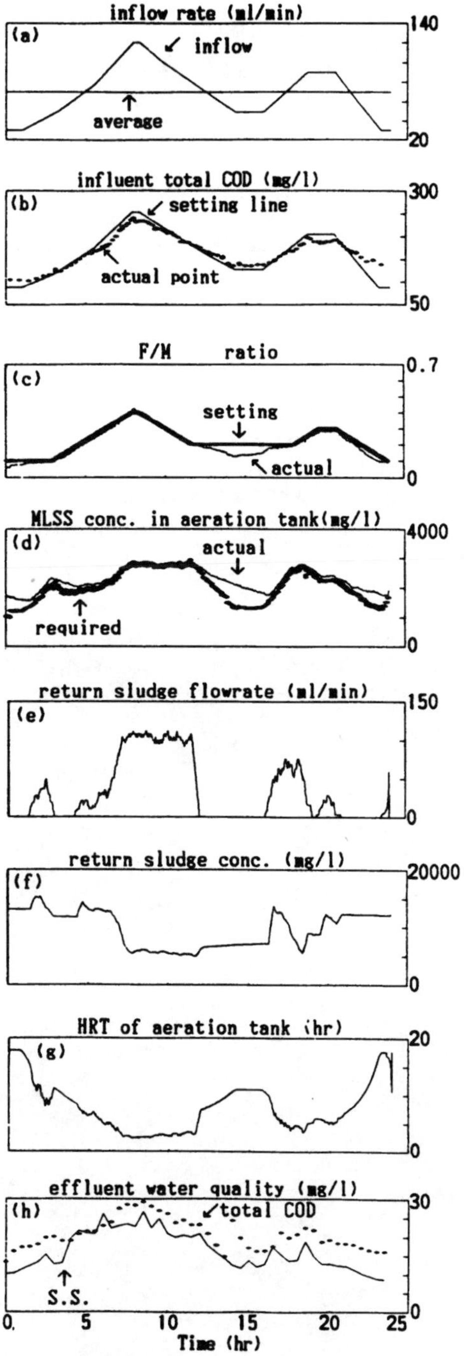

Fig. 4. Experimental data.

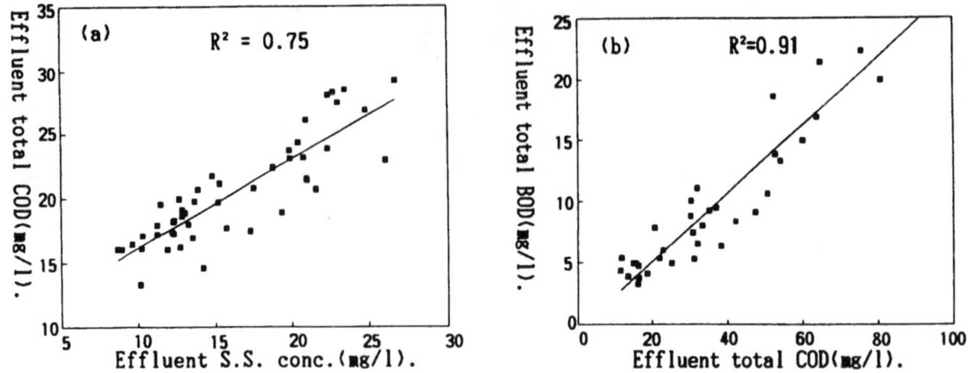

Fig. 5. Relationship between effluent S.S., total COD and total BOD.

Forecast Model Building

The factors affecting MLSS concentration in the aeration tank are influent flow rate (Q_{in}), return sludge flow rate (Q_r) and concentration (X_r) as seen in Eq.(10). The variation of X_r is small enough, except during peak flow time, to ignore for simplification of the model, so the MLSS forecasting model contains only two factors, namely Q_{in} and Q_r. The best model with the lowest performance index is shown in Eq.(12).

(12)

The time delays of Q_r and Q_{in} are 1.5 and 0 h respectively as seen in Eq.(12). In other words, MLSS concentration in the aeration tank at time t is affected by Q_r at time t-1.5 and the Q_{in} at time t. There are no clear physical meanings for the time delays obtained from experimental data. The results, as shown in Fig.6a, indicate good prediction for MLSS concentration.

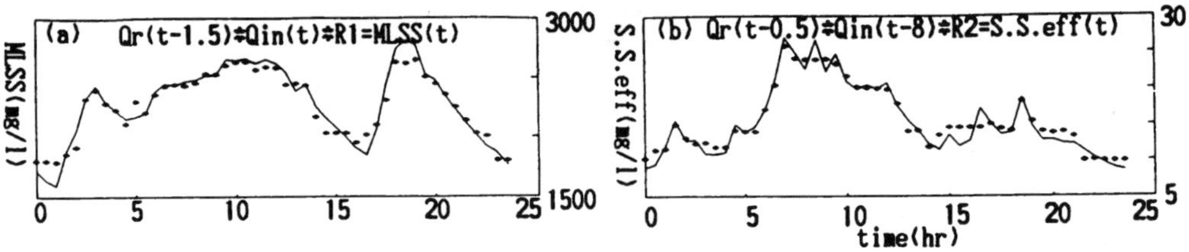

Fig. 6. Results of forecasting model.

Supposing the settling characteristics of sludge flowing into the final settler do not change with time, Q_r and Q_{in} are then also two important factors affecting effluent S.S. concentration and the best model is shown in Eq.(13).

(13)

The forecasting results indicate that effluent S.S. concentration at time t is affected by Q_r at time t-0.5 and Q_{in} at time t-8 and Fig.6b shows that the prediction is good.

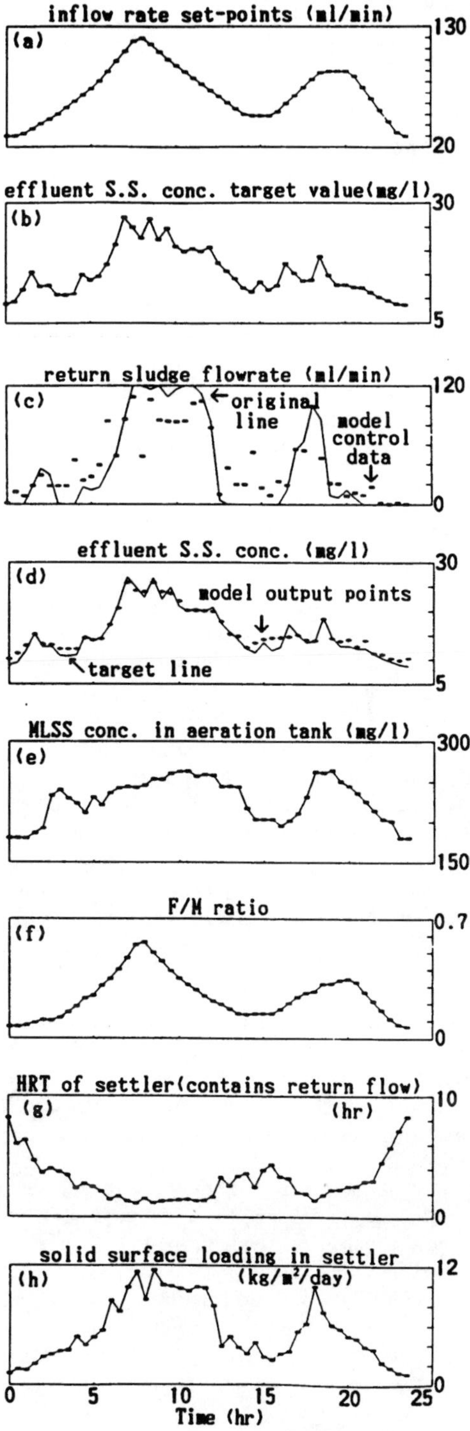

Fig. 7. Testing results of control model.

Control Model Building and Evaluating

Eq.(13) demonstrates that effluent S.S concentration is controlled by Q_{in} and Q_r. However, the amount of influent flow varies with the size of sewer network connected to the plant and is difficult to control manually. The return sludge flow rate is therefore the only variable to control effluent S.S. concentration. On the other hand, if $S.S._{eff}(t)$ and $Q_{in}(t-8)$ are given, the optimum control variable $Q_r(t-0.5)$ can be obtained. That is to say, in addition to forecasting, Eq.(12) can be used for control purposes. As for the target value of $S.S._{eff}(t)$, it is predetermined by decision makers.

To evaluate the control result of Eq.(13), the experimental data discussed in the above section are used to obtain the control variable, namely $Q_r(t)$. The input data sets of the model are influent flow rate and effluent S.S. concentration as shown in Fig.7a and 7b respectively. The output of the model, the theoretical return sludge flow rate, concides with raw control data in the experiment (Fig.7c, dotted and solid line respectively). Figure 7d is the comparison between the theoretical model output and actual experimental data of effluent S.S. concentration (dotted and solid line respectively). Based on these results, it indicates that the model is adequate to forecast and control effluent S.S. concentration. Here MLSS concentration in aeration tank, F/M ration, HRT and solid surface loadings of secondary clarifier are shown in Fig.7e, 7f, 7g and 7h respectively.

Minimization of Effluent S.S. Concentration

This study considers firstly 9 mg/l as the target value of effluent S.S. concentration because it is the minimum output in the experiment, and the total COD and BOD can be consequently reduced to about 15 and 4 mg/l respectively (obtained from Fig.5). The inputs of the control model are shown in Fig.8a and 8b and the output in Fig.8c. The control strategies shown in Fig.8c indicate that taking no action during peak flow time and pumping a large amount of sludge to the aeration tank during lower loading time to store sludge required during peak flow time will make effluent S.S. concentrations closest to the target value. Figure 8d shows the result that the derived control actions reduce overall effluent S.S. concentration significantly, especially during peak flow time and it is closer to the target value than that uncontrolled. MLSS concentration in aeration tank is maintained at 1700 mg/l to 2400 mg/l (Fig.8e) and F/M ratio at 0.08 to 0.5 (Fig.8f). These two are approximately equal to those of the uncontrolled. From Fig. 8g and 8h, the HRT of the final settler under control is shown to be more stable than that uncontrolled, so as to the shorter time of low HRT and the lower solid surface loadings. All these results indicate that the increase of HRT and decrease of solid surface loading are the two best strategies for reducing effluent S.S. concentration.

Minimization of Effluent S.S. Variation

If stable effluent S.S. concentration is required, the average value of experimental data can be used as target value (Fig.9b) to investigate the optimum control strategies at the same influent conditions (Fig.9a). The theoretical control measure (Q_r) derived from fuzzy control theory is stable (Fig.9c). The average return sludge flow rate is about 40 ml/min (0.33 Q_{max}). The theoretical effluent S.S. concentration is close to the target value except during peak flow time (still lower than that uncontrolled, see Fig.9d). The reason for higher effluent S.S. concentration during peak flow time is because the solid surface loading is higher (Fig.9h) and the HRT of final settler is shorter (Fig.9g) than any other time. The MLSS concentration in the aeration tank has been maintained at about 1900 mg/l to 2500 mg/l at all control time interval (Fig.9e) and F/M ratio (Fig.9f) is approximately equal to that uncontrolled. The HRT of the final settler is more stable and the solid surface loading is not only more stable but also lower than that uncontrolled. All these results show that the stabilities of Q_r, HRT and solid surface loading of the final settler will benefit reducing effluent S.S. concentration variation.

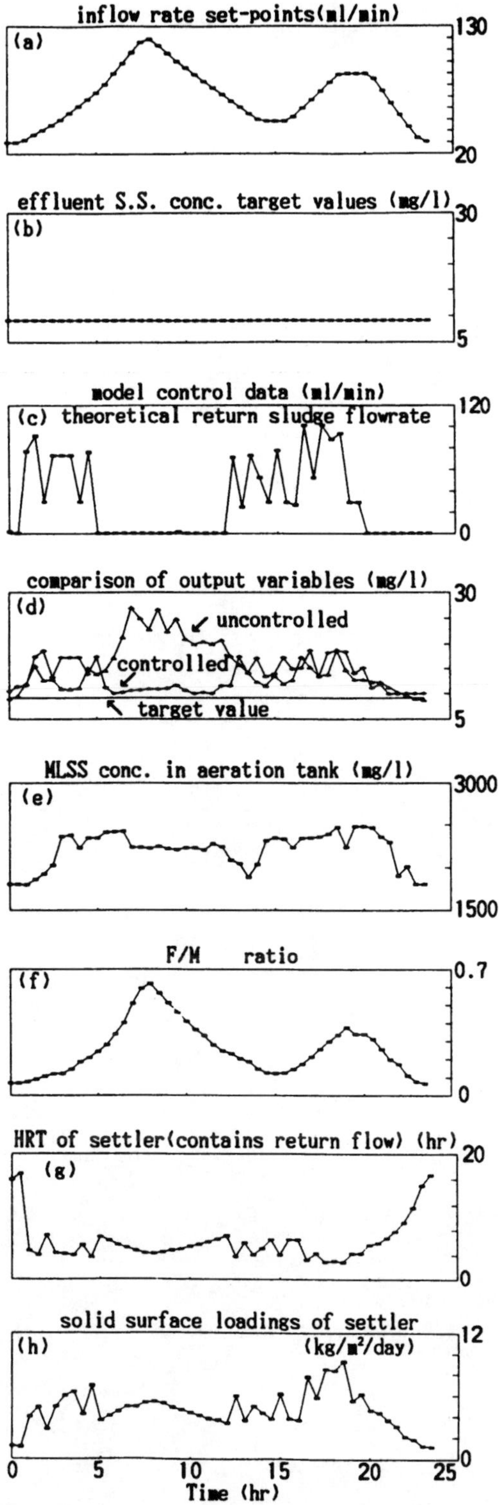

Fig. 8. Controlling results of minimum effluent S.S. concentration.

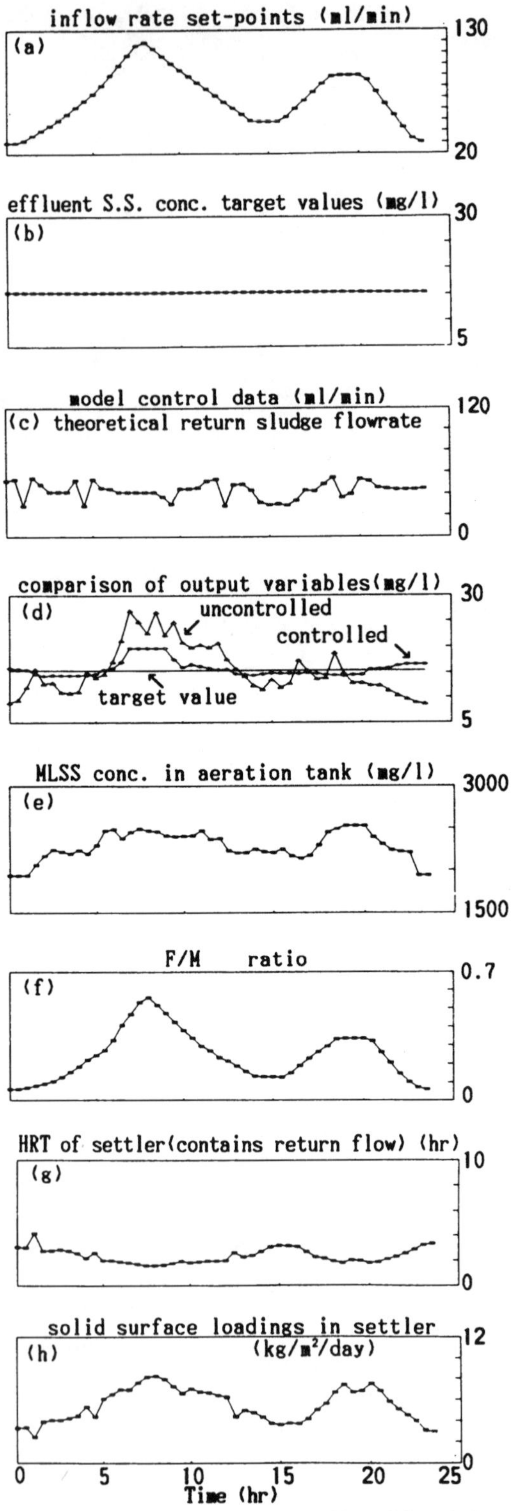

Fig. 9. Controlling results of minimum effluent S.S. variation.

CONCLUDING REMARKS

This paper described an application of fuzzy control techniques to the regulation of the activated sludge process. The process was simulated by the dynamic inputs and automatic control pilot plant. The resulting optimization problem of the fuzzy relational equation was solved by a modified Newton's approach. The major research conclusions are as follows. (1) The proposed fuzzy control techniques can properly handle nonlinearity, high dimensionality and uncertainty of the process. (2) The proposed approach can predict the outputs of the process and obtain the optimum control strategies, only by using experimental data instead of expert's knowledge. (3) The optimum control strategies for reducing effluent S.S. concentration include the increase of HRT, decrease of solid surface loading of settler, the returning of large amounts of sludge to the aeration tank at low influent loading time to store sludge for peak flow time and no returning action at the same time. (4) The optimum control strategies for minimizing effluent S.S. variation are the stabilization of HRT, solid surface loading of the settler and return sludge flow rate. (5) The decrease of effluent S.S. concentration can also reduce the effluent total BOD and COD at the same time. (6) Fuzzy control theory can assist the understanding of a complicated system for which there might exist no expert who can control it. Further extensions of this work include (1) considerations of other influent patterns simulating the large, medium and small scale city (2) considerations of the removal of soluble organic compounds (3) computational efficiency improvement.

ACKNOWLEDGEMENT

The gratitude is given to the NATIONAL SCIENCE COUNCIL in Taiwan for their financial support. (report N.O.: NSC81-0421-E-008-15-Z).

REFERENCES

A.P.H.A., A.W.W.A. and W.P.C.F. (1985). *Standard Methods for the Examination of Water and Wastewater*. 16th Ed., A.P.H.A, Washington.

Boscolo, A., Mangiavacchi, C., Drius, F., Rongione, F., Pavan, P. and Cecchi, F. (1993). Fuzzy control of an anaerobic digester for the treatment of the organic fraction of the MSW. *Wat. Sci. Tech.*, **27**(2), 57-68.

Cashion, B. S. and Keinath, T. M. (1983). Influence of three factors on clarification in the activated sludge process. *J. Wat. Pollut. Control Fed.*, **55**(11), 1331-1337.

Chapman, D. T. (1983). The influence of process variables on secondary clarification. *J. Wat. Pollut. Control Fed.*, **55**(12), 1425-1434.

Henze, M., Grady, C. P. L., Gujer, W., Marais, G. V. R. and Matsuo, T. (1986). Activated sludge model No.1. IAWPRC. Scientific and Technical Report No.1.

Mamdani, E. H. and Assilian, S. (1975). An experiment in linguistic synthesis with a fuzzy logic Controller. *Int. J. Man-Machine Studies*, **7**, 1-13.

Parker, D. S. (1983). Assessment of secondary clarification design concepts. *J. Wat. Pollut. Control Fed.*, **55**(4), 349-359.

Pedrycz, W. (1983). Numerical and applicational aspects of fuzzy relational equations. *J. Fuzzy Sets and Systems*, **11**, 1-18.

Severin, B. F. and Poduska, R. A. (1985). Prediction of clarifier sludge blanket failure. *J. Wat. Pollut. Control Fed.*, **57**(4), 285-290.

Tekippe, R. J. and Bender, J. H. (1987). Activated sludge clarifiers:design requirements and research priorities. *J. Wat. Pollut. Control Fed.*, **59**(10), 865-870.

Tong, R. M., Beck, M. B. and Latten, A. (1980). Fuzzy control of the activated sludge wastewater treatment process. *Automatica* **16**, 659-701.

Tuntoolavest, M., Miller, E. and Grady, C. P. L., Jr. (1983). Factors affecting the clarification performance of activated sludge final settlers. *J. Wat. Pollut. Control Fed.*, **55**(3), 234-248.

Zadeh, L. A. (1965). Fuzzy sets. *Information and Control*, **8**, 338-353.

RESPIROMETRY BASED PROCESS CONTROL

Pergamon

Wat. Sci. Tech. Vol. 28, No. 11-12, pp. 369–376, 1993.

0273–1223/93 $7·00 + 0·00

CONTROL OF ACTIVATED SLUDGE PLANTS BASED ON MEASUREMENT OF RESPIRATION RATES

Abraham Klapwijk, Henri Spanjers and Hardy Temmink

Department of Environmental Technology, Wageningen Agricultural University, P.O. Box 8129, NL-6700 EV Wageningen, The Netherlands

ABSTRACT

An integral control strategy for activated sludge plants presented in this paper consists of three low-level control tasks. The first is the actual respiration rate control by manipulating the influent flow rate, whereby the aerators are at a fixed position for each setpoint. The second task is the maximum respiration rate control by manipulating the waste activated sludge flow rate. The third task of the integral control strategy is the instantaneous respiration rate monitoring, which informs the operator about the performance of the process.

KEYWORDS

Respirometry; activated sludge; continuous respiration meter; process control.

INTRODUCTION

During the past few decades many municipal and industrial biological wastewater treatment plants have been constructed. However a significant number of the treatment plants in the USA (1) and in other countries are not meeting the effluent standards. Besides, in several plants the energy consumption and the operating costs are higher than necessary.

One of the origins of inadequacies in the performance of these plants is the lack of good operational control. Some variables are measured on-line in practice. Instruments currently in use are liquid level detectors, fluid flow rate meters and oxygen sensors. For other information the operator is using time-consuming laboratory analysis.

Instead of expanding wastewater treatment plants that do not meet the standards it would be better to improve the operational control. As there is a lack of on-line instruments, our research was focused on the design of reliable on-line respiration meters that provide the operator of the wastewater treatment plant with useful information on the activated sludge process. We developed a continuous respiration meter* (8,2,6) which can play an important role in measurement and control of activated sludge plants.

* This respirationmeter (RA-1000) is commercially available by Manotherm bv, P.O.B. 7050, 3000 HB Rotterdam, The Netherlands.

The respiration meter is not only developed for control purposes [3]. The meter was also used in the laboratory for kinetic studies [6,7], for toxicity experiments [10] and for evaluation of biodegradability. Further research to improve the meter is still going on [14].

In this paper we shall first present the respiration measurements by which the activated sludge process can be characterised, followed by a discussion of some controlled respiration variables and related manipulated variables. Finally we shall present an integral control strategy based on a continuous respiration meter.

MEASUREMENT OF RESPIRATION RATE

Respiration rates can be measured with a meter (Figure 1), which consisting of a closed, completely mixed respiration chamber (R) through which activated sludge is continuously pumped (P). Dissolved oxygen concentration is periodically measured with one and the same oxygen probe (Z) at the inlet as well as at the outlet of the respiration chamber. This is achieved by alternating the flow direction. The dissolved oxygen concentration in the outlet, which is equal to the dissolved oxygen concentration in the respiration chamber, should not be rate limiting. If necessary the activated sludge should be aerated before it enters the respiration chamber. The respiration rate is calculated every minute from the mass balance of dissolved oxygen over the respiration chamber [11,12].

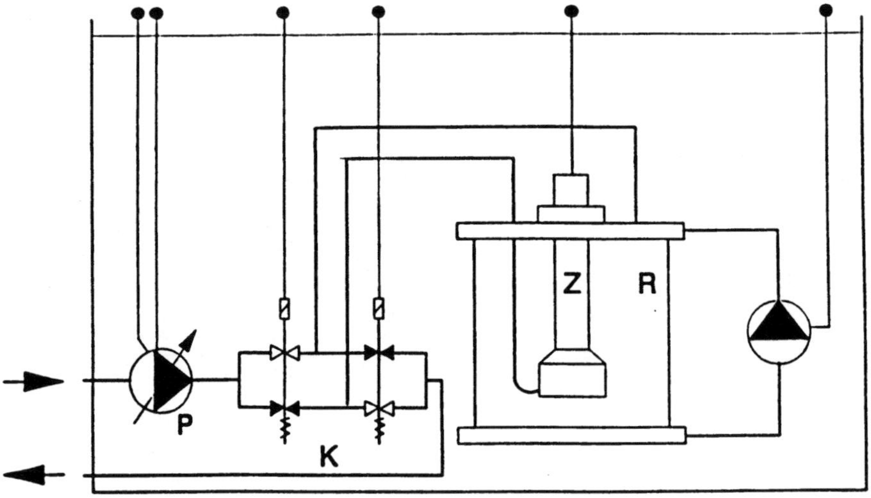

Fig. 1. Flowsheet RA-1000.

To characterize the activated sludge process, four different respiration rates are measured [11,12,13,5] (Figure 2).

The actual respiration rate is defined as the oxygen uptake rate in the aeration tank. To measure this rate, activated sludge from the aeration tank is continuously pumped into the respiration chamber. In addition wastewater is added to the sludge before it enters the respiration chamber. The respiration rate in the respiration chamber is equal to the (mean) actual respiration rate provided that the sludge loading in the respiration chamber equals the loading in the aeration tank. Therefore influent is continuously added to the sludge flowing into the respiration chamber in the proportion [9,11]:

$$Q_{sam} = Q_{in} \cdot V_{res} / V_{at} \tag{1}$$

Q_{sam} = influent sample flow to respiration chamber (m^3h^{-1}).
Q_{in} = influent flow (m^3h^{-1}).
V_{res} = volume respiration chamber (m^3).
V_{at} = volume aeration tank (m^3).

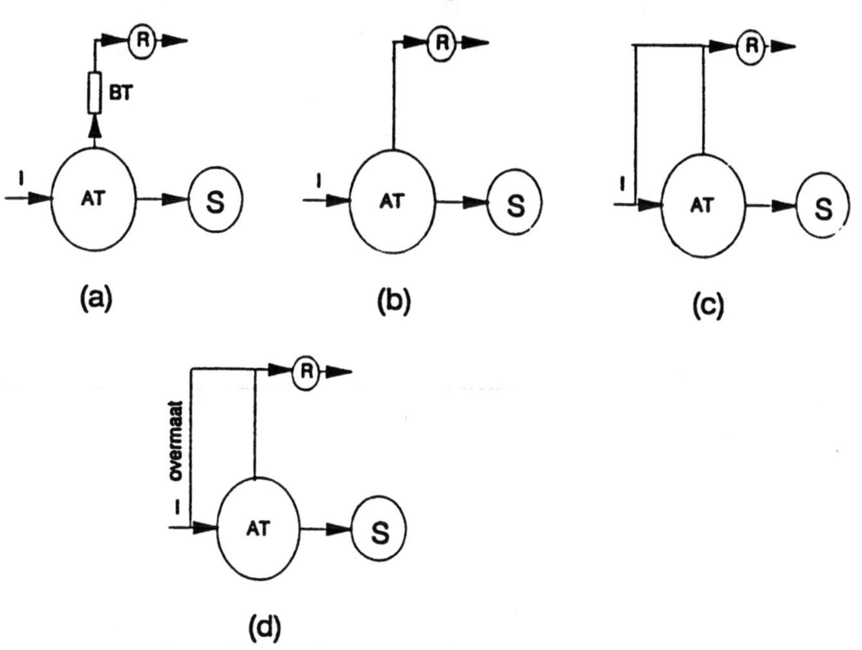

Fig. 2. Measurement of four different respiration rates:
a. basic or endogenous respiration rate (r_{bas})
b. instantaneous respiration rate (r_{ins})
c. actual respiration rate (r_{act})
d. maximum respiration rate (r_{max})
AT = aeration tank, I = influent, BT = bypass tank,
R = respiration chamber, S = settler

The maximum respiration rate is defined as the oxygen uptake rate of activated sludge with an excess of Readily-Biodegradable-Matter. This rate is measured when an excess of influent is continuously introduced to the sludge flowing into the respiration chamber. It was established experimentally for an activated sludge plant treating mainly domestic wastewater that, under normal operating conditions, an excess of influent is guaranteed when the ratio of influent flow to sludge flow is at least 0.03 [12]. However at other plants it is necessary to verify which ratio has to be used.

The endogenous or basic respiration rate is defined as the oxygen uptake rate of activated sludge that has been aerated for 1.5 hours without feeding. Most of the time this retention time is long enough to eliminate the Readily-Biodegradable-Matter completely. In practice the measurement is practically realised by measuring the respiration rate of activated sludge from a plug-flow bypass tank with a hydraulic retention time of 1.5 h, through which sludge from the aeration tank is continuously recirculated.

The instantaneous respiration rate is defined as the oxygen uptake rate of activated sludge flowing directly from the completely mixed aeration tank through the respiration chamber. This rate is lower than the oxygen uptake rate in the aeration tank: the actual respiration rate. The absolute value of the instantaneous respiration rate depends on the detention time in the respiration chamber.

CONTROL OF ACTIVATED SLUDGE PLANTS

The objectives of an aerobic activated sludge plant are:

- BOD removal,
- production of sludge with good settleability,
- keeping minimum energy costs.

These aims can be achieved by manual or automatic control.

Control is achieved by comparing the measured value of a controlled variable with a setpoint. Deviations of this setpoint will be adjusted by manipulation of manipulated variables. The most important manipulated variables in an activated sludge plant are:

1. the air flow rate;
2. the influent flow rate in case wastewater can be stored;
3. the waste activated sludge flow rate.

Several variables, for example dissolved oxygen concentration, biomass concentration and sludge loading can be used as controlled variables. In this paper our attention is focused on the use of respiration rates as controlled variables:

1. the actual respiration rate;
2. the maximum respiration rate.
3. the instantaneous respiration rate,

In the next paragraphs we discuss the different controlled variables and the control-strategies to be used. A synopsis of the control-strategies is given in Table 1.

TABLE 1. Synopsis of control strategies based on respiration rates

Controlled variables	Manipulated variables	
	Fixed	Continuous
actual respiration rate	automatic control	
setpoint 1	air flow rate 1	influent flow rate
setpoint 2	air flow rate 2	influent flow rate
setpoint 3	air flow rate 3	influent flow rate
-	-	-
-	-	-
maximum respiration rate	automatic control	
	flow rate	waste activated sludge
instantaneous respiration rate	manual control	
	variables to be chosen by the operator after evaluation of the process	

Actual Respiration Rate Control

The actual respiration rate, that is the (mean) oxygen uptake rate in the aeration tank of an activated sludge plant, is the resultant of two processes. The first process is the endogenous or basic respiration and the second is the oxygen uptake for oxidation of Readily- Biodegradable-Matter. A change in the loading of the aeration tank with Readily- Biodegradable-Matter has almost no effect on the first process. The rate of the second process however responds to a changing loading with Readily-Biodegradable-Matter. The actual respiration rate consequently is an indication of the loading of an aeration tank with Readily-Biodegradable-Matter.

The actual respiration rate is equal to the basic respiration rate if the loading with Readily- Biodegradable-Matter is zero. The actual respiration increases as the loading with Readily- Biodegradable-Matter increases till the maximum respiration rate is reached. At maximum respiration rate the activated sludge is overloaded, resulting in incomplete removal of Readily-Biodegradable-Matter. Before the maximum respiration rate is reached, most of the time the removal of Readily-Biodegradable-Matter becomes incomplete. This means that there is a critical actual respiration rate in between maximum and basic respiration rate, and at this rate the effluent quality meets the requirements and the removal of Readily-Biodegradable- Matter is satisfactory.

It is likely that the ratio between the critical actual respiration rate and the maximum respiration rate is constant for a given activated sludge plant as long as the oxygenation capacity is not rate limiting. This ratio can be found experimentally by evaluation of the removal efficiency at different ratios for actual/maximum respiration rates. If the oxygenation capacity is rate limiting the critical actual respiration rate is equal to the maximum oxygen intake rate of the aeration tank. Also in this case the critical actual respiration should be found experimentally. The critical actual respiration rate can be found by running the plant at different loadings and measuring the actual respiration rate and the dissolved oxygen concentration in the aeration tank.

There is an advantage when the actual respiration rate is always equal to the critical actual respiration rate. In this situation the activated sludge is never overloaded and works at the maximum acceptable rate. Therefore the total amount of activated sludge needed for the treatment process is as low as possible and the volume of the aeration tank can also be as small as possible.

A system can be installed to control the actual respiration rate by measuring this rate with a continuous respiration meter as described in the section on Measurement of Respiration Rate. The setpoint is the critical actual respiration rate and the influent flow rate is the manipulated variable. This is only possible with a storage tank for the wastewater and/or with use of the storage capacity of the sewer.

In a plant where the actual respiration rate is controlled by manipulating the influent flow, the uptake rate of oxygen in the aeration tank is constant. Therefore the aerators have to be designed for the expected oxygen uptake rate and such that overcapacity is not needed. In addition the aerators can be set in a fixed position with an optimal energy consumption/oxygen input ratio. In this strategy the air flow rate can also be seen as manipulated variable. In contrast to the influent flow rate the air flow rate is fixed for a given setpoint.

The ideal constant actual respiration rate can not always be met because the storage capacity is too low or because the wastewater is temporarily weak. Therefore an alternative for the proposed actual respiration rate control has been developed. In this alternative a series of setpoints for the actual respiration rate is introduced. Each setpoint is coupled with another oxygenation capacity (OC) of the aerators, for example:

setpoint 1 (50 g m^{-3} h^{-1}): air flow 1 (OC: 55 g m^{-3} h^{-1})
setpoint 2 (100 g m^{-3} h^{-1}): air flow 2 (OC: 110 g m^{-3} h^{-1})
setpoint 3 (150 g m^{-3} h^{-1}): air flow 3 (OC: 165 g m^{-3} h^{-1})

The highest setpoint should be the critical actual respiration rate by which the activated sludge works at full capacity. At the other setpoints, the oxygenation capacity is the limiting factor for the actual respiration rate and the activated sludge does not work at full capacity. In this approach with a series of setpoints the activated sludge plant does not always work at full capacity, but the aeration is performed in the most economic way. An economic evaluation should be executed to know what is cheaper: storage capacity and a small plant working at one actual respiration rate or a larger plant without or with smaller storage capacity working with a series of setpoints for the actual respiration rate.

We can conclude that actual respiration rate control can be performed by manipulating the influent flow whereby the aerators are at a fixed position. The advantages are that the loading of the activated sludge plant can be increased in comparison with a non-controlled plant and that aeration is optimised with respect to energy consumption.

Maximum Respiration Rate Control

The maximum respiration rate is defined as the respiration rate with excess of Readily- Biodegradable-Matter. In the experimental setup for the maximum respiration rate described in the section on Measurement of Respiration Rate the biomass concentration in the respiration chamber is equal to the biomass concentration in the aeration tank. Also environmental factors such as pH and temperature are the same in aeration tank and respiration chamber. Therefore the maximum respiration rate is equal to the maximum oxygen uptake rate in the aeration tank and is a measure of the concentration and quality of the biomass in the aeration tank.

To meet the effluent standards it is necessary to prevent overloading of the plant. Therefore a minimum amount of biomass should be kept in the aeration tank. The minimum amount of biomass is related to peak-loadings with Readily-Biodegradable-Matter. There should be just enough biomass to have satisfactory treatment at peak-loadings. Biomass control by manipulating the waste activated sludge flow is a possibility. A drawback of biomass control is that the relation between biomass and activity is not clear. A better approach therefore is maximum respiration rate control.

A historical data set of loadings with Readily-Biodegradable-Matter provides information about the desirable maximum respiration rate in the aeration tank. The maximum respiration rate should be somewhat higher than the actual respiration rate at the peak-loadings.

Control of the maximum respiration rate provides an opportunity to control the maximum treatment capacity. A setpoint for the maximum respiration rate is chosen as discussed and the maximum respiration rate is controlled by manipulating the waste activated sludge flow rate.

A combination of actual respiration rate control and maximum respiration rate control is most effective. However maximum respiration rate control can also be used without actual respiration rate control.

Instantaneous Respiration Rate Monitoring

One objective of the treatment process is a low value for Readily- Biodegradable-Matter in the effluent. For a completely mixed aeration tank, the concentration of the Readily- Biodegradable-Matter in the effluent is more or less equal to the Readily-Biodegradable- Matter-concentration in the aeration tank. Thus the Readily-Biodegradable-Matter- concentration in the tank itself should be very low too.

The respiration rate of the activated sludge is affected by the concentration of the dissolved Readily - Biodegradable-Matter. The respiration rate is equal to the endogenous or basic respiration rate if the concentration of dissolved Readily- Biodegradable-Matter is zero. The respiration rate increases with increasing concentration of dissolved Readily- Biodegradable-Matter until the maximum respiration rate is reached.

The respiration rate of activated sludge sampled from an aeration tank indicates the concentration of dissolved Readily-Biodegradable-Matter in the aeration tank provided that the respiration rate is measured just after sampling. A respiration rate almost equal to the basic respiration rate indicates a low concentration of dissolved Readily-Biodegradable- Matter.

The instantaneous respiration rate described in the section on Measurement of Respiration Rate is a continuous measurement of the respiration rate of activated sludge from the aeration tank just after sampling. This rate gives in combination with the basic respiration rate a good indication of the efficiency of the treatment plant. As long as both rates are more or less equal there is no problem. The concentration of Readily- Biodegradable-Matter in the effluent is low. If the difference between the two rates increases, the process deteriorates and the concentration of Readily- Biodegradable-Matter in effluent increases.

The question is in which way the instantaneous respiration rate has to be controlled. It is of course possible to use instantaneous respiration rate control as an alternative for actual respiration rate control. In this approach the loading is seen as the critical factor and the loading is influenced by manipulating the influent flow. However in our opinion a better approach is to use the instantaneous respiration rate measurements to inform if the plant works properly or not.

Overloading is one of possibilities of a plant which is not properly working. A change in aeration capacity is another. A shock-loading with toxic material also has a detrimental effect on the removal capacity of the plant. In addition other factors have an influence on the activated sludge process.

An increase of the instantaneous respiration rate informs the operator of the plant that the process is not working properly. He has to evaluate the plant at several points and based on this evaluation a decision is made which variable to manipulate.

An Integral Control Strategy

An integral control strategy presented in this paper for activated sludge plants consists of three low-level control tasks. The first is the actual respiration rate control by manipulating the influent flow rate, whereby the aerators are at a fixed position for a particular setpoint. The second task is the maximum respiration rate control by manipulating the waste activated sludge flow rate. The last task of the integral control strategy is the instantaneous respiration rate monitoring which informs the operator about the performance of the process.

In this paper not all aspects of an integral control strategy based on continuous measurement of respiration rates are discussed. For example the strategy to detect toxic loading [15] is not presented here, nor is the use of respiration rate measurement to control the efficiency of the aerator (4) and in addition other aspects of respiration rate control could be discussed.

The control strategy discussed in this paper can be used for activated sludge plants:

- that are or will be continuously overloaded in the near future,
- that are temporary overloaded,
- with high energy costs for aeration.

CONCLUSION

The continuous respiration meter described in this paper is able to continuously measure important variables in the activated sludge process.

An integral control strategy based on respiration rate measurements can be used:

- to have an optimal biomass concentration,
- to treat the maximum wastewater per volume aeration tank,
- with a minimum amount of air,
- while at every moment the effluent quality should meet the standards.

In this control strategy:

- the actual respiration rate is controlled by manipulating the influent flow,
- the maximum respiration rate is controlled by manipulating the waste activated sludge flow, and
- the instantaneous respiration informs the operator about the performance of the process.

REFERENCES

1. Hegg, B. A., Rakness, K. L. and Schultz (1978). Evaluation of operation and maintenance factors limiting municipal wastewater treatment plant performance. *Journ. Wat. Poll. Contr. Fed.*, **50**, pp. 419-426.
2. Klapwijk, A., Spanjers, H. and Van den Akker, D. (1987). Control of an activated sludge plant with a Wazu-respiration meter. *7th European Sewage and Refuse Symposium EWPCA*, Munich, May 19-22.
3. Klapwijk, A., Spanjers, H. and Temmink, H. (1991). Regeling van aktiefslibinstallaties op basis van respiratiemetingen. *H_2O*, **24**, 548-545.
4. Kim, C-W., Spanjers, H. and Klapwijk, A. (1991). Continuous determination of oxygen transfer coefficient with an on-line respiration meter. Wat. Sci. Tech., **28**(7) (this volume).
5. Kim, C-W., Spanjers, H. and Klapwijk, A. (1992). Validation of an on-line respiration meter and its applications by computer simulation. *Wat. Sci. Tech.*, **26**(5/6), 1355-1363.
6. Ossenbruggen, P., Spanjers, H., Aspegren, H. and Klapwijk, A. (1991). Designing experiments for model indentification of the nitrification process. *Wat. Sci. Tech.*, **24**(6), 9-16.
7. Ossenbruggen, P., Spanjers, H. and Klapwijk, A. (1992). Design and analysis of experiments for a two-step nitrification process model, to be published.
8. Spanjers, H. and Klapwijk, A. (1986). Een continue respiratiemeter - Nieuwe perspectieven voor het beheersen van het actiefslibproces. *H_2O*, **19**, 432-434.
9. Spanjers, H. and Klapwijk, A. (1987). The Wazu-respiration meter - A breakthrough in Automation and control. *Proceedings 10th Symposium on Wastewater Treatment*, Montreal.
10. Spanjers, H. and Klapwijk, A. (1987). Measurement of the toxicity of KCN and some organic compounds for the activated sludge using the Wazu-respiration meter. In: *Management of Hazardous and Toxic Wastes in the Process Industries*. (Edited by Kolaczkowski S. T. and Crittenden B. O.). Elsevier Amsterdam.
11. Spanjers, H. and Klapwijk, A. (1990a). On-line meter for respiration rate and short-term biochemical oxygen demand in the control of the activated sludge process. *Instrumentation, Control and Automation of Water and Wastewater Treatment and Transport Systems (Adv. Wat. Pollut. Control* no. 10) ed. R. Briggs, Pergamon, Oxford, pp. 67-77.
12. Spanjers, H. and Klapwijk, A. (1990b). Dynamiek van het zuurstofverbruik in aktiefslibsystemen gemeten met de Wazu-respiratiemeter. Vakgroep Milieutechnologie, Landbouwuniversiteit Wageningen, 35 pp.
13. Spanjers, H. and Klapwijk, A. (1991). Continuous Estimation of short term oxygen demand from respiration measurements. *Wat. Sci. Tech.* **24**(7), 29-32.
14. Spanjers, H. and Olsson G. (1992). Modelling of the dissolved oxygen probe response in the improvement of the performance of a continuous respiration meter. *Water Research* **26**, 945-954.
15. Temmink, H., Spanjers, H. and Klapwijk, A. (1990). Toepassing van een continue respiratiemeter bij toxiciteitstesten voor actief slib. *H_2O*, **23**, p538-545.

Pergamon

Wat. Sci. Tech. Vol. 28, No. 11-12, pp. 377–387, 1993.

0273–1223/93 $7·00 + 0·00

SIMULTANEOUS BIOKINETIC CHARACTERIZATION OF HETEROTROPHIC AND NITRIFYING POPULATIONS OF ACTIVATED SLUDGE WITH AN ON-LINE RESPIROGRAPHIC BIOSENSOR

P. A. Vanrolleghem and W. Verstraete

Laboratory for Microbial Ecology, Faculty of Agricultural and Applied Biological Sciences, University of Gent, Coupure Links 653, B-9000 Gent, Belgium

ABSTRACT

The more stringent requirements imposed on wastewater treatment systems result in an increasing demand for on-line monitoring equipment for these processes. The paper presents the modification of a fully operational respirographic biosensor to allow simultaneous and on-line biokinetic characterization of both BOD-removal and nitrification capacities of activated sludge.

For this characterization, advantage is taken of the calibration measurements that are performed regularly (every 2-3 h) during normal operation of this biosensor. The injection of a well chosen calibration mixture of readily biodegradable BOD and ammonia to a small activated sludge filled bioreactor results in oxygen uptake rate profiles that are interpreted in the framework of a mathematical model that describes the combined oxygen consumption for BOD removal and nitrification. Model identification gives rise to biokinetic parameters (max. conversion rates and affinity constants) that are indicative of the different metabolic capacities of the activated sludge. It is shown that this information can be used to assess the effect of injected wastewater samples on the sludge condition, e.g. adaptation or inhibition, and allows one to follow the change in relative abundance of heterotrophic and nitrifying populations.

Special attention is paid to the propei composition of the calibration solution. Especially the ratio between NOD and BOD is found to be critical. With the aid of simulation, sub-optimal experimental designs are proposed that allow reliable estimation of the biokinetic parameters. Finally, validation of the proposed design is given with experimental data.

KEYWORDS

Activated sludge process; kinetics; BOD removal; nitrification; oxygen uptake rate; biosensors; on-line monitoring; modelling, optimal experimental design; model identification.

INTRODUCTION

Wastewater treatment processes are becoming increasingly complex due to the more stringent effluent quality requirements that have to be met. Nowadays, not only suspended solids and residual carbon, but also nitrogen and phosphorus concentrations of the effluent must be taken into account for the design and operation of treatment plants.

During the last decades, large investments have been made in building wastewater treatment plants. A lot of interest exists to increase the capacity of or to add nutrient removal capacity to existing installations (Ermel *et al.*, 1992; von der Emde *et al.*, 1992). In the absence of space constraints, the construction of additional aeration tanks and clarifiers can be proposed. Alternatively, more elaborate control strategies that rely on new monitoring equipment can be installed allowing for an increase of the treatment capacity of existing plants. Indeed, in a lot of cases, plants have been largely overdimensioned to reach effluent quality criteria (Allsop *et al.*, 1990). Partly this is due to government policy that has provided grants for investment costs rather than operating costs.

Recent progress in the field of instrumentation is substantial. In addition to a number of physico-chemical sensors that have reached fully operational status, e.g. suspended solids (Stephenson *et al.*, 1981) and nitrate monitors (Isaacs *et al.*, in press), sensors for the direct monitoring of the biological processes involved have also been developed. Among these, respirographic sensors have already reached the status of practical implementation on site (Kalte, 1990; Spanjers and Klapwijk, 1990; Vanrolleghem *et al.*, 1990). The data provided by these instruments are especially valuable for control purposes because they are directly related to the central processes of aerobic biodegradation: carbon oxidation and nitrification.

In this paper, an extension of the capabilities of the RODTOX (acronym for Rapid Oxygen Demand and TOXicity tester) respirographic biosensor is proposed. First, the principles of operation of the RODTOX device in its current configuration are given. Second, the way oxygen uptake rates can be calculated from the raw sensor data is briefly introduced. Next, the model and the methods used to obtain biokinetic sludge characteristics from these oxygen uptake rate data are presented. In a fifth part, simulations based on the model are used to find an experimental design that allows one to obtain reliable estimates. Finally, experimental results obtained with nitrifying sludge are presented to show the potential of the approach.

DESCRIPTION OF THE RODTOX RESPIROGRAPHIC BIOSENSOR

Initially, the RODTOX has been developed for the monitoring of the BOD-load and potential toxicity of wastewaters (Vanrolleghem *et al.*, 1990). The sensor obtains this information by performing pulse additions of wastewater to a small bioreactor that contains activated sludge under controlled temperature and pH conditions.

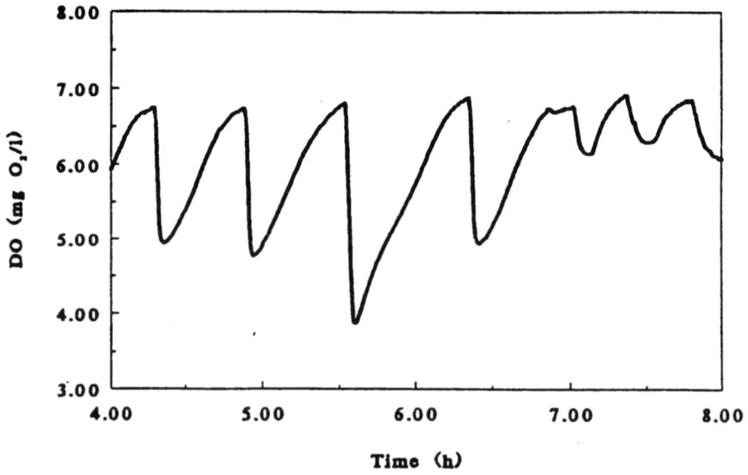

Fig. 1. Typical run of respirograms obtained in the RODTOX biosensor.

The dissolved oxygen concentration in the sensor's down-scaled aeration tank is determined by two competing processes: oxygen supply by continuous aeration and microbial oxygen uptake for endogenous and exogenous respiration. In the absence of substrate, the former type of respiration gives rise to a steady state oxygen concentration, the so-called baseline oxygen level. Pulse addition of substrate to the reactor immediately results in increased respiration. This disturbs the steady state in the bioreactor: dissolved oxygen decreases rapidly. Since the BOD load of the reactor is restricted, the exogenous respiration readily decreases due to substrate limitation. As a result of the oxygen mass transfer, the oxygen level will increase and will return to the baseline. Then the device is ready for the injection of a new pulse of wastewater. The described dissolved oxygen profile is called a respirogram (figure 1) and is used to calculate the short-term BOD and potential toxicity of the wastewater.

Essentially these short-term experiments of approx. 30 minutes reflect the impulse response of activated sludge on wastewater addition. Elsewhere, it is shown how model-based interpretation of such respirograms can yield information on the biodegradation kinetics of wastewaters. These include hydrolysis constants, max. growth rates, saturation constants, etc. (Vanrolleghem *et al.*, in press).

In this work we will focus only on the interpretation of respirograms obtained after pulse addition of a reference substrate. In view of this, it is important to note that in the current RODTOX configuration calibrations are performed regularly (every 2-3 hr). At present, the usefulness of these calibration measurements is rather restricted. Their main purpose is to check the proper operation of the system and to allow the elimination of oxygen mass transfer changes. The goal of the presented work is to extend the usefulness of these calibrations by extracting biokinetic characteristics of the two main microbial populations of nitrifying sludge. It will be shown that this is achievable by proper choice of the reference substrate.

CALCULATION OF OXYGEN UPTAKE RATES

As described above, the reactor dissolved oxygen concentration is the result of two processes. Most interest is given to the biological oxygen uptake and, therefore, the mass transfer is regarded as a parasite process that has to be eliminated from the raw electrode data. Several methods have been developed to assess the mass transfer characteristics without interruption of normal operation of the sensor (Vanrolleghem *et al.*, in press). All are based on the principle of reaeration after the disturbance of the steady state oxygen concentration. In one method this is accomplished by temporarily interrupting the aeration, in the other method advantage is taken of the increased respiration due to substrate oxidation. As exemplified in figure 1, the final part of a respirogram looks very similar to a normal reaeration curve. However, it is essential to determine the time at which the respiration is back at the endogenous level. Only when this condition is fulfilled, is the traditional interpretation of reaeration curves, resulting in K_La estimates allowed. Elsewhere, it is shown how this is accomplished (Vanrolleghem *et al.*, in press). Typical K_La-values are situated between 6 and 10 hr^{-1}. These are rather low, but acceptable values in view of the sub-optimal aeration equipment installed.

Once the mass transfer parameters are identified, oxygen uptake rates can be calculated from the dissolved oxygen mass balance. Since the endogenous respiration can be assumed constant in the (short) time interval of one respirogram, one can easily eliminate it from total respiration. As a result, exogenous respiration rates are available for further interpretation.

MATHEMATICAL MODEL AND BIOKINETIC CHARACTERIZATION PRINCIPLE

The exogenous oxygen uptake rate curves reflect the kinetics of aerobic biodegradation of substrates by the sludge. In the application developed in this paper, two aerobic processes are involved: nitrification and aerobic C-elimination. These processes are independent and their oxygen uptake rates add up. Different periods of exogenous respiration can be imposed by proper design of the batch experiments. This allows one to differentiate both processes on the basis of their respective oxygen uptake rates.

The extraction of biokinetic parameters from the respiration data requires an accurate model. The model used (built in the framework of the IAWPRC model n° 1, Henze *et al.*, 1987) is summarized in Table 1. A description of all symbols used can be found in the nomenclature at the end of the paper. The model describes the growth of the two aerobic populations on ammonia and readily biodegradable BOD. Thus, the degradative capacity of both groups is characterized in this model by their maximal growth rate μ_H and μ_A, resp.) and substrate affinity (K_S and K_{NH}, resp.). It is important to note that no hydrolysis of substrates is incorporated. The choice for this simplification is justified by the fact that there is no need for the injected calibration mixture to contain particulate substrates to assess the sludge biokinetics. On the contrary, such complex calibration solutions would only decrease the reliability of the parameter estimates obtained. However, if in a particular application, one is interested in the characterization of the hydrolytic capacity of activated sludge, the model can be extended in a straightforward way and used to interpret respirographic profiles obtained after the injection of a modified calibration mixture.

TABLE 1. Nitrifying Sludge Process Kinetics and Stochiometry.

Component →	i	1	2	3	4	5	6	Process Rate
j	Process ↓	S_I	S_S	X_H	X_A	S_{NH}	S_{O2}	ρ_j $[ML^{-3}T^{-1}]$
1	Aerobic growth of heterotrophs		$-\frac{1}{Y_H}$	1		$-i_{XB}$	$-\frac{1-Y_H}{Y_H}$	$\mu_H * \frac{S_S}{K_S+S_S} * X_{BH}$
2	Aerobic growth of autotrophs				1	$-i_{XB}-\frac{1}{Y_A}$	$-\frac{4.57-Y_A}{Y_A}$	$\mu_A * \frac{S_{NH}}{K_{NH}+S_{NH}} * X_{BA}$
3	Decay of heterotrophs	f_{SI}		-1			$-1+f_{SI}$	$b_H * X_H$
4	Decay of autotrophs	f_{SI}			-1		$-1+f_{SI}$	$b_A * X_A$

Also, biomass decay is not modelled as lysis to particulate material as in the IAWPRC model, but the description of this process is simplified to a model in which part of the biomass is respired, while a small biomass fraction f_{SI} is recalcitrant and is set free as inert matter. The same approach has been proposed before (Sollfrank and Gujer, 1991).

PARAMETER ESTIMATION

The model identification involves the estimation of parameters Y_H, Y_A, i_{XB}, mu_H, mu_A, K_S, K_{NH}, f_{SI}, b_H, b_A) and initial states ($S_S(0)$, $S_{NH}(0)$, $S_I(0)$, $X_{BH}(0)$, $X_{BA}(0)$). The estimation was performed with the aid of the MoSiFit (Model Simulator and Fitter) program developed at the Laboratory of Microbial Ecology. It incorporates, among others, a variable step 4th order Runge-Kutta routine to solve the stiff differential equations and a conjugative gradient optimization technique (Brent, 1973) was found to be the most adapted non-linear parameter estimation scheme. The objective function to be minimized was the sum of squared (absolute) errors. A good introduction in the field of model identification of ecological models can be found in Robinson (1985).

Saez and Rittmann (1992) have demonstrated the importance of diagnostic checking of the residuals when least squares is used for model identification. The residual errors must comply with the following conditions: independence, randomness and homoscedascity. The first and last conditions were checked on a substantial number of experimental data sets and were shown to be met (results not shown). A check for randomness is made for each model identification separately with a non-parametric runs test (Dixon and Massey, 1957) that attributes a level of confidence to the randomness of the model fit.

Global identifiability (as discussed by Chappell and Godfrey, 1992) of all parameters in the model cannot be achieved on the basis of oxygen uptake rate profiles only. Consequently, only combinations of certain parameters can be determined with the available data.

This is the case for the parameter couple $\mu * X_B$ (for both autotrophs and heterotrophs), which is estimated as such. Eventually, independent measurement of the biomass concentration allows calculation of max. growth rates.

Another complication concerns the estimation of the yield coefficients. It is only thanks to the fact that known amounts of BOD and ammonia are injected that yield coefficients theoretically can be estimated independently of the other parameters. Vanrolleghem *et al.* (in press) were confronted with the problem of non-identifiability of substrate concentrations and yields when assessing the characteristics of wastewaters with unknown composition. They found that some initial state and parameter values could only be estimated up to the factor $(1-Y_H)$. In this work, the approach to express the substrate concentration up to the yield factor was maintained, mainly because the calculation of the yield coefficient requires the exact concentration of substrate injected in the reactor. If the estimated initial substrate concentration is set equal to the injected amount, then all errors will accumulate in the estimate of the yield coefficient. This should be avoided.

In order to decrease the number of parameters and, consequent, make the identification process faster and more reliable, some reasonable assumptions were made. First, since the purpose of the work was not to determine any changes in decay rate and because biomass decay has only negligible effect on the extraneous respiration kinetics within the period of one respirogram, $(-1 + f_{SI}) * b_{H:A}$ was not estimated. Moreover, the endogenous respiration is eliminated from the oxygen uptake rate curves prior to model identification.

Since it is not possible to estimate the amounts of biomass, both types of biomass were lumped in a term X_B, which for its part is then lumped in other parameter combinations as shown above.
Finally, since the net growth of biomass in the RODTOX test vessel is rather low, the amount of ammonia incorporated in the cells i_{XB} is set to zero.

The resulting combinations of parameters that were estimated in this work from real-life data are given in Table 2.

SUBOPTIMAL EXPERIMENTAL DESIGN

Once the principle was established to use the combined oxygen uptake rates to characterize the two aerobic processes involved, it was essential to find the proper operating conditions that allow reliable estimation of the different biokinetic characteristics. Although statistical techniques are available to devise optimal experimental designs (e.g. Beck and Arnold, 1977), their use is not straightforward in the case of nonlinear models.

Instead, a heuristic approach on the basis of simulations was taken to find suboptimal experiments. The mathematical model introduced above was used for simulation, all variable and parameter values being taken from preliminary experiments where the two groups of organisms were characterized separately. In figure 2a, simulated oxygen uptake rates of an optimized experiment are shown. One clearly distinguishes the two aerobic processes, one being terminated halfway before the other's exogenous oxygen uptake ceases. The oxygen uptake rate attributed to nitrification (lower line) is included in the figure for clarity.

The objective of the proposed design process for suboptimal experiments is to obtain a profile where half of the respirographic data can unequivocally be attributed to the slowest process, i.e. nitrification in most sludges, while the other half, through subtraction, yields information on the faster oxidation process. Under such conditions, one expects that parameter estimation is reliable. The variables that can be manipulated to obtain such a profile are the substrate concentrations of the calibration solution: readily biodegradable BOD and ammonia. It is quite straight-forward to find a calibration mixture that satisfies this design objective.

Another design criterion is motivated by the requirement for on-line acquisition of the data: the time it takes to obtain the required data must be minimized. The length of a respirogram is deter-mined by the amount of substrate injected and the respective degradation rates. Consequently, one must restrict the injected amounts.

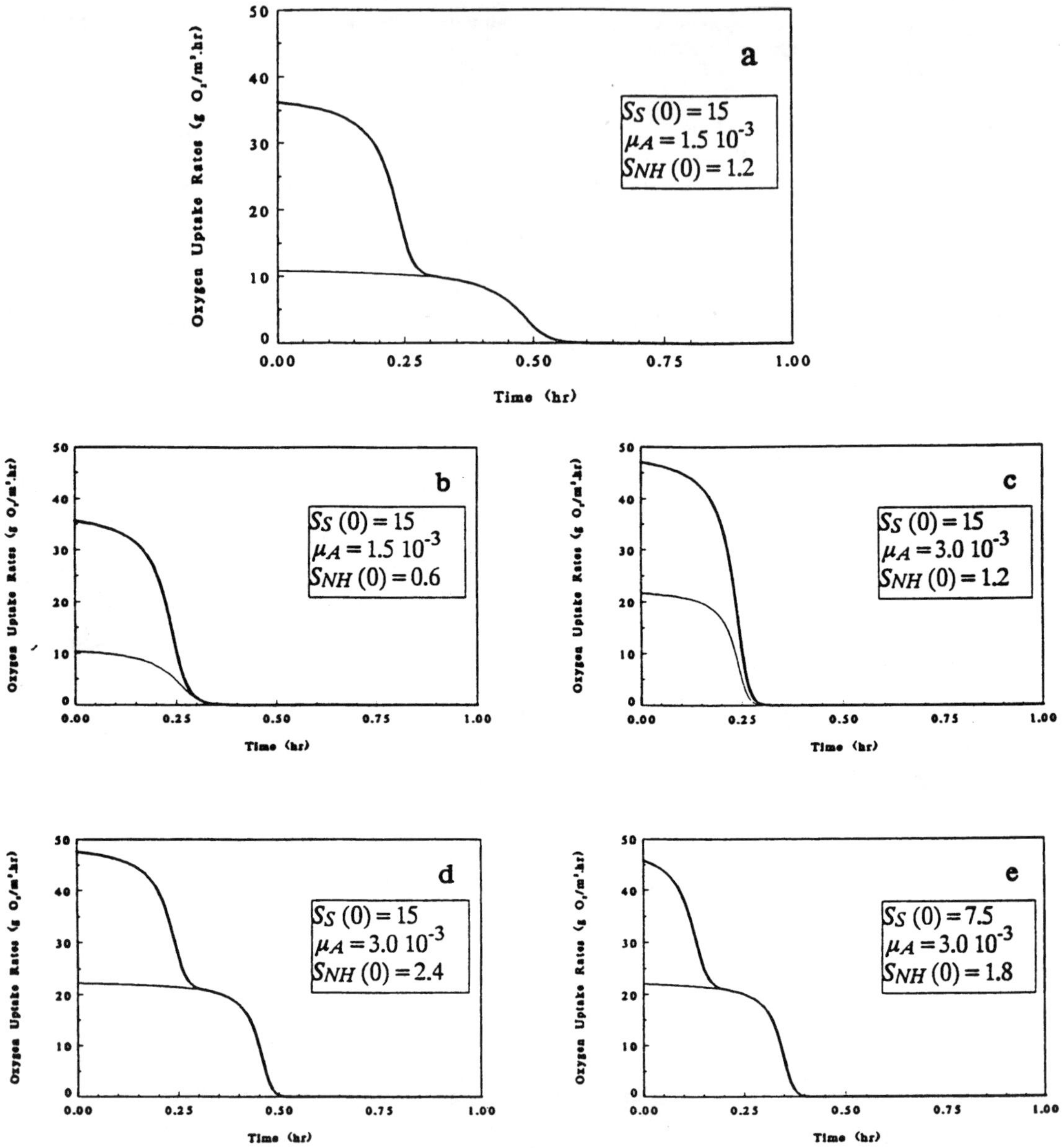

Fig. 2. Simulations of oxygen uptake rate profiles (overall, upper line; nitrification, lower line) obtained from different experimental designs using different parameter combinations and initial substrate concentrations. Parameters common to all simulations are: $_{H} = 0.012$; $K_S = 1.0$; $Y_H = 0.64$; $K_{NH} = 0.05$; $Y_A = 0.25$

In figure 2b, the effect of a wrong choice of substrate ratio is illustrated for worst case conditions. It is clear that it is not possible to differentiate the two processes from the overall oxygen uptake rates. The parameter estimation algorithm may therefore attribute different parameter sets to the data, each set still giving an equal fit: the model is no longer identifiable. Figure 2c shows an analogous case. Here the times of termination of exogenous respiration coincide due to an increased nitrification capacity. Note that the same amount of ammonia was injected as in the reference respirogram (figure 2a).

From these (worst case) examples it can be concluded that the optimal experimental design is dependent on the biokinetic characteristics of the sludge. Consequently, it is not possible to propose here an experimental design that is generally valid. The main purpose of this paper is to present and illustrate the principle of the method so that it can be applied under very diverse conditions.

In figures 2d and 2e, it is shown how such problems of non-identifiability can be solved by minor modification of the calibration mixture. In the first, extra ammonia was incorporated in the calibration solution to change the substrate rate, leading to a longer respirogram but identifiable parameters. The second solution, better in view of the on-line requirement, consists of decreasing the substrate concentrations and aiming for a new optimal substrate ratio. The result of such optimization exercises is that the two oxygen uptake rate profiles again allow reliable biokinetic characterization of the nitrifying sludge.

The cases of non-identifiability that have been illustrated with these examples can occasionally occur during on-line operation. It is therefore essential that the approaching of non-identifiable respirograms can be detected in time so that the proposed solutions can be applied. On-line detection of changing biokinetics is evident and, thus, an alarm signal can be given to the plant operator to change the calibration mixture so that subsequent characterizations are no longer endangered.

In order to eliminate the need for an intervention of the operator, on-line adaptation of the experiment to the new situation is required. However, this asks for a method which can yield optimal experimental designs independent of user interaction (non-heuristically). Therefore, currently, our research is aimed at finding optimal experimental designs using the statistical approach mentioned above. However, our efforts are not directed to the establishment of optimal batch experiments but instead optimal fed--batch experiments (with extra pulse additions in the course of a respirogram) will be looked for. Munack (1989) has demonstrated that the data from such experiments have an increased information content that improves the quality of the parameter estimates.

EXPERIMENTAL RESULTS

All experiments described were carried out with sludge taken from the municipal wastewater treatment plant situated at the hospital Maria Middelares in Gent. Nitrifying capacity was checked before the validation experiments started: a clear respirogram was recorded after the addition of NH_4Cl.

Validation of the proposed method for biokinetic characterization of nitrifying sludge is based on 6 typical respirograms. The experiments were carried out under the following practical conditions. The calibration pump was set to inject a volume of 2 (experiment 1), 3 (experiments 2,4 and 6) or 4 ml (experiments 3 and 5) of a calibration solution containing 40 g COD/l of an equimolar HAc/NaAc solution to the 10 l RODTOX reactor. In addition, the calibration mixture contained different amounts of ammonia nitrogen as NH_4Cl so that the initial reactor concentrations were: 0 (experiments 1 and 6), 0.5 (experiment 4), 1.0 (experiments 2 and 3) and 3.0 mg N/l (experiment 6).

The recorded dissolved oxygen profiles were processed to obtain oxygen uptake rates after elimination of the mass transfer. The estimated K_La-values had a mean value of 7.53 hr^{-1} and a coefficient of variation of 1.4%. The oxygen uptake rate profiles of the experiments in which ammonia was injected are given in figure 3. In three of the four experiments, one can easily detect the time after which the oxygen consumption is only governed by nitrification. The run with only 0.5 mg N/l as initial substrate concentration is a clear example of an experiment where the model is non-identifiable, while the run with 3.0 mg N/l is illustrative of an experimental design where the on-line requirement is not taken into ac-count.

The model identification results are summarized in Table 2 and simulations with the estimates illustrate the fit in figure 3. For sake of clarity, the oxygen uptake for nitrification is included in the figure. It is clear that the fit is acceptable.

TABLE 2. Model Identification Results.

Initial State/	Experiment					
Parameter	1	2	3	4	5	6
K_La	7.63	7.44	7.50	7.42	7.68	7.51
$\mu_H.X_B$	58.4	50.8	53.2	50.0	55.8	70.3
$K_S*(1-Y_H)$	0.182	0.116	0.112	0.096	0.125	0.179
$S_S(0)*(1-Y_H)$	2.82	4.64	6.11	4.60	6.31	5.29
$S_S(0)$real	8.0	12.00	16.00	12.00	16.00	12.00
Y_H#	0.65	0.61	0.62	0.62	0.61	0.56
$\mu_A.X_B$	-	0.786	0.708	0.486	0.870	-
$K_{NH}*(4.57-Y_A)$	-	0.435	0.466	0.103	0.501	-
$S_{NH}(0)*(4.57-Y_A)$	-	4.54	4.24	1.60	11.57	-
$S_{NH}(0)$real	-	1.00	1.00	0.50	3.00	-
Y_A#	-	0.03	0.33	1.37	0.71	-
Residual Error	148.4	73.0	53.9	30.1	100.7	101.1
N	160	241	264	220	337	144
Runs	23	40	43	27	34	25

Calculated under the assumption of exact knowledge of the real initial concentrations S.(0) real.

Conclusions that can be drawn from the results are the following. The data seem to contain ample information for model identification since the parameters are reproducible between the experiments. A considerable, but quite acceptable, deviation in parameter estimates is found for the nitrification process in experiment 4. It is obvious that this is the result of the non-optimality of the injected calibration mixture. The deduced yield coefficients are reasonable in view of the values reported in literature (Henze *et al.*, 1987; Sollfrank and Gujer, 1991). This is a strong indication that the proposed method is valid.

The runs test shows that the residuals are not distributed at random. This means some kinetics of the signal are not incorporated in the model. A closer look at the data, especially in figure 3b and 3d, suggests an oscillation with a period of approx. 2 minutes is superimposed on the data. Such oscillations give rise to significant sizes of separate runs and this is reflected in the low number of runs encountered. A reason for these oscillations is not clear, but it is unlikely that its origin is biological. Rather an artefact caused by hardware must be expected.

From an implementation point of view, it is important to note that the model identification, requiring the estimation of 8 para-meters in a non-linear model, takes about 10 minutes computation time per respirogram on a PC-486/33. Consequently, computation is not a bottleneck for a respirographic sensor carry-ing out this kind of on-line sludge characterization.

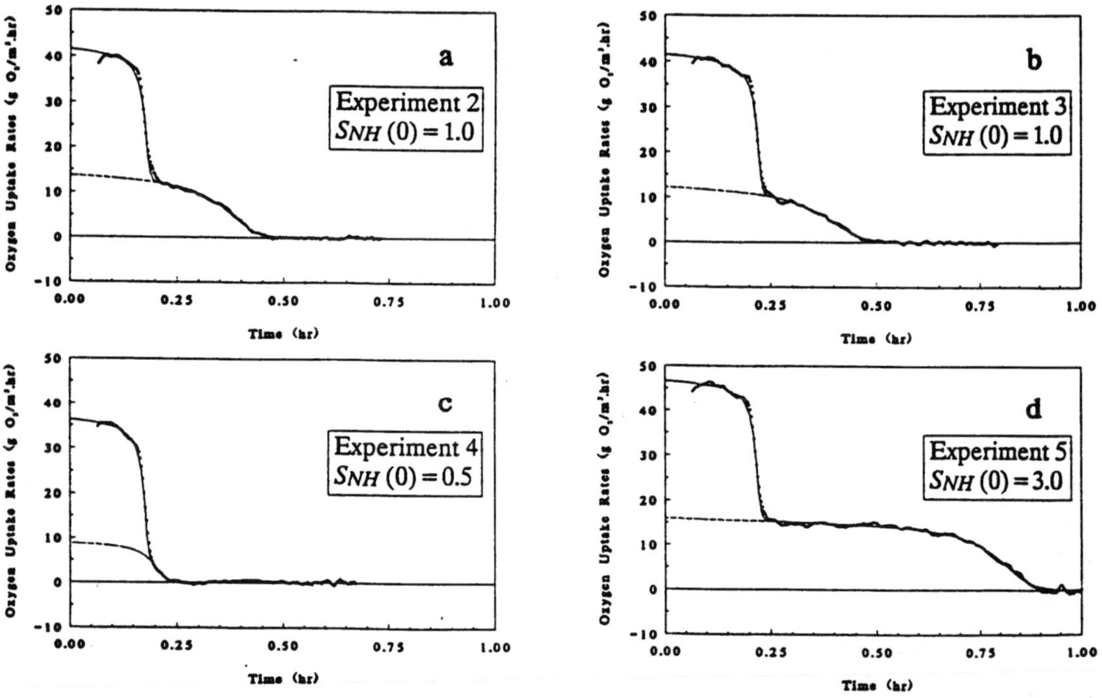

Fig. 3. Overall oxygen uptake rate profiles obtained with different BOD/NH<MV>4Cl-mixtures: Experimental data (dots) and simulations performed with the parameter estimates (full line); Simulated nitrification oxygen uptake rates (dashed line). For more details, see text.

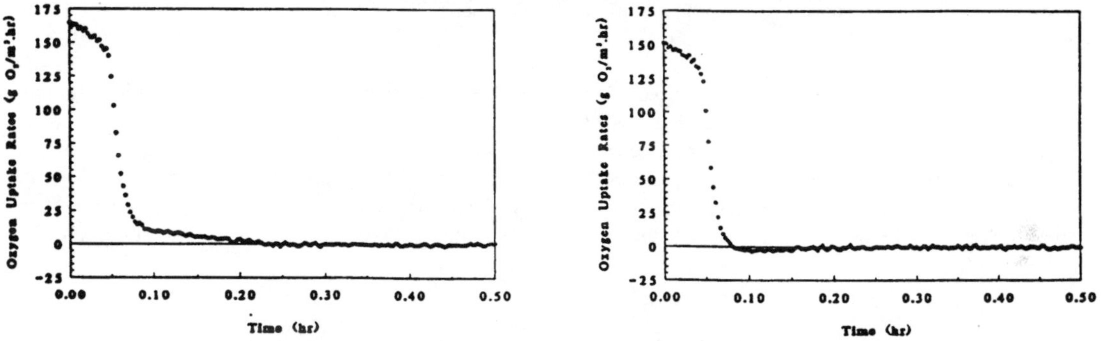

Fig. 4. Respirograms obtained with sludge with increased heterotrophic degradation capacity. The respirogram on the left is obtained from a BOD/NH_4Cl-mixture injection (20 mg BOD/l; 0.5 mg N/l); the one on the right is the result of an addition of 20 mg BOD/l only

Applications of this biokinetic sludge characterization are numerous. As an example, one can follow the activity of both populations when subjected to wastewater entering the plant where the RODTOX sensor is situated. Indeed, under normal operation, the sludge in the sensor's reactor is subject to a F/M load equivalent as the sludge in the aeration tank of the treatment plant it is supposed to monitor. There-fore, the monitoring of the sludge in the reactor allows one to follow the effect of a changing wastewater composition on the heterotrophic and nitrifying populations of the sludge. Especially adaptation and inhibitory effects seem important from a process control point of view. To illustrate this, a new batch of sludge was collected from the Maria Middelares treatment plant a few months after the first batch (used in the validation experiments) was collected. Again this sludge was subjected to injections of a BOD/NH_4Cl mixture. The resulting oxygen profiles are illustrated in figure 4a. In figure 4b, the respirographic profile resulting from an injection of BOD only is given for comparison. It is clear that the sludge has completely different biokinetic parameters: The heterotrophic capacity has increased dramatically, while the nitrifying rates have remained constant.

The latter example also indicates the usefulness of such a sensor in a lab environment for off-line biokinetic characterization of sludges. This may be particularly applicable to firms specializing in supervision of treatment plants because it allows them to rapidly gain information on the condition of sludge collected from the wastewater treatment works they are responsible for.

CONCLUSIONS

The operation of an existing respirographic biosensor was modified in order to take more advantage of the regular calibrations that are included in the system's operation. The injection of an appropriate calibration mixture containing readily biodegradable BOD and ammonia to a nitrifying sludge containing bioreactor results in oxygen uptake rate profiles rich in kinetic information. It was shown that model-based interpretation of these data allows biokinetic characterization of the heterotrophic and nitrify-ing populations in the sludge.

From some worst case examples it was deduced that the identifiability of the parameters is dependent on the experiment performed. Especially, the composition of the calibration solution was found to be essential.

As a result, some attention was paid to the design of a suboptimal experiment. It was concluded that the ratio between the two substrate concentrations is determined by the respective degradation kinetics, while the absolute amounts are governed by the allowable response time of the measurement.

The experiments carried out with nitrifying sludge collected at different time instants clearly showed the potential of the method to continuously follow any changes in a sludge's capacities for heterotrophic carbon elimination and nitrification. This on-line technique therefore allows one to monitor population shifts caused by changing conditions.

The possibility that the kinetics change in such a way that the experiment no longer results in respirograms that allow model identification is small, but this non-identifiability problem may not be neglected. It is quite straightforward to detect such a problem in time so that a modification can be made to the experiment, i.e. by changing the calibration mixture. At the time of this writing, this should be handled by the operator with the aid of the design principles given in the paper. Current research is however aimed at developing a method based on the optimal experimental design theory, that allows to do this without user intervention.

NOMENCLATURE

b_A : Decay rate for autotrophs (h^{-1})
b_H : Decay rate for heterotrophs (h^{-1})
f_{SI} : Fraction of biomass leading to inert material (g COD S_I/g COD X_B)
i_{XB} : Mass of nitrogen per mass of COD in biomass (g N/g COD)
K_La : Mass transfer coefficient (h^{-1})
K_{NH} : Monod half-saturation coefficient for S_{NH} (g N/m^3)
K_S : Monod half-saturation coefficient for S_S (g COD S_S/m^3)
N : Number of data points
S_I : Inert matter (g COD/m^3)
S_{O2} : Dissolved oxygen (g O_2 /m^3)
S_{NH} : NH_4 nitrogen (g N/m^3)
S_S : Readily biodegradable material (g COD/m^3)
t : Time (hr)
X : Biomass (g COD/m^3)

X_{BA} : Autotrophic biomass (g COD/m^3)
X_{BH} : Heterotrophic biomass (g COD/m^3)
Y_A : Yield coefficient for autot-rophic biomass (g COD X-$_A$/g N S_{NH})
Y_H : Yield coefficient for heterotrophic biomass (g COD X_H/g COD S_S)
μ_A : Maximum specific growth rate for autotrophic biomass (h-^{1})
μ_H : Maximum specific growth rate for heterotrophic biomass (h-^{1})

REFERENCES

Allsop, P. J., Moo-Young, M. and Sullivan, G. R. (1990). The dynamics and control of substrate inhibition in activated sludge. *CRC Crit. Rev. Environ. Control,* **20**, 115-167.

Beck, J. V. and Arnold, K. J. (1977). *Parameter Estimation in Engineering and Science.* Wiley, New York. pp. 501.

Brent, R. P. (1973). *Algorithms for Minimization without Derivatives.* Prentice-Hall, Englewood Cliffs, New Jersey. pp. 195.

Chappell, M. J. and Godfrey, K. R. (1992). Structural identifiability of a nonlinear batch reactor model. *Math. Biosci.*, **108**, 241-251.

Daigger, G. T., Buttz, J. A. and Stephenson, J. P. (1992). Analysis of techniques for evaluating and optimizing existing full-scale wastewater treatment plants. *Wat. Sci. Tech.*, **25**(4-5)., 103-118.

Dixon, W. J. and Massey, F. J. Jr. (1957). Nonparametric statistics. In: *Introduction to Statistical Analysis.* McGraw-Hill, New York. 280-303.

Ermel, G., Mennerich, A. and Höper G. (1992). Extension of the Seehausen sewage works to meet the new requirements on nitrogen and phosphorus removal. *Wat. Sci. Tech.*, **25**(4-5)., 43-50.

Fleckseder, H., Prendl, L. and Meulenbroek, H. (1992). The staging of large wastewater treatment plants – A "modular" approach. *Wat. Sci. Tech.*, **25**(4-5)., 67-73.

Henze, M., Grady, C. P. L. Jr., Gujer, W., Marais, G. v. R. and Matsuo, T. (1987). *Activated Sludge Model No. 1.* Scientific and Technical Reports No. 1, IAWPRC, London.

Isaacs, S. H., Soeberg, H. and Kummel, M. (in press). Monitoring and control of a biological nutrient removal process: Rate data as a source of information. In: *Proceedings IFAC-BIO2, Symposium on Modeling and Control of Biotechnical Processes.* Keystone, Colorado, March 29–April 2 1992.

Kalte, P. (1990). The continuous measurement of short-time BOD. In: *Advances in Water Pollution Control.* Ed. Briggs R., Pergamon Press, London. 59-65.

Munack, A. (1989). Optimal feeding strategy for identification of Monod-type models by fed-batch experiments. In: *Computer Applications in Fermentation Technology: Modern Control of Biotechnological Processes.* Eds. Fish, N., Fox, R. and Thornhill, N., Elsevier, Amsterdam. 195-204.

Robinson, J. A. (1985). Determining microbial parameters using nonlinear regression analysis: Advantages and limitations in microbial ecology. *Adv. Microb. Ecol.*, **8**, 61-114.

Romanelli, M. J. (1960). *Mathematical Methods for Digital Computers.* Ralston, A. & Willy, H.S.W., New York. pp. 110-120.

Saez, P. B. and Rittmann, B. E. (1992). Model-parameter estimation using least squares. *Wat. Res.*, **26**, 789-796.

Sollfrank, U., and Gujer, W. (1991). Characterisation of domestic wastewater for mathematical modelling of the activated sludge process. *Wat. Sci. Tech.*, **23**(4-6), 1057-1066.

Spanjers, H. and Klapwijk, A. (1990). On-line meter for respiration rate and short-term biochemical oxygen demand in the control of the activated sludge process. In: *Advances in Water Pollution Control.* Ed. Briggs R., Pergamon Press, London. pp. 67-77.

Stephenson, J. P., Luxon, E. G., Monaghan, B. A. and Gillespie, R. G. (1985). Evaluation of instruments for continuous activated sludge monitoring. *Wat. Sci. Tech.*, **13**(11-12), 721-728.

Vanrolleghem, P., Dries, D. and Verstraete, W. (1990). RODTOX: Biosensor for rapid determination of the biochemical oxygen demand and the on-line monitoring of the toxicity of wastewaters. In: *Proceedings Fifth European Congress on Biotechnology.* Copenhagen, Denmark, July 8-13 1990. Vol. 1, 161-164.

Vanrolleghem, P. A., Van Impe, J. F., Vandewalle, J. and Verstraete, W. (in press). Advanced monitoring and control of the activated sludge process: On-line estimation of crucial biological variables in a structured model with the RODTOX biosensor. In: *Proceedings IFAC-BIO2, Symposium on Modeling and Control of Bio-technical Processes.* Keystone Colorado, USA, March 29–April 2 1992.

von der Emde, W., Kroiss, H., Nagel, G. and Dornhofer, K. (1992). Design procedure and pilot investigations for the extension of the main treatment plant of Vienna. *Wat. Sci. Tec.*, **25**(4-5), 9-22.

Pergamon

Wat. Sci. Tech. Vol. 28, No. 11-12, pp. 389–399, 1993.

0273–1223/93 $7·00 + 0·00

ON LINE RESPIROMETRY: A POWERFUL TOOL FOR ACTIVATED SLUDGE PLANT OPERATION AND DESIGN

J. B. Watts* and W. F. Garber**

** Minworth Systems Ltd., Unit 8G, Castle Vale Industrial Site, Maybrook Road, Minworth, Sutton Coldfield, West Midlands B76 8AL, UK*
*** Civil Engineering and Engineering Science, Loyola Marymount University, Loyola Boulevard at West 80th Street, Los Angeles, CA 90045-2699, USA*

ABSTRACT

This paper will describe, with minimal reference to mathematically derived formulae, the importance and use of respirometry as a design and process operating tool. The authors will report on the latest findings of their team working in the United Kingdom. They will present case studies of events that were rapidly identified by the use of on-line respirometry.

KEYWORDS

Activated Sludge; Respirometry; Readily-Biodegradable Substrate; Oxygen Uptake Rate; Specific Oxygen Uptake Rate; Biological Nutrient Removal

INTRODUCTION

It has been recognised for many years that the aerobic biological treatment of wastewaters could be more efficient in energy use and the removal of organic contaminants than available alternative methods of treatment if good process control could be ensured. For larger and stronger flows this has come to mean the activated sludge process and its variants. In as much as growth of bacterial organisms required for such aerobic biological treatment is logarithmic in character, the best and most efficient treatment, from organic removal and energy-saving bases, would be when the rate of growth of the organisms was at its maximum. Gould (4) developed the so-called "High Rate" activated sludge process in the 1940-50's to utilize this relationship. It worked well where the incoming organic load had small variations and where temperature variations were relatively small. However, where there were larger variations in load and temperature the available process-control instrumentation systems were simply not good enough for the rapid changes which occurred in this process. It was recognised early that oxygen demand (respirometry) would be an excellent control parameter, since it represented a direct measure of the viability and activity of the biological organisms. However, no really usable on-line instrument was available. The Odeometer (5) marketed by the Chicago Pump Company was one of the first attempts to produce an instrument which was specific for activated sludge control and which would provide oxygen demand readings on samples brought to it. Later a chemical method (3) using grab samples taken from various points in the process was developed and used to provide control information (2).

Arthur (1) also built oxygen-demand instruments, but not readily usable on-line. These were, therefore, not adequate for the close process control necessary for minimization of energy use. Subsequently, dissolved oxygen measurements at various points in aerobic treatment were used and the process operated in the organism mature growth phase, where very close control was not as critical.

This paper describes an on-line respirometer developed within an operating treatment plant. Most detectors are self-calibrating and the process control software can feed information to a mathematical model of activated sludge where the effect of control changes can be analyzed to maximize operations. The effect of changes such as temperature and tankage changes can also be predicted to aid operation, maintenance and design tasks.

DESIGN PARAMETERS

Respirometry, or the measurement of oxygen uptake rates (OUR) of a living biomass, has been used in a basically laboratory mode for many years as an indicator of microbial activity in an aerobic biomass. Such measurements on-line could therefore be of significant value in controlling such aerobic biological wastewater processes as activated sludge. Additionally, on-line respirometry can provide a measure of the organic strength of a wastewater, and of how much of that strength must be handled at various points in the process (2).

The activity of an aerobic biomass is affected by variables such as:

a) The types of organisms present
b) Cell viability, or the actual concentration of living cells available to respire
c) Food strengths and its nature (amounts of readily and less-readily biodegradable materials)
d) Toxic or inhibitory components
e) Temperature
f) Availability of sufficient molecular oxygen.

Successful measurement of this activity in on-line respirometry must take into account these six factors as well as considerations of variables such as sampling method and size; instrument location; tank geometry; activated sludge process variation: and the reliability of basic sensors.

On-line respirometry might appear at first sight to be so complex as to suggest that a simpler measurement of the success of the process might be better, but in fact the strength of the on-line respirometric measurement is the fact that it represents a microcosm of the treatment facility and of what really affects process success. It shows in the influent what the strength of the organics needing immediate and longer term oxygen and biomass supplies are and allows evaluation of how oxygen should be supplied throughout the aerobic process.

The average age of the wastewater reaching the Hyperion facility of the City of Los Angeles is 9 hours, and wastewater temperature varies from 24 to 30 degrees Celsius. This means that partial digestion of organic materials has occurred and there is a high concentration of easily-oxidizable volatile organic acids and sulphides. An immediate oxygen demand of from 150 to 250 mg per litre per hour (2) results, requiring a steeply tapered oxygen supply in the plug flow tanks.

The Terminal Island Plant receives periodic inputs of blood from fish cannery operations, with resultant widely varying high oxygen demands. Successful operation of this facility requires flow/strength equalization and carrying of the highest possible active biomass. Manual grab samples analysed for oxygen demand by chemical means allowed critical process factors to be identified, but on-line respirometry is necessary for process maximization.

BOD measures oxygen demand under carefully-maintained artificial conditions and Total Organic Carbon does not necessarily pertain to biomass activity, so they are not really good operational parameters.

Active organisms in the biomass in an activated sludge facility are contained in the solid phase and are dependent for growth upon the carbohydrates, proteins, nucleic acids and lipids in the influent to the plant. Where there is primary treatment, primary effluent is the feed material. Molecular oxygen is needed for the continuing growth of the active organisms although oxygen-containing compounds such as nitrates can provide oxygen to keep them alive where anoxic zones exist. The organisms utilize the oxygen to oxidize the organic food in the wastewater to provide organic carbon to synthesize the compounds necessary for their reproduction. Respirometers measure the rate of the utilization of oxygen for this growth process. Reduction of the BOD and the growth of the organisms are interrelated. They occur concurrently and represent the rate of removal of the carbonaceous material from the wastewaters or the treatment process.

After the initial high rate of oxygen use, organism growth rate and organic material removal, a slower and relatively constant phase of oxidation occurs - the endogenous phase - corresponding to auto digestion of the organisms. This is clearly shown in Figure 1.

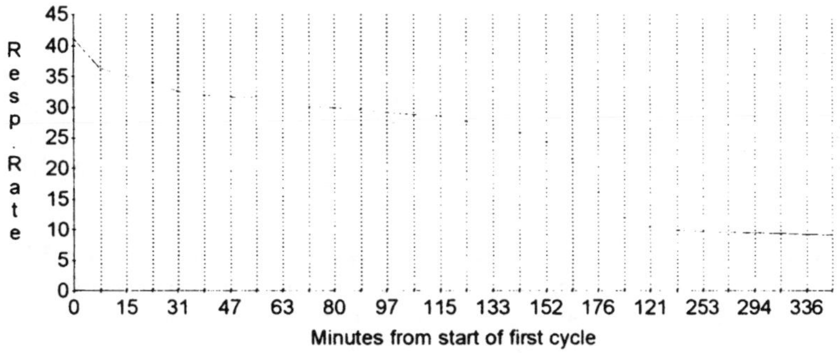

Figure 1: Variation in respirometric rate from initial high rate through to endogenous phase.

To better measure the viability of organisms in the solid phase or activated sludge it is desirable to separate the oxygen used by the organisms from the oxygen required by the whole sample. Oxygen demand from the whole sample can result from the presence of easily oxidizable chemical materials as well as the activity of the biological organisms. To this end, two types of oxygen uptake rates are referred to in this paper: the **total** oxygen uptake rate, designated "OUR" in this paper, and the **specific** oxygen uptake rate, designated "SOUR", of the viable organisms. SOUR is related to the amount of biomass present and can be expressed as mg oxygen used per gram of mixed liquor solids per hour. Figure 2 illustrates the readout obtained over a 24 hour period by the operating on-line respirometer to be described later in this paper. Readouts are for:

a) Dissolved oxygen in air prior to each sampling
b) Reading in air for the suspended solids probe
c) Value in the sample for this probe. (a & b permit re-calibration by re-scaling)
d) The Oxygen Uptake Rate (OUR) in mg/litre/hour
e) Temperature of the bioreactor
f) The Specific Oxygen Uptake Rate (SOUR) in mg/litre/gram suspended solids.

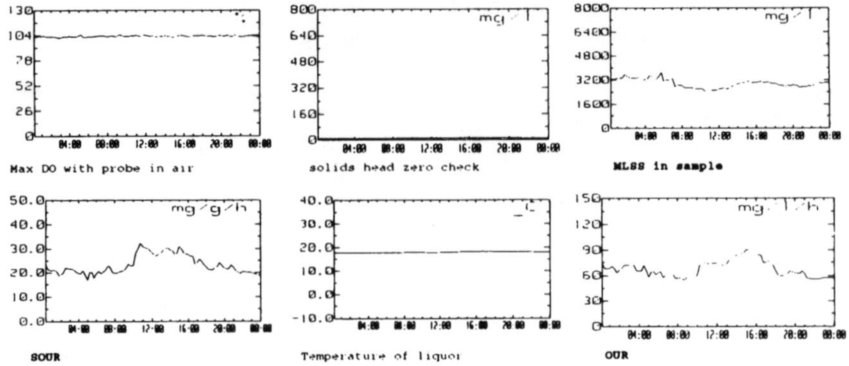

Figure 2: Typical key data from on-line respirometer for a 24 hour period.

The dissolved oxygen probe is self-calibrating, but the suspended solids probe must be re-calibrated at intervals. This is accomplished using the serial communications port in the instrument. A grab sample of the mixed liquor is taken next to the instrument, with the time of sampling recorded and the suspended solids is determined in the laboratory. The difference between the laboratory results and the instrument's suspended solids reading at the time of sampling is used to reset the scaling factor in the memory of the instrument. This is accomplished via the serial port of the instrument, so it can be made on-site via a small portable PC plugged into the serial port, or more conveniently the adjustment can be made remotely via a telephone line and modem. This procedure has been found to be much superior to the usual need to manually adjust potentiometers within the instrument.

Figure 3 illustrates how the differentiation between OUR (the total oxygen demand) and SOUR (the oxygen demand of the viable biomass) is of value operationally. It shows what happened to a plant when the returned activated sludge pumps failed and the viable biomass dropped.

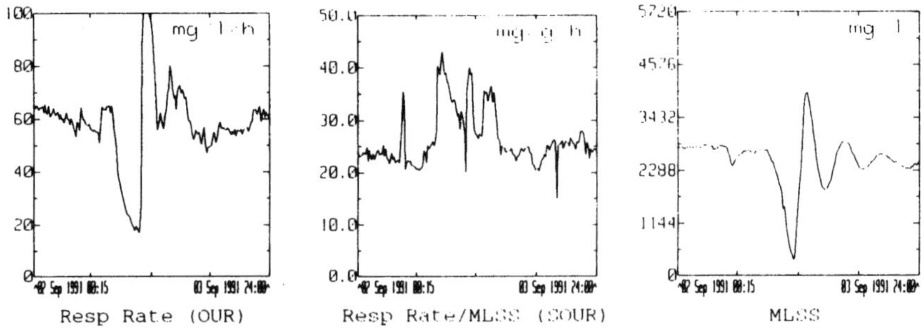

Figure 3: 2 Days during which RAS pumps failed.

On this occasion the OUR suggested that the oxygen requirement was **less than normal** because there were few viable organisms present, whereas SOUR showed that the organic loading on the viable organisms present was **very much higher than normal**. A rain storm in the facility where the on-line respirometer was in service would have given a similar picture, but in either case it would have indicated to the operator that action was necessary.

FACTORS IMPORTANT IN INSTRUMENT DESIGN

The Sampling Cycle

Development of the on-line respirometer was carried out at the Minworth Treatment Plant in Birmingham in the United Kingdom by personnel who had at one time been responsible for the operation of the plant, but were now a privatized Company called Minworth Systems Limited. During their operational experience they had noted the importance of reliable repeat sampling and of obtaining samples truly characteristic of the process stream. They were aware of the accurate sampling required for respirometry and over a 6 year period developed an in-situ floating ball device containing the sampling vessel, all necessary sensors and other equipment. Since the ball floats on the surface at the sampling point, its temperature is essentially the same as that of the bioreactor at that point, so the respirometric rate under actual plant conditions is measured. The ball with its sampling apparatus is allowed to sink into the process liquor to a depth where a portion of the liquor is sealed within the chamber in the ball. This portion is aerated to a set dissolved oxygen level and then the decay in the dissolved oxygen level in the sample with time is measured. The microprocessor controlling the instrument compares the rate of decrease of dissolved oxygen caused by the biomass with the usual characteristic curve. Major deviations from the usual decay curve require explanation and/or possible rejection of the sample results. Figure 4 illustrates a typical respirometer sample cycle of filling the chamber in the ball, aeration and dissolved oxygen utilization by the biomass.

Graphs A and B show the dissolved oxygen through the cycle and graph C shows the suspended solids content of the sample during the cycle. These data were obtained from the logging element in the instrument memory, which can store 14 days of 15-minute averages plus the last 5 individual sample cycles to the detail shown in the graphs.

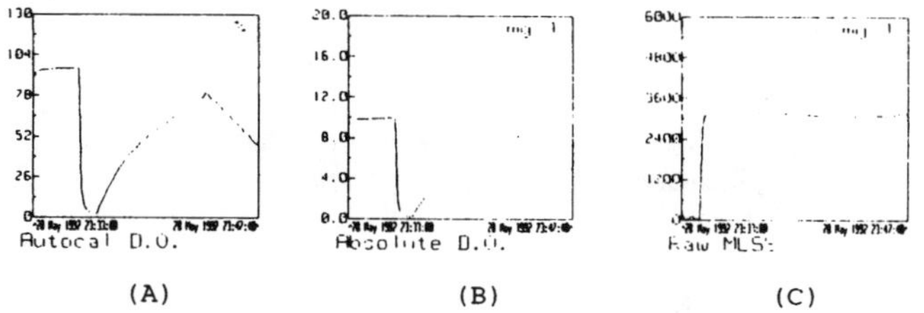

Figure 4: Typical on-line respirometer sample cycle.

When the sample is automatically ejected the dissolved oxygen and suspended solids probes are in air, so self-calibration of the oxygen probe and checking the extent of fouling of the suspended solids probe is possible. This allows analysis of the instrument's operation remotely through the telephone lines. The dissolved oxygen and level sensor (which closes the sample chamber at the sampling depth) are auto-cleaned during each sample cycle. The sample chamber is usually manually cleaned very irregularly. The suspended solids sensor requires the most operation attention depending upon the physical characteristics of the biomass. Weekly manual cleaning of this probe may be necessary in particularly fouling environments. Internal compensation software minimizes this problem by correcting the reading in the sample for that in air (zero suspended solids). The need for cleaning is usually indicated when the reading in air exceeds 10 per cent of that in the sample. In spite of this seeming complexity the respirometer has been functioning reliably with minimal operational attention on an on-line basis in several facilities in the United Kingdom.

INFORMATION AVAILABLE FROM THE ON-LINE RESPIROMETER

Development of a reliable on-line instrument allowed accurate data to be gathered which could in turn be used to obtain information of considerable value in both operation and design of activated sludge systems. In an earlier paper Bargman, Betz and Garber (4) showed how grab samples and chemical method respirometry had been used to both detect up-sewer trouble points (and measure in-sewer organic digestion) and to taper oxygen application rates within a plug flow reactor as well as to provide better operational control. However their method was crude and labour-intensive; could only be used at intervals by skilled chemists and could not provide the detailed information possible with the on-line respirometer whose utilization is described here.

During the development of the on-line respirometer it became evident that the location or locations had to be carefully planned if the information and control were to be maximized. The location where the primary effluent and return activated sludge initially mixed provided information as to the 24-hour variation of the wastewater oxygen demand, but did not allow good information as to whether biomass activity or wastewater strength were limiting. Thus it could be assumed that the maximum respirometric rate observed in a 24-hour period was limited by biomass viability and that the rest of the period accurately reflects the organic strength of the primary effluent. Figure 5 illustrates this by showing what happened to SOUR when the organic loading on a plant suddenly increased.

OUR and SOUR doubled in a few days and the final effluent started to exceed the prescribed ammoniacal nitrogen consent limit. The operator increased the biomass by limiting wasting and the SOUR gradually returned to normal, but the sludge mass increased as did the measure of poor sludge settleability, the Sludge Volume Index (SVI) and solids were carried from the final settling tanks into the receiving river waters. The respirometer had almost immediately noted the increased loading on the viable biomass and, if it had been put as part of a control loop, could have allowed a constant ratio of food to biomass to be obtained so that violations in the ammoniacal nitrogen and suspended solids of the effluent could have been avoided.

Analysis of figure 5 shows the value of the in-line respirometer when placed where the incoming primary effluent strength is measured. However in this location, where the return organisms and the primary effluent are mixed, it is difficult to obtain a measure of the activity and viability of the activated sludge biomass itself. This measurement is important since the continuing active growth of these organisms is what the process seeks to encourage to maximize organic removal. The question then becomes what is a proper level of return sludge activity and how can its determination aid in process control. Returning to Figure 1, some information can be obtained.

The data in figure 1 was obtained by treating a single sample within the respirometer to 20 repeated aerations and utilizations by the organisms of the oxygen to oxidize and utilize the organic foods. The initial slope represents the readily biodegradable COD with the curve gradually flattening as carbonaceous and nitrogenous materials are oxidized. The slope after about 220 minutes is believed to be the endogenous respiration rate which in the case of this plant at the time approximated to 4.5 mg of oxygen per gram of biomass per hour.

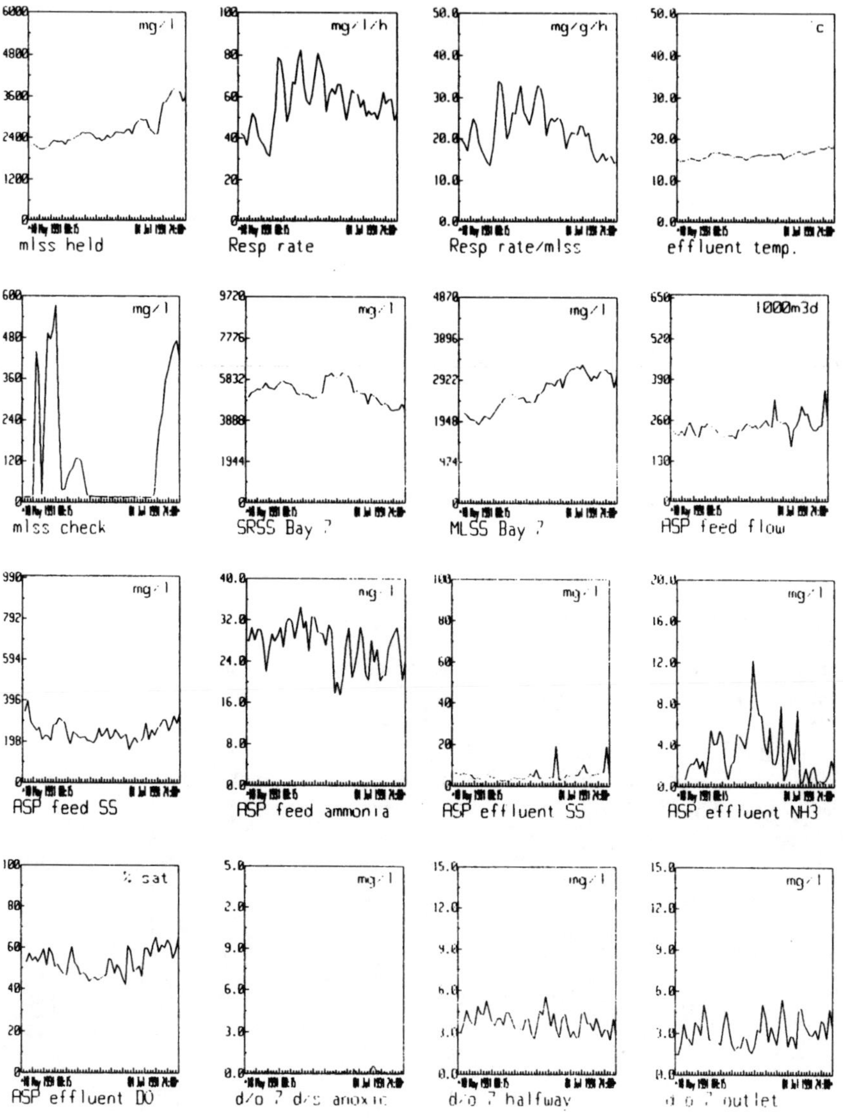

Figure 5: Effect of sudden organic loading on a plant in the UK (55 days' data shown).

The respirometer can be remotely re-programmed, via its serial port and modems and phone lines, to make whatever tests operational analysis and experience, or simple curiosity suggests. The number of aeration/utilization cycles carried out on the same sample before it is ejected, the dissolved oxygen level to which it is raised and the level to which the dissolved oxygen is allowed to decay before re-aeration can all be adjusted remotely as described. In just such an experiment, the respirometer was programmed to take three consecutive samples over a twelve hour period at the Minworth facility in Birmingham, England. The results of this experiment are presented in Figure 6.

An important observation in figure 6 is the time variance in the onset of endogenous respiration throughout the 12 hours. Continuing to add oxygen over the time and at the rate necessary for the top (solid) would represent a significant waste for the wastewater of the second (dotted) and third (dashed)curves.

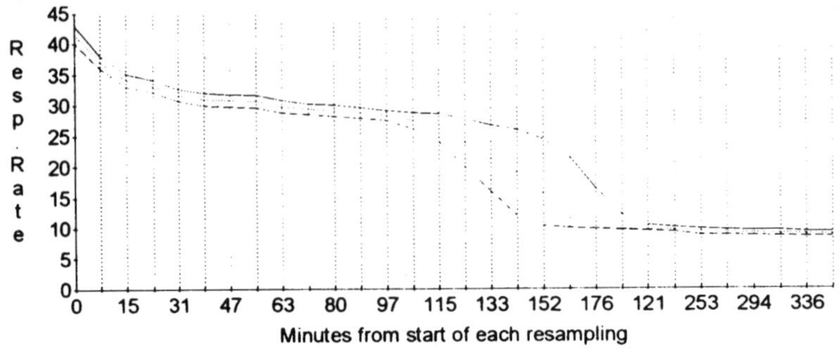

Figure 6: Multiple aeration/utilize cycles for 3 separate samples over a 12-hour period.

As energy costs continue to climb the importance of such careful control becomes obvious. Also in terms of process control a slope on the curve can be taken as the best place for process control - that is, the point in organism activity desired. For standard rate activated sludge with minimal chance of upset it might be at the point where endogenous respiration starts. For high rate activated sludge and maximum energy and tankage economy it would be before the steep drop to endogenous respiration or even further upslope. Probably such high rate operation is only suitable where ocean diffusion is possible and nitrogen and phosphorus represent nutrients beneficial to the marine environment.

Figure 7 illustrates the data obtained over a 24 hour period when an on-line respirometer was remotely directed (with a radio link in this particular facility) to separately analyse the change in oxygen demand of a single sample subjected to 5 aerations/decays before being ejected and a new sample being taken and similarly analysed.

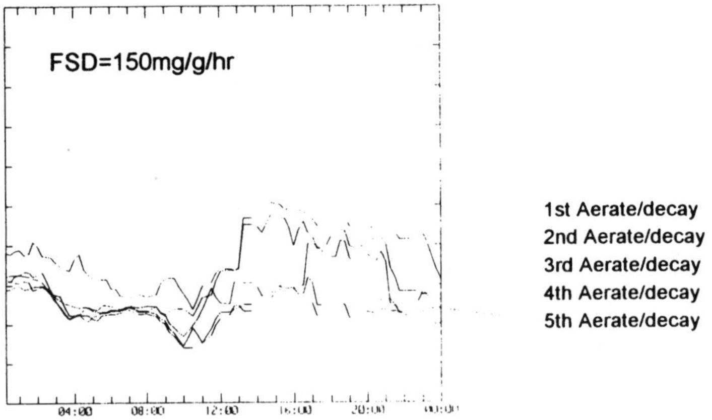

Figure 7: 24 Hours data from sequential samples each subjected to 5 aerate/decay cycles.

Each separate sample's 5-cycle measurement takes about 25 minutes to accomplish. Figure 7 presents these measurements in the form of 15-minute averaged data over a 24 hour period. The calculation, logging and averaging of the different respirometric rates is accomplished within the instrument's built-in microprocessor/data logger.

Figure 8 illustrates the detailed data for just one of these cycles from which the instrument calculates the 5 separate SOURs.

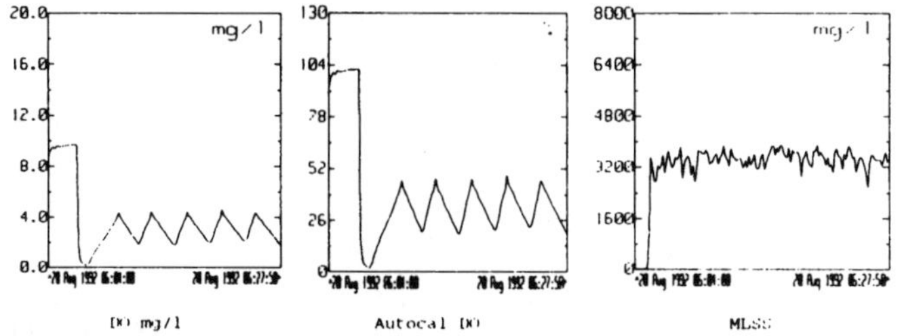

Figure 8: From the instrument's detailed area for 5 aerate/decays on a single sample.

In figure 7 the readily-biodegradable COD - perhaps due to volatile fatty acids and sulphides appears in the first cycle. The amounts of less-readily biodegradable components then show in the succeeding re-aerations. From these immediate air-oxygen needs can be evaluated. Figure 7 is fairly typical of the pattern observed in dry weather for the particular plant this on-line respirometer was monitoring. The differences between consecutive cycles varies through the day, indicating variable amounts of readily-biodegradable substrate at different times.

At times during the day the first and second cycles have very similar SOURs, and these are higher than for cycles 3,4 and 5. This suggests that there was still some readily biodegradable COD remaining after the initial aeration and that this continued to be readily utilized by the biomass so that its capacity to oxidize organics had not been exceeded. It is postulated that the mean of these two similar and higher SOUR readings represent a good index of the maximum rate at which the biomass could utilize the incoming organic food (substrate). This could then provide information as to the maximum rates at which the organic wastes can be fed to the biomass. This will be an interesting avenue of further research.

Figure 9 illustrates what happened when the flow from a rain storm in a combined sewer system reached the Minworth Treatment Plant.

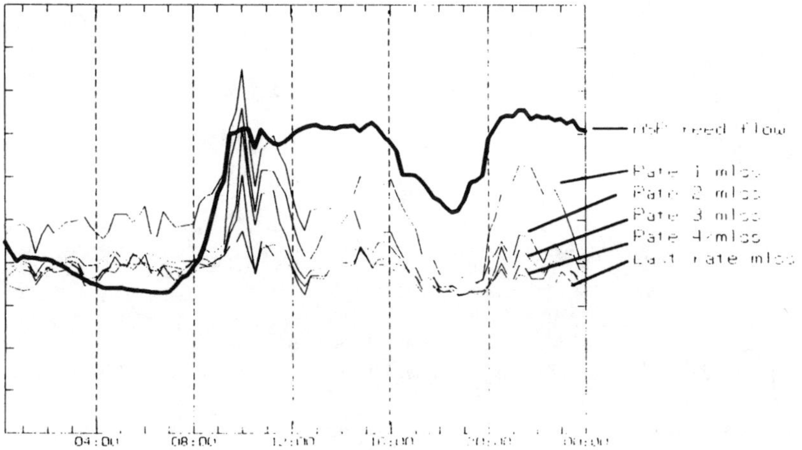

Figure 9: Effect of two consecutive rain storms on the SOURs (Minworth Treatment Facility).

About 30 minutes after the initial high flow was received the SOUR climbed to a maximum because of washout of readily biodegradable COD. largely from the primary tanks. About 4 hours later a second high SOUR was recorded as a result of material flushed from the collection system. A second storm some 10 hours after the first also resulted in a steep increase in SOUR. Figure 9 does not show the subsequent 24 hours, but interestingly there was no similar peak 4 hours later from the collection system after this second storm. This is not surprising since presumably the first storm rapidly flushed any readily-biodegradable material that had been deposited in the collection system through to the treatment plant. The sort of information illustrated in this case shows how the respirometer can provide better process management opportunities when storm flows are a problem.

The ability to discriminate between readily-biodegradable and less-readily-biodegradable components in the primary effluent suggests that the respirometer could be used to monitor and control volatile fatty acid injection into Biological Nutrient Removal facilities. This must currently be done by manual spot checks. Since the process involves passing primary effluent through thickened sludge or cold digested sludge to increase the concentration of volatile acids before entering the aeration basins, on-line control by the respirometer would be a significant step forward.

SUMMARY AND CONCLUSIONS

The use of an accurate and reliable on-line respirometer which has operated in an automatic mode in wastewater treatment plants has been described. The instrument can be instructed to change its operational mode and to produce different types of data remotely by telephone or radio. The acquired data is also received by telephone or radio remotely from the instrument. It can be placed to analyze the incoming wastewater and/or in the return activated sludge to examine its viability and organic handling capacity on a continuous basis. Other possible placements for the control of other operations such as Biological Nutrient Removal are suggested. Historically the value of respirometry for the control of aerobic biological processes has long been recognized as desirable. The Minworth Systems' instrument described here is believed to be the first on-line instrument shown to be reliably usable on a long term-basis.

The difference and significance in terms of process control of the Oxygen Uptake Rate (OUR) and Specific Oxygen Uptake Rate (SOUR) has been described. Maximization of process effectiveness and minimization of energy use is one possible benefit of using the instrument and the parameters suggested. By utilizing repeated re-aerations the difference between readily and less-readily biodegradable organic wastes can be identified and the rate at which oxygen is needed in the process determined.

Points at which endogenous respiration is reached can also be determined and air addition varied according to such data. Figure 1 shows such data. In this type of measurement, the area under the curve represents the total organic load. The time to endogenous respiration is also an index to the necessary contact time within the bioreactor, assuming that the oxygen needs identified by the respirometer are satisfied in the actual plant.

From the experience with the instrument in wastewater treatment plants to date the following appear possible.

1) Taking a sample in the respirometer to the endogenous rate (see figure 1) gives a measure of the retention time needed in the plant for very complete or less complete (high-rate) treatment. It also aids. control in these modes.
2) Joint analysis of the SOUR in mixed liquor and the returned activated sludge can aid in the maximization of whatever operational mode is chosen. The examples cited are for bioreactors with theoretically plug flow configuration.
3) The actual organic loading and the biomass activity can be quickly checked and plant adjustments can be made before difficulties occur.

4) The SOUR for the carbonaceous and nitrification organisms can be detected so that changes to maintain their viability can be made as necessary.
5) Close analysis of the process on a continuous basis is possible, so the aerobic process can be maximized in terms of needed goals and the use of energy to achieve those goals.

Instrumentation in any wastewater treatment facility should be designed to withstand the usually hostile environment while producing data significant to operational personnel. Data from reliable respirometry has been a desired goal of operational personnel for many years. Instruments of the type described have been operating in working treatment plants for a period long enough to demonstrate that reliable on-line respirometric information is now available.

ACKNOWLEDGEMENTS

The authors wish to thank the staff of the wastewater facilities used to provide the data for the case studies presented in this paper, and the staff of Minworth Systems, particularly Des McGarr, for assistance in the production of this paper.

REFERENCES

(1) Arthur, R. - Arthur Instruments, Wisconsin USA
(2) Bargman, R.D., Betz, J.M. and Garber, W.F. . "Aeration requirements of a high oxygen demand sewage", Sewage And Industrial Wastes, 27,7,768-790 (1957)
(3) Garber, W.F. , "The Activated Sludge Process", Operator's Work Book, California Water Pollution Control Association, Annual Conference 1962 (1962).
(4) Gould, R.H. "Wards Island Plant Changes", Sewage and Industrial Wastes, 22, 8, 997 (1950)
(5) Spiegel,M, "Oxygen demand measurements using the Odeometer", Chicago Pump Company, Chicago Illinois, USA (1948)

Pergamon

Wat. Sci. Tech. Vol. 28, No. 11-12, pp. 401-414, 1993.

0273-1223/93 $7·00 + 0·00

DETERMINING INFLUENT SHORT-TERM BIOCHEMICAL OXYGEN DEMAND BY COMBINED RESPIROMETRY AND ESTIMATION

Henri Spanjers*, Gustaf Olsson** and Abraham Klapwijk*

** Department of Enviornmental Technology, Wageningen Agricultural University, P.O. Box 8129, NL-6700 EV Wageningen, The Netherlands*
*** Department of Industrial Electrical Engineering and Automation, Lund Institute of Technology, P.O. Box 118, S-221 00 Lund, Sweden*

ABSTRACT

This paper describes a method for the continuous estimation of the influent short-term biochemical oxygen demand (BOD_{st}) of a wastewater treatment plant. The method uses the dissolved oxygen (DO) concentration in the inlet and outlet of a closed, completely mixed respiration chamber during the transients between two modes of respiration measurement which are alternately executed. In one mode, sludge having the endogenous respiration rate is directed through the respiration chamber. In the other mode, sludge from the aeration tank directly flows through the respiration chamber and, in addition, a sample of influent is continuously introduced in the chamber. The influent BOD_{st} is estimated for each cycle of both modes by integrating the mass balances of BOD_{st} and DO over the respiration chamber. Knowledge of the degradation kinetic of readily biodegradable matter is not required. The proposed procedure is verified using both simulated and experimental data from a pilot plant. It is concluded that the procedure enables the estimation of the influent BOD_{st} with a relative error ranging from 2 - 6%.

KEYWORDS

activated sludge process, biochemical oxygen demand, BOD, influent, oxygen uptake rate, respirometry

NOMENCLATURE

BOD_{st} = short-term biochemical oxygen demand
c = DO concentration in aeration tank (mass volume^{-1})
c_a = DO concentration in respiration chamber (mass volume^{-1})
DO = dissolved oxygen
h = sampling interval DO concentration (inlet or outlet) (time)
$p = c - s$ = help variable (mass volume^{-1})
q = sludge flow through respiration chamber (volume time^{-1})
q_i = influent sample flow through respiration chamber (volume time^{-1})
$r = r_s + r_{end}$ respiration rate (mass volume^{-1} time^{-1})
r_{end} = endogenous respiration rate (mass volume^{-1} time^{-1})
r_s = substrate respiration rate (mass volume^{-1} time^{-1})
s = BOD_{st} in the respiration chamber (mass volume^{-1})
s_0 = BOD_{st} in the respiration chamber at the beginning of a measuring mode (mass volume^{-1})
s_a = BOD_{st} in the aeration tank (mass volume^{-1})
s_i = BOD_{st} of the influent (mass volume^{-1})
s_n = BOD_{st} in the respiration chamber at the end of a measuring mode (mass volume^{-1})
v = volume respiration chamber (volume^{-1})
$\alpha = q/v$ dilution rate (time^{-1})
$\alpha_i = q_i/v$ dilution rate (time^{-1})
$\alpha_r = (q+q_r)/v$ dilution rate (time^{-1})

INTRODUCTION

Because activated sludge plants receive wastewater with varying flow and waste concentration, the process is seldom in steady-state. In order to operate the activated sludge process (have a sufficient degree of waste removal and production of sludge with good settling properties) control is needed. By measuring the influent waste concentration, useful information for control is obtained. Biochemical Oxygen Demand (BOD) is the most widely used measure for the waste concentration, because it is related to the biochemical process of the activated sludge itself.

Unfortunately, the BOD is unsuitable for control purposes, as it is mostly expressed in BOD_5, which requires a five days analysis. This time span is too long for real-time control purposes. To overcome this problem other measures like chemical oxygen demand (COD) and total organic carbon (TOC) are used, however these do not predict how the activated sludge would respond to changes in the waste concentration. Another possibility is to seek a correlation between the BOD_5 and variables, like influent flow, that can be measured continuously.

Considerable efforts have been spent to develop methods which enable the continuous or batchwise measurement of the BOD for monitoring purposes. All these methods are based on a respirometric principle and they yield a short-term BOD (BOD_{st}) that is not equivalent with the BOD_5. Several authors correlated the BOD_{st} with the conventional BOD_5, with more or less success (Vernimmen *et al.*, 1967; Riegler, 1984; Strand and Carlson, 1984; Vasel *et al.*, 1991). However, the BOD_{st} is, in contrast to the BOD_5, a more valuable variable for control purposes, because it represents the oxygen demand within the time constraints of the activated sludge process and it is, in most cases, intimately related to the specific organisms of the activated sludge plant. The BOD_5 merely characterizes the impact on the receiving water.

We propose to classify the methods for monitoring the BOD_{st} into three categories: state estimation, BOD probe method and respirometer method.

Holmberg (1982) proposed a method to estimate the influent BOD from a simplified model of the activated sludge process. In fact this method uses the aeration tank as a respiration chamber. No further experimental verification of the approach has been found.

A BOD probe consists of immobilized living cells, a membrane and a dissolved oxygen electrode. The signal of the probe is a measure of the activity of the cells, which is, in turn, a measure of the substrate concentration in the water. Probes with both activated sludge organisms (Strand and Carlson, 1984; Princz and Oláh, 1990) and yeast cultures (Harita *et al.*, 1985; Riedel *et al.*, 1988) are reported in the literature. The performance of the probe, calibrated in a certain solution, is dependent on the composition of wastewater. Another disadvantage is that the linear range of BOD probes is below the BOD_{st} of most wastewaters, so that samples have to be diluted, which is unsuitable for continuous measurement.

Respirometric BOD procedures can be divided into batch methods and continuous methods. Typical of batch methods is that the BOD is calculated from a respirogram obtained after the addition of a wastewater sample to a respirometer. Closed respirometers, which are not aerated, use a DO-probe or a manometer to record the respirogram (Vernimmen *et al.*, 1967; Tebbutt and Berkun, 1976; Cadena *et al.*, 1988). These meters are laborious to operate and unsuited for continuous measurement. Open aerated respirometers need either a calibration, using a substrate with known BOD (Farkas, 1981; Vanrolleghem *et al.*, 1990), the aeration constants need to be known (Suschka and Ferreira, 1986; Ros *et al.*, 1990; Ciaccio, 1992) or two DO measurements at a separated closed respiration chamber are needed (Spanjers and Klapwijk, 1990). When applying a sampling device for sludge and wastewater, most batch methods can be automated to obtain a discontinuous monitor of the influent BOD.

In continuous respirometric BOD methods, the wastewater sample is continuously fed into a flow-through reactor having a volume of maximum some litres. Heckershoff and Wiesmann (1986) suggested a method which measures the difference of the oxygen consumption between two parallel reactors, both continuously fed with sludge. Additionally, one is fed with influent sample while the other is fed with tap water. The influent BOD is calculated from DO and BOD mass balances over the reactors. For this calculation it is necessary to assume that the BOD in the activated sludge supplied is equal to zero and that the DO concentrations in both reactors are equal. Riegler (1984) described a method which is based on calibration with a standard with known BOD. The sample is continuously diluted with tap water and the mixture is directed through a reactor with immobilized microorganisms. The DO concentration difference of the flow entering the reactor and the flow leaving the reactor is measured and kept constant by manipulating the dilution ratio. From the comparison with a standard with known BOD_5, the BOD_5 of the wastewater can be determined. In fact, the measurement uses the correlation between BOD_{st} and BOD_5 so that the result is dependent on the type of wastewater. Moreover, since the bacteria are separately grown, the population of microorganisms in the reactor may not have the same properties as the activated sludge in the plant.

Spanjers and Klapwijk (1990) proposed a method to calculate the influent BOD_{st} from three types of respiration rate: the endogenous, the instantaneous and the actual rate. The endogenous respiration rate is the oxygen uptake rate of sludge free of readily biodegradable matter. The instantaneous rate is the oxygen uptake rate measured when sludge directly flows from the aeration tank through a respiration meter. The actual respiration rate, defined as the real oxygen uptake rate in the aeration tank, is measured if the respiration chamber is equally loaded with BOD_{st} as the aeration tank. The authors showed, using a mass balance over the respiration chamber, that the influent BOD can be calculated from these three rates. The model is based on the assumptions that the degradation of readily biodegradable matter can be described by a half order kinetic and that there is steady state during the respiration measurements.

In this paper, we propose and test a procedure for the continuous estimation of the influent BOD_{st}. The method uses the DO concentration in the inlet and outlet of a closed respiration chamber during the transients between two modes of respiration measurement. Knowledge of the degradation kinetic of readily biodegradable matter is not required. Other advantages are that a BOD calibration is not needed and that the aeration constants need not to be known. The proposed procedure is verified using both simulated and experimental data from a pilot plant. It is concluded that the procedure enables the estimation of the influent BOD_{st}.

PRINCIPLE OF THE MEASUREMENT

For the continuous estimation of the influent BOD_{st}, a measuring system is developed which is fed with sludge of the plant under consideration. This means that the measuring system is closely linked to the activated sludge plant. Figure 1 shows a schematic of the measuring system along with the activated sludge plant.

The measuring system consists of a closed, completely mixed respiration chamber, a DO probe, sample pumps for activated sludge and influent and six solenoid valves. By using four valves that switch the flow through the respiration chamber, typically every 30 seconds, the DO concentration is successively measured in the inlet and outlet of the chamber. From the DO-concentrations, the

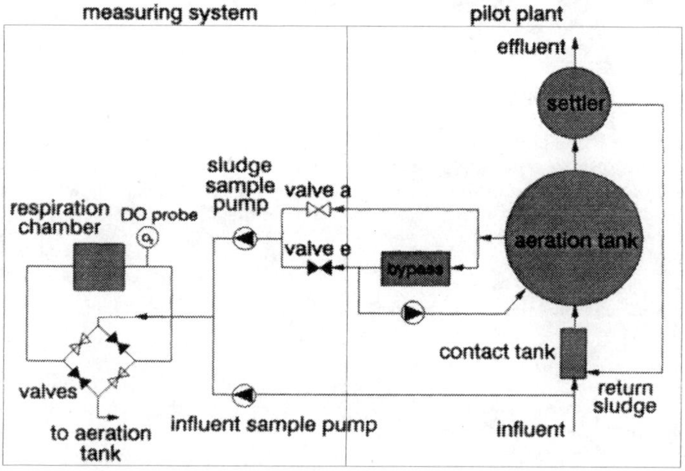

Fig. 1. Schematic of the activated sludge plant and the measuring system.

respiration rate can be calculated using a numerical approximation of the mass balance over the respiration chamber (Spanjers and Klapwijk, 1990).

In the present work, we calculate the respiration rate for the assessment of the endogenous respiration rate. Furthermore, the influent BOD_{st} is estimated from the DO concentration in the inlet and outlet of the respiration chamber during the transients between two modes of operation. In the endogenous mode (e-mode), sludge having the endogenous respiration rate is directed through the respiration chamber. This sludge is produced in an aerated bypass tank (Spanjers and Klapwijk, 1990). In the mode, where the actual respiration rate is measured (a-mode), sludge from the aeration tank directly flows through the respiration chamber and, in addition, a sample of influent is continuously introduced in the chamber. It can be demonstrated (Spanjers and Klapwijk, 1987; Sollfrank, 1988) that the BOD_{st} in the chamber equals the BOD_{st} in the aeration tank if the ratio of the influent sample flow to the chamber volume equals the ratio of the influent flow to the aeration tank volume. The measuring period for one mode is typically 15 minutes and is sufficiently long to let the BOD_{st} and the respiration rate in the chamber converge to an ultimate value.

At the end of each a-mode, the measurement is changed to the e-mode and the sludge in the respiration chamber is gradually replaced by endogenous sludge. The endogenous sludge contains no readily biodegradable matter, so the BOD_{st} of the endogenous sludge is assumed to be zero. Hence, in the e-mode, the BOD_{st} in the respiration chamber remaining from the a-mode will be washed out and the respiration rate will decrease. At the end of each e-mode the BOD_{st} in the respiration chamber is zero and the rate measured is the endogenous respiration rate.

At the end of each e-mode, the measurement is changed to the a-mode and the sludge in the chamber is gradually replaced by sludge from the aeration tank, supplied with influent sample. Since, in the a-mode, the respiration chamber is equally loaded as the aeration tank, the BOD_{st} in the chamber will converge to a value equal to the BOD_{st} in the aeration tank (s_a).

In the next section we shall show how the DO measurements in the inlet and outlet of the respiration chamber during the transients can be used to estimate the influent BOD_{st}.

ESTIMATION OF THE INFLUENT BOD_{st}

The influent BOD_{st} (s_i) is calculated from cycles, each of them consisting of two successive modes: one a-mode and one e-mode. Because each mode takes 15 minutes, the calculation interval of s_i is typically 30 minutes. Figure 2 schematically illustrates the estimation procedure. This involves three steps, executed for each cycle:

1. assessment of the endogenous respiration rate (r_{end});
2. calculation of the BOD_{st} in the aeration tank (s_a);
3. calculation of the influent BOD_{st} (s_i).

1. Endogenous respiration rate

It is assumed that, at the end of each e-mode, the BOD_{st} in the respiration chamber is zero, so that the endogenous respiration rate (r_{end}) is measured. It is further assumed that r_{end} is constant during the whole cycle of 30 minutes. To obtain an operational r_{end} for the whole cycle, the last measured rates of the current and previous cycle are averaged [Fig. 2(b)].

2. BOD_{st} in the aeration tank (s_a)

In the e-mode, the mass balances of DO and BOD_{st} over the respiration chamber are respectively:

$$\frac{dc}{dt} = \alpha\,(c_a - c) - r_s - r_{end} \tag{1}$$

$$\frac{ds}{dt} = -\alpha s - r_s \tag{2}$$

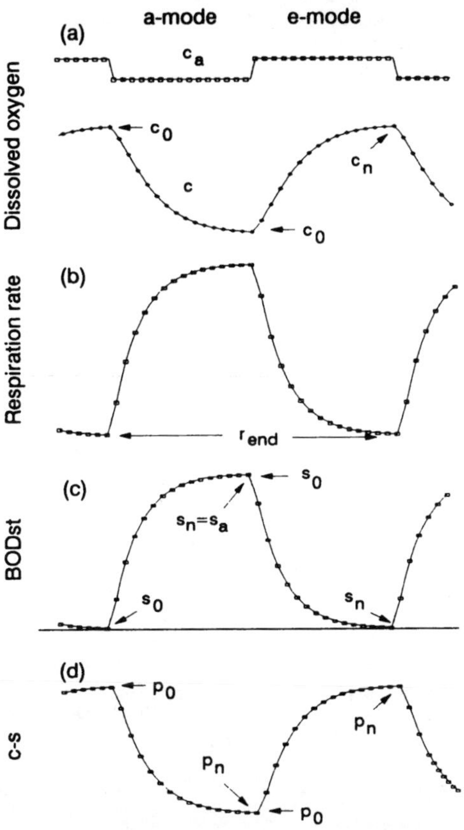

Fig. 2. Schematic representation of the procedure for the estimation of s_i. One cycle, consisting of one a-mode and one e-mode is shown. (a) Measured DO concentration at the inlet, c_a, and outlet, c, of the respiration chamber. (b) Respiration rate calculated from the DO concentrations. (c) Pattern of the BOD_{st} in the respiration chamber. (d) Difference between c and s.

The BOD_{st} in the respiration chamber is known as being zero at the end of the e-mode [Fig. 2(c)]. The BOD_{st} during the e-mode can now be calculated directly from the DO concentrations. From the observation that r_s can be eliminated from (1) and (2), we introduce a help variable p=c-s [Fig. 2(d)]:

$$\frac{dp}{dt} = \alpha (c_a - p) - r_{end} \tag{3}$$

The variable p is known at the end of the e-mode, so that its value at the beginning of the e-mode can be calculated. Given the discrete time measurements, equation (3) has to be approximated by a difference equation. Approximating the derivative with the trapezoidal rule and solving for p_{k-1} we get for the e-mode:

$$p_{k-1} = \frac{1}{\frac{1}{2}\alpha h - 1}\left[-\left(\frac{1}{2}\alpha h + 1\right)p_k + \alpha h c_{a,k-\frac{1}{2}} - h r_{end}\right] \tag{4}$$

where k=1, 2,..., n (n typically 15).

Using the end condition $s_n=0$, so that $p_n=c_n-s_n=c_n$, p_0 can be calculated. This value is used to find the influent BOD_{st} from the a-mode.

Knowing that s at the beginning of the e-mode is equal to s at the end of the a-mode, and recalling that the BOD_{st} in the respiration chamber at the end of the a-mode is equal to s_a, the latter can be calculated as an intermediate:

$$s_a = s_0 = c_0 - p_0 \tag{5}$$

3. Influent BOD_{st} (s_i)

In the a-mode, the mass balances of DO and BOD_{st} over the respiration chamber are respectively:

$$\frac{dc}{dt} = \alpha c_a - \alpha_r c - r_s - r_{end} \tag{6}$$

$$\frac{ds}{dt} = \alpha s_a + \alpha_i s_i - \alpha_r s - r_s \tag{7}$$

The BOD_{st} in the respiration chamber is known as being zero at the beginning of the a-mode. Using the same help variable, p, to eliminate r_s, we obtain:

$$\frac{dp}{dt} = \alpha (c_a - s_a) - \alpha_r p - \alpha_i s_i - r_{end} \tag{8}$$

The variable p is known at the beginning of the a-mode, and its value at the end of the a-mode can be calculated. Approximating the derivative and solving for p_k we get for the a-mode:

$$p_k = \frac{1}{\frac{1}{2}\alpha_r h + 1}\left[-\left(\frac{1}{2}\alpha_r h - 1\right)p_{k-1} + \alpha h c_{a,k-\frac{1}{2}} - \alpha h s_a - \alpha_i h s_i - h r_{end}\right] \tag{9}$$

Using the initial condition s_0=0, so that $p_0=c_0-s_0=c_0$, p_n can be calculated. Finally, s_i can be found explicitly from equation (9) so that p_n (a-mode) = p_0 (e-mode).

The procedure for the estimation of s_i was verified using both simulated data and experimental data from a pilot plant.

MATERIALS AND METHODS

In the simulations, executed with SIMNON (Elmqvist *et al.*, 1986), measuring data were generated using a model of the pilot plant and the measuring system (Fig. 1). This model simulates, for a known s_i, the DO at the inlet and at the outlet of the respiration chamber during the alternating modes of operation (a-mode and e-mode). The DO was given a noise having a normal distribution with a standard deviation of 0.005 mg l^{-1}. The values of the parameters and variables (Table 1) were in accordance with the experimental conditions of the pilot plant. First order kinetic for BOD_{st} was employed. The DO measurements generated with the simulation model were saved in a file and processed according to the procedure proposed in this paper to calculate s_i. The result was compared with the simulated, known s_i.

In the experiments, the measurements were carried out at a pilot plant equipped with a respiration measuring system (Fig. 1). This system was capable of measuring consecutively the respiration rate of sludge in the a-mode and in the e-mode. The control of valves and pumps, collection of data and the basic calculations were performed by a modular I/O processor (μMac 6000, Analog Devices).

The pilot plant, fed with presettled domestic sewage, consisted of an unaerated contact tank (four 2 litre compartments in series), a completely mixed aeration tank and a settling tank. The air flow rate was maintained constant using a flow controller. Sludge with the endogenous respiration rate was produced in a bypass tank, consisting of five 5 litre aerated compartments in series, through which activated sludge from the aeration tank was continuously recirculated. It was verified that, under normal loading conditions, the sludge leaving this tank was free of readily biodegradable matter.

The measuring system consisted of a respiration meter (prototype of RA-1000, Manotherm) having a respiration chamber, an influent sample pump and six valves. To operate alternately in the a-mode and in the e-mode, two valves were used to select sludge from the aeration tank and the bypass respectively. During the a-mode, a sample of influent was continuously added to the sludge flowing through the respiration chamber. Table 1 lists the average process conditions during the measuring period.

To verify the proposed estimation procedure, the following strategy was used. During the period 12:00 h to 24:00 h, the influent was taken from a slowly stirred 500 litre storage tank. Because the concentration of readily biodegradable matter of the stored wastewater was assumed constant, it was expected that the estimated s_i during this period was constant. In addition, the BOD_{st} of the wastewater was independently determined using a batch respiration measurement system (Spanjers and Klapwijk, 1987), equipped with a RA-1000 respiration meter. During the period 0:00 h to 12:00 h the storage tank was refilled with wastewater from the sewer while the reactor was simultaneously fed with the same water. During this period, there was a varying concentration of readily biodegradable matter in the influent and consequently s_i was expected to vary.

Table 1. Conditions of simulations and experiments for estimation s_i. Irrelevant condition indicated with "-".

	simulation	experiment
sampling interval c_a, c and r (s)	60	60
duration one mode (min)	15	15
volume aeration tank (l)	475	475
influent flow ($l\ h^{-1}$)	40	40-42
return sludge flow ($l\ h^{-1}$)	-	41-43
waste sludge flow from AT ($l\ h^{-1}$)	-	1.43
air flow ($m^3\ h^{-1}$)	-	2.4
volume bypass tank (l)	-	25
recirculation flow bypass ($l\ h^{-1}$)	-	26
volume respiration chamber (l)	0.731	0.731
sludge flow through resp. cham. ($l\ h^{-1}$)	21	21.78
influent sample flow ($l\ h^{-1}$)	0.0616	0.06160
influent BOD_{st} ($mg\ l^{-1}$)	100 if 0<time<6 h 250 if 6≤time<12 h	100-350
influent ammonium-N on 17 October ($mg\ l^{-1}$)	-	23.5
influent total nitrogen ($mg\ l^{-1}$)	-	34.5
influent COD on 17 October ($mg\ l^{-1}$)	-	243
MLSS ($g\ l^{-1}$)	-	2.42-2.52
MLVSS ($g\ l^{-1}$)	-	1.92-1.99
temperature sludge (°C)	-	17.8-19.2
pH	-	7.2-7.3
endogenous respiration rate ($mg\ l^{-1}h^{-1}$)	20	17-21
endogenous sludge BOD_{st} ($mg\ l^{-1}$)	0	assumed 0
biodegradation in settler	none	assumed 0
biomass growth	none	unknown
kinetics	1st-order 10 h^{-1}	unknown

For the batch experiments, a sludge sample from the aeration tank was transferred to the batch system and the respiration rate was recorded. When the rate reached the endogenous level, a known volume of wastewater from the storage tank was added. The rate was recorded until the endogenous level was reached again. The BOD_{st} of the wastewater sample was calculated from the total amount of oxygen additionally used to the endogenous oxygen consumption. This procedure was repeated several times with another addition of the same water sample. During the batch experiments, pH and temperature were maintained at 7.5 and 20 °C.

ESTIMATION OF s_i FROM SIMULATED MEASUREMENTS

To verify the proposed procedure and to examine the error in the estimated s_i, this variable was estimated from simulated DO data. These data were generated from a mathematical model which was given a known input step function of s_i. Normally distributed noise was added to the simulated DO. In this simulation, any deviation of the estimated values from the original ones resulted from the noise added to the DO and the assumptions associated with the estimation procedure. Figure 3 shows the original and estimated values. Since r_{end} and s_a, as intermediates in the estimation procedure, have an influence on s_i, these variables are also given in the figure.

As required by the method, the estimated r_{end} did not depend on the influent BOD_{st}, was close to the original value of 20 mg $l^{-1}h^{-1}$ and showed a small standard deviation [Fig. 3(a)]. As dictated by the model, the simulated s_a was a function of the influent BOD_{st}. The estimated s_a [Fig. 3(b)] also showed the step response. However, the average values of this variable (calculated from repeated simulations) deviated from the simulated ones. Moreover, the estimated s_a showed a substantial standard deviation. Before the step increase, when the influent BOD_{st} was 100 mg l^{-1}, the error was even greater than estimated value itself. When the influent BOD_{st} was 250 mg l^{-1}, considered to be a moderate value for the wastewater, the relative error was 53%.

Because of its large value the estimated s_i was much less sensitive to noise on the DO. The average s_i before and after the step change, 99±3.8 mg l^{-1} and 250±3.4 mg l^{-1} respectively, were close to the simulated values.

The simulation leads to the conclusion that the procedure proposed here allows the estimation of the influent BOD_{st} (s_i). The BOD_{st} in the aeration tank (s_a), an intermediate in the estimation of s_i, shows a substantial relative error. This is due to the low value of s_a.

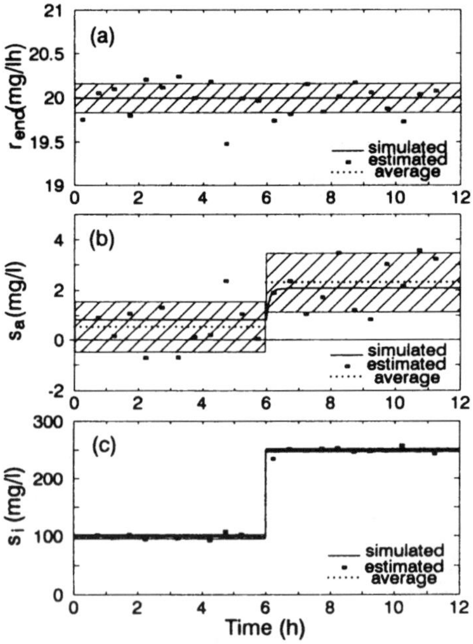

Fig. 3. Result of the estimation of (a) r_{end}, (b) s_a and (c) s_i from simulated DO measurements. The simulated s_i was given a step change from 100 mg l^{-1} to 250 mg l^{-1} at 6 h. Noise with standard deviation 0.005 mg l^{-1} was added to the simulated DO. Solid lines: simulated value, points: estimated values; dotted lines: averaged estimate; dashed zones: standard deviation estimation.

ESTIMATION OF s_i FROM EXPERIMENTAL DATA

During a period of several days, the pilot plant was fed alternately with wastewater from a storage tank and wastewater directly from the sewer system. The respiration measuring system was operated alternately in the two modes of operation. The DO concentration, meanwhile measured at the inlet and outlet of the respiration chamber, was used to estimate the BOD_{st} of the influent (s_i). An illustrative example of the DO measured is given in Figure 4(a). It is recalled that the measurement is done with one and the same probe. Figure 4(b) shows the respiration rate calculated from the DO measurements. The peaks at the end of each mode are a consequence of the numerical approximation used in the calculation of r. The magnitude of these errors was dependent on the extend of the jump in c_a, being the difference between the DO in the bypass tank and the aeration tank. The peaks have been excluded from the calculation of r_{end} in order to minimize the error.

The estimated variables, for a measuring period of 53 hours, are given in Figure 5. r_{end}, having relatively little noise, shows a gradual variation, probably caused by varying activity and biomass concentration. The decrease in the night of 18 October, is partly caused by a loss of biomass. The MLSS of grab samples of activated sludge, taken at about noon from the aeration tank was 2.50 g l^{-1} and 2.42 g l^{-1} on 18 and 19 October respectively.

As expected from the simulation, s_a showed considerable variation [Fig. 5(b)]. Assuming that the real s_a was constant during the periods with influent from the storage tank, the standard deviation of the estimated s_a in these periods would be a indication of the estimation error. For the first and the second period we got 2.51 ± 1.12 mg l^{-1} or $\pm 45\%$ and 4.40 ± 0.80 mg l^{-1} or $\pm 18\%$ respectively.

Figure 5(c) shows that the estimated s_i was relatively constant during the periods with influent from the storage tank. For the first and second period, assuming constant BOD_{st} during storage, we obtained for s_i: 104 ± 6 mg l^{-1} or $\pm 5.8\%$ and 287 ± 7 mg l^{-1} or $\pm 2.6\%$ respectively. To verify the estimated s_i during the periods with influent from the storage tank, the BOD_{st} of samples of the stored water was also determined using the batch respirometric method. These BOD_{st} values are also shown in Figure 5(c).

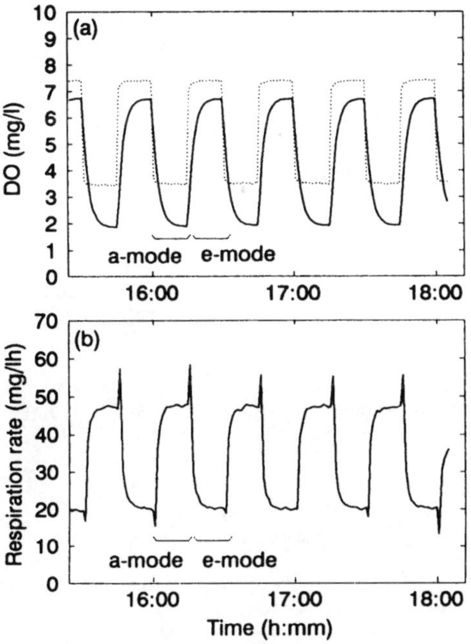

Fig. 4. Part of (a) the measured DO concentration and (b) the calculated respiration rate. Calculation interval respiration rate: 60 seconds.

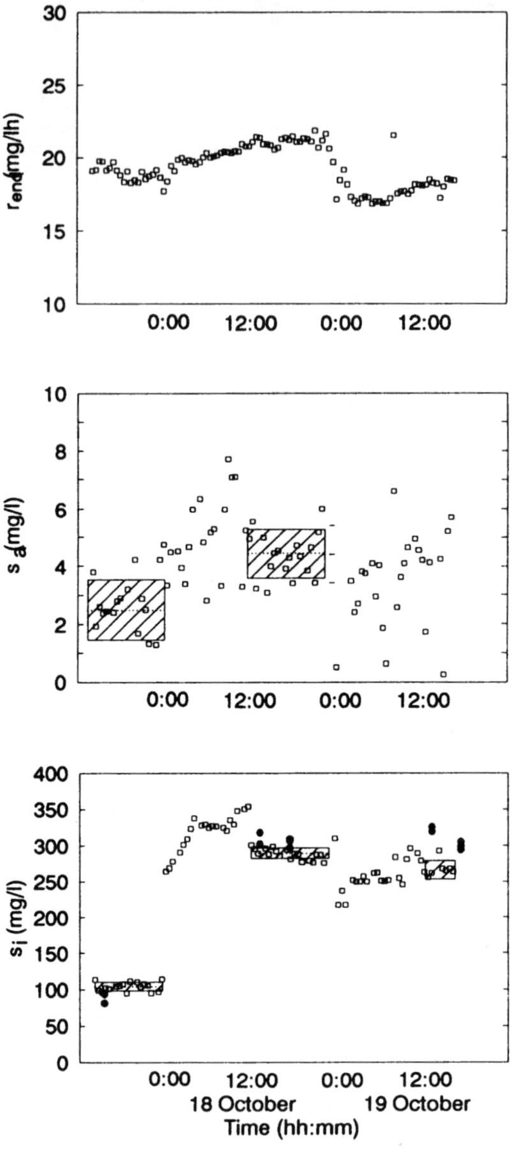

Fig. 5. Results of the estimation procedure applied on experimental measurements. (a) r_{end}, (b) s_a and (c) s_i. Open markers: estimated values; dotted lines: average; dashed zones: standard deviation; filled markers: BOD_{st} determined in batch experiments.

Both methods matched reasonable well, indicating that the method proposed enables the estimation of s_i. Table 2 summarises the values of both methods. The wastewater usually shows a diurnal peak in the waste concentration at about noon (Spanjers and Klapwijk, 1991). A part of this peak can be seen in Figure 5(c), on 18 October at the end of the period with influent from the sewer system. On 18 October, both methods indicated a decrease in the influent BOD_{st} during storage. An explanation is that there was biochemical oxidation in the storage tank. A higher biochemical activity during storage of the wastewater in the second period, compared to the first one, was not unlikely, because of higher wastewater temperature (1.0 to 1.5 °C) and waste concentration.

Table 2. Influent BOD_{st} from estimation procedure and batch experiments compared. avg = averaged, std = standard deviation.

period	BOD_{st} estimation average	std	batch time				avg
Oct 17, 12:00 h - 24:00 h	104	6	15:00	94	82	80	85
Oct 18, 12:00 h - 24:00 h	287	7	13:00	303	317		310
			16:00	310	298	311	306
Oct 19, 12:00 h - 17:00 h	264	11	12:30	319	313	318	317
			16:00	300	294	293	296

Examples of respirograms obtained with the batch method are given in figure 6. In agreement with earlier results (Spanjers and Klapwijk, 1990), each respirogram shows an initial peak, a first shoulder and, at the end, a second shoulder. During these additions, the respiration rate at the baseline only decreased from 22 to 20 mg $l^{-1}h^{-1}$ in 110 minutes, indicating that the endogenous respiration rate, as defined in this work, was reached. However this rate was still higher than r_{end} measured at the pilot plant: 18.5 mg $l^{-1}h^{-1}$. This is probably caused by the higher temperature in the batch experiments: 20 •C compared to 18.7 •C in the pilot plant.

Further experimental evidence, using ammonium concentration in the wastewater (table 1), has confirmed that the BOD_{st} of the wastewater considered is mainly caused by ammonium being oxidised by nitrifiers.

From the experimental results it is concluded that the proposed method enables the estimation of the influent BOD_{st}. The BOD_{st} in the aeration tank, estimated as an intermediate variable, however, show substantial lack of precision.

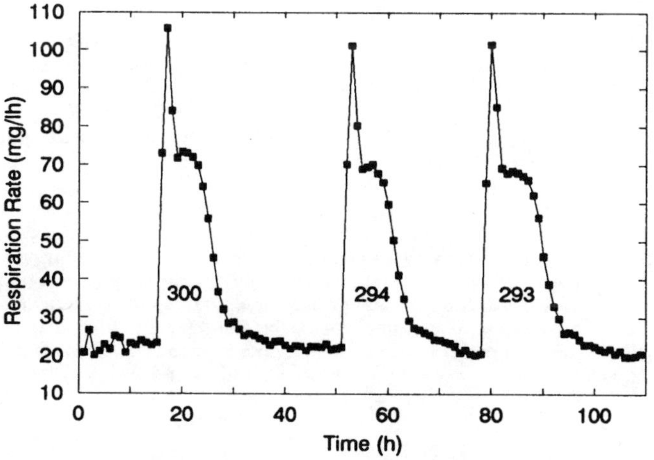

Fig. 6. Batch respirograms of one wastewater sample from the storage tank in the afternoon of 19 October. The activated sludge and the wastewater were sampled at 15:30 h and 16:00 h respectively. The volume of wastewater added was 50 ml for the first and second respirogram and 60 ml for the third one. The wastewater BOD_{st} obtained is indicated in the figures.

DISCUSSION

The purpose of our investigation was to develop and test a procedure for the continuous estimation of the influent BOD_{st}. The proposed procedure used the DO concentration in the inflow and outflow of a respiration chamber during the transients of two modes of operation: a-mode and e-mode. Advantage of the procedure, as compared to a previously published method (Spanjers and Klapwijk, 1990), was that an assumption on the biodegradation kinetics of readily biodegradable matter was not needed. However, an important condition was that, at the end of the e-mode, the BOD_{st} in the respiration chamber was zero. The procedure was tested in a simulation and verified in an experiment.

The simulation showed that the estimation procedure for the model considered was correct and the assumptions were reasonable. A typical error of the simulated DO led to an acceptable relative error of the simulated s_i. However, the relative error of the estimated s_a, an intermediate in the estimation procedure, was unacceptable.

The measurements showed similar results with regard to the errors of the estimated variables. Assuming that, during the periods with influent from the storage tank, s_a and s_i were constant, their relative errors could be determined as being 18-84 % and 2 - 6 % respectively. Under the plant conditions in this work, s_a was only 1.5 to 2.5% of s_i, meaning that s_a could have been neglected in the estimation procedure. However, in a higher loaded plant, this may not be allowed.

Because the calculation of s_a started at the end of the e-mode, with the last c, the numerical integration was much affected by the inaccuracy of this DO value. This may be considered in the algorithm, and features can be added to the integration to minimize this sensitivity. One possibility would be to start the calculation where the slope of p [Fig. 2(d)] becomes significantly different from zero. A critical part of the estimation is the filtering and integration of the measurement values. This may be subject for further investigations.

Because, in the proposed method, the respiration rate is measured as a function of a continuously varying BOD_{st} in the respiration chamber, the kinetic relationship between these variables may be determined for each cycle. Steady state with regard to the transport of substrate must then be assumed. From equations (1) and (6), r_s can be calculated. Then, from equations (2) and (7), s is calculated using the boundary conditions $s_0=0$ (a-mode) and $s_n=0$ (e-mode). Finally, the kinetic relationship can be fitted on the corresponding points r_s and s.

Our experience was that the proposed method was easily applicable in practice. Most attention had to be paid to the correct measurement of the a-mode, since here it was necessary that the ratio of the influent sample flow to the chamber volume matched the ratio of the influent flow to the aeration tank volume. Since the influent flow was constant in the experiment described here, the sampled flow could be fixed. A varying influent flow would require an accurate control of the sample flowrate.

The experiments showed that the BOD_{st}, as defined in this work, is an important variable for characterising the dynamics of the oxygen demand in the activated sludge process. In this work the BOD_{st} was defined as the oxygen demand in excess of the demand for endogenous oxygen consumption. The endogenous respiration rate was practically defined as the oxygen uptake rate of sludge after a mean residence time of 1 hour in an aerated reactor consisting of five 5 litre compartments in series. The batch experiments clearly illustrated (Fig. 6) that this was a reasonable definition since the r_{end} level changed only slightly during a period of 2 hours after sampling and, after the addition of wastewater, the rate rapidly returned to this level. The endogenous rate includes the oxidation of slowly biodegradable matter. The gradual increase of r_{end} on 18 October (Fig. 5) partly indicates an enlarged oxygen uptake rate for the degradation of slowly biodegradable matter caused by the higher loading compared to 17 october.

CONCLUSIONS

The method proposed here allowed the estimation of the influent BOD_{st}. When it was assumed that, during the measurement with influent from the storage tank, the concentration of readily biodegradable matter was constant, the relative error of the estimated BOD_{st} could be determined: 2 to 6%. There was a reasonable agreement between the estimated BOD_{st} and the BOD_{st} determined in batch experiments.

ACKNOWLEDGEMENT

We acknowledge Professor Lambertus Lijklema and Hardy Temmink for reviewing the manuscript. We would like to thank the Netherlands Organization for Scientific Research (NWO) for financial support.

REFERENCES

Cadena F., Drohobyczer A., Beach M.I., Barnes D. (1988) A novel approach to simplified respirometric oxygen demand determinations. *43rd Purdue Industrial Waste Conference Proceedings*, 459-467.

Ciaccio L.L. (1992) Instrumental determination of the energy oxygen and BOD_5. *Wat. Sci. Tech.* **26**, No. 5-6, 1345-1353.

Elmqvist H., Åström K.J., Schönthal T. and Wittenmark B. (1986) *Simnon, User's Guide for MS-DOS Computers*. SSPA Systems, Sweden.

Farkas P.A. (1981) The use of respirography in biological treatment plant control. *Wat. Sci. Technol.* **13**, 125-131.

Harita K., Otani Y., Hikuma M. and Yasuda T. (1985) BOD quick estimating system utilizing a microbial electrode. In *Advances in Water Pollution Control, Proc. 4th IAWPRC Workshop on Instrumentation and Control of Water and Wastewater Treatment and Transport Systems*, 27 April- 4 May, Houston and Denver, U.S.A., pp. 529-532.

Heckershoff H. and Wiesmann U. (1986) Ein neues Messgerät für die Betriebskontrolle und Reglung von Belebtschlammanlagen. Teil 1: Messung der Substratkonzentration. *Korrespondenz Abwasser* **33**, 508-513.

Holmberg A. (1982) Modelling of the activated sludge process for microprocessor-based state estimation and control. *Wat. Res.* **16**, 1233-1246.

Kalte P. (1990) The continuous measurement of short-time BOD. In *Advances in Water Pollution Control, Proc. 5th IAWPRC Workshop on Instrumentation, Control and Automation of Water and Wastewater Treatment and Transport Systems*, 26 July - 3 August, Yokohama and Kyoto, Japan, pp. 59-65.

Princz P. and Oláh J. (1990) BOD measurement of non-toxic wastewaters with an improved microbial probe. In *Advances in Water Pollution Control, Proc. 5th IAWPRC Workshop on Instrumentation, Control and Automation of Water and Wastewater Treatment and Transport Systems*, 26 July - 3 August, Yokohama and Kyoto, Japan, pp. 79-87.

Riedel K., Renneberg R., Kühn M. and Scheller F. (1988) A fast estimation of biochemical oxygen demand using microbial sensors. *Appl. Microbiol. Biotechnol.* **28**, 316-318.

Riegler G. (1984) Kontinuierliche Kurzzeit-BSB-Messung. *Korrespondenz Abwasser* **5**, 369-377.

Ros M., Zagorc-Koncan J., Dular M. and Pompe M. (1990) Respirometric determination of wastewater biodegradation in a biological treatment plant. *Z. Wasser, Abwasserforschung* **23**, 236-240.

Sollfrank U. (1988) Bedeutung organischer Fraktionen in kommunalem Abwasser im Hinblick auf die mathematische Modellierung von Belebtschlammsystemen. Ph.D. thesis, Technical University, Zurich, Switzerland.

Spanjers H. and Klapwijk A. (1987) The Wazu-respiration meter. A breakthrough in automation and control for a wastewater treatment plant. In *10th Symposium on Wastewater Treatment*, November 1987, Montréal, Québec, Canada.

Spanjers H. and Klapwijk A. (1990) On-line meter for respiration rate and short-term biochemical oxygen demand in the control of the activated sludge process. In *Advances in Water Pollution Control, Proc. 5th IAWPRC Workshop on Instrumentation, Control and Automation of Water and Wastewater Treatment and Transport Systems*, 26 July - 3 August, Yokohama and Kyoto, Japan, pp 67-77.

Spanjers H. and Klapwijk A. (1990) Dynamiek van het zuurstofverbruik in aktiefslibsystemen gemeten met de Wazu-respiratiemeter. Report, Wageningen Agricultural Universisty.

Spanjers H. and Klapwijk A. (1991) Continuous estimation of short term oxygen demand from respiration measurements. *Wat. Sci. Tech.*, **24**, 29-32.

Strand S.E., Carlson D.A. (1984) Rapid BOD measurement for municipal wastewater samples using a biofilm electrode. *J. Wat. Pollut. Control Fed.* **56**, 464-467.

Suschka J. and Ferreira E. (1986) Activated sludge respirometric measurements. *Wat. Res.* **20**, 137-144.

Tebbutt T.H.Y. and Berkun M. (1976) Respirometric determination of BOD. *Wat. Res.* **10**, 613-617.

Vanrolleghem P. Dries D. and Verstraete W. (1990) RODTOX: biosensor for rapid determination of the biochemical oxygen demand and the on-line monitoring of the toxicity of wastewaters.

In *Proceedings Fifth European Congress on Biotechnology.* 8-13 July, Copenhagen, Denmark, Vol. 1, pp 161-164.

Vasel J.L., Warnier P., Jupsin H. and Schrobiltgen P. (1991) Automatisation et comparaison de differentes méthodes respirométriques d'estimation rapide de la DBO. *Revue des Sciences de l'Eau* **4**, 415-435.

Vernimmen A.P., Henken E.R. and Lamb J.C. (1967) A short-term biochemical oxygen demand test. *J. Wat. Pollut. Control Fed.* **39**, 1006-1020.

Pergamon

Wat. Sci. Tech. Vol. 28, No. 11-12, pp. 415–425, 1993.

0273–1223/93 $7·00 + 0·00

BIOLOGICAL EARLY WARNING SYSTEMS FOR TOXICITY BASED ON ACTIVATED SLUDGE RESPIROMETRY

H. Temmink*, P. Vanrolleghem**, A. Klapwijk* and W. Verstraete**

* *Department of Environmental Technology, Wageningen Agricultural University, P.O. Box 8129, NL-6700 EV Wageningen, The Netherlands*
** *Laboratory of Microbial Ecology, University of Gent, Coupure Links 653, B-9000 Gent, Belgium*

ABSTRACT

The proper operation of the activated sludge process is endangered by toxic spills in the wastewater that has to be treated. Therefore there is a need for so called biological early warning systems (BEWSs) that can detect toxicity for the activated sludge process and allow operators to take protective actions.

Two independently developed BEWSs are described and evaluated. Both systems are based on the on-line measurement of the respiration rate of activated sludge taken from the plant to be protected. Their practical implementation however is different with respect to the way the suspicious wastewater is brought into contact with the sludge. It is illustrated what the consequences of the design choices are with respect to the sensitivity, the response time and the positioning at the treatment plant of these BEWSs.

KEYWORDS

Activated sludge; biological early warning systems (BEWS); oxygen uptake rate; respirometry; toxicity.

INTRODUCTION

The proper operation of activated sludge plants treating industrial wastewaters is endangered by toxic spills. This may result in a deteriorating effluent quality and higher costs. Therefore there is a need for so called biological early warning systems (BEWS) which can detect toxic spills and leave sufficient time to take protective actions (Solymon *et al.*, 1976). To indicate toxicity, a BEWS measures a variable which reflects the viability of the sludge. The respiration rate is a direct measure of the primary function of the activated sludge process and therefore preeminently is suitable (Vargaz-Lopez, 1988). This is in contrast to a concentration value produced by a peak in a GC-fingerprint or other time consuming and elaborate analytical techniques (Herricks *et al.*, 1991). In the sequel we speak of a toxic wastewater if it affects the respiration rate of the sludge with which the wastewater has to be treated.

The basic principle of a BEWS based on the measurement of respiration variables is illustrated in Figure 1. An aerated test vessel, which is a physical model of the aeration tank that has to be protected, is fed with the suspicious wastewater and sludge. Optionally, calibration water or synthetic wastewater is supplied. In the test vessel, one or more respiration variables are measured to indicate the toxicity of the wastewater.

When converting this basic principle into a practical implementation, two considerations are important:

1) Toxic wastewater must be detected before it inhibits the activated sludge process itself. This can be achieved by a different operation of the test vessel as compared to the treatment plant:

- the test vessel is fed with wastewater that is sampled upstream of the treatment plant or
- the test vessel is exposed to a higher wastewater loading meaning that the concentration of a toxic compound will increase more rapidly in the test vessel than it will in the treatment plant or
- the contact time between sludge and the suspicious wastewater in the test vessel is longer resulting in an increased sensitivity of the sludge to toxic wastewater.

2) The contact time between the sludge and the wastewater in the test vessel must be long enough to allow toxic compounds to actually affect the sludge.

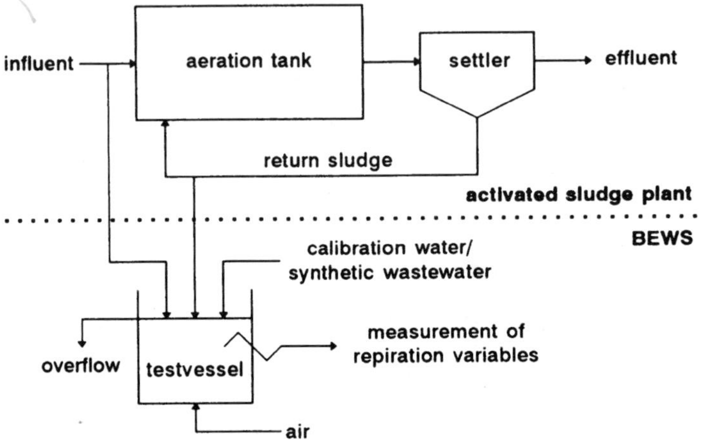

Fig. 1. Schematic representation of a BEWS to detect toxicity of wastewaters for the activated sludge process, based on the measurement of respiration variables.

This paper presents and evaluates two different BEWSs that are based on the measurement of respiration variables to detect the toxicity of wastewaters to the activated sludge process. The first system is called the RODTOX (Rapid Oxygen Demand and TOXicity tester). The second system is based on a continuous respirometer called the RA-1000 (Respiration Analyzer). First, the two respirometers are described. Next, in two separate cases, the practical implementation of these respirometers in a BEWS is described. Finally, in an evaluation these BEWSs are compared. Because the performances of the two BEWSs have not been compared in an actual side-by-side study, the evaluation will be concentrated on their practical implementation.

RESPIROMETERS

RODTOX

The RODTOX was developed at the Laboratory of Microbial Ecology of the University of Gent, Belgium (Vandebroek, 1986). It consists (Figure 2) of an aerated vessel filled with sludge originating from the treatment plant. The aerated vessel is operated as a batch system. Typically, every 30 min the sludge in the vessel is fed with a known volume of wastewater. Periodically, wastewater with known BOD is added to perform a calibration. The oxidation of substrate is monitored with a DO-probe. Figure 3 gives an illustration of the resulting DO-profiles, called respirograms, in case of a toxic spill. Each respirogram is characterized by three variables: peak slope (PS), peak height (PH) and peak area (PA). These variables can be used to indicate toxicity (see below). Further details on the measurement and applications other than toxicity detection can be found elsewhere (Vanrolleghem *et al.*, 1990; Vanrolleghem *et al.*, 1992).

RA-1000

The RA-1000 was developed at the Department of Environmental Technology of the Agricultural University of Wageningen, The Netherlands. This continuous respirometer is schematically presented in Figure 4. A sample pump continuously feeds a mixture of sludge and wastewater to a closed and completely mixed respiration chamber. A solenoid system periodically reverses the flow direction through this chamber. Thus, with one DO-probe, both the DO of the incoming flow as well as the DO of the flow leaving the respiration chamber are measured. A typical recording of raw DO-data is given in Figure 5. From the mass balance for DO over the respiration chamber the respiration rate is calculated every minute. It can be used to detect toxicity. Further details on the measurement and applications other than toxicity detection can be found elsewhere (Spanjers and Klapwijk, 1990; Spanjers, 1992).

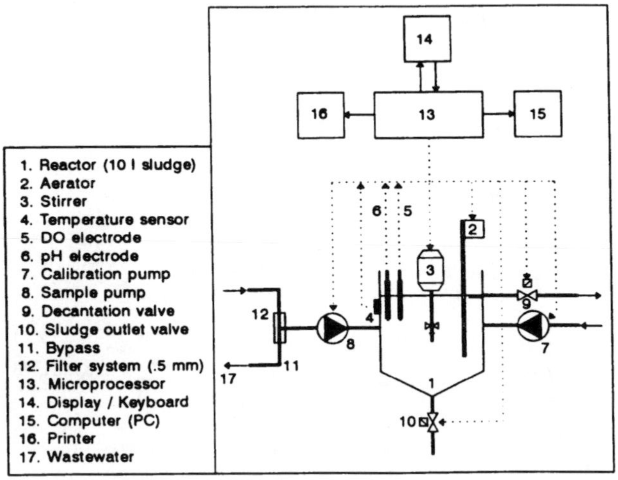

Fig. 2. RODTOX.

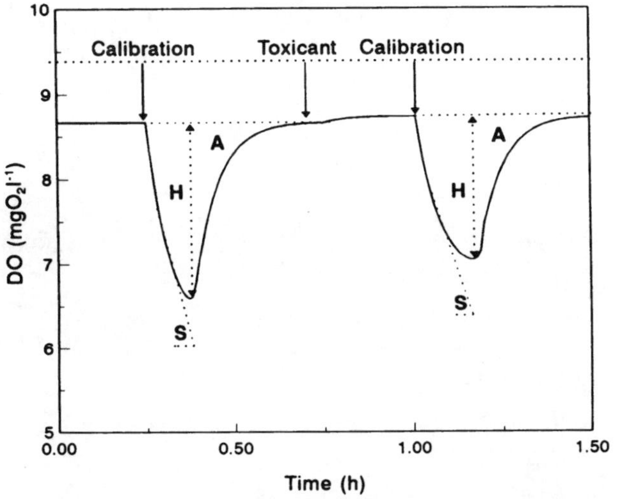

Fig. 3. DO-profiles (respirograms) as recorded with the RODTOX.

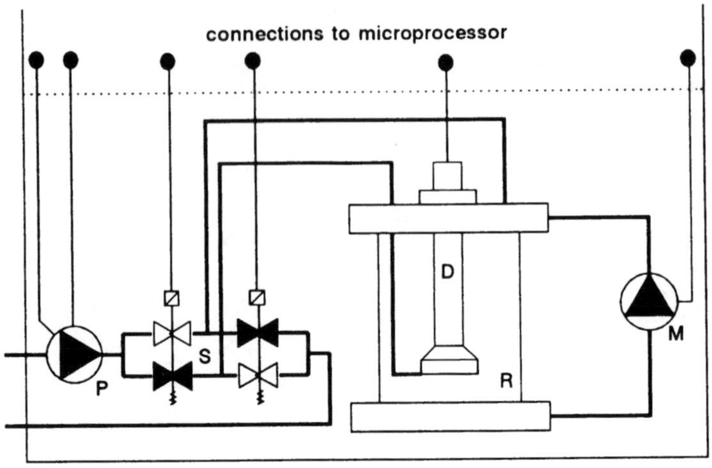

Fig. 4. Continuous respirometer RA-1000 (P=sample pump, S=solenoid system, R=respiration chamber, D=DO-probe, M=mixing pump).

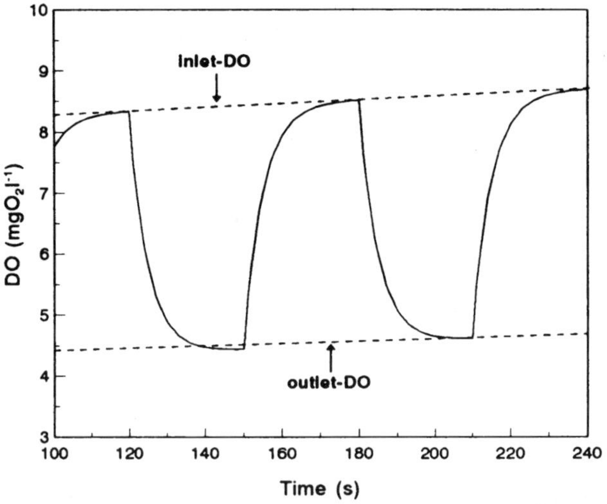

Fig. 5. DO-profile as recorded with the RA-1000.

Comparison of the Two Respirometers

Table 1 compares the most important characteristics of both respirometers.

TABLE 1 Comparison of the Respirometers RODTOX and RA-1000

characteristic	RODTOX	RA-1000
respiration chamber	open and aerated	closed
measured variable	DO	
sensor	one DO-probe	
derived variable	PS, PH and PA of respirogram	respiration rate
frequency	every 30 min.	every min.
data processing	on-line	
sludge supply	batch , every week	continuous
wastewater supply	batch, every 30 min.	continuous

CASE RODTOX

Practical Implementation

The open and aerated respiration chamber of the RODTOX corresponds to the test vessel of Figure 1. Environmental variables such as pH and temperature can be controlled. Initially the vessel is filled with 10 l of sludge originating from the treatment plant. Every 30 min a known volume of wastewater is fed to the sludge until a total volume of 13 l is reached. Then a 30 min decantation phase is initiated and 3 l of supernatant is withdrawn. Every 2-3 h, addition of wastewater from the treatment plant is replaced by addition of calibration water to obtain a reference respirogram. The calibration water has a known composition which is representative for the real wastewater. The toxicity of the wastewater is tested by comparing the respiration variables peak slope (PS) and peak area (PA) of the reference respirogram with the PS and PA of a previous reference respirogram (see also Figure 3). A decrease in PS indicates an inhibited capacity of the sludge to degrade substrate. PA can yield additional information on the toxicity of wastewaters to the sludge. If a specific group of microorganisms, for instance the group of nitrifying bacteria, is inhibited completely, this will be reflected in a decrease of PA. The level of toxicity is calculated as:

$$\%I = 100 \frac{P_0 - P_1}{P_0}$$

with:

%I : percentage of inhibition (%)
P_0 : PS or PA of reference respirogram before addition of wastewater (mg $O_2 l^{-1} min^{-1}$; mg $O_2 l^{-1}$)
P_1 : PS or PA of reference respirogram after addition of wastewater (mg $O_2 l^{-1} min^{-1}$; mg $O_2 l^{-1}$)

The response time for toxicity detection is dependent on the imposed calibration frequency. As a worst case, a toxic spill enters the measurement just after a calibration and will not be detected until a new calibration is performed. Apart from a higher calibration frequency (once every hour) the response time for toxicity detection can also be shortened by monitoring the baseline DO between two successive additions of wastewater. Because this baseline DO is a measure of the endogenous respiration rate of the sludge, an increasing baseline indicates toxicity. In this case, a calibration cycle is initiated immediately to check for toxicity.

The wastewater loading of the test vessel is determined by the volume of the wastewater additions. Normally, a volume is selected such that the wastewater loading of the test vessel corresponds to the wastewater loading of the activated sludge plant. Therefore, the concentration of a toxic compound in the test vessel will also increase at the same rate as the concentration in the aeration tank. However, in contrast to the aeration tank, the sludge in the test vessel is not continuously replaced by fresh (return) sludge. The resulting cumulative exposure to the toxic wastewater makes the sludge in the test vessel more sensitive compared to the sludge in the treatment plant and consequently a shorter response time is obtained. Still, to leave sufficient time for protective measures, it is advisable to sample wastewater upstream of the treatment plant, e.g. from equalization or primary sedimentation tanks.

Testing the Measurement

Some data illustrating the performance of the RODTOX as a BEWS are given below.

Figure 6 shows the response of sludge which was not adapted to phenol to a pulse of wastewater containing this compound in considerable amounts. The sequence of respirograms consists of calibration peaks (CAL) for every three influent sample measurements. As the first 7 sample respirograms are nearly identical, this shows the composition of the wastewater was almost constant. At the eighth sample respirogram the wastewater is polluted with phenol (dotted line) and this drastically affects the condition of the sludge. At the fourth calibration respirogram of the depicted sequence, a significant decrease in peak slope was detected. A 37% inhibition was calculated using PS-values. As a result, an alarm was toggled.

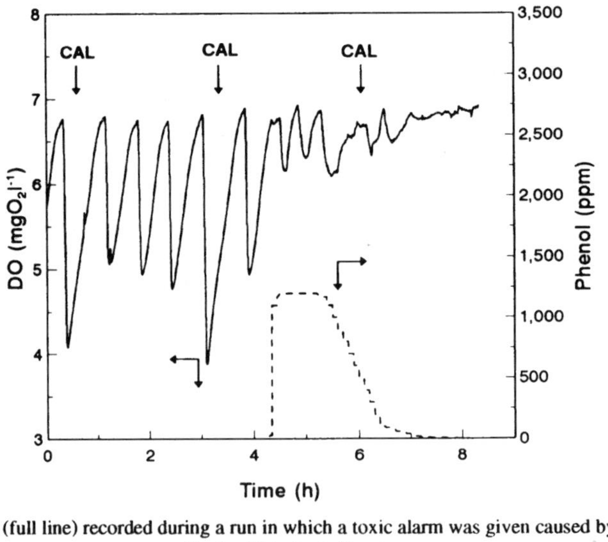

Fig. 6. Respirograms (full line) recorded during a run in which a toxic alarm was given caused by a phenol (dotted line) spill in the influent.

In case the toxicant concentration increases with time, PS gradually decreases (Figure 7). All respirograms are the result of injections with wastewater doped with increasing concentrations of copper. From the third respirogram onwards, copper is entering the treatment plant. The copper concentration in the influent was increased logarithmically from 25 ppm to 200 ppm during the experiment. The sludge showed 30 % inhibition of PS from the second addition onwards. When the copper concentration in the influent had reached the 200 ppm level, the sludge in the test vessel had been inactivated completely.

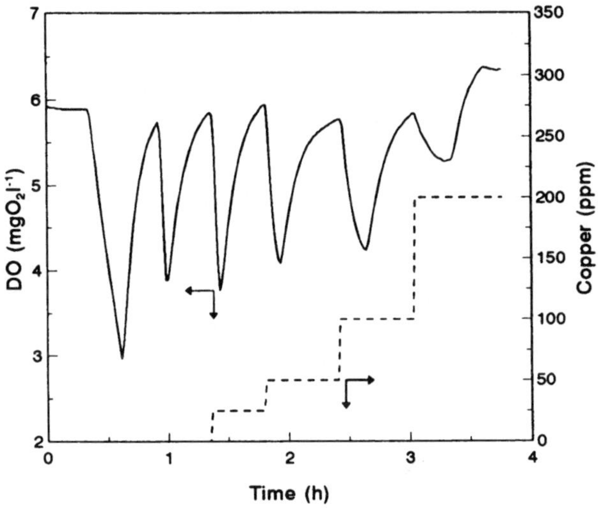

Fig. 7. Respirograms (full line) taken from a pilot experiment in which the influent was doped with a logarithmically increasing copper concentration (dotted line).

CASE RA-1000

Practical Implementation

The continuous respirometer is connected to an aerated test vessel (see also Figure 1). This vessel is designed as a small and highly loaded aeration tank which is continuously fed with the suspicious wastewater and return sludge originating from the treatment plant. In the absence of toxicity a maximum respiration rate is measured which is an indication of the potential sludge activity. A (peak) loading of toxic compounds in the wastewater is detected as a diminishing maximum respiration rate. For proper operation, the respiration rate must be at a constant reference level in the absence of toxicity. Therefore, the influence of factors other than toxicity must be eliminated. At high substrate concentrations the respiration is at the maximal rate and, consequently, is relatively insensitive to variations in substrate concentration. Hence, the ratio between wastewater and (return) sludge flow into the test vessel must guarantee a certain critical level of substrate. Apart from the substrate concentration the substrate composition is also an important factor to be considered. If a stable reference respiration rate is not feasible because these factors fluctuate too much, synthetic wastewater has to be fed to the test vessel. If necessary, environmental variables such as temperature and pH can be controlled, preferably at the average level at which they prevail in the aeration tank.

The retention time of the wastewater in the test vessel determines the response time to a toxic spill. Because the test vessel is designed as a highly loaded aeration tank the response time is very short compared to the response time in the aeration tank. Consequently, the test vessel can be situated close to the treatment plant and toxicity will still be detected before the activated sludge process itself is affected.

On the other hand, too short a retention time is not acceptable because a certain contact time between sludge and wastewater is required to allow toxic compounds in the wastewater to affect the respiration rate. The minimum required contact time depends on the toxic compounds which can be expected.

Testing the Measurement at AKZO-Botlek

The chemical company of AKZO-Botlek in Rotterdam, The Netherlands, discharges wastewater which is biologically treated on-site in an activated sludge plant of the type Carrousel. Spills of toxic compounds originating from a herbicide plant and a vinyl chloride plant may appear in the wastewater. In general, two types of wastewater may occur: a mixture of vinyl chloride effluent and herbicide effluent or just vinyl chloride effluent.

In order to apply the RA-1000 at the AKZO-Botlek treatment plant, preliminary experiments were performed to find the operating conditions which fulfil the prerequisites of the measurement. The critical loading above which a maximum respiration rate is measured was determined for both types of wastewater at average operating conditions (Figure 8). The specific respiration rate (per unit of mass sludge) as a function of the loading of the test vessel with shortterm-BOD (BOD_{st}) (Spanjers and Klapwijk, 1990) shows that the respiration rate is independent of the loading above approximately 70 $mgO_2g^{-1}h^{-1}$.

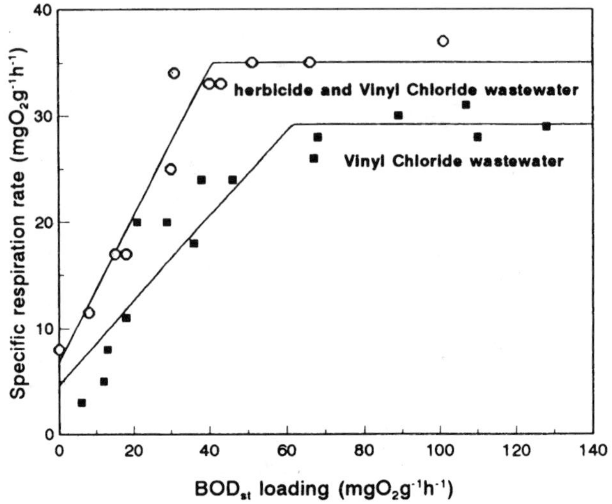

Fig. 8. Specific respiration rate in test vessel as a function of BOD_{st}-loading.

The retention time of the wastewater in the test vessel was set at 15 to 30 min, which is very short compared to the retention time in the aeration tank of approximately 20 h. This leaves sufficient time to take protective actions. For the most suspicious compounds, the minimum required contact time before they actually affect the respiration rate was determined in batch. A contact time of 15 min proved to be adequate.

Both the behaviour of the measurement at non-toxic conditions as well as the behaviour in case of toxic spills must be judged. Figure 9 gives two examples of the behaviour of the RA-1000 installed at AKZO-Botlek under non-toxic conditions. The upper line of Figure 9 (A) is taken from a period in which both the vinyl chloride plant and the herbicide plant were discharging wastewater at normal rates. The respiration rate shows only a gradual variation in time and never gives rise to an alarm. The other line (B) illustrates a period of measurement during which the herbicide plant was out of production and did not contribute to the substrate composition of the wastewater. In addition, the vinyl chloride plant also periodically stopped discharging wastewater. The corresponding dips in the respiration rate are caused by lack of substrate. If the operator is not informed about the prevailing situation, toxicity would be unjustly detected. However, it is also possible to temporarily increase the wastewater flow rate. If not toxicity but lack of substrate was responsible, the respiration rate would recover.

The behaviour in case of toxic spills is illustrated in Figure 10. It demonstrates the adverse effect of a spill of 4-chloro-cresol on the respiration rate of the sludge taken from the AKZO-Botlek plant. The respiration rate immediately responds to the spill. Interrupting the spill results in a recovery of the respiration rate thanks to wasting of 4-chloro-cresol from the test vessel and supplying with fresh (unaffected) sludge.

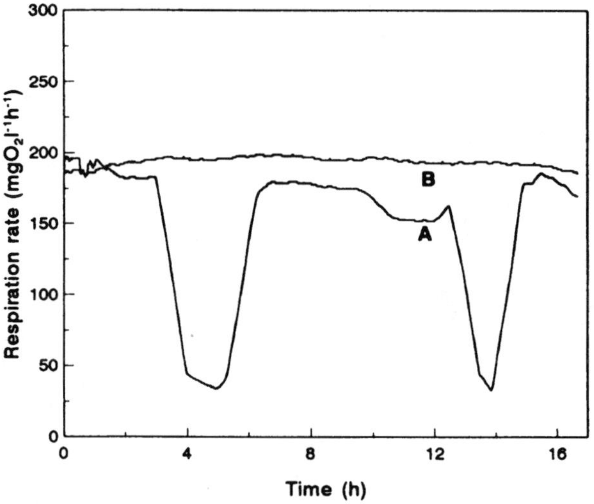

Fig. 9. Respiration rate at non-toxic conditions (A=normal wastewater discharge, B=extremely low wastewater discharge).

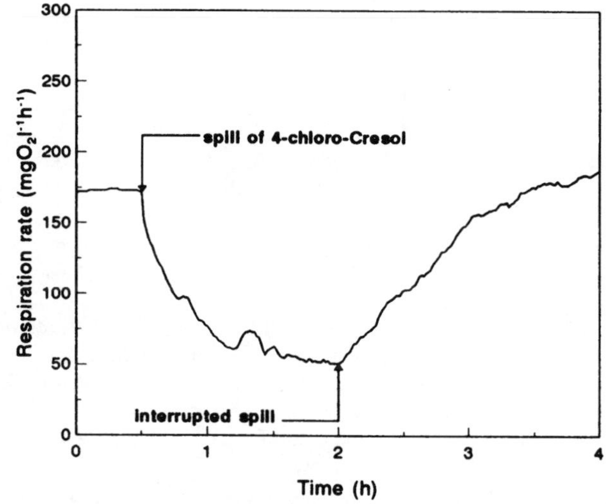

Fig. 10. Response of respiration rate to a (toxic) spill of 4-chloro-cresol.

EVALUATION

In this paper, two independently developed BEWSs have been presented. Both systems monitor the respiration rate of activated sludge to detect toxicity. Examples clearly showed that both systems can give rapid indication of the potential toxicity of influents.

TABLE 2. Comparison of the RODTOX and RA-1000 for Toxicity Detection

characteristic	RODTOX	RA-1000
testvessel	built-in	additional aerated vessel
environmental conditions	can be controlled	
sludge origin	treatment plant	
sludge supply	periodic refill	continuous
wastewater supply	batch	continuous
reference substrate	wastewater with known BOD-content	wastewater from treatment plant and optionally synthetic wastewater
reference variable	PS, PA of reference respirogram	maximum respiration rate in testvessel
interpretation	acute inhibition of degradative capacity	
	complete loss of degradation of specific substrates	
contact time	equivalent to treatment plant (cumulative exposure)	retention time of testvessel (10-15 min.)
response time	in case of inhibition of endogenous respiration : 30 min. else 2-3 hours	15 min
action after sludge intoxication	sludge withdrawal and fresh sludge supply	continuous sludge refreshment

Since both systems operate with sludge originating from the treatment plant that has to be protected, their output can be regarded as being representative of this sludge. Therefore, few false alarms are to be expected.

The practical implementation of the two systems is different resulting from a different approach to obtain the necessary early warning character. This has consequences for the way the suspicious wastewater is brought into contact with the sludge that has to be protected. The RODTOX operates in batch mode whereby the loading of the sludge with wastewater is comparable to the treatment plant. In the RA-1000 approach sludge and wastewater renewal are continuous and the loading of the sludge with wastewater is much higher than the treatment plant.

The cumulative exposure of the sludge in the RODTOX to toxic wastewater makes the sludge more sensitive compared to the sludge in the treatment plant and consequently results in a shorter response time. A higher loading of the sludge in the RA-1000 approach results in a short response time to a toxic spill compared to the response time in the treatment plant.

Although the measurement frequency of the RODTOX is small compared to the measurement frequency of the RA-1000, the information content of the raw data is such that the complete loss of degradative capacity can also be detected (e.g. nitrification capacity).

Since the RODTOX does not require continuous sludge supply, the exact position with respect to the treatment plant is not important. However, to leave sufficient time for protective actions, it is advisable to sample the suspicious wastewater upstream of the treatment plant. In practice, samples may be taken from the equalization or sedimentation tanks. The short response time in the RA-1000 approach makes it possible to situate the measurement close to the treatment plant. This position is also preferable because continuous sludge supply must be provided.

Table 2 summarizes the most important characteristics of the evaluated BEWSs.

CONCLUSIONS

Although the two BEWSs described both monitor respiration variables to indicate toxicity of wastewaters for activated sludge, their practical implementation is different with respect to the way the suspicious wastewater is brought into contact with the sludge that has to be protected. This has consequences for the sensitivity, response time and position of the measurement.

ACKNOWLEDGEMENT

The authors gratefully acknowledge Henri Spanjers and Frans Houwen for reviewing the manuscript.

REFERENCES

Herricks, E., Rittmann, B., Grady, C., Pascoe, D., Somlyody, L., Fleit, E., Olah, J., Van der Gaag, M., Van De Guchte, C., Verstraete, W., Dries, D. and Schowanek, D. (1991). Advancements in toxicity testing applied to design and control of biological processes. *Wat. Sci. Technol.* **23** (1-3), 271-282.

Solymon, P., Boman, B., Björndal, H. (1976). Continuous monitoring of acute-toxic substances in wastewater. *Prog. Wat. Tech.*, **8**, (2/3), 417-422.

Spanjers, H. and Klapwijk, A. (1990). On-line meter for respiration rate and short-term biochemical oxygen demand in the control of the activated sludge process. *Advances in Water Pollution Control*, no 10, ed. R Briggs, Pergamon, Oxford, 67-77.

Spanjers, H. (1992). PhD Thesis, Department of Environmental Technology, The Agricultural University of Wageningen, The Netherlands.

Vandebroek, R. (1986). Study and development of a microcomputer controlled sensor for the determination of the biodegradibility and the toxicity of wastewaters: the RODTOX. PhD Thesis, Laboratory of Microbial Ecology, The University of Gent, Belgium.

Vanrolleghem, P., Dries, D. and Verstraete, W. (1990). RODTOX: Biosensor for rapid determination of the biochemical oxygen demand and the on-line monitoring of the toxicity of wastewaters. *Proceedings Fifth European Congress on Biotechnology*, 8-13 July, Copenhagen, Denmark. Vol. 1, 161-164.

Vanrolleghem, P.A., Van Impe, J.F., Vandewalle, J. and Verstraete, W. (1992). Advanced monitoring and control of the activated sludge process: On-line estimation of crucial biological variables in a structured model with the RODTOX biosensor. *Proceedings 2nd IFAC Symposium on Modelling and Control of Biotechnical Processes*, 29 March - 2 April, Keystone Colorado, USA.

Vargaz-Lopez, C.E. (1988). The relationship between respiration rate and microbial activity and kinetic modelling of the activated sludge extended aeration process. PhD Thesis, Department of Civil Engineering, The University of Leeds, England.

Pergamon

Wat. Sci. Tech. Vol. 28, No. 11-12, pp. 427–434, 1993.

0273–1223/93 $7·00 + 0·00

ON-LINE ESTIMATION OF THE RESPIRATION RATE IN AN ACTIVATED SLUDGE PROCESS

Bengt Carlsson

Systems and Control Group, Uppsala University, P.O. Box 27, S-751 03 Uppsala, Sweden

ABSTRACT

In this paper, a novel approach to estimate the respiration rate in an activated sludge process is developed. The proposed estimator for the respiration rate is simple to implement and can be used during normal closed loop control of the dissolved oxygen. The basic idea is to use a two step procedure where in the first step, a model of the oxygen transfer rate K_La is estimated from a data set. The nonlinear K_La function is modelled with a static, constrained, piecewise linear model. In the second step, the respiration rate is estimated from measurements of dissolved oxygen concentration and air flow rate using the K_La model obtained in step 1.

KEYWORDS

Activated sludge; dissolved oxygen; Kalman filters; oxygen transfer rate; oxygen utilization rate; parameter estimation; respiration rate.

INTRODUCTION

The estimation of the dissolved oxygen concentration (DOC) dynamics in an activated sludge process has become a subject of increasing interest because of its impact on process diagnosis (see Olsson and Andrews, 1978; Olsson, 1985), and DOC control (see Ko *et al.*, 1982; Rundqwist, 1986; Holmberg *et al.*, 1989). A common assumption in many earlier identification schemes is that the respiration rate, R, is constant or that the oxygen transfer rate, commonly denoted K_La, is directly proportional to the air flow rate (e.g. Ko *et al.*, 1982; Holmberg, 1982; Holmberg and Olsson, 1985). The respiration rate has, however, normally a diurnal variation and the oxygen transfer rate is nonlinear, see for example, Holmberg *et al.* (1989). One simple approach (Bocken *et al.*, 1989), to account for the nonlinearity of the K_La function is to approximate it with a linear trend. In Holmberg (1990) it was argued that a linear trend model gives identifiability problems in the sense that the absolute level of the estimated respiration rate may be biased. A review of oxygen transfer rate in biological wastewater treatment can be found in Bennett (1980).

The objective of this paper is the development of a method for on-line estimation of the respiration rate. The basic idea is to use a two step procedure. In the first step, a pre-estimation is made where both the respiration rate and a global model of the oxygen transfer rate K_La are estimated. The nonlinear K_La function is modelled with a static, constrained, piecewise linear model. This model reduces the bias problem discussed in Holmberg (1990) since no bias term is needed. In order to achieve an accurate estimate of the K_La function in step 1, an exciting signal is added to the control input signal (the air flow rate). The estimation procedure in step 1 is performed on a relatively short data set, typically a few hundred data points. In the second step, the respiration rate is estimated from measurements of the dissolved oxygen concentration and air flow rate using the K_La model obtained in step 1. A benefit with this procedure is that the respiration rate can be estimated during normal closed loop control except for the initial estimation phase (step 1). It is also easy to implement the estimator for the respiration rate in step 2 in, for example, the plant supervision computer system.

THE DISSOLVED OXYGEN DYNAMICS

The aerobic basin in an activated sludge process is often divided into a number of zones where, in each zone, the DOC is measured and controlled individually. Applying a DOC mass balance to one zone, regarding the zone as completely mixed, yields

$$\frac{dy(t)}{dt} = D(t)(y_{in}(t) - y(t)) + K_La(u(t))(y_{sat} - y(t)) - R(t) \tag{1}$$

where $y(t)$ is the DOC in the zone, $y_{in}(t)$ is the DOC of the input flow rate, y_{sat} is the saturated value of the DOC, $D(t)=Q(t)/V$ is the dilution rate, $Q(t)$ is the wastewater flow rate, V is the volume, $K_La(u(t))$ is the oxygen transfer rate, $u(t)$ is the air flow rate into the zone from the air production system and $R(t)$ is the respiration rate (oxygen utilization rate). In general, it will be assumed that $y_{in}(t)$ and $D(t)$ are measured and that y_{sat} is known.

The respiration rate $R(t)$ depends on the biomass concentration and describes the oxygen consumption. It consists of two terms, oxygen needed for growth and oxygen needed for maintenance energy. The respiration is time varying. Typically, it has a daily variation around a nonzero mean value. Abrupt changes in respiratory rate may, however, also occur, for example, if a toxic disturbance occurs. The variation of the oxygen transfer rate, K_La, depends mainly on the air flow rate u

The model (1) can be discretized by, for example, a Euler approximation which gives

$$y(k) = y(k-1) + hD(k-1)(y_{in} - y(k-1)) + hK_La(u(k-1))(y_{sat} - y(k-1)) - hR(k-1) + v(k) \tag{2}$$

where h is the sampling interval. In equation (2), a term $v(k)$ has been added to describe measurement noise and model misfit. For sampled values of a signal, the notation $x(k)=x(t)|_{t=kh}$, where k is a running index, is used. For simplicity, we have not indicated any time argument in y_{in}.

THE PIECEWISE LINEAR MODEL OF THE K_La FUNCTION

In this section we describe a model of the K_La function which will be used in the first step in the estimation procedure given in the next section.

The oxygen transfer rate can be expected to be a nonlinear function of the air flow rate into the zone. Several theories and empirical relations for describing the OXYGEN TRANSFER RATE have been suggested, see Bennett (1980). Here, the K_La function will be modeled with a static piecewise linear model. Some general applications of modeling static nonlinearites with piecewise linear approximation can be found in Wigren (1993). The model of $K_La(u)$ is constrained so that

$$K_La(0)=0. \tag{3}$$

This makes physical sense since when the air flow rate is zero the OXYGEN TRANSFER RATE vanishes (neglecting the OXYGEN TRANSFER RATE from the surrounding air). Note that the constraint (3) means that the K_La model does not have a bias parameter. This means that the problem of estimating a bias parameter reported in Holmberg (1990) will not occur here.

A piecewise linear model can be uniquely described by the values of K_La in a set of user chosen grid points, denoted by

$$\text{grid}=[u_1\ u_2 \ldots u_n]. \tag{4}$$

The values of K_La in the grid points are described with the parameter vector

$$\theta_\kappa=[\theta_{\kappa,1}\ \theta_{\kappa,2} \ldots \theta_{\kappa,v}]^T \tag{5}$$

where T denotes transpose. The piecewise linear model of the OXYGEN TRANSFER RATE can now be determined from the conditions

$$K_La(\theta_\kappa,u_j) = \theta_{\kappa,j},\ j=1,2\ldots n. \tag{6}$$

$$K_La(\theta_\kappa,0) = 0. \tag{7}$$

Straightforward calculations, utilizing (6) and (7), then give the model

$$K_La(\theta_\kappa,u) = \frac{u}{u_1}\theta_{\kappa,1},\quad 0 \le u \le u_1 \tag{8}$$

$$K_La(\theta_\kappa,u) = \frac{u_{j+1}-u}{u_{j+1}-u_j}\theta_{\kappa,j} + \frac{u-u_j}{u_{j+1}-u_j}\theta_{\kappa,j+1},\quad u_j < u \le u_{j+1},\quad j=1 \ldots n-1. \tag{9}$$

The piecewise linear model (8)-(9) is illustrated in Fig. 1. The model can be written in linear regression form.

$$K_La(\theta_\kappa,u) = \theta_\kappa^T\phi_\kappa(u) \tag{10}$$

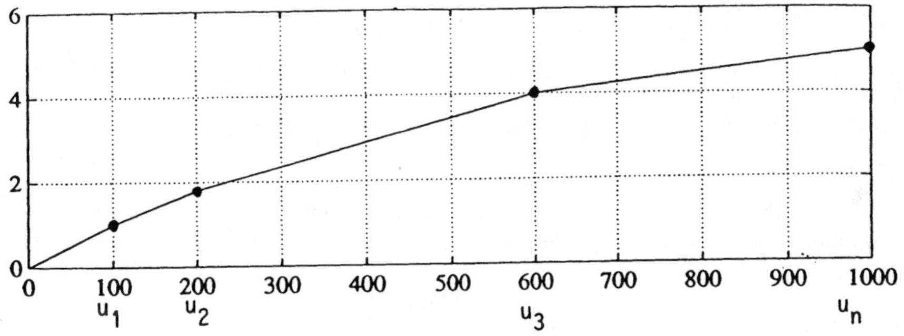

Figure 1. A piecewise linear model of the K_La function with the constraint (3).

with

$$\phi_\kappa(u) = [\frac{u}{u_1}\ 0 \ldots 0]^T,\quad 0 \le u \le u_1 \tag{11}$$

$$\phi_\kappa(u) = [0 \ldots 0\ \frac{u_{j+1}-u}{u_{j+1}-u_j}\ \frac{u-u_j}{u_{j+1}-u_j}\ 0 \ldots 0]^T,\quad u_j < u \le u_{j+1},\quad j=1 \ldots n-1. \tag{12}$$

In (12), the components of the row vector $\mathsf{l}_{\kappa}(u)$ are zero except for the positions j and j+1. This completes the description of the K_La model.

The model (10) can now be utilized in order to transform (2) into

$$y(k) = y(k-1)+hD(k-1)(y_{in}-y(k-1))+h\theta_{\kappa}^T\phi_{\kappa}(u(k-1))(y_{sat}-y(k-1))-hR(k-1)+v(k). \quad (13)$$

It is easily seen that this model can be written as

$$y(k) = \theta^T(k)\varphi(k) + d(k) + v(k) \quad (14)$$

with

$$\begin{aligned} d(k) &= (1 - hD(k-1))y(k-1) + hD(k-1)y_{in} & (15)\\ \theta(k) &= [\theta_{\kappa}^T(k) \quad R(k-1)]^T & (16)\\ \varphi(k) &= [h\phi_{\kappa}(u(k-1))^T(y_{sat} - y(k-1)) \quad -h]^T & (17) \end{aligned}$$

In (16), possible time variations of the K_La parameters have been indicated by using $\theta_{\kappa}(k)$.

THE TWO STEP ESTIMATION PROCEDURE

In this section, a two step estimation procedure is presented. In the first step, both the parameters in the K_La model and the respiration rate are estimated. The objective is to obtain an accurate estimate of the (nonlinear) K_La function. In the second step, the respiration rate is estimated using the estimated K_La model obtained from step 1. Note that the procedure in step 1 is performed during a fixed period of time, typically a number of hours while step 2 can be implemented for permanent operation.

Step 1

A Kalman filter is used to estimate the parameter vector $\theta(k)$ in (14). It is then possible to tune the estimation gain for the parameters individually. A simple choice (see Ljung and Soderstrom, 1983), that will be adopted here, is to model the time variations as a random walk

$$\theta(k) = \theta(k-1) + w(k). \quad (18)$$

By regarding $w(k)$ as white Gaussian noise with zero mean and covariance matrix $Ew(k)w(k)^T = R_1$, a formal Bayesian approach to the identification procedure is possible and a Kalman filter interpretation can be made. Let the measurement noise $v(k)$ in (14) be white with zero mean and variance r_2. The following recursive identification procedure then results (Ljung and Söderström, 1983):

$$\begin{aligned} \hat{\theta}(k) &= \hat{\theta}(k-1) + L(k)\,(y(k) - \varphi^T(k)\,\hat{\theta}(k-1) - d(k)\,)\\ L(k) &= \frac{P(k-1)\,\varphi(k)}{r_2 + \varphi^T(k)\,P(k-1)\,\varphi(k)} \\ P(k) &= P(k-1) - \frac{P(k-1)\,\varphi(k)\,\varphi^T(k)\,P(k-1)}{r_2 + \varphi^T(k)\,P(k-1)\,\varphi(k)} + R_1. \end{aligned} \quad (19)$$

In (19), $d(k)$ and $\vartheta(k)$ are obtained from (15) and (17), respectively. The parameter estimate $\hat{\theta}(k) = [\hat{\theta}_{\kappa}^T(k) \; \hat{R}(k-1)]^T$ in (19), gives the values of the K_La model at the grid points as well as the respiration rate.

Step 2

In this step, the respiration rate is estimated based on the estimated K_La model obtained in step 1. Denote the estimated parameters of the K_La model obtained in step 1 with $\hat{\theta}_{\kappa}^{step1}$. Now, using this K_La model, the following estimator for the respiration rate is obtained from (13).

$$\hat{R}(k-1) = -\frac{y(k)-y(k-1)}{h} + D(k-1)(y_{in} - y(k-1)) + (\hat{\theta}_{\kappa}^{step1})^T \phi_{\kappa}(u(k-1))(y_{sat} - y(k-1)) \quad (20)$$

To reduce the noise sensitivity of the estimated respiration rate, it can be useful to lowpass filter the estimate, see further the numerical illustration.

Remarks

In step 1, it is important to have a sufficiently exciting air flow rate to obtain an accurate estimate. This is not guaranteed during normal closed loop control of the DOC but can be achieved by adding an external signal to the control signal, see Carlsson (1992). A typical choice of external signal is a white noise sequence. This signal deteriorates the control of the DOC and, if used continuously, may damage the control valve and air production system. These problems are the main motivation for step 2 where no additional signal is needed.

The tracking performance of the algorithm in step 1 is determined by the ratio R_1/r_2 which is treated as a user choice. In the following we set r_2=1. A typical choice is to choose R_1 diagonal with $R_1(j,j)=\lambda_1$ for j=1 ... n and $R_1(n+1,n+1)=\lambda_2$ with $\lambda_2 >> \lambda_1$ in order to reflect that the Respiratory is expected to change considerably faster than the K_La function.

The number of grid points (n) determine the number of parameters of the static nonlinear model of the K_La function. A low value on n should be adequate in most cases. A typical choice may be in the interval 2-4. A priori knowledge of the K_La function should be used in selecting the location of the grid points. If no a priori knowledge exists, the grid points may be located at equivalent distances from each other.

Since no matrix computations etc. are needed in step 2 it is easy to implement this procedure in the plant supervision computer system.

A drawback with the proposed two step procedure is that a change in the K_La function is not detected in step 2. Periodic utilization of step 1 (once a month, say) will, however, reduce this problem.

An alternative in step 2 is to use the Kalman filter (19). Then, to keep the K_La model obtained from step 1 fixed, set the first n rows and columns of the initial P matrix to zero and set λ_1 in R_1 to zero. The parameter λ_2 in R_1 can be used to trade off noise sensitivity and trackability to changes in the respiratory rate.

Under some simplified assumptions on the true system, the model and the experimental conditions, it is possible to prove parameter identifiable, see Carlsson and Wigren (1993).

NUMERICAL ILLUSTRATION

In this section, the performance of the suggested estimation procedure will be numerically illustrated. In the simulations, the data were generated from (1) with $y_{in}(t)$ = 2mg/l, y_{sat} = 10mg/l and $D(t)=Q(t)/V = 0.04^{-1}$. The K_La function used to generate the data was taken as $K_La(u)$=3 arctan($u\pi$/1000) h^{-1}. This is a monotone function with a decaying local gain (see Fig. 2d below) which corresponds to a typical behaviour of the OXYGEN TRANSFER RATE. The respiration was taken as a sinusoid (with 24 h period time) added to a constant level (see Fig. 2e). Closed loop operation was considered, where the DOC was controlled by a standard PI controller with set point equal to 2 mg/l. The data was collected with the sampling interval 0.1 h.

In step 1, the number of data points was N=200, corresponding to an experiment length of 20 h. The grid points in the K_La model were grid=[200 600 1000]. Hence, the K_La model had three parameters. The structure of R_1 was taken as described above, with $\lambda_1=0$ and $\lambda_2=100$. The estimation algorithm (19) was initialized with $\theta(0)$ and $\vartheta(0)$ set to zero vectors and the matrix $P(0)=10^6 I$ were I is the identity matrix. This means that practically no prior information was used. Correct values were used for $D=Q/V$, y_{in} and y_{sat}. White Gaussian noise with zero mean and standard deviation 200 was added to the input signal u. The estimation experiment is summarized in Fig. 2a-e. It is seen from Fig. 2e that already after a few hours of estimation, a fairly good tracking of the respiration rate is obtained. The estimated piecewise linear K_La model gives a good description of the true nonlinear K_La function as shown in Fig. 2d. This estimate will be used in step 2 of the estimation procedure.

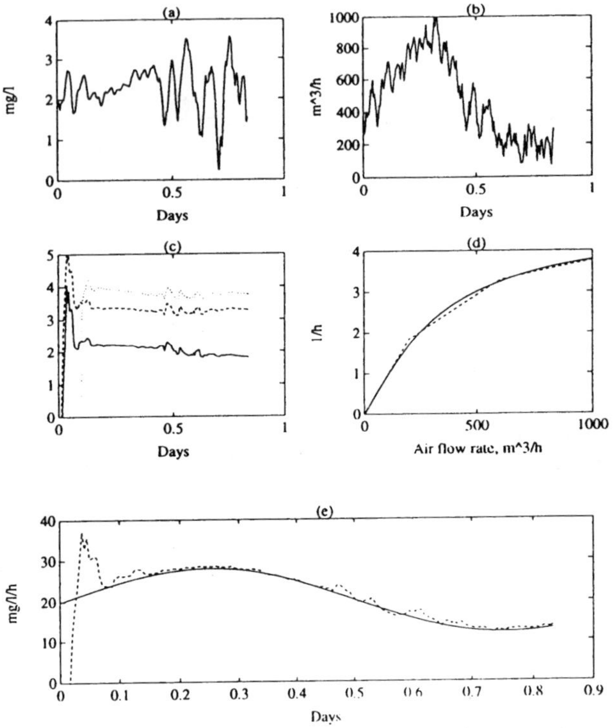

Figure 2. Estimation experiment, step 1. In (a) and (b), the DOC and air flow rate are shown. The estimated parameters of the K_La model are shown in (c). The true (solid line) and estimated K_La function (dashed line) obtained at the end of the run are shown in (d). The true (solid line) and estimated (dashed line) respiration rate are shown in (e).

In Fig. 3a-e, the estimation procedure in step 2 is illustrated. The respiration is estimated using (20). In this case, no additional input signal is used as shown in Fig 3a-b. In Fig. 3c, the true and estimated respiration rates are shown. Note that the true respiration rate makes an abrupt change after about 3.2 d, simulating a process disturbance. As seen from the figure, the estimated respiration is very accurate. This is due to the good model of the K_La function obtained in step 1. In Fig. 3d, the estimated respiration rate is shown when the measured DOC is corrupted with white measurement noise with standard deviation 0.1 mg/l. Fig. 3e shows the same situation but where the estimated respiration rate is low pass filtered. This reduces the noise in the estimate. The choice of filter bandwidth is a trade off between high noise sensitivity and good trackability. This trade off can be simplified by using an adaptive filter, see for example, Carlsson (1989).

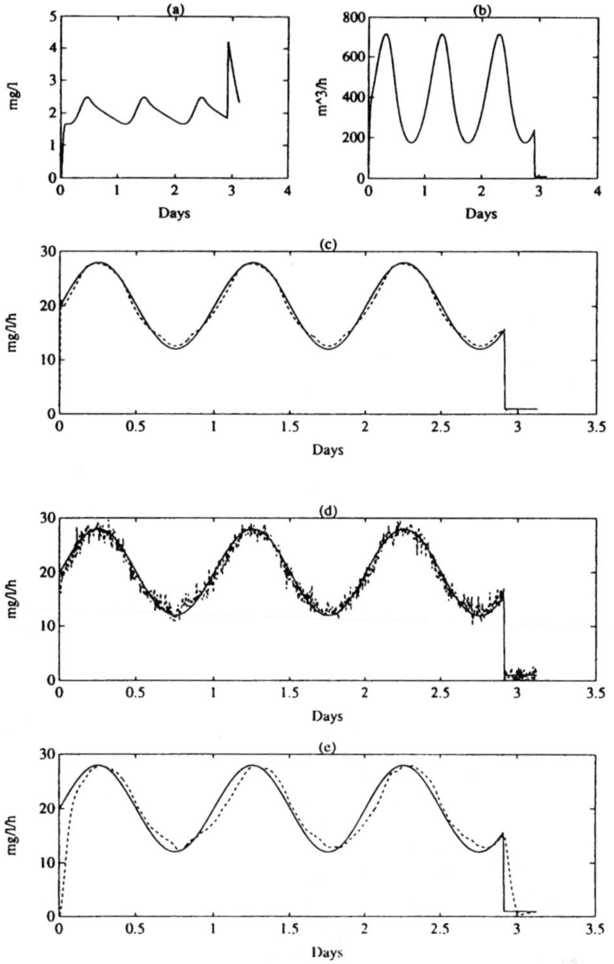

Figure 3. Estimation experiment, step 2. The DOC is shown in (a) and the air flow rate is shown in (b). The true (solid line) and estimated (dashed line) respiration rate are shown in (c) for the case of noise free measurements and in (d) for the case when the DOC is corrupted with white noise with zero mean and standard deviation 0.1 mg/l. A low pass filtered estimate is shown in (e).

CONCLUSIONS

The problem of on-line estimation of the respiration in an activated sludge process has been considered. A two step procedure was suggested. In the first step, a pre-estimation is made where a global constrained piecewise linear model of the K_La function is estimated. In the second step, the respiration rate is estimated using the K_La model obtained in step 1. A benefit with this procedure is that no external exciting of the airflow rate is needed except for the initial estimation phase (step 1). It is also easy to implement the estimator for the respiration rate in step 2 in, for example, the plant supervision computer system. The identification procedure was illustrated in a simulation study. Current research is focused on applying the suggested approach to real data. Some initial tests have shown very promising results, see Carlsson and Wigren (1993).

ACKNOWLEDGEMENTS

This work has been financially supported by the Swedish National Board for Technical Development (contract 91-00019P) within the STAMP (Control and Operation of Wastewater Treatment Systems) project.

REFERENCES

Bennett, G. F. (1980). Oxygen transfer rates, mechanisms, and applications in biological wastewater treatment. *CRC Critical Reviews in Environmental Control*, May, 301-391.

Bocken, S. M., Braae M. and Dold P. L., (1989). Dissolved oxygen control and oxygen utilization rate estimation: Extension of the Holmberg/Olsson Method. *Wat. Sci. Technol.*, **21**, 1197-1208.

Carlsson, B. (1989). Digital differentiating filters and model based fault detection. PhD Thesis, Acta Universitatis Upsaliensis, **28**,Uppsala University, Uppsala, Sweden.

Carlsson, B. (1992). On-line estimation of the oxygen transfer rate and respiration rate in an activated sludge process. Report UPTEC 92078R, Department of Technology, Uppsala University, Uppsala, Sweden.

Carlsson, B. and T. Wigren, (1993). On-line identification of the dissolved oxygen dynamics in an activated sludge process. Accepted for publication in the *12th IFAC World Congress*, Sydney, Australia, July 19-23.

Holmberg, A. (1982). Modelling of the activated sludge process for microprocessor-based state estimation and control. *Wat. Res.*, **16**, 1233-1246.

Holmberg, U. (1990). On the identifiability of dissolved oxygen concentration dynamics. Instrumentation, Control and Automation of Water and Wastewater Treatment and Transport System (*Adv. Wat. Pollut. Control* no 10) ed R Biggs, Pergamon, Oxford, 113-119.

Holmberg, U. (1991). Relay feedback of simple systems. PhD Thesis, TFRT-1034, Department of Automatic Control, Lund Institute of Technology, Lund, Sweden.

Holmberg, U. and Olsson G. (1985). Simultaneous on-line estimation of the oxygen transfer rate and respiration rate. *IFAC Symposium on Modelling and Control of Biotechnological Processes*, 185-189.

Holmberg, U., Olsson G. and Andersson B. (1989). Simultaneous DO control and respiration estimation. *Wat. Sci. Technol.*, **21**, 1185-1195.

Ko, K. Y., McInnis B. C. and Goodwin G. C. (1982). Adaptive control and identification of the dissolved oxygen process. *Automatica*, **18**, 727-730.

Ljung, L. and Söderström T. (1983). *Theory and Practice of Recursive Identification*. MIT Press, Cambridge, MA.

Olsson, G. (1985). Control strategies for the activated sludge process. *In Comprehensive Biotechnology* , Pergamon Press, 1107-1119.

Olsson, G. and Andrews J. F. (1978). The dissolved oxygen profile - A valuable tool for control of the activated sludge process. *Wat. Res.*, **12**, 985-1004.

Rundqwist, L. (1986). Self-tuning control of the dissolved oxygen concentration in an activated sludge process. Licentiate Thesis, TFRT-3180, Dept. of Automatic Control, Lund Institute of Technology, Lund, Sweden.

Wigren, T. (1993). Recursive prediction error identification using the nonlinear Wiener Model. *Automatica*, **12**.

INSTRUMENTATION APPLICATIONS AND MAINTENANCE

Pergamon

Wat. Sci. Tech. Vol. 28, No. 11-12, pp. 435–445, 1993.

0273–1223/93 $7·00 + 0·00

DEVELOPMENT OF THE AMMONIA BIOSENSOR MONITORING SYSTEM

H. Tanaka*, E. Nakamura*, H. Hoshikawa** and Y. Tanaka**

** Public Works Research Institute, Ministry of Construction, 1 Asahi, Tsukuba City, Ibaraki 305, Japan*
*** Fuji Electric Corporate Research and Development, Ltd., 2-2-1, Nagasaka, Yokosuka City, 240-01, Japan*

ABSTRACT

The ammonia biosensor system was newly developed for monitoring ammonia in a river and in secondary effluent from a wastewater treatment plant. The detector of the ammonia biosensor is nitrifying bacteria immobilized by two acetic cellulose membranes and the transducer of the sensor is a dissolved oxygen probe. The microbial probe performs excellent selectivity and quick response to ammonia although control of pH with a phosphate buffer solution and control of temperature with an isothermal water bath are required. The developed ammonia biosensor system processes automatic calibration, sampling, and measurement as a monitor in the field. Accuracy and reliability of the biosensor system were also demonstrated by field tests accomplished at a wastewater treatment plant and at a river quality monitoring station.

KEYWORDS

Ammonia biosensor; *Nitrosomonas europaea*; Immobilized nitrifying bacteria; DO probe; Automatic water quality measurement; Monitoring system; Secondary effluent; River water.

INTRODUCTION

Necessity of ammonia monitoring is increasing in natural waters, wastewater treatment, and water treatment. Existing ammonia in natural waters is undesirable for fish and other aquatic organisms (Russo, 1985). Monitoring of ammonia in the rivers receiving municipal and industry wastewater is important for protection of the aquatic environment. Nitrification of ammonia is required in wastewater treatment if the receiving water body has an insufficient dilution capacity. Denitrification is also required in wastewater treatment plants discharging their effluents into stagnant water bodies, such as lakes, reservoirs, and enclosed bays. In these cases, monitoring of ammonia is valuable for control of the nitrification process and for saving energy of aeration. Ammonia in raw water for water supply consumes chlorine in pre-chlorination and post-chlorination processes. Dirunal variation of the ammonia concentration in the raw water may cause insufficient removal of pathogenic, coloring and odorous agents due to insufficient free available chlorine in finished water, or pipe corrosion and generation of undesirable odor due to excessive free available chlorine (Tamai, 1987). In such cases, monitoring of ammonia and control of chlorine are effective for water treatment.

Existing ammonia monitors are classified into an ion polarography type and a colorimetric auto-analyzer type. Although polarographic sensors are operated as automatic monitors for rivers in Japan, deterioration of the membranes potentially causes maintenance problems (Murakami, 1977). The colorimetric ammonia analyzers are also applied to the process monitors for wastewater treatment in Japan. The reagents for the colorimetric analyzer need frequent replacement due to instability (Annaka, 1979). The analyzers are inappropriate for the river monitoring stations situated far from maintenance stations or for the small-scale treatment plants where a small number of operators reside. Sensors generally consist of "receptors" detecting specific chemical materials and "transducers" changing reactions of the receptors into electric signals. Biosensors apply biomaterials such as enzymes and microorganisms as their receptors (Suzuki, 1985). This paper presents the development of the ammonia biosensor system and its application to an ammonia monitor for a river and secondary effluent from a wastewater treatment plant.

MICROBIAL PROBE

Principle of the Ammonia Biosensor

A chemical sensor generally consists of a receptor detecting a specific chemical material and a transducer converting the reaction of the receptor into an electric signal caused by the magnitude of the chemical concentration. The ammonia biosensor applies *Nitrosomonas europaea* (ATCC 25978), nitrifying bacteria, to its receptor for ammonia and a dissolved oxygen (DO) probe to its transducer. The nitrifying bacteria selectively utilize ammonia for their energy source and change ammonia into nitrite, which results in consumption of oxygen in water. The above reaction is expressed by the following chemical equation (Grady and Lim, 1980):

$$\underset{\text{Consumption of ammonia}}{55\,NH_4^+} + \underset{\text{Consumption of oxygen}}{76\,0_2} + 109\,HCO_3^- \rightarrow \underset{\text{Growth of } \textit{Nitrosomonas}}{C_5H_7O_2N} + 54\,NO_2O + 104\,H_2CO_3$$

The growth yield of the nitrifying bacteria is so small that change of the biomass is negligibly small, which is suitable to a microbial biosensor because of stability. The nitrifying bacteria can be cultured on a a large scale using the Pramer's broth listed in Table 1 (Lewis and Pramer, 1958). The cultured nitrifying bacteria are immobilized by acetic cellulose membranes depicted in Fig.1. The nitrifying bacteria are sandwiched between the two membranes to form a single microbial film. The microbial film is attached to the end of the DO probe.

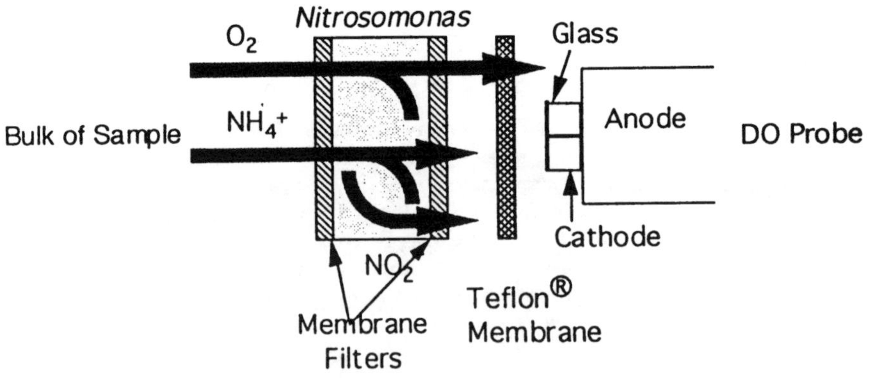

Fig.1. Principle and structure of the microbial probe.

When the microbial probe is immersed in a pH buffer solution without ammonia, the current of the DO probe becomes constant because the rate of oxygen diffusing into the microbial film from the bulk of the solution reaches equilibrium with the rate of oxygen consumed by endogenous respiration of the nitrifying bacteria. This current level coincident with the DO at the end of the DO probe is called a "base current". When the microbial probe is immersed into an ammonia solution, ammonia diffuses into the microbial film from the bulk of the solution. Then, the current of the DO probe reaches a constant current level, which is smaller than the base current, because the rate of oxygen diffusing into the microbial film achieves equilibrium with the enhanced respiration rate of the microbial film by increasing ammonia. The current level of the DO probe is called a "peak current". The difference between the base current and the peak current is defined as an "output". Because the magnitude of the output is proportional to the ammonia concentration in a sample in a range, the unknown sample concentration is predictable based on measurement of the output of the microbial probe.

Characteristics of the Microbial Probe

Because activities of nitrifying bacteria depend on pH and temperature (Grady and Lim, 1980), characteristics of the microbial probe were investigated before assembling the ammonia biosensor system.

TABLE 1 Pramer's Culture Medium for *Nitrosomonas*

Chemicals	
$(NH_4)_2SO_4$	2.5 g
Na_2HPO_4	13.5 g
KH_2PO_4	0.7 g
$NaHCO_3$	0.5 g
$MgSO_4 \cdot 7H_2O$	0.1 g
$CaCl_2 \cdot 2H_2O$	5 mg
$FESO_4 \cdot 7H_2O$	30 mg
Distilled water	1,000 ml

Selectivity

The microbial probe responded to ammonia solutions, but showed no response to a nitrite solution or to organic compound solutions such as glucose and peptone. Thus, the microbial probe is highly selective to ammonia. The output of the microbial probe was quite consistent. Reproductivity of the output was 4% in terms of a CV (coefficient of variation) value when an ammonia solution of 2 mg/L as NH_4-N was used.

pH

The output of the microbial probe was measured when the pH value in a sample varied between 6 and 11. As illustrated in Fig.2, the output increased with increasing pH, and became maximum at the pH value of 9. However, the output decreased sharply when the pH value exceeded 10. Therefore, the pH of the microbial probe should be maintained at a constant level by mixing a sample solution with the buffer solution which has an efficient capability of pH adjustment. Several trials were made to keep the microbial probe active for a long period. The results showed that a phosphate buffer, pH 8, including magnesium ion could maintain the microbial probe in active condition over 50 d.

Temperature

The output of the microbial probe was also studied when the temperature of the microbial probe varied between 14 and 51°C. As illustrated in Fig.3, the output increased with the increasing temperature, and became maximum between 35 and 40°C. However, the output dramatically dropped when the temperature exceeded 40°C. Thus, the temperature of the microbial probe should be controlled at 30°C considering a margin of safety. The ammonia biosensor requires an isothermal bath keeping the microbial probe and sample at 30°C.

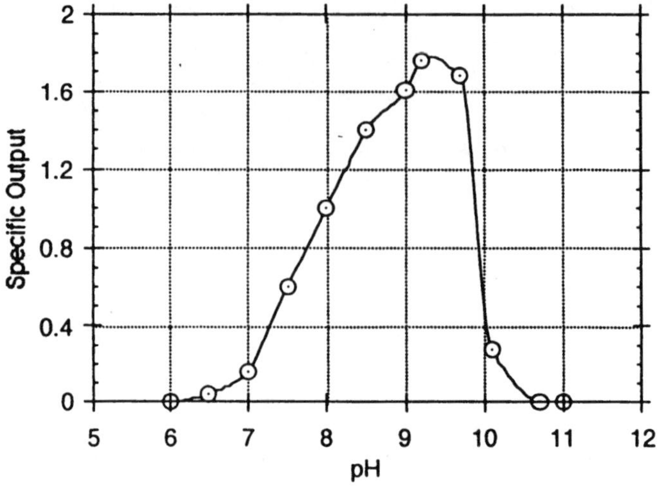

Fig.2. Relationship between the outputs of the microbial probe and pH values.

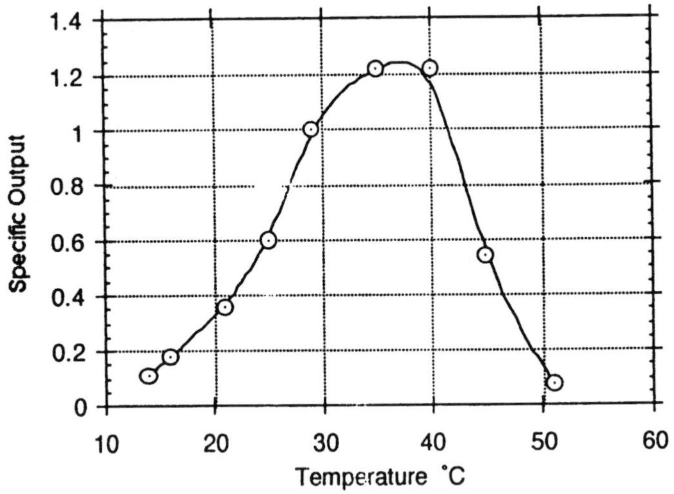

Fig.3. Relationship between the outputs of the microbial probe and temperature.

Amount of the immobilized nitrifying bacteria

The amount of the immobilized nitrifying bacteria affects the DO consumption rate. Fig.4 illustrates the relationships between the outputs of the microbial probe and the amounts of the immobilized nitrifying bacteria in terms of adenosine triphosphate (ATP) density. The microbial film consisting of the high density nitrifying bacteria has a narrow range where the output of the microbial probe is linear and sensitive to ammonia concentration. On the contrary, the microbial film consisting of the low density nitrifying bacteria has a wide range where the output of the microbial probe is linear and insensitive to ammonia concentration. These results imply that an upper detection limit of the microbial probe is controllable to a degree by selecting the amount of the immobilized nitrifying bacteria.

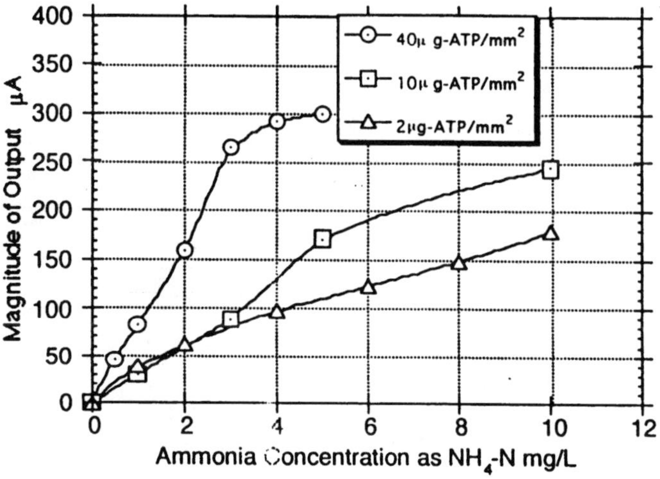

Fig.4. Effects of immobilized biomass on the relationships between the outputs of the microbial probe and ammonia concentrations.

MONITORING SYSTEM

Description of the Monitoring System

The developed microbial probe is supported by several components to establish an automatic monitoring system. Three standard solutions of ammonia and the buffer solution are prepared for automatic calibration of the microbial probe. The microbial probe is attached to a flow cell which the buffer solution or a sample solution flows through. A microcomputer controls the valves and the pumps supplying the standard solutions, the buffer solution and a sample into the flow cell. The microcomputer digitally records the amplified outputs of the microbial probe in time series. The flow rates of the buffer solution and a sample solution are 6 ml per min. After start of operation, the microcomputer orders the buffer pump to supply the buffer solution and records a base current. Next, the microcomputer orders the supply of a standard solution and records its peak current. After the microcomputer orders the pump to stop sending the standard solution, it orders the buffer pump to send the buffer solution and identifies recovery of the DO current to the previously recorded base current. Then, the microcomputer orders measurement of the peak current for another standard solution in the same way. The same process proceeds for the other standard solution. Using the base current and the three peak currents, the microcomputer can make a calibration line between ammonia concentrations and outputs of the microbial probe. After sending the buffer solution and identifying recovery to the base current of the microbial probe, the microcomputers orders supply of a sample solution to the system and measures the output of the microbial probe. The microcomputer converts the output for the sample into an ammonia concentration based on the calibration line and orders display of the result on the monitor and a printout. Samplings are repeated according to the sampling period given by users until sampling times reach the number set by users.

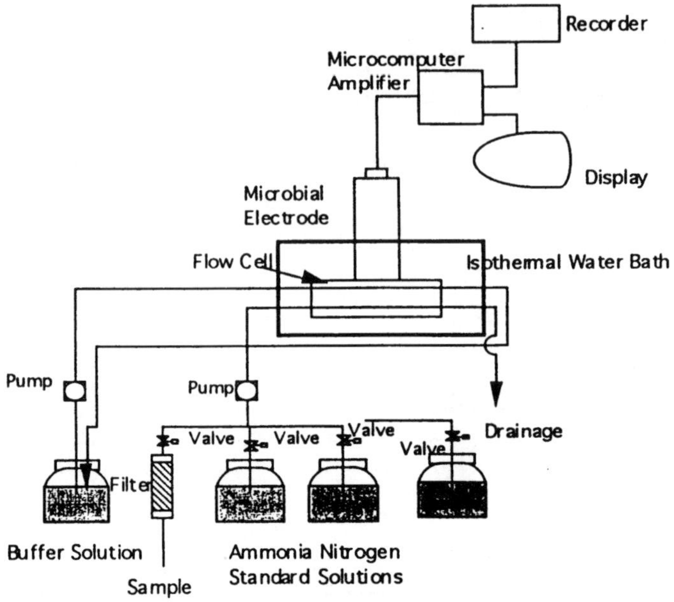

Fig.5. Schematic diagram of the ammonia biosensor system.

Improvement of the Flow Cell Structure

In the prototype ammonia biosensor system (Nakamura and Tamaka, 1988), the buffer solution was directly mixed with a sample solution ↑ to adjust pH of a sample automatically. The buffer solution was sent into the flow cell and was discarded at any time even while no measurements of samples were made. However, it is quite important for the monitor in the field to reduce the consumption rate of the buffer solution because the amount of the buffer solution may be critical to improvement of maintenance frequency and cost. The modified system incorporates a circulation line for the buffer solution illustrated in Fig.6. The flow cell consists of two lines. One is for a sample solution and the other is for the buffer solution. The buffer solution flows through a space between the microbial film and the DO probe at any time, while the sample solution flows outside of the microbial film only during measurement of the sample. A sample solution fed into the system is drained after measurement, but the buffer solution recycles in the system. Both the solutions fed into the system are aerated using air pumps to maintain the DO concentration in the flow cell at a saturated level. The output of the sensor becomes stable 10 min after a sample solution is fed into the system. The output of the sensor is stable due to the buffer solution even if the pH value of a sample varies between 6 and 9. In an ammonia concentration range of 0 to 1 mg/L as NH_4-N, the output of the sensor is proportional to the ammonia concentration in the samples. However, because the relationship between outputs of the microbial probe and ammonia concentrations is nonlinear in the range of more than 1 mg/L of NH_4-N, the ammonia biosensor system facilitates an automatic dilutor to control the ammonia concentration within the above linear range.

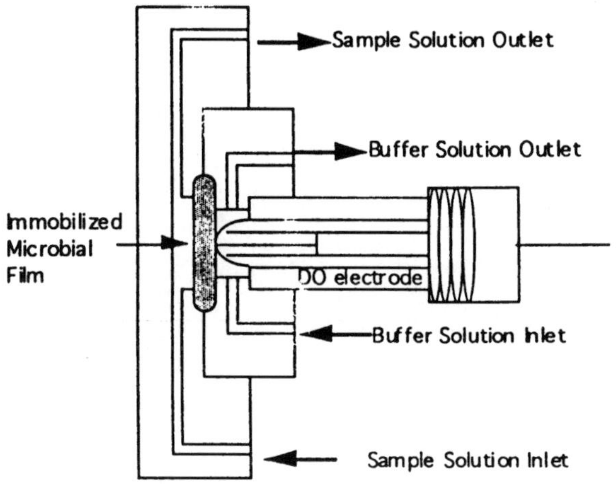

Fig.6. Structure of the improved flow cell of the ammonia biosensor.

Filtration

Effluent from wastewater treatment plants and water in a river include suspended solids. The suspended solids in such a sample should be removed to avoid clogging of the transmission tube. A sample flows inward through a screen made of stainless steel and the filtrate is fed to the ammonia biosensor system. The collected solids on the screen are regularly backwashed by the supplied water flowing outward through the screen.

FIELD TESTS

Application to Wastewater Treatment Plants

Several secondary effluents from wastewater treatment plants were measured to verify the accuracy of the ammonia biosensor system. As shown in Fig.7, the ammonia concentrations measured with the ammonia biosensor system showed good coincidence with the ammonia concentrations measured with the analytical method based on the distillation method in the Sewage Test Methods (Japan Sewage Works Association, 1984).

Verification of the ammonia biosensor system was demonstrated in a field experiment at a wastewater treatment plant. The secondary effluent from the plant was monitored over three weeks. Selectivity of the ammonia biosensor was verified because ammonia concentrations in samples measured with the ammonia sensor system coincided with those in the same sample to which glutamate was added at 100 mg/L three weeks after start of operation. The only observed problem was control of temperature in the isothermal bath. Because the developed monitoring system was set outdoors without any cooling units, the atmospheric temperature in the system exceeded 50°C in summer day time, which decreased activity of the microbial probe.

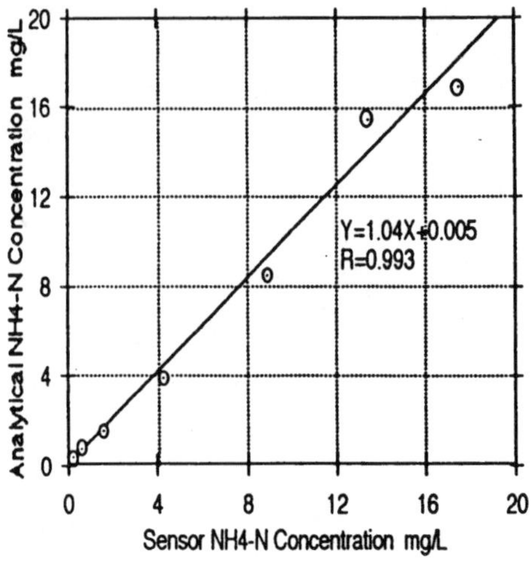

Fig.7. Verification of the measured ammonia concentrations by the ammonia biosensor system.

Application of the Monitoring System to a River

Over 11 months the system was also applied to a river monitoring station in the Tama River at the Tamagawara Bridge in Tokyo. The microbial film was exchanged once a month to avoid deterioration of the film. The standard solutions and the buffer solution were supplied twice a month. The stainless steel screen for solid removal required an exchange in 2 months. Sensor ammonia concentrations were regularly compared to the analytical ammonia concentrations and were found to coincide with analytical values as illustrated in Fig.8. Because no samples exceeded more than 7 mg/L of NH_4-N, the ammonia standard solution was added to the samples for verification of the sensor at higher concentrations. These results demonstrate that an ammonia sensor system is also applicable to ammonia monitors for rivers.

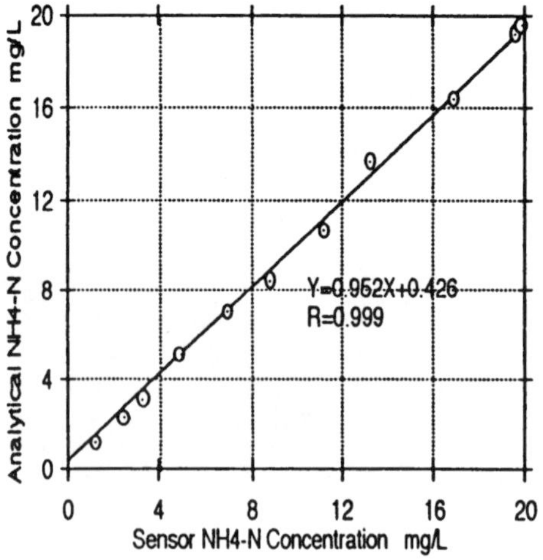

Fig.8. Verification of the measured ammonia concentrations in the Tama River by the ammonia biosensor system.

Monthly and daily variations of ammonia concentrations in the Tama River at the Tamagawara Bridge are illustrated in Fig.9 and Fig.10. Ammonia concentrations were high in winter and low in summer, which suggests that temperature of the river affects the nitrification rate to a large extent. Anyway, the sensor ammonia concentrations coincided with the analytical concentrations very well all the year round.

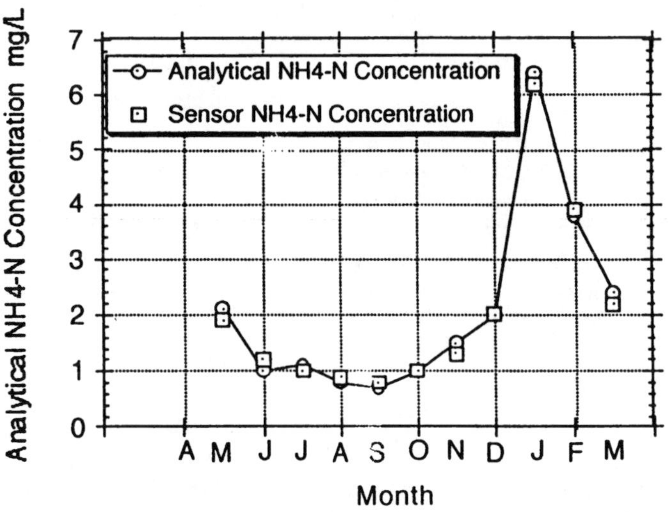

Fig.9. Monthly variation of the ammonia concentrations in the Tama River at the Tamagawara Bridge using the ammonia biosensor system.

Typical daily variation of the ammonia concentration is illustrated in Fig.10. The ammonia concentration became maximum around 3 am and between 1 and 3 pm. Compared with a typical discharge pattern of domestic wastewater (Metcalf and Eddy, 1991), the diurnal variation pattern of the ammonia concentrations shows a time lag of approximately six hours. The time lag is presumably caused by a hydraulic detention time of domestic wastewater in an upstream wastewater treatment plant close to the monitoring station.

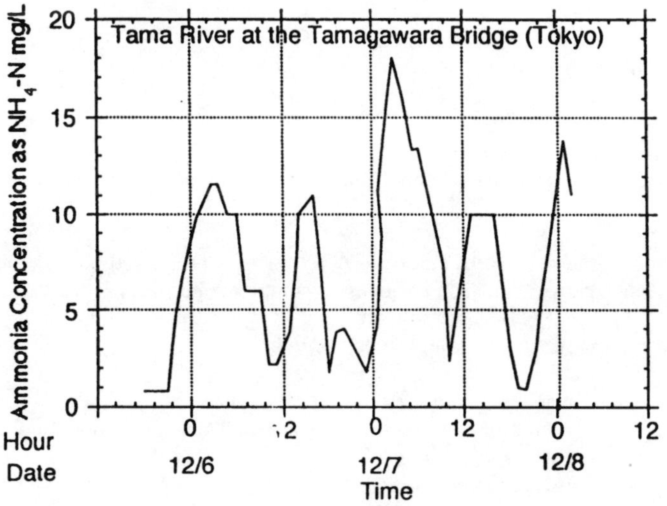

Fig.10. Example of diurnal variation of ammonia concentrations in the Tama River at the Tamagawara Bridge using the ammonia biosensor system.

TABLE 2 Threshold Concentrations [a] of Selected Pollutants Inhibitory to the Ammonia Biosensor System

Toxic Materials	Critical Concentrations [a]
Arsenic	>0.05 mg/L
Lead	>0.1 mg/L
Cyanide	0.04 mg/L
Ortho-chlorophenol	0.06 mg/L
Trichloroethylene	0.06 mg/L
Tetrachloroethylene	0.04 mg/L

a. Threshold concentrations or critical concentrations are defined as the concentrations of the toxic pollutants which inhibit the ammonia biosensor by 5% depression of the output of the microbial probe when an ammonia solution of 1 mg/L as NH_4-N is supplied.

INHIBITORS

Several inhibitors to the ammonia biosensor system were studied because nitrifying bacteria are described to be considerably susceptible to toxic materials (Grady and Lim, 1980). The results of selected toxic materials are described in Table 2. Inhibitory effects of these materials were measured in terms of decreases of the ammonia sensor output. The output of the ammonia biosensor system may be inhibited by high concentrations of toxic pollutants. However, such a disadvantage also implies that the ammonia sensor system can be used for detection of toxic materials in water and for measurement of their toxicities as Nakamura and Tamaka pointed out (1988).

CONCLUSION

The developed ammonia biosensor system demonstrated satisfying performance in the field tests at a wastewater treatment plant and at a river monitoring station. The advantages of the ammonia biosensor system can be summarized as follows:

- The sensor has excellent selectivity to ammonia.
- Required sample volume is relatively small.
- Structure of the sensor and the monitoring system are quite simple.
- Required period to analysis is quite short.
- Little pre-treatment of a sample is required.
- Consumption of reagents is relatively small and the reagents are stable, so maintenance frequency is relatively small.

Moreover, if the other biosensors, for instance the BOD sensor, are applied to the monitoring system in the future, significant parts of the system, such as the microcomputer and the amplifier, can be mutually utilized.

However, the following disadvantages should be considered:

- Replacement of the microbial film is required on a regular basis.
- The measurable range of ammonia concentrations is small. Thus, an automatic dilution process may be required if a sample varies over a wide range.
- Several inhibitory pollutants may affect the sensor output.
- In an atmospheric temperature exceeding 30°C the ammonia biosensor system requires a cooling unit.

ACKNOWLEDGEMENT

The research was jointly promoted by the Public Works Research Institute and Fuji Electric Co., Ltd. under the Biofocus Project of the Ministry of Construction.

REFERENCES

Annaka, T. (1979). Development of automated continuous P-D, Hydro and NH_3 Analyzers. In: *Papers presented at the Sixth U.S.-Japan Conference on Sewage Treatment Technology*, Ministry of Construction, Japan.

Grady, C. P. L. and Lim, H. C. (1980). *Biological Wastewater Treatment*, Theory and Applications, Marcel Dekker, Inc.

Japan Sewage Works Association (1984). *Sewage Test Methods*, Japan Sewage Works Association, Tokyo, Japan (*in Japanese*).

Lewis, R. F. and Pramer, J. (1958). Isolation of *Nitrosomonas* in pure culture, *J. Bacteriology*, **76**, 524.

Metcalf & Eddy (1991). *Wastewater Engineering; Treatment, Disposal, and Reuse*, Third Edition, McGraw-Hill, Inc.

Murakami, K. (1975). Automatic Water Quality Measurement for Wastewater Treatment, In: *Automation and Instrumentation for Wastewater Treatment Plant presented at Fourth US/Japan Conference on Sewage Treatment Technology*, Ministry of Construction, Japan.

Nakamura, E. and Tanaka, H. (1989). Toxicity monitoring by biosensor, *Proceedings of 12th US/Japan Conference on Sewage Treatment Technology*, USEPA.

Russo, R. C. (1985). Ammonia, nitrite, and nitrate. In; *Fundamentals of aquatic toxicology*, M. R. Gray and S. R. Petrocelli (ed.), Hemisphere Publishing Co., New York

Suzuki, S. (1985). *Biosensor*, Kohdansha Publishing Co., Tokyo, Japan (*in Japanese*).

Tamai, Y. (1987). Water Quality Control in the Yodo River System, *Proceedings of 24th Annual Technical Conference: International Session*, Japan Sewage Works Association.

Pergamon

Wat. Sci. Tech. Vol. 28, No. 11-12, pp. 447-456, 1993.

0273-1223/93 $7·00 + 0·00

INSTRUMENT AND SYSTEM MAINTENANCE: A DESIGN AND OPERATIONAL NECESSITY

J. B. Watts* and W. F. Garber**

** Minworth Systems Ltd., Unit 8G, Castle Vale Industrial Site, Maybrook Road, Minworth, Sutton Coldfield, West Midlands B76 8AL, UK*
*** Civil Engineering and Engineering Science, Loyola Marymount University, Loyola Boulevard at West 80th Street, Los Angeles, CA 90045-2699, USA*

ABSTRACT

This paper will present a novel approach to the maintenance of instrumentation and systems. It will show that the true cost of ownership can be quantified by careful thought in the conception and design of systems and instrumentation. The authors will try to identify the true cost of inaccurate process information caused by poorly conceived instrumentation and maintenance.

KEYWORDS

Wastewater Treatment; Maintenance; Instrumentation; Systems; ICA.

INTRODUCTION

In any large industrial enterprise organised to produce a product, the necessary components include people (at management, planning, engineering, research and production levels), production machinery; product, and sales or product disposal mechanisms. The water industry, which in this paper is considered the continuum of water supply to wastewater treatment and disposal, includes all of these elements. However, it is also unique in several ways. For example, its final product often has little value and is discarded, although this same product is also hedged with many difficult and expensive-to-achieve requirements. The activity is also often looked upon as a necessary evil within society and consequently can find capital and operation and maintenance funds difficult to obtain. It is frequently a function of government rather than of a private industrial concern, so its managers may be chosen for reasons other than for their capability within the industry they are responsible for, and their goals may not favour operations. Operational personnel are also often rather low on the decision tree, and have difficulty in putting information into decisions in which they may in fact be the actual experts. Because this function was and/or is in large part a function of government, it additionally became a function of the public works activity. This was usually the responsibility of the civil/sanitary engineering specialists within the municipality served, so facility design was most frequently contracted to civil/sanitary engineering firms.

The water/wastewater facilities and operations within a governmental unit were often their most capital-intensive permanent structures in terms of facilities and equipment and later in operation and maintenance.

As a result of this the instrumentation portions of facilities construction contracts were often considered as relatively small and received little attention. This in part is because the dynamics of the biological treatment systems were often little understood, and in part because both the civil engineering designers and the plant operators had little knowledge of instrumentation systems and computer control systems. As a consequence, systems which were very expensive were installed and quickly became non-functional and mistrusted by operators. Some of these difficulties have been reported in the literature (1),(2). Many operators, after such an expensive and unsatisfactory experience, became unwilling to further test instrument control systems, although they recognised that they might in theory significantly aid them in operational control.

Analysis of the problem with its many and expensive failures indicated several factors were present. For example :

1. Instrumentation specialists did not know the necessary and important operating parameters.
2. Operational specialists did not understand how to use computer control and instrumentation.
3. The dynamics of biological systems were not understood in terms of control.
4. Probes were developed for industries where the variations in organic content and characteristics and the hostile environment found in wastewaters were not factors.
5. One-off computer control systems were designed and installed. These were obsolete at the start of their service and within 1-3 years could not be repaired.
6. The whole process was perpetrated by designers and instrument specialists as a 'mystery', presumably to protect their reputation as 'experts' and to protect development under way and possible patents. Treatment plant operators and engineers consequently learned little of what was really necessary for a successful system.

Fortunately some people within operations recognized the value of a good control system to process maximization when very tight discharge standards and operational cost criteria existed. In addition these operational specialists had become computer and control literate and became aware of the potential power of the new generation of computers and PLCs, of their basic ruggedness, lower cost, relative ease of programming and resistance to early obsolescence.

In summary, conditions became such that operators who understood treatment could now have a very real input into the design and operation of control systems, instead of the previous situation where specialists in computers and instrumentation made some attempt to understand operations and translate such knowledge into a usable system (3).

At the Minworth Treatment Plant in the U.K. serving the City of Birmingham and its environs, a group of people responsible for operations realized the value of close control and developed a computer control system and special probes and methods for ongoing system maintenance. Their group later became a private company, Minworth Systems Ltd. This group's experiences form the basis for the work to be presented in the rest of this paper.

Monitoring of treatment plant operations and of effluent quality has become a major expense within the Developed World Nations where environmental concerns are high and high-technology solutions are affordable. In Southern California in the U.S.A. the monitoring of effluents and receiving waters costs around $20,000,000 per year (4), since laboratory analyses of both the plant processes and of effluents and receiving waters must be carried on. Much of the in-plant process and analytical control could be carried on by presently available computers, PLCs, probes and control software if maintenance such that reliable data was consistently obtained could be assured. Monitoring and maintenance of major capital installations and facilities within a treatment plant is well understood and can be planned for in terms of budget and personnel. However this is not true of the instrument control systems required to maximize treatment and minimize energy usage. Personnel whose understanding of the dynamics of biological systems, process operations and control system individual units is equalled by their understanding of systems are needed so as to make sure that such systems are designed properly and continue to perform their design functions accurately. The ways in which these tasks are accomplished are discussed in the paper.

PRESENT CONTROL SYSTEM MAINTENANCE REGIMES

As requirements for wastewater treatment become more restrictive in terms of air, land and water impacts; and as costs in terms of energy and personnel increase it becomes more and more necessary that instrument systems which include computers, detection probes, software and certain operating units be available and working well. To work well it is essential that good design based upon parameters needing control, good quality instrumentation and effective maintenance be present. Among the difficulties arising from poor process control are:

1. Higher than necessary electrical energy costs because of over-aeration, poor organism viability, insufficient biomass control or similar factors.
2. Inaccurate knowledge of parameters such as flows to the plant and unit processes, actual air amounts delivered and other properties which bear upon the actual processing capabilities of the plant.
3. Poor information regarding the efficiency of operation of the plant and unit facilities making supervision and planning difficult and cost control hard to accomplish.
4. Difficulty in predicting maintenance needs and in planning and providing for proper unit process and control system maintenance.
5. Difficulty in planning for and obtaining the types of personnel necessary to keep facilities at maximum levels. For control systems maintenance this would mean instrument technicians and engineers as well as computer and software specialists.

Maintenance has always been one of the most important functions of wastewater treatment operations. The increasing complexity of treatment, required standards and necessary control systems with the resultant possible heavy costs make proper maintenance increasingly important.

Observation of procedures used in various large and small treatment facilities in various parts of the world over the years suggests that there are four general categories. These could be designated as:

1. Total neglect - reacting only to breakdowns.
2. More-or-less planned breakdown maintenance.
3. Planned maintenance by the facility staff.
4. A combination of planned staff maintenance and total care maintenance of special portions of the facility by contract to specialists.

The application of each of these categories in the various units in a treatment plant will positively or negatively affect the way the required standards are met. This paper is restricted to discussion of maintenance of the overall instrument control systems for critical wastewater processes because of the importance of careful process control.

It should be obvious to anybody that neglect and breakdown maintenance are totally unsatisfactory, and, even though they might be present, should be changed to plant staff and/or contract maintenance. For the complex systems now required to enable maximum economy and treatment, completely effective maintenance by plant staff is difficult. Some very large facilities may have the skills, but many of these also depend upon contract maintenance by skilled instrument companies or manufacturers. Such contract specialist groups and/or must however really know both treatment process details and the philosophy of the system design and its critical components, including software, in order to be effective. A significant variant of the usual contract procedure would be for the contract organisation to be involved in the original system design and then to take total responsibility for maintenance of the system, including upgrades. Under such a contract the specifier (the facility) would require the contractor to provide Total Care Maintenance (TCM).

To ensure that a computerised instrumentation control system will function continuously in the manner it was designed to do, the following requirements should be part of the specifications for every installation.

1. The system should be hardware independent and be able to work on any hardware platform if this platform is maintained as part of the total system.
2. The system maintenance package should be the responsibility of a **single** entity, so that the operator will not have to become involved in resolving disputes between component suppliers as to responsibility for malfunction.
3. The system should be "future-proofed". That is, it should be amenable to upgrading of hardware and software and should be designed such that replacement with advanced hardware and software would be simple.
4. Under a total care maintenance contract the supplier-maintainer of the system should supply the user software upgrades without cost as they become available. If the user requests software from another party, the system should be able to integrate it, although an additional cost should be expected for such service.
5. The user and supplier-maintainer should actively explore other control possibilities and use for the system. These should be furnished at no cost if it results in an enhancement of the system which will also aid other users.

The kind of situation where contractual maintenance by a single supplier/maintainer of computer-instrumentation monitoring and control systems is practiced is very rare throughout the world. However, the authors believe that the needs and complexities are now so specific that such a procedure is now a necessity for the maximum efficiency of operations. General knowledge and general skills are **not** sufficient and both capital investments in systems and the costs and risks of poor operation are such that a knowledgeable approach must be adopted.

WHY A SINGLE DESIGNER-SUPPLIER-MAINTAINER IS NECESSARY

A major problem with the present ICA systems, including instruments, computers and other hardware and software is one of the commitment to the need to know and accept the **real** cost of ownership. Most maintenance regimes now used **hide** the real cost of not owning guaranteed fully functioning equipment. They fail to include the cost of not operating the facilities they control with energy efficiently so as to meet the tight control standards. Inefficient operation also translates to loss in capacity and difficulties with control authorities. In effect, the costs arising from the misuse of capital investment in both control systems and treatment by allowing poor systems maintenance are ignored - that is, people are often ignorant of these costs because they do not appreciate the true cost penalties of 'saving money' by neglecting maintenance, specifying cheap, unreliable equipment, and contractually obliging suppliers down to providing continual total support. The costs arising from the piecemeal maintenance of the instrumentation monitoring and control system and of replacing and updating hardware and software in the system on a regular basis is not included in most cost calculations. Many factors contribute to this problem, including internal organisation of operations and maintenance by owners of systems. For example, centralised purchasing systems may not be the right platform to allow proper evaluation of the sophisticated control system needs of different plants. Also, although they may have good instrument repair groups, group responsibilities for many differing in-plant systems may not allow the degree of concentration on the basic treatment process that is necessary.

Control of biological processes is very specialised since the knowledge of the dynamics of the growth of biological systems must be a principal factor in design, system maintenance and close control. Further, wastewaters and wastewater plant atmospheres are very hostile to the hardware components of an ICA system. Design and maintenance must be predicated upon this fact to allow reliable and predictable date from the monitoring and control system. The fact that a system produces **consistent** data does not mean the data is **reliable** and is suitable for control. It is therefore important that a process instrumentation, monitoring and control system be considered as a single entity in terms of design, operation maintenance and upgrading.

For many years the manager of the operation of a treatment plant has had to rely upon laboratory results obtained from grab and/or composite samples which can only at best tell what has happened in the recent past. His experience, aided by manual plotting of trends for certain parameters were then used to make decisions on alterations in plant operation. Sometimes this system provided for reasonable control where conditions did not vary too greatly. Unfortunately a wastewater treatment facility has very limited control over its influent, so the possibilities perceived in installing an instrumentation control and automation system were accepted with cautious optimism. However, the computers, probes and on-line analysers first used were adapted from laboratory uses or from clean streams where the problems resulting from the hostile wastewater treatment environment were not present.

Early wastewater control systems suffered because:

1. Parameters necessary for process evaluation were not really measured or were not even fully known.
2. The data from the probes and the instrument were at best consistent, but users had no way to check their accuracy and to recalibrate them on a continuing basis, over-riding inbuilt calibration frequencies if necessary.
3. The widely varying characteristics of wastewater, such as the content of hydrogen sulphide and dissolved sulphides had not been taken into account in terms of service life.
4. Responsibility for the system was distributed between designers, operators, manufacturers and data transfer specialists so that assessing the reason for difficulties became a multi-party controversy and therefore often very hard to pin down which party had responsibility for correction of the problem.
5. Proper maintenance (often consequently from item 4) was not done and many systems failed within a short time (1). It became apparent that design, construction, operation and ongoing maintenance (including the possibility of upgrading)) were all necessary.

The full package of design, construction, data accuracy and upgrading could have been accomplished internally by the operational authority if it had possessed the organisation and the varied skills required. Another possible procedure would have been to use a contract group specialised in the performance of the varied skills necessary to ensure design, construction and continuous operation and upgrade of such a system - if such an organisation could be found. One of the ways in which this is now achieved in the U.K. is described later in the paper. If an operating agency has a skilled instrumentation group functioning, the contract operation could mesh with them to complete the necessary maintenance team.

Most of the instrumentation, monitoring and control systems available are designed to accommodate the instrumentation control and automation experts. To answer questions as to the accuracy and calibration of probes requires specialised knowledge and special instruments which a plant manager or his staff do not usually have. The plant influent is normally beyond the control of the manager in terms of flows, loads and contaminants which can very rapidly change the biological population he must use for treatment.

For instruments and probes to be of value and really utilised by plant managers and operators they need to have features such as the following.

1. They should be self-cleaning
2. They should be self-calibrating.
3. The instrument itself should obviously supply local output analogue signals proportional to the measured values, but it should also be able to store the data obtained in a suitable form (we suggest 15-minute averaged data) for a suitable length of time (we suggest 14 days). It should also store in detail the last few calibration cycles, so these can be examined in detail retrospectively should the need arise to check the accuracy. This would simplify the system and allow non-dedicated telephone lines or radios to access the instrument to check accuracy and performance.
4. The instrument should be capable of interrogation and re-configuration remotely to allow for checks on its accuracy and performance and for its method of operation to be changed if necessary.
5. Instruments and probes should be able to handle the hostile environments found in wastewater systems.
6. In addition to a local display, which may be a digital readout, instruments should provide continuous analogue information to allow non-specialist users to evaluate trends and be confident in the accuracy of results.
7. A local display should allow the operator on the plant to examine the status of the instrument at any time.
8. Instruments should be of modular design and easy for regular employees to carry out any non-specialist service needed.
9. They should be **FAIL SAFE** - that is, they should give accurate readings or none at all!
10. The should allow future advances to be readily incorporated.
11. They should be such that supply and maintenance costs, inclusive of labour materials and upgrades could be quoted on an annual basis with some guarantee of a similar cost for a longer period such as five years. This allows the TRUE cost of ownership to be established from the outset.

TOTAL CARE MAINTENANCE OF INSTRUMENTS

Total care requires that the designer/supplier/maintainer be able to monitor the system installed at a user's site. This involves linking the systems and systems components in a loop by way of a modem to a maintainer centre using regular dial-up telephone lines. In some sites simple radio links are also part of the system. On a regular daily basis the central maintenance centre communicates with each system and its components, interrogating them individually to establish that they are working properly. If the local operator contacted the maintainer, querying the operation of an instrument, the same initial response would occur. The maintainer would contact and interrogate the instrument to diagnose its performance. In case there are any problems or potential problems of which the local user may not be aware, the central computer obtains detailed information from the remote memory in the instruments, including the calibration cycles. Various diagnostic techniques are then utilised to determine whether corrections can be made remotely or whether on-site service is necessary. Data obtained over periods from minutes to years can be called up from the maintainers archive to aid diagnosis.

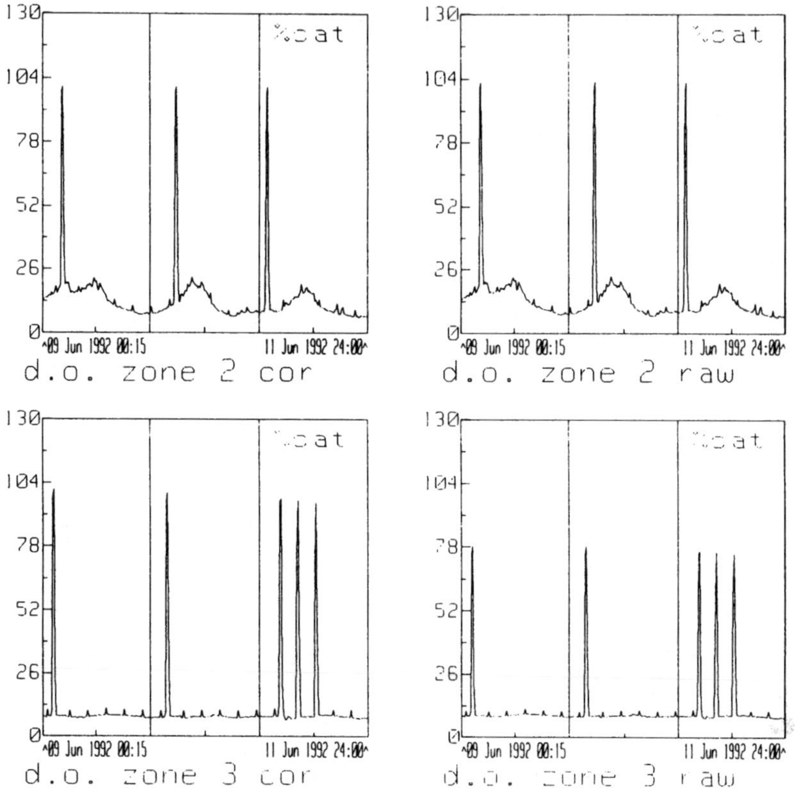

Figure 1: Example of repeated re-calibrations triggered by the inbuilt software of an MSL 'intelligent' dissolved oxygen instrument.

The lower 2 graphs in Figure 1 illustrate a three-day plot in which automatic repeated calibrations have been triggered on a Minworth Systems Auto-Calibrating dissolved oxygen probe in zone 3 of the particular plant on the third day. The (upper two) graphs for the instrument in zone 2 are provided for comparison. Usually (as in zone 2, and zone 3 for the first 2 days) there would be only one calibration per day if the calibration sequence was accepted by the software on the instrument that statistically checks its acceptability, although in some particularly fouling environments more frequent self-cleans and re-calibrations may be deliberately remotely programmed into the instrument. In this example the probe was beginning to foul and the software was rejecting the calibration and triggering another clean/auto-calibrate four hours later. Note that this did not mean the measurement from the instrument was not usable - the software checks are made very strict to show up any signs of impending inaccuracy. A further feature of the diagnostic system identified the problem as fouling by allowing comparison of the uncorrected ('RAW') signal from the sensor with the corrected ('AUTOCAL') signal, which is that automatically corrected for any deviation from 100% saturation in air.

Also visible in Figure 1 are small spikes at 4-hourly intervals on the trace. These result from the 4-hourly self-clean performed by the instrument in addition to the daily clean/auto-calibrate cycle. Both these spikes are, of course, suppressed on the sampled-and -held output signal that is fed from the instrument into the DO control system.

Figure 2 illustrates the slow fouling of a dissolved oxygen probe. It shows the RAW and Autocal data as the fouling progresses over a period of 11 days. It should be recognised that fouling in any system or probe can occur eventually even though carefully designed self-cleaning features are fitted. The important thing is to make the self-cleaning as efficient and reliable as possible and make the system recognise when manual intervention is required by the operator or contractor.

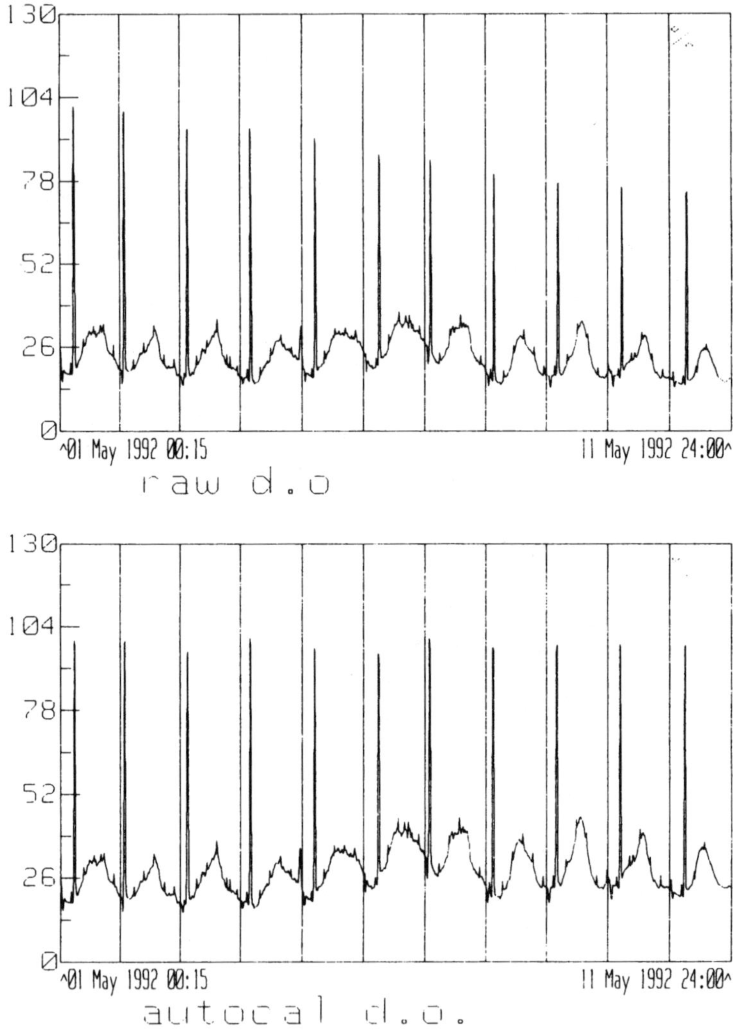

Figure 2: Illustrating the slow progressive fouling of a dissolved oxygen probe.

Note in Figure 2 that although the **raw** DO signal, as illustrated by the daily calibration spikes, is progressively falling, the **auto-calibrated** signal is compensated and still maintained at an accurate 100% saturation in air by the software built-in to the instrument.

Figure 3 shows an example of the data permitting remote diagnosis of a problem which had developed. There had been a drop in the pressure of the air supply which the instrument requires for its self-cleaning and auto-calibration. The central maintenance computer system operators picked up the problem and notified the operator with instructions as to how to rectify the fault.

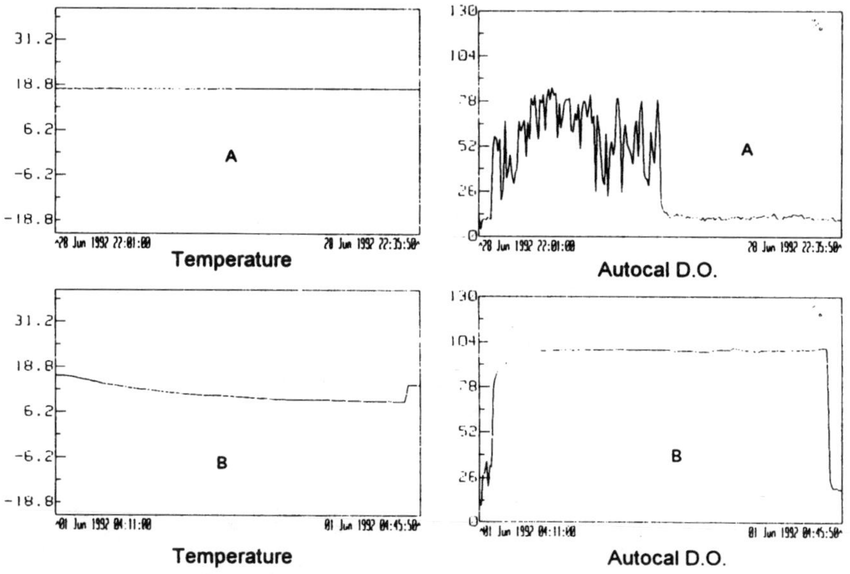

Figure 3: Detailed area from dissolved oxygen instrument, which permits remote fault diagnosis

The top graphs, labelled 'A', show the logged detailed area from the instrument for the day the fault developed. The bottom graphs, labelled 'B' show the usual type of detailed area for this particular instrument in this particular plant. The instrument in question is unusual in that its air supply for cleaning/calibration comes from a compressed air cylinder, instead of the more normal line-voltage powered air compressor and reservoir system. (Plant conditions dictate that a mains(line)-powered compressor cannot be used in this particular installation.) The Raw DO trace in 'A' is simply caused by the cylinder being practically empty, and not being able to supply enough air for the calibration cycle. Note also the temperature trace. With this particular instrument, the sort of trace in 'B' is normal. There is a significant drop in temperature of the air used during calibration, because it is supplied from a high-pressure air cylinder, and it cools significantly as it expands from this cylinder. Note the virtual absence of this effect in the 'A' graph (because the air cylinder supply ran out), and compare the 'normal' trace in the 'B' graph.

The maintenance system central computer interrogates each site instrument computer and transfers a copy of all their data to the central archives where it is stored for backup, and for treatment and process diagnosis as necessary. The confidentiality of any such data is assured by the integrity of the contractor and by the terms of the total care maintenance agreement. The client operator therefore does not require expensive on-site archives or backup archival systems. The central system diagnostics and its operators analyse the transferred data and any fault or potential fault (which the operator may not be aware of) is reported within 24 hours. Plant managers are accordingly assured that correct data is provided, so that the state of the treatment process can be continually evaluated.

Remember, all this central diagnosis and maintenance service is additional to local features of the instrument, which themselves display a message and supply an alarm signal if the integrity of the measurement is in doubt. The plant manager can concentrate his attention on fine-tuning the plant and does not have to worry whether he is trying to make decisions based on erroneous data!

A centralised archive system which continuously queries intelligent instruments and analyses their performance is central to the Total Care Maintenance Package. The ability to use normal telephone lines and the ability to ensure continuing upgrades make the system valuable and economical.

In the UK plant systems with a replacement cost of well over $5,500,000 are now using total care maintenance.

SUMMARY AND CONCLUSIONS

The wastewater treatment industry faces very severe challenges in that it must accept a continuously variable influent and maximise the economy of its treatment while meeting effluent standards which are increasingly more difficult. Biological systems of treatment such as activated sludge are the most satisfactory in terms of economy and the ability to handle most organic contaminants. However biological systems, like every system of living organisms, are sensitive to changes which are too severe. Manual control of biological systems by humans is possible, but becomes more difficult as discharge requirements increase in severity and energy costs increase.

The use of instrumentation, control and automation therefore becomes essential. As these instrumentation systems become more complex, it becomes less satisfactory to deal separately with designers, instrument manufacturers, information transfer specialists and operators separately. A total care approach by a group which can accept complete responsibility is required. This group needs to be based in operations to really know the dynamics of the biological systems used as well as being competent in computerised instrumentation, monitoring and process control design, implementation and maintenance.

ACKNOWLEDGEMENTS

The Authors wish to thank the staff of Minworth Systems, particularly Des McGarr, for their assistance in the production of this paper.

REFERENCES

1. Watts, J. B. and Evans, K. (1985). An operational users approach to the introduction and implementation of ICA to a large inland sewage works. *Instrumentation and Control of Water and Wastewater Treatment and Transport Systems (Adv. Wat. Pollut. Control* no. 2) ed. R. A. R. Briggs, Pergamon, Oxford, pp. 311-322.
2. Garber, W. F. and Anderson, J. J. (1985). From the stand-point of an operator, what is really needed in the automation of a wastewater treatment plant? *Instrumentation and Control of Water and Wastewater Treatment and Transport Systems (Adv. Wat. Pollut. Control* no. 2) ed. R. A. R. Briggs, Pergamon, Oxford, pp. 429-442.
3. Cardona, M. (1992). What do you do when a single design mainframe computer system controlling complex wastewater treatment processes irretrievably crashes? Newsletter, Specilaist Group on Instrumentation and Control and Automation of Water and Wastewater Treatment and Transport Systems, Number 2.
4. Marine Board (1990). Managing troubled waters – the role of marine environmental monitoring, Natural Research Council, National Academy of Sciences, Washington, D.C., USA.

Pergamon

Wat. Sci. Tech. Vol. 28, No. 11-12, pp. 457–463, 1993.

0273–1223/93 $7·00 + 0·00

CROSS COMPARISON OF TECHNIQUES FOR THE MONITORING OF TOTAL ORGANIC CARBON (TOC) IN WATER SOURCES AND SUPPLIES

B. MacCraith*, K. T. V. Grattan**, D. Connolly*, R. Briggs**, W. J. O. Boyle** and M. Avis***

** Department of Physics, Dublin City University, Dublin 9, Ireland*
*** Department of Electrical Electronic and Information Engineering, City University, Northampton Square, London, EC1V OHB, UK*
**** Resource Consultants Cambridge (RCC), The Mount, Slater Road, Longstanton, Cambridge CB4 5DS, UK*

ABSTRACT

A comparison of UV (254nm) absorption spectrometry, optical fluorescence (excitation at 340nm, emission measurement at 420nm) and established chemical methods for estimating Total Organic Carbon (TOC) in water samples obtained from a variety of sources has been made. In addition, a proprietary UV in-line absorption monitor commonly used in industry is assessed relative to these measurements. The work has shown a strong correlation between Total Organic Carbon (TOC) measurements and both UV absorption and fluorescence measurements. These measurements indicate the feasibility of both approaches in the design of new field based instruments for TOC determination using optoelectronic devices.

KEYWORDS

Water; total organic carbon; TOC; UV absorption; fluorescence; cross comparison.

INTRODUCTION

The development of optical fibre based sensors for the water industry has been a major research interest for some years and sensors for dissolved oxygen, chlorine, ammonium and nitrate have been developed [1]. In this work the effort is being extended to include the determination of Total Organic Carbon (TOC). There is considerable interest in the development of field based instruments for the measurement of TOC as it is an extremely important gross indicator of the viability of a water course to sustain life. Previous work by one of the authors [2] has led to the development of an instrument for measurement of TOC based on UV absorption and which is now marketed commercially. An investigation of this instrument is made in this study as a benchmark of currently used technology.

BACKGROUND AND MOTIVATION

The TOC value of natural waters is composed of naturally occurring and man-made organic compounds. Many of the naturally occurring organic compounds are not implicated as health hazards. In contrast, many man-made compounds have serious implications for the use of natural waters as amenities and as sources of drinking water. The nature of the potential health hazard depends on the source of the polluting material; influxes of slurry or silage effluent can lead to eutrophication whereas influxes of agri-chemicals or fuel oils can lead to poisoning. Therefore the estimation of TOC is an extremely important gross indicator of the health of a water course.

Most methods for the determination of TOC content in water are laboratory based. In general these methods operate on the same principle; the organic carbon is converted to a gas by oxidation which is then detected and quantified as a measure of TOC content. Oxidation is performed by high temperature catalytic, chemical or photochemical methods and produces carbon dioxide. Detection of the gas may be performed by flame ionization, non-dispersive infra-red, titrimetric, nephelometric, electrical conductive, spectrophotometric and resistivity methods [3].

A high TOC content in natural waters is not necessarily an indication of pollution by harmful organic compounds as many organic substances may be present naturally. Although normal levels of TOC content are unlikely to represent a health hazard an increase [4] however may indicate the introduction of something harmful and would suggest that further analysis is prudent. Therefore rapid detection of potential pollution is a higher priority than accuracy and for rapid detection, continuous monitoring is necessary. Laboratory based systems are not capable of effective continuous monitoring as they rely on grab sampling, the analysis is time consuming and the process is destructive of the sample. A field-based system should be capable of continuous monitoring to be of realistic value. Analysis must be almost instantaneous and optical methods such as absorption and fluorescence fulfil this criterion.

BENEFITS OF OPTICAL SYSTEMS

The concept of using optical methods for the examination of waters is a long established one [2,4 to 18]. The benefits of optical methods for the monitoring of TOC are many. Optical methods such as absorption and fluorescence do not require reagents or any special preparation of the sample and the processes are almost instantaneous. Optical methods can utilise fibre optics giving them further advantages. Fibre optic technology allows remote sensing to be carried out giving greater flexibility as to the location of monitoring equipment, making multi-point analysis possible.

In this study optical methods are investigated as methods for the effective determination of TOC. Previous work [2] led to the use of absorption for the estimation of TOC and other research [4,5] has suggested that fluorescence could be used as a method of determining TOC in waters. Fluorescence has some advantages over UV absorption in that generally it has higher sensitivity, and it is also considerably less affected by sample turbidity [4].

INSTRUMENTAL TECHNIQUES

The measurement techniques used in this study are familiar to the analytical chemist but are chosen because they can be configured to allow for the use of modern optical and fibre optic technology, the basic of a new generation of field instruments.

The water samples are from river, treated sewage and landfill seepage sources. These samples are representative of the variation in waters routinely found in measurement laboratories. Samples were supplied with chemically determined measurements of TOC and were stored at approximately 4°C, while awaiting further investigation.

Chemical determination of TOC content in samples was made using a Dohrmann DC-80 TOC analyzer. Here TOC is measured by ultra-violet promoted persulphate oxidation and the evolved CO_2 is measured by infra-red absorption. UV absorption measurements were made using a Hewett Packard HP8452A Diode Array Spectrophotometer. This instrument is capable of measuring absorbance over the range 190nm to 820nm. In addition a commercial device, a pHOX Organic Pollution Monitor (OPM) was used on the same samples. The principle of operation of this instrument is based on absorption at 254nm with a correction for sample turbidity based on absorption in the visible. Figure 1 shows a diagram the set-up of the pHOX OPM.

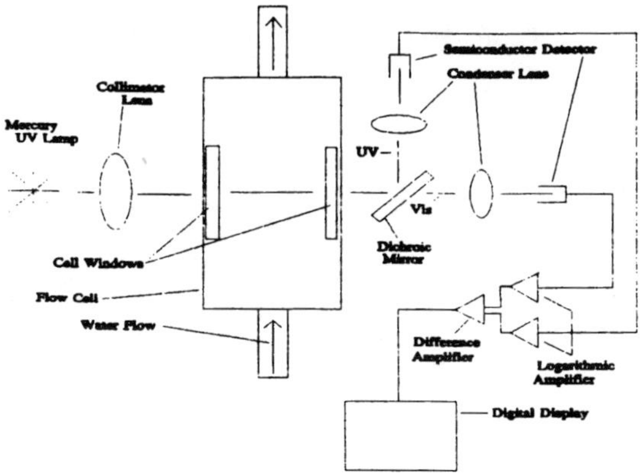

Fig. 1. Schematic of the pHOX OPM device.

Fluorescence measurements were made using a Perkin Elmer LS-50 Spectrofluorimeter. The wavelengths at which the samples were excited and at which the fluorescence emission was measured were determined initially by the examination of the excitation and emission spectra of samples of contaminated water. Based on these results a fibre optic based sensor system was constructed. Interference filters were used to select the excitation and emission wavelengths. Figure 2 is a diagram of the fibre based fluorimetric sensor.

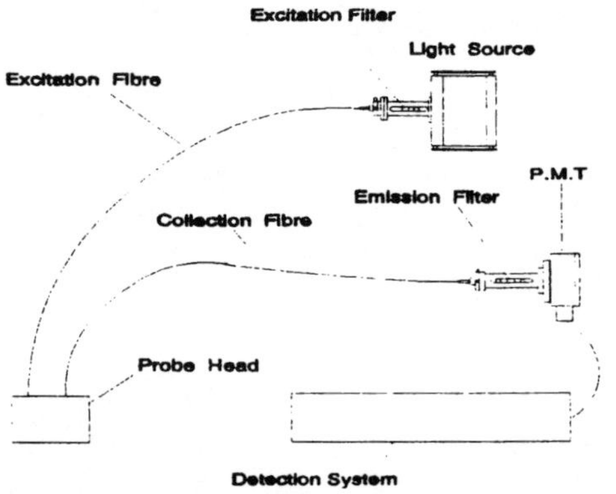

Fig. 2. Schematic of the fibre-based fluorimetric sensor system.

Figure 3 shows the schedule for the measurements and the exchange of samples between the participating laboratories. The involvement of several laboratories in different countries in the work enabled a range of samples and a range of techniques to be evaluated, where each laboratory employed different expertise, derived from its interest in using and developing different appropriate field instruments.

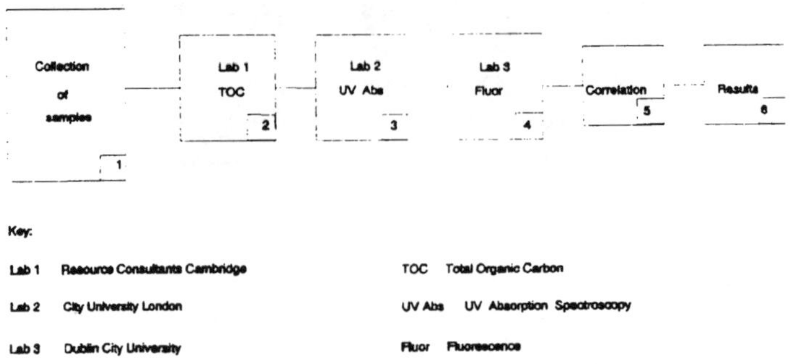

Fig. 3. Schedule for the laboratory measurements.

RESULTS

Strong linear dependencies were found between the chemical analysis of TOC and the absorption and fluorescence measurements in the region of 0mg/l to 15mg/l of TOC for the RCC samples.

Absorption measurements were made at step 3 in Figure 3 (the measurement schedule). An absorption spectrum of each of the RCC samples was taken from 190nm to 780nm. The absorbance value at 254nm was obtained from each of the spectra. Figure 4 shows a plot of this absorption at 254nm against TOC concentration. The correlation coefficient for this graph is 0.991. This illustrates that the optical absorption method provides an excellent means of identification of TOC concentration, over the range shown, with light from a conventional 254 nm source used. Thus this gives confidence that a dedicated instrument using such an approach has potential for further development.

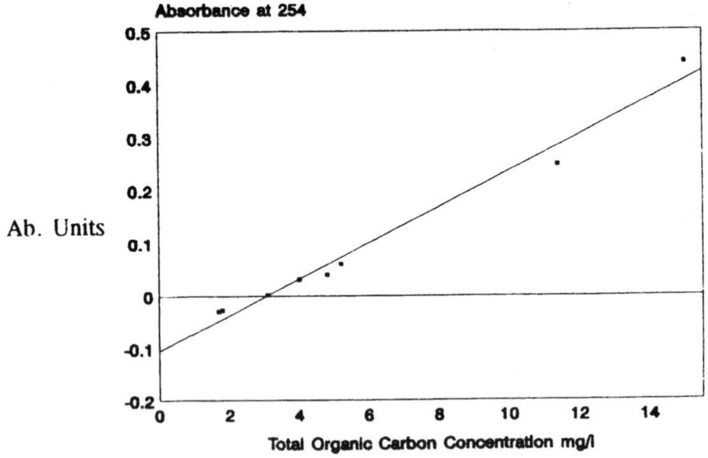

Fig. 4. Plot of absorbance verses TOC for samples from laboratory 1, obtained with a spectrophotometer (HP8452A).

Figure 5 shows a graph of the measurements obtained with the pHOX meter against TOC. The correlation coefficient for this graph, over the same TOC range, is 0.988. The result of this investigation also shows a high correlation coefficient for the linear relationship between absorption and TOC concentration, using a proprietary piece of equipment. This is also based upon the absorbance at 254 nm, this being the corrected value with sample turbidity being taken into account. The agreement between the two graphs (Figures 4 &

5) arises due to the absence of suspended matter in the sample, which could otherwise lead to disagreement in the values obtained.

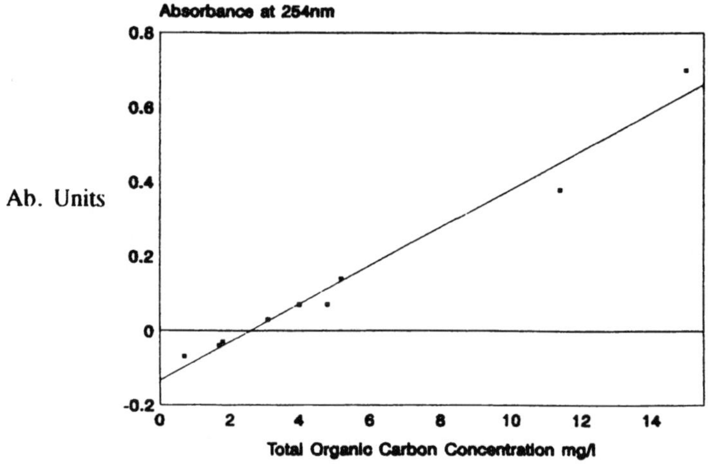

Fig. 5. Plot of absorbance versus TOC for samples from Laboratory 1, obtained with the pHOX OPM.

Fluorimetric measurements were obtained at step 4 in Figure 3 after the samples had been transported from Laboratory 2 to Laboratory 3 (London to Dublin). Figure 6 shows a plot of the fluorescence signal from the samples against the TOC concentration. The correlation coefficient for the graph is 0.994. This result, obtained with a linear relationship between the count of fluorescent photons and the TOC concentration also shows a high value of correlation coefficient and within experimental error, the graph passes through the origin, as would be expected. This, in addition to the relative insensitivity of a fluorescence method to interference from the turbidity of the sample, gives confidence in the efficacy of the method. This is particularly so as it was undertaken with the same samples as were used for the results of Figures 4 & 5, and the agreement obtained with the use of an alternative optical method is very satisfactory.

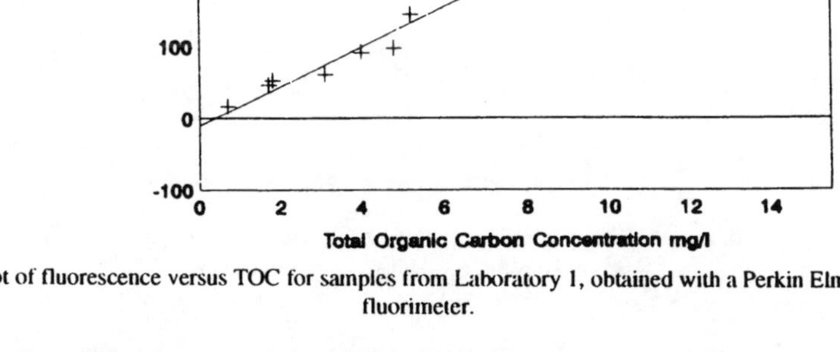

Fig. 6. Plot of fluorescence versus TOC for samples from Laboratory 1, obtained with a Perkin Elmer LS-50 fluorimeter.

Figure 7 shows a plot of fluorescence signal against TOC using the fibre optic based sensor system. The correlation coefficient for this graph is 0.993. This illustrates the results of the use of a fibre optic version of the open air-path instrument, and the same good agreement, both in terms of linearity and correlation coefficient, is seen. This gives confidence that a fibre optic version could be constructed as a viable instrument, with no loss of fidelity and a more convenient optical configuration being employed. That such a conclusion can be reached is highly satisfactory.

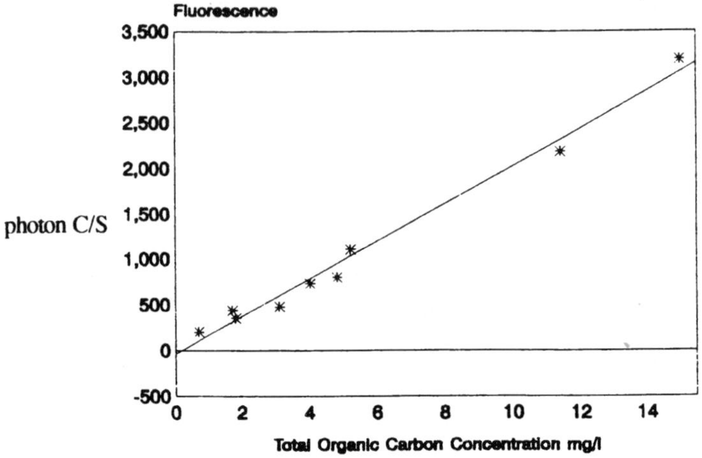

Fig. 7. Plot of fluorescence versus TOC for samples from Laboratory 1, obtained with the fibre optic sensor device shown in figure 2.

Figure 8 shows a plot of the fluorescence signals obtained using the Perkin Elmer LS-50 against the absorption at 254nm using the HP8452A spectrophotometer for the samples from Laboratory 1. The correlation coefficient for this graph is 0.995. This Figure is used to illustrate the expected relationship between the fluorescence and absorption measurements, in light of the individual results obtained and shown in Figures 4 - 7. This would not, of course, be used as part of an instrumental procedure to determine TOC in a sample, but it does give confidence to the suggestion that both the sample and the instrumental system have not changed during the course of this set of experiments (carried out in two countries over several weeks). In any instrument which is used, *in situ*, for continuous monitoring, this is a very important factor. In further work, an investigation of more highly turbid samples could be undertaken to investigate the effect, and the degree of maintainence of the degree of correlation.

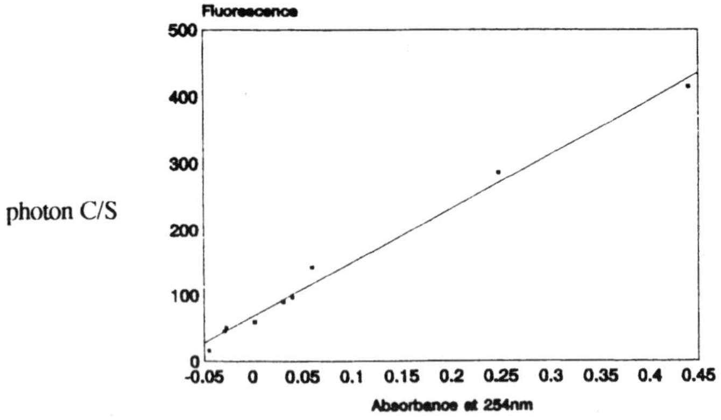

Fig. 8. Plot of fluorescence using the Perkin Elmer LS- 50 Fluorimeter versus absorbance using the HP 8452A spectrophotometer for samples from Laboratory 1.

CONCLUSIONS AND DISCUSSION

The strong correlation between the UV absorption measurements and the chemically determined TOC values confirm the value of absorption as a method of measuring TOC.

In addition the positive correlation between fluorescence and the chemically determined TOC concentration indicates that fluorescence is a viable method for measuring TOC from 0 up to at least 15 mg/l.

The close agreement between fluorescence and absorption at 254 nm indicates that it is unlikely that samples changed significantly during transportation from London to Dublin and also point to fluorescence as a method of measuring TOC.

In particular, the conclusions reached and shown in Fig. 8, show the close correlation between the use of fluorescence and absorption techniques for the monitoring of TOC. This provides encouragement towards the development of field instruments based on either (or both) of these instruments.

ACKNOWLEDGEMENTS

The authors are pleased to acknowledge the support of EOLAS and the British Council in facilitating this work, and SERC for funding for WJOB.

REFERENCES

Crompton, T. R. (1991). *Analytical Instrumentation for the Water Industry*, Publ.Butterworth-Heinemann, Oxford.

Briggs, R., Schofield, J. W. and Gordon, P. A. (1976). *Instrumental Methods of Monitoring Organic Pollution, Wat. Pollut.Control.*

Her Majesty's Stationary Office (1979). *The instrumental Determination of Total Organic Carbon, Total Oxygen Demand and Related Determinands.*

Smart, P. L. Finlayson, B. L., Rylands, W. D. and Ball, C. M. (1976). The Relation of Fluorescence to Dissolved Organic Carbon in Surface Waters, *Wat. Res.* **10**, 805-811.

Lakshman, G., (1975). Monitoring Agricultural Pollution Using Natural Fluorescence', *Wat. Resources Res.*, October 1975, **11**, No 5.

Christman, R. F. and Ghassemi, M. Chemical Nature of Organic Colour in Water, *J. of the Am. Wat. Wks Ass.*, **58**, 723-741.

Larson, R. A. (1978). Dissolved Organic Matter of a Low-Coloured Stream, *Freshwat. Biol.* **8**, 91-104.

Susic, M. and Boto, K. G.,(1989). High-Performance Liquid Chromatographic Determination of Humic Acid in Enviromental Samples at the Nanogram Level using Fluorescence Detection, *J. Chromatog.*, **482**, 175-187.

Camagni, P., Colombo, A., Koechler, C., Ormennetta, N., Qi, P. and Rossi, G. (1991). Fluorescence Response of Mineral Oils: Spectral Yield vs Absorption and Decay Time, *Appl. Optics* **30**, No 1.

Power, J. F. and Langford, C. H. (1988). Optical Absorbance of Dissolved Organic Matter in Natural Water Using the Thermal Lens Effect, *Analyt. Chem.*, **60**, 842-846.

Baumgartner, D. G., Feldman, M. H. and Gibbons, C. L. (1971). A Procedure for Tracing Kraft Mill Effluent from an Ocean Outfall by Constituent Fluorescence, *Wat. Res.*, **5**, 533-544.

Evans, K. M., Gill, R. A. and Robotham, P. W. (1990). The Source, Composition and Flux of Polycyclic Aromatic Hydrocarbons in Sediments of the River Derwent, Derbyshire, U.K., *Wat., Air Soil Pollut.*, **51**, 1-12.

Hirschfield, T., Haugen, G. and Milanovich, F. (1983). Instrumentation for Remote Sensing over Fibre Optics, *Proceedings of the 26th ORNL-DOE Conference on Analytical Chemistry in Energy Technology*, Knoxville,Tennessee,October 11-13.

Dobbs, R. A., Wise, R. H. and Dean, R. B. (1972). The Use of Ultra-Violet Absorbance for Monitoring the Total Organic Carbon Content of Water and Wastewater, *Wat. Res.*, **6**, 1173-1180.

Tipping, E., Hilton, J. and James, B. (1988). Dissolved Organic Matter in Cumbrian Lakes and Streams, *Freshwat. Biol.*, **19**, 371-378.

Mrkva, M. (1975). Automatic U.V.-Control System for Relative Evaluation of Organic Water Pollution, *Wat. Res.*, **9**, 587-589.

Moore, T. R. (1987). An Assessment of a Simple Spectrophotometric Method for the Determination of Dissolved Organic Carbon in Freshwaters, *NZ J. Mar.Freshwat. Res.*, **21**, 585-589.

Carpenter, P. D. and Smith, J. D. (1984). Simultaneous Spectrophotometric Determination of Humic Acid and Iron in Water, *Analytica Chim Acta*, **159**, 299-308.

Pergamon

Wat. Sci. Tech. Vol. 28, No. 11-12, pp. 465-471 1993.

0273-1223/93 $7·00 + 0·00

MEASUREMENT OF TRIHALOMETHANES (THM) IN DRINKING WATER WITH AN AUTOMATIC THM ANALYZER

K. Kawakami, Y. Tanaka, Y. Zaitsu and H. Hoshikawa

Water Treatment and Bioelectronic Laboratory, Fuji Electric Corporate Research and Development Ltd., 2-2-1 Nagasaka, Yokosuka City, 240-01, Japan

ABSTRACT

For waterworks the most fundamental subject is to maintain the safety of drinking water. It has become important every year to monitor the concentrations of trihalomethanes (THM). We developed a new automatic THM analyzer based on fluorometry with membrane separation, and investigated a correlation between the gas chromatographic method and our THM analyzing method on drinking waters at 10 cities in Japan. The investigation resulted in good correlation (r=0.9807) between both methods. Good results were obtained when continuous measurements of THM in drinking water were performed by the analyzer. Consequently, it was found that the present analyzer could measure THM in drinking waters simply, rapidly, with good precision and could be applied to continuous monitoring of THM.

KEYWORDS

Trihalomethane; Fluorometry with membrane-separation; Automatic THM analyzer; Gas chromatography; Correlation; Continuous measurement

INTRODUCTION

The THM (trihalomethanes: chloroform, bromodichloromethane, chlorodibromomethane and bromoform) are formed during the chlorination of waters containing precursor compounds (Rook, 1974). The major THM precursors appear to be aquatic humic substances (Rook, 1976; Oliver and Visser, 1980; Urano *et al.*, 1983; El-Rehaili and Weber, 1987). THM were discovered by the EPA National Organics Reconnaissance Survey in the drinking waters of 79 out of 80 US cities (Symons *et al.*, 1975). Following this survey, toxicological investigations assessed certain trihalomethanes, chloroform in particular, as suspected carcinogens (Dowty *et al.*, 1975; Harris and Breecher,1974; National Cancer Institute, 1976). The USEPA, in November 1979, and the Japanese Ministry of Health and Welfare, in March 1981, established a maximum contaminant level (MCL) of 100 μg/l THM in drinking water (NIPDWRCTDW, 1979). The MCL was established as 350 μg/l in Canada and was recommended to be 25 μg/l in West Germany.

For the purpose of solving the above THM problem and ensuring safety of drinking water, it is important to develop not only a system of water treatment which causes no THM formation, but also a method which analyzes the THM precisely, rapidly and continuously. The gas chromatographic method (GC method) is usually used as a standard method for the above analysis. However, in case of application of the GC method to continuous monitoring of THM in the water treatment process, the method has several shortfalls. That is to say, it is time consuming and a batch method which require pretreatment and preconcentration, and requires skill for precise analysis.

It is our purpose to develop a new automatic THM analyzer, based on fluorometry with membrane-separation, and which is able to measure THM in drinking water simply, rapidly, with good precision and continuously. In this paper, following previous basic studies (Aoki and Kawakami, 1989; Kawakami *et al.*, 1990), we report on a correlation between the GC method and our THM analyzing method for THM in drinking waters at 10 cities in Japan, and on an application to continuous measurements of THM in drinking water with the automatic THM analyzer.

EXPERIMENTAL

Reagents

The distilled water that was used was boiled for more than 30 minutes and cooled so as to minimize absorption of THM from the atmosphere, and all reagents were of analytical reagent grade.

Apparatus and procedure

A flow diagram of the automatic THM analyzer used in this study is shown in Fig.1. The THM analyzer consists of five main units, that is, a solution carrier unit, a separation unit, a reaction unit, a detection unit (fluorometer) and display, a record unit. The analyzer, for example, automatically opened and shut electromagnetic valves in a solution carrier unit, and was able to perform from calibration by a standard solution to calculation, display and record of the measured values after the analysis of samples according to predetermined sequences and conditions. A standard solution (chloroform) or a sample was mixed with a solution of acidic reducing agent (1% hydrazine sulfate) to decompose the free chlorine and chloramines in the sample. The mixed solution was called a sample solution, which was conveyed to a separation unit and was drained out of the unit. In the separation unit maintained at 70℃, the volatile THM permeated through the microporous PTFE membrane and was separated from the sample solution. On the other hand, a solution of 0.3M sodium hydroxide was mixed with a 30% nicotinamide solution and propelled to the separation unit. This

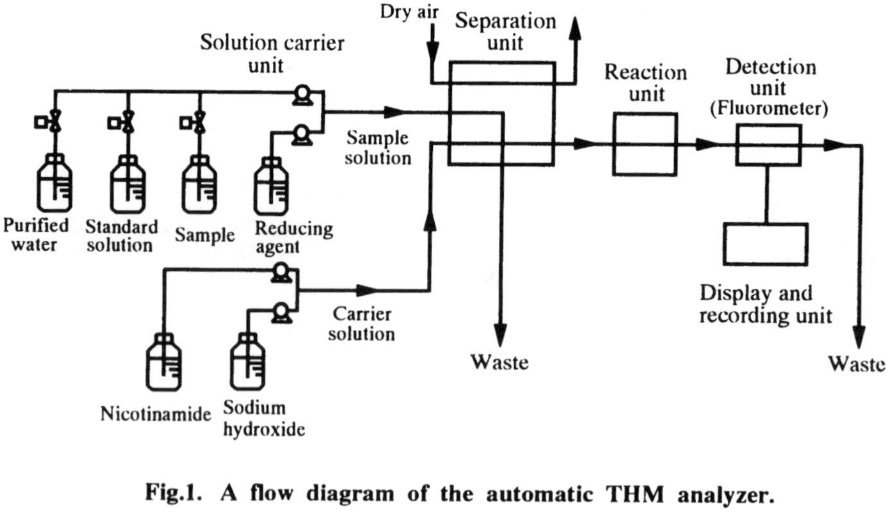

Fig.1. A flow diagram of the automatic THM analyzer.

mixed solution was called a carrier solution. In the separation unit, the separated THM dissolved into the carrier solution through the microporous PTFE tube. Then the carrier solution was passed into a reaction unit, where it was heated to 90℃, and the reaction product was led to the detection unit. After the fluorescence intensity excited at 369 nm was measured at 458 nm, the carrier solution was drained out of the detection unit. Since the fluorescence intensity was proportional to the concentration of THM, it was capable to determine the THM in the sample. After the determination finished, purified water and dry air were passed into the separation unit in order to remove the remaining THM inside of the unit. Flow rates of purified water, standard solution and the sample were each and all 4.0 ml/min, and those of the solution of 1% hydrazine sulfate, the 30% nicotinamide solution and the solution of 0.3M sodium hydroxide solutions were each and all 0.5 ml/min. The measurement time was within about 40 minutes. There were two measurement modes for the automatic THM analyzer, that is, a manual measurement mode (MANU mode) and an automatic measurement mode (AUTO mode). The MANU mode is that the analyzer measures THM in different samples one by one, and the AUTO mode is that the analyzer automatically performs the continuous measurement of THM in the same sample at fixed hours.

Experimental method

Correlation to the conventional GC method. In order to investigate a correlation between the GC method and our THM analyzing method, the THM in tap waters at 10 cities in Japan was determined with both methods. Analyses for the THM in the tap waters by the GC method were performed by the head space method according to the Standard Methods (Japan Water Works Association,1985). In this study, the MANU mode was used in the THM analyzer.

Continuous monitoring of drinking water. The concentration of THM in certain tap water was continuously measured by using the AUTO mode in the analyzer.

RESULTS AND DISCUSSION

Preliminary experiments

Precision of measurements. Table 1 shows the concentrations of THM standard solutions and the results of the repeated 5 analyses at each solution. It was found that the analyzer could measure precisely enough for the sample of lower THM concentration since the coefficients of variation (C.V.) were below 6.4% except a standard solution of 2 μg/l.

TABLE 1 Precision of measurements in the automatic THM analyzer

Standard Solution (μg/l)	Analytical Value(N=5) (μg/l)	Mean Value (μg/l)	Standard Deviation	Coefficient of Variation (%)
50	50,50,51,49,50	50	0.63	1.3
10	9.9,9.3,9.9,10.4,10.5	10	0.43	4.3
5	5.5,4.7,5.1,4.6,5.1	5.0	0.32	6.4
2	2.9,1.9,1.4,1.9,1.9	2.0	0.49	25

Storage of samples. In the Japanese standard method for THM measurements, it is prescribed that fixation reagents of samples are sodium sulfite as the reducing agent and phosphoric acid as the acid, and that THM of the fixed samples are immediately determined by the GC method after cooled storage by ice. But, in case of the present analyzer, these preparations have two problems as follow. One is that sulfur dioxide, formed by the reaction of the sodium sulfite with the acidic reducing agent used for the THM analyzer, led to a positive error of measurements when the sodium sulfite was used as the fixation reagent. The other is that it was relatively difficult to transport samples cooled by ice when drinking waters were obtained as the samples at various places. In this study, therefore, it was investigated whether the following two conditions were available as the conditions at fixation and storage of samples. The first was to use the same acidic reducing agent (hydrazine sulfate) as in the case of the THM analyzer instead of the sodium sulfite as the fixation agent. The second condition was to transport the samples at ordinary temperature after fixation. As a method of the above study, it was investigated to observe a change of THM concentrations when both the fixed tap waters and those unfixed by the fixation reagents (hydrazine sulfate and phosphoric acid) were stored for days in incubators maintained at 5℃, 20℃ and 37℃, respectively, and the TTHM (total THM; sum of THM concentration) in the samples were determined by the GC method. The results are shown in Fig.2.

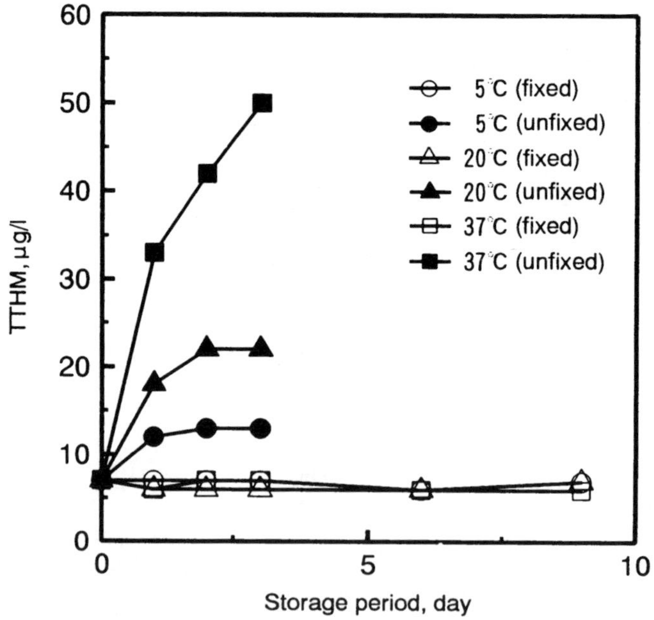

Fig.2. Change of TTHM in fixed and unfixed samples.

The TTHM in the unfixed samples increased with the storage period and as the storage temperature to the initial TTHM (7 μg/l). On the other hand, the TTHM in the fixed samples was almost constant at the initial TTHM in all conditions.
Therefore, it was found that it was possible to store samples for at least 9 days on the above conditions.

Correlation to the conventional GC method

Table 2 shows a result of measurements of THM in tap waters, which were sampled from 10 cities in Japan between July 6 and July 17 in 1992, by both the GC method and the THM analyzer between July 14 and July 23 in 1992. Bromoform was not detected (N.D.) in all samples, and the TTHM by the GC method ranged from 17 to 32 μg/l except a special case where it was N.D. in the Toyama sample.

TABLE 2 Comparison of measurements in 10 cities by both methods (values in μg/l)

Sampling city	GC method					THM analyzer
	$CHCl_3$	$CHCl_2Br$	$CHClBr_2$	$CHBr_3$	TTHM*	
Sapporo	6	8	8	N.D.	22	17
Sendai	16	11	5	N.D.	32	31
Tokyo	15	7	4	N.D.	26	27
Yokosuka	14	8	4	N.D.	26	23
Toyama†	N.D.‡	N.D.	N.D.	N.D.	N.D.	N.D.
Nagoya	12	4	1	N.D.	17	16
Osaka	17	8	3	N.D.	28	26
Hiroshima	13	7	3	N.D.	23	22
Takamatsu	15	4	1	N.D.	20	18
Fukuoka	10	8	4	N.D.	22	19

* Sum of concentration of THM ($CHCl_3$, $CHCl_2Br$, $CHClBr_2$, and $CHBr_3$).
† Ground water is used as drinking water.
‡ Not detected.

As the result, the measured values with the THM analyzer were nearly equal to the TTHMs by the GC method. Furthermore, the correlation diagram between both methods is shown in Fig.3. As shown in Fig.3, good correlation was obtained, that is, the correlation coefficient (r) between both methods was 0.9807.

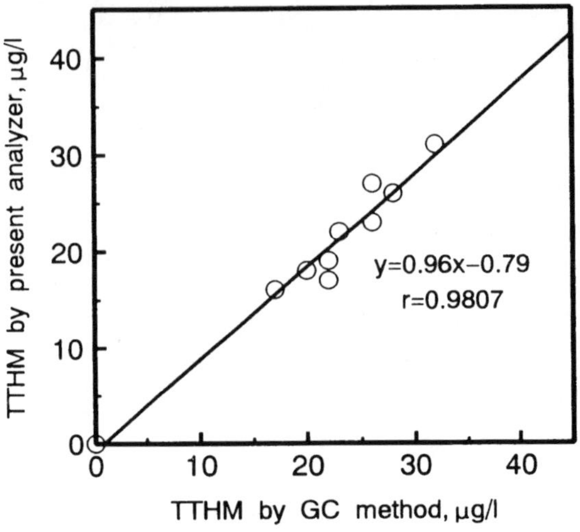

Fig.3. Correlation diagram between GC method and present THM analyzer.

Continuous monitoring of drinking water

Fig.4 shows a result of automatic measurements of THM in tap water at Yokosuka city, drained continuously off in our Laboratory, by using the AUTO mode in the THM analyzer every hour for a day. Also, a result of measurements of the same water by the GC method every 6 hours are plotted in Fig.4. The measured values with the THM analyzer were nearly equal to those by the GC method every 6 hours. Therefore, it was found that the present analyzer could be applied to continuous monitoring of THM.

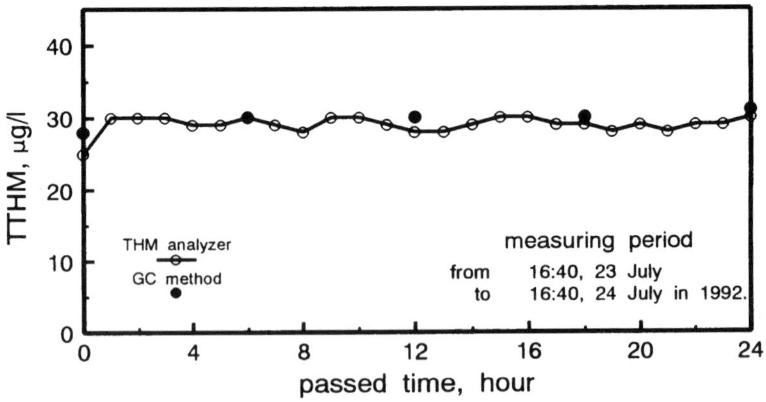

Fig.4. A example of continuous measurements with present THM analyzer.

CONCLUSION

The developed automatic THM analyzer based on fluorometry with membrane separation demonstrated satisfying performance in the determinations of drinking water. The following two points of the THM analyzer could be found:

- the measured values with our THM analyzing method on tap waters at 10 various cities in Japan were nearly equal to those by the conventional GC method of the Standard Method, and a good correlation was obtained between both methods;
- the analyzer could be applied to continuous monitoring of THM.

REFERENCES

Aoki T. and Kawakami K. (1989). Continuous flow method for determination of total trihalomethanes in drinking water with membrane-separation. *Wat. Res.* **23,** 739–742.

Dowty B., Carlisle D., Laseter J. L. and Storer J. (1975). Halogenated hydrocarbons in New Orleans drinking water and blood plasmas. *Science* **187,** 75–80.

El-Rehaili A. M. and Weber W. J. Jr (1987). Correlation of humic substance trihalomethane formation potential and adsorption behavior to molecular weight distribution in raw and chemically treated waters. *Wat. Res.* **21,** 573–582.

Harris R. H. and Breecher E. M. (1974). Is the water safe to drink? *Consumer Rep.* **23,** 436–452.

Japan Water Works Association (1985). *Standard Methods for the Examination of Drinking Water*, pp. 482–486 (*in Japanese*).

Kawakami K., Tanaka Y., Hoshikawa H., Aoki T., Ito T., Kaji S., Fujiyoshi K., Yamamoto M. and Yamanaka Y. (1990). Examinations of a new trihalomethane analyzer. *Proceedings of the 41st Congress of Jap. Water Works Assoc.*, 599–601 (*in Japanese*).

National Cancer Institute (1976) Report on carcinogenesis bioassay of chloroform, Bethesda, Md.

National Interim Primary Drinking Water Regulations for Control of Trihalomethanes in Drinking Water (1979). *Final Rule Fed. Reg.*, **44,** 231, 68624.

Oliver B. G. and Visser S. A. (1980). Chloroform production from the chlorination of aquatic humic material. *Wat. Res.* **14,** 1137–1141.

Rook J. J. (1974). Formation of haloforms during chlorination of natural waters. *Wat. Treat. Exam.* **23,** 234–243.

Rook J. J. (1976). Haloforms in drinking water. *J. Am. Water Works Assoc.* **68,** 168–172.

Symons J. M., Bellar T. A., Carswell J. K., DeMarco J., Kropp K. C., Robeck G. G., Seeger D. R., Slocum C. J., Smith B. L. and Stevens A. A. (1975). National organics reconnaissance survey for halogenated organics. *J. Am. Water Works Assoc.* **67,** 634–647.

Urano K., Wada H. and Takemasa T. (1983). Empirical rate equation for trihalomethane formation with chlorination of human substances in water. *Wat. Res.* **17,** 1797–1802.

Pergamon

Wat. Sci. Tech. Vol. 28, No. 11-12, pp. 473-480, 1993.

0273-1223/93 $7·00 + 0·00

COMPARISON OF GOLD AND PLATINUM ELECTRODE RESPONSES IN ACTIVATED SLUDGE

A. Heduit*, B. Martin*, I. Duchamp* and D. R. Thevenot**

** Centre National du Machinisme Agricole, du Génie Rural, des Eaux et des Forêts (CEMAGREF), Division Qualité des Eaux, 14 avenue de Saint-Mandé, 75012 PARIS, France*
*** Université Paris XII, Val de Marne (U.P.V.M.), U.F.R. de Sciences et de Technologie, Laboratoire de Bioélectrochimie et Analyse du Milieu (LABAM), Avenue du Général de Gaulle, 94010 Créteil Cedex, France*

ABSTRACT

Gold and platinum were compared to ascertain how they expressed a stabilized potential in activated sludge. The comparison was based on electrochemical determination of the electron transfer rate (i.e. equilibrium exchange current density) and recording of potentials against time.

When both metals are treated in the same way, platinum gives equilibrium exchange current densities approx. 10 times higher than gold, both in aerated activated sludge and in treated water. For platinum, the equilibrium exchange current densities range from 0.1 to 0.25 $\mu A/cm^2$ immediately after polishing and decrease during prolonged contact with activated sludge subjected to alternating aeration/anoxia sequences.

The lower kinetics of electron transfer on gold go together with significant differences in response :

- In an aerobic medium a gold electrode potential is lower than that of a platinum electrode. In a strongly anaerobic medium, the reverse is true. Consequently, the amplitude of the potential variation between aerobic and anaerobic media is smaller for gold than for platinum. Under our experimental conditions this amplitude was approx 350 mV for gold and 850 mV for platinum.

- The slopes of the linear relationships between potential and pH or potential and the logarithm of the dissolved oxygen concentration are two or three times greater for platinum than for gold.

Although the values obtained with platinum electrodes cannot represent a veritable equilibrium state, the platinum electrode zero-current potential would seem to be far more sensitive to variations in the medium than that of the gold electrode; it is, therefore, more suitable for use in activated sludge.

KEYWORDS

Redox potential; platinum electrode; gold electrode; activated sludge; equilibrium exchange current density; zero-current potential.

INTRODUCTION

The zero-current platinum electrode potential is an attractive parameter for controlling the nitrification and denitrification processes in activated sludge (Charpentier *et al.*, 1989; Audic, 1992; Deguin, 1992).

Thermodynamically, gold is "nobler" than platinum. Nonetheless, bibliographic studies show that platinum is almost always used to measure potential in both water and activated sludge (Heduit, 1989). Authors rarely explain this choice.

Whitfield (1972) obtained estimations of the equilibrium exchange current densities on platinum (0.02 to 0.38 μA/cm^2) and on gold (0.003 to 0.008 μA/cm^2) by discharging natural redox cells through an external resistance. Buhler and Galster (1980) confirmed that platinum gave considerably higher equilibrium exchange current densities than gold in water. Coquery (1986) stopped using gold electrodes in activated sludge due to "their poor sensitivity to variations in the medium".

The aim of this project was to compare gold and platinum to ascertain how they expressed a stabilized potential in activated sludge: the comparison was based on electrochemical determination of the electron transfer rate (i.e. equilibrium exchange current density) and recording of potentials against time.

We therefore performed measurements of:

equilibrium exchange current densities (after polishing and after one week immersion in activated sludge)

zero current potentials in aerobic media (change after polishing, relationships potential vs pH and potential vs log [O_2])

zero current potentials in an anaerobic medium (change after polishing).

MATERIAL AND METHODS

Determinations of the Equilibrium Exchange Current Densities

Procedure. The equilibrium exchange current densities (J_0) were determined by recording low speed current potential curves (anode and cathode scans on either side of the equilibrium potential). The logarithm of the current densities was then plotted against the corresponding overpotential. The equilibrium exchange current densities were then determined by extrapolating these curves to zero overpotential. This technique has already been described in a previous paper (Heduit and Thevenot, 1989).

Electrochemical curve plotting equipment. The current-potential curves were plotted at low speed (1.25 mV/s) on a Tacussel PRG3 assembly comprising:

- a potential control unit (PRT 20-2X)
- an adaptor unit
- a voltage recorder (EPL 2) connected to the potential control unit.

Electrodes. The characteristics of the metal electrodes used are presented in Table 1.

TABLE 1 Metal Electrode Characteristics

TRADEMARK	TYPE	SHAPE	SIZE (mm) height	SIZE (mm) diameter	AREA (cm^2)
INGOLD	Pt 805	ring	3	6	0.56
TACUSSEL	Pt 102	rod	12	1	0.38
INGOLD	Au 800/5	rod	5	2	0.34
INGOLD	Au 800/7	rod	7	2	0.47

The reference electrodes were Xerolyt Ingold electrodes (potential = + 210 mV/NHE at 20°C).

A Ponselle 02P oxygen-meter and a Schott CG 817 pH meter coupled to combined Ingold/Xerolyt pH electrodes, were used.

Preparatory treatment of electrodes. For the first series of tests, the metal electrodes were polished using bands of powdered abrasive (granularity 5 and 0.3 µm) fixed onto a plastic support (Tacussel BSC 3 and BAO 3 abrasive bands). The electrodes were then rinsed with distilled water. For the second series of tests, the electrodes were first polished and then left in alternately aerated sludge for 7 d.

Test cell. A 250 ml glass test cell was used. The cover contained 5 orifices for the measurement electrode, the reference electrode, the working electrode, the oxygen electrode and an air feed pipe. The air was provided by an aquarium compressor (adjustable flow) connected to a diffuser.

Activated sludge and treated water. The activated sludge and treated water were taken from an extended-aeration treatment plant (Maincy - Seine et Marne). The sludge was intermittently aerated (15 min aeration followed by 15 min with no aeration) for 2 to 4 d in a laboratory. Throughout this period, approx 100 ml of waste water per litre of sludge was added each day.

Measurement of Zero-Current Potentials in Aerobic Media

Procedure. The metal electrodes and the common reference electrode were connected:

either to high input impedance millivoltmeters (Knick 646), themselves connected to a multi-channel graphic recorder (Goerz Servogor 460 06), including variable-sensitivity racks (experiments on Maincy sludge),

or a Hewlett Packard HP 3421 A automatic data acquisition system (Central processor unit equipped with a 10-channel multiplexer card, internal resistance 10^{10} Ω, accuracy at the 3 V range: 0.3 mV) + HP 41 CV calculator + 82162 A thermal printer + 82161 A microcassette recorder (experiments on the Valenton sludge).

The electrodes were polished prior to each series of tests and then placed in the medium until the response stabilized.

A Schott CG 817 pH meter, coupled to a combined pH/Xerolyt Ingold electrode and Ponselle O2P and YS1 57 oxygen-meters were used. The dissolved oxygen concentration in the sludge was varied by decrementing the air pump flow and then increasing it in the same sequence. The pH was varied by adding sulphuric acid (Merck pro analysi) then soda (Prolabo Normapur). The potential was measured 20 to 30 min after changing the conditions in the medium.

Laboratory test cell. This is a 2-litre transparent PVC rectangular cell, capable of accommodating 10 electrodes. It was placed on a magnetic agitator and the liquid was stirred by a Teflon rod. It was thermostat-controlled for certain experiments (see results).

Activated sludge and treated water. The activated sludge and treated water were taken from biological treatment plants at Maincy (Seine et Marne) and Valenton (a low-loaded plant, Val de Marne). The sludge was stored in the laboratory as described previously.

Measurement of Zero-Current Potentials in an Anaerobic Medium

The sludge taken from the Valenton treatment plant was stored in the laboratory, under anaerobic conditions, for several days and periodically activated with wastewater taken at the station inlet. The electrodes were mounted on the above cell, containing this sludge plus 50 mg/l of potassium acetate (Prolabo Normapur) until the potential stabilized. The electrodes were then again polished and the potential from each was monitored using the data acquisition system.

RESULTS AND DISCUSSION

Equilibrium exchange current densities

Results. Figure 1 shows the equilibrium exchange current densities obtained, at various stabilized dissolved oxygen concentrations, from 2 platinum and 2 gold electrodes in the Maincy activated sludge, only a few minutes after polishing.

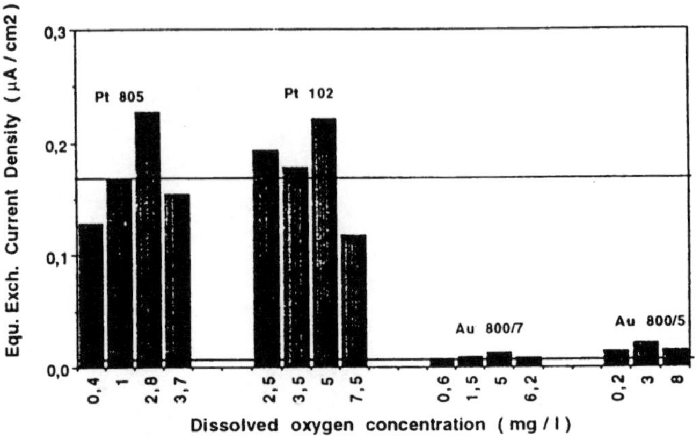

Fig.1. Equilibrium exchange current densities against dissolved oxygen concentration, for platinum and gold electrodes a few minutes after polishing (Maincy activated sludge).

The values obtained for platinum are between 0.12 and 0.24 $\mu A/cm^2$ (mean: 0.17 $\mu A/cm^2$). The results for gold are between 0.008 and 0.021 $\mu A/cm^2$ (mean : 0.013 $\mu A/cm^2$). The dissolved oxygen content does not seem to affect the measured values for either metal.

Table 2 shows the equilibrium current densities obtained from platinum and gold electrodes in treated water saturated with dissolved oxygen. The electrodes were first polished and then left a week in sludge alternately aerated and supplied with wastewater, before being placed in aerated treated water.

TABLE 2 Equilibrium Exchange Current Densities (in mA/cm2) from Platinum and Gold Electrodes in Aerated Treated Water, Immersed Directly after Polishing or after Polishing + 7 days in Maincy sludge.

Type of electrode / Treatment before voltametry in treated water	Pt 805	Pt 102	Au 800/7	Au 800/5
Polishing	0.22	0.24	0.018	0.013
Polishing and 7 days in activated sludge	0.042	0.052	0.006	0.005

With both metals, the current densities are considerably lower after 7 d submersion.

Discussion. The current densities we measured during these two series of experiments were 0.04 to 0.24 $\mu A/cm^2$ for platinum and 0.005 to 0.018 $\mu A/cm^2$ for gold. These values must be compared with those Whitfield obtained: 0.02 to 0.38 $\mu A/cm^2$ for platinum and 0.003 to 0.008 $\mu A/cm^2$ for gold.

Heduit and Thevenot (1989) using unpolished platinum electrodes which were very probably contaminated by sulphur obtained $J_o = 0.04\ \mu A/cm^2$ in aerated sludge and $J_o = 0.02\ \mu A/cm^2$ in sludge held under anoxic conditions.

Consequently, provided both metals are treated identically, platinum gives equilibrium current densities approx 10 times higher than gold. In both cases, the density decreases after prolonged contact with activated sludge alternately subjected to aeration and anoxia.

Measurement of Zero-Current Voltages in an Aerobic Medium

Evolution in potential after polishing. Figure 2 represents changes in the electrode potentials, a few minutes after polishing, in sludge which was previously aerated and then no longer supplied with wastewater for several hours. For electrodes in the same metal and treated in exactly the same way, the potentials vary very little from one electrode to another.

The platinum potentials are far higher than those for gold. Under the experimental conditions (pH = 7.8, (O_2) = 8 mg/l), the difference is approx 200 mV. After polishing, the potential varies in a similar way for both metals. This development is probably not representative of the large differences in current densities mentioned previously.

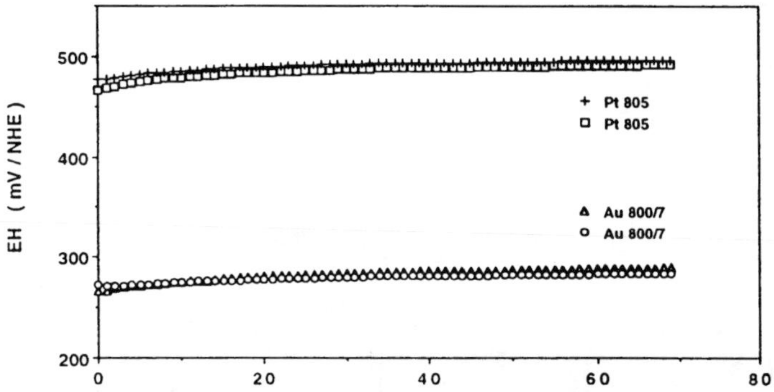

Fig. 2. Changes in the potential of polished platinum and gold electrodes (EH) a few minutes after polishing in aerated activated sludge (VALENTON activated sludge).

Effect of pH on potential. Figure 3 shows the effect of pH on the potential of platinum and gold electrodes in treated water from the Valenton treatment plant, at 27 Â 0.1°C. The pH was varied by adding sulfuric acid or soda.

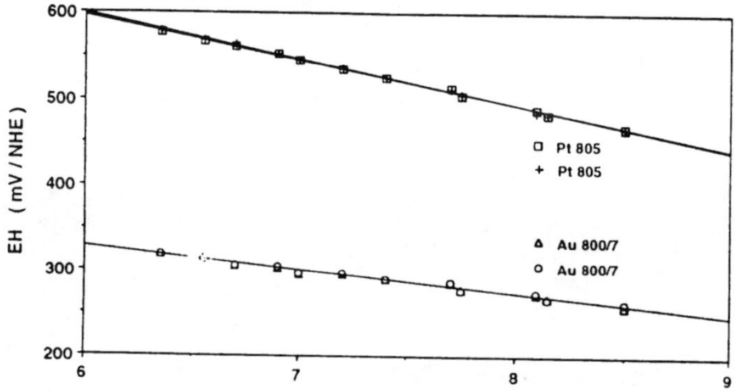

Fig. 3. Potential against pH at 27 +/- 0.1°C in aerated treated water from the Valenton plant.

Table 3 summarizes the results obtained during this experiment and those obtained from a sample of treated water from the Maincy plant whose pH was varied in the same way.

TABLE 3 Results Pertaining to the Relationships Between Potential and pH in Treated Water from the VALENTON and MAINCY Plants when the pH is Varied Between 6 and 9.

Origin of water	Type of electrode	Equations of linear regression lines EH = A - B pH (mV/NHE)	EH at pH 7 (mV/NHE)	r^2	n	effluent content (mg/l)		
						NH_4N	NO_2N	NO_3N
VALENTON	Pt 805	EH = 903 - 51.1 pH	545	0.998	12	0.4	0.1	27
	Pt 805	EH = 915 - 52.7 pH	546	0.999	12			
	Au 800/7	EH = 489 - 26.8 pH	301	0.984	12			
	Au 800/7	EH = 487 - 26.5 pH	301	0.980	12			
MAINCY	Pt 102	EH = 884 - 58.5 pH	474	0.994	11	0.5	< 0.1	< 0.1
	Pt 805	EH = 885 - 58.6 pH	475	0.996	11			
	Au 800/7	EH = 524 - 32.8 pH	294	0.980	11			
	Au 800/7	EH = 482 - 28.6 pH	282	0.950	11			

These results confirm that in treated water the zero current potential is highly dependent on the metal of the electrode.

For platinum, although the experimental values of coefficients A and B in the EH = A - B pH equations are lower than the thermodynamic values corresponding to the Nernst equation for the Pt/PtO system (EH (mV) = 980 - 59 pH), this is still the most plausible electroactive system to explain the variation in potential with pH (Heduit and Thevenot, 1992). The correlation factors exceed 0.994 for both samples of treated water.

The slopes of the EH against pH lines obtained for gold are roughly half those obtained with platinum. This could reflect the low current densities with gold or the mixed nature of the potentials measured. The correlation factors for gold are not so good, possibly due to poorer stabilization of the potentials.

If the "sensitivity" of a sensor is considered as the strength of its response to changes in the medium, the polished platinum electrode is more sensitive to variations in pH than the gold electrode.

Effects of oxygen concentration on potential. As demonstrated previously (Heduit and Thevenot, 1989) the stabilized potentials generated by platinum electrodes when the pH is held constant is related to the dissolved oxygen concentration by a semi-logarithmic equation of the form EH = C + D log (O_2). The same result was demonstrated for the gold electrode.

Figure 4 gives the curves of EH against log (O_2) obtained from Valeton sludge. The oxygen content was stabilized at various levels by progressively reducing the flow from the air pump and then increasing it in the same sequence.

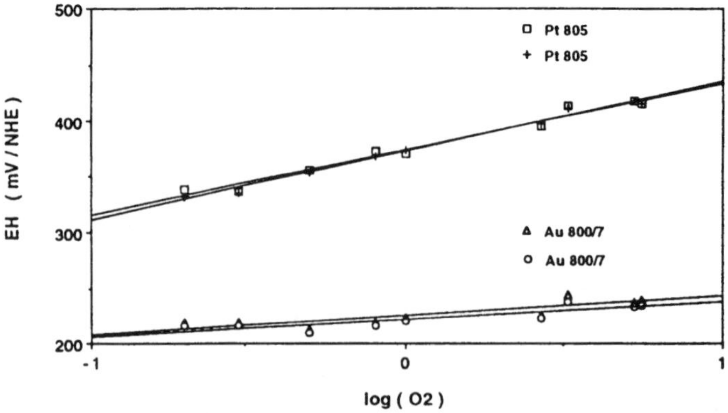

Fig. 4. Potential (EH) against log (O_2) (Valenton activated sludge).

Table 4 shows the results obtained from this experiment and those from the mean of 7 tests carried out on Maincy activated sludge. The pH varied between 7.6 and 8.15 during these experiments and, consequently, the measured potentials were corrected back to pH = 7.9 using the equations determined previously.

TABLE 4 Potential (EH) Against Log (O_2), Results Obtained from Valenton and Maincy Activated Sludge.

Origin of water	Type of electrode	Equations of linear regression lines EH = C - D log (O_2) (mV/NHE)	EH at 1 mg O_2/l (mV/NHE)	r^2	n
VALENTON	Pt 805	EH = 375 + 59.2 log (O_2)	375	0.975	9
	Pt 805	EH = 373 + 62.3 log (O_2)	373	0.990	9
	Au 800/7	EH = 225 + 17.6 log (O_2)	225	0.736	9
	Au 800/7	EH = 221 + 15.7 log (O_2)	221	0.746	9
MAINCY	Pt 805	EH = 482 + 68 log (O_2)	482	0.95	7
	Au 800	EH = 316 + 29 log (O_2)	316	0.85	7

Except for factor D obtained for gold electrodes in Valenton activated sludge, the experimental coefficients are very different from the thermodynamic coefficients which represent the electrochemical reduction of dissolved oxygen:

$$EH = 730 + 15 \log (O_2) \qquad pH = 7.9 \text{ at } 27°C$$
(mV / NHE) (mg/l)

The values of factor D obtained for gold during the various experiments confirmed the very low current densities and, therefore, the mixed nature of the potentials measured.

Platinum is more sensitive to variations in dissolved oxygen concentration than gold. The correlation factors are generally better for platinum.

Measurement of Zero-Current Voltages in an Anaerobic Medium

Figure 5 shows the variation in the potential taken by mechanically polished electrodes in anaerobic sludge. Unlike the results from an aerobic medium, the platinum potentials are lower than gold (approx. 350 mV). For a given metal, the values obtained vary little from one electrode to another (differences are less than 10 mV). Under the experimental conditions, the platinum electrode potentials stabilized roughly 30 min after polishing while the gold electrode potentials initially stabilized and then continued to decrease slowly.

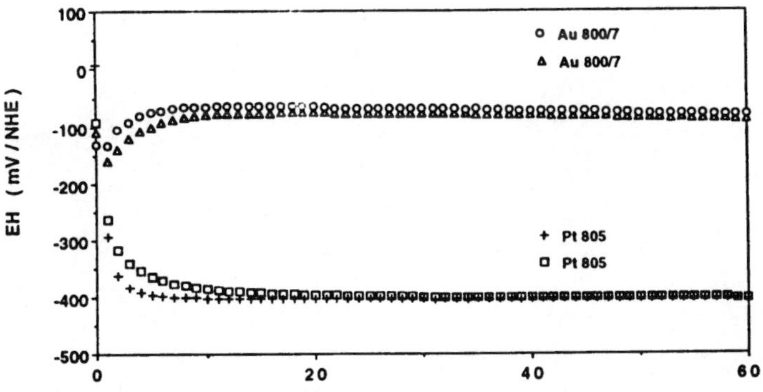

Fig. 5. Changes in the potential of polished platinum and gold electrodes (EH) a few minutes after polishing in anaerobic sludge (Valenton activated sludge).

CONCLUSIONS

The equilibrium exchange current densities and potential values obtained from a given type of electrode are highly repeatable after identical mechanical treatment or submersion in activated sludge. Platinum electrodes give current densities approx 10 times higher than gold.

As shown by the experiments conducted in various conditions of pH and dissolved oxygen concentrations, the lower kinetics of electron transfer on gold go together with significant differences in behaviour.

The potential of a gold electrode in an aerobic medium can be 200 mV less than that from a platinum electrode. In a strongly anaerobic medium, this situation reverses and the gold potential can be up to 350 mV higher than platinum. The difference between the potential in an aerobic medium and that in an anaerobic medium was, under our experimental conditions, 300 to 400 mV for gold and 800 to 900 mV for platinum.

The slopes of the linear relationships between potential and pH or potential and the logarithm of the dissolved oxygen concentration are lower for gold (29 ± 3 mV/pH and 22 ± 7 mV/decade (O_2)) than for platinum (55 ± 4 mV/pH and 60 ± 4 mV/decade (O_2)).

Although the values obtained with platinum electrodes do not represent a veritable equilibrium state, since they differ from those suggested by, for example, a Pt/PtO or O_2/H_2O system the platinum electrode nonetheless seems more sensitive to changes in the medium than the gold electrode and, therefore, more suitable for use in activated sludge.

REFERENCES

Audic J.M. (1992). L'exemple d'une station entièrement automatisée. TSM 4, 195-199.

Bühler et GASTER (1980). Mesures redox : bases et problèmes. INGOLD *Technique Ed.* 23 p.

Charpentier J.,Godart H.,Martin G. and Mogno Y. (1989). Oxidation reduction potential (ORP) regulation as a way to optimize aeration and C, N and P removal : experimental basin and various full-scale examples. *Wat. Sci. Tech.* 21 (10/11). 1209-1223.

Coquery, M. (1986). Limites et applications des mesures de potentiels d'équilibre dans les boues activées. *Mémoire de D.E.A. Sciences et Techniques de l'Environnement.* Université Paris Val de Marne - Agence Financière de Bassin Seine-Normandie - CEMAGREF - 63 p.

Deguin A. (1992).Intérêt pratique de la mesure du REDOX pour la conduite de l'élimination de l'azote en boues activées. TSM 4, 187-194.

Heduit A. (1990).Potentiel d'électrode de platine en épuration biologique. Thèse de Doctorat de l'Université Paris Val de Marne. CEMAGREF Collection Etudes nÁ 1 - 163 p.

Heduit A. and THEVENOT D.R. (1989). Relation between redox potential and oxygen levels in activated sludge reactors. *Wat Sci. Tech.* 21 (8/9), 947-956.

Heduit A. and Thevenot D.R., (1992). Elements in the interpretation of platinum electrode potentials in biological treatment. *Wat. Sci. Tech.* 26 (5/6), 1335-1344.

Whitfield M. (1972). The electrochemical characteristics of natural redox cells. *Limn. Oceanogr.*, 17, 383-393.

Pergamon

Wat. Sci. Tech. Vol. 28, No. 11-12, pp. 481–488, 1993.

0273–1223/93 $7·00 + 0·00

A NEW SENSING TECHNIQUE FOR OPTIMUM SLUDGE CONDITIONING FOR WASTE SLUDGE DEWATERING

Y. Yasukawa, T. Totoki and H. Shigemi

Central Laboratories, Kurita Water Industries Ltd., 7-1, Wakamiya, Morinosato, Atsugi, Kanagawa 243-01, Japan

ABSTRACT

A new sensing technique using a Hot-Film sensor for determining the optimum polymer dosage for sludge dewatering has been developed. Correlations between the behavior of the Hot-Film sensor and dewaterability of sewage and night soil sludge were investigated. The results of the study revealed the following: the Hot-Film sensor values depended on the filtrate characteristics related to the sludge dewaterability which is measured by the filterability of the conditioned sludge and by the moisture content of the dewatered sludge cake; the optimum polymer dosage for sludge dewatering corresponded with the minimum output value on the characteristics curve of the Hot-Film sensor. This paper describes the observations of the measurement test by the Hot-Film sensor as well as the results of the laboratory test of sludge dewatering for optimum sludge conditioning.

KEYWORDS

Sludge dewatering; sludge conditioning; polymer; polyelectrolyte; belt press filter; centrifuge; hot-film sensor; residual polymer concentration of filtrate; moisture content of dewatered sludge cake.

INTRODUCTION

In recent years, the polyelectrolyte has become a primary choice as a conditioning agent for sludge dewatering operations. The availability of polyelectrolyte has allowed different types of dewatering machines, such as belt press filter and centrifuge. The performance of such dewatering equipment critically depends on the sludge characteristics.

Many factors affecting sludge dewaterability have been suggested including pH, suspended solids, organic content, viscosity, yield stress, torque, biopolymer, colloidal particles and colloid equivalent. Using such factors several methods have been suggested to determine an optimum dosage of polyelectrolyte. These methods use the following techniques: (1) measurement of unconditioned sludge by ultrasonic wave densitometer, (2) measurement of conditioned sludge by torque meter, (3) measurement of filtrate by colloid equivalent meter, and so forth. However, these methods still have certain limitations in their application to different cases.

The objectives of this study include testing and confirmation of practical uses of a new method determining an optimum dosage of polyelectrolyte for sludge dewatering using a so-called Hot-Film sensor.

The Hot-Film sensor is a new and unique device for measuring the potential of sludge dewaterability. When a Hot-Film sensor chip in the filtrate is heated electrically with a constant direct current under a constant flow over the film, the sensor temperature reaches a steady state at a certain level depending on the sludge characteristics, seemingly representing the potential of sludge dewaterability.

This paper describes the progress of studies where we tried to establish a method of determining an optimum polymer dosage for sludge dewatering using a Hot-Film sensor. Here we consider an optimum dosage such that the highest dewaterability as expressed by the gravity filtration is achieved along with the minimum moisture content in the sludge cake.

MATERIALS AND METHODS

An overall schematic block diagram of the test procedure for the study is shown in Figure 1.

Hot-Film Sensor Apparatus

Hot-Film sensor apparatus consists of a small sensor chip and a stirrer located in a measuring vessel, a power supply, a voltmeter, an A/D converter, a personal computer and a printer. A schematic diagram of the apparatus is shown in Figure 2. The thermal sensor is made with a platinum thin film coated on a square chip 10 mm x 10 mm. The stirrer agitates the filtrate to generate a constant flow over the sensor chip. The power supply causes a constant direct electric current through the sensor film. And the voltmeter indicates the temperature of the sensor.

We started the test by supplying a constant direct electric current to the sensor chip placed in the filtrate of dewatered sludge under constant agitation. When the sensor chip was heated with the electric current, the temperature of the sensor chip rose to a steady state level according to the filtrate characteristics depending upon sludge dewaterability, which was measured by the voltmeter.

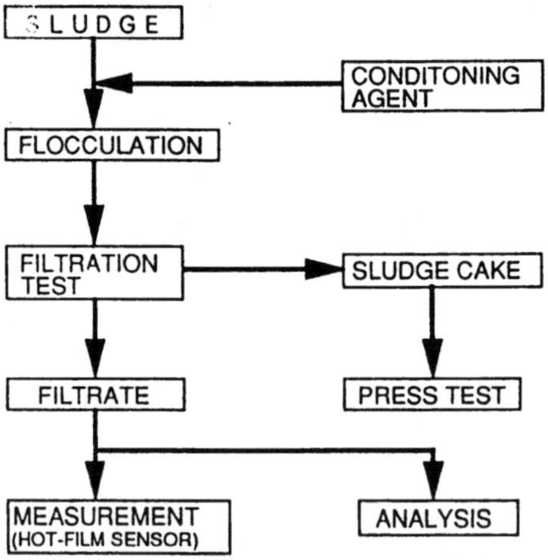

Fig. 1. Flow diagram of experiment

Characteristics of Sludge

Characteristics of sludge samples taken from three different full scale sludge treatment plants are shown in Table 1. One of the samples was taken from a primary settled sludge of a sewage plant and the others from excess activated sludge out of two different night soil plants respectively.

TABLE 1 Characterists of Sludges and Sludge Conditioning Agent

Sample	Sludge Type	Properties of Sludge pH (–)	Cond. (μS/cm)	SS (g/dl)	VSS/SS (%)	Sludge Conditioning Agent
Sludge -A	Primary settled sludge of sewage plant	5.5	3190	2.23	83.9	Kurifix CP604
Sludge -B	Excess activated sludge of night soil plant	6.9	2200	1.39	85.6	Kurifix CP604
Sludge -C	Excess activated sludge of night soil plant	6.7	2500	2.52	75.3	PFS + Kurifix CP542

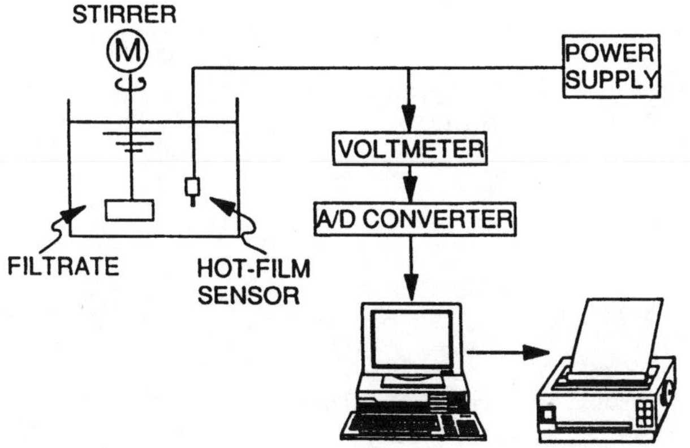

Fig. 2. Schematic diagram of Hot-Film sensor apparatus.

Sludge Conditioning Agent

Descriptions of conditioning agents studied in this work are presented in Table 2. Two different types of sludge conditioning methods were tested: namely, conditioning by cationic polymer alone and by ampholytic co-polymer with PFS (poly ferric sulfate) at the same time.

Measurement of Dewatered Sludge Characteristics

Each of the 200 ml sludge samples was pre-conditioned by the flocculant shown in Table 2, and filtered through a Buchner funnel with polyester filter cloth as used in a belt press filter. The filtrate was collected in a graduated cylinder for 30 secs and measured. This result was used to calculate the gravity filtration velocity of flocculated sludge. Colloid equivalent of the filtrate was measured by colloidal titration method. The residual polymer concentration of filtrate was then calculated from the measurement. The sludge sample left on the Buchner funnel was placed between two pieces of polyester filter cloth, and pressed at 1.0 kg/cm^2 for 1 min, and the moisture content of the dewatered sludge cake was measured.

TABLE 2 Description of Conditioning Agent

Sample	Description
Kurifix CP604	Cationic Polymer High molecular weight High cationic charge
Kurifix CP542	Ampholytic Co-polymer High molecular weight High cationic charge and low anionic charge
PFS	Inorganic Coagulant (Poly ferric sulfate)

Measurements of Hot-Film Sensor

A 1000 ml sample of filtrate was poured into the measuring vessel of the Hot-Film sensor apparatus shown in Figure 2. A development of temperature was observed in an electrically heated sensor chip. The sensor chip was heated with a constant direct electric current under constant agitation. The measured data was automatically transmitted to a personal computer at a regular interval (0.1 s) during the heating period (30 s). These data were averaged, converted to percentages of a full span value previously set up, and presented on the display.

RESULTS AND DISCUSSION

Water Solution of Polyelectrolyte

The output values of the Hot-Film sensor turned out to be sensitive to the polymer concentration in water. Figure 3 shows the correlation between the concentration of cationic polymer in distilled water and the output values of the sensor. As shown in the figure the sensor reading increases with the increase of the polymer concentration. The result of this experiment seems to support the assumption as follows: as polymer concentration increases, the heat transfer out of the sensor film slows down, as a result, the temperature of the sensor chip rises.

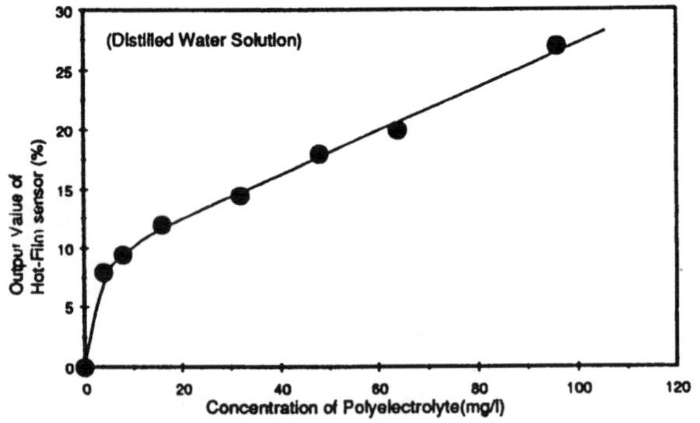

Fig. 3. Correlation between concentration of polyelectrolyte and output value of Hot-Film sensor (Kurifix CP604).

Sludge of Sewage Plant Conditioned by Cationic Polyelectrolyte

Results of dewatering experiment with Sludge-A sample (See Table 1) are shown in Figure 4. The Sludge-A sample was taken from primary settled sludge of a sewage plant, and the flocculant used for conditioning this sludge was cationic polymer (Kurifix CP604). In the figure, the filterability (gravity filtration velocity of flocculated sludge) and the moisture content of the sludge cake are plotted versus polymer dosage. The dosage that yielded the maximum filterability and the minimum moisture content at the same time is clearly the optimum dosage for sludge conditioning. (Generally these two points do not coincide but our experiments with polyelectrolyte flocculation yielded such results.) The optimum polymer dosage here fell to ca. 6.7 mg/g-ds for Sludge-A.

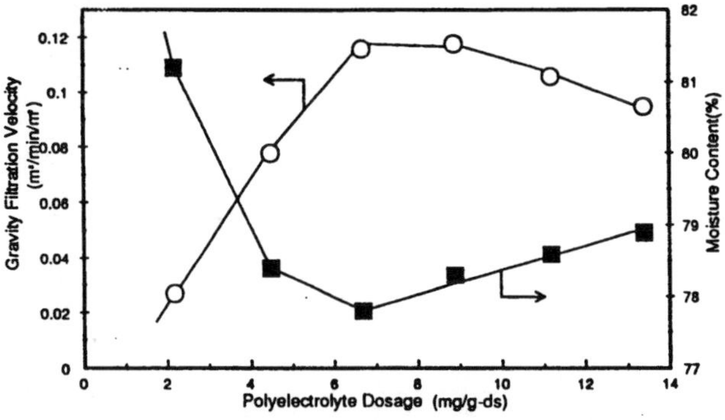

Fig. 4. Effects of polyelectrolyte dosage on dewatering experiment Isludge-A; Kurifix CP604).

Figure 5 shows the output values of Hot-Film sensor and the residual polymer concentration in the filtrate plotted against polymer dosage. The sensor output value seems to decrease with increasing polymer dosage up to 6.7 mg/g-ds, and to increase again with further increase of polymer dosage. The residual polymer rapidly increases beyond 6.7 mg/g-ds in a similar way. That is, the polymer dosage ranging above 6.7 mg/g-ds is considered overdose. With these results combined, an optimum polymer dosage seems to correspond to the minimum output value of the Hot-Film sensor.

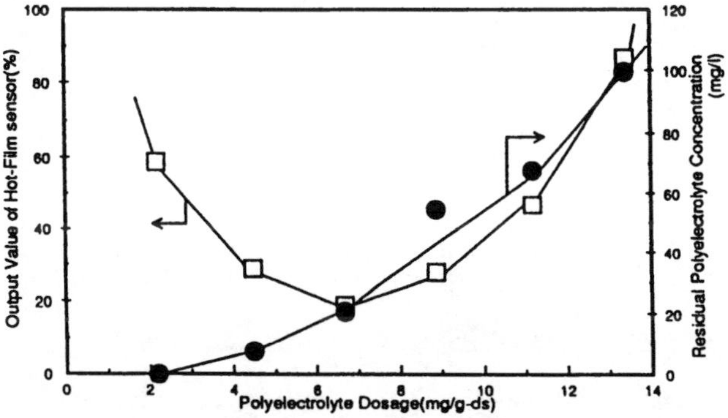

Fig. 5. Correlation betwen output value of Hot-Film sensor residual polyelectrolyte concentration and polyelectrolyte dosage (sludge; Kurifix CP604).

Summarizing the results, a good correlation between the behavior of the Hot-Film sensor and the performance of the dewatering experiments was obtained. It can be said that the output value has presented itself as a possible index of sludge conditioning characteristics for sludge dewatering.

Sludge of Night Soil Plant Conditioned by Cationic Polyelectrolyte

Figure 6 shows the results of dewatering experiments with Sludge-B sample (See Table 1). The Sludge-B sample was taken from excess sludge of a night soil plant, and the flocculant used for conditioning this sludge was cationic polymer (Kurifix CP604). The results of the dewatering experiments revealed that the optimum polymer dosage falls to around 14.5 mg/g-ds for this sludge. Figure 7 shows a good correlation between the output values of the Hot-Film sensor and the polymer dosage. The output values of the sensor decreased with increasing polymer dosage up to 14.5 mg/g-ds, and their increased again. It can be seen that the optimum polymer dosage coincided with the minimum output value on the characteristic curve of the Hot-Film sensor, a similar agreement being seen with Sludge-A even though the particular value was different for the different sludge.

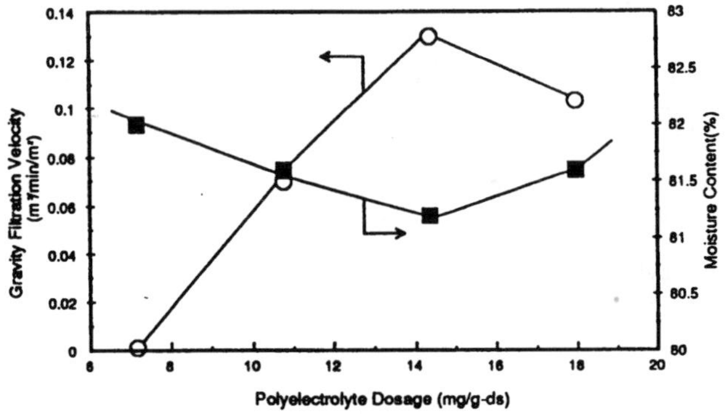

Fig. 6. Effects of polyelectrolyte dosage on dewatering experiment (sludge-B; Kurifix CP604).

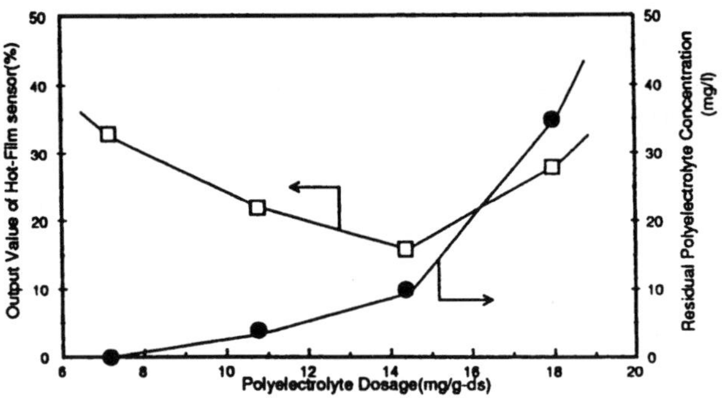

Fig. 7. Correlation between output value of Hot-Film sensor residual polyelectrolyte concentration and polyelectrolyte dosage (sludge-B; Kurifix CP604).

Sludge of Night Soil Plant Conditioned by PFS & Ampholytic Co-polymer

Figure 8 shows the results of the dewatering experiments with Sludge-C (See Table 1 for sludge characteristics). The Sludge-C sample is from excess sludge of another night soil plant. The flocculants used here were PFS (poly ferric sulfate) and ampholytic co-polymer (Kurifix CP542) together. Data of the dewatering experiments in Figure 8 reveal that the optimum polymer dosage was 20 mg/g-ds for this sludge. Figure 9 shows a good correlation between the sensor output values and the polymer dosage. The output value decreases with increase of the polymer dosage up to 20 mg/g-ds, and their increases again. It can also be seen that the optimum polymer dosage here coincides with the minimum output value on the sensor characteristic curve.

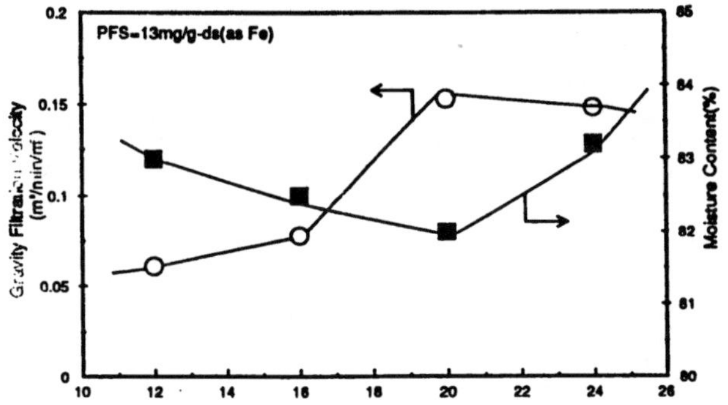

Fig. 8. Effects of polyelectrolyte dosage on dewatering experiments (sludge-C; PFS+Kurifix CP542).

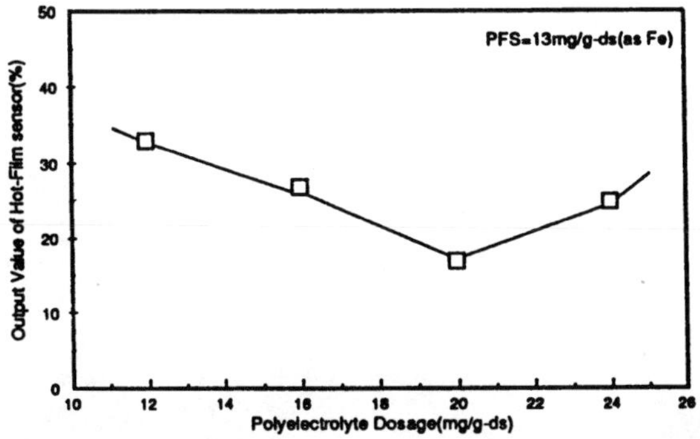

Fig. 9. Correlation between output value of Hot-Film sensor and polyelectrolyte dosage (sludge-C; PFS+ Kurifix CP 542).

CONCLUSIONS

The Hot-Film sensor apparatus provided a possible, simple, rapid and inexpensive means for measuring characteristics to find an optimum sludge conditioning for dewatering of various kinds of sludge. The major points of the investigation are summarized as follows.

1. The output value of the Hot-Film sensor was found to depend on the characteristics of filtrate which has a strong relation with the sludge conditioning prior to sludge dewatering.

2. The optimum polymer dosage for sludge dewatering was found to be in good agreement with the minimum output value on the characteristics curve of the Hot-Film sensor.

Measurement with a Hot-Film sensor seems capable of monitoring the potential of sludge conditioning for dewatering. It is encouraging that these results show the possibilities of the Hot-Film sensor as a means to find an optimum sludge conditioning parameter. We would like to further confirm the possibility in the course of future development. We are going to develop our study towards on-line measurements with the Hot-Film sensor based on the assumption that the minimum output value on the characteristic curve of a Hot-Film sensor indicates the optimum sludge conditioning. We are hoping that this study will eventually lead us to develop an automatic control device for a future sludge dewatering process.

REFERENCES

Hashimoto, M. and Hiraoka, M. (1990). Characteristics of Sewage Sludge Affecting Dewatering by Belt Press Filter. *Wat. Sci. Tech.*, **22**(12), 143-152.

Campbell, H. W. and Crescuolo, P. J. (1982). The Use of Rheology for Sludge Characterization. *Wat. Sci.Tech.*, **14**(6/7), 475-489.

Crawford, P. M. and Shigemi, H. (1990). Control of Polymer Consumption for Sludge Dewatering: The Japanese Experience. *Instrumentation, Control and Automation of Water and Wastewater Treatment and Transport Systems, (Adv. Wat. Pollut. Control* no 10) ed. R. Briggs, Pergamon, Oxford, pp. 351-358.

Pergamon

Wat. Sci. Tech. Vol. 28, No. 11-12, pp. 489-498, 1993.

0273-1223/93 $7·00 + 0·00

INVENTORY AND ASSESSMENT OF AUTOMATIC NITRATE ANALYZERS FOR URBAN SEWAGE WORKS

Hervé Wacheux*, Sylvain Da Silva* and Jacques Lesavre

* *Anjou Recherche, Centre de Recherche de la Compagnie des Eaux, Chemin de la Digue, BP 76, 78603 Maisons-Laffitte Cedex, France*
** *Agence de l'Eau Seine Normandie, 51 Rue Salvador Allende, 92027 Nanterre Cedex, France*

ABSTRACT

An inventory of the European market showed that 12 automatic nitrate analyzers were available using measuring techniques based on ion specific electrodes, colorimetry and UV absorption.

Among these analyzers the following six - Data Link, Dr Lange, M.E, Phox, Polymetron, Process Styrning - were put through two series of tests :

- metrological characteristics of the instruments,
- behaviour in waste water and comparison with analyses of samples in laboratory.

The metrological quality of all the analyzers is sufficient, but all the ion specific electrode analyzers showed contamination and rapid drift when used in waste water.

The others show greater resistance. UV absorption systems are simpler and quicker to use. They are also less demanding with regard to sample conditioning than colorimetric apparatus.

KEYWORDS

Monitoring, On-line sensors, Automatic nitrate analyzers, Ion specific analyzers, UV absorption, Colorimetry, Sensor testing.

INTRODUCTION

The application of Directive 91/271/EEC of 21 May 1991 concerning the treatment of waste water will entail in Europe, France and in works operated by the Compagnie Générale des Eaux Group, the renovation and extension of several sewage works in sensitive areas where nitrogen and phosphorus must be removed. This intensive treatment, performance monitoring and the need for an almost perfect reliability makes it essential to use new operating tools.

Sensors, more than PLCs and supervisors remain the key to success. They permit local or remote monitoring, the regulation of various elements (flow rates, aeration, return sludge, and removal, etc...), the compensation of the dynamic effects of incoming loads, and the development of control strategies to prepare for changing operating setpoints in advance.

Automatic NH4, NO3 and PO4 analyzers are beginning to be used on French sewage works, hence the need to proceed with a preliminary evaluation of the equipment proposed by various manufacturers.

This paper deals exclusively with nitrate analyzers for which we were able to classify 12 suppliers in Europe according to three analytical techniques.

Specific electrode analyzers

SUPPLIERS	ABB KENT Great-Britain	BRAN ET LUBBE Germany	Hydro-Environnement France	PHOX Great-Britain
Model	Model 8086	Ionmeter 90	IGEA-ES	Model1802
Electrode	E.I.L.	Phillips or Orion	Orion	Phillips or Russel
Flow	continuous	continuous	sequential	continuous
Measuring range	0-1100 mg/l (2-3 decades)	0-1000 mg/l	0-200 mg/l	0-100 mg/l
Detection threshold	1 mg/l	0.05 mg/l	0.1 mg/l	0.05 mg/l
Accuracy	5 %	3 %	5 %	3 %
Response time	4-5 minutes	7 minutes	10 minutes (sequential)	< 4 minutes
Analog-digital output	0/4-20 mA RS 232	0/4-20 mA RS 232	0/4-20 mA RS 232	4-20 mA RS 232
Reagents autonomy	15 days	3 weeks	10 days min.	3 weeks
Price FF (tax excl.) basic service	98,000	79,000	170,80	82,000

Specific electrode analyzers (cont.)

SUPPLIERS	POLY-METRON France	PROC. STYRNING Sweden	SERES France	TACUSSEL France
Model	POLYSTAT	CONTRONIC	IGEA-ES	AMA 410S
Electrode	Ingold	EDT	Potentio-oxidation of nitrate	Tacussel
Flow	sequential	sequential	sequential	continuous
Measuring range	0-100 mg/l	2-40 mg/l	Nitrates : 0-22 mg/l max Nitrites : 0.3 mg/l	1 0-1000 mg/l
Detection threshold		0.1 mg/l	0.1 mg/l	1 mg/l
Accuracy	4 %	5%	2 %	5 %
Response time	5 minutes (sequential)	15 minutes (sequential)	10 minutes (sequential)	3 minutes
Analog-digital output	0/4-20 mA	0/4-20 mA	4-20 mA RS 232	4-20 mA RS 232
Reagents autonomy	1-3 weeks	1-3 weeks	1 month	
Price FF (tax. excl) basic service	≈ 90,000	≈ 41,000 in Sweden	100,000	60,000

Colorimetric analyzers and UV absorption

SUPPLIERS	SKALAR Netherlands	Meerestechnik Elektronik (ME) Germany	DATA LINK France	DR LANGE Germany
Model	SA 9000	APP	Nitral 100	NITRAX
Principle	Diazototation	Diazototation	UV absorption 2 λ	UV absorption 2 λ
Measuring range	5-100 mg/l	0-100 mg/l	0-22.6 mg/l	0-22.6 mg/l
Detection threshold	10 ppb	0.1 mg/l	0.1 mg/l	0.1 mg/l (0-25)
Accuracy	4 %	2%		4%
Response time flow	7 minutes	10 minutes (sequential)	immediate	immediate
Analog-digital output	4-20 mA RS 232	0/4-20 mA RS 232	4-20 mA RS 232	0/4-20 mA
Reagents autonomy	10 days	14 days	varies according to cleaning	varies according to cleaning
Price FF (tax excl) basic sercice	160,000	90,000	52,000	108,000

SELECTION OF INSTRUMENTS AND TEST PROCEDURES

After contacting suppliers, six agreed to lend us a complete system, free of charge, to carry out these tests lasting about one month at the Compagnie Générale des Eaux Research Centre at Maisons-Laffitte.

The firms concerned are Data Link, Dr Lange, M.E, Phox, Polymetron, Process Styrning.

The systems were installed by our technicians on a test platform, then examined by a representative of each of the above firms before being put through two series of tests, one metrological and the other an in-situ test on waste water.

The procedure for metrological tests follows mainly the AFNOR T90-550 standard for the determination of :

- response time
- reproducibility
- response linearity
- short-term stability
- interference of the measurement with other components
- response hysteresis.

Tests on waste water consisted in following the response of the different systems exposed to the same variable sample, versus time.Since all the instruments were not delivered on the same date, these tests were carried out in two series.

First series

The test was designed to follow a Biostyr pilot plant. As we only disposed of a small flow rate we tried cross-flow filtration, using a paper filtering cartridge. The diameter of the filter pores was about 8 µm. This size clogged too fast and so the next pore size was tried. We finally opted for filter cartridges made of sintered polyethylene, pore size 15 µm.

The instruments were fed (see fig 1) from an overflow pot receiving filtered water.

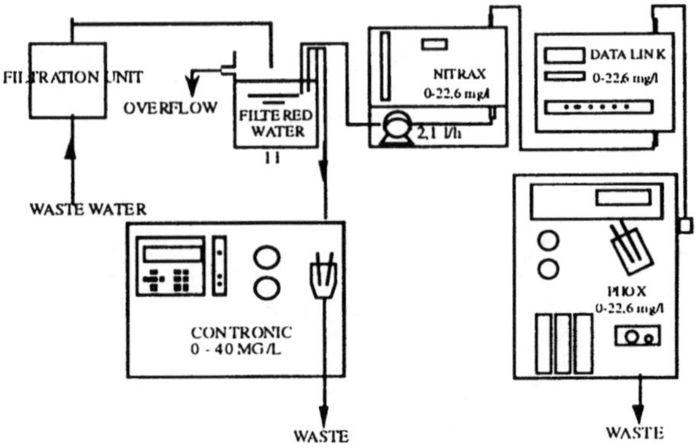

Figure 1 : Analyzers installation layout diagram (1st series of tests)

The sensors operated on-line and their outputs were registered on a Grant data-logger. After compensating transfer time due to assembly, and the response time specific to each analyzer, the results were compared one with the other, as well as with the measurements obtained on a spectrometer S1000 and with the sample analyses performed in the central laboratory (according to AFNOR NFT 90-012). Samples were collected either by hand or with an automatic sampler during the two series of tests on waste water.

Second series of profiles

The feed flow diagram (fig 2) was different from the one described above for technical reasons.

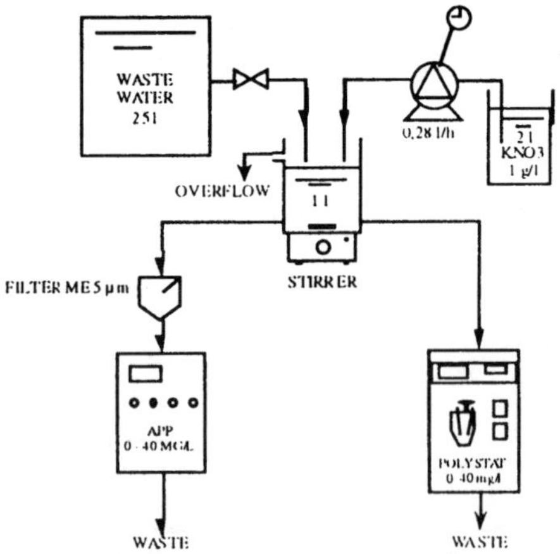

Figure 2

RESULTS AND DISCUSSION

Metrological tests

Due to a late delivery, Phox and Polystat analyzers were installed directly on waste water so the metrological tests were performed after those on waste water. In the case of the Phox system, results were affected by the contamination of the electrode.

Determination of response time

The measuring sequence was as follows : 5-10-5-20 or 40 mg/l (depending on the instrument's measuring range) then 5 as mg/L of N-NO_3.

Analyzer	Time required to reach 90 %
Nitrax 1 min	10 s (± 6-10 sec)
Nitral 100	1 min 15 sec (± 10 sec)
Contronic	1 o r 2 measurements (15-30 min)
Phox	4 min 30 sec (± 30 sec)
APP	1 or 2 measurements (20 to 40 min)
Polystat	1 measurement (5 min)

Response times for UV systems are the shortest (≈1 minute). This is due to the principle of the measuring technique.

Reproducibility

The reproducibility of these instruments was measured on a solution containing 5 mg/l of N-NO_3. We presented successively to the analyzer a solution at 5 mg/l then, after stabilization, a solution without nitrate (distilled water) and that several times. As the metrological trial was following the trials on waste water, the electrodes were unstable.

	Average value (mg/l)	Standard deviation (mg/l)	Coefficient of variation
Nitrax	8.44	0.17	2.01 %
Nitral	9.61	0.61	6.35 %
Contronic	10.97	0.52	4.74 %
APP-ME	5.12	0.24	4.69 %

Interference with other constituents

Summary table

	NO_2	Cl^-	HCO_3^-	DCO	Detergent	ClO^-
Nitrax	++	0	0	+	+++	0
Nitral 100	++	0	0	+	+++	0
Contronic	+	+	+	NT	NT	NT
Phox	unstable	NT	NT	NT	NT	NT
APP	++	-	NT	-	NT	NT
Polystat	+	+	NT	NT	NT	NT

NT : not tested
- : weak negative interference
0 : zero interference
+ : weak positive interference
++ : strong positive interference
+++ : very strong positive interference

The UV analyzers measure the sum of nitrates + nitrites. This is not a drawback for controlling denitrification, since the aim is the reduction of nitrogen in all its forms.

They are very sensitive to detergents but since we do not know precisely what the concentrations in the waste water will be, we cannot assess interferences from that source.

The most inconvenient problem is the UV absorbant organic matter in UV tests. Our experiments were carried out on treated effluent, hence in the most favourable case. If we transpose the operation upstream, after the anoxic tank for instance, the problem may become crucial.

For the other instruments, our tests did not reveal any major source of interference among the substances contained in the waste water, at the concentrations usually found.

Response hysteresis

We started off with a solution of 5 mg/l of $N\text{-}NO_3$ and we examined the back flow after saturation.

	Nitrax	Nitral 100	Contronic	Phox	APP	Polystat
N°of cycles	5	5	5	5	5	5
Average	4.06	4.58	5.8	5.369	5.44	3.95
Coefficient of variation	0.50 %	6.76 %	1.75 %	6.79 %	6.25 %	3.67 %

Profiles on waste water

First test series

The profile of 4 February 1992 (see figure 3 below) shows the data recorded by the various instruments working together, after readjustment to allow for the transfer times. We note that the UV systems and the Phox analyzer are in perfect correlation.

The Nitral 100 analyzer varies in steps of 1.4 mg/l of $N\text{-}NO_3$. With 15 minute analyzing periods, the Contronic sensor is not quite quick enough to sense the rapid variations in concentration.

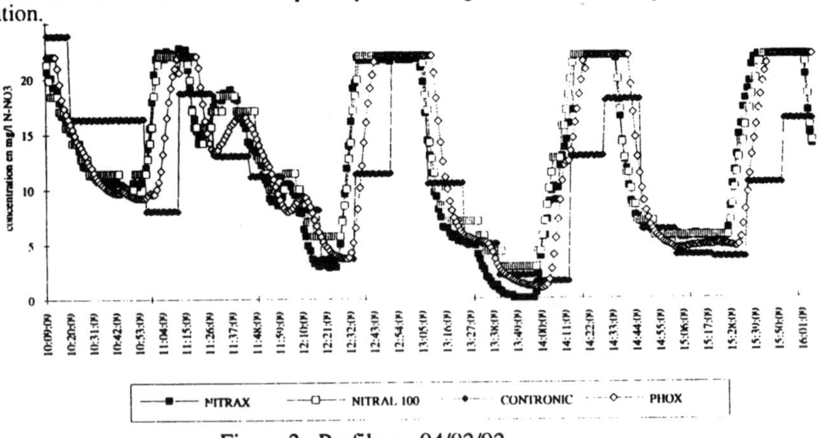

Figure 3 : Profile on 04/02/92

In the same conditions results are not so good the following day.

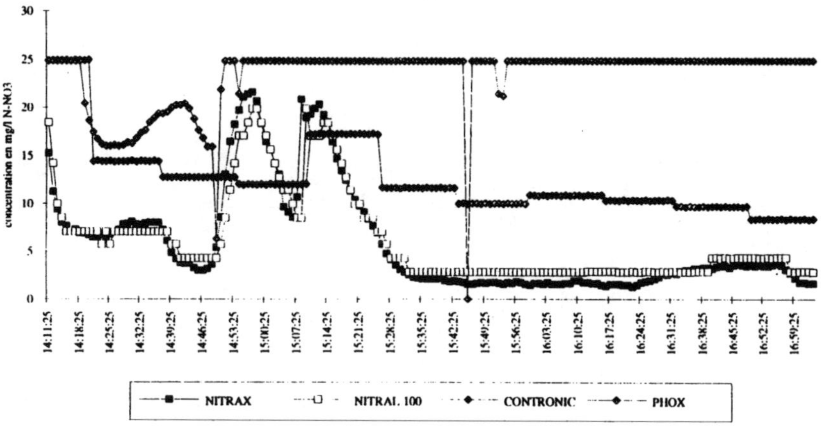

Figure 4 : Profile on 05/02/92

Nitrax and Nitral 100 are always in accordance but Contronic and Phox supply incorrect data. The amplitude of the signals fades, due to a decrease in the gain factor of these analyzers. The Phox system quickly reaches saturation point and stays jammed at full scale. The cause is contamination of the electrode by a dissolved substance contained in the sample.

M. Fleet, General Manager of Phox, told us that this problem sometimes occurred on certain types of sewage. The contaminant is not yet identified, in these cases he proposes a system with a spare electrode. While one electrode is in operation, the other is regenerated by leaving it to soak in distilled water.

The representative of Contronic said electrode drift is a well known phenomenon and proposed regeneration in a concentrated nitrate solution. That is why the Contronic analyzer operates sequentially (shorter time in contact with the sample). The electrode plunges into the high calibration solution after each measurement (regeneration of the ion exchange substances in the membrane).

The Contronic analyser does not saturate because its measuring range is twice that of the Phox model. Both manufacturers could only propose changing the electrodes for our tests. This remedy is not good enough since the drift appeared very quickly (within a few days) and several times.

- Comparison of analyzer results with laboratory analyses

We observe an average correlation factor between the analyzers'output and the laboratory analysis of samples; R = 0.878 for Nitrax and 0.879 for Nitral 100 calculated on 53 measurements. However, it is important to realize that a part of the difference in values is attributable to the sampling method.

In fact, one of the major problems in assessing the analyzers is the representativeness of the sample when it varies quickly.

- Comparison of data obtained by Nitrax and Nitral 100

Since both are based on UV absorption, it was interesting to compare their performance.

Using all the data from the different profiles established, we performed a correlation of the data concerning Nitral 100 with that of Nitrax. We obtained a correlation factor R = 0.975 based on 80 measurements.

Notwithstanding the step by step mode of the Nitral 100, both of the systems are very comparable in their results.

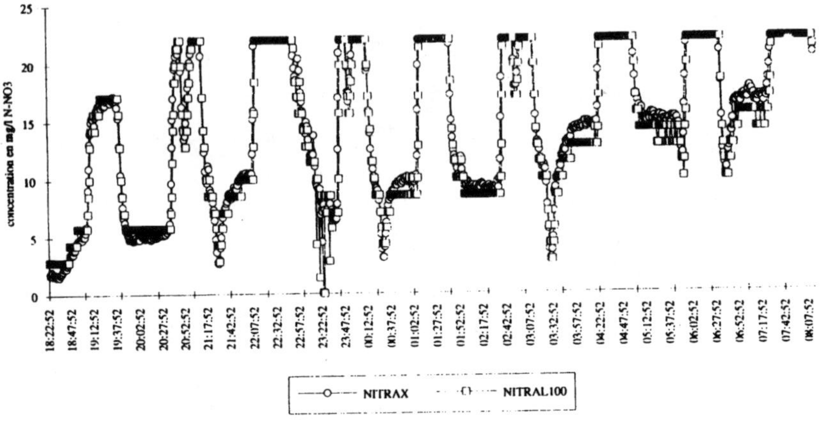

Figure 5 : Profile of 06/02/92

- Comparison of measurements performed by Contronic and Phox

The measurements in the profile, effected by the two electrode analyzers respectively, were compared. We have eliminated the pairs of points which did not correspond to the range of the Phox analyzer.

The correlation factor is R = 0.851 for a set of 30 points, in other words, quite a good correlation. Wide discrepancies in the points were noted in the higher values.

Second series of tests

For this second series of assessments, we established one profile in relation to the laboratory results. This profile is given hereunder.

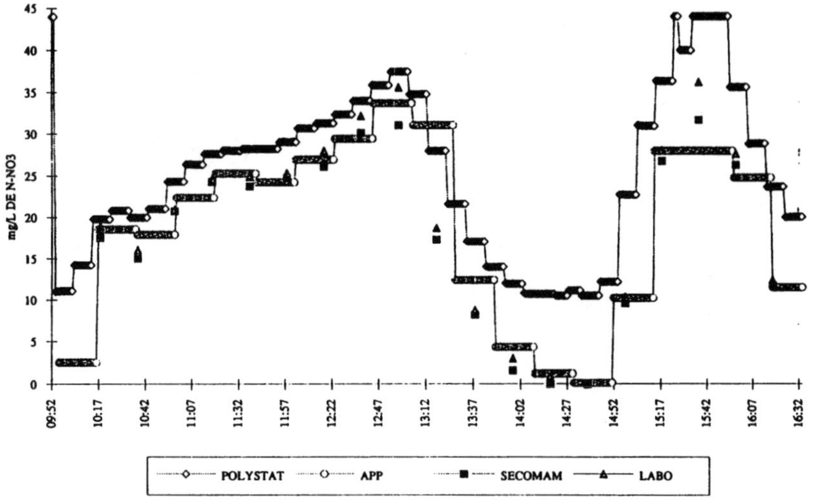

Data from the analyzers are in accordance with each other and with those from the laboratory analyses.

For the APP colorimeter, the correlation factor was R = 0.938 obtained with 18 samples. We note a shift of the zero point due to zero instability of the analyzers. When the colorimeter is fed with a solution of distilled water, the output fluctuate between 0 and 2 mg/l.

The Polystat apparatus, having worked properly for several days, started to drift. In particular, it was incapable of descending below a certain value (approx. 10 mg/l). It seems that despite a number of arrangements, such as using a buffer against interfering elements, washing the electrodes after each measurement, the electrode was either contaminated or fouled.

This phenomenon does not result in poor correlation of the measurements compared with the laboratory analyses, since the correlation factor was R = 0.952. The zero shift was, on the contrary, very big (10 mg/l) and the gain factor is rather weak (0.8 instead of 1).

CONCLUSION

There are quite a lot of manufacturers producing nitrate analyzers. Three basic techniques are used :

- ion specific electrodes (Phox, Polymetron, Process styrning)
- absorption by UV (Data Link and Dr Lange)
- colorimetry (ME).

The tests performed were partly metrological, based on the AFNOR T 90-550 standard "Evaluation of water analyzers", and partly on-stream waste water tests compared with the standardized analyses of samples in Central Laboratory of Compagnie Générale des Eaux.

The UV systems tested (Nitrax and Nitral 100) are very similar. They quickly measure the total nitrate + nitrite content, and require very little maintenance.

This method is not selective enough in the case of UV absorbant organics. The interference caused is limited with regard to treated water but may prove to be too much along the treatment line.

Out of the 3 electrode type analyzers tested, the Contronic analyzer has neither automatic calibration nor temperature compensation. Moreover, the response time is slow (15 minutes).

The Phox and Polystat systems are more complete and better suited to industrial measuring. We did not find any interfering elements worth mentioning among those tested, but the electrodes very soon showed a sharp drift in the waste water used (a few days). This risk, of which manufacturers are aware, does not makes electrode analyzers for measuring nitrates in waste water an attractive proposition.

The APP colorimeter behaves as well as the UV systems and does not show interference as far as those tested are concerned. The drawback remains the time taken by the analytical process, 21 minutes. Sudden changes in concentration are also beyond its perception, which can be inconvenient for regulation use.

REFERENCES

AFNOR (1987). Norme T 90-550 Essais des eaux. Evaluations des caractéristiques des analyseurs d'eaux. Définitions. Méthodes générales, 20 pp.

Andersen, K.L. and Wagner, P. (1990). Testing on-line monitors for ammonia, nitrate, orthophosphate and suspended solids in activated sludge sewage treatments plants. *Instrumentation, Control and Automation of Water and Wastewater Treatment and Transport Systems (Adv. Wat. Pollut. Control* no. 10) ed. R. Briggs, Oxford, Pergamon, pp. 667-675.

Fraboul, P. (1991). Automatisation des stations d'épuration, évaluation d'analyseurs industriels (ammonium, nitrate, phosphates) Rapport de D.E.A, Université de Créteil.

Million, J.L. (1991), Mesure des nitrates en station d'épuration. Evaluation de l'électrode spécifique nitrates Philips IS 561. Rapport Anjou Recherche JLM/mf 91/447, 20 pp.

Sakamoto, T. and Kameyama, K. (1990). A simple method for simultaneously measuring nitrate nitrogen and nitrite nitrogen in water using ultraviolet absorbance. ICA Preprint of posters paper of 5 th IAWPRC Worshop. Yokohama and Kyoto. Japan. 26 July - 3 August 1990 pp 9-12.

CONTROL STRATEGIES APPLICATION

Pergamon

Wat. Sci. Tech. Vol. 28, No. 11-12, pp. 499-512, 1993.

0273-1223/93 $7·00 + 0·00

EFFECT OF SUPPLEMENTAL CARBON SOURCE ON PHOSPHATE REMOVAL IN AN ALTERNATING ACTIVATED SLUDGE PROCESS

S. H. Isaacs*, M. Henze**, H. Søeberg* and M. Kümmel*

* *Department of Chemical Engineering, Building 229, Technical University of Denmark, DK-2800 Lyngby, Denmark*
* *Department of Environmental Engineering, Building 115, Technical University of Denmark, DK-2800 Lyngby, Denmark*

Abstract

Pilot plant studies indicate that the addition of COD in the form of acetate to an alternating nutrient removal activated sludge process as a means to improve denitrification rates is not detrimental to biological phosphate removal. This is in spite of the fact that the added acetate can cause a significant release of phosphate during the anoxic phases in the aeration tanks. The added COD also appears to result in increased rates of aerobic phosphate uptake, thus improving the phosphate removal ability of the process. These observations are presented and discussed in the paper.

Keywords

Activated sludge, biological nitrogen and phosphate removal, nutrient removal, organic substrate supplement, process control, waste water treatment

Introduction

Activated sludge processes which, in addition to organic removal, include the biologically mediated removal of nutrients require newer and more complex strategies for their control and efficient operation. Here the goal is to simultaneously and continually maintain effluent water criteria with respect to organic, nitrogen and phosphorus content in the face of the daily, hourly, weekly and seasonal variations inherent to the input and environment of these processes. One of the difficulties in the development of new control strategies is the relatively low number of variables which can be manipulated. The traditional control variables for activated sludge processes of aeration intensity, internal recycle rates, and sludge wasting rates (Olsson, 1976) do not afford adequate operational flexibility when the additional goals of nitrogen and phosphorus removal are considered.

A means by which the nitrate effluent quality of an alternating type nutrient removal process can be improved has been discussed and demonstrated on a pilot plant scale by Isaacs, et al. (1993). This entails the addition of a readily degradable organic substrate to that point in the process at which denitrification occurs. This serves to increase the denitrification rate on demand, thus compensating for factors such as diurnal increases in nitrogen loading and winter time operating temperatures. In the study, acetate had been used as model carbon source, which exhibits an immediate and strong effect on the denitrification rate. However, acetate also promotes phosphate release from phosphate accumulating bacteria which, at a first glance, would appear to be undesirable for processes which are designed for biological phosphate removal. The pilot plant behaviour nevertheless indicates that the contrary may be true. The acetate induced phosphate release occurring during denitrification appears not to adversely effect effluent quality. Furthermore, a general improvement in phosphate uptake ability has been observed, which may be due to the additional supply of organics in the system. Whereas the above cited paper focused on the aspect of nitrogen removal, this paper discusses the implications of the proposed control strategy from the point of view of phosphate removal.

The Alternating Process

The experimental investigations presented in this work have been performed on a pilot plant process of the BIO-DENIPHO[1] type. This is an activated sludge nutrient removal process employing an alternating mode of operation (Arvin, 1985; Einfeldt, 1992). The two subprocesses involved with biological nitrogen removal, nitrification and denitrification, are performed in a semi-batch manner by periodically changing the path of flow through two parallel aeration tanks which are aerated according to a fixed (or controlled) strategy. An anaerobic zone at the front of the process promotes the growth of a type of bacteria having the ability to take up and store excessive amounts of phosphate under aerobic conditions (Wentzel, et al., 1992). By removing excess sludge at the point in the process at which the phosphorous content of these bacteria is highest, the biological removal of phosphorous is achieved.

The pilot plant is a part of the experimental facilities of the Department of Environmental Engineering at the Technical University of Denmark and is located approximately 14 km north of Copenhagen in Lundtofte. The key components of the plant are shown schematically in Fig 1. The plant feed is preclarified municipal waste water with a flowrate on the order of 100 l hr^{-1}. The anaerobic zone, ANA, consists of a plexiglass column with a holding volume of 200 liters and provided with mild agitation. The feed to this zone is a mixture of the preclarified waste water and return sludge. The flow rates of the inlet water and return sludge are held to a 1:1 ratio and are controlled with the two pumps P1 and P2. The two aeration tanks, T1 and T2, each have a holding volume of 800 liters and are provided with an agitator and an air diffuser. The holding volume of the settling tank, SED, is 1100 liters. A flow injection analysis system measures the aqueous concentrations of

---

1 BIO-DENIPHO process is a patented process developed by I. Krüger Systems in cooperation with the Department of Environmental Engineering at the Technical University of Denmark.

ammonia (NH_4-N), nitrate plus nitrite (NO_x-N) and phosphate (PO_4-P) at the four locations indicated as SP1 to SP4 in Fig. 1: the inlet waste water, the top of the anaerobic column, the aeration tank T2, and the settler effluent (Pedersen, et al. 1990). These measurements are performed for tank T2 at a rate of every 3 minutes and for the remaining three locations plus a standard solution at a rate of every 12 minutes.

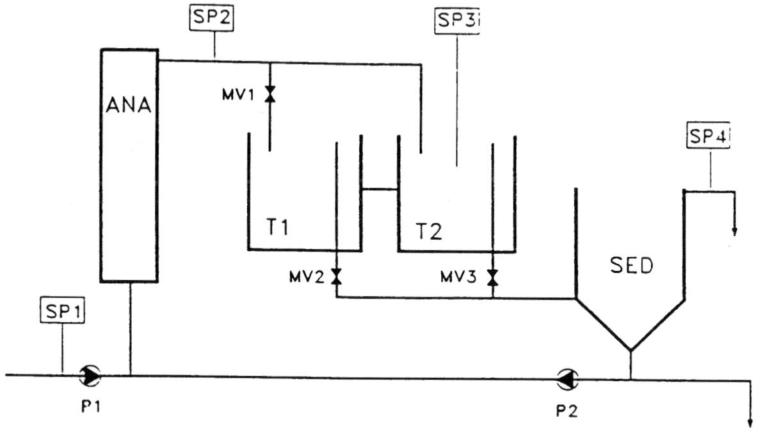

Fig. 1. Schematic diagram of the pilot plant. ANA: Anaerobic column; T1, T2: aeration tanks; SED: settler; P1, P2: pumps; MV1-MV3: motor valves; SP1-SP4: sample points.

The alternating process calls for a cyclic manner of operation by which each cycle is divided into a number of phases representing different flow and aeration patterns. At the time of this investigation, the six phase schedule illustrated in Fig 2 was employed (shown here for a two hour cycle length). This fixed schedule provides, however, an excessively long aeration period each cycle, as nitrification occurs at a higher rate than denitrification in the pilot plant. For this reason, an operational strategy has been implemented by which aeration is terminated after nitrification has come to completion, based on the measured NH_4-N in tank T2. The unmeasured tank T1 is presumed to mirror that of tank T2, so that tank T1 is aerated for the same length of time as that received by tank T2 during the preceding half cycle.

The Control Strategy

One of the problems commonly encountered with the pilot plant process is an inability to completely convert all the nitrate and nitrite to nitrogen gas within the time allotted to denitrification, resulting in an accumulation of NO_x-N in the aeration tanks and an associated increase in effluent water NO_x-N concentration. This is illustrated in Fig. 3 which shows an excerpt of the measurement data over a three day period. The entire data

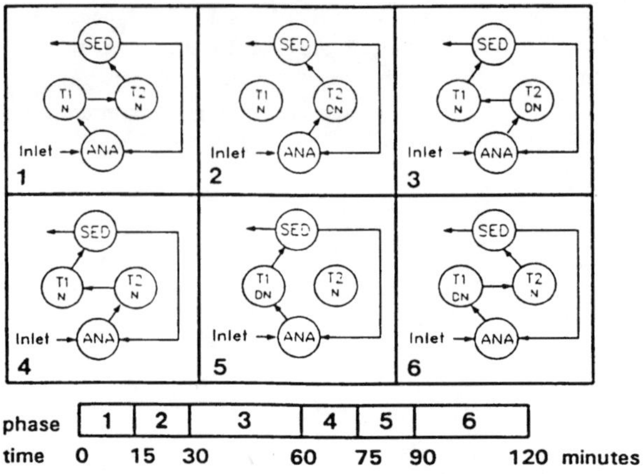

Fig. 2. Phase scheduling of the alternating process employed during the course of the presented investigations.

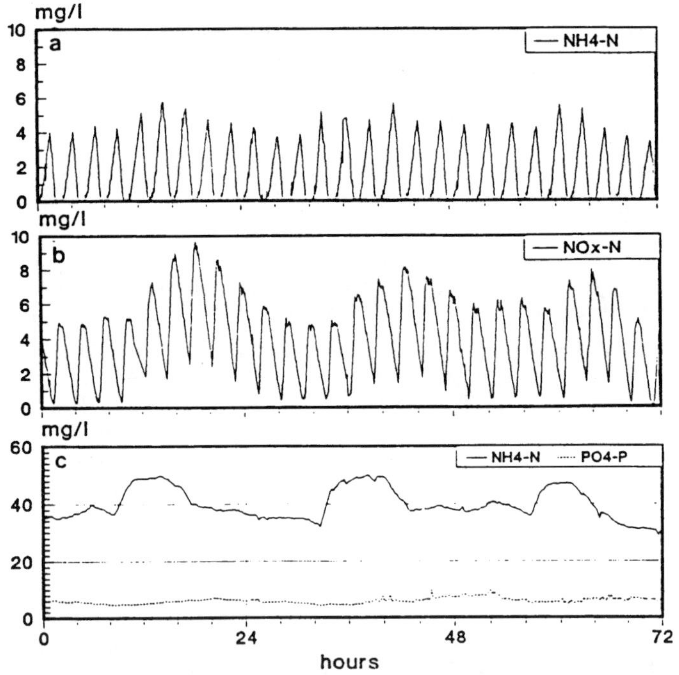

Fig. 3. Concentrations of NH_4-N and NO_x-N in tank T2 (a and b) and of NH_4-N and PO_4-P in the inlet waste water (c) recorded in the pilot plant over a three day period.

set including a discussion of pilot plant performance can be found elsewhere (Isaacs, et al. 1992a). Whereas nitrification comes to completion during each process cycle (Fig. 3a), NO_x-N accumulates over the course of several periods during each day, causing a corresponding rise in the effluent NO_x-N concentration (Fig 3b). This situation represents a temporary nitrogen overload caused by the characteristic diurnal increase in the incoming waste water's ammonia content (Fig. 3c).

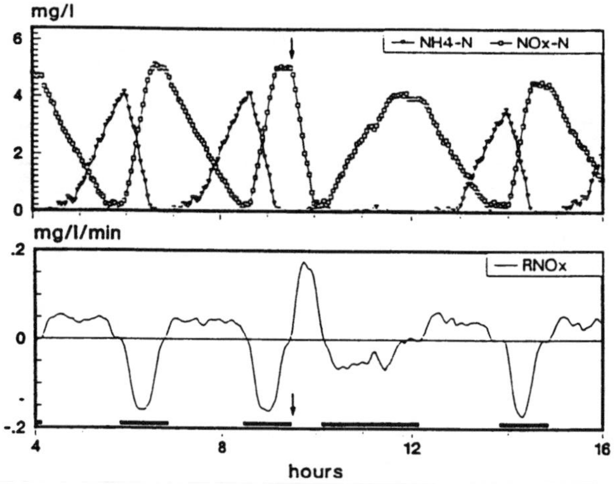

Fig. 4. NH_4-N and NO_x-N measurements in tank T2 (upper plot) and the rate of nitrification and denitrification, R_{NOx} (lower plot). The arrow denotes the time of addition of 200 g sodium acetate.

In order to improve this situation, a strategy has been examined involving the controlled addition of a readily metabolizable carbon source to tanks T1 and T2 during their respective denitrification phases (Isaacs, et al. 1993). This strategy is based on the results of a study which shows that the denitrification rate can be significantly and immediately increased in this manner (Nielsen et al. 1992). This is illustrated in Fig. 4. The top portion of the figure shows the measurements of NH_4-N and NO_x-N in tank T2 of the pilot plant. The curve in the bottom portion of the figure labelled R_{NOx} shows the course of the nitrification rate (negative y-axis) and denitrification rate (positive y-axis) which have been calculated using the NO_x-N measurements and a mass balance over tank T2 (Isaacs, et al. 1992b). At the point in time indicated by the arrow, 200 g of sodium acetate ($CH_3COONa{\cdot}3H_20$) was added to tank T2, which is equivalent to a supplement of approximately 116 mg l^{-1} COD. Immediately following acetate addition, the denitrification rate increased to about 0.18 mg $(l{\cdot}min)^{-1}$ which is more than a threefold increase over the general rate of about 0.05 mg $(l{\cdot}min)^{-1}$ seen during the periods before and after acetate addition.[2]

2 Reaction rates are reported in this work on a per volume rather than a per VSS basis. During the course of these investigations, the pilot plant VSS has been held relatively constant at about 3.8 g l^{-1}.

To demonstrate this strategy, a temporary overload with respect to nitrogen was induced by feeding a solution of ammonia just after sample point SP2 at the top of the anaerobic column between the 7th and 13th hours on four separate days. The feed rate of ammonia was occasionally adjusted in order to compensate for the ammonia content of the incoming waste water. In this manner, a similar overload was obtained for each day. On three of the days, a solution of sodium acetate (50, 100 or 200 g l^{-1} $CH_3COONa{\cdot}3H_2O$ in tap water) was added to both tanks T1 and T2 during their respective denitrification phases at a flow rate of 12.5 ml min^{-1}. This addition was performed only from the time at which the first high NO_x-N peak appeared in tank T2 at approximately the 10th hour until the 15th hour. For each denitrification phase of each tank, the addition of acetate was begun as soon as the dissolved oxygen concentration in that tank dropped below 0.5 mg l^{-1}. Acetate addition was terminated in tank T2 at the first occurrence of either a drop in the measured NO_x-N concentration below 0.3 mg l^{-1} or at the start of the next aerated phase. For tank T1, the same duration of acetate addition was applied as that for tank T2 during the preceding half period.

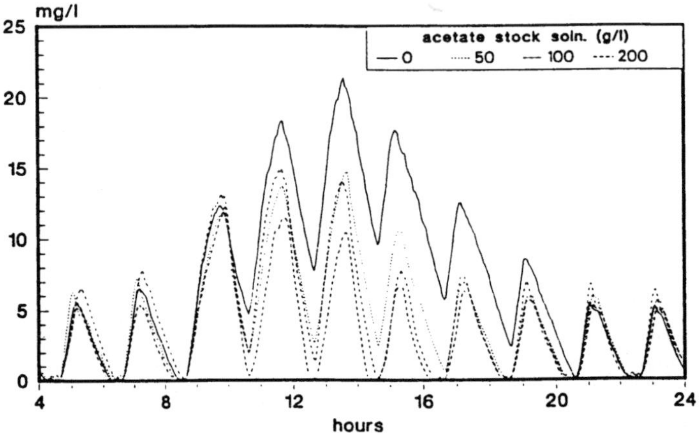

Fig. 5. The course of NO_x-N in tank T2 recorded over four days of the control strategy demonstration.

Fig. 5 shows the course of the NO_x-N measurements of tank T2 recorded during the demonstration. As compared to the day which did not receive supplemental COD (labelled here and in the following as 0 g l^{-1} stock solution), the addition of acetate to denitrification resulted in a reduction in NO_x-N accumulation. NO_x-N accumulation was in fact completely avoided during the trial employing 200 g l^{-1} acetate stock solution. The corresponding improvements in the process effluent NO_x-N content can be seen in Fig. 6. The higher rates of denitrification during those cycles in which acetate was added is responsible for the reduced nitrate levels (cf. Isaacs, et al., 1993).

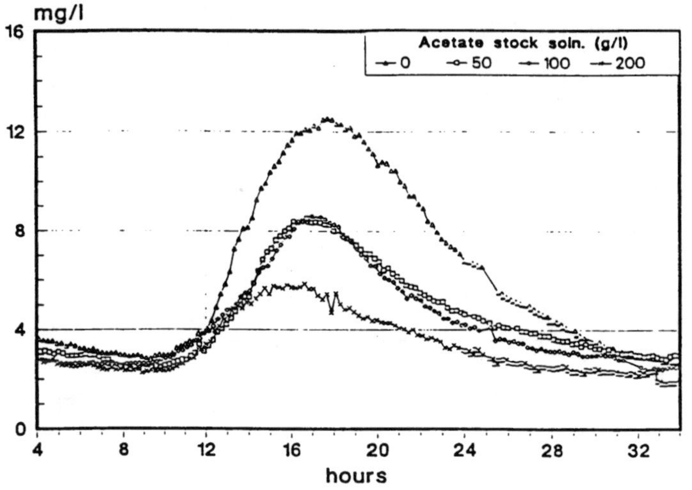

Fig. 6. The NO_x-N measurements of the process efflent which were recorded over four days of the control strategy demonstration.

Phosphate Behaviour

The application of this control strategy to an alternating process designed for biological phosphate removal has associated with it a side effect, namely the release of phosphate from the phosphate-accumulating bacteria within the (non-aerated) aeration tanks. Particularly the use of a volatile fatty acid such as acetate as supplemental carbon source could be expected to result in a high level of phosphate release, since this type of carbon substrate is readily sequestered by these organisms (Wentzel, et al. 1992). This situation is illustrated in Fig. 7 which shows the PO_4-P measurements of tank T2 corresponding to the experiment of Fig. 3 involving batchwise acetate addition. Also shown as a continuous curve is the rate of phosphate uptake or release, R_{PO4}, which had been calculated from the PO_4-P measurements (Isaacs, et al. 1992b). Positive values of R_{PO4} signify phosphate uptake and negative values signify phosphate release. Shortly after acetate addition, the PO_4-P concentration rose rapidly, reaching a level of about 24 mg l^{-1} within 40 minutes. For this data set, the peak PO_4-P concentration attributable to inflow from the anaerobic zone alone lies in the range of from 3 to 5 mg l^{-1}. Hence this PO_4-P peak represents an acetate induced phosphate release in the range of from 19 to 21 mg l^{-1}.

The solid bars under the curves in Fig. 7 indicate periods of aerobic conditions. The aeration time for the cycle following acetate addition had been prolonged in order to see if all of the released phosphate would be re-absorbed by the phosphate-accumulating bacteria. Indeed this was the case. However, had the normal operational strategy been followed, a significant amount of phosphate would still have been present when aeration was terminated. This would then have resulted in an increased effluent PO_4-P concentration.

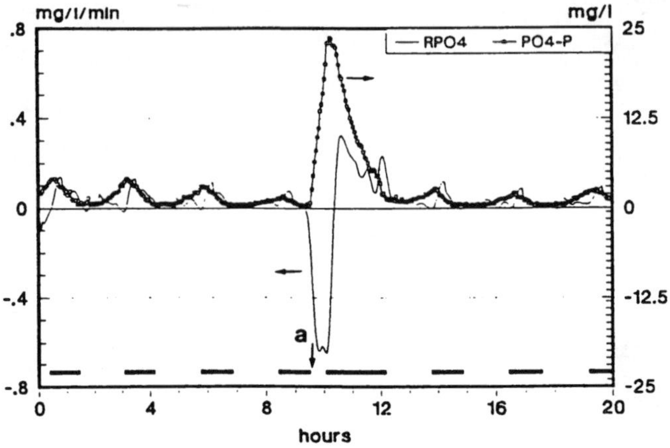

Fig. 7. The PO_4-P measurements in tank T2 and the calculated rate of phosphate uptake/release, R_{PO4}, which correspond to the batchwise acetate addition experiment of Fig. 4. The solid bars denote times of aerobic conditions. The arrow labelled a denotes the time of acetate addition.

The course of the PO_4-P measurements of tank T2 recorded during the four days of the control demonstration discussed above are shown in Fig. 8. The excessive peaks which appear between the 10th and 16th hours for the trials employing 100 and 200 g l^{-1} acetate stock solution are again due to acetate induced phosphate release. What occurs during the trial employing 200 g l^{-1} acetate stock solution is illustrated in more detail with the help of Fig. 9. Shown in the main plot to the left are the PO_4-P measurements as well as the course of the NO_x-N measurements and the phosphate uptake/release rate RPO_4 for this trial. The two sets of solid bars under the curves denote when aerobic conditions existed (upper set) and when the process was either in the second or third of the six phases shown in Fig. 2 (lower set). During these two phases, tank T2 is not aerated and receives inlet flow from the anaerobic zone. The solid curve segments represent approximately the course that the PO_4-P concentration would have taken during these two phases of each cycle had there been neither utilization nor release of phosphate within tank T2. These segments were calculated assuming that the tank behaved like an ideally well stirred vessel according to

$$C(t)=C_0(t)\exp(-Dt)+C_{in}\cdot[1-\exp(-Dt)] \quad , \qquad (1)$$

where C symbolizes PO_4-P concentration (mg l^{-1}), t is the time (min), D is the dilution rate (0.005 min^{-1}), C_0 is the PO_4-P concentration in tank T2 at the start of each calculation, and C_{in} is the PO_4-P concentration at the top of the anaerobic zone averaged over the time span of each calculation.

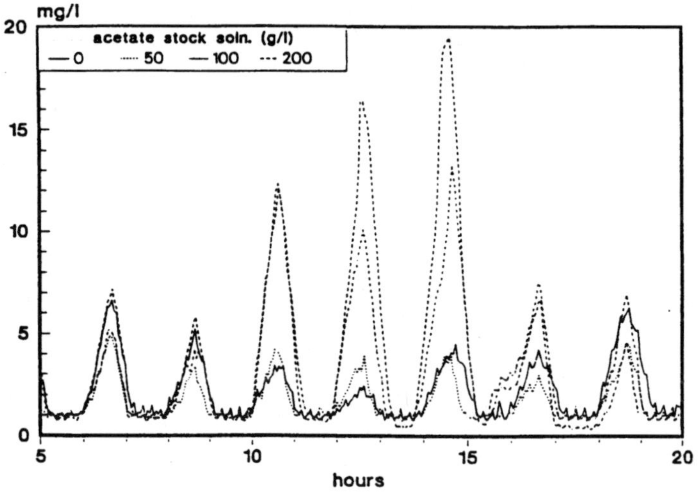

Fig. 8. Course of PO_4-P in tank T2 recorded during the control strategy demonstration.

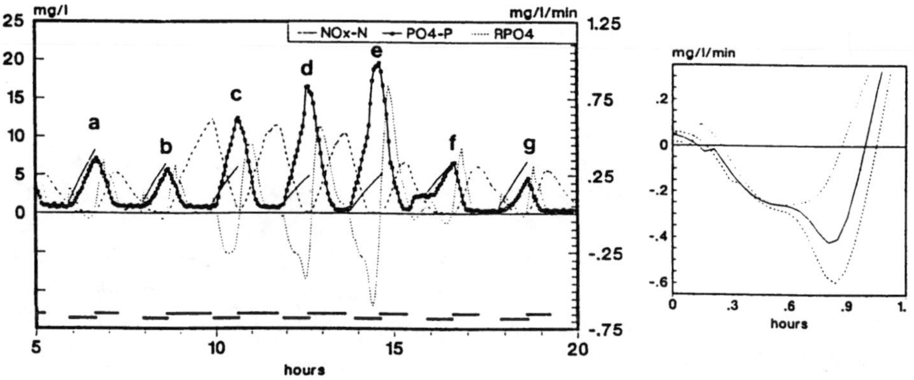

Fig. 9. Measurements and derived quantities pertaining to the control demonstration trial for which 200 g l^{-1} acetate stock solution was employed (see text for more details).

The slopes of the rising edges of the peaks labelled **c**, **d** and **e** are much greater than those determined with Eq. (1). This indicates that free phosphate has appeared in the tank to a greater extent than that which could be attributed to inflow alone. The rising edges of the three peaks labelled **a**, **b**, and **g**, which occurred entirely outside the time period of acetate addition, lie slightly under but otherwise agree fairly well with those calculated from Eq. (1). The deviations which do occur for these peaks can be explained on the basis of phosphate uptake under anoxic conditions and phosphate release under anaerobic conditions. This behaviour is reflected in the course of R_{PO4}. During phases 2 and 3 for the peaks **a**, **b**, and **g**, R_{PO4} takes on positive values when NO_x-N is present and negative values after NO_x-N has been consumed.

A close look at the course of R_{PO4} for peaks **c**, **d** and **e** leads to some interesting observations. R_{PO4} remains negative throughout phases 2 and 3 for all three peaks, signifying that phosphate is released under anoxic as well as anaerobic conditions. The negative courses of R_{PO4} for these three peaks have been overlaid and magnified in the plot to the right. For peak **c**, anoxic conditions prevail for the entire duration of phases 2 and 3. The course of R_{PO4} indicates a maximum phosphate release rate of about 0.25 mg $l^{-1}min^{-1}$. For peaks **d** and **e**, a maximum release rate of approximately the same magnitude appears also to be approached under anoxic conditions just until NO_x-N is completely consumed, after which a dramatic increase in the release rate occurs. The switch to anaerobic conditions occurs earlier for peak **e** which may explain why a larger rise in the phosphate release rate was observed for this peak. One interpretation of these results is that phosphate is taken up and released simultaneously under anoxic conditions when acetate is added. The net effect of these two opposing reactions is the observed release of phosphate at a rate of about 0.25 mg $l^{-1}min^{-1}$. After NO_x-N is consumed, those phosphate accumulating organisms which can denitrify terminate their anoxic phosphate uptake activity (and possibly also begin to release phosphate), resulting in the observed increase in release rate (Kerrn-Jespersen and Henze, 1992).

Referring once again to Fig. 8, it might be anticipated that the phosphate released in the aeration tanks for the trials employing the two highest acetate stock solutions will find its way to the settler, resulting in an increased effluent phosphate concentration. This has not occurred, however, as evidenced by the PO_4-P measurements of the process effluent which appear in Fig. 10. In fact, the lowest concentrations have been recorded for the trial employing 200 g l^{-1} acetate stock solution. The high initial PO_4-P content of the trial employing 100 g l^{-1} acetate stock solution is due to a process disturbance which occurred on the day prior to performing this trial. Except for this disturbance, the effluent PO_4-P concentrations of all four trials lie close to each other and remain under 1.25 mg l^{-1}.

These results can be explained with the help of Fig. 11. The course of the PO_4-P measurements in tank T2 for the trial employing 200 g l^{-1} acetate stock solution are shown as a broken curve. The thick solid curve segments denote those times when only water had been sent directly from tank T2 to the settler (phases 1, 2 and 6, cf. Fig. 2). Between these times, water enters the settler directly from tank T1, where here it is reminded that the dynamics of this tank approximately mirror those of tank T2 with a half cycle delay. According to this figure, all the phosphate which was released as a result of acetate addition had been once again taken up during the ensuing aerobic period and before the flow of water was directed to the settler. Consequently, the high phosphate peaks appearing in the aeration tanks were effectively invisible to the process effluent.

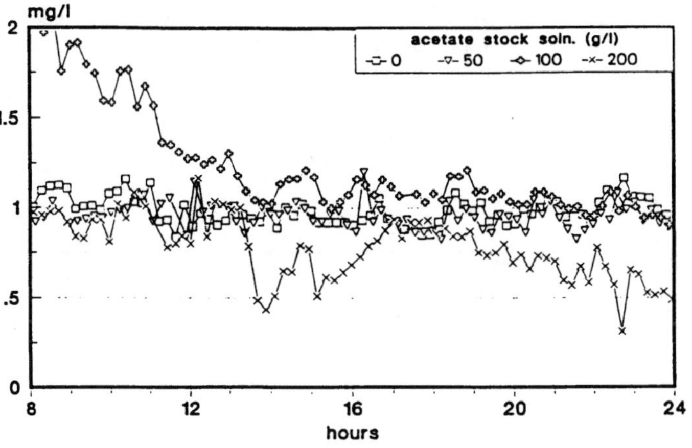

Fig. 10. Effluent PO_4-P concentration measurements recorded during the four days of the control strategy demonstration.

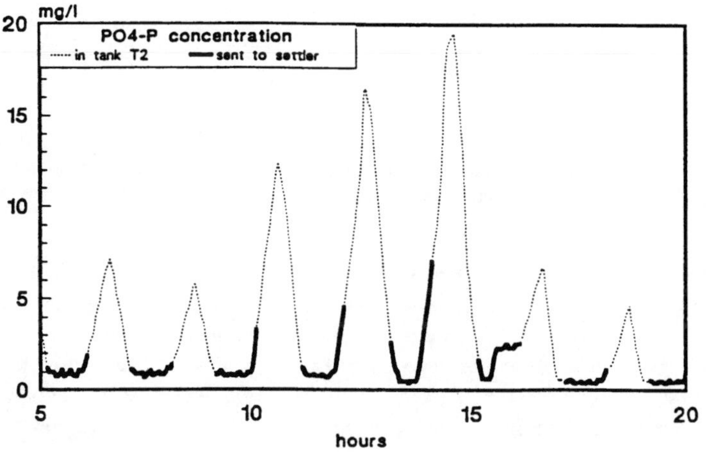

Fig. 11. Course of PO_4-P in Tank T2 pertaining to the control demonstration trial for which 200 g l^{-1} acetate stock solution was employed. The thick curve segments denote those times at which water was directed from tank T2 to the settler.

The rates of phosphate uptake/release, R_{PO4}, calculated from the PO_4-P measurements of all four trial days appear together in Fig. 12. Both the rate of uptake as well as release exhibit a distinct dependency on the acetate stock solution strength. Consider first the course of R_{PO4} during phases 2 and 3 between the 10th and 15th hours when acetate was added. In this figure, the approximate times of phases 2 and 3 are denoted by the bars under the curves. R_{PO4} exhibits a definite downward trend, i.e. in the direction of a greater net phosphate release rate, with increasing stock solution strength. For the trial receiving no COD supplement, R_{PO4} is positive over these time periods, indicating the occurrence of phosphate uptake under anoxic conditions. The general level of R_{PO4} is still positive but to a lesser extent for the trial employing 50 g l^{-1}, suggesting once again that a competition between phosphate uptake and release occurs. R_{PO4} is distinctly negative for the trials employing the two strongest stock solutions.

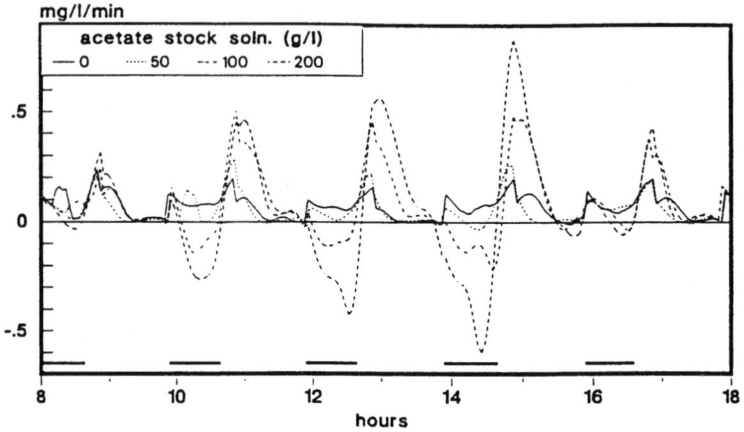

Fig. 12. Rate of phosphate uptake/release, R_{PO4}, derived from the PO_4-P measurements of the four days of the control demonstration. The solid bars denote the times of phases 2 and 3.

The factor of significance which is revealed by this figure is the generally high level of the phosphate uptake rate following phosphate release for the trials employing 100 and 200 g l^{-1} acetate stock solution. The uptake rate is greater than the net rate of phosphate release, and it is for this reason that the phosphate concentration can return to low levels before aeration is terminated and before flow is directed to the settler. This was not the case for the batchwise experiment shown in Fig. 7, for which the maximum uptake rate attains only 50% of the maximum release rate of 0.64 mg $l^{-1}min^{-1}$. The higher uptake rates appearing in Fig. 12 may have to do with the fact that experimentation involving the addition of acetate as well as hydrolysed sludge had been undertaken over the course of several weeks prior to performing the control demonstration. Hence one explanation is an enhancement in the percentage of phosphate-accumulating bacteria in the activated sludge due to the increase in readily degradable organics within the system. A second possibility is that the repeated supplement of COD has caused a general increase in the amount of organics stored by the phosphate-accumulating bacteria throughout the process. Therefore, all the phosphate-accumulating bacteria, including that which entered the tank after the acetate addition during a cycle had been terminated, could contribute to phosphate uptake at a relatively high rate. In the one-shot experiment of Fig. 7, on the other hand, the sludge which entered tank T2 after acetate addition had not been exposed to the extra COD.

Regardless of whether or not one or both of these explanations apply, the fact of the matter is that during and immediately after the time period of experimentation involving supplemental COD, very low levels of residual free phosphate were consistently attained in the aeration tanks at the conclusion of the aerated phase. This resulted in a considerable improvement in effluent phosphate content and in the phosphate removal ability of the pilot plant.

Conclusions

The noteworthy conclusions that can be drawn from the observed pilot plant behaviour presented and discussed above can be summarized as follows:

- The release of phosphate which occurs when adding organic substrate to enhance the denitrification rate of the alternating process will not necessarily lead to a higher effluent phosphate content. The phosphate which is released during a denitrification phase will not appear as free phosphate in the plant effluent as long as it can be re-taken up quickly enough during the ensuing aerated period, as witnessed here in the control strategy demonstration.
- Plant behaviour indicates that the addition of a readily metabolizable organic substrate to the denitrification phase may in fact also improve the biological removal of phosphate. Higher aerobic phosphate uptake rates and lower residual phosphate levels in the aeration tanks were observed during and immediately after the few week time period of experimentation involving COD supplementation.

Finally it should be mentioned that the basis for this presentation has been the experience gained when acetate was used as model carbon source for increasing the denitrification rate. Acetate may not, however, represent the ideal organic substrate from an economic point of view, and alternatives might be found which do not promote phosphate release as strongly. In reference to the results shown in Fig. 8 using acetate stock solution at the weaker strength of 50 g l^{-1}, a satisfactory compromise might also exist whereby sufficient improvement in denitrification rate is gained but with an insignificant release of phosphate.

Acknowledgements

Funding for this investigation is provided by the Center for Biotechnological Process Research of Denmark. The project environmental technician, Ms. A. Nielsen, is gratefully acknowledged for her aid in performing the experimental studies and maintaining the plant and measurement system operational.

REFERENCES

Arvin, E. (1985). Biological Removal of Phosphorus from Wastewater. **CRC Critical Rev. Environ. Control** 15, pp. 25-64.

Einfeldt, J. (1992). The Implementation of Biological Phosphorus and Nitrogen Removal with the Bio-Denipho Process on a 265,000 PE Treatment Plant, **Wat. Sci. Tech.** 25, (4/5), 161-168.

Isaacs, S.H., Søeberg, Kümmel, M. (1992a). On the Monitoring and Control of a Biological Nutrient Removal Plant. Proceedings of the 6th Forum for Applied Biotechnology, Brugge, Belgium, September 24-25, 1992.

Isaacs, S.H., Søeberg, H., Kümmel, M. (1992b). Monitoring and Control of a Biological Nutrient Removal Process: Rate Data as a Source of Information. Proceedings of the ICCAFT 5/IFAC-BIO 2 conference, Keystone, Colorado, March 30-April 1, 1992.

Isaacs, S.H., Henze, M., Søeberg, H., Kümmel, M (1993). Control of an Activated Sludge Nutrient Removal Process through Supplemental Carbon Source Addition. Submitted to IFAC World Congress 1993, 18-23 July, Sydney, Australia.

Kerrn-Jespersen, J.P., Henze, M. (1992). Biological phosphorus uptake under anoxic and aerobic conditions. **Water Research**, **27**, 4, 617-624.

Pedersen, K.M., M. Kümmel, and H. Søeberg (1990). Monitoring and control of biological removal of phosphorus and nitrogen by flow-injection analysers in a municipal pilot-scale waste-water treatment plant. **Analytical Chimica Acta** 238, pp. 191-199.

Nielsen, E., Jørgensen, B., Isaacs, S., Henze, M. (1992). Controlled carbon source addition to an alternating nitrification-denitrification wastewater process. Department of Environmental. Engineering, Technical University of Denmark.

Wentzel, M.C., Ekama, G.A., Marais, GvR. (1992) Processes and Modelling of Nitrification Denitrification Biological Excess Phosphorus Removal Systems - A Review. **Water Sci. Techn.** 25 (6), 59-82

Pergamon

Wat. Sci. Tech. Vol. 28, No. 11-12, pp. 513-521, 1993.

0273-1223/93 $7·00 + 0·00

SIMULTANEOUS REMOVAL OF NITROGEN AND PHOSPHORUS IN INTERMITTENTLY AERATED 2-TANK ACTIVATED SLUDGE PROCESS USING DO AND ORP-BENDING-POINT CONTROL

Kousei Sasaki*, Yasuji Yamamoto**, Kazushi Tsumura***, Shigeru Hatsumata† and Masahiro Tatewaki‡

* *Water Treatment and Bioelectronic Laboratory, Fuji Electric Corporate Research and Development Ltd., 2-2-1 Nagasaka, Yokosuka City, Kanagawa, Japan*
** *Department of Environmental Sanitation, Osaka Prefectural Institute of Public Health, 1-3 Nakamichi, Higashinari-ku, Osaka City, Osaka, Japan*
*** *Division of Global Environment Engineering, Faculty of Engineering, Kyoto University, Yosida, Sakyo-ku, Kyoto, Japan*
† *Public Works Division, Fuji Electric Company Ltd., 12-1, Yurakucho 1-chome, Chiyoda-ku, Tokyo, Japan*
‡ *Central Research Laboratory, Unitika Ltd., 23 Ujikozakura, Uji City, Kyoto, Japan*

ABSTRACT

The 2–tank intermittent aeration method is an anaerobic–aerobic activated sludge process of time–sharing type in which 2 complete mixing reaction tanks are connected in series, and aeration and agitation are periodically repeated in each tank. We have developed a new control system for the process which can secure anaerobic, anoxic and aerobic conditions through a combination of DO and ORP–bending–point (corresponding to termination of denitrification) emergence time control. In the 1st tank, nitrification and phosphorus uptake occur in the aeration period, followed by denitrification and phosphorus release in the agitation. The 2nd tank performs nitrification and phosphorus uptake in the aeration and denitrification in the agitation. One cycle of aeration and agitation is approximately 2 hours. This control system was applied to the test plant (influent flow rate: 225 l/day) for two months under the conditions of HRT 16 hours and temperature 20±2 ℃. We achieved stable and high removal ratios: TOC 94.9 %, T–N 89.4 %, and T–P 95.5 %. We also investigated the mechanisms of nitrogen and phosphorus removal and their material balance.

KEYWORDS

Nitrogen and phosphorus removal; intermittent aeration; DO control; ORP–bending–point control; time sharing; 2–tank activated sludge process; nitrification; denitrification; phosphorus release; phosphorus uptake.

INTRODUCTION

In order to cope with the recent serious problem of eutrophication, various processes for simultaneous removal of nitrogen and phosphorus have been developed. Typical examples of the new processes include the A_2O process, the sequencing batch reactor (SBR) and the intermittent aeration activated sludge process. However, difficulties have been reported in stabilizing the removal efficiency of these two nutrients in these processes. In the A_2O process, while nitrogen removal efficiency was approximately 60% and remained stable during the summer, it was unstable in winter (Minagawa *et al.*, 1989). Mean phosphorus concentration of the effluent was 0.8 mg/l, but it fluctuated from 0.2 to 1.2 mg/l depending on DO concentration in the aeration period during a pilot plant experiment of 7 months in SBR (Hamamoto *et al.*, 1984). The intermittent aeration method has also been considered a suitable process for removing nitrogen only.

To solve the problems related to the instability of nutrients removal, we have been developing a new process in which the introduction of automatic control is effective. An initial experiment of the automatic control using DO and ORP meters as sensors was carried out at an extended aeration plant having two aeration tanks connected in series (design capacity 625 m^3/day). In that experiment, the two aeration tanks were operated using the intermittent aeration method; concentrations of nitrogen and phosphorus over a year were kept under 5 and 1 mg/l respectively (Yamamoto *et al.*, 1992). Thus, the 2-tank intermittent aeration activated sludge process was found to be a feasible means of nutrient removal. In this paper, we will describe the control method highly improved after the first experiment in order to achieve more stable and efficient simultaneous removal of nitrogen and phosphorus, and will then present the results of experiments performed at a test plant.

CONTROL METHOD

Figure 1 shows the DO and ORP variation for each tank per cycle (comprising both the aeration and agitation) and the biological reactions which accompany nitrogen and phosphorus removal. Figure 2 exemplifies by the experimental apparatus the system configuration for the 2-tank intermittent aeration activated sludge process. As Figure 1 indicates, nitrification and phosphorus uptake occur in the aeration period at the 1st tank, followed by denitrification and phosphorus release during agitation. The 2nd tank performs nitrification and phosphorus uptake in the aeration and denitrification in the agitation. The aeration time in the 1st tank is different from that of the 2nd tank, though the time of one cycle (2 hours) is the same for both tanks.

Control method of the 1st tank. The main role of the 1st tank is to secure the phosphorus release. In general, a certain portion of the entire amount of activated sludge must be placed under anaerobic conditions for phosphorus release in an enhanced biological phosphorus removal. We assign 25 % (this set value can be changed depending upon the influent quality) of the activated sludge for that purpose, and allow it 60 minutes in the 1st tank. Consequently, since one cycle is 120 minutes, the aeration and denitrification in the agitation phase must be performed in the remaining 60 minutes. We set the cycle length at 2 hours for the following reasons. If the cycle was longer than 2 hours, the variation of water quality in the tanks would become larger, thereby deteriorating effluent quality, and the response of the automatic controls to changes in influent quality would also become slow. On the other hand, detection of the bending point P1 (described later) would be difficult in a shorter cycle.

A detailed control method of the 1st tank is as follows. A combination of the DO and ORP-bending-point (corresponding to the completion of denitrification) emergence time control is performed in this tank. DO1 is controlled so that the sum T_{P1} of the aeration period T_{N1} which is preliminarily set at 30 minutes, and the denitrification period T_{D1} of the agitation phase becomes equal to a predetermined period T_{P1S}. After 30 minutes have elapsed, the aeration is stopped and the agitation begins, i.e. denitrification starts. The denitrification period T_{D1} is measured as a period from the start of the agitation to the appearance of the bending point P1. Phosphorus is then released until an aeration restart signal is delivered from the 2nd tank. Since the period T_{D1} is proportional to the concentration of NO_3-N that has been generated by the nitrification, the denitrification period T_{D1} can be controlled by controlling nitrification during the aeration period T_{N1}; as a result, T_{P1} can be controlled to T_{P1S}. DO1 is thus controlled during the aeration period because the activity of nitrifying bacteria depends on the DO concentration. DO1 is adjusted according to Equation (1):

$$DO1_n = DO1_{n-1} + K1(T_{P1S} - T_{P1n-1}) \quad (1)$$

where

$DO1_n$ = set DO in the current cycle,
$DO1_{n-1}$ = set DO in the previous cycle,
T_{P1S} = preset value of the sum of the aeration period and the denitrification period,
T_{P1n-1} = sum of the aeration period and the denitrification period in the previous cycle,
$K1$ = constant.

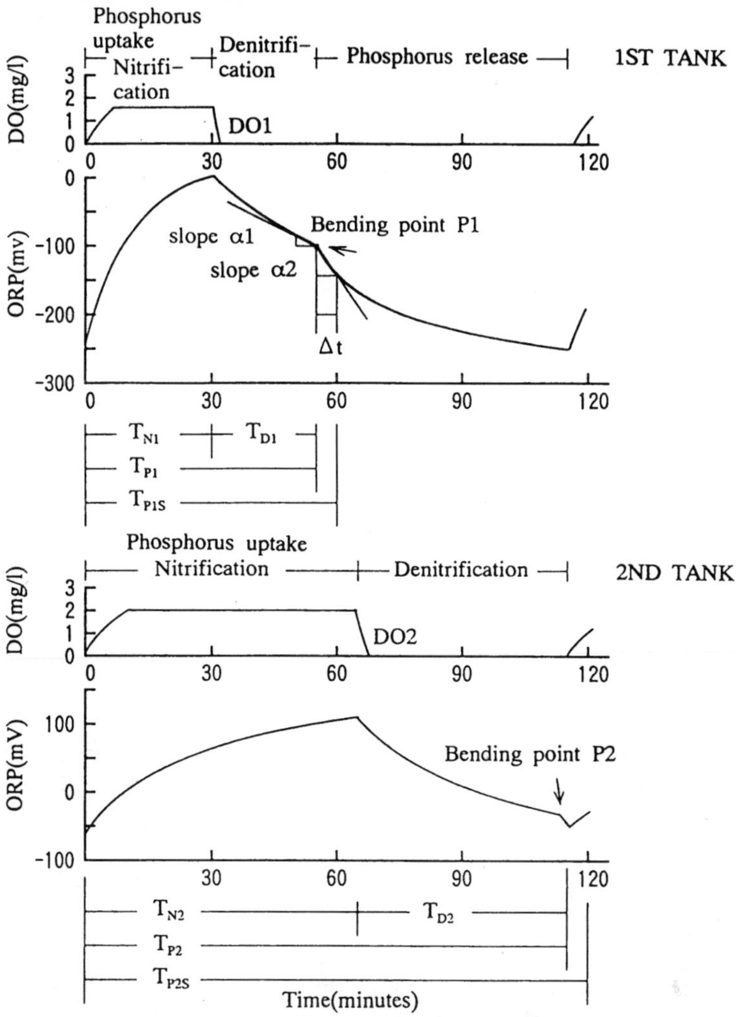

Fig. 1. DO and ORP variation during one cycle

Control method of the 2nd tank. The main role of the 2nd tank is to remove residual nitrogen and phosphorus, and to maintain the length of one cycle at the preset value. In this tank, the sum T_{P2} of the aeration period T_{N2} and the agitation period T_{D2} is controlled to assure that sufficient phosphorus release has occurred in the 1st tank when the denitrification in the 2nd tank is completed. That is, the sum period T_{P2} is controlled to the period T_{P2S} that is preset at one cycle period. The agitation period T_{D2} is obtained through the same procedure as described above. The aeration period T_{N2} is adjusted according to Equation (2):

$$T_{N2n} = T_{N2n-1} + K2(T_{P2S} - T_{P2n-1}) \quad (2)$$

where

T_{N2n} = aeration period of the current cycle,
T_{N2n-1} = aeration period of the previous cycle,
T_{P2S} = preset value of the sum of the aeration period and the agitation period,
T_{P2n-1} = sum of the aeration period and the agitation period of the previous cycle,
K2 = constant.

DO2 is set at 2 mg/l so as not to impede the nitrification. When the bending point P2 is detected, the 1st and 2nd tanks return to aeration mode at the same time and, thereafter, the above control is repeated.

Detection of ORP-bending-point. The bending point P1 shown in Figure 1 is detected in the following manner. The slope of the ORP curve is calculated with a time increment Δt. Then, $\alpha 2/\alpha 1$ is calculated, where $\alpha 1$ and $\alpha 2$ represent the respective slopes. The ratio $\alpha 2/\alpha 1$ is close to unity while the slope of the ORP curve does not vary, taking on a larger value when the slope suddenly changes at the bending point P1. Therefore, employing a threshold of β, the bending point P1 can be detected when the condition $\alpha 2/\alpha 1 > \beta$ is satisfied. P2 is detected in the same manner.

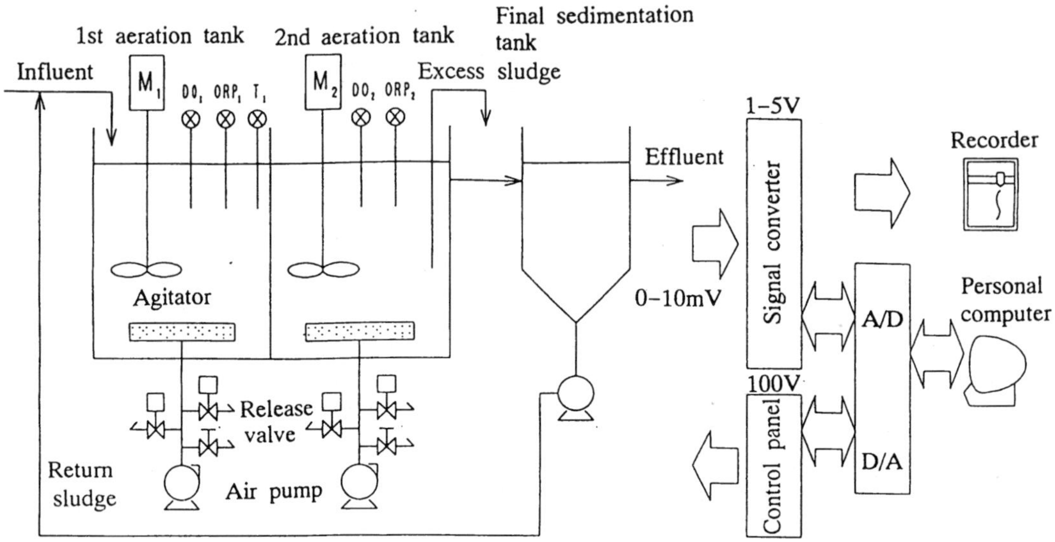

Fig. 2. Schematic diagram of the test plant

EXPERIMENTAL METHODS

Apparatus. As shown in Figure 2, the test plant primarily consisted of 2 aeration tanks (76,74 l) and a final sedimentation tank (36 l). Each of the aeration tanks was equipped with an independently operable agitator (30 rpm) and an air pump having electromagnetic air release valves for DO control. We used DO and ORP electrodes for laboratory use, which have no cleaning mechanism. The operation and control of the test plant were performed by a control panel linked to a personal computer, which was used for calculation.

Conditions. The experimental conditions are listed in Table 1. Hydraulic retention time was 16 hours with constant influent flow rate at 225 l/day, and the water temperature was maintained at 20±2 ℃ by cooling and heating equipment for laboratory use (not shown in Figure 2). We set one cycle T_{P2S} at 120 minutes and the sum T_{P1S} of aeration period T_{N1} and denitrification period T_{D1} at 60 minutes securing 60 minutes of phosphorus release time. MLSS was 4730 mg/l on the average. SRT was 24.0 days, and consequently A-SRT became 8.5 days. Excess sludge was wasted directly from the 2nd tank once every one or two days. The feed water was artificial sewage prepared automatically once a day by mixing tap-water, sugar, sodium acetate, soap, detergent, human feces and cafeteria wastewater. Nitrogen and phosphorus concentrations were adjusted by adding ammonium chloride and trisodium phosphate to the feed water. Sodium hydrogen carbonate was also added to increase alkalinity. For the water quality, grab samples were analyzed according to the Sewage Testing Methods of the Japan Sewage Works Association, and ion chromatography method was used to analyze NO_3-N and PO_4-P.

TABLE 1 Experimental Conditions

Tank	Items	Unit	Value
Aeration tank	Hydraulic retention time	(h)	16.0
	Temperature	(℃)	20±2
	1st aeration tank		
	Aeration T_{N1}	(min)	30
	Denitrification T_{D1}	(min)	30
	Phosphorus release	(min)	60
	2nd aeration tank		
	Sum of aeration and denitrification T_{P2S} (set period for one cycle)	(min)	120
	MLSS	(mg/l)	4730
	SVI	(-)	128
	SRT	(day)	24.0
	A-SRT	(day)	8.5
	T-N loading	(kg/kg.ss.day)	0.012
	T-P loading	(kg/kg.ss.day)	0.0013
Final sedimentation tank	Hydraulic retention time	(h)	3.9
	Surface-loading rate	(m^3/m^2.day)	4.5
	Recirculation ratio	(%)	100

RESULTS AND DISCUSSION

Variation of water quality in one cycle. Figure 3 shows the variation of water quality in the 1st and 2nd aeration tanks during one cycle. In the 1st tank, D-TOC decreased during the aeration period and increased as organic matter was supplied from the influent in the agitation period. NH_4-N was reduced from 11.5 to 7.4 mg/l by nitrification accompanied by an increase of NO_2-N+NO_3-N during the aeration period, and was increased due to the nitrogen in the influent during the next phase. NO_2-N+NO_3-N disappeared in 30 minutes by denitrification. PO_4-P sharply decreased in the aeration period from 9.2 to 0 mg/l, indicating phosphorus uptake by the activated sludge, and was maintained at 0 mg/l until the denitrification was completed. After that, PO_4-P increased because of the phosphorus release and the amount supplied in the influent. Further, the ORP-bending-point P1 appeared clearly at the end of denitrification.

In the 2nd tank, D-TOC was lower and more stable than that in the 1st tank. NH_4-N became 0 mg/l before the end of the aeration period owing to the high nitrification capacity of the process and increased to only 3.2 mg/l by the end of the agitation. NO_2-N+NO_3-N increased during aeration, and then decreased slowly as a result of endogenous denitrification until the end of the cycle when the ORP-bending-point P2 was detected. PO_4-P remained at 0 mg/l the entire time. We think this phenomenon is due to the occurrence of phosphorus uptake under aerobic conditions and limited uptake under anoxic conditions (Kerrn-Jespersen *et al.*, 1993). We calculated the material balance and found that the activated sludge in the 2nd tank had absorbed phosphorus at the rate of 0.30 mg PO_4-P/g.mlss.h. This figure was identical to the phosphorus uptake rate at 0.29~0.32 mgPO_4-P/g.mlss.h obtained under anoxic conditions by Kerrn-Jespersen *et al.* In fact, we confirmed this type of phosphorus uptake at a different experiment later. Therefore, these phosphorus uptake mechanisms were considered to be contributing the high phosphorus removal efficiency.

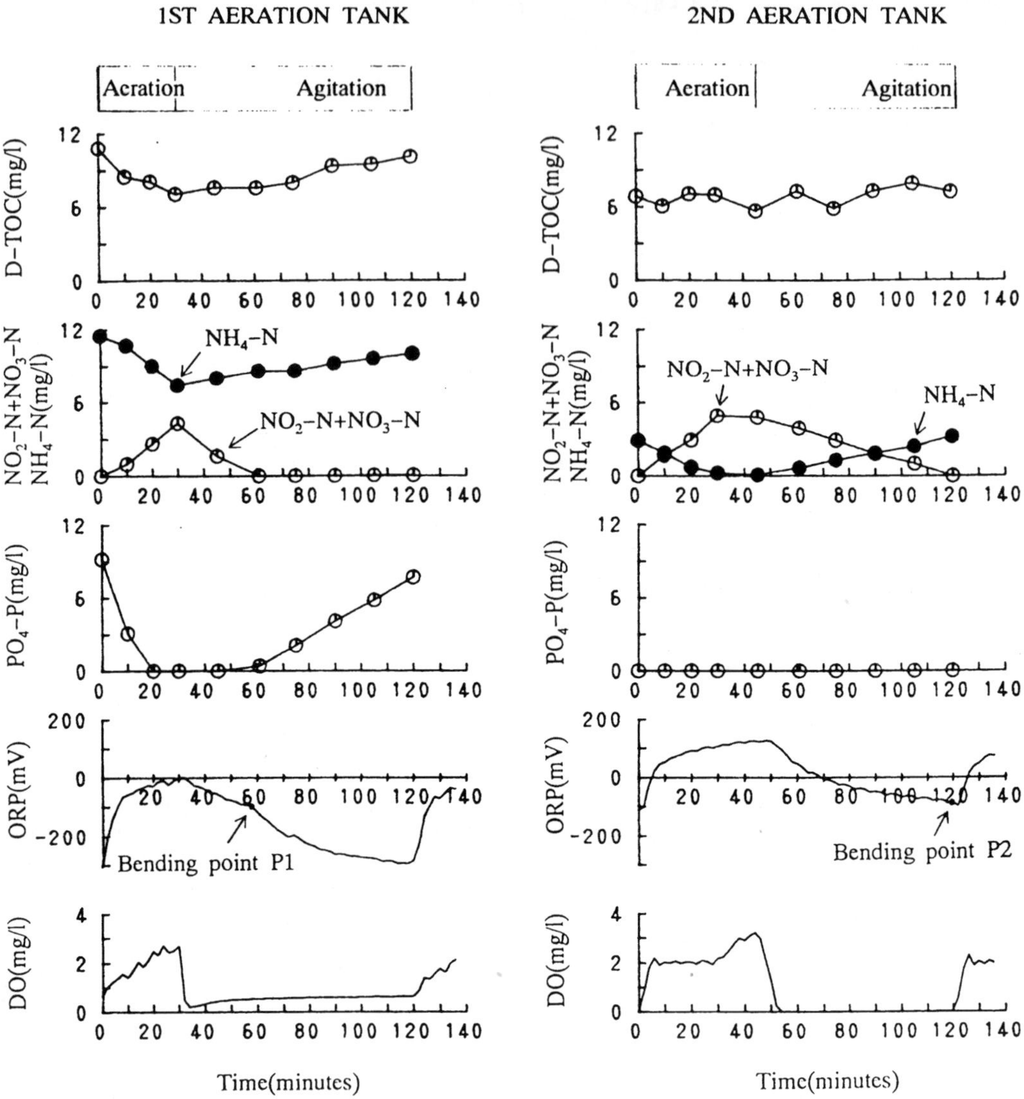

Fig. 3. Variation in water quality over one cycle

Results of automatic control. Figure 4 shows the concentrations of TOC, T–N and T–P in influent and effluent during the 2 months that the automatic control experiment was conducted. Detailed water qualities are listed in Table 2. As shown in Figure 4, the concentrations of TOC, T–N and T–P in effluent were low and quite stable despite their variable levels in the influent. TOC concentration was 126.7 mg/l in the feed water and decreased to 6.5 mg/l in effluent (removal ratio 94.9 %). Similarly, T–N at 37.8 mg/l went down to 4.0 mg/l (89.4 %), while T–P went from 3.97 mg/l to a very low 0.18 mg/l in effluent (95.5 %). These results indicate that, when there is no diurnal change of the influent, simultaneous removal of nitrogen and phosphorus at high removal ratios is possible in the 2–tank intermittent aeration activated sludge process through this control method.

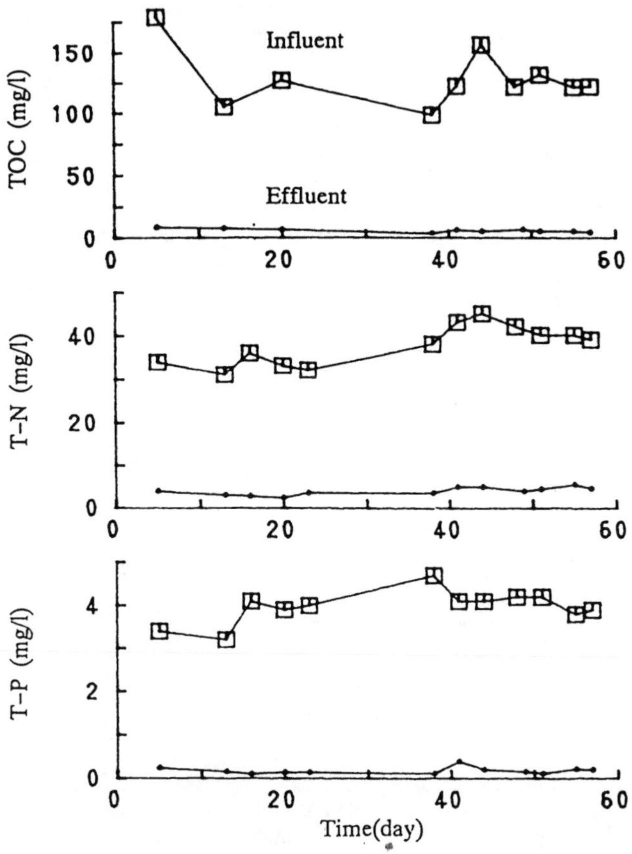

Fig. 4. Results of the control experiment

TABLE 2 Average Quality of Influent and Effluent

		Influent	Effluent	% removal
PH	(-)	7.1	7.0	-
COD	(mg/l)	97.2	10.6	89.1
D-COD	(mg/l)	30.8	-	-
TOC	(mg/l)	126.7	6.5	94.9
D-TOC	(mg/l)	38.8	-	-
SS	(mg/l)	145.1	3.8	97.4
T-N	(mg/l)	37.8	4.0	89.4
NH_4-N	(mg/l)	28.9	1.4	-
NO_2-N+ NO_3-N	(mg/l)	0.0	2.1	-
T-P	(mg/l)	3.97	0.18	95.5
PO_4-P	(mg/l)	1.51	0.04	-

Material balance of nitrogen and phosphorus. Figure 5 shows the relative material balance of nitrogen and phosphorus assuming the amount of those entered to the process as 100 %. In the 1st tank, where the hydraulic retention time is 8 hours and the predominant growth of activated sludge takes place, 56.9 % of the nitrogen was removed. We subtracted 24.6 %, the amount of nitrogen contained in the excess sludge (nitrogen content 8.0 %), from that figure and obtained 32.3 % as the amount of nitrogen removed through denitrification in the 1st tank. The result showed that almost the same amount of nitrogen was removed by denitrification in both tanks, although the denitrification period of 30 minutes in the 1st tank was about half that in the 2nd tank, as shown in Figure 3. We measured the denitrification rate in both tanks to explain this and found that the rate in the 1st tank was 1.5 mg/g.mlss.h, which was twice as large as that of 0.7 mg /g.mlss.h in the 2nd tank. Entire nitrogen removal by denitrification was 62.8 %. Phosphorus was removed by 94.4 % as excess sludge in which phosphorus content was 3.1 %.

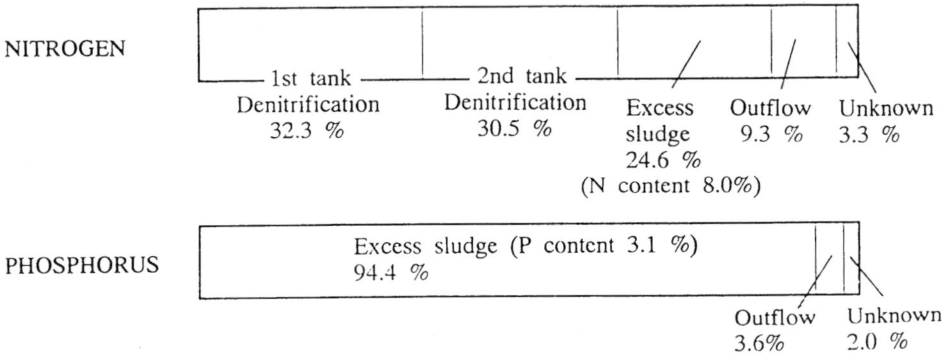

Fig. 5. Material balance of nitrogen and phosphorus in the process

Maintenance of electrodes. We had to wash slime from the top of the electrodes with water or soft papers an average of once a week for the DO meters and once every two weeks for the ORP meters. This was necessitated by our choice of electrodes for laboratory use, which have no cleaning devices. The DO meters were calibrated about once a month. If DO meters for process control with automatic cleaning devices were applied, once-a-month calibration would be enough. ORP does not require strict maintenance because the control method depends on the shape of the ORP curve (i.e. the bending point) rather than measured values. If ORP meters for process control with automatic cleaning devices were used, the maintenance interval could be extended beyond once a month.

SUMMARY

As described above, we have developed a new control method for simultaneous removal of nitrogen and phosphorus, and demonstrated the stable and high removal efficiency of the process through the test plant experiment of the intermittently aerated 2–tank activated sludge process. Four major biological reactions for nutrients removal, nitrification, denitrification, phosphorus release, and phosphorus uptake, clearly took place in the two aeration tanks during the course of one operation cycle of aeration and agitation; moreover, set reaction periods could be changed in accordance with the change of operation conditions. In that sense, we regard the 2–tank intermittent aeration process using this control as a time–sharing type A_2O process which is more flexible to variable operating conditions, including water quality and temperature, than the ordinary A_2O process in which the volume for major reactions is fixed. Another important characteristic of this method is its applicability to existing activated sludge processes with longer hydraulic retention times in which only minor structural changes are required.

We are currently developing a further improved control method in which the aeration period T_{N1} of the 1st tank varies under DO control of 2.0~2.5 mg/l maintaining maximum nitrification rate (Sasaki *et al.*, 1993). Moreover, we are planning a demonstration experiment at a larger plant using sewage to acquire detailed operation and design data. We believe this process is suited for small– or medium–scale sewage treatments.

REFERENCES

Minagawa, T., Yuasa, K., Takagi, T. (1989). The influence of rain fall on A_2O process and utilization of raw sludge as a counterplan for rain (in Japanese). *J. Japan Sewage Works Association*, **26**, 301, 43–53.

Hamamoto, Y., Okubo, Y., Hayakawa, N. (1984). Simultaneous removal of nitrogen and phosphorus in sequencing batch reactor with continuous feed (in Japanese). *J. Water and Waste*, **26**, 12, 1279–1283.

Yamamoto, Y., Tsumura, K., Nakano, J. (1992). Two stage intermittently aerated activated sludge system for simultaneous removal of nitrogen and phosphorus (in Japanese). *J. Japan Society on Water Environment*, **15**, 8, 541–553.

Kerrn–Jespersen, J., Henze, M. (1993). Biological phosphorus uptake under anoxic and aerobic conditions. *Water Research*, **27**, 4, 617–624

Sasaki, K., Yamamoto, Y., Tsumura, K., Tatewaki, M. (1993). Simultaneous removal of nitrogen and phosphorus in intermittently aerated 2–tank activated sludge process using DO and ORP –bending–point control (in Japanese). *Proceedings of the 30th Japan Sewage Works Associations annual technical conference (in press).*

Pergamon

Wat. Sci. Tech. Vol. 28, No. 11-12, pp. 523–530, 1993.

0273–1223/93 $7·00 + 0·00

ALTERNATIVE METHOD FOR THE CONTROL OF BALANCING TANK AT A WASTEWATER TREATMENT PLANT

A. J. Mather* and I. S. Shaw**

* *Watson Edwards Incorporated, P.O. Box 98867, Sloane Park, 2152, South Africa*
** *Faculty of Engineering, Rand Afrikaans University, South Africa*

ABSTRACT

Wastewater Treatment Plants are often designed with a balancing tank upstream of the bioreactors. The purpose of the balancing tank is to act as a buffer against large fluctuations in inflow volume and pollutant concentration. This paper presents the options available for control of the balancing tank. The paper also presents the results of simulation tests conducted to determine the effectiveness of a modelfree, neural-network-based controller, when used for balancing tank control.

KEYWORDS

Biological nutrient removal; balancing tanks; wastewater treatment; design; aeration; neural networks; dald algorithm.

BACKGROUND AND PROCESS THEORY

Modern Wastewater Treatment Plants utilise the biological nutrient removal process for treating domestic wastewater. The heart of the process is the biological reactor where aerobic bacteria remove biodegradable organic matter, nitrogen and phosphorus from the wastewater to produce an unpolluted effluent. The key control parameter is the dissolved oxygen concentration in the bioreactor. The pollutant load in the influent to the bioreactor determines the oxygen demand of the bacterial mass. To maintain the correct dissolved oxygen concentration the aeration capacity has to be adjusted accordingly.

Two methods of aeration are used for aerating the sludge within the bioreactor module. The first method, known as FBDA or "Fine Bubble Diffused Aeration", utilises a network of pipes with diffusers laid at the bottom of the reactor. This network of pipes is connected to an air header fed by a number of large low-pressure high-volume air blowers. The reactor is divided into a number of aeration zones and the flow of air into each zone is controlled by a control valve feeding that zone. The level of dissolved oxygen in the zone is measured by means of a dissolved oxygen sensor and this signal is fed into a PID loop controller which in turn modulates the air control valve in such a way as to maintain a fairly constant dissolved oxygen level within the basin.

The second method of aeration, known as "Surface Aeration", utilises a number of surface- mounted mechanical aerators which fling the biomass into the air allowing them to entrain oxygen. The control of the level of dissolved oxygen in an aeration control zone is achieved by the switching of aerators and mixed liquor recycle pumps on and off.

PROBLEM STATEMENT

The inflow of wastewater or sewage into a treatment works is characterised by large variations in both flow and concentration (of pollutant). In the absence of any load equalisation upstream of the reactors, the oxygen demand in the reactors would undergo large fluctuations. These large fluctuations would make control of the oxygen supply to the reactors extremely difficult. In the case of surface aeration, where control is achieved by switching on and off of the aerators, control is not only difficult but would be harmful to the mechanical equipment and would lead to significant increases in power consumption. As a result of these difficulties, it has become common practice to construct a balancing or equalisation tank upstream of the reactors. This tank is used as a buffer tank to dampen out the effect on the bioreactors of large changes in inflow to the plant. It is necessary to optimise the use of the tank so that it provides a fairly constant outflow to the reactor, whilst at the same time not running dry or overflowing. In view of the non-linearity of the control function a simple PID loop controller cannot be used in isolation. There is also an inverse relationship between volume and pollutant load and generally the load decreases as the volumetric flow into the works increases. Too large a tank would result in premature biological reactions and the settling of solids before reaching the bioreactors.

PRACTICAL PROBLEMS ASSOCIATED WITH BALANCING TANK CONTROL

Inflow variations

The problem which has already been discussed is the large variations in inflow. Not only does the inflow vary during the day, but the weekday daily inflow pattern differs from the inflow pattern over a weekend. In addition, Summer inflow patterns are different to Winter inflow patterns. Storm conditions also have to be catered for by the control algorithm.

Outlet control valves

As the outlet flow control valves are normally fairly large (600 mm - 1000 mm), their selection is problematical. Also a butterfly type valve cannot be used as rags or plastic bags can accumulate around its shaft. The correct valve from a process and mechanical point of view is therefore a gate valve or a diaphragm valve. In contrast, both gate valves and diaphragm valves have very poor control characteristics in that they are extremely non-linear. To aggravate matters, these valves need to operate over a fairly large range and therefore it is not possible to operate them over a linear part of the control curve. Any control system has to cater for the non-linear response from the control valves.

Environmental considerations

A wastewater treatment works is a fairly harsh environment characterised by high quantities of dust and a corrosive atmosphere. In addition, a high incidence of lightning is experienced at most plants in South Africa. Electronic instruments and controllers therefore have to be carefully selected, housed and protected to ensure the long-term reliability of the control system.

CONTROL METHODS PREVIOUSLY USED

Manual control

The first method of control used, and the most common method in use in Southern Africa at present, was manual control. The plant operator would control the outflow from the balancing tank by manually adjusting the outlet control valve. He would utilise his past experience in terms of knowing trends in daily inflow patterns, as well as the current level in the balancing tank, to determine the required outlet flow.

Although an experienced operator can achieve very good results with this method, an inexperienced operator can lead to disaster. This method of control also requires almost constant monitoring by the operator and is therefore undesirable.

Dold algorithm

In an attempt to automate the functions performed by the operator, Dold *et al* developed an algorithm to predict the required outflow from the balancing tank. This algorithm uses historical trends of daily inflows and the current level in the balancing tank to predict the required outflow from the balancing tank. An error function is then used to correct the historical data for the next iteration of the programme. The algorithm was tested on a simulator and favourable results were achieved. However, owing to practical problems with the microprocessor hardware and lightning strikes, the algorithm was not successfully implemented on a wastewater treatment plant.

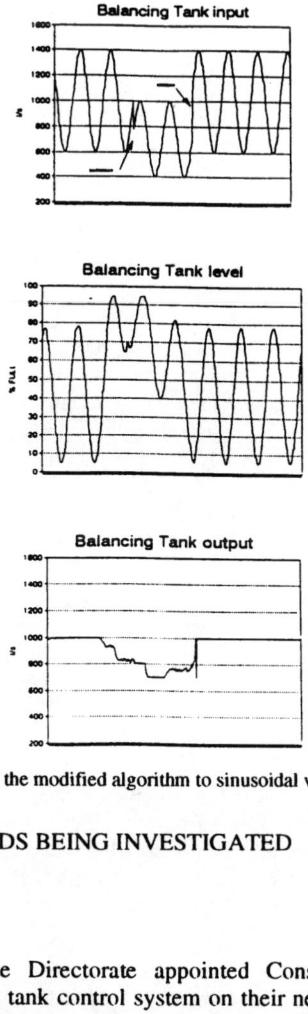

Fig. 1. Response of the modified algorithm to sinusoidal variations in input.

CONTROL METHODS BEING INVESTIGATED

Modified Dold algorithm

The Johannesburg Water and Waste Directorate appointed Consulting Engineers Watson Edwards Incorporated to implement a balancing tank control system on their new Northern Works Unit 4 Treatment Plant. Their planned approach is loosely based on the Dold Algorithmic Model (Dold 1992). Additional checks have been implemented to prevent tank overflow and the balancing of pollutant concentration has been given a reduced weighting. The modified algorithm has been linked to a supervisory control and data acquisition software package to allow operator interfacing, graphic representation of data and the trending of critical parameters. This system is scheduled to be installed on site in early 1993. Figure 1 shows the response of the algorithm to a sinusoidal variation of the inlet flow.

Neutral network controllers

Whilst Watson Edwards Incorporated were in the early design stage with their balancing tank control system, they provided data to the Intelligent Control Research Group, Faculty of Engineering, of the Rand Afrikaans University. Based on this data, the Research Group launched a feasibility study to determine whether or not an intelligent, model-free, neural-network-based controller could control the balancing tank as well as or better than either the expert human operator or the Dold Algorithm (Dold 1992). In this context, model-free means the absence of a fixed mathematical model. In this study, only steady-state operation was considered.

What are neural networks?

As the name indicates, a neural network resembles biological nervous systems as we currently understand them. To distinguish them from their biological counterparts, they are often referred to as "artificial" neural networks. In contrast to traditional computing machines which rely on one or more relatively complex processors, neural networks consist of a large number of simple computational units, so-called neurons, highly interconnected with communication pathways. In essence, a neural network is a massively parallel self-adaptive dynamic system which can alter its inner structure incrementally in time, until it achieves the desirable performance. This structural change occurs during the so-called training or learning phase, whereby the network is exposed to known patterns of input-output data rather than following programmed rules. In turn, the thus trained network can recognize and classify unknown or incomplete input data patterns. While conventional computers are good in performing complex calculations accurately and handle data sequentially, neural networks are good at shape recognition, speech synthesis, clean-up of noisy signals, perceiving partially hidden objects in a complicated scene, or recognizing subtle relationships which are not obvious to humans. Another advantage is that they are inherently fault and noise tolerant. In a conventional computer, if one part fails, usually the whole system breaks down, while in a neural network fault tolerance is due to a distributed processing and memory structure: if one neural element fails, its erroneous output would be "swamped" by the correct outputs of its neighbouring elements. Noise tolerance is attributable to the averaging effect of the parallel processing of input data. These favourable characteristics prompted new applications in adaptive control systems for model identification: when trained with a set of acceptable known input output data pairs, neural networks can accurately estimate the correct outputs to inputs not contained in the original input training set. Since this estimate is based on the actual data, no assumption of linearity for the process being controlled is necessary, in contrast to mathematical models where linearity is often assumed to make the mathematics tractable. Kaya (1992), Lawrence and Lawrence (1990), Nguyen and Widrow (1990) and Willis *et al* (1991) provide additional information on neural networks and their recent application in model identification and control theory.

Figure 2 illustrates the basic architecture of a simple neuron having a number of inputs and a single output. The neuron calculates its output by finding the weighted sum of its inputs. Each input connection has a weight w in the net total effect of the inputs. In a network of such neurons, at any point in time t, the ith neuron adds up the weighted inputs to produce an *activation value* $a_2(t)$, whose value is the algebraic sum of the weighted inputs at that time (excitatory inputs are positive while inhibitory inputs are negative). The connection from the jth neuron's output to the kth neuron's input is represented by w_{kj}. Thus in general, the throughput is expressed as:

$$net_i = \sum_{j=1}^{N} w_{ij} o_j$$

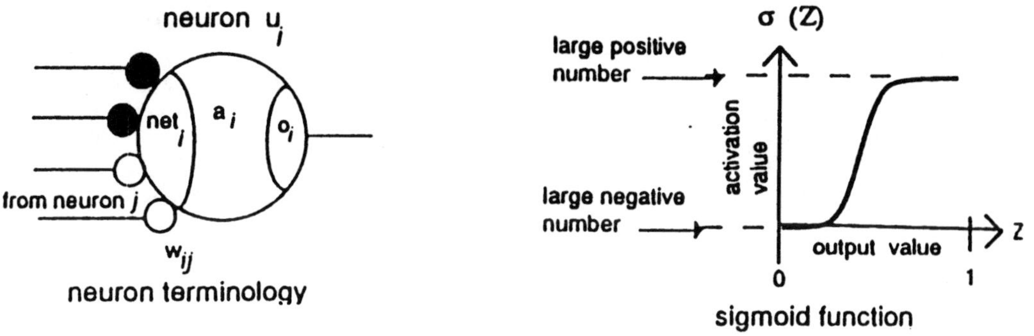

Fig. 2. Simple neuron architecture.

where o_j is the jth output of N outputs. In turn, the activation value is passed through a transfer function f_i which produces the actual output for the neuron for that time. For example, the transfer function may be a *sigmoid function* which has a low and a high saturation limit and a proportional range in between them. The function is "0" when the activation value is a large negative number. The sigmoid function is "1" when the activation value is a large positive number, and makes a smooth transition in between. A neural network may have many attributes which define its architecture, method of learning, means of handling data, and its use of time. There are many different neural network models described in the literature with different attributes. The most popular network is the so-called *backpropagation model* (Lawrence and Lawrence, 1990) representing a specific learning scheme. While training this network with known input-output patterns, the *weights* are changed so that the network will eventually produce the matching output pattern when given a corresponding input pattern of that pair.

REALIZATION OF NEURAL NETWORKS

Most neural networks in use today are software simulations with neurons represented by memory locations in a conventional computer. For example, commercially available neural network program packages for PC-type computers give the user the appearance of running normal computer programs. If the neural network is used in a control application, the network inputs and outputs can be received or transmitted respectively by means of appropriate computer interfaces.

The salient feature of neural-network-based models is that an accurate mathematical model of the plant to be controlled is not necessary, thus plant non-linearity is of no consequence. The model is built up during the training phase by observing known input-output data pairs; thus the model structure and parameters are "customized", as it were, to be optimal for the specific application involved.

FEASIBILITY STUDY FOR BALANCING TANK CONTROL USING A NEURAL NETWORK

The use of neural-network technology is thus far limited to a few selected fields such as speech and voice recognition. The specific application as a basic controller as suggested above may open new doors for use in other more complex applications.

As stated earlier, the conditions for control are: the tank may not run dry; it may not overflow; the outflow rate is to be kept as constant as possible; sharp changes in rates of outflow are to be avoided.

Measurements were taken at the Goudkoppies Wastewater Treatment Plant near Johannesburg where the balancing tank had a volume 22,500 m^3, a length of 150 m, width of 50 m, and a depth of 3 m.

The feasibility study set out first to answer the questions: **what should the optimum outflow rate be on a specific day and how should the balancing tank function to ensure this optimum?**

The tank achieves balancing by storing those inflow rates that exceed the average daily inflow rate. In turn, the stored volume is released during periods whenever the inflow rate falls below the daily average. This average inflow rate is used in setting up a mass-flow hydrogram in order to determine the minimum wastewater volume in the tank (referred to as the tank volume) needed for flow balancing. The optimum outflow rate can then be determined from the mass-flow hydrogram where the cumulative volume is plotted against time. In this way the connection between the three state variables (inflow rate, outflow rate, wastewater volume in the tank) was established. This data was later used for "training" of the neural network.

TRAINING AND RUNNING THE NEURAL NETWORK

The commercial software package simulating the neural network employed in this study uses the backpropagation training algorithm (Lawrence and Lawrence, 1990). Training the neural network with the three state variables as inputs required measured data. Various networks were built using combinations and differing amounts of data. The trained networks were then tested.

The most reliable network trained from the above data was exposed or tested with data used by Dold *et al* in 1979. Typical responses of a human operator and the neural network are shown in Figure 3, while those of the Dold Algorithm and the neural network are presented in Figure 4.

These preliminary results show that the outflow provided by the neural-network controller is as good as or better than the output provided by a human operator and the Dold Algorithm.

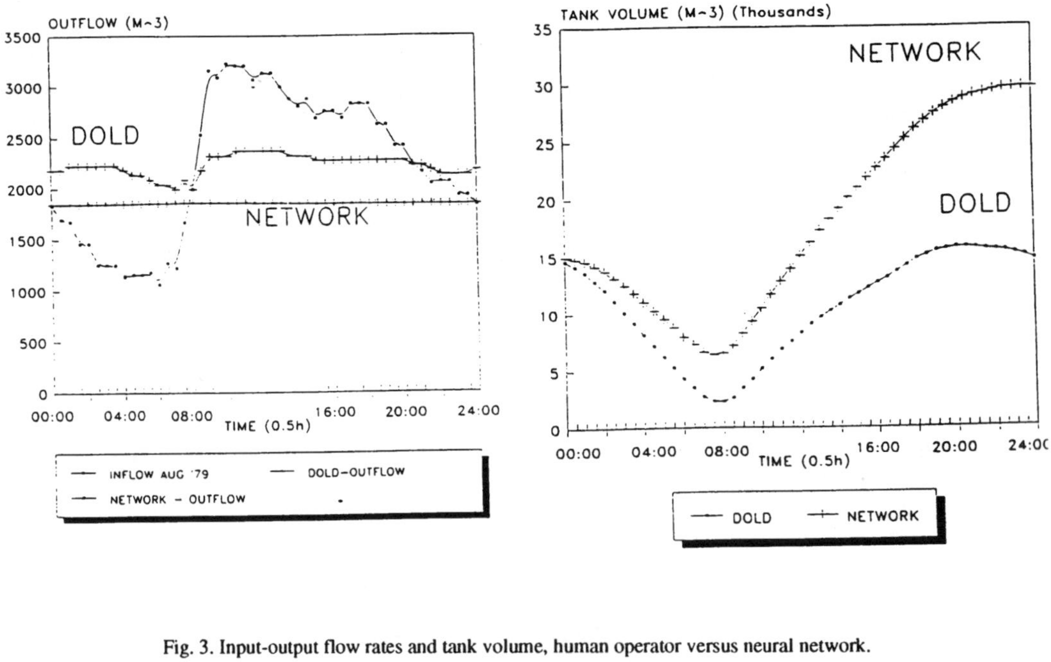

Fig. 3. Input-output flow rates and tank volume, human operator versus neural network.

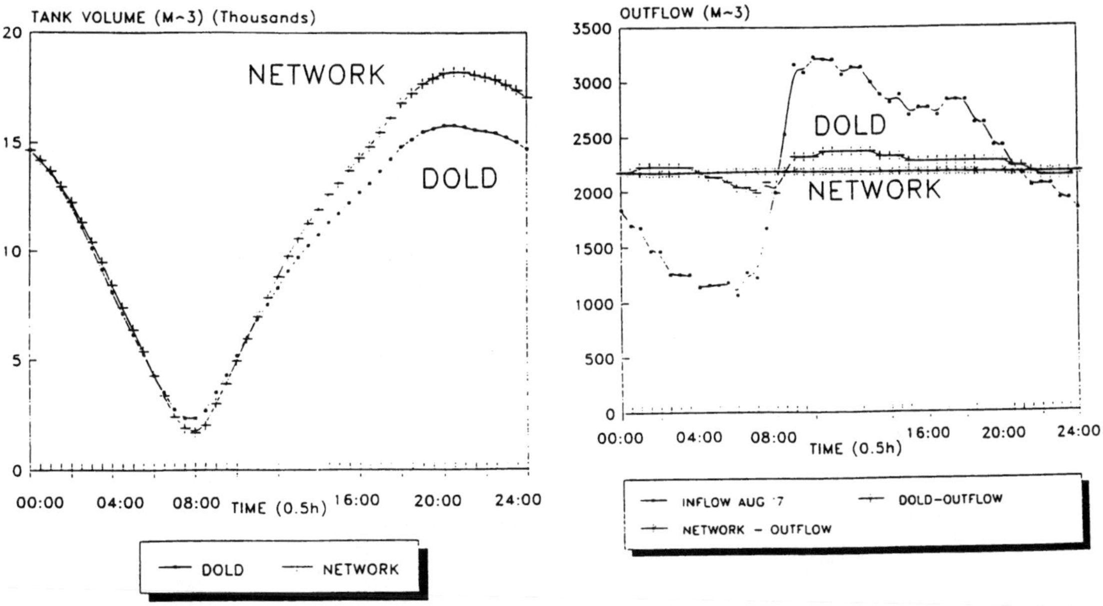

Fig. 4. Input-output flow rates and tank volume, mathematical model versus neural network.

CONCLUSION

Although experienced plant operators can achieve satisfactory results in controlling the outflow from a balancing tank, this method of control is labour intensive and prone to poor repeatability.

Mathematical algorithms have been developed to replace the operator and favourable results have been achieved during simulation tests with these algorithms.

Preliminary results from a feasibility study have however shown that using an intelligent model-free neural-network-based controller for the control of a balancing tank at a wastewater treatment plant can yield results as good as or better than a human operator or a mathematical algorithm. From the simulations all criteria for tank operation were satisfied: the simulated outflow of the neural network model tended to be more constant than that which was achieved with the other two methods, with no under- or overflow of the tank.

ACKNOWLEDGEMENTS

The authors wish to extend their gratitude to the staff of the following organisations who supported and/or assisted with this research: (1) The City of Johannesburg's Water and Waste Directorate; (2) Rand Afrikaans University; and (3) Watson Edwards Incorporated.

REFERENCES

Dold, P. L. (1992). Design and Control of Equalisation Tanks. Research Report W42, University of Cape Town.

Kaya, A. (1992). An intelligent PID Control Algorithm by Neural Networks". *IFAC Symposium on Intelligent Components and Instruments for Control Applications*, Malaga, Spain, May 1992, pp 417-420.

Lawrence, M. and Lawrence, J. (1990). *Brainmaker User's Guide and Reference Manual*. 5th Edition, (California Scientific Software Inc., Grass Valley, CA., USA).

Nguyen, D. H. and Widrow, B. (1990). Neural Networks for Self-Learning Systems. *IEEE Control Systems Magazine*, April 1990, pp 18-23.

Willis, M. J., Dimassimo, C., Montague, G. A., Tham, M. T. and Morris, A. J. (1991). Artificial Neural Network in Process Engineering". *IEE Proceedings - D*, **138**(3) 256-266.

Pergamon

Wat. Sci. Tech. Vol. 28, No. 11-12, pp. 531–538, 1993.

0273–1223/93 $7·00 + 0·00

CONTROL STRATEGIES FOR A HIGHLY LOADED BIOLOGICAL AMMONIA ELIMINATION PROCESS

B. Teichgräber

Emschergenossenschaft/Lippeverband, Kronprinzenstr. 24, D-45128 Essen, Germany

ABSTRACT

A nitrification/denitrification process was applied to reject water treatment from sludge dewatering at Bottrop central sludge treatment facilities of the Emschergenossenschaft. On-line monitoring of influent and effluent turbidity, closed loop control of DO and pH, and on-line monitoring of nitrogen compounds were combined to a three level control pattern. Though on-line measurement of substrate and product showed substantial response time it could be used to operate nitrification/denitrification within process boundaries.

KEYWORDS

Reject water; nitrification/denitrification; on-line monitoring; response time.

INTRODUCTION

The Emschergenossenschaft is operating three wastewater treatment plants serving more than 5 million population equivalents. Sewage sludge is transported via pipeline from the WWTPs to the central sludge treatment facilities (CSTF) located in Bottrop.

Sewage sludge is treated by digestion thermal conditioning, mixing with coal, dewatering and combustion. About 10 000 m^3/d water carrying 11 t TKN/d and 60 t COD/d are rejected from sludge dewatering to sewage treatment.

To meet rising standards for environmental quality the Emscher system will be ecologically improved by constructing piped sewers and dividing the catchment of the Emscher Mouth WWTP into six subcatchments with separate WWTPs. The largest of these WWTPs in Bottrop (Teichgräber, 1991) is under construction now.

Today reject water is fed to the influent of Emscher Mouth WWTP. When the Bottrop WWTP starts operating reject water from the Bottrop CSTF is bound to be treated in this WWTP. Due to high nitrogen load of the reject water problems dealing with pH depression, insufficient sludge properties and exceeding German minimum requirements are expected. Moreover refractory COD is suspected to be produced by the combined thermal conditioning and digesting steps.

To remedy those predicted problems described above several methods to treat reject water are investigated. Objectives are:

- elimination of TKN to a degree of at least 90 %,
- reducing refractory COD,
- minimization of odour problems.

Two alternatives to treat reject water were tested in pilot scale:

- steam stripping process and
- nitrification/denitrification in a low loaded activated sludge process which is described below.

PROCESS SCHEME

Biological nitrogen elimination even from high strength waste water can be achieved by a single stage activated sludge system. (Mennerich, Albers 1988). Stockholm Water and Wastewater works tested nitrification/denitrification of reject water from digested sludge in a pilot plant using brewery waste as external carbon source for denitrification (Hultgren *et al.* 1989). Pilot tests are performed with a pre-denitrification system in half-technical scale (1:4000) which treats the side stream containing the highest COD and TKN loads (from thermal conditioning of secondary sludge).

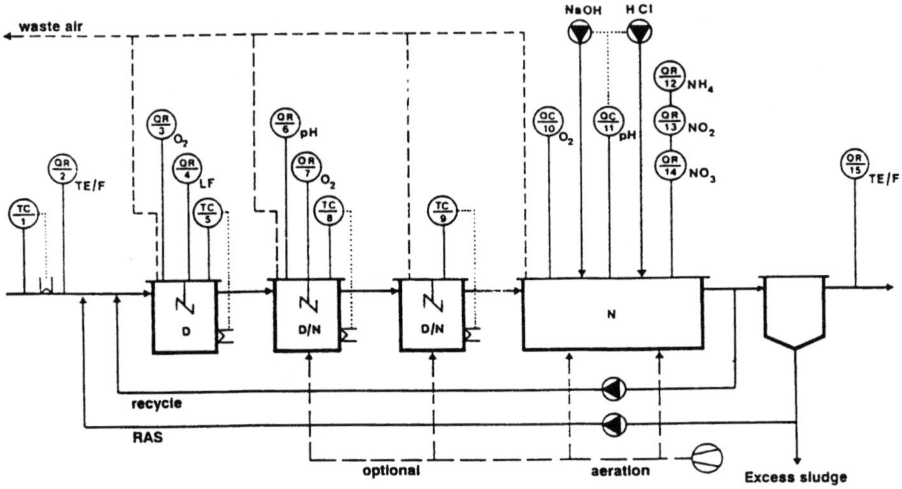

D denitrification
D/N facultative denitrification/nitrification
N nitrification
RAS return activated sludge

Fig. 1. Flow scheme of nitrification/denitrification process.

Figure 1 shows the flow pattern of the anoxic/aerobic process. The denitrification zone is divided into three compartments. Depending on process results two chambers can be shifted to the nitrification zone. All tanks contain devices to destroy foam which otherwise causes severe problems.

Due to the high efficiency required recycle rates can be varied between 1000 and 5000%. Return activated sludge ratio varies between 50 and 150%. Reaction rates for nitrification and denitrification therefore are similar to those of simultaneous nitrification/denitrification.

As ammonia gas (NH_3) and nitrous acid (HNO_2) inhibit nitrification these concentrations must be kept low. Especially the startup of the process must be controlled carefully.

Nitrification and denitrification shift pH and buffering capacity sharply. On the other hand the pH heavily influences nitrification rates. Therefore continuous pH-control is essential.

The temperature of the reject water varies between 30 and 50°C. Nitrification reates directly depend on temperature (Downing, Hopwood 1964). The optimum temperature ranges between 30 and 35°C (Rheinheimer *et al* 1988). Due to this the biological reject water treatment is operated at about 30°C. Cooling the influent and electric heating of tanks guarantees this temperature for the whole year.

SENSORS

An overview on controlled process steps is shown in Fig. 1. Table 1 presents parameters and monitors for this purpose.

TABLE 1 List of monitors

Nr.	Parameter	Manufacturer	Monitor	control loop
1	Temperature	Winkler	WRK 82/60	infl. cooling
2	turbidity	BTG	MEX 3	-
3	diss. Oxygen	Danfoss	Emco	-
4	conductivity	WTW	LF 161 T	-
5	temperature	Winkler	WRK 82/60	tank heating
6	pH	WTW	pH 96	-
7	diss. oxygen	WTW	Oxi 96	-
8	temperature	Winkler	WRK 82/60	tank heating
9	temperature	Winkler	WRK 82/60	tank heating
10	diss. oxygen	Danfoss	air supply	
11	pH	WTW	pH 161 TR	acid/base feed
12	ammonia	TeCom	MSR Contitest	-
13	nitrite	Bran & Lübbe	Monitor 90	-
14	nitrate	KOWA	Photometer	-
15	turbidity	BTG	MET 3000	-

From the practical point of view monitors can be divided into two groups:

1. measurement of physical properties or primary chemical composition, such as temperature, turbidity, conductivity, pH and DO and

2. measurement of sewage constituents such as ammonia, nitrite and nitrate.

Parameters of the first group usually are measured in sewage treatment plants. Measuring devices are developed for this purpose. Mostly they can be installed close to the control point.

Continuous analyzers for the second group of parameters are much more complicated to install and to maintain (Kayser, Ermel 1985). Those analyzers quoted in Table 1 require a sample free of suspended solids. To produce this sample an ultrafiltration unit, manufactured by TeCom, was installed (see Fig. 2).

The ammonia monitor (MSR Contitest AM by TeCom) is based on a gas sensitive probe. Sample and caustic soda are mixed in a thermostat cell containing 10 ml total volume. The frequency of measurement cycles can be varied between 4 min and 24 h. For most applications this kind of measurement can be regarded as semi-continuous. The range of measurement is specified between 0.1 and 1000 mg/l NH_4-N.

Nitrate measurement in the KOWA analyzer is performed by UV absorption. The extinction of a UV flashlight, filtered at 232 nm and 253 nm, is transformed to the nitrate concentration signal. Due to short flash lengths the UV flash bulb lasts for long time. The measurement range is 0 - 45 mg NO_3-N/l. Interferences are reported for high organic contents. The effluent of the pilot plant contains between 100 and 500 mg/l of soluble COD. Transmission is reduced by this blank absorbance to 30 - 50 %. To remove this absorbance a granular activated carbon (GAC) vessel containing 2 l was located between diverting chamber and nitrate monitor. Up to now no activated carbon could be selected which does not retain any nitrate but removes COD to a sufficient extent. Therefore the GAC vessel influences the nitrate measurement by adsorbing nitrate during startup of GAC treatment and by mixing in the vessel.

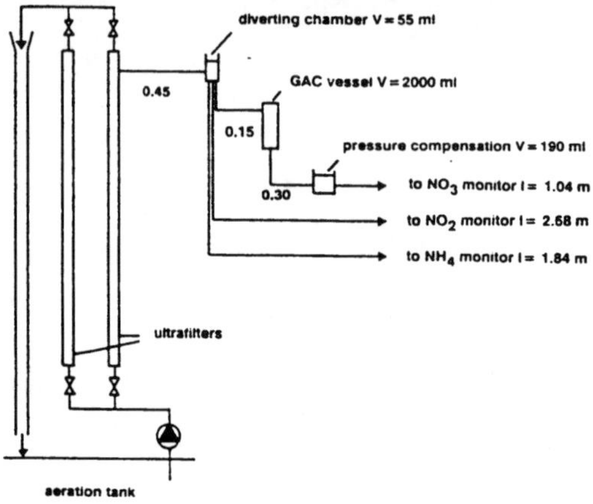

Fig. 2. Ultrafiltration unit.

Nitrite measurement is based on a modified EPA Standard method, developed by Griess. The sample is acidified and mixed with sulfanilamide to produce an azo compound. The extinction is measured at 565 nm. Originally the measurement ranges between 0 and 3 mg NO_2-N/l. By installing a dilution device this range could be enlarged to 0 - 10 mg NO_2-N/l.

CONTROL STRATEGIES

Control loops 1, 5, 8 and 9 refer to temperature control. The tanks of the pilot plant lose much energy due to their large specific surface. This control will not be necessary in a full scale reject water treatment plant.

Turbidity (No. 2) and conductivity (No. 4) indicate variations of the influent composition caused by the preceding sludge treatment steps. During pilot tests these parameters cannot be influenced. When reject water treatment is installed in full scale the operation of the sludge treatment facilities must be coordinated with the needs of the reject water treatment.

Dissolved oxygen concentration in the first denitrification reactor (No. 3) shows whether recirculation rates are set properly. Measurable DO concentrations indicate that the anoxic respiration rate is too low and that recirculation ratio can be diminished. This control loop is operated by hand. Probably it will not be automated as further information might be important which is difficult to introduce into closed loop control (e.g. minimum reaction times, effluent concentrations). This control loop might be a typical application for control by expert systems.

DO concentration in the second tank (No. 7) can serve both to indicate proper denitrification as well as to control aeration when this tank is used for nitrification. PH monitoring (No. 6) also indicates extent of denitrification.

The air supply of the aeration tank (N) is controlled by continous DO measurement (No. 10) in a closed loop control. Set points usually are 2.0 and 2.5 mg/l for minimum and maximum concentration. During startup phases or when inhibition of nitrification is observed DO concentration setpoints are elevated to 3.5 or 4.5 mg/l. In parallel the aeration of two facultative tanks is tunen in to achieve nitrification in these tanks. Decisions are made by operators and not automatically.

PH control in the nitrification tank is performed in a closed loop control. A pH monitor manufactured by WTW and a controller with chemical feed pumps manufactured by proMinent guarantee a stable pH-value. During startup phases the pH decreases due to nitrification and caustic soda must be dosed. When nitrification is depressed while denitrification still works for short periods hydrochloric acid must be dosed. According to the recommendations of Anthonisen *et al.* [1976] pH was kept between 6.5 and 6.7.

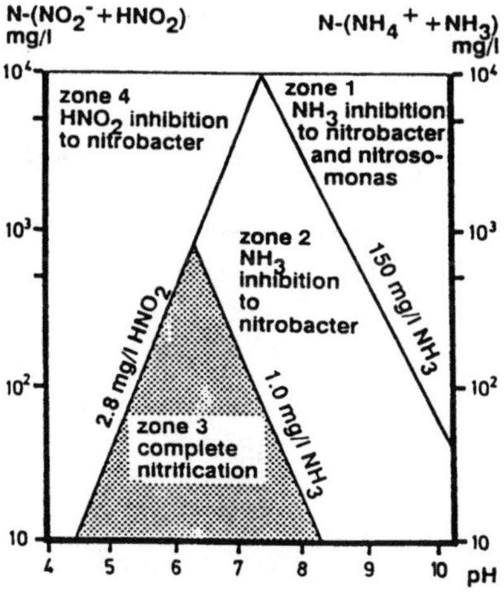

Fig. 3. Inhibition of nitrification due to free ammonia and free nitrous acid (from Anthonisen *et al.* [1976]).

Ammonia, nitrite and nitrate are monitored to remedy inhibitions due to ammonia or nitrous acid. Figure 3 gives the limits which must be kept to avoid inhibitions. On the other hand this is quite difficult because there are no parameters to influence these inhibiting substances by a kind of step forward control. During startup phase influent might be reduced. Moreover facultative anoxic/oxic tanks might be shifted to aeration to enhance nitrification. These decisions only can be made by operators, not by automatic control.

The turbidity of the effluent is monitored by a photometer. This indicates loss of biomass, caused by insufficient settling properties or high biomass concentrations in the activated sludge system.

RESULTS

The startup phase for the nitrification/denitrification process was quite difficult to manage, especially to avoid nitrite and ammonia accumulation. Moreover excess sludge production is very small when loads are kept low. The import of sludge via influent or the loss of biomass via effluent may disturb the whole startup of the process. Turbidity monitors were used to control these paths.

During proper operation the process parameters given in Table 2 could be kept.

TABLE 2 Process Parameters

retention time	d	7
sludge age (MCRT)	d	30
$B_{X,BOD5}$	kg BOD_5/kg MLSS*d	0.1
$B_{X,COD}$	kg COD/kg MLSS*d	0.2
$B_{X,TKN}$	g TKN/kg MLSS*d	20

Operational results (see Table 3) show the nitrification/ denitrification process being suitable for reject water treatment though operational targets could not be reached fully up to now.

TABLE 3 Operational Results of Nitrification/Denitrification Process to Treat Reject Water

		influent	effluent	operational target
COD	mg/l	15000 [1)]	1500	500
TKN	mg/l	2000	30	100
NH_4-N	mg/l	1500	20	90
SS	mg/l	200	20	50

[1)] Digestion of reject water has been bypassed to enhance denitrification capacity

DISCUSSION

Practical experience with the measuring equipment described above was fairly acceptable. Without these devices to indicate several parameters startup and proper operation could not have been performed.

Precision and response time of monitors measuring directly the controlled media are well known and described. Special problems arise when a sample preparation becomes necessary, which can be shown by analyzing the signals produced by control loops 12, 13 and 14. Figure 4 presents two response curves, of the ammonia monitor and of the complete analyzing system including ultrafiltration, to a simulated step load with defined concentrations.

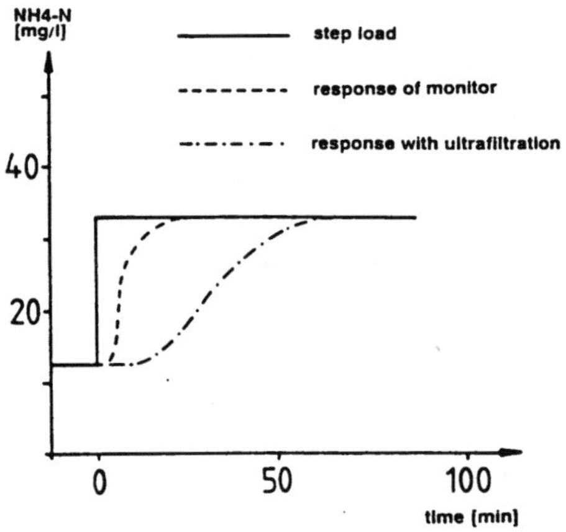

Fig. 4. Response curve of the ammonia analyzing system to a step load.

The tranformation of the input signal shows two components:

- linear time shift and

- a relax time until the response signal is constant.

The time shift can be explained by transport in tubes. Relax time is caused by mixing in the ultrafilter, in the diverting chamber and partially in the tube system and by reaction time of the potentiometric probe. It is very important that the ammonia monitor itself shows a time shift of 5 min and a relax time of about 20 min. The ultrafiltration system enlarges these times substantially.

Figure 5 shows the response times of the nitrate measurement. Time shift and relax time of the nitrate monitor are small (< 5 min). The preparation of the sample causes important transformations of the signal. In the first example the shift of the input signal was transformed to a continuous increase lasting 100 min. The final signal corresponds to the input signal. The second graph shows a short time shift of the input signal. The output signal starts with a deficit of 10 % and shows after setting back the input signal a remarkable memory effect. This indicated that the activated carbon affects the nitrate content and that it is not completely loaded.

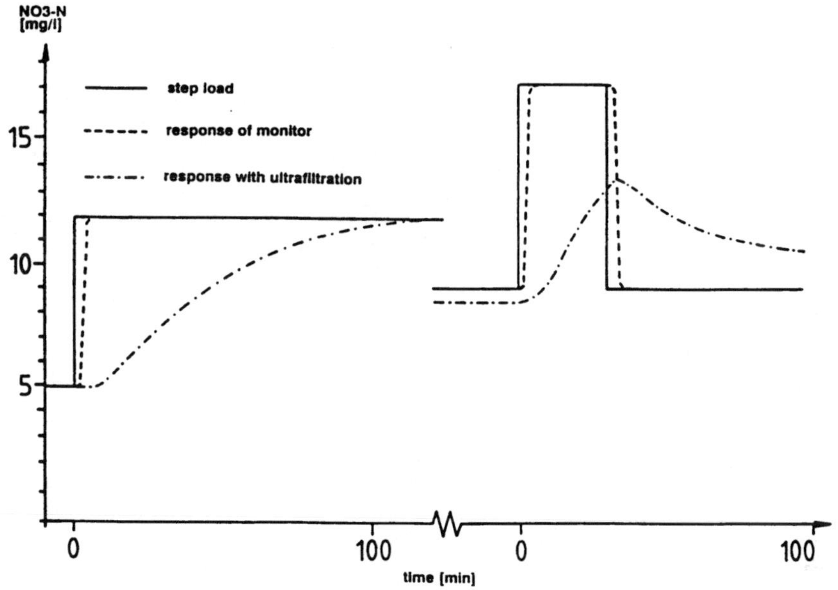

Fig. 5. Response curve of the nitrate analyzing system to a step load.

The output signals of the ammonia monitor (see Fig. 6) and of the nitrate monitor fit very well into results of laboratory measurements based on German standard methods. The ultrafiltration partially influences nitrate outputs as has been explained above. Nitrite analysis responds very fast and is not influenced by ultrafiltration. The nitrite monitor produces nonlinear deviations from laboratory analysis.

CONCLUSIONS

The biological treatment process for high strength reject water from sewage sludge treatment must be controlled intensively to identify operational problems, especially due to interactions between sludge treatment and reject water composition.

Three levels of control should be regarded:

a. Influent and effluent control by turbidity and conductivity monitoring,

b. closed loop control for important boundaries of biological treatment (temperature, DO and pH control) and

c. direct process control by monitoring substrate or product concentrations.

Measuring devices for level a and b are well proven for their application in sewage treatment. Their information is accurate and reproducible, but sometimes it may be difficult to derive orders for other control loops from this information. This problem arises from process configuration and lack of knowledge about the process, not from the measurement itself.

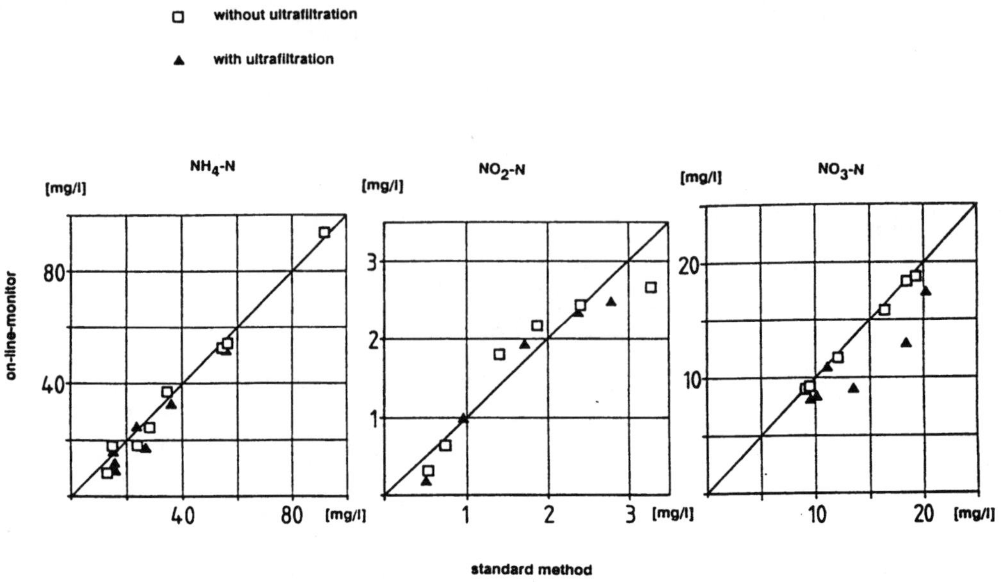

Fig. 6. Comparison between on-line and laboratory measurement.

Level c monitors are more difficult to install and to operate. The accuracy of these measurements depends on various factors and often is not investigated. In spite of this direct measurement of sewage constituents often can be applied successfully to control loops when changes in wastewater composition pass slowly and if the parameters of closed loop control are set properly. For the pilot plant described above the continuous measurement of nitrogen compounds delivers all necessary information to operate the process according to Anthonisen's diagram (Fig. 3).

Response times of total analyzing system including ultrafiltration show that this system must be optimized. General advices (short tubes etc.) is evident, but the combination of several sample streams, further sample treatment and sample flow must be improved.

REFERENCES

Anthonisen, A. *et al.* (1976). Inhibition of nitrification by ammonia and nitrous acid. *Journal WPCF*, **48**(5), 835-852.

Downing, A. and Hopwood, A. (1964). Some Observations on the Kinetics of Nytrifying Activated-Sludge Plants. Schweizerische Zeitschift für Hydrologie, 26, pp. 271-288.

Hultgren *et al.* (1989). Upgrading of the Treatment Plants in Stockholm to meet more stringent Requirements. Proceedings of International Specialized Conference on Upgrading of Wastewater Treatment Plants, Munich, September 3rd - 7th, 1989, pp. 73-80.

Kayser, R. and Ermel, G. (1985). Control of Simultaneous Nitrification-Denitrification. Advances in Water Pollution Control - Instrumentation and Control of Water and Wastewater Treatment and Transport Systems. Oxford: Pergamon, pp.481 - 488.

Mennerich, A. and Albers, H. (1988). Nitrification/Denitrification of Landfill Leachates. *Water Supply*, **6**, 157-166.

Rheinheimer, G. *et al.* (1988). Stickstoffkreislauf im Wasser. Wien: Oldenbourg, pp. 109 f.

Teichgräber, B. (1991). Decentralisation of the Emscher Sewerage System – Planning of the Bottrop biological Sewage Treatment Plant. *Wat. Sci. Tech.* **25**, (4/5), 51-58.

Pergamon

Wat. Sci. Tech. Vol. 28, No. 11-12, pp. 539-548, 1993.

0273-1223/93 $7·00 + 0·00

THE USEFULNESS OF ON-LINE MONITORING IN EFFECTING SAVINGS IN COMBINED BIOLOGICAL AND CHEMICAL REMOVAL OF PHOSPHORUS IN ACTIVATED SLUDGE PLANTS

L. H. Lotter and A. R. Pitman

City Council of Johannesburg, Water and Waste Directorate, P.O. Box 4323, Johannesburg 2000, South Africa

ABSTRACT

On-line analyzers to monitor the effluent levels of ammonia and phosphate have been installed at all four nutrient removal activated sludge plants in Johannesburg. The data from these instruments is transferred to a SCADA system for control purposes. The on-line system is supplemented by laboratory data, which is stored on a separate computer system. A combination of this data is used for routine operational control.

Dissolved oxygen probes are installed in the aerobic reactors of all plants and are used for aeration control. Automatic aeration control has recently been installed at one of the plants. In this paper some experience with on-line monitoring is discussed in terms of its usefulness and cost savings achieved. Savings of 1 million US $ in the purchase of phosphate precipitation chemicals have been achieved in the current financial year as a result of more efficient operational control.

KEYWORDS

Activated sludge; on-line measurement; nutrient removal; chemical dosing.

INTRODUCTION

It is expected that the future demand for water in the Republic of South Africa will exceed the economically exploitable sources within two to three decades. Because of erratic runoff conditions the construction of large impoundments with long retention periods has become necessary. Eutrophication in storage dams adds considerably to the cost of potable water treatment (Viljoen, 1984; Taylor et al., 1984). Phosphate is the limiting nutrient in most South African impoundments (Toerien et al., 1975), and in 1980 legislation was introduced to restrict the amount of phosphate entering river systems via sewage effluents. The promulgated limit of 1 mg P/l for certain catchments was however only enforced from 1985.

Johannesburg finds itself in the unenviable position of being situated in the middle of two drainage systems; namely, Southwards to the Vaal River and Northwards to the Hartebeespoort Dam. Both catchments are subject to the 1 mg P/l standard at all times. Approximately 600 Ml sewage are treated daily at the four works in Johannesburg, approximately 85% of which is treated by the activated sludge process.

During the last fifteen years researchers in Johannesburg have been striving to remove phosphate from sewage effluent in the most cost-effective manner. This has resulted in the optimisation of the biological process wherever possible. Osborn et al. (1986) showed that the additional cost of phosphate removal by chemical precipitation would be 4,94 US c/kl in comparison to 3,86 US c/kl for the biological option.

In spite of the disadvantage of high cost, the use of chemical precipitation cannot be avoided in older extended aeration plants not designed for nutrient removal and the early 5 stage BNR plants.

In a plant designed for nutrient removal, operational optimisation of the biological phosphorus removal process is directly dependent on meeting the oxygen demand in the aerobic zone and providing the correct substrate in the anaerobic zone (Pitman, 1991). Reliable continuous measurement of the dissolved oxygen levels in the aerobic zone forms an integral part of any aeration control strategy (Pitman, 1992).

Investigations into combined chemical and biological phosphorus removal have revealed that one of the biochemical processes, crucial to successful biological phosphorus removal, namely polyphosphate storage is inhibited by chemical dosing (Lotter, 1991). During further investigations into this phenomenon, the necessity of estimating the levels of chemical and biological removal became apparent (Lotter and Pitman, 1992).

On-line monitoring of activated sludge plants is used throughout the world to optimize plant operation (Buttz and Daigger, 1990; Spanjers and Klapwyk, 1990; Watts et al., 1990). The usefulness of continuous measurements of effluent phosphate levels by on-line analyzers was first evaluated in Johannesburg in 1986 (Osborn et al., 1986) and by 1991 on-line monitoring of all activated sludge plant effluents had been introduced.

By 1990 in situ dissolved oxygen probes had also become an integral feature of all activated sludge plants in Johannesburg.

In this paper, the use of on-line effluent monitoring and continuous measurement of dissolved oxygen in the operation of activated sludge plants in Johannesburg is discussed. Experience at three plants will be used to illustrate the usefulness of reliable on-line measurement in effecting cost savings. The importance of performance indicators in operator motivation is also discussed.

DESCRIPTION OF PLANTS AND EQUIPMENT

Olifantsvlei Works

The Olifantsvlei activated sludge plant was designed to treat 80 Ml of unsettled sewage in four parallel modules of 20 Ml each. The aeration tanks are divided into four compartments separated by baffles providing semi-plug flow operation. The plant was designed to nitrify completely and has a nominal hydraulic retention time of 27 hours.

By switching off aerators in the first and third compartments, biological phosphate removal was achieved (Venter et al., 1978). Effluent phosphate levels were reduced although not to consistently below 1 mg P/l. The creation of anaerobic conditions by switching off aerators is only possible when the carbon load is sufficiently low so as not to require the full aeration capacity. The implementation of the 1 mg P/l standard and increased carbon load necessitated the use of chemicals at this plant.

Since August 1985 ferric chloride and more recently a ferrous/ferric chloride mixture has been used to reduce phosphorus to less than 1 mg P/l in the effluent. The chemical was added at various positions, before finally deciding on a point downstream of the primary sedimentation tanks.

During 1991 two two channel Bran and Luebbe (Kolorimeter AC570) phosphate analyzers were installed to monitor the phosphate levels in the effluent. Effluent from the clarifier outlets of each module is pumped via a multi stage centrifugal pump to a monitoring room between 50 and 100 m away. The sample stream is pumped through a filter system to the analyzers.

At the same time two two channel ammonia analyzers were also installed. During September 1991 the single chemical tank from which all modules were dosed was replaced with four individual tanks.

Northern Works

The Northern Works activated sludge plant was designed to treat 150 Ml of settled sewage in three identical modules of 50 Ml each. One of the three modules was subsequently modified to incorporate endogenous denitrification ahead of the anaerobic zone. Fermentation and subsequent elutriation of primary sludge in the primary sedimentation tanks has allowed fairly consistent biological phosphate removal to be achieved at this plant (Osborn et al., 1986; Osborn et al., 1989; Lotter and Pitman, 1992).

Unlike the Olifantsvlei extended aeration plant, which was not designed for biological phosphorus removal, this plant achieved intermittent biological removal as a result of primary sludge fermentation. Although recent commissioning of specially designed primary sludge fermentation and elutriation facilities have reduced the chemical dose requirement significantly, supplementary chemical removal is still required in order to guarantee compliance with the standard.

On-line analyzers were first installed at this plant and were extensively evaluated. The second set of Bran and Luebbe instruments was installed in 1991. Effluent is pumped from the clarifier outlets via a multi-stage centrifugal pump to a monitoring room approximately 100 m away. As in the case of Olifantsvlei the sample stream is pumped through a filter system to the analyzers.

Bushkoppie Works

The Bushkoppie plant was originally designed as a 150 Ml/d biological nutrient removal plant. Each of the 4 parallel reactors comprises 3 zones, in series, anaerobic, anoxic and aerobic. Subsequently full primary sedimentation was retrofitted thus increasing the treatment capacity to 200 Ml/d.

In contrast to previous designs in Johannesburg, aeration is by means diffused air instead of surface aerators. The aeration system has 3 blowers, (2 operating; 1 standby) and an aeration system with 5 aerator zones per module, each with a different diffuser density and air supply from the main air header. The aeration system is tapered in 2 of the 3 operational modules.

The DO control system was upgraded to provide a separate blower control based on an air pressure set point in the main air header and DO control in the individual aeration zones of each reactor. The DO probes and controllers for zones 1 to 3 control the air flows to zones 1 to 2 respectively, while the DO probe and controller for zone 3 control air supply to zones 3 and 4. Zone 5 has an uncontrolled minimal fixed air input.

As the plant has a more modern configuration and favourable feed characteristics, iron salt addition is only used infrequently. It is dosed manually to the 4 reactors. Bran and Luebbe (Kolorimeters) phosphate and ammonia analyzers were installed in May 1992.

Date Processing

The signals from the analyzers at all the plants are transmitted to a SCADA system (Turbolink). This system has been configured to provide mimic screens showing realtime concentrations of the analytes, and archived trends. The data is trended every 30 minutes and the previous 31 days data may be viewed.

A daily production report showing the minimum average and maximum for each analyte in a reactor is printed daily. Alarms are triggered if concentrations exceed pre-set limits.

The signals from the DO probes are transmitted to the same SCADA system via programmable logic controllers at the Bushkoppie and Northern Works Plants.

RESULTS AND DISCUSSION

Olifantsvlei Works

Initially ferric chloride was dosed at an average rate of approximately 300 kg Fe/d per module. Continuous dosing of ferric chloride eventually resulted in tripling of the iron dose being used, with only intermittent compliance with the 1 mg P/l being achieved. (See Figure 1)

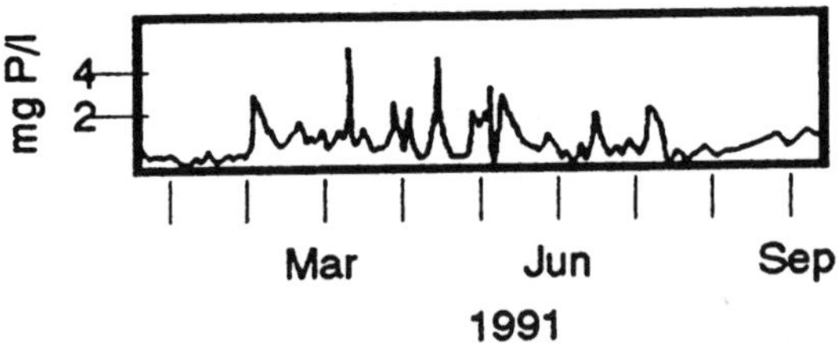

Figure 1 : Effluent phosphate levels : Olifantsvlei combined effluent. March to June 1991.

The phosphate levels in the combined effluent regularly exceeded the 1 mg P/l standard. Dosing continued on the same basis from a single tank, with intermittent excedances of the effluent phosphate standard until 1991 when the on-line analyzers were installed. At the same time improved measuring of the iron dose was introduced. (See Figure 2).

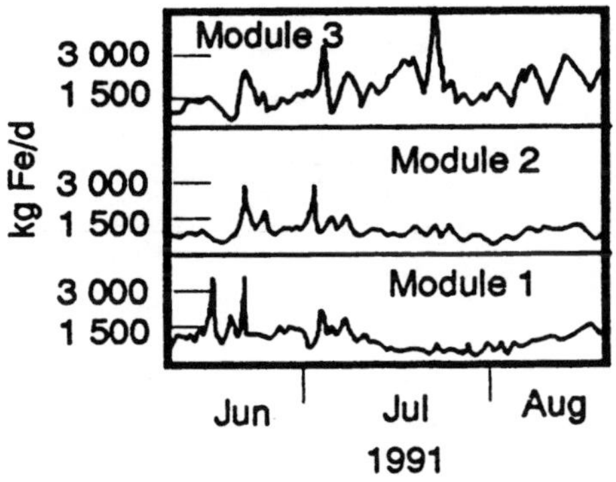

Figure 2 : Daily iron dose to each module : Olifantsvlei. June to August 1991.

Compliance with the standard improved significantly. (See Figure 3).

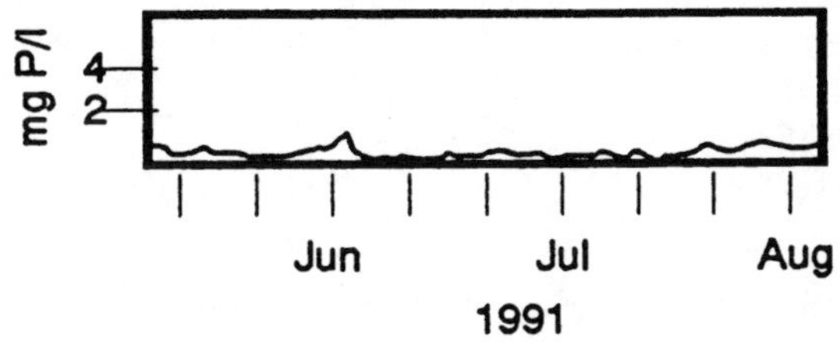

Figure 3 : Effluent phosphate levels : Olifantsvlei combined effluent. June to August 1991.

Better dosage control was achieved by operators checking the on-line phosphate value on an hourly basis then adjusting the iron dose accordingly. The daily iron dose was further reduced as shown in Figure 4, while still achieving very good effluent compliance. (See Figure 5)

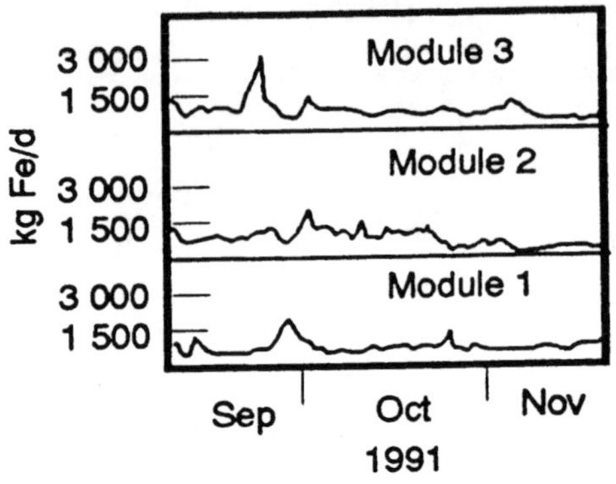

Figure 4 : Daily iron dose to each module : Olifantsvlei. September to November 1991.

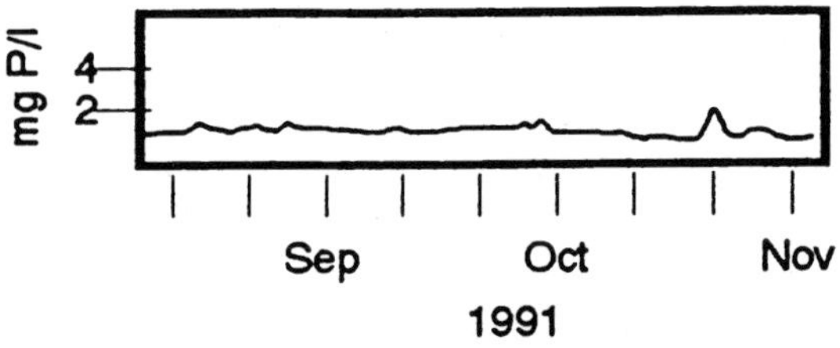

Figure 5 : Effluent phosphate levels. Olifantsvlei combined effluent. September to November 1991.

Northern Works

During research into combined biological and chemical phosphorus removal (Lotter and Pitman, 1992), it became clear that a parameter was required which would provide an indication of the amount of chemical precipitation taking place in a combined system. In order for a parameter to be useful in this context, it should provide at least some measure of the amount of phosphorus removed by this route. The daily iron (Fe) dose and phosphorus removed were measured and fed into the laboratory computer. The Fe/P ratio was then calculated daily from this data.

Evaluation of the Fe/P ratio on a plant where negligible biological phosphorus removal was being achieved indicated that Fe/P ratios of between 1.8 and 2.0 could readily be achieved with careful dosing control. The stoichiometric ratio for ferric phosphate precipitation is known to be 1.8. Achievement of ratios lower than 1,8 indicates the degree of biological phosphate removal while higher ratios indicate overdosing.

This parameter has proved useful in reducing the iron dose as well as evaluating biological removal under conditions of combined chemical and biological removal. The use of this parameter for two of the modules at Northern Works is shown in Figures 6 and 7.

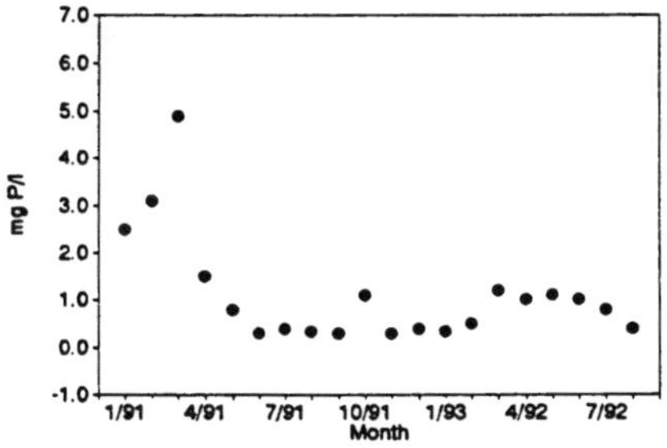

Figure 6 : Monthly average Fe/P ratio. Northern Works Module 1.

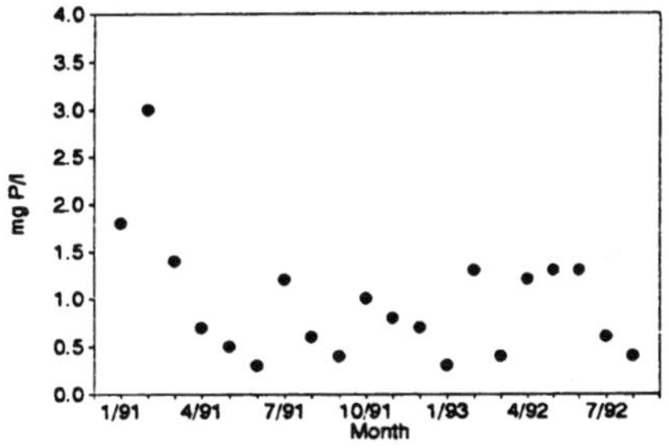

Figure 7 : Monthly Fe/P ratio. Northern Works Module 2.

During September 1990 operational problems with the old primary sedimentation tanks resulted in a decrease in biological removal. The greater requirement for precipitation is shown in the increased Fe/P ratios. The usefulness of this parameter in estimating biological response to operational changes is clearly evident in Figures 6 and 7.

During March 1991 new primary fermentation and elutriation facilities were commissioned resulting in improved biological removal as borne out by the decrease in Fe/P values from May 1991. Mechanical problems with these facilities resulted in larger chemical doses being required periodically. Adjustment of the iron dose in response to on-line effluent values ensured that chemical usage was kept to a minimum.

Efficient biological phosphorus removal is mainly dependent on three operational factors: correct substrate obtained by primary fermentation, meeting oxygen demand in the aerobic zone and prevention of nitrate ingress to the anaerobic zone.

Aeration control was achieved by manually switching aerators on and off as directed by the DO level. Satisfactory control was achieved during the week with this method but worked less efficiently during weekends with less skilled operators.

It was therefore decided to automate the aeration control system. This was done via a PLC SCADA system controlling the switching on and off of aerators to maintain a DO level of 2-3 mg/l throughout the aerobic zone.

Preliminary results with this system, which has only recently been commissioned, have shown that oxygen levels are controlled within narrower limits with this technique thus reducing unnecessary aeration and preventing under-areation.

Bushkoppie Works

As a result of the favourable feed characteristics this plant was dosed infrequently to achieve the 1 mg P/l limit. Consequently accurate measurement of the dose was only introduced during May 1992 as a result of research carried out on other plants. In June 1992 the on-line analyzers were fully commissioned and operational. Dosing was then adjusted according to the effluent phosphate levels indicated by the on-line analyzers. The beneficial effect of the analyzers on the iron dose is shown in Table 1.

Table 1 : Monthly variation in average Fe/P ratio.

	Monthly variation in average Fe/P ratio
Month	Fe/P
June	0,9
July	0,5
August	0,1
September	0,1

The aeration control system described above is currently being installed in all three modules. The beneficial effect of satisfactory aeration control is highlighted by the example of a week's operation in October 1992. See Figure 8.

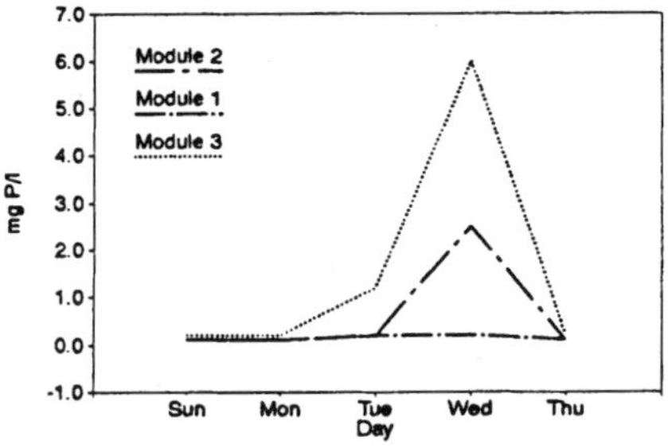

Figure 8 : Bushkoppie effluent phosphate levels.

An unusually high feed COD load resulted in a higher than normal oxygen demand. In module 1, with tapered aeration and fully automated aeration control, aeration was increased to match oxygen demand and no loss in biological phosphorus removal was observed.

In module 3, also with tapered aeration, but with aeration control not fully operational the oxygen demand was not completely satisfied resulting in high effluent phosphate levels.

In module 2 the old conventional aeration system (i.e. not tapered) has not yet been replaced thus making it more difficult to satisfy high oxygen demands at the start of the aerobic reactor.

In addition the aeration control system is not fully operational. The significantly high effluent phosphate level in this module provides support for the advantages of both automatic aeration control and tapered aeration.

Performance Targets

A series of monthly performance targets in addition to percentage compliance have been developed for each plant based on some of the parameters mentioned above. Operators are thus able to evaluate their performance on a regular basis. In an attempt to further optimize power consumption the automatic aeration control is supplemented by the setting of a power consumption target, which is kWh/kgO_2 demand. A specific target is determined for each plant. The theoretical oxygen demand is determined on a monthly basis for each plant using the UCT activated sludge model. A comparison of the values for Bushkoppie and Northern Works is given in Table 2.

Table 2 : Comparison of power usage between Bushkoppie and Northern Works.

Month	Northern Works	Bushkoppie
June	1,60	0,80
July	1,65	0,84
August	1,84	0,73
September	1,68	0,71

Specific targets are also set for the Fe/P ratio for each plants.

CONCLUSIONS

It is estimated that the optimisation of chemical dosing which has been achieved by more acurate dosing control using on-line analyzers and the setting of performance indicators has resulted in an estimated 1 million US $ saving in chemical purchases in the current financial year.

Automatic aeration control has been shown to be advantageous not only in respect of power saving but also in achieving compliance with the phosphate standard.

In order to further improve operational control, the current system needs to be further refined by;

- introducing direct SCADA linked measurement of iron dose

- direct daily feedback to SCADA of Fe/P ratio
- automatic iron dosing control based on effluent phosphate levels
- extension of automatic aeration control to all plants.

ACKNOWLEDGEMENTS

This paper is published with the permission of the Executive Director : Water and Waste. The authors wish to acknowledge the staff in the Wastewater and Scientific Services Departments who contributed to this work.

REFERENCES

BUTTZ, J. and DAIGGER, G. (1990) Real-time biological process optimisation using on-line monitoring. Adv. in Wat. Poll. Contr., (ed) R Briggs, 343-349.

LOTTER, L.H. (1991) Combined chemical and biological removal of phosphate in activated sludge plants. Wat. Sci. Tech., 23(4-6), 611-621.

LOTTER, L.H. and PITMAN, A.R. (1992) Chemical augmentation of biological phosphorus removal. Report to the Water Research Commission. In press.

OSBORN, D.W., LOTTER, L.H., PITMAN, A.R. and NICHOLLS, H.A. (1986) Enhancement of biological phosphate removal by altering process feed composition. Report to the Water Research Commission, South Africa, WRC 137/1/86.

OSBORN, D.W., LOTTER, L.H., PITMAN, A.R. and NICHOLLS, H.A. (1989) Enhancement of biological phosphate removal by altering process feed composition : Plant Studies. Report to the Water Research Commission, South Africa, WRC 137/2/89.

PITMAN, A.R. (1991) Design considerations for nutrient removal activated sludge plants. Wat. Sci. Tech., 23(4-6), 781-790.

PITMAN, A.R. (1992) Process design of new biological nutrient removal extensions to Northern Works, Johannesburg. Proceedings of European Conference on Nutrient Removal from Wastewater, Leeds, September 1992.

SPANJERS, H. and KLAPWIJK, A. (1990) On-line meter for the respiration rate and short-term biochemical oxygen demand in the control of the activated sludge process. Adv. in Wat. Poll. Contr., (ed) R Briggs, 67-77.

TAYLOR, R., BEST, H.J. and WIECHERS, H.N.S. (1984) The effluent phosphate standard in perspective : Part 1, Impact, control and management of eutrophication, IMIESA, October, 43-56.

TOERIEN, D.F., HYMAN, K.L. and BRUMER, M.J. (1975) A preliminary trophic status classification of some South African impoundments. Water S.A., 1, 15-23.

VENTER, S.L.V., HALLIDAY, J. and PITMAN, A.R. (1978) Optimisation of the Johannesburg Olifantsvlei extended aeration plant for phosphorus removal. Prog. Water Technol. 10(1/2), 279-292.

VILJOEN, S.C. (1984) The necessity of phosphate restrictions within the Vaal River Barrage catchment. IMIESA, September, 11, 3-5.

WATTS, J.B., EVANS, K. and MOLLOY, A. (1990) A wastewater/water process monitoring system allowing operational control - Developed by an operational Manager. Adv. in Wat. Poll. Contr., (ed) R Briggs, 143-154.

Pergamon

Wat. Sci. Tech. Vol. 28, No. 11-12, pp. 549–560, 1993.

0273–1223/93 $7·00 + 0·00

NUTRIENT REMOVAL: ON-LINE MEASUREMENTS AND CONTROL STRATEGIES

Dines E. Thornberg*, Marinus K. Nielsen* and Klaus L. Andersen**

* *Krüger Systems AS, 363 Gladsaxevej, DK-2860 Soeborg, Denmark*
** *Krüger Inc, 401 Harrison Oaks Blvd, Suite 100, Cary, NC 27513, USA*

ABSTRACT

New control strategy principles with respect to nitrogen and phosphorus removal in activated sludge plants are described. The principles include control of phase length and oxygen set point based on criteria functions and continuous on-line measurements of ammonium, nitrate, phosphate and pH. Five case studies are stated which describe improvements in biological nitrogen and phosphorus removal and in chemical phosphorus removal at activated sludge plants ranging from 5000 to 220000 population equivalents (0.25 to 11 MGD). The improvements due to the new strategies are basically a reduction of nutrients in the effluent and a reduction in the amount of chemicals and electricity consumed.

KEYWORDS

Activated sludge; Nutrient removal; Nitrogen; Phosphorus; On-line measurements; Control strategies; Criteria functions.

INTRODUCTION

The new strict demands on Danish wastewater treatment plants have led to the development of a new approach to control. The new demands mainly concern nitrogen and phosphorus removal. The demands can be met by a traditional extension of the existing plants by a factor 1.5 to 2. Another, and often more advantageous possibility, is to establish on-line measurements in the aeration tanks, measure the compounds that are to be removed and use these measurements to control the plant operation either in open or closed control loops.

There are two major obstacles to be considered:

1) that on-line measurements of ammonium, nitrate, phosphate, suspended solids and sludge blanket are too unstable to be used for on-line control;

2) that there are no efficient control strategies yet documented in the literature.

Obstacle #1 is now overcome as these measurements have been established at several plants in Denmark and are used for direct control.

This paper describes one successful attempt to solve obstacle #2 and deals with case studies at 5 different full scale activated sludge plants.

CONTROL IDEAS

The On-Line Measurement Systems

I. Krüger has carried out a two year project investigating the applicability of different on-line monitors for measuring ammonium (as ammonia), nitrate, phosphate and suspended solids in aeration tanks and effluents. This project led to the conclusion that with little, but responsible, surveillance, some of the monitors can indeed be used for on-line control (Thomsen and Nielsen, 1992).

I. Krüger has installed permanent on-line monitors at some plants. A mobile monitoring station, containing monitors for ammonium, nitrate, phosphate and suspended solids plus a PC able to monitor and control the plant and to transmit the acquired data by modem, has also been constructed. The mobile station has been monitoring various plants for about 14 at each plant. This period of time is sufficient to get a good understanding of the dynamics in the load pattern and performance of the plant. A few control strategies can be tested off-line which make it possible to decide whether permanent on-line control or the existing but adjusted simple control is optimal in terms of both effluent quality and economy.

Description of the BioDenipho Concept

The control experiments described in this paper have all, except one, been carried out at activated sludge plants operated according to the Biodenitro or BioDenipho method.†

In figure 1 the principle of BioDenipho operation is illustrated. The figure only shows the biological part of the plant.

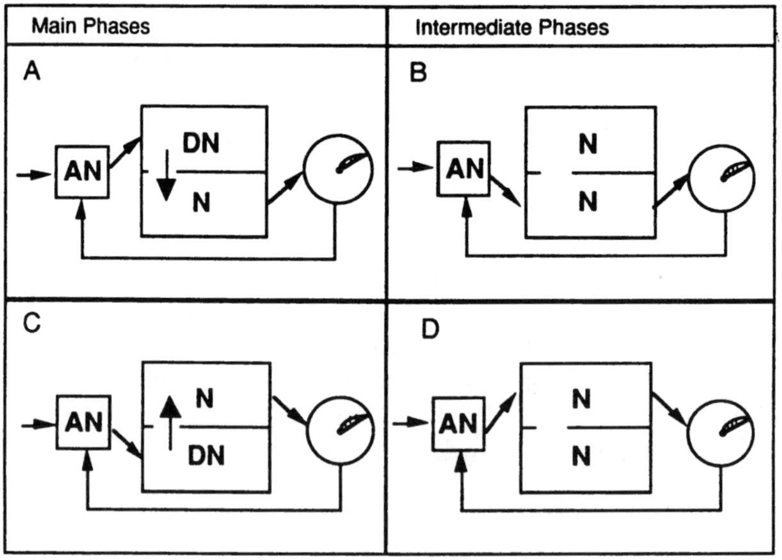

Fig. 1. The BioDenipho Cycle

During one cycle, the phases A to D are performed. The duration of one cycle is typically 3 h. The intermediate phases last from 15 to 30 min and the main phases from 60 to 90 min. The first tank is the anaerobic selector inducing P-release, and the nitrification/denitrification (N/DN) processes are performed in the two succeeding reactors.

It is important to understand that in the BioDenipho (or BioDenitro) operation the ammonium and nitrate time series will display a zig-zag pattern. In the nitrification phase the ammonium will be converted to nitrate and consequently the concentration of ammonium will decrease while the nitrate concentration will increase. In the denitrification phase the nitrate concentration will decrease and the ammonium concentration will increase due to influent ammonium. This is illustrated schematically in figure 2.

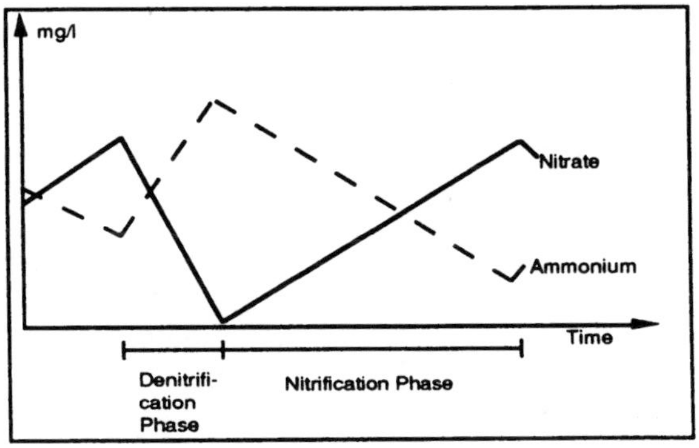

Fig. 2. The characteristic zig-zag pattern of the BioDenipho method.

One of the advantages of the BioDenipho concept compared to normal recirculation processes (Bardenpho and others) is that no recirculation is needed. Instead of the nitrate being recirculated back to the carbon source, which in this case is raw wastewater, the carbon source is directed to the nitrate.

The BioDenitro concept is similar to the BioDenipho concept but does not incorporate an anaerobic selector (Bundgaard and Nielsen, 1989).

Phase Length Control

The most common way of designing a plant to perform biological nitrogen and phosphorus removal is to "distribute" a fixed volume to the anaerobic, anoxic and aerobic processes, respectively.

In the BioDenipho concept, the distribution of anoxic and oxic capacity is based on phase lengths. Normally, the phase lengths are controlled by a software time scheme in the Supervision, Control And Data Acquisition (SCADA) computer with variations on workdays/weekends and during summer/winter.

These time schemes do not take into account sudden increases in the loading of the sludge plant, e.g. during rain or a factory start-up. The time schemes are often planned with a safety margin inducing "waste of process time" in some periods.

From a control point of view, the BioDenipho process has the capability to change from 100% denitrification to 100% nitrification in only 5 min. This powerful control possibility opens up the prospect of automatic compensation of dynamic differences in load, BOD/N ratios and degradability of the wastewater during diurnal variations or industrial holidays.

N-removal. By installing on-line measurements of ammonium and nitrate in continuous samples from the aeration tanks, it is possible to get the required information to introduce a total dynamic control of the phase lengths. The basic principle is that a nitrification phase is ended when ammonium is low, and the denitrification is ended when the nitrate is low.

The control of the phase lengths is based on the criteria functions that are shown in figures 3 and 4.

In the nitrification phase the computer checks the (NO_3-N, NH_4-N) point in the above diagram every 5 or 6 min. As the phase proceeds, the point will move from the upper left down towards the lower right as indicated by the arrow (less ammonium, more nitrate). When the line is crossed the phase is ended.

The same principle is used in the denitrification phase with the criteria function shown in figure 4.

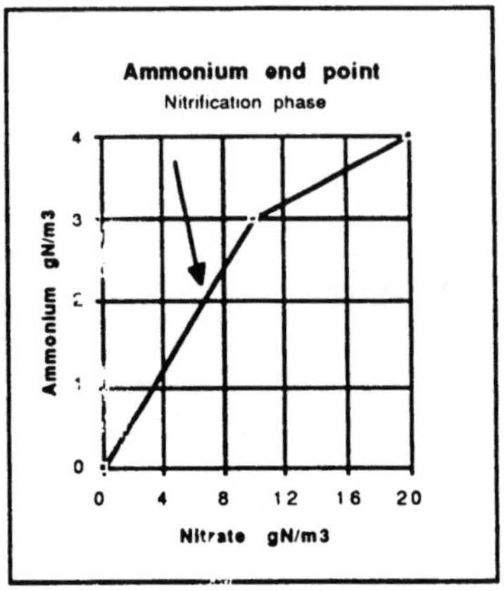

Fig. 3 . Criteria function for shift from nitrification phase to denitrification phase.

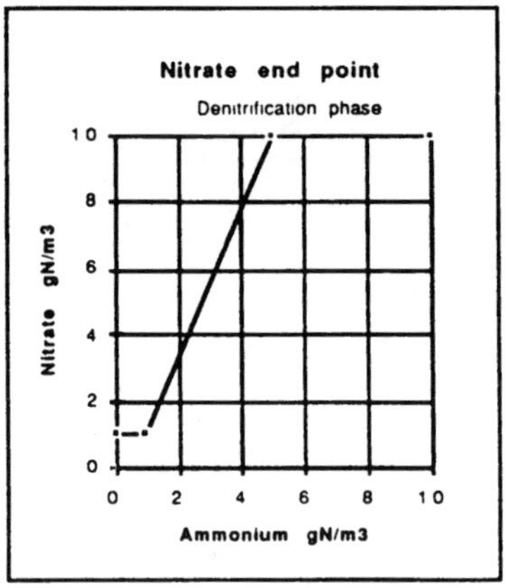

Fig. 4. Criteria function for shift from denitrification phase to nitrification phase.

If the loading at the plant increases as it typically does in the morning, the criteria functions allow both ammonium and nitrate to rise. This leads to an increase in the process rates which are presumably limited in a Monod kinetic fashion.

The shape of the criteria functions can be changed according to the effluent demands and the characteristics of the treatment plant and wastewater.

P-removal. At one plant not furnished with an anaerobic selector, Bio-P removal was obtained by extending an anoxic phase into anaerobic conditions once a day. This is described below under Frederikssund WWTP.

The Control of Oxygen Set Point

Nitrification and denitrification processes are both sensitive to the concentration of dissolved oxygen. By changing the set point in the oxygen control it is possible to affect the process rates, thereby following the dynamic variations in the nitrogen load.

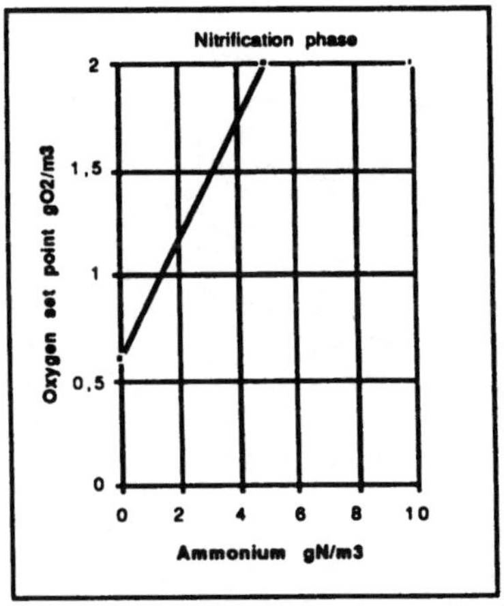

Fig. 5. Oxygen set point as a function of ammonium concentration.

Figure 5 shows how a higher ammonium concentration induces a higher oxygen set point, thereby increasing the process rate. When the ammonium concentration is lowered during nitrification, a gradual conversion of the process conditions into simultaneous denitrification and energy saving is obtained. If the plant loading is low, for instance in weekends, the oxygen set point is lowered automatically. This reduces the need for carbon source, which is the limiting factor for biological nutrient removal at many plants.

CASE STUDIES

Below, five examples of different implementations of on-line control are described. Each description is divided into 3 sections: background, measurements and results.

STENLOSE WWTP

Background

The treatment plant in Stenlose municipality was originally a conventional nitrifying activated sludge plant with extended aeration in a single aeration tank. The wastewater is almost entirely domestic with practically no industrial load, resulting in a normal degradability, and the BOD/N ratio is 5-8. The sludge loading is 0.03-5 kg BOD/kg MLSS*d and the temperature 7-15°C. In 1987, equipment for on-line measurement was installed, and during the next 3 years, control experiments were conducted as phase 1. The nitrogen removal was obtained by intermittent aeration controlled by criteria functions as described above. Stenlose WWTP was the first plant to be controlled in this way (Thornberg, 1988).

In phase 2, the plant was extended to 18 000 pe and included P removal by the addition of an anaerobic selector and the complementing of the processes with simultaneous precipitation. This has now been in operation for 2 years .

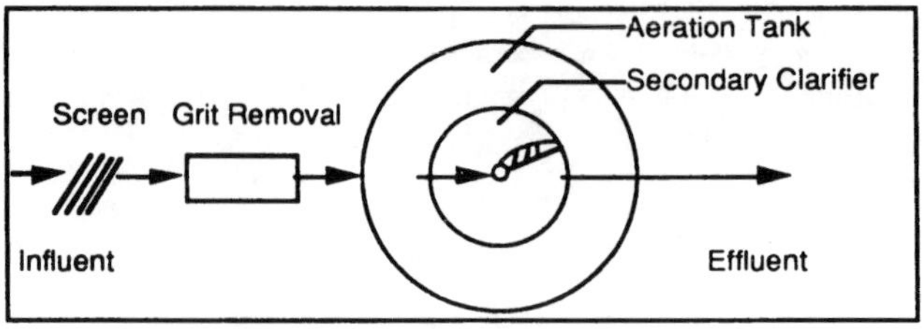

Fig. 6. Flow sheet of Stenlose WWTP

Measurements

The measurement system is based on ion-selective electrodes, which measures ammonium and nitrate directly in a continuous sample stream of MLSS from the aeration tanks. The system has a sensitivity check every 30 minutes, is auto-calibrated twice a day, and auto-cleaned every 24 h. The measurement system is controlled by the same computer that controls the plant, thus facilitating the synchronization of calibration and cleaning procedures with periods where the signal is needed less. The response time in this system is below 3 mins.

Two hours of maintenance are required a week for 3 x 2 measurements. The yearly measurement cost is approx. US$10 000 incl. manpower, but excluding depreciation on the monitor costs which is about 50 000 US$. The system is one of three prototypes, and it is not available as a commercial product.

Results

The results from phase 1 demonstrated that control of oxygen supply is sufficient to remove more than 80% nitrogen, even though 1/2 of the raw wastewater is supplied during nitrification periods. The concentration of total nitrogen in the effluent dropped from about 20 mg/l to 6-8 mg/l. Further, the energy consumption was reduced by 15-20%. In phase 2 the intermittent aeration operation mode was maintained even though the plant is now able to operate in a recirculation mode.

STEGHOLT WWTP

Background

This plant consists of pre-clarification followed by a BioDenipho plant performing biological nitrogen and phosphorus removal. There is no chemical support for the biological P-removal, although it is widely used in Denmark. The plant treats wastewater from households and industry. In the beginning of 1992 the wastewater had changed due to a milk powder factory s alteration of its operation. This resulted in problems with the sludge dewatering and thereby the biological P-removal.

The plant operators asked Krüger for help and in June 1992 the plant was adjusted during a short-term visit by the mobile monitoring station from the 10 to the 24 of June, 1992.

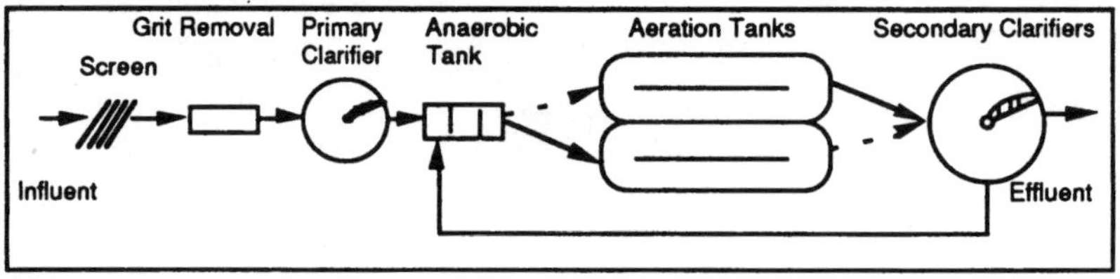

Fig. 7. Flow sheet of Stegholt WWTP

Measurements

There was no closed loop on-line control based on the nutrient monitors in the mobile monitoring station. Instead the standard control systems were adjusted manually according to visual inspection of the time series of ammonium, nitrate, phosphate, suspended solids and redox potential measured in the aeration tank. These graphs were generated along with graphs based on the normal logging of dissolved oxygen and operation of the aeration equipment and pumps.

Results

The phase shift time scheme program had been adjusted so that 57% of the cycle-time was allocated to nitrification and 43% to denitrification, during the startup of the plant. The investigations based on the nutrient measurements showed that the denitrification only lasted about one third of the time assigned for this process. This led to anaerobic conditions in the aeration tanks, followed by extended release of phosphorus, which could not be taken up again in the following nitrification phase. Consequently the phosphorus concentration in the effluent was too high.

The phase shift program was changed manually once a day, for 5d, reaching an optimum of 67% nitrification time and an effluent P concentration lower than the standards.

TABLE 1 Results of Phase Time Adjustment. Concentrations are in mg/l

Date	NH_3-N	NO_3-N	PO_4-P	% Nitrification time
9 June 92	0.74	0.9	7.7	57
16 June 92	0.24	1.6	1.2	75
23 June 92	0.77	1.1	1.2	67

As is seen from Table 1 the concentration of phosphorus in the effluent decreased, while ammonium and nitrate concentrations remained fairly constant.

MARSELISBORG WWTP

Background

Marselisborg treatment plant was retrofitted in 1986-90 and thus became a nitrogen and phosphorus removing plant instead of a primary treatment plant. It was designed to remove nitrate in the summer only. The plant is designed for 220 000 population equivalents (approx. 11 MGD) and comprises an AB process, i.e.biosorption followed by a BioDenitro plant.

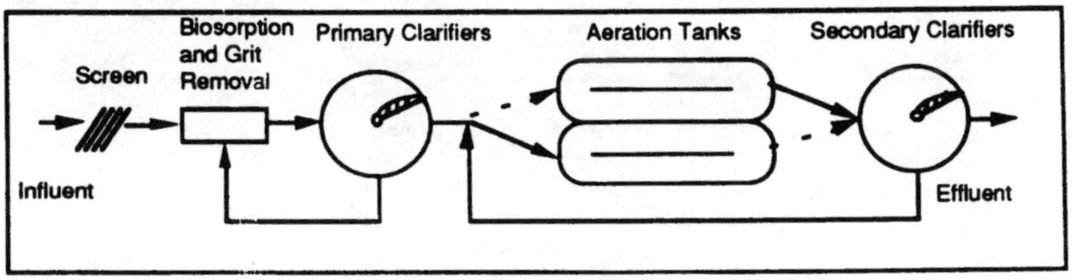

Fig. 8. Flow sheet of Marselisborg WWTP.

Measurements

In June 1991, permanent on-line monitors for nitrate and ammonium, were installed. There are continuous sampling strings from 2 (a couple) out of 4 aeration tanks but the filtrate is multiplexed to the one ammonium and the one nitrate monitor. The multiplexing is operated so that it is always the decreasing concentrations that are measured. These slopes provide information about nitrification and denitrification rates.

Results

Along with the installation of on-line monitors the SCADA computer was reprogrammed to enable dynamic phase length and oxygen set point control. The control strategies were based on criteria functions as described above.

TABLE 2 shows a comparison of the removal percentage before and after the control system was installed.

	SS	COD	BOD	Tot N	Tot P
Jan-May 91	96	91	97	80	95
Jan-May 92	98	91	98	87	96

It should be added that between the two periods the plant COD loading drops about 10% and the '92 consumption of chemicals in step B is reduced to about 25% of the '91 consumption. The important point in Table 2 is that the nitrogen removal is increased from 80 to 87% while removal of organics is constant.

The concentration of total nitrogen in the effluent decreased from 6.8 to 4.5 mg/l. It was mainly the nitrate fraction of the total nitrogen that decreased. The oxygen set point control has moved a part of the aeration time from the night time to the day time. The ammonium peaks in the day time have thereby been lowered, and thereby the ammonium concentration in the effluent. This result has not been quantified. The total energy consumption is constant. Sorensen *et al.*(1992) describe the results in more detail.

SOUTH OMME WWTP

Background

This plant was constructed as an SBR with two parallel tanks and no clarifier. Phosphorus was removed by simultaneous precipitation with lime and ferric chloride dosed in the influent. The incoming nitrogen was nitrified. From 1989 to 1991 Krüger tested several on-line monitors at this plant. At the same time, tests were conducted to reduce the consumption of chemicals and to introduce denitrification.

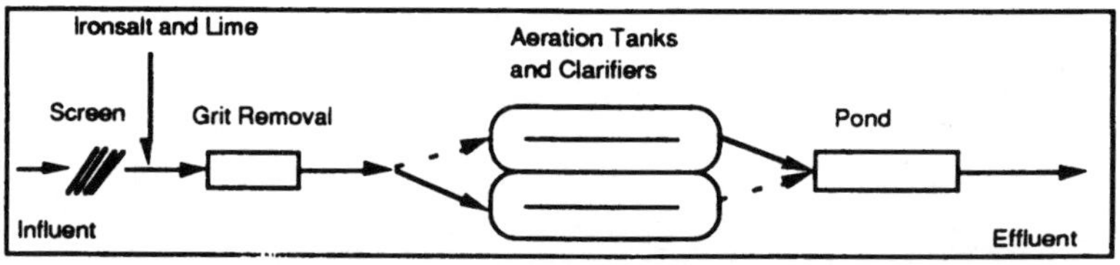

Fig. 9. Flow sheet of South Omme WWTP

Measurements

The plant was initially provided with flow, oxygen and pH meters. The monitors, temporarily installed at the plant, measured phosphate, ammonium and nitrate multiplexing between samples from the two SBRs and the effluent.

Results

The strategy for the dosing of ferric chloride was changed in March 1991. Once a day the dosing was adjusted according to the on-line measurement of phosphate. Before that the dosing was time proportional.

This control has reduced the consumption of $FeCl_3$ by 25-40%. The denitrification was introduced as an extra phase in the fixed time scheme only. The phase lengths were adjusted off-line on the basis of the N-monitors. The total nitrogen concentration in the effluent of the plant dropped about 25%.

As a consequence of the reduced dosing of ferric chloride and the denitrification, the drop in alkalinity in the relative soft water was not as extensive as before. That is one of the reasons why the dosing of lime could be reduced about 75%. The main reason though, is probably the new control based on continuous pH measurements instead of time proportional control. The reduced dosing of chemicals results in an increased ratio of biologically active sludge.

TABLE 3. Results from closed loop control at South Omme WWTP. The concentrations were measured in the effluent.

	Before mg/l	After mg/l	Relative Change in %
NH_4-N	1.8	1.2	-25
NO_3-N	4-6	2-4	-40
Tot-N	7-10	5-7	-25
Tot-P	0.17-0.34	0.34	+20
	Consumption of		
	Electricity		-5
	Lime		-75
	Iron Salt		-25

In spite of the simultaneous precipitation it could be seen from the on-line PO_4 measurement that there was biological activity connected to the phosphate. There was a sharp increase in the PO_4 concentration when the nitrate fell below 0.5 mg/l. It is not possible to estimate the amount of phosphate removed biologically.

FREDERIKSSUND WWTP

Background

The plant was originally built for nitrification only. In the nineteen-seventies the plant was changed to include full nitrogen removal just by the installation of motors on the weirs. This was one of the first plants at which the BioDenitro method was tested. The plant has no anaerobic selector in front of the aeration tanks but early experiments back in 1978 had shown that biological P removal could be obtained to a large extent in this plant. The wastewater composition is very favourable for Bio-P removal, since much of the carbon is easily degradable due to a juice factory discharging to the plant.

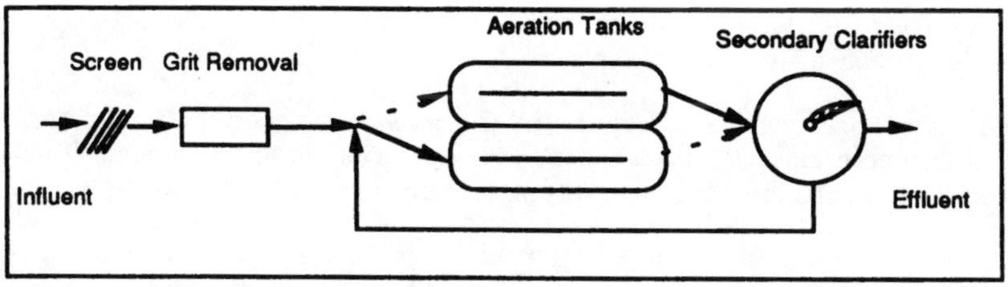

Fig. 10. Flow sheet of Frederikssund WWTP.

Measurements

In 1989, Krüger installed equipment for permanent measurements of ammonium, nitrate and phosphate. The nitrate measurement did not function very well and instead a redox potential electrode (ORP measurement) was installed.

The phase shift control was based on the phosphate, ammonium and ORP measurements. The anaerobic phase was diagnosed and ended either by a sharp rise in the phosphate concentration or by a drop in the ORP.

Results

In the autumn of 1991, it was shown that to enhance the growth of phosphorus accumulating bacteria it was sufficient to expose the sludge to anaerobic conditions once a day. The effluent contained less than 1.5 mgPO_4-P/l. The influent concentration is 8 mg PO_4-P/l on the average.

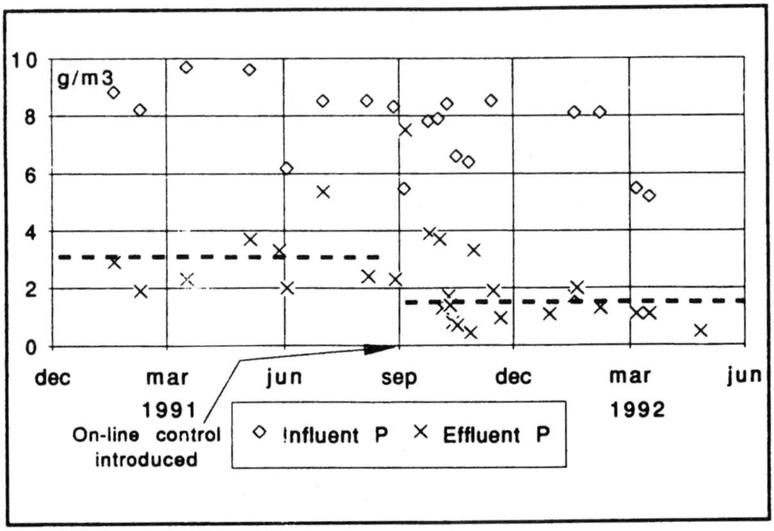

Fig. 11. Effluent concentrations of phosphate before and during the on-line phase shift control. The dotted horizontal lines indicate average effluent concentrations.

DISCUSSION

The on-line measurement of nutrients, sludge blanket and suspended solids, reveals a whole new world of information, and a new understanding of the processes in the wastewater treatment plants. The above examples describe some of the phenomena that have been seen in the enormous amount of data one receives with on-line monitors for several parameters. The investigations described herein show that it is possible to develop efficient control strategies for an enhanced removal of nitrogen and phosphorus by means of on-line measurements of ammonia, nitrate and phosphorus.

It is not always necessarily the closed loop control that leads to the cleaner effluents or reduced operation costs. Some of the improvements in these experiments were achieved just by looking closely at the way the plant is operated, and combining this with the detailed information available from the on-line measurements.

The investigations described in this paper have been carried out at plants removing nitrogen by the BioDenitro or BioDenipho methods. The improvements obtained by closed loop control with ammonia and nitrate can be obtained at other types of plants. Kayser and Ermel (1983, 1984, 1987) have shown that on-line measurements can be used to introduce denitrification in carrousels and oxidation ditches.

Process Stability

One result that is certainly obtained by installing on-line control is a greater process stability in the effluent concentrations of nutrients. This was shown at Marselisborg WWTP (Sorensen *et al.*, 1992). If the effluent standards are based on a maximum concentration, the new control will decrease the risk of not meeting the standards. This is due to the quick response to sudden load changes like for instance stormwater.

Cost/Benefit

There is a running debate on the cost/benefit in installing on-line monitors for nutrients permanently at a wastewater plant, and it is not possible to give a general answer to this.

The cost of installing two on-line monitors, for each of the compounds ammonia, nitrate and phosphorus, is approximately 100 000 US$. To this should be added the programming cost related to the control loops.

If a plant larger than 20 000 pe has capacity problems, there is no doubt that the installation of on-line control will be less costly than a physical extension of the plant. The question is how much the plant capacity can be increased by installing on-line control. The results from these investigations show that the increase in the N and P removal capacity is in the range from 5 to 30%. At plants performing nitrification at a sludge load of 0.03-0.08 kg BOD/kg MLSS *d, the extension of the plant to perform denitrification might be done with a smaller safety margin than the one usually used. The planned aeration tank and clarifier volume can be reduced by up to 30%.

The Control of Data Quality

The next step is to develop better data quality control by using experience about the way in which the measurements normally behave and by comparing trends in different parameters that are known to be interlinked. The on-line signals are still unstable to a certain degree. Therefore, it is necessary to include a check procedure in the software to sort out the bad data. So far, only checking on high, low and slope limits has been done automatically. New concepts of data quality check are made. The basic principle is that the parameters must show the expected tendencies according to the operation situation and the known variance, in order to be accepted. One example is the link between ammonium load and the oxygen uptake rate (OUR). OUR can be calculated automatically based on monitoring of the aeration system and the oxygen concentration. If a sudden unexpected increase in ammonium occurs this should normally be seen in OUR estimates as well. If no increase is detected in the OUR, the measurement of ammonium is likely to be erroneous. At present principles of forecasting nutrient concentrations in the aeration tanks to obtain a better control even when the on-line signals are unstable, are being developed (Carstensen *et al.*, 1992).

The STAR Program

The control strategies described in this paper have been incorporated into one single software package together with new strategies for return sludge control and sludge control during stormwater flow. The package called STAR (Superior Tuning And Reporting) has been implemented on a UNIX based computer system at Aalborg WWTP in the north of Denmark. The results of this are described in (Lynggaard and Nielsen,1993).

CONCLUSIONS

- Continuous on-line measurements of nutrients in the aeration tanks can be used for closed loop control, provided that the monitors are maintained in a responsible way.

- Use of on-line measurements for control optimization is advantageous at most large plants with biological nutrient removal. The control can be either open or closed loop.

- Control strategies for the closed loop control have been tested at several Danish wastewater treatment plants. These tests show that the suggested control strategies are efficient and robust for practical implementation.

REFERENCES

Bundgaard, E. and Nielsen, F. M. (1989). Stability of Effluents from Biological Nutrient Removal Plants - Danish Long-Term Operating and Optimization Experience. WCPF 62nd Annual Conference, San Francisco.

Carstensen, J., Madsen, H., Poulsen, N. K. and Nielsen, M. K. (1992). Identification of wastewater processes using non-linear grey box models. Internal Report from the Institute of Mathematical and Statistical Analysis at the Technical University of Denmark.

Ermel, G. (1983). Steuerung der simultanen Denitrifikation über den Nitratgehalt des Belebtschlammes. GWF-wasser / Abwasser, 124, H 10, pp. 484-487.

Kayser, R. and Ermel, G. (1984). Control of simultaneous Denitrification by a Nitrate Controller. *Wat.Sci.Tech.* **16**(10/11), 143-150.

Kayser, R. (1987). Biologishe Stickstoff - und Phosphor elimination in Abwasserreinigungsanlagen. Veröffentlichungen des Instituts für Stadtbauwesen, Heft 42, Technishe Universität Braunschweig.

Lynggaard, A. and Nielsen, M. K. (1993). Superior Tuning And Reporting (STAR) - a new concept for On-line Process Control of Wastewater Treatment Plants. Paper submitted for the 6th IAWPRC conference on Instrumentation and Control in Hamilton, Canada 1993.

Sørensen, J., Nielsen, M. K. and Thornberg, D. E. (1992). Optimisation of a Nutrient Removing Plant using On-line Measurements. Paper submitted for publication in *Research Journal of WPCF*.

Thomsen, H. A. and Nielsen, M. K. (1992). Practical Experience with On-line Measurements of NH_4, NO_3, PO_4, Redox, MLSS, and SS in Advanced Activated Sludge Plants. HYDROTOP Conference, Colloquium "The city and the Water", Marseilles.

Thornberg, D..E (1988). Nitrogen Removal by Computer Control of a Simple Activated Sludge Plant. *Wat. Supply*, **6**, pp 361-369

KEYWORD INDEX

Printed in Dunstable, United Kingdom

91897606R10335